*Surveying/Construction/
Transportation/Energy/
Economics & Government/
Computers*

CIVIL ENGINEERING PRACTICE

4/Surveying
Construction
Transportation
Energy
Economics & Government
Computers

EDITED BY

PAUL N. CHEREMISINOFF
NICHOLAS P. CHEREMISINOFF
SU LING CHENG

IN COLLABORATION WITH

F. A. Ahmed	N. F. Danial	R. Janardhanam	Y. Makigami	E. F. Roof
G. J. Alexander	E. G. Dauenheimer	W. D. Kahn	W. W. Mann	N. M. Rouphail
R. F. Astrack	G. Dioguardi	B. E. Keenan	M. Morsi	K. Seil
K. W. Bauer	J. J. Degnan	J. D. Keenan	M. Nakajima	W. S. Smith
N. A. Baumann	R. W. Eck	K. Kinose	Z. A. Nemeth	C.-N. Sun
P. R. Bhave	T. S. Englar, Jr.	W. Konon	D. L. Olson	D. Teodorović
R. K. Block	R. P. Guenthner	N. Kumaki	A. Polus	M. F. Tewfik
N. Bodnar	B. F. Hobbs	C. F. Lam	D. D. Pfaffinger	C. D. Tockstein
J. D. Bossler	R. Hussein	S. M. Lee	A. K. Rathi	M. Vaziri
P. D. Cady	M. C. Ircha	H. Lunenfeld	G. L. Reynolds	

TECHNOMIC
PUBLISHING CO., INC.

LANCASTER · BASEL

Published in the Western Hemisphere by
Technomic Publishing Company, Inc.
851 New Holland Avenue
Box 3535
Lancaster, Pennsylvania 17604 U.S.A.

Distributed in the Rest of the World by
Technomic Publishing AG

©1988 by Technomic Publishing Company, Inc.
All rights reserved

No part of this publication may be reproduced, stored in a
retrieval system, or transmitted, in any form or by any means,
electronic, mechanical, photocopying, recording, or otherwise,
without the prior written permission of the publisher.

Printed in the United States of America
10 9 8 7 6 5 4 3 2 1

Main entry under title:
 Civil Engineering Practice 4—Surveying/Construction/Transportation/
 Energy/Economics & Government/Computers

A Technomic Publishing Company book
Bibliography: p.
Includes index p. 681

Library of Congress Card No. 87-50629
ISBN No. 87762-537-9

TABLE OF CONTENTS

Preface   vii

Contributors to Volume 4   ix

SECTION ONE: **SURVEYING**

1 A Geometric Framework for Land Data Systems 3
   K. W. Bauer, *Southeastern Wisconsin Regional Planning Commission, Waukesha, WI*

2 Plane Trilateration Adjustment by Finite Elements 13
   N. F. Danial and M. F. Tewfik, *University of Petroleum and Minerals, Dhahran, Saudi Arabia*

3 Area Computation Using Salient Boundary Points 39
   F. A. Ahmed, *University of Petroleum & Minerals, Dhahran, Saudi Arabia*

4 Redefinition of the North American Geodetic Networks 47
   J. D. Bossler and N. Bodnar, *National Oceanic and Atmospheric Administration, Rockville, MD*

5 Centimeter Precision Airborne Laser Ranging System for Rapid, Large-Scale Surveying and Land Control ... 71
   J. J. Degnan and W. D. Kahn, *NASA/Goodard Space Flight Center, Greenbelt, MD*
   T. S. Englar, Jr., *The Johns Hopkins Applied Physics Laboratory, Laurel, MD*

6 Inertial Surveying 95
   E. F. Roof, *USAETL, Fort Belvoir, VA*

SECTION TWO: **CONSTRUCTION**

7 An Organizational Model for Building Construction Firms in the Computer Age 125
   G. Dioguardi, *Universita degli Studi di Bari, Bari, Italy*

8 CPM in Construction 189
   E. G. Dauenheimer, *New Jersey Institute of Technology, Newark, NJ*
   R. K. Block, *Conti Construction Company, South Plainfield, NJ*

9 Environmental Concerns During Construction 197
   W. Konon, *New Jersey Institute of Technology, Newark, NJ*

10 Construction: Traffic Control at Freeway Work Sites 211
   Z. A. Nemeth, *The Ohio State University, Columbus, OH*
   A. K. Rathi, *KLD Associates, Huntington Station, NY*
   N. M. Rouphail, *The University of Illinois at Chicago, Chicago, IL*

11 Determination of Power Spectral Density Functions from Design Spectra 231
   D. D. Pfaffinger, *P + W Engineering, Zurich, Switzerland*

12 Chance Constrained Aggregate Blending 241
   S. M. Lee, *University of Nebraska, Lincoln, NE*
   D. L. Olson, *Texas A&M University, College Station, TX*

13 Constitutive Modelling of Ballast 257
    R. Janardhanam, *University of North Carolina at Charlotte, Charlotte, NC*

SECTION THREE: **TRANSPORTATION**

14 Public Transportation in Urban America 273
    W. S. Smith, *Columbia, SC*

15 Mass Transportation: Carpool Assignment Technique Application 335
    W. W. Mann, *Metropolitan Washington Council of Governments, Washington, DC*

16 Transit Performance Evaluation 345
    R. P. Guenthner, *Marquette University, Milwaukee, WI*

17 Matching of Transportation Capacities and Passenger Demands in Air Transportation 365
    D. Teodorović, *University of Belgrade, Belgrade, Yugoslavia*

18 Pedestrian Flow Characteristics 393
    A. Polus, *Technion — Israel Institute of Technology, Haifa, Israel*

19 Simulation Model Applied to Japanese Expressway 407
    Y. Makigami, *Ritsumeikan University, Kyoto, Japan*
    N. Kumaki, *Japan Highway Public Corporation, Hiroshima, Japan*
    M. Nakajima, *Institute of Systems Science Research, Kyoto, Japan*
    K. Seil, *Kyoto University, Kyoto, Japan*

20 The Role of Driver Expectancy in Highway Design and Traffic Control 429
    G. J. Alexander, *Positive Guidance Applications, Inc., Rockville, MD*
    Lunenfeld, *Federal Highway Administration, Washington, DC*

21 Technique for Identifying Problem Downgrades 457
    R. W. Eck, *West Virginia University, Morgantown, WV*

22 Transportation Planning and Network Geometry: Optimal Layout for Branching Distribution Networks 473
    P. R. Bhave, *Visvesvaraya Regional College of Engineering, Nagpur, India*
    C. F. Lam, *Medical University of South Carolina, Charleston, SC*

23 Factors Affecting Driver Route Decisions 499
    M. Vaziri, *University of Kentucky, Lexington, KY*

SECTION FOUR: **ENERGY**

24 Cooling Water Supply for Energy Production 513
    B. F. Hobbs, *Case Western Reserve University, Cleveland, OH*

25 Storage Treatment of Salt-Gradient Solar Pond 535
    K. Kinose, *National Research Institute of Agricultural Engineering, Yatabe, Tsukuba, Ibaraki, Japan*

26 Economics of Industrial Fluidized Bed Boilers 557
    J. D. Keenan, *University of Pennsylvania, Philadelphia, PA*
    B. E. Keenan, *Schlusser & Reivers, Wilmington, DE*

SECTION FIVE: **ECONOMICS/GOVERNMENT/DATA ACQUISITION**

27 Managing Public Involvement 571
    R. F. Astrack, *U.S. Army Corps of Engineers, St. Louis, MO*
    N. A. Baumann, *Planning and Managing Consultants, Ltd., Carbondale, IL*
    G. L. Reynolds, *University of Central Arkansas, Conway, AR*

28 Municipal Service Distribution: Equity Concerns 589
    M. C. Ircha, *University of New Brunswick, Fredericton, New Brunswick, Canada*

29 Engineering Economic Evaluation 601
    P. D. Cady, *The Pennsylvania State University, University Park, PA*

30 Microprocessor-Based Data Acquisition Systems 651
    C.-N. Sun and C. D. Tockstein, *Tennessee Valley Authority, Knoxville, TN*

31 BASIC Programming for Civil Engineers on Micro-Computers 669
    R. Hussein, *University of the District of Columbia, Washington, DC*
    M. Morsi, *Arab Bureau for Design and Technical Consulting, Abbasseya, Cairo, Egypt*

Index   681

PREFACE

While the designation civil engineering dates back only two centuries, the profession of civil engineering is as old as civilized life. Through ancient times it formed a broader profession, best described as master builder, which included what is now known as architecture and both civil and military engineering. The field of civil engineering was once defined as including all branches of engineering and has come to include established aspects of construction, structures and emerging and newer sub-disciplines (e.g., environmental, water resources, etc.). The civil engineer is engaged in planning, design of works connected with transportation, water and air pollution as well as canals, rivers, piers, harbors, etc. The hydraulic field covers water supply/power, flood control, drainage and irrigation, as well as sewerage and waste disposal.

The civil engineer may also specialize in various stages of projects such as investigation, design, construction, operation, etc. Civil engineers today as well as engineers in all branches have become highly specialized, as well as requiring a multiplicity of skills in methods and procedures. Various civil engineering specialties have led to the requirement of a wide array of knowledge.

Civil engineers today find themselves in a broad range of applications and it was to this end that the concept of putting this series of volumes together was made. The tremendous increase of information and knowledge all over the world has resulted in proliferation of new ideas and concepts as well as a large increase in available information and data in civil engineering. The treatises presented are divided into five volumes for the convenience of reference and the reader:

VOLUME 1 Structures
VOLUME 2 Hydraulics/Mechanics
VOLUME 3 Geotechnical/Ocean Engineering
VOLUME 4 Surveying/Transportation/Energy/Economics & Government/Computers
VOLUME 5 Water Resources/Environmental

A serious effort has been made by each of the contributing specialists to this series to present information that will have enduring value. The intent is to supply the practitioner with an authoritative reference work in the field of civil engineering. References and citations are given to the extensive literature as well as comprehensive, detailed, up-to-date coverage.

To insure the highest degree of reliability in the selected subject matter presented, the collaboration of a large number of specialists was enlisted, and this book presents their efforts. Heartfelt thanks go to these contributors, each of whom has endeavored to present an up-to-date section in their area of expertise and has given willingly of valuable time and knowledge.

PAUL N. CHEREMISINOFF
NICHOLAS P. CHEREMISINOFF
SU LING CHENG

CONTRIBUTORS TO VOLUME 4

F. A. AHMED, University of Petroleum and Minerals, Dhahran, Saudi Arabia

G. J. ALEXANDER, Positive Guidance Applications, Inc., Rockville, MD

R. F. ASTRACK, U.S. Army Corps of Engineers, St. Louis, MO

K. W. BAUER, Southeastern Wisconsin Regional Planning Commission, Waukesha, WI

N. A. BAUMANN, Planning and Managing Consultants, Ltd., Carbondale, IL

P. R. BHAVE, Visvesvaraya Regional College of Engineering, Nagpur, India

R. K. BLOCK, Conti Construction Company, South Plainfield, NJ

N. BODNAR, National Oceanic and Atmospheric Administration, Rockville, MD

J. D. BOSSLER, National Oceanic and Atmospheric Administration, Rockville, MD

P. D. CADY, The Pennsylvania State University, University Park, PA

N. F. DANIAL, University of Petroleum and Minerals, Dhahran, Saudi Arabia

E. G. DAUENHEIMER, New Jersey Institute of Technology, Newark, NJ

J. J. DEGNAN, NASA/Goddard Space Flight Center, Greenbelt, MD

G. DIOGUARDI, Universita degli Studi di Bari, Bari, Italy

R. W. ECK, West Virginia University, Morgantown, WV

T. S. ENGLAR, JR., The Johns Hopkins Applied Physics Laboratory, Laurel, MD

R. P. GUENTHNER, Marquette University, Milwaukee, WI

B. F. HOBBS, Case Western Reserve University, Cleveland, OH

R. HUSSEIN, University of the District of Columbia, Washington, DC

M. C. IRCHA, University of New Brunswick, Fredericton, New Brunswick, Canada

R. JANARDHANAM, University of North Carolina at Charlotte, Charlotte, NC

W. D. KAHN, NASA/Goddard Space Flight Center, Greenbelt, MD

B. E. KEENAN, Schlusser & Reivers, Wilmington, DE

J. D. KEENAN, University of Pennsylvania, Philadelphia, PA

K. KINOSE, National Research Institute of Agricultural Engineering, Yatabe, Tsukuba, Ibaraki, Japan

W. KONON, New Jersey Institute of Technology, Newark, NJ

N. KUMAKI, Japan Highway Public Corporation, Hiroshima, Japan

C. F. LAM, Medical University of South Carolina, Charleston, SC

S. M. LEE, University of Nebraska, Lincoln, NE

H. LUNENFELD, Federal Highway Administration, Washington, DC

Y. MAKIGAMI, Ritsumeikan University, Kyoto, Japan

W. W. MANN, Metropolitan Washington Council of Governments, Washington, DC

M. MORSI, Arab Bureau for Design and Technical Consulting, Abbasseya, Cairo, Egypt

M. NAKAJIMA, Institute of Systems Science Research, Kyoto, Japan

Z. A. NEMETH, The Ohio State University, Columbus, OH

D. L. OLSON, Texas A&M University, College Station, TX

D. D. PFAFFINGER, P + W Engineering, Zurich, Switzerland

A. POLUS, Technion—Israel Institute of Technology, Haifa, Israel

A. K. RATHI, KLD Associates, Huntington Station, NY

G. L. REYNOLDS, University of Central Arkansas, Conway, AR

E. F. ROOF, Fort Belvoir, VA

N. M. Rouphail, The University of Illinois at Chicago, Chicago, IL
K. Seil, Kyoto University, Kyoto, Japan
W. S. Smith, Columbia, SC
C.-N. Sun, Tennessee Valley Authority, Knoxville, TN
D. Teodorović, Univerity of Belgrade, Belgrade, Yugoslavia

M. F. Tewfik, University of Petroleum and Minerals, Dhahran, Saudi Arabia
C. D. Tockstein, Tennessee Valley Authority, Knoxville, TN
M. Vaziri, University of Kentucky, Lexington, KY

CIVIL ENGINEERING PRACTICE

VOLUME 1
Structures

SECTION 1   Reinforced Concrete Structures
SECTION 2   Structural Analysis
SECTION 3   Stability
SECTION 4   Pavement Design
SECTION 5   Wood Structures
SECTION 6   Composites

VOLUME 2
Hydraulics/Mechanics

SECTION 1   Hydraulics/Open Channel Flow
SECTION 2   Flow in Pipes
SECTION 3   Flow With Bed Load
SECTION 4   Mechanics/Solid Mechanics
SECTION 5   Fluid Mechanics
SECTION 6   Solid-Fluid Interaction

VOLUME 3
Geotechnical/Ocean Engineering

SECTION 1   Soil Mechanics
SECTION 2   Stability
SECTION 3   Bearing Capacity
SECTION 4   Buried Structures
SECTION 5   Waves and Wave Action
SECTION 6   Coastal Structures

VOLUME 4
Surveying/Construction/ Transportation/Energy/ Economics & Goverment/Computers

SECTION 1   Surveying
SECTION 2   Construction
SECTION 3   Transportation
SECTION 4   Energy
SECTION 5   Economics/Government/Data Acquisition

VOLUME 5
Water Resources/Environmental

SECTION 1   Water Supply and Management
SECTION 2   Irrigation
SECTION 3   Environmental

SECTION ONE
Surveying

CHAPTER 1	A Geometric Framework for Land Data Systems	3
CHAPTER 2	Plane Trilateration Adjustment by Finite Elements	13
CHAPTER 3	Area Computation Using Salient Boundary Points	39
CHAPTER 4	Redefinition of the North American Geodetic Networks	47
CHAPTER 5	Centimeter Precision Airborne Laser Ranging System for Rapid, Large-Scale Surveying and Land Control	71
CHAPTER 6	Inertial Surveying	95

CHAPTER 1

A Geometric Framework for Land Data Systems

KURT W. BAUER*

INTRODUCTION

There is a growing interest in the United States today in land data systems. This interest ranges from a relatively narrow concern about the need to modernize land title recordation systems to the relatively broad concern about the need to create entirely new land-related data banks for multipurpose application. This growing interest has involved many disciplines, ranging from surveyors, abstractors, assessors, and attorneys narrowly concerned with the fiscal and legal administration of real property to planners, engineers, and public administrators broadly concerned with community development and resource management. Much of the interest has centered around the use of electronic computers for the storage, manipulation, and retrieval of the data and, more recently, the use of graphic display hardware for the reproduction of the data in mapped, as well as tabular, form.

For practical reasons the development of automated land data systems may have to begin with development of single-purpose cadastres relating to the registration of land ownership and perhaps to the value of real property as a basis for taxation. Such cadastres should, however, be amenable to evolutionary development into true data systems that provide information on the characteristics, capabilities, and existing and potential uses of land for planning and management purposes, as well as on the ownership and value of land.

Any land data system requires some method of spatial reference for the data. Indeed, the National Research Council has identified the basic components of a land data system as: 1) a spatial reference framework consisting of geometric control points; 2) a series of accurate large-scale base maps; 3) a cadastral overlay that delineates all cadastral parcels and displays a unique identifying number for each; and 4) a series of compatible registers of interest in, and data about, the land parcels keyed to the parcel identifier.[1] As indicated, an adequate geometric framework for such spatial reference must, if it is to serve even the narrowest purposes of a land data system, permit identification of land areas by coordinates down to the individual ownership parcel level. The provision of a geometric framework of adequate accuracy and precision to permit system operation at a highly disaggregate parcel level is the most demanding specification possible and permits ready aggregation of information from the more intensive and detailed level to the more extensive and general level as may be necessary.

The decision concerning the type of geometric framework to be provided for any new land data system will be one of the key determinations affecting the long-term utility and efficiency of the system. Any error in this determination should be on the side of potential utility. A determination to provide a geometric framework more precise and accurate than may be required ultimately would mean that a part of the capital required to implement the system may be wasted. A determination, however, to provide a geometric framework less precise and accurate than may be required ultimately will mean that much, if not all, of the capital investment required to implement the system will have been essentially wasted. Further, this capital investment itself may form an insurmountable impediment to later evolutionary development of the system, since the committed decision will with time make it increasingly difficult and costly to effect any required reforms.

Because of the importance of the geometric framework

---

*Executive Director, Southeastern Wisconsin Regional Planning Commission, Waukesha, WI

[1] *Procedures and Standards for a Multi-Purpose Cadastre,* Panel on a Multi-Purpose Cadastre, Committee on Geodesy; Comission on Physical Science, Mathematics, and Resources; National Research Council; National Academy Press; Washington, DC (1983).

for spatial reference of data to the long-term success of any multipurpose land data bank, and because that importance is apt to be overlooked by planners and decision-makers in their deliberations of other important issues involved in the creation of land data systems, this chapter discusses certain basic concepts that should be applied in the design of the geometric framework for any land data bank system. It also describes one kind of geometric framework that is based on these concepts and that can serve as a sound foundation for the evolutionary development of a multipurpose land data system.

SOME BASIC CONCEPTS

A good multipurpose land data system must be able to store in coordinated, machine-readable form a wealth of data essential to sound land use planning and management. Historically, such data would have been typically stored on maps. Consequently, certain concepts which apply to the design and preparation of good maps also apply to the design and implementation of the geometric framework for a land data system.

Any accurate mapping project requires the establishment of a system of survey control. This survey control consists of a framework of points whose horizontal and vertical positions and interrelationships have been accurately established by field surveys, to which the map details are adjusted and against which such details can be checked. The survey control system should be carefully designed to fit the specific needs of the particular map being created. For multipurpose application, it is essential that this survey control system meet two basic criteria if the maps are to be effective planning and management tools. First, it must permit the accurate correlation of real property boundary line data with topographic, earth science, and other land and land-related data. Second, it must be permanently monumented on the ground so that lines on the maps may be accurately reproduced in the field when land use development and management projects reach the regulatory or construction stage. That is, the survey control system must be such as to provide finished maps, the points and lines of which not only accurately reflect both cadastral and earth science field conditions but also points and lines which can be readily and accurately reproduced upon the ground as well. This property is important not only to the use of the maps but to the maintenance of the maps in a current condition. Conceptually, the geometric framework for a land data system is the equivalent of the survey control system for a map; and the same principles apply to its design and implementation.

Unfortunately, in the United States two different, and heretofore largely uncoordinated, systems of survey control have evolved. One—the State Plane Coordinate System—is founded in the science of measurement and is utilized as a basis for the collection of earth science data and the preparation of earth science maps, such as topographic, geologic, soils, and hydrographic maps. The other—the U.S. Public Land Survey System—is founded in the principles of property law, as well as in the science of measurement, and is utilized for the collection of cadastral data and the preparation of cadastral maps—such as real boundary line maps.

U.S. Public Land Survey System

For most of the United States, the federal government has provided the basic survey control system for cadastral mapping in the form of the U.S. Public Land Survey System.[2] This system is founded in the best features of the English common law of boundaries, superimposing on that body of law systematic land survey procedures under which the original public domain is surveyed, monumented, and platted before patents are issued; legal descriptions are by reference to a plat; lines actually run and marked on the ground control boundaries; adjoiners are respected; and the body of law in effect at the time the deed is issued is controlling and forever a part of the deed. Unlike scientific surveys, which are made for the collection of information and which can be amended to meet improved standards or changing conditions, the original government land survey in an area cannot be ignored, repudiated, altered, or corrected so long as it controls rights vested in lands affected.

The U.S. Public Land Survey System is one of the finest systems ever devised for describing and marking land. It provides a basis for a clear, unambiguous title to land, together with the physcial means by which that title can be related to the land which it describes. The system is ingenious, yet simple, easy to comprehend and administer; and without it, the nation would be unquestionably poorer. The "rectangular" land survey system, however, has one serious flaw. Its use requires the perpetuation of monuments set by the original government surveyors, monuments the positions of which are not precisely related to the surface of the earth through a scientifically established map projection.

State Plane Coordinate System

A strictly scientific survey control system designed to provide the basic control for all federal—and most private—topographic and other earth science mapping operations exists separately from the U.S. Public Land Survey System in

---

[2]Under the regulations imposed by the Congress, the U.S. Public Land Survey System has been extended into 30 of the 50 states, covering all but Connecticut, Delaware, Georgia, Hawaii, Kentucky, Maine, Maryland, Massachusetts, New Hampshire, New Jersey, New York, North Carolina, Pennsylvania, Rhode Island, South Carolina, Tennessee, Texas, Vermont, Virginia, and West Virginia.

the triangulation and traverse stations established by the National Geodetic Survey (formerly U.S. Coast and Geodetic Survey). The triangulation and traverse stations established by this agency comprise a nationwide network connecting thousands of monumented points whose geodetic positions in terms of latitude and longitude are known. In order to make the National Geodetic Survey control network more readily available for local use, the U.S. Coast and Geodetic Survey in 1933 devised the State Plane Coordinate System. This system transforms the spherical coordinates—latitudes and longitudes—of the stations established in the national geodetic survey into rectangular coordinates—eastings and northings—on a plane surface. This plane surface is mathematically related to the spheroid on which the spherical coordinates of latitude and longitude have been determined. The mutual relationship which makes it practicable to pass with mathematical precision from a spherical to a plane coordinate system makes it also practicable to utilize the precise scientific data of the national geodetic survey control network for the reference and control of local surveying and mapping operations. A limitation on such uses, however, is imposed by the relatively widespread location of the basic triangulation and traverse stations and the difficulties often encountered in the recovery and use of these stations.

A RECOMMENDED GEOMETRIC CONTROL SYSTEM

From the foregoing brief discussion of the U.S. Public Land Survey and State Plane Coordinate Systems, it is apparent that two essentially unrelated survey control systems have been established in the United States by the federal government. One of these—the U.S. Public Land Survey System—is founded in the legal principles of the real property description and location and was designed primarily to provide a basis for the accurate location and conveyance of ownership rights in land. The other—the State Plane Coordinate System—is founded in the science of geodesy and was designed primarily to provide a basis for earth science mapping operations and for the conduct of high precision scientific and engineering surveys over large areas of the earth's surface. Both systems have severe inherent limitations for use as a geographic framework for a land data system. By combining these two separate survey systems into one integrated system, however, an ideal system for the geometric control required for land data systems is created. This ideal system requires the relocation and monumentation of all U.S. Public Land Survey section and quarter-section corners, including the centers of sections, within the geographic area for which the land data system is to be created and the utilization of these corners as stations in second order traverse and spirit level nets, both nets being tied to the National Geodetic Data. The traverse net establishes the true geographic positions of the U.S. Public Land Survey corners in the form of state plane coordinates, while the spirit level net establishes the exact elevation above mean sea level of the monuments marking the corners.

Such a system of survey control has the following advantages as a geographic framework for a multipurpose land data system:

1. It provides an accurate system of control for the collection and coordination of cadastral data since the boundaries of the original government land subdivision form the basis for all subsequent property divisions and boundaries. Thus, all subsequent legal descriptions and plats must be tied to the U.S. Public Land Survey System; and the accurate reestablishment and monumentation of the quarter-section lines and corners permit the ready compilation of accurate property boundary line data and the ready maintenance of these data in a current form over time. The data can be readily and accurately updated and extended since all new land subdivisions must by law be tied to corners established in the U.S. Public Land Survey and since the accuracy of the surveys for these subdivisions can be readily controlled by state and local land subdivision regulations. The recommended survey control system thus fully meets the needs of a narrowly defined cadastre for the fiscal and legal administration of real property, yet, a cadastre which can be developed readily and soundly into a multipurpose land data system.

2. It provides a common system of control for the collection and mapping of both cadastral and earth science data. By relocating the U.S. Public Land Survey corners and accurately placing them on the State Plane Coordinate System, it becomes possible to accurately correlate real property boundary line information with earth science data. This placement of property boundary and earth science data on a common datum is absolutely essential to the sound development of any multipurpose land data system. Yet, such a common control datum is rarely used. The establishment of state plane coordinates for the U.S. Public Land Survey corners permits the correlation with mathematical precision of data supplied by aerial and other forms of earth science mapping with property boundary line data compiled through the usual land surveying methods. Only through such a common geometric control system can all of the information required for a multipurpose land data system be accurately collected and correlated.

3. It permits lines and areas entered into the data base—whether these lines represent the limits of land to be reserved for future public uses, the limits of land to be taken for immediate public use, the limits of districts to which the public regulations are to be applied, or the location and alignment of proposed new property boundary lines or of proposed constructed works—to be accurately and precisely reproduced upon the ground.

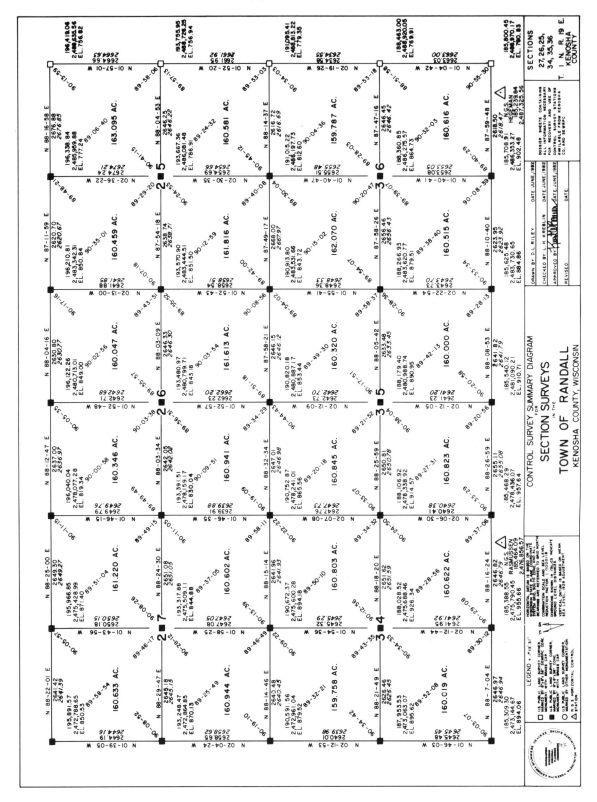

FIGURE 1. Control survey summary diagram. A geometric framework for land data systems.

U.S. PUBLIC LAND SURVEY CORNER $\frac{24|19}{24|19}$ T $\underline{1}$ N, R $\underline{18/19}$ E, KENOSHA COUNTY, WIS.
GEODETIC SURVEY BY: AERO-METRIC ENGINEERING, INC.   YEAR: 1980
STATE PLANE COORDINATES OF: QUARTER SECTION CORNER
  NORTH   198,197.35
  EAST   2,456,906.93
ELEVATION OF STATION: 830.066
HORIZONTAL DATUM: WISCONSIN STATE PLANE COORDINATE SYSTEM, SOUTH ZONE
VERTICAL DATUM: NATIONAL GEODETIC VERTICAL DATUM OF 1929
CONTROL ACCURACY: THETA ANGLE: +1-09-53
  HORIZONTAL: THIRD ORDER, CLASS I   VERTICAL: SECOND ORDER, CLASS II

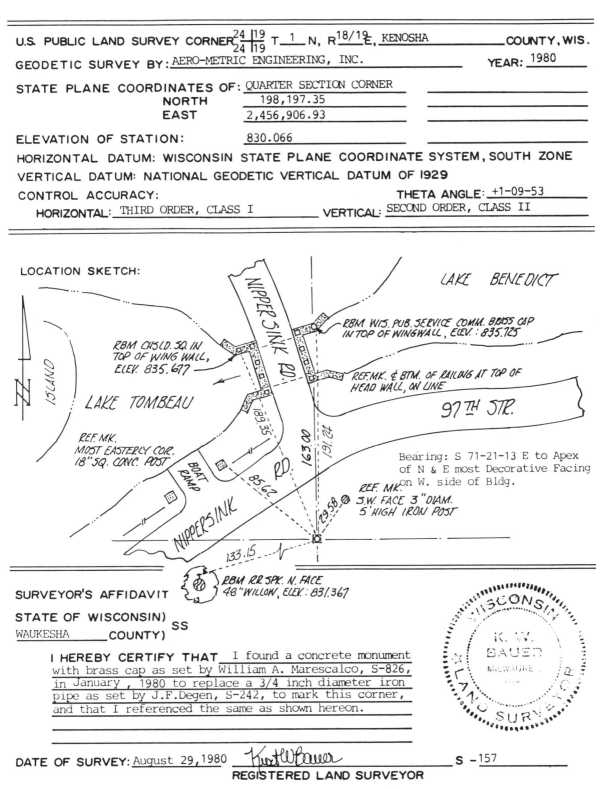

LOCATION SKETCH:

SURVEYOR'S AFFIDAVIT
STATE OF WISCONSIN) SS
WAUKESHA COUNTY)

I HEREBY CERTIFY THAT I found a concrete monument with brass cap as set by William A. Marescalco, S-826, in January, 1980 to replace a 3/4 inch diameter iron pipe as set by J.F.Degen, S-242, to mark this corner, and that I referenced the same as shown hereon.

DATE OF SURVEY: August 29, 1980   S-157
REGISTERED LAND SURVEYOR

FORM PREPARED BY SOUTHEASTERN WISCONSIN REGIONAL PLANNING COMMISSION

FIGURE 2. Record of U.S. Public Land Survey Control Station.

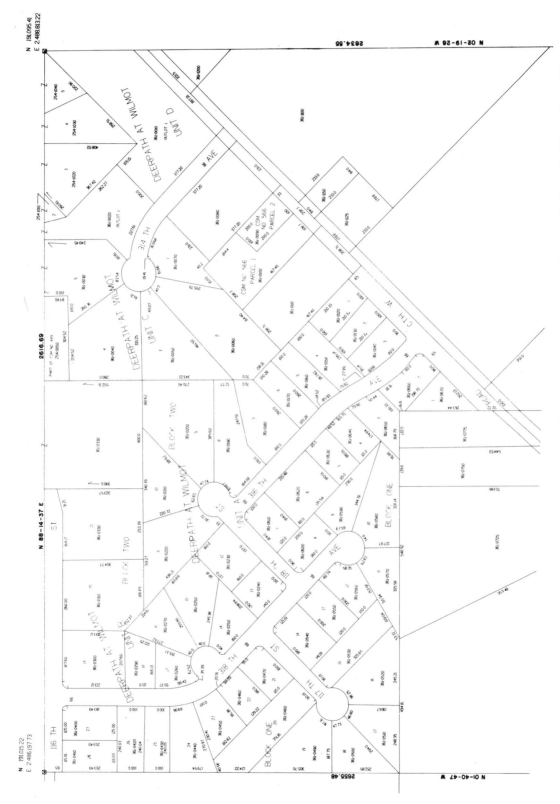

FIGURE 3. Machine plotted cadastral map properly referenced to geometric framework. Northeast one-quarter of section 36, Township 1 North, Range 19 East.

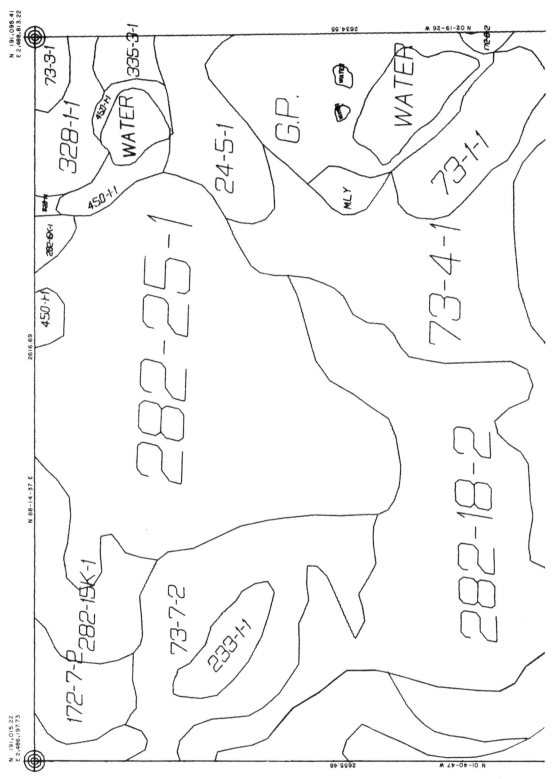

FIGURE 4. Machine plotted soils map properly referenced to geometric framework. Northeast one-quarter of section 36, Township 1 North, Range 19 East.

4. The system is readily adaptable to the latest survey techniques and is of relatively low cost as compared to alternate survey control systems that meet the criteria for use with a multipurpose land data system.

The specific geometric framework described herein is, of course, applciable only to those parts of the United States which have been covered by the U.S. Public Land Survey System. The fundamental concept involved—that is, the need to place both cadastral and earth science data on a common geometric base—is, however, applicable to any area. In those portions of the United States that have not been covered by the U.S. Public Land Survey System, the application of this concept may well be more difficult and costly, requiring the incremental placement of the corners of the individual real property boundaries on the State Plane Coordinate System, but is just as essential if a comprehensive land data system is to be created over time.

RECOMMENDED ACCURACY

Coordinates of the geometric framework should be entered into the computer readable land data base numerically through direct keypunch entry rather than graphically through any mechanical digitization procedures. If the coordinates are so entered to the nearest 1/100th of a foot, as provided by the recommended control surveys, the geometric framework becomes independent of any map scales and thereby provides a sound basis for the evolution of a multipurpose land data bank over time, being able to accept for accurate reference digitized data not only from mapped sources at any scale down to one to one but numeric data from direct field measurements. The recommendation is as important as it is subtle and should not be overlooked in the design of any land data system.

APPLICATION

The recommended system of survey control has been to date widely utilized within the Southeastern Wisconsin Region, a seven-county, 2,689 square mile planning area. One of the important reasons advanced for the application of this control system within the Region was to prepare the way for the eventual development of an automated land data system which would have multipurpose application. As of the end of 1986, 6,632, or about 56 percent, of the U.S. Public Land Survey corners in the Region have been relocated, monumented, and placed upon the State Plane Coordinate System in order to permit the development of large-scale topographic and cadastral maps for 1,281 square miles, or about 48 percent, of the total area of the Region.

Figure 1 illustrates a control survey summary diagram produced from the field surveys. This diagram, when numerically entered into the land data bank, forms the geometric framework for the spatial reference of all data in that bank, whether earth science or cadastral based. The stations in the framework are marked by monuments in the field and can, therefore, be recovered for use in all subsequent data collection and verification operations. Figure 2 illustrates a station dossier sheet, which provides the information necessary for the recovery and use of the control stations in subsequent data collection and verification efforts. Figure 3 illustrates a portion of a machine plotted cadastral map prepared from land data base records referenced to the recommended geometric framework. Figure 4 illustrates the machine produced detailed soils map also referenced to the same geometric framework. Because both the cadastral and earth science data illustrated in these two machine produced maps are referenced to the same geometric framework, the data can be fully coordinated graphically and numerically for use in analysis. Similar examples could be provided with other forms of data, including hydrographic, topographic, land use and cover, flood hazard, and wetland data, among others.

This survey network already forms the basis for a regional planning data bank in which all forms of comprehensive planning data—both socioeconomic and geophysical—are related to a common geographic framework at the one-quarter section level. It permits the ready and economic creation of a modern land title recordation system readily adaptable for the use of data, all of which can be overlaid on a parcel level map. Once the cadastral and earth science data are stored in a computer, the data are scale free; and maps can be plotted by machine from the data at any desired scale. The coordinate values can be readily transformed from one system to another, if necessary; and metrication can be accomplished by computer programming.

SUMMARY AND CONCLUSION

The development of dynamic land data systems that can be practically implemented initially as single-purpose cadastres providing information for the fiscal and legal administration of real property, but which can also be evolved efficiently into multipurpose land data banks providing information essential for comprehensive land use planning and management, will require careful planning and design. Such land data systems all ultimately depend for successful application upon some method of spatial reference for the various kinds of data involved. Indeed, the geometric framework for spatial reference is one of the essential factors upon which the ultimate success or failure of any land data system will depend. The necessary geometric framework should permit identification of land areas by coordinates down to the individual parcel level, while permitting the precise mathematical correlation of real property boundary and earth science data. By combining the two essentially unrelated survey conrol systems heretofore established in the United States by the federal government for real property boundary and earth science mapping, the necessary framework can be provided. This requires the relocation and monumentation of all the U.S. Public Land Survey corners within the geographic area for which the land data system is to be created and the utilization of these corners as stations in second order traverse and spirit level nets tied to the National Geodetic Datum. The traverse nets establish the true geographic positions of the U.S. Public Land Survey corners in the form of state plane coordinates, thereby providing a common system of control for the collection and coordination of both cadastral and earth science data. The monumented, coordinated corners, in turn, provide the basis for readily maintaining the data base in current condition, since all future surveys can be tied to these corners.

The specific geometric framework described herein is, of course, applicable only to those parts of the United States which have been covered by the U.S. Public Land Survey System. The fundamental concept involved—that is, the need to place both cadastral and earth science data on a common geometric basis—is, however, applicable to any area. In those portions of the United States that have not been covered by the U.S. Public Land Survey System, the application of this concept may well be more difficult and costly, requiring the incremental placement of the corners of the individual real property boundaries on the State Plane Coordinate System, but is just as essential if a comprehensive land data system is to be created over time.

The importance of the establishment of a sound geometric framework for land data systems is apt to be overlooked by planners and decision-makers as a technical detail in their deliberations of other important issues involved in the creation of such systems. The establishment of a sound geometric framework for land data systems is, however, a fundamental, as well as a major, undertaking which clearly will require much understanding, foresight, and commitment on the part of the technicians and decision-makers concerned. Failure to make the proper decisions concerning this basic foundation of any land data system during the formative period will jeopardize the future utility of the system, for reform will become increasingly costly and difficult over time.

CHAPTER 2

Plane Trilateration Adjustment by Finite Elements

NAGUIB F. DANIAL* AND MONEER F. TEWFIK*

INTRODUCTION

A plane trilateration net can be likened to a plane structural framework. In the former, distances are measured between ponts, while in the latter, elastic members are joined together by pins. Redundant measurements generally do not fit in the trilateration net due to measuring errors. Similarly redundant members do not fit in the framework due to fabrication errors. Forcing them into position will cause strains to develop in members, causing changes in their lengths. The corrections to the distances and the changes in length of the members are governed by minimum conditions, i.e., the weighted sum of the squares of distance corrections and the work done in connecting the redundant members to the framework are minimum. It is possible therefore to adjust trilateration nets by structural methods. Energy methods of structural analysis such as virtual work and finite element methods, can be adopted for this purpose. Structural analysis approach would provide the forces and distortions in members and the displacements of nodes for the equivalent structural framework. These are directly related to the corrections in the corresponding distances and the changes of coordinates of the points for the trilateration network.

In order to understand the applicability of the energy methods in the trilateration adjustments, the analogy between the virtual work method and the least squares method is explained in details [2].

METHOD OF VIRTUAL WORK

Case of One Redundant Measurement

One imagines that elastic members equal in length to the distances measured have been put together to form the

*University of Petroleum and Minerals, Dhahran 31261, Saudi Arabia

$n - 1$ triangles shown in Figure 1. Since the $(2n - 1)$ members are just sufficient to define the shape of the measured net, they will not carry any force. If line $P_1 P_n$ has been measured and found equal to L_{2n}, which is slightly different than the calculated distance $P_1 P_n$ difficulty will arise in placing this member in the previously built framework. To fit it between points P_1 and P_n a force, either tensile or compressive, must be applied depending on whether this member is shorter or longer than the calculated distance $P_1 P_n$. As a result, all members, including L_{2n} itself, will be stressed axially and will suffer changes in lengths. Said force must be exactly sufficient to make the deformed length of the redundant L_{2n} equals exactly the same value as obtained by calculations from all other deformed members.

The virtual work method can be conveniently applied to find this necessary force. The method is based on cutting the redundant member in the middle and applying a unit virtual (imaginary) load in both cut segments. The unit load will undergo a virtual external work W due to the displacement D in its direction. The external work done will therefore be

$$W_{ext} = 1\ D \tag{1a}$$

Due to this unit load, an internal axial force will be developed in each member m, which will undergo a virtual internal work due to the resulting change in its length.

$$W_{int} = \Sigma\ S_m\ \delta L_m \tag{1b}$$

in which, W_{int} = the internal work done by all members; S_m = the axial force in member m due to a unit load acting in the redundant; and δL_m = the change in length in member m caused by the force S:

$$\delta L_m = \frac{S_m\ L_m}{A_m\ E_m} \tag{2}$$

in which L_m = the length of member m as mentioned previ-

13

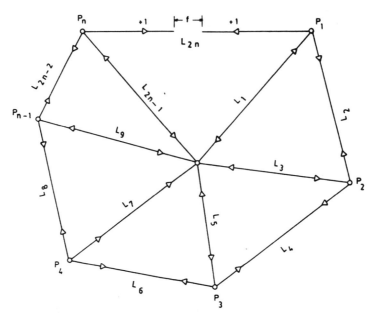

FIGURE 1. P_1P_n is chosen as redundant member in central figure $P_1P_2 \ldots P_n$: f is the closing error.

ously; A_m = its cross-sectional area; and E_m = its modulus of elasticity. According to the law of conservation of energy $W_{int} = W_{ext}$. Equating Equation (1a) with Equation (1b) and considering Equation (2) gives

$$D = \sum_m \frac{S^2 L}{AE} = \sum_m \frac{S^2}{k} \tag{3}$$

in which $k_m = A_m E_m / L_m$ = the axial stiffness of the member m.

In fitting the measured member L_{2n} in the calculated distance $P_1 P_n$ a gap (or overlap) f, which is the closing error, is created. If X is the necessary force needed to close this gap, the following equation may be written

$$XD + f = 0 \tag{4}$$

X may be obtained by combining Equations (3) and (4), from

$$X \sum_m \frac{S^2}{k} + f = 0 \tag{5}$$

Finally every member m will carry a force F_m which is the final force acting in it after closing the gap.

$$F_m = X S_m \tag{6}$$

Similar to Equation (2) one gets

$$\Delta L_m = \frac{F_m L_m}{A_m E_m} = X \frac{S_m L_m}{A_m E_m} = X \frac{S_m}{k_m} \tag{7}$$

in which ΔL_m = the final change in length of member m after closing the gap.

Case of More than One Redundant Observation

The previous review was limited to the case of one redundant only. The same procedure can be applied to any number r of redundant observations. In this case all redundant members are cut and unit virtual loads are applied in each of them. The forces $S_{1m}, S_{2m}, \ldots S_{rm}$ due to these unit loads are calculated for all members m. A set of r displacement condition equations corresponding to the r redundants can be formulated similar to Equation (4):

$$\begin{aligned} D_{11} X_1 + D_{12} X_2 + \ldots D_{1r} X_r + f_1 &= 0 \\ D_{21} X_1 + D_{22} X_2 + \ldots D_{2r} X_r + f_2 &= 0 \\ &\cdots\cdots\cdots\cdots\cdots\cdots \\ D_{r1} X_1 + D_{r2} X_2 + \ldots D_{rr} X_r + f_r &= 0 \end{aligned} \tag{8}$$

in which D_{ij} = the change in length in redundant i due to a unit virtual force in redundant j:

$$D_{ij} \sum_m \frac{S_{im} S_{jm}}{k_m} \qquad (9)$$

where S_{im} = the force in the uncut member m due to a unit virtual load in redundant i; $f_1, f_2, \ldots f_r$ = the gaps created in the cut members $1, 2, \ldots r$; $X_1, X_2, \ldots X_r$ = the forces in the redundants after all gaps have been closed. They can be found by solving the set of linear simultaneous algebraic Equations (8).

The final force F_m in any member can be computed from

$$F_m = S_{1m} X_1 + S_{2m} X_2 + \ldots + S_{rm} X_r \qquad (10)$$

which will cause a final change in length

$$\Delta L_m = \frac{F_m L_m}{A_m E_m} = \frac{S_{1m}}{k_m} X_1 + \frac{S_{2m}}{k_m} X_2 + \ldots + \frac{S_{rm}}{k_m} X_r \qquad (11)$$

SIMILARITY BETWEEN THE LEAST SQUARES AND THE VIRTUAL WORK METHODS

As mentioned before there is an apparent similarity between a framework of members and joints and a trilateration net of measured sides and points. The relationships can be established as follows.

Relationship between Minimum Work and Minimum Sum of Squares of Corrections

The work W done in fitting the redundant members in their position is minimum. Therefore $W = \Sigma F_m \Delta L_m$ = min. Substituting ΔL_m by their values as given in Equation (7) and (11) gives

$$W = \Sigma \frac{F^2 L}{AE} \qquad (12)$$

which can be written

$$W = \Sigma \left[\left(\frac{FL}{AE} \right)^2 \left(\frac{AE}{L} \right) \right] = \Sigma [(\Delta L_m)^2 k_m] \qquad (13)$$

If the change in length ΔL_m is replaced by the correction v_m and the stiffness k_m of the member m is replaced by the weight w_m of the observation L_m, Equation (13) will become

$$\Sigma(v^2 w) = \text{min.} \qquad (14)$$

which is the basic formula for the method of least squares.

Relationship between Forces in Members Due to Unit Loads Corresponding to Redundants and Coefficients of Condition Equations

Substituting Equation (6) into Equation (4) and considering Equation (3) gives

$$\Sigma S \frac{FL}{AE} + f = 0 \qquad (15)$$

for the case of one redundant, which may be written

$$\Sigma S \, \Delta L + f = 0 \qquad (16)$$

In the case of r redundants one obtains

$$\begin{aligned} S_{11} \Delta L_1 + S_{12} \Delta L_2 + \ldots + S_{1n} \Delta L_n + f_1 &= 0 \\ S_{21} \Delta L_1 + S_{22} \Delta L_2 + \ldots + S_{2n} \Delta L_n + f_2 &= 0 \\ &\cdots \\ S_{r1} \Delta L_1 + S_{r2} \Delta L_2 + \ldots + S_{rn} \Delta L_n + f_r &= 0 \end{aligned} \qquad (17)$$

Replacing ΔL_m by the corrections v_m as mentioned previously, and the forces S_{jm} by coefficients a_{jm} gives

$$\begin{aligned} a_{11} v_1 + a_{12} v_2 + \ldots + a_{1n} v_n + f_1 &= 0 \\ a_{21} v_1 + a_{22} v_2 + \ldots + a_{2n} v_n + f_2 &= 0 \\ &\cdots \\ a_{r1} v_1 + a_{r2} v_2 + \ldots + a_{rn} v_n + f_r &= 0 \end{aligned} \qquad (18)$$

which are the known condition equations.

In conclusion, the force S_{jm} developed in the member m due to a unit virtual load in the redundant member j is the coefficient a_{jm} of the correction of the corresponding measured line m due to the distance condition j.

Relationship between Forces in Redundant Members and Correlates

If the observations $L_1, L_2, \ldots L_n$ have been measured with the standard deviations $\sigma_1, \sigma_2, \ldots, \sigma_n$, weights $w_1, w_2, \ldots, w_n$ could be assigned to them such that

$$w_m = \frac{1}{\sigma_m^2} \qquad (19)$$

If the stiffness k_m of the members m are replaced by the weights w_m of the measured distances and the forces in the redundants $X_1, X_2, \ldots X_r$, are replaced by the correlates

$C_1, C_2, \ldots, C_r$, Equations (8) become

$$\left[\frac{a_1a_1}{w}\right]C_1 + \left[\frac{a_1a_2}{w}\right]C_2 + \ldots + \left[\frac{a_1a_r}{w}\right]C_r + f_1 = 0$$

$$\left[\frac{a_1a_2}{w}\right]C_1 + \left[\frac{a_2a_2}{w}\right]C_2 + \ldots + \left[\frac{a_2a_r}{w}\right]C_r + f_2 = 0$$

$$\left[\frac{a_1a_r}{w}\right]C_1 + \left[\frac{a_2a_r}{w}\right]C_2 + \ldots + \left[\frac{a_ra_r}{w}\right]C_r + f_r = 0$$

(20)

which are the known normal equations. It is therefore clear that the correlates $C_1, C_2, \ldots, C_r$ are exactly the axial forces $X_1, X_2, \ldots, X_r$ need to be applied on the redundant members to close the gaps created in them by the erroneous measurements.

Relationship Between Deformation ΔL_m and Corrections v_m

The correction v_m of a member m is the final change ΔL_m in its length. Substituting the previously shown relationships in Equation (11) results in

$$v_m = \frac{a_{1m}}{w_m}C_1 + \frac{a_{2m}}{w_m}C_2 + \ldots + \frac{a_{rm}}{w_m}C_r \quad (21)$$

Equivalent expressions relating structural terms to surveying terms are provided in Table 1.

The virtual work method of structural analysis has led to

TABLE 1. Equivalent Expressions.

Structure (1)	Surveying (2)
Framework	Net
Plane Truss	Plane Trilateration Net
Node, Joint	Point
Member, Rod Element	Distance
Distortion, Misfit	Misclosure, f
Stiffness, k	Weight, w
X Coordinate	E Coordinate, Easting
Y Coordinate	N Coordinate, Northing
Modulus of Elasticity, E	Distance, Length L
Cross-section Area, A	Weight, w
Support, S	Fixed Point
Member distortion, ΔL	Length Correction, v
Joint Displacements ΔX and ΔY	Coordinate Corrections ΔE and ΔN

Equations (14), (17), (20), (21) which are the basic formulas in the method of least squares. In general, energy methods based on the minimum principle can be applied to any adjustment problem. One such method, which is readily available and very commonly used, is the finite element method. This method can handle axially loaded as well as bending members and elements for two and three dimensional structures. The part that includes axially loaded planer frameworks is briefly presented in the following.

FINITE ELEMENT METHOD (FEM)

The finite element method, like many other numerical methods, is a discretization procedure. A continuum with infinite number of degrees of freedom is replaced by a model that simulates the real case, but has only a finite number of degrees of freedom. The model is an assemblage of discrete elements which are connected to each other only at node points.

Consider, e.g., a rod element of length L, cross section area A, and elastic modulus E, as shown in Figure 2. The axial forces S_1 and S_2 that are acting at the nodal points 1 and 2, respectively, will cause axial displacements u_1 and u_2 of these points. Employing the principle of minimum potential energy (or any other variational principle like virtual work, etc.), will result in the following force-displacement relationship for the element (6):

$$\begin{Bmatrix} S_1 \\ S_2 \end{Bmatrix} = \frac{AE}{L}\begin{bmatrix} 1 & -1 \\ -1 & 1 \end{bmatrix}\begin{Bmatrix} u_1 \\ u_2 \end{Bmatrix} \quad (22)$$

If the forces S_1 and S_2 applied to the rod are known, then one can calculate the resulting displacements u_1 and u_2 from Equation (22) and vice versa. Equation (22) can be written in a general form as follows:

$$\{S\} = [k]\{\delta\} \quad (23)$$

in which $\{S\}$ = element force vector, all forces acting on the element nodes; $\{\delta\}$ = element displacements vector, displacements of the element nodes; and $[k]$ = element stiffness matrix, incorporating the material and geometrical properties of the element. The assembly of all elements to form the particular model under consideration, results in obtaining a set of equations which have the following form:

$$\{F\} = [K]\{\Delta\} \quad (24)$$

in which $\{F\}$ = generalized nodal force vector for entire model; $\{\Delta\}$ = generalized nodal displacement vector for entire model; and $[K]$ = assembled stiffness matrix for en-

FIGURE 2. Forces and displacements of nodes of rod element.

tire model. Equation (24) is lengthy, especially for large frameworks. Therefore, for such cases, the manual solution is not recommended. A tremendous gain in time and accuracy is achieved when a finite element program is employed for this purpose [4]. Many of these programs are available commercially, such as STRUDL, NASTRAN, ASKA, SAP, MARK, FINITE, and others. The program STRUDL is chosen here to demonstrate the application of this method in adjusting trilateration nets, because it is versatile and quite commonly used by civil engineers.

STRUDL

STRUDL is a problem oriented STRUctural Design Language. It is one of the most commonly used structural analysis and design computer packages. The program can perform several types of analysis on many types of structures like plane and space trusses and frames, plane grids, plate bending and shells [5]. Each input has a certain syntax and its order in the program is usually free, but in some parts the order should be logical. The input consists of the type of the structure, geometry of the structure (in terms of joint coordinates, and member/element incidences), member/element properties, and loading (forces, temperature and distortions).

The output commands are simple expressions. They may instruct the computer to print out one, several or all member forces, member distortions, and joint displacements.

DATA PREPARATION FOR STRUDL

There are several computations to be performed before the measurements can be introduced as input data. All these computations are routine, and familiar to a person with an average knowledge in surveying. The computational steps are as follows.

Number of Redundants

The first step in preparing the data is to determine the number of the redundant measurements. This is reached by calculating the minimum number of distances needed to locate the position of the different points. Since the first three distances in a trilateration net define the location of three points, and any additional pair of distances locate a new point, the following relationship can be used:

$$M = 2I - 3 \quad (25)$$

in which M = the minimum number of distances; and I = the number of points. The number of the redundants r, is the difference between the number of all measurements n, and the minimum number M. Thus:

$$r = n - M = n - 2I + 3 \quad (26)$$

Supports

In order to apply STRUDL program the plane trilateration net must be replaced by a plane truss with members equal in length to the measured distances. The plane truss needs to be stable. This can be provided if the truss is made of triangles, which is the case of a trilateration net. Also at least two joints (points) are restrained with one or both of them prevented from movement (fixed point). If fixed points are given they will provide the necessary supports. Otherwise any two points, preferably but not necessarily the end points of one member, can be chosen for this purpose. Taking the end points of one member as supports will facilitate the calculation of the initial coordinates as will be shown in the numerical examples.

Choice of Redundants and Calculation of Initial Coordinates

The application of the finite element method requires the calculation of the initial coordinates of all points. If no known points are given then the coordinates of one support and the azimuth of the line joining both supports must be assumed.

Unlike the method of observation equations, in which approximate coordinates are used, the initial coordinates in the finite element method must be calculated exactly from an initial figure which is based on the measured distances, excluding the redundants. Any distance, or distances, can be chosen as redundant. It is preferable however, to choose the redundants in such a way that the calculation of the initial coordinates is convenient. This can best be demonstrated for the case of the central figure, as examined herein.

Figures 3a and 3b show a central figure $P_1P_2P_3P_4P_5$, with the center at P_0, in which the perimeter side $P_1 P_5$ is chosen as a redundant. The coordinates of the central point P_0, and the azimuth of the line P_0P_1, are either given or assumed. The initial coordinates of all points can be calculated either from the radial lines (Figure 3a) or from a closed traverse $P_0P_1P_2P_3P_4P_5P_0$ (Figure 3b). The first solution requires the computation of the four central angles, γ_1, γ_2, γ_3, and γ_4. The initial coordinates of all points can be checked by

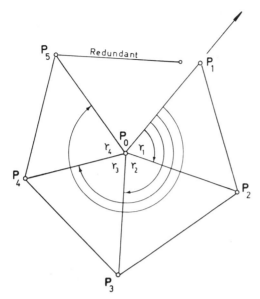

FIGURE 3a. Central figure with perimeter side as redundant: initial coordinates of points $P_1, P_2, \ldots P_5$ are calculated from lengths and directions of radial lines $P_0P_1, P_0P_2, \ldots, P_0P_5$.

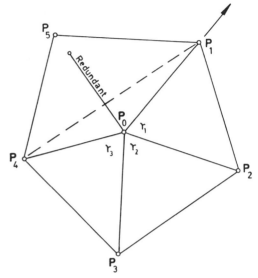

FIGURE 4. Central figure with radial side as redundant: initial coordinates of point P_5 are calculated by intersection of lines—P_4P_5 and P_1P_5—redundant radial line P_0P_5 is not used in these calculations.

calculating the perimeter distances P_1P_2, P_2P_3, P_3P_4, and P_4P_5. They must be exactly equal to their measured values. The second solution requires the calculation of the eight perimeter angles α_1, β_1, α_2, β_2, α_3, β_3, α_4, and β_4. The mathematical check is obtained directly if the calculated

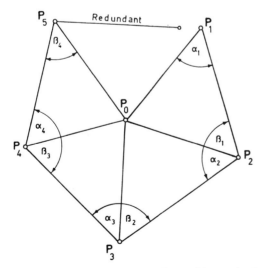

FIGURE 3b. Central figure with perimeter side as redundant: initial coordinates of points $P_1, P_2, \ldots, P_5$ are calculated from closed traverse $P_0P_1P_2P_3P_4P_5P_0$.

coordinates of point P_0 are the same as the assumed coordinates of this point at the beginning. It should be noticed that the triangle $P_5P_0P_1$, which has the redundant member P_5P_1 as a side, is not used for the computation of any of the aforementioned angles.

Figure 4 shows the same central figure, but the radial member P_0P_5 is chosen as redundant. The initial coordinates of points P_1, P_2, P_3, and P_4 can be obtained by either one of the previously mentioned procedures. To determine the coordinates of point P_5, however, neither of the triangles $P_4P_0P_5$ and $P_5P_0P_1$ can be used since they contain the redundant member P_0P_5 as a side. To overcome this difficulty, the distance P_1P_4 must be calculated from triangle $P_4P_0P_1$ as follows:

$$P_1P_4 = \left[\overline{P_1P_0}^2 + \overline{P_4P_0}^2 - 2 \times \overline{P_1P_0} \times \overline{P_4P_0} \right.$$
$$\left. \times \cos\left(360 - \sum_1^3 \gamma\right) \right]^{1/2} \quad (27)$$

Triangle $P_1P_4P_5$ can now be solved and the coordinates of point P_5 can be obtained by intersection of the two distances P_4P_5 and P_1P_5.

The calculation of the initial coordinates is apparently longer when a radial member is taken as a redundant, instead of a perimeter member in a central figure. The same principle applies to complicated nets. A proper choice of the redundants will reduce the amount of calculations.

Misclosures

The distance between any two points at which a redundant member is to be connected can be calculated from the already computed initial coordinates of these points. The calculated length generally differs slightly from the measured length. The difference, f, computed in the sense "measured minus calculated," is called misclosure or misfit.

Weights

From the similarity between the method of least squares and the method of minimum work, as explained previously, the following relationships are obtained:

$$w = k = \frac{AE}{L} \quad (28)$$

and

$$v = \Delta L = \frac{FL}{AE} \quad (29)$$

The stiffness of each member k is calculated by the computer program from the given values of the modulus of elasticity E, the cross section area A, and the length L, which is computed from the initial coordinates of the nodal points. The orientation of each member with respect to X (East) and Y (North) axes, is also taken into consideration. To be able to introduce the weight with the data, the modulus of elasticity of every member is assumed to be the value of its length. Thus:

$$E = L \quad (30)$$

Substituting in Equation (28) gives

$$A = w \quad (31)$$

In other words, the weight w can be introduced in the data instead of the cross section area A, if the modulus of elasticity E is replaced by the length of the member L.

For cases in which the weights of all distances are equal, it is assumed that

$$A = 1 \quad (32)$$

For the cases when the weights are inversely proportional to the measured lengths, i.e., $w = 1/L$, the following relationship is obtained from Equation (28):

$$AE = 1 \quad (33)$$

For programs which do not give member distortions directly but give member forces instead, care should be taken when Equation (33) is used. By substituting Equation (33) into Equation (29) one gets

$$F = \frac{\Delta L}{L} \quad (34)$$

Since ΔL is in the order of millimeters, and L in the order of multiples of 10 m, the ratio $\Delta L/L$ is quite small. Therefore, the application of Equation (33), as shown, may lead to the loss of significant figures in the printout for the axial forces F, from which the corrections v are calculated. If, however, the ratio AE/L is chosen to be in the order of 1, then the forces F will become of the same order as the distortions (see Example 1). It is therefore recommended to take

$$AE = 10^n \quad (35)$$

in which n = an integer such that the product AE will be of the same order as the distances.

INPUT

The input format of STRUDL program is presented in Tables 4, 7, 12, and 14. The sequence of data entry is as follows.

TYPE PLANE TRUSS

JOINT COORDINATES

$$\begin{array}{cccc} & X & Y & \\ i & E_i & N_i & S \end{array} \quad (36)$$

in which E_i and N_i = initial or given easting and northing of point i, for $i = 1, 2, \ldots I$, and S is typed only if point i is a support.

MEMBER INCIDENCES

$$j \quad G_j \quad H_j \quad (37)$$

in which G_j and H_j = numbers of first and second end points of distance j, and are also defined as node 1 and node 2 of the member. $j = 1, 2, \ldots n$.

MEMBER PROPERTIES

If all distances have the same weight then the command is

$$1\ 2\ 3 \ldots n \text{ PRISMATIC AX } 1 \quad (38a)$$

or

MEMBER 1 TO n PROPERTIES PRISMATIC AX 1

(38b)

If the distances have different weights, the property of each member must be entered on a separate line

$$j \quad \text{PRI} \quad \text{AX} \quad (w_j) \quad (39)$$

$j = 1, 2, \ldots, n$, and w_j is the weight of member j.

CONSTANTS

$$E \quad L_j \quad j \quad (40)$$

in which L_j is the measured length of member j. $j = 1, 2, \ldots n$. It is also possible to write several distances in one line like, for example, for members 1, 2, and 3:

$$E \quad 227.644 \quad 1 \quad 154.965 \quad 2 \quad 205.732 \quad 3$$

Every new line must start with E.

If the weights are inversely proportional to the lengths of the distances the MEMBER PROPERTIES command should be the same as in Equation (38) and the CONSTANTS command can be written as follows

$$E \quad 10^n \quad \text{ALL} \quad (41)$$

in which n is an integer such that AE will be of the same order as the distances.

LOADING 1 "MISFIT"

The framework is loaded by applying either forces and/or displacements (i.e., distortions of rod elements). In trilateration nets, however, distortions occur only along element axes. The input information is therefore provided for each redundant as follows.

MEMBER DISTORTIONS

$$(r) \text{ CONCENTRATED FRACTIONAL } (N) \text{ DISPLACEMENT } (f_r) \quad (42)$$

in which r = the number of the redundant member; N = location along rod where distortion is imposed; f_r = magnitude of distortion (misclosure) of member r. $r = 1, 2, \ldots, r$. Since the distortion is axially in pure trilateration, N can be imposed at any point along the redundant. It can therefore assume any value between zero and one.

JOINT RELEASES

Supports representing fixed points are not allowed to move. In free trilateration nets, however, one of the two chosen supports must be allowed to move in the direction of the line joining them so that their relative location can also be adjusted. If this direction coincides with the X-axis the command is

$$i \quad \text{FORCE} \quad X \quad (43a)$$

where i is the number of support to be released. It takes the following form if the direction of movement is along the Y-axis.

$$i \quad \text{FORCE} \quad Y \quad (43b)$$

If the direction of movement makes a counterclockwise angle ϑ with respect to the X-axis (see Figure 5), then the command is

$$i \quad \text{FORCE} \quad X \quad \text{TH1} \quad (\vartheta) \quad (43c)$$

The same result is obtained if the direction is given counterclockwise with respect to the Y-axis.

$$i \quad \text{FORCE} \quad Y \quad \text{TH1} \quad (270 + \vartheta) \quad (43d)$$

Since the azimuths (Az.) in surveying are measured clockwise from the north (Y-axis) direction, the command may be written

$$i \quad \text{Force} \quad Y \quad \text{TH1} \quad (-\text{Az.}) \quad (43e)$$

It should be noted here that the azimuth is introduced

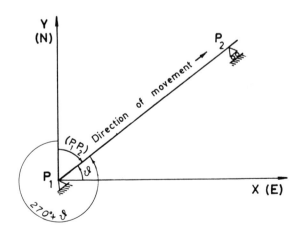

FIGURE 5. Geometrical representation of directional constraint on joint releases.

negative as calculated from the fixed support to the movable one. Also STRUDL program reads any angle in radians unless other units are defined earlier. To be able to introduce the azimuth in decimal degrees the following command

$$\text{UNITS DEGREES} \qquad (44)$$

must be written before the command of JOINT RELEASES.

OUTPUT INFORMATION

Member Distortions

Member distortions can be obtained directly by using the following command

$$\text{LIST DISTORTIONS ALL} \qquad (45)$$

For the case of a truss the only distortion is the axial change in length of the member. This would give the correction of the measurement v_j = axial distortion of member j = ΔL_j.

Member Forces

It is possible and simple to calculate the distortions of the members from the axial forces in them. These can be listed in the output by using the following command:

$$\text{LIST FORCES ALL} \qquad (46)$$

These axial forces can either be positive (tensile) or negative (compressive) as they act on the joint printed opposite to them. The change ΔL in the length of each element j can be obtained from the force acting in it. Considering this change in length as the correction v_j of the distance j, it is possible to obtain from Equations (28) and (29)

$$v_j = \frac{F_j}{A_j E_j / L_j} = \frac{F_j}{w_j} \qquad (47)$$

In other words, the correction of a distance and its algebraic sign is obtained by dividing the force in the corresponding member, as obtained from the output, by its weight.

Resultant Joint Displacements

The nodal displacements ΔX_i and ΔY_i in the X (East) and Y (North) directions, respectively, are listed for each nodal point in the model by using the command

$$\text{LIST DISPLACEMENTS ALL} \qquad (48)$$

They represent the corrections ΔE_i, and ΔN_i of the point i and are of course zeros if point i is a fixed point. Adding these corrections to the initial coordinates will give the final adjusted eastings and northings of all points.

$$E_i = E_i' + \Delta E_i = X_i + \Delta X_i \qquad (49a)$$

$$N_i = N_i' + \Delta N_i = Y_i + \Delta Y_i \qquad (49b)$$

NUMERICAL EXAMPLES

The application of the finite element method to adjust trilateration nets is demonstrated by the following four examples, which practically cover all types of trilateration problems. No references to the foot or meter are given, since the computations are valid for any linear unit.

Example 1: Multiple Section of Arc.

The four distances L_1, L_2, L_3, and L_4 are measured from the known and fixed points P_1, P_2, P_3, and P_4, respectively, to the new point P_5 (Figure 6a). The results of the measurements are recorded in Table 2. The coordinates of the given points are presented in Table 3. It is required to find the best coordinates of point P_5, knowing that the weights of the measured distances are inversely proportional to their lengths.

Solution: (computer input and output in Table 4).

NUMBER OF REDUNDANTS

The minimum number of observations needed to determine the location of point P_5 is two by inspection. The two additional measurements are redundants.

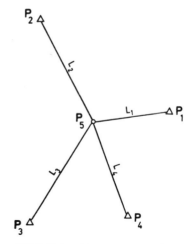

FIGURE 6a. Multiple section of arc.

TABLE 2. Measured and Adjusted Distances (Example 1 and Table 4).

	Inut		Output		Further Calculations	
No. (1)	Length $E = L$ (2)	Weight $A = w$ (3)	Axial Force F (4)	Member Distortion $\Delta L = v$ (5)	Correction $v = F/A$ (6)	Corrected Length $L + v$ (7)
1	111.08	0.90	+0.0046	+0.0051	+0.0051	111.085
2	158.20	0.63	−0.0184	−0.0292	−0.0292	158.171
3	171.40	0.58	+0.0063	+0.0108	+0.0109	171.411
4	124.49	0.80	−0.0230	−0.0287	−0.0287	124.461

Output values are rounded.

TABLE 3. Initial or Given Coordinates and Displacements of Points after Adjustment (Example 1 and Table 4).

	Input			Output	
	Given and Initial Coordinates			Joint Displacements	
Point (1)	X (E') (2)	Y (N') (3)	Support (4)	ΔX (ΔE) (5)	ΔY (ΔN) (6)
1	262.270	306.080	S	0.0	0.0
2	91.880	435.890	S	0.0	0.0
3	56.830	147.450	S	0.0	0.0
4	196.320	173.300	S	0.0	0.0
5	152.401	289.724		−0.0093	0.0277

Output values are rounded.

TABLE 4.

```
STRUDL 'TRILATERATION', 'EXAMPLE 1'

*************************************************
*                                               *
*              ICES STRUDL-II                   *
*         THE STRUCTURAL DESIGN LANGUAGE        *
*                                               *
*           IUG VERSION V3M1,  MAY  1977        *
*             SIZE OF POOL  8K BYTES            *
*                 8:47:44     5/04/85           *
*                                               *
*************************************************

TYPE PLANE TRUSS

JOINT COORDINATES

1   262.270   306.080   S

2    91.880   435.890   S

3    56.830   147.450   S

4   196.320   173.300   S

5   152.401   289.724

MEMBER INCIDENCES

1   1   5

2   2   5

3   3   5

4   4   5
```

(continued)

TABLE 4. (continued).

```
        MEMBER PROPERTIES         ┌ ─ ─ ─ ─ ─      alternate input
                                  │
        1  PRISMATIC  AX 0.90     │
                                  │  MEMBER 1 TO 4 PROPERTIES PRISMATIC AX 1
        2  PRI        AX 0.63     │  CONSTANTS E 130  ALL
                                  │
        3  PRI        AX 0.58     │
                                  │
        4  PRI        AX 0.80     │
                                  │
        CONSTANTS                 │
                                  │
        E  111.08  1    156.20  2 │
                                  │
        E  171.40  3    124.49  4 │
                                  └ ─ ─ ─ ─ ─
        LOADING 1 'MISFIT'
        MEMBER DISTORTIONS
        3  CONCENTRATED FRACTIONAL 1.0 DISPLACEMENT +0.007
        4  CONCENTRATED FRACTIONAL 1.0 DISPLACEMENT +0.058
        STIFFNESS ANALYSIS
        LIST DISTORTIONS ALL
        LIST DISPLACEMENTS ALL
        LIST FORCES ALL
******************************
*RESULTS OF LATEST ANALYSES*
******************************
JOB ID -  TRILATER     JOB TITLE - EXAMPLE 1
ACTIVE UNITS -  LENGTH        FORCE         ANGLE      TEMPERATURE      TIME
                INCH          LB            RAD        DEGF             SEC
ACTIVE STRUCTURE TYPE - PLANE    TRUSS
ACTIVE COORDINATES AXES  X Y

***************************************************************************
*   LOADING - 1            MISFIT
***************************************************************************

     MEMBER  DISTORTIONS

MEMBER            /—————————————— DISTORTION L——————————————//——————————
                     AXIAL           SHEAR Y         SHEAR Z         TORSION

1                  0.0051487
2                 -0.0292038
3                  0.0108238
4                 -0.0287477

        RESULTANT JOINT DISPLACEMENTS - SUPPORTS

   JOINT            /—————————————— DISPLACEMENT ——————————/
                     X DISP.         Y DISP.         Z DISP.

1        GLOBAL       0.0             0.0
2        GLOBAL       0.0             0.0
3        GLOBAL       0.0             0.0
4        GLOBAL       0.0             0.0

        RESULTANT JOINT DISPLACEMENTS - FREE JOINTS

   JOINT            /—————————————— DISPLACEMENT ——————————/
                     X DISP.         Y DISP.         Z DISP.

5        GLOBAL      -0.0093354       0.0277428

        MEMBER  FORCES

MEMBER   JOINT     /—————————————— FORCE —————————————//——————————
                     AXIAL           SHEAR Y         SHEAR Z         TORSIONAL

1        5          0.0046338
2        5         -0.0183984
3        5          0.0062781
4        5         -0.0230088
```

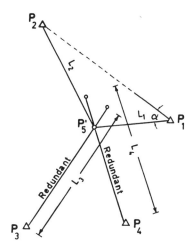

FIGURE 6b. Chosen redundants in an equivalent framework.

INITIAL COORDINATES

To compute the initial coordinates for P_5, any two of the measured four distances can be used. The distances L_1 and L_2 are chosen here for this purpose. The angle $P'_5 P_1 P_2 = \alpha$ (Figure 6b) can be calculated by the cosine law.

$$\alpha = \cos^{-1} \frac{\overline{P_1 P_2}^2 + L_1^2 - L_2^2}{2 \times \overline{P_1 P_2} \times L_1} = 45.7689°$$

The azimuth of line $P_1 P_2 = (P_1 P_2) = \tan^{-1}[(E_2 - E_1)/(N_2 - N_1)] = 307.3016°$.

The azimuth of line $P_1 P_5 = (P_1 P_5) = (P_1 P_2) - \alpha = 261.5327$.

The initial coordinates of point P'_5 are $X_5 = E'_5 = E_1 + L_1 \sin(P_1 P_5) = 262.27 + 111.08 \sin(261.5327) = 152.401$. $Y_5 = N'_5 = N_1 + L_1 \cos(P_1 P_5) = 306.08 + 111.08 \cos(261.5327) = 289.724$.

MISCLOSURES

The distances between this initial location (P'_5) and both points P_3 and P_4 are calculated from the coordinates and are found to be 171.393 and 124.432, respectively. The differences between the measured and the calculated lengths are the misclosures $f_1 = L_3 - P_3 P'_5 = 171.40 - 171.393 = +0.007$; $f_2 = L_4 - P_4 P'_5 = 124.49 - 124.432 = +0.058$.

No further calculations are required for the input. The input data for the program is introduced as mentioned previously and shown in Table 4.

DISTANCE CORRECTIONS

The output gives both the axial forces and member distortions. The corrections of the measured distances can be obtained either by dividing the resulting axial forces by the corresponding weights or directly as the obtained member distortions. They are listed together with the corrected lengths in Columns 5 and 6 of Table 2.

FINAL COORDINATES

The final coordinates of point P_5 are obtained by adding to the initial coordinates $X'_5 (= E'_5)$ and $Y'_5 (= N'_5)$ the joint displacements $\Delta X_5 = 0.009$ and $\Delta Y_5 = +0.028$, respectively, as obtained from the output and rounded to three decimals.

$$E_5 = X_5 + \Delta X_5 = 152.401 - 0.009 = 152.392$$

$$N_5 = Y_5 + \Delta Y_5 = 289.724 + 0.028 = 289.752$$

Note, since the weights in this problem are inversely proportional to the lengths, the MEMBER PROPERTIES and CONSTANTS commands and values may be replaced by the commands (see "alternate input" in Table 4).

MEMBER 1 TO 4 PROPERTIES PRISMATIC AX 1

CONSTANTS E 100 ALL

Example 2: Trilateration Net with a Fixed Point and a Given Direction

The nine distances shown in Figure 7a have been measured with the same accuracy and are provided in Table 5. Approximate coordinates were also given for all points, but it was required to maintain the coordinates of point P_1 and the azimuth $(P_1 P_2)$, as calculated from the approximate coordinates of points P_1 and P_2, unchanged during the process of adjustment. The given approximate coordinates are:

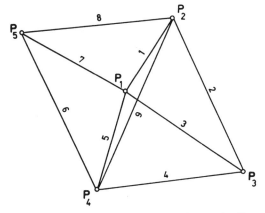

FIGURE 7a. Trilateration net with one fixed point P_1 and one fixed direction $(P_1 P_2)$.

TABLE 5. Measured and Adjusted Distances (Example 2 and Table 7).

	Input		Output	Computations
No. (1)	Length $E = L$ (2)	Weight $A = w$ (3)	Member Distortion $\Delta L = v$ (4)	Adjusted Length $L + v$ (5)
1	11,679.756	1	−0.0086	11,679.747
2	22,182.161	1	−0.0126	22,182.148
3	17,164.398	1	+0.0179	17,164.416
4	18,802.410	1	−0.0097	18,802.400
5	13,169.814	1	−0.0102	13,169.804
6	22,451.497	1	−0.0106	22,451.486
7	16,025.191	1	+0.0167	16,025.208
8	19,134.584	1	−0.0101	19,134.574
9	24,816.476	1	+0.0223	24,816.498

Output values are rounded.

for P_1: $E_1 = 3550295.83$, $N_1 = 5804022.95$; for P_2: $E_2 = 3555548.31$, $N_2 = 5814454.98$.

Solution: (Computer input and output in Table 7)

NUMBER OF REDUNDANTS

By Equation (25), the minimum number of distances required to form the basic figure is

$$M = 2 \times 5 - 3 = 7$$

Since the total number of measurements is nine, the number of redundant measurements is therefore two.

INITIAL COORDINATES

Members 8 and 9 are chosen as redundant members and are removed (Figure 7b). Accordingly, the basic figure consists of three triangles: $P_1P_2P_3$, $P_1P_3P_4$, and $P_1P_4P_5$. The central angles γ_1, γ_2, and γ_3 in these triangles are calculated by the cosine law. They are found to be $\gamma_1 = 98.753013°$, $\gamma_2 = 75.325548°$, and $\gamma_3 = 100.072026°$.

The azimuth of line P_1P_2 is $(P_1P_2) = \tan^{-1}[(E_2 - E_1)/(N_2 - N_1)] = 26.725048°$.

The azimuths of lines P_1P_3, P_1P_4, and P_1P_5 can now be calculated by adding the corresponding central angle or angles to (P_1P_2) as follows:

$$(P_1P_3) = (P_1P_2) + \gamma_1 = 125.478061°$$

$$(P_1P_4) = (P_1P_2) + \gamma_1 + \gamma_2 = 200.803609°$$

$$(P_1P_5) = (P_1P_2) + \gamma_1 + \gamma_2 + \gamma_3 = 300.875635°$$

Substituting these azimuths and the corresponding measured distances in the following equations gives the required initial coordinates for points P_2, P_3, P_4, and P_5.

$$X_i = E_i' = E_1 + P_1P_i \sin(P_1P_i)$$

$$Y_i = N_i' = N_1 + P_1P_i \cos(P_1P_i)$$

where $i = 2, 3, 4,$ and 5. To check the calculations, the distances P_2P_3, P_3P_4, and P_4P_5 may be evaluated from the initial coordinates. They must be exactly equal to the measured distances. The initial coordinates are listed in Table 6.

MISCLOSURES

Calculating the redundant distances from the initial coordinates gives $P_2P_4 = 24816.5197$ and $P_2P_5 = 19134.5079$. The misclosures are therefore

$$f_1 = L_9 - P_2P_4 = 24816.476 - 24816.5197 = -0.0437$$

$$f_2 = L_8 - P_2P_5 = 19134.584 - 19134.5079 = +0.0761$$

CONSTRAINTS

To keep point P_1 fixed, and to maintain the direction from P_1 to P_2 unchanged, joint 2 must be released in the direction P_1P_2. This will allow member L_1 (distance P_1P_2) to change only its length during the process of adjustment. The commands are:

> UNITS DEGREES
> JOINT RELEASES
> 2 FORCE Y TH1 − 26.7250

Note that the azimuth (P_1P_2) is introduced in decimal degrees as a negative value with respect to the Y-axis and that four decimals are sufficient.

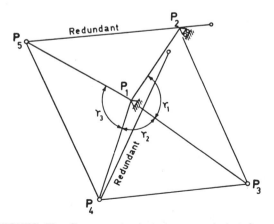

FIGURE 7b. Chosen redundants in an equivalent framework.

TABLE 6. Initial and Given Coordinates, Joint Displacements and Adjusted Coordinates (Example 2 and Table 7).

	Input			Output		Computations	
	Given and Initial Coordinates			Joint Displacements		Adjusted Coordinates	
Point (1)	X (E') (2)	Y (N') (3)	Support (4)	ΔX (ΔE) (5)	ΔY (ΔN) (6)	E (7)	N (8)
1	3550295.830	5804022.950	S	0.0	0.0	. .5.830	. .2.950
2	3555548.327	5814455.015	S	−0.0039	−0.0077	. .8.323	. .5.007
3	3564273.448	5794060.885		+0.0395	+0.0246	. .3.488	. .0.910
4	3545618.362	5791711.764		+0.0536	−0.0095	. .8.416	. .1.755
5	3536541.678	5812246.699		−0.0628	−0.0725	. .1.615	. .6.627

Output values are rounded.

TABLE 7.

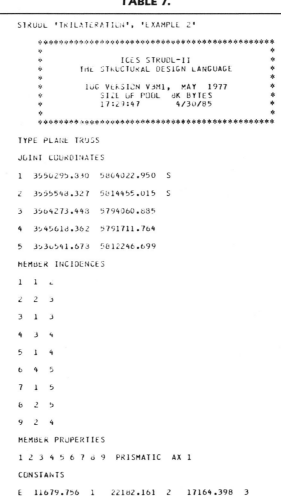

(continued)

TABLE 7. (continued).

```
      E  16802.410  4   13169.814  5   22451.497  6
      E  16025.191  7   19134.584  8   24816.476  9
      LOADING 1 'MISFIT'
      MEMBER DISTORTIONS
      6  CONCENTRATED FRACTIONAL 1.0 DISPLACEMENT +0.0761
      9  CONCENTRATED FRACTIONAL 1.0 DISPLACEMENT -0.0437
      UNITS DEGREES
      JOINT RELEASES
      2 FORCE Y THI -26.7250
      STIFFNESS ANALYSIS
      LIST DISTORTIONS ALL
      LIST DISPLACEMENTS ALL
   ****************************
   *RESULTS OF LATEST ANALYSES*
   ****************************
```

JOB ID - TRILATER   JOB TITLE - EXAMPLE 2

ACTIVE UNITS - LENGTH   FORCE   ANGLE   TEMPERATURE   TIME
                INCH   LB   DEG   DEGF   SEC

ACTIVE STRUCTURE TYPE - PLANE   TRUSS

ACTIVE COORDINATES AXES X Y

```
******************************************************************************
*  LOADING - 1              MISFIT
******************************************************************************
```

MEMBER DISTORTIONS

| MEMBER | /----------------- DISTORTION ------------------//--------------- |
	AXIAL	SHEAR Y	SHEAR Z	TORSION
1	-0.0086171			
2	-0.0126178			
3	0.0179018			
4	-0.0096913			
5	-0.0101863			
6	-0.0106356			
7	0.0166857			
8	-0.0100887			
9	0.0223267			

RESULTANT JOINT DISPLACEMENTS - SUPPORTS

| JOINT | | /---------------- DISPLACEMENT ---------------/ |
		X DISP.	Y DISP.	Z DISP.
1	GLOBAL	0.0	0.0	
2	GLOBAL	-0.0038752	-0.0076966	

RESULTANT JOINT DISPLACEMENTS - FREE JOINTS

| JOINT | | /---------------- DISPLACEMENT ---------------/ |
		X DISP.	Y DISP.	Z DISP.
3	GLOBAL	0.0395048	0.0245854	
4	GLOBAL	0.0535584	-0.0094504	
5	GLOBAL	-0.0627980	-0.0725113	

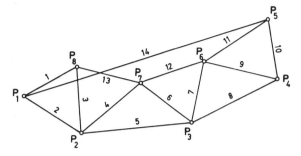

FIGURE 8a. Free trilateration net with a long diagonal.

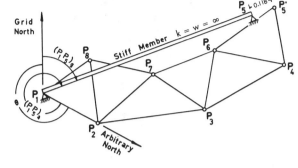

FIGURE 8c. Trilateration chain between two fixed points (P_1 and P_5).

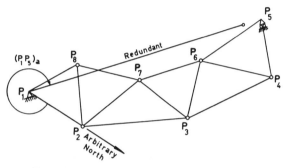

FIGURE 8b. Long diagonal as redundant in an equivalent framework.

DISTANCE CORRECTIONS AND FINAL COORDINATES

The results obtained from the computer analysis are used to evaluate the adjusted lengths of the measured distances and the final coordinates of the points, as presented in Tables 5 and 6, respectively.

Example 3: Free Trilateration Chain with Long Diagonal

In Figure 8a, a trilateration chain is presented in which the long diagonal between points P_1 and P_5 was measured in addition to all other distances. The results of the measure-

TABLE 8. Measured and Adjusted Distances (Example 3 and Table 12).

	Input		Output	Computations
No. (1)	Length $E = L$ (2)	Weight $A = w$ (3)	Member Distortion $L = v$ (4)	Adjusted Length $L + v$ (5)
1	464.61	1	−0.0140	464.596
2	530.15	1	−0.0037	530.146
3	496.82	1	+0.0095	496.829
4	607.19	1	−0.0132	607.177
5	875.16	1	+0.0079	875.168
6	500.24	1	−0.0024	500.238
7	470.37	1	−0.0007	470.369
8	764.75	1	+0.0068	764.757
9	608.40	1	−0.0070	608.393
10	460.03	1	+0.0045	460.034
11	609.48	1	−0.0173	609.463
12	523.26	1	−0.0225	523.238
13	525.55	1	−0.0135	525.537
14	2026.97	1	+0.0162	2026.986

Output values are rounded.

ments are listed in Table 8. It is required to adjust the measured distances.
Solution: (Computer input and output in Table 12)

NUMBER OF REDUNDANTS

The total number of measurements is 14 and that of the points is 8. By Equation (26) the number of redundant measurements is $r = 14 - 2 \times 8 + 3 = 1$.

INITIAL COORDINATES

The easiest way to solve such problem is to choose the long diagonal as the redundant measurement [1]. Since no coordinates are given, an arbitrary coordinate system is chosen with point P_1 as origin and the direction from point P_1 to point P_2 as pointing north (see Figure 8b). All angles of the measured triangles are calculated by the cosine law and listed in Table 9. The angles of the closed traverse $P_1P_2P_3P_4P_5P_6P_7P_8P_1$ are then formed and are given also in Table 9. The solution of said closed traverse is shown in Table 10 in which the initial coordinates of all points are given in the last two columns.

MISCLOSURE

The distance P_1P_5 is calculated from the initial coordinates of points P_1, P_5 and is found to be 2027.088. The misclosure is therefore $f = 2026.97 - 2027.088 = -0.118$.

CONSTRAINTS

Points P_1 and P_5 are chosen to act as supports to the framework. The first point is taken fixed and the second is released in the direction P_1P_5 during the process of adjustment. The arbitrary azimuth $(P_1P_5)_a$ is calculated from the coordinates of both points $(P_1P_5)_a = \tan^{-1} [(-1512.282 - 0.0)/(1349.847 - 0.0)] = 311.751765°$. The commands are similar to those shown in Example 2.

UNITS DEGREES

5 FORCE Y TH1 −311.7518

ADJUSTED DISTANCES AND FINAL COORDINATES

The member distortions and joint displacements are added to the measured lengths and to the initial coordinates, respectively, to obtain the adjusted values. These are listed in Tables 8 and 11.

Example 4: Trilateration Chain between Two Fixed Points

This example is basically identical to Example 3 except that here the points P_1 and P_5 are fixed, and have the following coordinates:

TABLE 9. Calculation of the Angles of Closed Traverse $P_1P_2P_3P_4P_5P_6P_7P_8P_1$.

Point	Angle		Point	Angle	
P_1	$P_8P_1P_2$	59.4912557°	P_5	$P_4P_5P_6$	67.6823547°
	$P_1P_2P_8$	53.6774562		$P_5P_6P_4$	44.3861973
	$P_8P_2P_7$	55.7786161		$P_4P_6P_3$	89.3436917
	$P_7P_2P_3$	33.6841468		$P_3P_6P_7$	60.1768363
P_2	$P_1P_2P_3$	143.1402191	P_6	$P_5P_6P_7$	193.9067253
	$P_2P_3P_7$	42.3136945		$P_6P_7P_3$	54.6626803
	$P_7P_3P_6$	65.1604834		$P_3P_7P_2$	104.0021587
	$P_6P_3P_4$	52.7026881		$P_2P_7P_8$	51.4136034
P_3	$P_2P_3P_4$	160.1768660	P_7	$P_6P_7P_8$	210.0784424
	$P_3P_4P_6$	37.9536202		$P_7P_8P_2$	72.8077800
	$P_6P_4P_5$	67.9314481		$P_2P_8P_1$	66.8312881
P_4	$P_3P_4P_5$	105.8850683	P_8	$P_7P_8P_1$	139.6390681

For P_1: $E_1 = 7163.690$, $N_1 = 2594.810$

For P_5: $E_5 = 8936.982$, $N_5 = 3576.667$

Solution: (Computer input and output in Table 14)

NUMBER OF REDUNDANTS

There is a condition here which must be satisfied in the process of adjustment, namely, the distance between points P_1 and P_5 as calculated from all measured distances must be equal to the distance between them as calculated from their given coordinates. This has the effect of having one redundant measurement.

INITIAL COORDINATES

In order to find the initial coordinates of all other points, it is necessary to orient the chain of triangles, so that the line joining the two fixed points assumes the given azimuth $(P_1P_5)_g$. $(P_1P_5)_g = \tan^{-1} [(8936.982 - 7163.690)/(3576.667 - 2594.810)] = 61.027066°$. This is done in two steps. First, point P_1 is assumed as an origin, and line P_1P_2 as the north direction in an arbitrary coordinate system (see Figure 8c). Second, the open traverse $P_1P_2P_3P_4P_5$ is solved and the azimuth (P_1P_5) is calculated. The whole framework is then rotated an angle ϑ, which is equal to the difference between the arbitrary and the given azimuths (the arbitrary azimuth in this example is the same as calculated in Example 3). Thus: $\vartheta = (P_1P_5)_a - (P_1P_5)_g = 311.751765° - 61.027066° = 250.724699°$. The closed traverse $P_1P_2P_3P_4P_5P_6P_7P_8P_1$ (see Figure 8c) can now be calculated in the correct north-

TABLE 10. Solution of Closed Traverse $P_1P_2P_3P_4P_5P_6P_7P_8P_1$ (Example 3).

Pt. (1)	Angle (2)	Azimuth (3)	Distance (4)	Latitude (5)	Departure (6)	N' (7)	E' (8)
1	59.491256°					0.0	0.0
		0.0	530.15	530.15	0.0		
2	143.140219					530.150	0.0
		323.140219°	875.16	700.221	−524.972		
3	160.176866					1230.371	−524.972
		303.317085	764.75	420.056	−639.058		
4	105.885068					1650.427	−1164.030
		229.202153	460.03	−300.580	−348.252		
5	67.682355					1349.847	−1512.282
		116.884508	609.48	−275.603	543.607		
6	193.906725					1074.244	−968.675
	+1	130.791233	523.26	−341.848	396.157		
7	210.078442					732.396	−572.518
		160.869676	525.55	−496.527	172.232		
8	139.639068					235.869	−400.286
		120.508744	464.61	−235.869	400.286		
1						0.0	0.0
Sum	1079.999999 +1			0.0	0.0		

Angle 7 is corrected so that all angles add up to 1080° exactly.

east system, thus providing the initial coordinates of all points. The solution is shown in Table 13.

MISCLOSURE

After rotating the traverse, point P'_5 will be on the line P_1P_5 or its extension. It will not coincide with the original point P_5 if there are measuring errors. The distance between both points is the misclosure:

$f = P_5P'_5 =$ known distance P_1P_5 − calculated distance $P_1P'_5$

$= [(8936.982 − 7163.690)^2 + (3576.667 − 2594.810)^2]^{1/2}$

$− [(8937.085 − 7163.690)^2 + (3576.725 − 2594.810)^2]^{1/2}$

$= 2026.97 − 2027.088 = −0.118$

Since the distance P_1P_5 is not measured but computed from fixed coordinates, it will not get any correction. To be

TABLE 11. Initial and Given Coordinates, Joint Displacements and Adjusted Coordinates (Example 3 and Table 12).

	Input			Output		Computations	
	Initial and Given Coordinates			Joint Displacements		Adjusted Coordinates	
Point (1)	X (E') (2)	Y (N') (3)	Support	ΔX (ΔE) (4)	ΔY (ΔN) (5)	E (6)	N (7)
1	0.0	0.0	S	0.0	0.0	0.0	0.0
2	0.0	530.150		+0.0590	−0.0037	+0.059	+530.146
3	−524.972	1230.371		+0.0822	+0.0237	−524.890	+1230.395
4	−1164.030	1650.427		+0.0433	−0.0231	−1163.987	+1650.404
5	−1512.282	1349.847	S	+0.0760	−0.0678	−1512.206	+1349.779
6	−968.675	1074.244		+0.0911	+0.0005	−968.584	+1074.244
7	−572.518	732.396		+0.0834	+0.0260	−572.435	+732.422
8	−400.286	235.869		+0.0288	+0.0213	−400.257	+235.890

Output values are rounded.

TABLE 12.

```
STRUDL 'TRILATERATION', 'EXAMPLE 3'

****************************************************
*                                                  *
*              ICES STRUDL-II                      *
*         THE STRUCTURAL DESIGN LANGUAGE           *
*                                                  *
*         IUG VERSION V3M1,  MAY  1977             *
*          SIZE OF POOL   8K BYTES                 *
*            17:19:46         4/30/85              *
*                                                  *
****************************************************

TYPE PLANE TRUSS
JOINT COORDINATES

 1      0.000        0.000   S
 2      0.000      530.150
 3   -524.972     1230.371
 4  -1164.030     1650.427
 5  -1512.282     1549.847   S
 6   -968.675     1074.244
 7   -572.518      732.396
 8   -400.286      235.869

MEMBER INCIDENCES

 1  1  8
 2  1  2
 3  2  8
 4  2  7
 5  2  3
 6  3  7
 7  3  6
 8  3  4
 9  4  6
10  4  5
11  5  6
12  6  7
13  7  8
14  1  5

MEMBER 1 TO 14 PROPERTIES PRISMATIC AX 1
CONSTANTS
E   464.61  1     530.15  2
E   496.82  3     607.19  4
E   875.16  5     500.24  6
E   470.37  7     764.75  8
E   608.40  9     460.03 10
E   609.48 11     523.26 12
E   525.55 13    2026.97 14
```

(continued)

TABLE 12. (continued).

```
    LOADING 1 'MISFIT'

    MEMBER DISTORTIONS

    14 CONCENTRATED FRACTIONAL 1.0 DISPLACEMENT -0.118

    UNITS  DEGREES

    JOINT RELEASES

    5 FORCE Y TH1 -311.7518

    STIFFNESS ANALYSIS

    LIST DISTORTIONS ALL

    LIST DISPLACEMENTS ALL

    ****************************
    *RESULTS OF LATEST ANALYSES*
    ****************************


JOB ID - TRILATER     JOB TITLE - EXAMPLE 3

ACTIVE UNITS -  LENGTH       FORCE         ANGLE        TEMPERATURE        TIME
                INCH          LB            DEG            DEGF             SEC

ACTIVE STRUCTURE TYPE - PLANE    TRUSS

ACTIVE COORDINATES AXES  X Y


***********************************************************************************
*   LOADING - 1              MISFIT
***********************************************************************************


    MEMBER  DISTORTIONS


MEMBER                /----------------- DISTORTION ------------------//---------------
                           AXIAL            SHEAR Y           SHEAR Z          TORSION

1                       -0.0140071
2                       -0.0036606
3                        0.0094953
4                       -0.0131605
5                        0.0079330
6                       -0.0024122
7                       -0.0007011
8                        0.0067604
9                       -0.0070163
10                       0.0044865
11                      -0.0173381
12                      -0.0225279
13                      -0.0134797
14                       0.0161769

    RESULTANT JOINT DISPLACEMENTS - SUPPORTS

JOINT                 /---------------- DISPLACEMENT ---------------/
                           X DISP.          Y DISP.          Z DISP.

1        GLOBAL          0.0              0.0
5        GLOBAL          0.0759640       -0.0678042


    RESULTANT JOINT DISPLACEMENTS - FREE JOINTS

JOINT                 /---------------- DISPLACEMENT ---------------/
                           X DISP.          Y DISP.          Z DISP.

2        GLOBAL          0.0589767       -0.0036606
3        GLOBAL          0.0822207        0.0236809
4        GLOBAL          0.0433487       -0.0231496
6        GLOBAL          0.0911344        0.0004607
7        GLOBAL          0.0834087        0.0259907
8        GLOBAL          0.0288233        0.0213241
```

TABLE 13. Solution of Closed Traverse $P_1P_2P_3P_4P_5P_6P_7P_8P_1$ (Example 4).

Pt. (1)	Angle (2)	Azimuth (3)	Distance (4)	Latitude (5)	Departure (6)	N' (Y) (7)	E' (X) (8)
1	59.491256°					2594.810	7163.690
		109.275301°	530.15	−175.006	500.432		
2	143.140219					2419.804	7664.122
		72.415520	875.16	264.396	834.266		
3	160.176866					2684.200	8498.388
		52.592386	764.75	464.571	607.467		
4	105.885068					3148.771	9105.855
		338.477454	460.03	427.954	−168.770		
5	67.682355				−1	3576.725	8937.085
		226.159809	609.48	−422.156	−439.602		
6	193.906725					3154.569	8497.482
		240.066534	523.26	−261.104	−453.460		
7	210.078443					2893.465	8044.022
		270.144977	525.55	1.330	−525.548		
8	139.639068					2894.795	7518.474
		229.784045	464.61	−299.985	−354.784		
1						2594.810	7163.690
Sum	1080.000000			0.000	0.001		
					−0.001		

Departure of line P_5P_6 is corrected so that all departures add up exactly to zero.

TABLE 14.

```
STRUDL 'TRILATERATION', 'EXAMPLE 4'

*******************************************
*                                         *
*           ICES STRUDL-II                *
*      THE STRUCTURAL DESIGN LANGUAGE     *
*                                         *
*       IUG VERSION V3M1,  MAY  1977      *
*         SIZE OF POOL   8K BYTES         *
*           13:32:18       5/02/85        *
*                                         *
*******************************************

TYPE PLANE TRUSS

JOINT COORDINATES

1    7163.690    2594.810   S
2    7664.122    2419.804
3    8498.388    2684.200
4    9105.855    3148.771
5    8937.085    3576.725   S
6    8497.482    3154.569
7    8044.022    2893.465
8    7518.474    2894.795

MEMBER INCIDENCES

1   1   8
2   1   2
3   2   8
```

(continued)

TABLE 14. (continued).

```
 4  2  7
 5  2  3
 6  3  7
 7  3  6
 8  3  4
 9  4  6
10  4  5
11  5  6
12  6  7
13  7  8
14  1  5
```

MEMBER PROPERTIES
MEMBER 1 TO 13 PROPERTIES   PRISMATIC   AX 1
MEMBER 14 PROPERTIES PRISMATIC AX 1000000.00
CONSTANTS
E   464.61   1   530.15   2
E   496.82   3   607.19   4
E   875.16   5   500.24   6
E   470.37   7   764.75   8
E   608.40   9   460.03 10
E   609.48 11   523.26 12
E   525.55 13   2026.97 14
LOADING 1 'MISFIT'

MEMBER DISTORTIONS
14 CONCENTRATED FRACTIONAL 1.0 DISPLACEMENT -0.118
UNITS   DEGREES
JOINT RELEASES
5 FORCE Y TH1 -61.0271
STIFFNESS ANALYSIS
LIST DISTORTIONS ALL

LIST DISPLACEMENTS ALL

(continued)

TABLE 14. (continued).

```
****************************
*RESULTS OF LATEST ANALYSES*
****************************


JOB ID -  TRILATER     JOB TITLE - EXAMPLE 4

ACTIVE UNITS -  LENGTH         FORCE          ANGLE         TEMPERATURE      TIME
                INCH           LB             DEG           DEGF             SEC

ACTIVE STRUCTURE TYPE - PLANE  TRUSS

ACTIVE COORDINATES AXES  X Y


*******************************************************************************
*  LOADING - 1              MISFIT
*******************************************************************************


   MEMBER  DISTORTIONS


MEMBER             /---------------- DISTORTION ----------------//-------------
                         AXIAL           SHEAR Y         SHEAR Z        TORSION
  1                   -0.0162325
  2                   -0.0042422
  3                    0.0110038
  4                   -0.0152513
  5                    0.0091933
  6                   -0.0027954
  7                   -0.0008125
  8                    0.0078344
  9                   -0.0081309
 10                    0.0051993
 11                   -0.0200926
 12                   -0.0261071
 13                   -0.0156213
 14                    0.0000000


       RESULTANT JOINT DISPLACEMENTS - SUPPORTS

  JOINT                    /--------------- DISPLACEMENT -------------/
                                X DISP.         Y DISP.         Z DISP.

   1          GLOBAL            0.0             0.0
   5          GLOBAL           -0.1032320      -0.0571588


       RESULTANT JOINT DISPLACEMENTS - FREE JOINTS

  JOINT                    /--------------- DISPLACEMENT -------------/
                                X DISP.         Y DISP.         Z DISP.

   2          GLOBAL           -0.0265659      -0.0631148
   3          GLOBAL           -0.0055489      -0.0990009
   4          GLOBAL           -0.0419068      -0.0385634
   6          GLOBAL           -0.0343598      -0.0998688
   7          GLOBAL           -0.0034768      -0.1011841
   8          GLOBAL            0.0123002      -0.0396877
```

TABLE 15. Measured and Adjusted Distances (Example 4 and Table 14).

	Input		Output	Computations
No. (1)	Length E = L (2)	Weight A = w (3)	Member Distortion ΔL = v (4)	Adjusted Distance L + v (5)
1	464.61	1	−0.0162	464.594
2	530.15	1	−0.0042	530.146
3	496.82	1	+0.0110	496.831
4	607.19	1	−0.0153	607.175
5	875.16	1	+0.0092	875.169
6	500.24	1	−0.0028	500.237
7	470.37	1	−0.0008	470.369
8	764.75	1	+0.0078	764.758
9	608.40	1	−0.0081	608.392
10	460.03	1	+0.0052	460.035
11	609.48	1	−0.0201	609.460
12	523.26	1	−0.0261	523.234
13	525.55	1	−0.0156	525.534
14	2026.97*	10^6	0.0	2026.970

*Distance was not measured but has been calculated from the known coordinates of the given fixed points.
Output values are rounded.

able to introduce it in the input it can be assumed as measured, but has an infinite weight associated with it. This situation is represented by a very stiff member in the corresponding framework (3 and 4). For the input data, a weight of 10^6 is given to it.

CONSTRAINTS

Joint 5 must be released in the given direction between the fixed points P_1 and P_5. The commands are similar as mentioned in the previous example.

UNITS DEGREES

5 FORCE Y TH1 −61.0271

ADJUSTED DISTANCES AND FINAL COORDINATES

Adjusted distances are obtained by adding the corrections (member distortions) to the measured distances (Table 15). Similarly the final coordinates (Table 16) are obtained by adding the joint displacements to the corresponding initial coordinates $X(E')$ and $Y(N')$.

TABLE 16. Initial and Given Coordinates, Nodal Displacements and Adjusted Coordinates (Example 4 and Table 14).

	Input			Output		Computations	
	Initial and Given Coordinates			Nodal Displacement		Adjusted Coordinates	
Point (1)	X (E') (2)	Y (N') (3)	Support (4)	ΔX (ΔE) (5)	ΔY (ΔN) (6)	E (7)	N (8)
1	7163.690	2594.810	S	0.0	0.0	7163.690	2594.810
2	7664.122	2419.804		−0.0266	−0.0631	7664.095	2419.741
3	8498.388	2684.200		−0.0055	−0.0990	8498.382	2684.101
4	9105.855	3148.771		−0.0419	−0.0386	9105.813	3148.732
5	8937.085	3576.725	S	−0.0032	−0.0572	8936.982	3576.668
6	8497.482	3154.569		−0.0344	−0.0999	8497.448	3154.469
7	8044.022	2893.465		−0.0035	−0.1012	8044.019	2893.364
8	7518.474	2894.795		+0.0123	−0.0397	7518.486	2894.755

Output values are rounded.

NOTATION

A = Cross-sectional area
Az = Azimuth
a = Coefficient of condition equation
C = Correlate
D = Displacement
E = Modulus of elasticity
E_i = Final or adjusted easting of point i
E'_i = Calculated initial easting of point i
F = Final force in member
F = Generalized nodal force vector for entire model
f = Linear closing error, misclosure or misfit
G_j = Number of first end point of distance j (STRUDL input)
H_j = Number of second end point of distance j (STRUDL input)
I = Total number of points or joints
K = stiffness matrix of entire model
k = stiffness of a member
L = Length of a distance or member
M = Minimum number of distances needed to locate all points
m = Member
N_i = Final or adjusted northing of point i
N'_i = Calculated initial northing of point i
n = Number of all measured distances
P = Point identification
(P_iP_j) = Azimuth of line P_iP_j from point P_i to point P_j
r = The number of redundants or conditions
S = Force in member due to unit virtual load in redundant
S = Support identification (STRUDL input)
u_i = Axial displacement of point i
v = Correction of a distance ($v = \Delta L$)
W = work done
w = Weight of a measured distance
X = Force in X-direction (STRUDL input)
X_i = X-coordinate (easting) of point i
X_r = Force in redundant r
Y_i = Y-coordinate (northing) of point i
α, β = Perimeter angles in a central figure
γ = Central angle
Δ = Generalized nodal displacement vector for entire model
ΔE_i = Correction to initial easting of point i ($\Delta E_i = \Delta X_i$)
ΔL = Change in length of member
ΔN_i = Correction to initial northing of point i ($\Delta N_i = \Delta Y_i$)
ΔX_i = Displacement, in the X-direction, of joint or node i
ΔY_i = Displacement, in the Y-direction, of joint or node i
δ = Member displacement vector
δL_m = Change in length in member m caused by force S
σ = Standard deviation
ϑ = Rotation angle between an arbitrary and a known azimuth

Subscripts

a = Arbitrary
g = Known or given
i = Current index of a point
j = Current index of a distance
m = Current index of a member
r = Current index of a redundant distance or a redundant member

REFERENCES

1. Danial, N. F., "Die Zentral- und Diagonalenbedingungen in ebenen Streckennetzen," *Vermessung, Photogrammetrie, Kulturtechnik*, Switzerland, pp. 249–257 (October 1979).
2. Danial, N. F., "Virtual Work Adjustment of Trilateration Nets," *Journal of the Surveying and Mapping Division*, ASCE, Vol. 105, No. SU1, Proc. Paper 14955, pp. 67–83 (November 1979).
3. Danial, N. F., "Adjustment of Trilateration Nets with Fixed Points," *Survey Review*, England, No. 197, pp. 66–83 (July 1980).
4. Danial, N. F. and T. Krauthammer, "Trilateration Adjustment by Finite Elements," *Journal of the Surveying and Mapping Division*, ASCE, Vol. 106, No. SU1, Proc. Paper 15812, pp. 73–93 (November 1980).
5. *ICES–STRUDL–II, The Structural Design Language, Engineering User's Manual*, Volume 1, R68-91, Department of Civil Engineering, M.I.T., Cambridge, Mass. (1968).
6. Zienkiewicz, O. C., *The Finite Element Method*, 3rd ed., McGraw-Hill Book Co., Inc., New York, N.Y. (1977).

CHAPTER 3

Area Computation Using Salient Boundary Points

FOUAD A. AHMED*

There are three sources of errors encountered when employing any of the methods used in the survey and computation of areas bounded by curved boundaries. Simpson's rule is the prime example, as this formula has proven to give better accuracy than any other. First, the discontinuity of the assumed parabolas results in sharp junction points between the parabolas which do not match the actual shape of the boundary line. Second, if the existing number of subdivisions is odd, it is difficult to choose the part which should be excluded from Simpson's rule. Third, since the intercepts have to be equal there is no possibility of choosing the governing points of the curvilinear boundary (such as points of maximum and minumum offsets, points of curvature inflection, and terminals of constant slopes). Accordingly, the computed area may differ dramatically from the actual figure, especially if the survey boundary line has several points of inflections as shown in Figure 1.

A theoretical example is given demonstrating this situation. In this example the area under the sine curve is computed as shown in Figure 2. The x-axis is shifted downward by 2 units and the exact area is computed from $x = \pi$ to $x = 10\pi$ and found to be 17.36338π. The same area is then computed using Simpson's rule with a constant intercept of 1.5π. The resulting points are mixed between maximum, minimum, and inflection and the computed area is 20π with a difference of 15%. If all the points of maximum, minimum, and inflection are taken into consideration in the area computation (in other words if the constant intercept was chosen as 0.5π), the computed area is 17.33333π with a difference of only 0.17%. This shows the importance of taking the offsets at all the maximum, minimum, and inflection points if an accurate estimation of the area is required. Now, to show that just points of maximum and minimum offsets are not sufficient for precise area determination, the area based on only the maximum and minimum points between 1.5π and 9.5π is computed by the Simpson's rule and found to be 18.66666π with a difference of 16.7% from actual area of 16π.

It is seen from this that if the offsets at the points of maximum, minimum, and inflection of a curvilinear boundary are measured, the accuracy of the computed area is maximized while the field work is minimized.

The method presented fulfills the need of a simple and practical formula while eliminating the drawbacks of Simpson's rule.

PRINCIPLES OF SALIENT BOUNDARY POINTS METHOD

Consider the curvilinear boundary from a to r of the ground surface as shown in Figure 3. It is assumed that the surveyor is able to locate in the field, within reasonable limits, the position of the points of maximum and minimum offset and the points of inflections. Points b, h, P, and r are points of maximum or minimum, M, and points a, c, e, g, and q are points of inflection, I, or terminals of a curve (other than M). If a sharp break, such as point g, exists at a certain location on the boundary, this point should be considered an inflection point I. Points d, f, and n are middle points on the straight lines ce and hP, respectively, so that $\ell_3 = \ell_4$ and $\ell_8 = \ell_9$. It is not required to determine their offsets y_4 and y_9 in the field but it is essential to calculate the values of these offsets to include them into the computations. A middle point on a straight line will be M if it lies between two I's and vice versa. As for the curve efg, which neither has a point of maximum nor a point of minimum, the middle point on this curve should be labeled M and its corresponding offset y_6 should be measured. It is therefore necessary for convenience of computation that points are labeled alternatively as I and M.

---

*University of Petroleum & Minerals, Dhahran 31261 Saudi Arabia

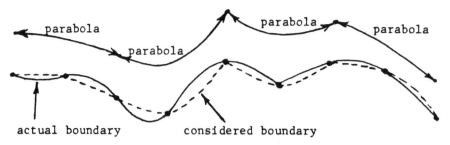

FIGURE 1. Difference between actual and boundary defined by Simpson's rule.

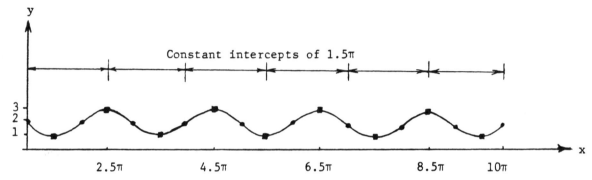

FIGURE 2. Simulated terrain described by a sine curve: • points of inflection; ■ points of maximum and minimum.

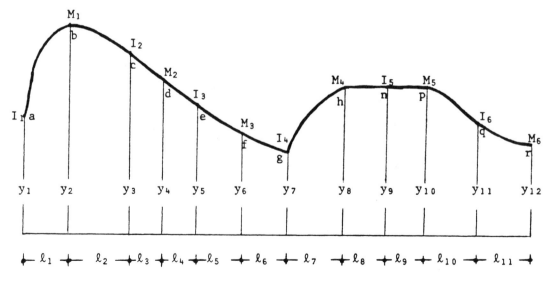

FIGURE 3. Change points on a curvilinear.

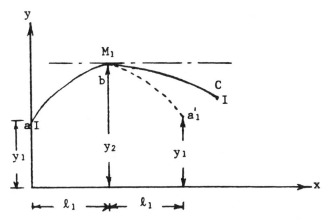

FIGURE 4. Calculation of area under curve ab.

The offsets y_1, y_2, y_3, y_5, y_6, y_7, y_8, y_{10}, y_{11}, and y_{12} are measured in the field as well as the spacing ℓ_1, ℓ_2, $(\ell_3 + \ell_4)$, $(\ell_5 + \ell_6)$, ℓ_7, $(\ell_8 + \ell_9)$, ℓ_{10}, and ℓ_{11}.

The area under *ab* is computed assuming that the curve *ab* is a parabola. Since *b* is a point of maximum the tangent at this point is perpendicular to the offset. If *ab* is extended to *a'* so that *ba'* is symmetrical to *ba*, as shown in Figure 4, the area under the curve *aba'* is given by

$$A_{aba'} = \frac{1}{3} \ell_1 (2y_1 + 4y_2)$$

Accordingly the area under *ab* is found from

$$A_{ab} = \frac{1}{3} \ell_1 (y_1 + 2y_2)$$

or

$$A_{ab} = \frac{1}{3} \ell_1 (y_{I_1} + 2y_{M_1}) \qquad (1)$$

Similarly,

$$A_{bc} = \frac{1}{3} \ell_2 (y_{I_2} + 2y_{M_2}) \qquad (2)$$

The area under the straight line *ce* is given by

$$A_{ce} = \frac{1}{2} \ell_{ce} (y_3 + y_5) \text{ where } \ell_{ce} = 2\ell_3 = 2\ell_4$$

or

$$A_{ce} = \frac{1}{6} \ell_{ce} (3y_3 + 3y_5)$$

Introduce

$$y_4 = \frac{1}{2} (y_3 + y_5)$$

Hence,

$$A_{ce} = \frac{1}{6} \ell_{ce} (4y_4 + y_3 + y_5)$$

Therefore the area under the straight line may be written in the form

$$A_{ce} = \frac{1}{3} [\ell_3(2y_4 + y_3)] + \frac{1}{3} [\ell_4(2y_4 + y_5)]$$

or

$$A_{ce} = \frac{1}{3} [\ell_3(2y_{M_2} + y_{I_2})] + \frac{1}{3} [\ell_4(2y_{M_2} + y_{I_3})] \qquad (3)$$

The area under the curve *efg* which neither has a point of maximum nor a point of minimum is given by

$$A_{efg} = \ell_{eg}(y_5 + y_7 + 4y_6)/6 \text{ where } \ell_{eg} = 2\ell_5 = 2\ell_6$$

This equation may take the form

$$A_{efg} = \frac{1}{3} [\ell_5(2y_6 + y_5)] + \frac{1}{3} [\ell_6(2y_6 + y_7)]$$

or

$$A_{efg} = \frac{1}{3} [\ell_5(2y_{M_3} + y_{I_3})] + \frac{1}{3} [\ell_6(2y_{M_3} + y_{I_4})] \qquad (4)$$

The area under the curve gh can be obtained from

$$A_{gh} = \frac{1}{3} \ell_7 (2y_8 + y_7)$$

or

$$A_{gh} = \frac{1}{3} \ell_7 (2y_{M_4} + y_{I_4}) \quad (5)$$

The straight line hP is considered horizontal. The area under this line is given by

$$A_{hP} = \ell_{hP}\, y,$$

where $y = y_8 = y_9 = y_{10}$ and $\ell_{hP} = 2\ell_8 = 2\ell_9$.

This equation may be written in the form

$$A_{hP} = \frac{1}{6} \ell_{hP} (2y_8 + 2y_9 + 2y_{10})$$

or

$$A_{hP} = \frac{1}{3} [\ell_8 (2y_8 + y_9)] + \frac{1}{3} [\ell_9 (2y_{10} + y_9)]$$

Therefore,

$$A_{hP} = \frac{1}{3} [\ell_8 (2y_{M_4} + y_{I_5})] + \frac{1}{3} [\ell_9 (2y_{M_5} + y_{I_5})] \quad (6)$$

The area under the reverse curve pgr can be obtained in a way similar to that of the curve ab. Therefore,

$$A_{pg} = \frac{1}{3} \ell_{10} (2y_{M_5} + y_{I_6}) \quad (7)$$

and

$$A_{gr} = \frac{1}{3} \ell_{11} (2y_{M_6} + y_{I_6}) \quad (8)$$

The total area A_{ar} is obtained by adding equations from (1) to (8).

$$A_{ar} = \frac{1}{3} [y_{I_1} \ell_1 + y_{I_2} (\ell_2 + \ell_3) + y_{I_3} (\ell_4 + \ell_5)$$
$$+ y_{I_4} (\ell_6 + \ell_7) + y_{I_5} (\ell_8 + \ell_9) + y_{I_6} (\ell_{10} + \ell_{11})$$
$$+ 2y_{M_1} (\ell_1 + \ell_2) + 2y_{M_2} (\ell_3 + \ell_4)$$
$$+ 2y_{M_3} (\ell_5 + \ell_6) + 2y_{M_4} (\ell_7 + \ell_8)$$
$$+ 2y_{M_5} (\ell_9 + \ell_{10}) + 2y_{M_6} \ell_{11}]$$

The generalized notation that covers all cases, whether the boundary starts and/or ends with an I point or M point, is

$$A = \frac{1}{3} \left[\sum_{i=1}^{k} y_{I_i}(\ell_{L,I_i} + \ell_{R,I_i}) + 2 \sum_{i=1}^{j} y_{M_i}(\ell_{L,M_i} + \ell_{R,M_i}) \right] \quad (9)$$

where

k is the number of I points
j is the number of M points
ℓ_{L,I_i} is the distance right to the ith I point
ℓ_{L,I_i} is the distance left to the ith I point
ℓ_{L,M_i} is the distance left to the ith M point
ℓ_{R,M_i} is the distance right to the ith M point

Note that one distance, ℓ_L, at the beginning, and one distance, ℓ_R, at the end, will be zero in Equation (9), which has to be considered from the sketch. Recall also that it is necessary for the points to be marked alternately as I and M.

AREA ENCLOSED BY CURVED BOUNDARY

The same approach can be applied when the area is enclosed by a curvilinear boundary. Polar coordinates may be easier to use in this case as shown in Figure 5. In the figure the area A_{IMP} is bounded by the curve IM where M is a point of maximum at which the tangent is perpendicular to the line MP and I is an inflection point. The measurements are the two distances d_I and d_M from the chosen pole station P and the included angle θ. If IQ and PQ are two orthogonal lines, then

$$A_{IMP} = A_{IMPQ} - A_{IPQ} = PQ(IQ + 2MP)/(3) - 0.5\ PQ.IQ \quad (10)$$

After introducing the angle θ in the equation and after some reduction the area can be computed from

$$A_{IMP} = d_I(8d_M \sin \theta - d_I \sin 2\theta)/12 \quad (11)$$

If the area is closed about the pole P, all M's are located and the corresponding d_M's are measured. The I's should also be found and their corresponding d_I's are measured. There should be an inflection point between every two M's. If, in some cases, it is difficult to find an inflection point, a point I may be assumed and its corresponding d_I is measured as

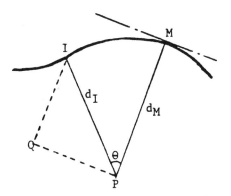

FIGURE 5. Area calculation with polar coordinates.

ERROR ANALYSIS

The main source of errors in the present techniques is the determination of the positions of the M and I points. There are two advantages when using the polar coordinates. The first advantage is that electromagnetic distance measuring (EDM) instruments can be used since all the measurements are conducted from one station, the pole, and only the reflector will be moving along the curved boundaries. The second advantage is that it is very easy to determine the location of points of maximum and minimum by making use of the tracking mode of the EDM instrument used. Continuous change of the read out is obtained at the instrument station when the reflector is moved slowly along the curved boundaries. The surveyor should record the readings at which their trend changes from increasing to decreasing or vice versa. The determination of the location of points of inflection in both techniques is simpler since their locations do not depend on the position of the traverse line (in the first technique) or the pole (in the second technique). The point at which the curvature changes its direction can be easily located even by the unaided eye of the observer along the boundary line itself. Points of maximum and minimum offsets in the first technique can easily be determined by the use of a tape pulled perpendicular to the traverse line. The effect of angle measurements when using the polar coordinates depends mainly on the kind of theodolite used. If a second-theodolite is used and if the standard error in angle measurement is ± 5 seconds, the corresponding error in determining an M or I point at an average distance from the pole of 50 m is about ± 20 mm. Such error is practically negligible in earth work computations.

shown in Figure 6. The total area is given by

$$A = \frac{1}{12} \sum_{i=1}^{n/2} [8d_{2i}(d_{2i-1} \sin \theta_{2i-1} + d_{2i+1} \sin \theta_{2i}) - d_{2i}^2 (\sin 2\theta_{2i-1} + \sin 2\theta_{2i})] \quad (12)$$

Where n is the total number of M and I points, with M points designated as odd numbers and I points designated as even numbers, thereby using the notation d_{2i-1} or d_{2i+1} for d_M; and d_{2i} for d_I.

It is recommended that this technique be used only when the curved boundary forms a closed loop. Otherwise, there will be only a few maximum and minimum points.

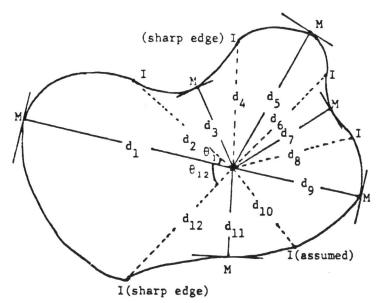

FIGURE 6. Polar coordinate system.

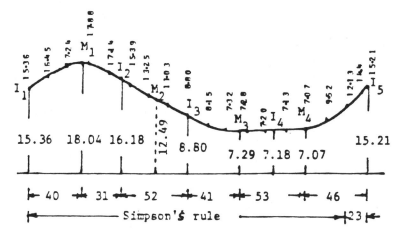

FIGURE 7. Boundary offsets in a practical example [1 yard (yd) = 0.914 meter (m)].

In case of area determination of vertical profile or cross sections, the location of the points of maximum is easily done by the instrument man, but the points of inflection are located by the rod man. The accuracy of locating these points is affected mainly by two factors: (1) the curvature and (2) the length of the curve. Such accuracy decreases with the decrease of the boundary curvature and with the length of the curve. In different experiments the standard deviation m_p, based on observation repeatability, of locating an I or M point between the two surrounding M or I points, respectively, was calculated. Also the distance, S, between the two surrounding I or M points was measured. Thus, the relative accuracy m_p/S was computed. It was found that, in more than 90% of the cases, the average relative accuracy was 0.05. To elevate the propagation of the variancies m_p^2 into the variance of the area Equation (9) is considered.

If an error, $d\ell$, is made in locating an M point between two I points so that $\ell_{L,M}$ is increased, $\ell_{R,M}$ will be decreased by the same value $d\ell$. If dy_M is the corresponding error in the determination of the offset y_M, the resulting error in the computation of the included area between the two I points, I_i and I_{i+1}, can be calculated from

$$dA_M = \frac{1}{3}[d\ell(y_{I_i} - y_{I_{i+1}}) + 2dy_M(\ell_{L,M} + \ell_{R,M})] \quad (13)$$

But since the actual value of the errors $d\ell$ and dy_M are not known, the variance of the area is considered.

$$m_{A_M}^2 = \frac{1}{9}[(y_{I_i} - y_{I_{i+1}})^2 m_{\ell_M}^2 + 4(\ell_{L,M} + \ell_{R,M})^2 m_{y_M}^2] \quad (14)$$

In a similar way the variance of the area between two successive M points, M_i and M_{i+1} can be obtained as a function of the variance of locating the included I point.

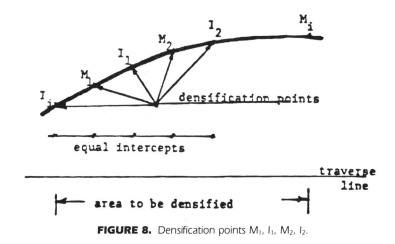

FIGURE 8. Densification points M_1, I_1, M_2, I_2.

$$m_{A_I}^2 = \frac{1}{9} [4(y_{M_i} - y_{M_{i+1}})^2 m_{\ell_I}^2 + (\ell_{L,I} + \ell_{R,I})^2 m_{y_I}^2] \quad (15)$$

m_y^2 is a function of the curve shape and of m_ℓ^2. If the equation of the curve is known, $y = f(\ell)$ and considering that $m_{\ell_M}^2 = m_{\ell_I}^2 = m_\ell^2$, therefore Equations (14) and (15) may be written in the form

$$m_{A_M}^2 = \frac{1}{9} \left[(y_{I_i} - y_{I_{i+1}})^2 + 4(\ell_{L,M} + \ell_{R,M})^2 \left(\frac{\partial y}{\partial \ell_M}\right)^2 \right] m_\ell^2 \quad (16)$$

$$m_{A_1}^2 = \frac{1}{9} \left[4(y_{M_i} - y_{M_{i+1}})^2 + (\ell_{L,I} + \ell_{R,I})^2 \left(\frac{\partial y}{\partial \ell_I}\right)^2 \right] m_\ell^2 \quad (17)$$

Equations (14) and (15) or (16) and (17) are applied for all the M and I points on a curvilinear boundary; the summation gives the variance of the whole area. Note that the applications of Equations (14) or (16) for all the M points gives the variance of the whole area due to standard errors in locating only these M points. As $\partial y/\partial \ell_M$ is zero, and under the simplifying assumption that points of inflection fall approximately midway between M points, Equation (16) is nearly zero. Thus, the summation of Equation (17) should be used to compute the variance of the whole area.

When recalling the example illustrated in Figure 2, the equation of the simulated terrain is $y = 2 + \sin x$. The evaluation of the derivative of this equation at the M points gives 0, and at the I points gives 1. The intercepts are $\ell_{L,M} = \ell_{R,M} = \ell_{L,I} = \ell_{R,I} = \pi/2$, $(y_{I_i} - y_{I_{i+1}}) = 0$ and $(y_{M_i} - y_{M_{i+1}}) = 2$.

The standard deviation m_ℓ is taken as $\pm 0.05\pi$. The first point (at $x = \pi$) and the last point (at $x = 10\pi$) are assumed to be errorless. Substituting with the previous values in Equation (17), the standard deviation of the total area, m_A, is found to be $\pm 0.24\pi$. The relative accuracy is therefore computed to be $\pm 1.5\%$. If a confidence limit of 95% is assumed the maximum expected error is 4.5%. This is less than the 15% error obtained when Simpson's rule was applied in the same example.

PRACTICAL EXAMPLE

The following is an illustration of a practical example of area computation using salient points in comparison with Simpson's rule. An offsetting operation was carried out three times along a 263 m long traverse line. In the first time the offsets determined every one meter in order to obtain the closest value to the area A_a. A_a was computed using numerical integration by Simpson's rule and found to be $A_a = 3022.7$ m². During the second time 16 offsets, with a spacing of 15 m, were then measured starting from I_1 as shown in Figure 7. Simpson's rule was applied in the calculation of the area. The last portion, which has an intercept of 23 m, was assumed to be trapezoid and its area was computed accordingly. The total area, A_s, was found to be 3091.7 m². In the third case three offsets were chosen at points of maximum and minimum offsets from the traverse line (M_1, M_3, and M_4) as shown in Figure 7. Also, four points of inflection were chosen (I_1, I_2, I_3, and I_5) and their corresponding offsets were measured. The offsets at M_2 and I_4 were computed by linear interpolation since it was noticed that the boundary about these points is almost straight. The area was then computed using Equation (9) and found to be $A_{new} = 3025.5$ m². It is seen that although the field work is reduced by more than 50% (based on the number of offsets measured) the computed area was even closer to the correct value A_a in the first case, i.e., within 2.2%.

Where the distance between two successive points, M_i and I_i, is too long, densification points can be chosen on the curve between them in order to improve the accuracy. It is proven (see Equation (4)) that if Equation (9) is applied, the answer will be identical to that given by Simpson's formula provided that the densification points are equally spaced and named alternatively M and I as shown in Figure 8.

REFERENCE

Kelly, L., *Handbook of Numerical Methods and Applications,* Addison-Wesley Publishing Company, pp. 54–56 (1967).

CHAPTER 4

Redefinition of the North American Geodetic Networks

JOHN D. BOSSLER* AND NICHOLAS BODNAR**

INTRODUCTION

New geodetic techniques and computer technology are providing Federal engineers and scientists in the United States and other countries of North America with the tools needed to redefine and readjust the North American horizontal and vertical geodetic control networks—a web of geodetic survey monuments that spans the continent. When the project is completed (1986 for the horizontal network and 1989 for the vertical) the datums will provide a greatly improved foundation for mapping and engineering projects, crustal motion studies, inventories of natural resources and land parcel recordation, and the launch and recovery of space vehicles. In addition, earth scientists will have available, in an instant, 100 years of historical geodetic data in computer-readable form, supplying valuable data in support of crustal motion and earthquake prediction studies.

The National Geodetic Reference System of the United States has evolved over the last 150 years. Built up year by year from the continuous addition of new surveying data, the horizontal, vertical, and gravity networks have extended into all parts of the nation. One consequence of this growth is that newer surveys were forced to fit into the existing framework, even though many of the newer surveys have been performed more accurately than those comprising the original framework. The result is a national reference system that varies in accuracy from area to area and is in need of major repair. The new adjustments, now underway, will remove these discrepancies and provide new datums for space-age surveying techniques.

This paper describes the salient features of the new adjustments of the North American datums and the tasks necessary to complete this monumental endeavor by the National Geodetic Survey (NGS). The new horizontal datum is referred to as the North American Datum of 1983 (NAD 83) and the new vertical datum is called the North American Vertical Datum of 1988 (NAVD 88).

The term "North American" describes the international scope of this project. All countries of North America, from Greenland (part of Kingdom of Denmark) to Panama and many Caribbean Islands are participating. Greenland and the Caribbean Islands, however, will not be part of the vertical network due to a lack of connecting observations.

Persons interested in obtaining additional information on the new adjustment or ordering geodetic data for the United States can contact:

National Geodetic Information Center, (N/CG174)
National Ocean Service
National Oceanic and Atmospheric Administration
U.S. Department of Commerce
Rockville, Maryland 20852

HORIZONTAL NETWORK

Need for New Adjustment

Figure 1 shows the extent of the geodetic observations used in the last general horizontal network adjustment of 1927. This framework included 41 loops ranging in circumference from several hundred to 3,000 km. The average closure of these large loops, after adjustment, was on the order of one part in 300,000, which was satisfactory for basic control for that period. However, considering the overall sparseness of control points, combined with each

---

*Rear Admiral, NOAA Corps, Office of Charting and Geodetic Services, National Oceanic and Atmospheric Administration, Rockville, MD
**Commander, NOAA Corps, Office of Charting and Geodetic Services, National Oceanic and Atmospheric Administration, Rockville, MD

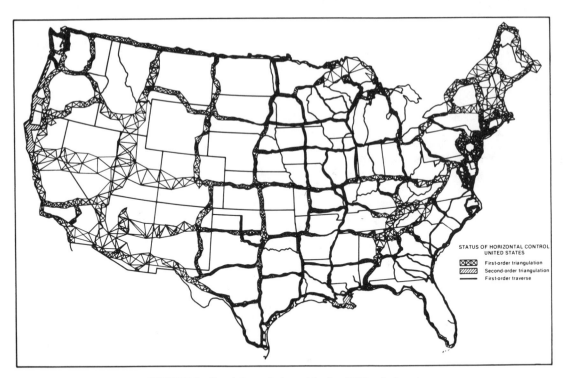

FIGURE 1. Geodetic control used for the 1927 adjustment.

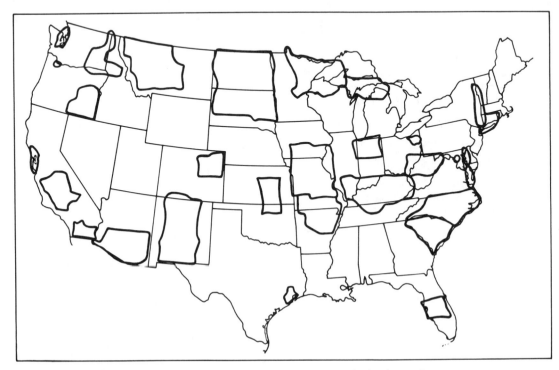

FIGURE 2. Regions of the United States that have required major readjustments.

loop's varying degrees of relative positional accuracies, this measure could easily mislead those expecting to achieve high-precision survey results. Shortly after the eastern portion of the adjustment was completed, and before the results were published, a discrepancy of approximately 10 m in latitude was discovered along the U.S. border in northern Michigan. A portion of the network in Wisconsin and Michigan was later readjusted to redistribute this discrepancy.

Other discrepancies went undetected until they were finally identified through the process of fitting and adjusting new survey data. Large regions of the network were then readjusted to distribute the discrepancies over larger areas, thereby minimizing the local effects of distortion (see Figure 2). These regional readjustments were a mending process which did not eliminate the errors, but rather distributed them and minimized their effects on local network applications. Unfortunately, this attempt to improve the existing reference system frequently changed control point coordinates for certain portions of the network.

The discovery of these weaknesses did not indicate careless planning or execution of the 1927 adjustment, but represented significant improvements in surveying methods of the following decades. One weakness was inadequate length control. Measuring precise base lines presented unique challenges to early surveyors. Coefficients of thermal expansion were difficult to determine and they used calibrated bars of dissimilar metal, usually 3 to 6 meters in length. In the 1800's the surveyors kept the measuring bars in ice water to minimize changes in bar temperature. Consider the effort needed to measure a 10 km base line in this manner, let alone to find a convenient and adequate supply of ice in the field! Azimuth control, based on astronomic observations and precise time measurements, was another weakness in the NAD 27 datum. Here again, advances in technology were largely responsible for uncovering the deficiency, as radio signals replaced telegraph chronometers in 1922 for measuring precise time intervals. Two other factors further weakened the system: the 1927 adjustment did not include survey observations from the Atlantic seaboard, and survey observations in Alaska were connected to the national network during World War II by only a single arc of triangulation along the Alaska Highway, providing marginally adequate positional control for Alaska.

A well-established engineering principle states that subordinate surveys must be started from and closed upon geodetic stations of higher-order accuracy [16]. NAD 27 readily provides geodetic control for one part in 25,000 but it attains one part in 100,000 only in limited portions of the network. It simply cannot provide modern users with the accuracies demanded today.

Changes Resulting from Redefinition and Readjustment

DATUM REDEFINITION

Most existing geodetic datums, including NAD 27, have been defined to minimize differences between the geoidal and ellipsoidal surfaces for the region of the world served by a particular datum. In effect, the geoid and the ellipsoid are usually taken to be synonymous for that region. For these datums, measured distances reduced to the geoid, are assumed to lie on the ellipsoid. Theodolites leveled according to the Earth's real gravity field, that is, perpendicular to the geoid, are correspondingly assumed to be oriented perpendicular to the ellipsoid [2].

Deflection of the Vertical

The NAD 83 project will not neglect the proper reduction of the surveying observations to the reference ellipsoid. To accomplish this task, both the geoid height and the deflection of the vertical must be determined for every control point occupied for horizontal observations—a significant effort considering the national network contains approximately 180,000 occupied points. The deflection of the vertical is computed for all occupied control points, with precise geoid heights being obtained as a by-product of the computational process. The computation method is a numerical quadrature using the classical Stokes and Vening-Meinesz equations. The method requires a detailed free-air anomaly field within 45 minutes of arc of the control point and mean gravity anomalies beyond.

Within the United States, astronomic positions are measured directly at nearly all NGS stations where there is an actual or potential line of sight with a vertical angle of 7° or greater. For lines with vertical angles of less than 7°, the maximum propagated error in a horizontal direction, as a result of error in the deflection of the verticle is 0.25″. This is well below the accidental error component of first-order directions [17].

Earth Centered Datum

The NAD of 1927 is based on the astronomic position of triangulation station MEADES RANCH in Kansas and the Clarke Ellipsoid of 1866. NAD 83, however, will not be based on a single point, such as MEADES RANCH, but rather on numerous stations whose positions have been determined from satellites or other super-precise methods [1]. The Clarke Ellipsoid of 1866 will be replaced by an Earth-centered datum called the Geodetic Reference System 1980 (GRS 80) [14]. (See Table 1 for the numerical values defining the datum.) The result is a globally best-fitting reference surface replacing one which fit the needs of North America alone.

Most *recent* datums have been geocentric, i.e., the center of the reference figure coincides, as closely as measurements allow, with the center of mass of the Earth. The Defense Mapping Agency (DMA) of the U.S. Department of Defense (DoD) defined such a datum in 1972, identified as WGS 72. This datum has been used extensively by the world mapping and charting community for global applications related to civilian and defense purposes.

There are only a few disadvantages of a globally best-fitting datum. For instance, in general, a larger separation of

TABLE 1. Geodetic Reference System 1980 Numerical Values.

Defining constants (exact)*
- $a = 6378\ 137$ m — semimajor axis
- $GM = 3\ 986\ 005 \times 10^8\ m^3 s^{-2}$ — geocentric gravitational constant
- $J_2 = 108\ 263 \times 10^{-8}$ — dynamical form factor
- $\omega = 7\ 292\ 115 \times 10^{-11}$ rad s^{-1} — angular velocity

Derived geometrical constants
- $b = 6\ 356\ 752.3141$ m — semiminor axis
- $E = 521\ 854.0097$ m — linear eccentricity
- $c = 6\ 399\ 593.6259$ m — polar radius of curvature
- $e^2 = 0.006\ 694\ 380\ 022\ 90$ — e = first eccentricity
- $e'^2 = 0.006\ 739\ 496\ 775\ 48$ — e' = second eccentricity
- $f = 0.003\ 352\ 810\ 681\ 18$ — flattening
- $f^{-1} = 298.257\ 222\ 101$ — reciprocal flattening
- $Q = 10\ 001\ 965.7293$ m — meridian quadrant
- $R_1 = 6\ 371\ 008.7714$ m — mean radius $R_1 = (2a + b)/3$
- $R_2 = 6\ 371\ 007.180$ m — radius of sphere of same surface
- $R_3 = 6\ 371\ 000.7900$ m — radius of sphere of same volume

Derived physical constants
- $U_0 = 6\ 263\ 686.0850 \times 10$ m^{2}s^{-2} — normal potential at ellipsoid
- $J_4 = 0.000\ 002\ 370\ 912\ 22$ ⎫
- $J_6 = 0.000\ 000\ 006\ 083\ 47$ ⎬ spherical-harmonic coefficients
- $J_8 = -0.000\ 000\ 000\ 014\ 27$ ⎭
- $m = 0.003\ 449\ 786\ 003\ 08$ — $m = \omega^2 a^2 b/GM$
- $\gamma_e = 9.780\ 326\ 7715$ ms^{-2} — normal gravity at equator
- $\gamma_p = 9.832\ 186\ 3685$ ms^{-2} — normal gravity at pole
- $f^* = 0.005\ 302\ 440\ 112$ — $f^* = (\gamma_p - \gamma_e)/\gamma_3$
- $k = 0.001\ 931\ 851\ 353$ — $k = (b\gamma_p - a\gamma_e)/a\gamma_e$

The Geodetic Reference System 1980 was adopted at the XVII General Assembly of the International Union of Geodesy and Geophysics in Canberra, December 1979 [14]. These values are accurate to the number of decimal places given.

the geoid from the ellipsoid will result from a geocentric datum. In the conterminous United States, the change in the separation of the geoid and ellipsoid for a global datum, as compared to the NAD 27, will be a maximum of about 35 m. Because this affects the reduction of distances by an amount approximately equal to 35/6,378,137, or 1/180,000, this quantity concerns only those of us involved in extremely accurate surveys. It is probably fair to speculate that at least 90 percent of the U.S. surveying community could ignore this error for nearly all of their surveys. Certainly, it is true for surveyors involved in lot surveys, subdivision layout, and similar activities.

Considering the global applications, and weighing all considerations carefully, the countries involved in the project formally have decided the NAD 83 would be a geocentric datum, or best fitting in a global sense. This decision means that the center of the ellipsoid must coincide with the origin of the coordinate system—the center of mass.

Datum Reference

The orbits of the Navy Transit satellites have been determined by the DoD, and Doppler signals from these satellites are used when one determines positions using Doppler receivers. If the orbit is considered to be known, then positions referred to the orbit are determined. These positions are, to a very close approximation, referred to the center of mass of the Earth because the satellite orbit is clearly referenced to a mass center. By including the geocentric Doppler positions in the adjustment, the datum will be referenced to the center of the mass of the Earth. The directions of the coordinate axis of the datum can be determined by defining a pole and reference for longitude.

In space, the direction of the rotational axis of the physical Earth is not invariant with time. Many reasons account for this condition. (See [15] for a complete discussion.) Here, we need to consider only the relationships between the instantaneous axis of rotation (pole) of the Earth; the axis of the reference figure, which is the semi-major axis of the ellipsoid; and a mean or established direction of the axis of rotation. A successful orientation of the datum would be accomplished if we make the axis of the ellipsoid coincide with the mean value of the axis of rotation (mean pole). This orientation has also been agreed upon by the countries involved in the NAD project.

The NGS has been observing polar motion, which is the difference in direction between the instantaneous axis of rotation and a mean of established directions of the axis of rotation for more than 75 years. These optically observed values have been combined with values from other countries by several international organizations that publish the positions of the mean pole. All pertinent data in the adjustment will be referred to the mean pole and, therefore, the datum will be so oriented. Defining the direction of the pole fixes two of the three required angles. The third is defined by adopting a zero longitude. NGS has agreed to orient the datum by referencing the zero longitude to the Greenwich meridian.

NEW ADJUSTMENT

In contrast to the new datum definition, the new adjustment requires a computational procedure to remove distortions in the coordinates that represent the existing network. It requires far more effort than defining a new datum. The new adjustment places all appropriate national network survey observations into a coherent reference system of known accuracy.

The new adjustment involves the largest computing task ever attempted—the solution of a system of equations with approximately 900,000 unknowns, the latitude and longitude unknowns of the network's 250,000 control points and certain "nuisance" parameters. Planning for such an undertaking has required some innovative approaches to data processing, quite aside from the effort needed to ensure the correctness and completeness of the 2.5 million survey observations included in the process. It is necessary to take a closer look at the individual steps required to complete the project to understand both the magnitude of effort involved and just what is needed to update a continental reference network.

Tasks Necessary to Complete the NAD 83 Project

CONVERT ARCHIVAL DATA TO READABLE FORM

Even with modern electronic computers, the size and difficulty of today's general network adjustment are several times greater than the one undertaken in 1927, considering the increased volume of data and the steps needed to satisfy modern accuracy requirements. All acceptable observational data in the NGS archives associated with conventional horizontal control surveys (horizontal directions, taped and electronically measured distances, vertical angle observations, and astronomic positions and azimuths) have been converted to computer-readable form—a massive effort by any standard. Each observation record includes a source document identifier, an observation date, an indicator of station intervisibility at ground level, the survey observation itself, and an observational standard error based on the type of instrument used and the number of repeat measurements.

DETERMINATION OF PRECISION OF OBSERVABLES

The observables (measured quantities such as horizontal directions or angles, distances, and astronomic azimuths) were obtained by employing a variety of instrumentation that will not be described here. However, to assure a high quality of the coordinates of the horizontal network stations after the new adjustment, it is extremely important to know the precision of the observations. Thus, we will describe the observables in terms of their standard errors. These standard errors always reflect the combined effect of both internal and external errors that affect the observation.

The internal error is obtained from within the sample. Its presence is readily apparent by analyzing the sample, whereas the external error can only be obtained by comparing the mean values of several samples obtained under different conditions. The internal error is basically instrument dependent. The external error, however, depends on factors such as the instrument operator, weather conditions, and observing techniques used [7].

Horizontal Directions

Most observables consist of horizontal directions. The horizontal network contains an average of 10 horizontal directions for each theodolite-occupied station, bringing the total estimated number of horizontal direction observations to 1.7 million.

The precision of horizontal directions, in general, is given

TABLE 2. The Precision of Direction Observations.

Order of Point	Standard Error of a Single Position (arc seconds)	Average Number of Telescope Observations
High Precision	2.0	32
First-Order	2.4	16
Second-Order	2.8	16
Third-Order	3.4	8
Intersection	6.0	4

as a function of the "order" of the two end-points of the line. Table 2 gives the standard error of a single observation and the average number of telescope positions for each order of station [7]. As an example, consider a line originating at a second-order station, going to a first-order station: The lower order station will take precedence, and thus, assuming 16 positions of the telescope, the standard error of the mean of 16 direction observations will be $2.8''/\sqrt{16} = 0.7''$.

Zenith Distances (Vertical Angles)

As distances became easier to measure with EDM (electronic distance measuring) equipment, the need for good elevations for points increased; and the NGS made it a policy to measure zenith distances at all projects. The standard error of the mean for zenith distances is $\sigma_z = 5.0'' + 1.0''/\text{km}$, where km is the distance in kilometers between stations [7].

Astronomic Latitudes, Longitudes, and Azimuths

The most important application of astronomic data is the establishment of Laplace stations—stations where astronomic latitude, longitude, and azimuth have been determined—to orient the horizontal network during adjustment. These corrections are particularly important when the two ends of a line are at substantially different elevations. In mountainous terrain, this correction may reach as much as $1''$; however, it is generally much smaller ($<0.5''$) for the conterminous United States [7].

Table 3 gives the estimated accuracy, in terms of standard error, of Laplace stations determined by the NGS during the last 130 years. Although the latitude and azimuth error

TABLE 3. Estimated Accuracy of Laplace Stations.

Type of Determination	Internal Precision (arc seconds)	External Precision (arc seconds)	Total Precision (arc seconds)
Latitude	±0.18	±0.26	±0.32
Longitude	±0.20	±0.45	±0.49
Azimuth	±0.3	±1.1	±1.1

TABLE 4. The Precision of Distance Observations.

	Standard Error	
Type of Measurement	Constant (Millimeters)	Distance Dependent (PPM)
Taped Distance Under 50 m	1	
Taped Base Lines	10	1.5
Lightwave EDM Instruments*	15	1.0
Microwave EDM Instruments*	30	3.0

*The above values for Electronic Distance Measurement (EDM) instruments are average values. They may change for specific instruments.

budgets may generally be considered time invariant for the 130-year period, this is not true for astronomic longitudes because of the continuous improvements in timing procedures. As a result, the tabulated longitude error budget is to be considered as a pooled (average) estimate over the 130-year period [18].

Distance Observations

The standard error of distance observations is composed of two parts: an instrument constant standard error and a distance-dependent standard error. Using the notation ppm (parts per million) of the distance, Table 4 lists mean values of standard errors in the four broad categories of distance measurements [7].

Position Observations

Preliminary investigations into the precision of Doppler point positioning have shown that the standard error of each coordinate (X, Y, Z rectangular Earth-centered coordinates) is of the order of 1.0 m [7]. However, in the conterminous United States, approximately 50 stations of the 150 Doppler-station network are located on the high precision Transcontinental Traverse (TCT). An analysis by NGS of the comparison of Doppler satellite positions to TCT positions showed that the estimated root mean square (rms) accuracies for the Doppler positions were 50 cm for latitude and 75 cm for longitude. Hothem [9] indicates that the estimated rms accuracy for height-above-ellipsoid is 50 cm.

DOPPLER SATELLITE POSITIONING PROGRAM

Two significant field observing programs were conducted for the new adjustment in terms of the effort involved and for their importance in establishing a precise framework for NAD 83. For one of these programs, NGS established a primary network of approximately 600 Doppler satellite stations. This network provides scale and orientation control for the new adjustment as well as the relationship of the new datum to the Earth's center of mass.

Using the Doppler principle, if the parameters of a satellite's orbit are well-known, signals from the satellite can be used to compute a position on the Earth's surface. As mentioned before, the satellite's orbit is closely referenced to the Earth's center of mass; therefore, the positions computed from the Doppler observations are also referenced to the Earth's center of mass. With the stations of the primary Doppler network spaced at intervals of up to 300 km, this network provides relative accuracies in azimuth and distance between stations on the order of one part in 200,000 or better [9].

TRANSCONTINENTAL TRAVERSE

The second of these field programs, called the high-precision Transcontinental Traverse (TCT), further improves scale and orientation of the National Horizontal Geodetic Network. TCT comprises super-precise length, angle, and azimuth measurements in somewhat rectangular loops covering the continental United States. Using a specially designed observing scheme of elongated polygons, TCT provides position control of approximately one part in 1,000,000 between connected stations. In this project, two different observers measured each line on different nights using at least two high-precision EDM instruments. The instrument frequencies were checked each week, and those recorded results became part of the observation record for that week.

All distance measurements for the TCT project were made from towers at least 10 m in height, to obtain a representative value for the refractive index along each line. Atmospheric pressure, temperature, and humidity values were recorded at the end-points of the lines, with mid-line temperatures obtained using a remote-reading thermometer supported by a balloon. In addition, first-order astronomic position and azimuth observations were made at the connecting stations of each polygon. The azimuth observations were also taken on two nights with a different observer using a different instrument each night. Observed horizontal directions to all adjacent stations were generally included along with the direction to Polaris.

This project remains a remarkable accomplishment, in terms of the accuracies achieved and the scope of the project (22,000 km of ultra-precise measurements). Combined with the Doppler satellite positioning program, it provides a uniform standard of accuracy for the network in all regions of the country and a rigidly accurate framework for the NAD 83 reference system.

DATA BASE MANAGEMENT SYSTEM

Organizing, storing, and manipulating the huge volume of heterogeneous data associated with more than 250,000 network stations that have accumulated over a century pose distinct challenges. Consider the data base information for just one of these stations: station name and date of establishment, latitude, longitude, height above vertical reference surface, height above the reference ellipsoid, state plane coordinate zone identifier, source number for the field project in which the station's position was computed, quad identifier which indicates the 7.5-minute quadrangle in which the station is located, quad sequence number to locate the station in that particular 7.5-minute quadrangle, a code indicating the order of accuracy of the main-scheme network containing the station, and a code to indicate the surveying method (triangulation, traverse, resection, other) used to determine the station's position.

All of the above information covers only the position record for a particular point. In addition to position records, the data base contains observational records, including taped and electronically measured distances, horizontal angle measurements, astronomic observations, and Doppler-derived coordinates. The data base also houses a descriptive record for each station, including instructions on how to reach the station; ownership of the property on which the station is located; the type of station mark and how it is set; the specific designation stamped on the primary station mark and its associated azimuth and reference marks, and the type of mark used for those associated points; and information on the location and condition of the survey marks obtained when the site was visited. All in all, the horizontal data base contains 2 gigabytes of data.

PROJECT-LEVEL VALIDATION

This first major element of the recomputation process necessitated a minimally constrained least-squares adjustment on each of the more than 5,000 field survey projects contained in the NGS archives. In this phase of the project, the preliminary adjustment of each field project verified the mathematical consistency of the observations within the project. These least-squares adjustments held the position of one station fixed and included one distance and azimuth to provide scale and orientation for the project.

BLOCK-LEVEL VALIDATION

The next step of the new adjustment involved block validation and data entry. Here, the automated and provisionally adjusted observational data from the individual field projects were partitioned into blocks, each containing roughly the same volume of data. Data from all automated projects containing geographic positions in a well-defined area were combined and checked for completeness and consistency among projects observed in different years. Observations were cross-checked, duplicates deleted, errors in preliminary positions resolved, geographic positions in the automated field data compared with those in the NGS data base to ensure all stations contained supporting observations, and station description and recovery notes were compared to verify the accuracy of the automation process.

These verified and provisionally adjusted field project data were then combined with all the observational data stored in the NGS data base for those stations located in that block. The combined data were reformatted into one data module and subjected to a least-squares adjustment to validate the mathematical consistency among data from different field projects. Any internal singularities (an insufficient number of observations to determine a point's position) or weaknesses were evaluated and resolved. Finally, the validated block of observational data was entered into the NGS data base, but not before making a thorough check of the data to ensure it would fit correctly into the scheme of neighboring blocks [11].

HELMERT BLOCK ADJUSTMENT

With modern computers, adjusting small-scale geodetic networks can be done using manageable, if not simple, computational procedures. As the network grows larger, however, the computational requirements grow rapidly to the degree that the system of equations describing the large-scale geodetic network is too large to fit into the memory of a single computer. The solutions to this challenge include exploiting the symmetry and sparseness of the equation matrices and using peripheral storage hardware to augment the storage capabilities of the main-frame computer.

The principle equations of the adjustment algorithm (normal equations) are matrices in which the elements above the main diagonal of the matrix are identical in value to the corresponding elements below the diagonal. NGS found that valuable computer storage space could be saved by forming, storing, and using only the upper triangular portion of each matrix. In fact, this tactic cuts storage requirements nearly in half [8].

Sparseness is another attribute of the normal equation matrix which has been exploited to facilitate network adjustment. In a typical horizontal geodetic network, a station may be connected by direction or distance observations to about ten other stations. These connections appear in a normal equation matrix as nonzero terms at the intersection of the rows and columns containing coefficients of connected stations. In no instance, however, is every station connected to every other station. The result is that a typical normal equation matrix contains many more zero elements than nonzero elements. By carefully rearranging the coefficients of the unknowns, the number of "fill-in" elements can be reduced drastically, with a corresponding reduction in storage requirements and computing time [8]. An additional benefit of reducing the number of computations in this way is a reduction in the accumulated error due to "rounding-off" in the literally billions of arithmetic operations undertaken to adjust a large geodetic network [12].

After taking advantage of these matrix characteristics and significantly reducing storage requirements for the normal equations, the storage space needed for the equations is still much too large to be accommodated in the central memory of the computer. Therefore, a third device must be used to make such large adjustments possible—the use of peripheral storage hardware such as random-access disks. With this technique, the system of equations needed for the network adjustment is partitioned and stored externally on tapes or disks. These can be moved into and out of the central memory on command when they are needed for their part in the reduction and solution of the entire system.

Even after exploiting all these opportunities to streamline the "number-crunching" process, modern computers would still not be able to adjust a large-scale network such as NAD 83 without applying a technique geodesists call Helmert blocking. As geodetic networks become larger, the computational requirements to adjust them grow immensely. For example, the final procedure in adjusting NAD 83 requires solving a system of equations with 900,000 unknowns. To do this without taking advantage of symmetry and sparseness, would require 900,000-squared, or 250 billion words of storage space for just the final form of the normal equations. The number of arithmetic operations (multiplications, divisions, subtractions, and additions) needed to solve such a matrix is on the order of 900,000-cubed separate computations. Beyond that, the number of computations needed to form and reorder the matrix is even greater than the quantity needed to solve it in its final form!

The sheer size of such a task is in itself a deterrent, and that is how the Helmert blocking strategy offers a solution. Helmert blocking has been routinely employed in satellite orbit analysis since the early 1960s. Using this technique, a large problem is divided into several manageable sub-problems. The result is that by combining the separate solutions to the subproblems we obtain exactly the same result as if we were solving the original problem in a single pass. For the Helmert blocking approach to the NAD 83 adjustment, the entire national network partitioned into approximately 161 computer-manageable sub-networks, each consisting of 1,500 to 2,000 stations.

The Helmert blocking phase of the NAD 83 adjustment began after completion of block-level validation. At this point the preliminary positions for the national network's control points and the observations constituting the network have been verified and entered into the NGS data base. Helmert blocking software extracts from the data base chunks of the national network called first-level Helmert blocks, each containing the positional and observational information for approximately 2,000 network stations. Using all of the observational data associated with those stations located inside the block boundary, the software creates a normal equation system for the stations in that block. The size of these blocks and the locations of their boundaries are determined in part by the availability of computer resources, since it requires a significant amount of internal memory to form and reorder the normal equations for networks of this size.

Another factor in determining the boundaries of these blocks includes limiting the number of stations connected by observations to control points located in another block. These stations are called junction points. When the Helmert blocking software creates a normal equation matrix for the first-level block, there will be insufficient information to form all of the necessary matrix coefficients for the junction points, since part of the observational information needed to form the coefficients must come from another block. Therefore, the Helmert blocking system must postpone completion of the accumulation of junction point coefficients until the area being considered contains all of the necessary information. If the block boundaries are defined in a way that decreases the number of junction point stations, then the software system will have to postpone the solution for a fewer number of stations until the higher levels of the scheme. The result is that minimizing the number of junction point stations reduces the total number of Helmert blocks and the levels of blocking, thus reducing computer storage and time.

The software system can, however, proceed to form and reorder a normal equation matrix for those points in the block which have the needed observational information contained totally within the block boundaries. These points are called interior points. After the system has fully formed and reordered the equation matrix for the interior points, and partially formed the equation matrix for the junction points, the process is halted. The set of equations for the interior points is then temporarily eliminated from the Helmert blocking strategy and stored on an external storage device (a numbered magnetic tape). The partially formed sets of equations from the junction points of adjacent blocks remain and are combined to form a series of second-level blocks.

The Helmert blocking software system then forms and reorders a matrix for the interior points of the second-level block, a block consisting of first-level junction points. As with first-level blocks, the fully formed matrix coefficients are temporarily eliminated and stored on a magnetic tape, while the partially formed sets of coefficients from the second-level junction point blocks remain to form a third-level block. This procedure is repeated through as many levels as necessary until all of the observations in the network are processed [10].

At the top of this hierarchy, the Doppler satellite observations plus other space geodesy data, e.g., Very Long Baseline Interferometry (VLBI) observations and the results of the reduced Canadian data for the border junction points are introduced into the solution process. This establishes the best three-dimensional control for the new reference system and relates it to the Earth's center of mass. The Doppler data will be introduced as observed values, assigned appropriate weights, and allowed to accept corrections in the solution process. The solution of the highest level block will result in a set of "unknowns" for all the highest level junction points and Doppler satellite points.

The Helmert blocking software system then uses a back-substitution procedure to solve for all the unknowns associated with the highest level of junction points. The resulting values are then introduced into the matrix for the next lower level and the back-substitution process continues, with the software system solving unknowns for each subsequent lower level.

This back-substitution process continues downward through the approximately nine levels of the Helmert blocking scheme. If a second or third iteration through the Helmert blocking scheme is needed, then the preliminary positions used for each of these later iterations will be updated by the solution results obtained from the previous iteration. Before any subsequent iteration is attempted, a thorough statistical analysis is conducted on the results produced from the previous iteration. A final satisfactory solution occurs when the corrections to the preliminary position of the control points lie within a specified tolerance.

The NAD Helmert blocking strategy and the software system designed for the Helmert block adjustment possess many distinct advantages. One advantage of applying this technique is that it provides the ability to detect and evaluate intermediate problems as the network adjustment progresses. This ability to intercede in the adjustment process furnishes distinct information pathways that can be used to solve coding and observational blunders or numerical singularities. Errors in the observational data, for example, are automatically traced back to the first level. This avoids the unfortunate possibility that a simultaneous adjustment of the entire national network would be unable to iterate to a satisfactory solution, and that identifying the causes of such a breakdown would be as difficult as trying to find a needle in a haystack.

Helmert blocking strategy and its related software system also offer the project management a flexibility in allocating personnel and computer resources needed for the readjustment project. The software system directing the Helmert blocking scheme operates automatically. Once all first-level blocks are defined, the system proceeds on its own up to the highest level, creating its own job runs and job control language, and maintaining a list of activities it has conducted and the results of those activities. The software system automatically calls for each of the approximately 700 sequentially numbered tapes containing the set of normal equations for each block in the scheme. It then schedules each for the necessary operations in the computer central processing unit.

DEVELOPING THE MATHEMATICAL MODEL

To reduce horizontal observations made from a particular point to the ellipsoid, we need to know the height of that point with respect to the ellipsoid (not to be confused with heights above or below the geoid or mean sea level) and the direction of gravity at that point.

This classical approach to computing positions by reducing observations to the ellipsoid can be simplified by merely computing in three dimensional space, without first reducing the observed values to a reference ellipsoid. Astronomic latitudes and longitudes (either observed or appropriately computed) are needed for all points where horizontal directions are observed, and these values are held fixed [21]. The adjusted observations are unoriented horizontal directions, astronomic azimuths, and spatial distances; and two coordinate unknowns are associated with each point. If we keep the heights of the points fixed, then we have a height-controlled, three dimensional system, the mathematical model chosen for the NAD 83 adjustment.

In this height-controlled, three-dimensional system, the distance between two points is a segment of a straight line and the horizontal direction between them is one that is measured in the plane perpendicular to the local direction of gravity at the standpoint. This method avoids the time-consuming computation of geodetic azimuths and distances reduced to the reference ellipsoid before the iterations of an adjustment. It uses the relatively simple equations of three-dimensional geodesy to produce X, Y, Z coordinates. The three-dimensional coordinates can later be referenced to the chosen ellipsoid to transform them into geodetic latitudes and longitudes.

The height-controlled adjustment method is conceptually much simpler than the classical approach. It is more efficient because it avoids numerous trigonometric functions and complicated computations on the ellipsoid. No restrictions are imposed on the lengths of the lines nor on the extent of the network.

GEODETIC DATA SHEET

A new Horizontal Control Data Sheet (Figure 3) was designed to support both the NAD project and automated processing and printing. Like the old form, the new data sheet will contain: the station name, the order and class of accuracy for the station and the surveying method (triangulation, traverse, or intersection) used to determine its position, the geodetic latitude and longitude, the state plane coordinates, the azimuth to one or more azimuth mark(s) (from north), the year established (and reestablished, if applicable), the source number for the project in which the station position is computed, and the narrative description of the station location along with the directions and distances to the azimuth and reference marks associated with the station. Table 5 summarizes the density of horizontal control stations by state.

The new form of the geodetic data sheet will contain all of the above information, plus the following elements useful for a large variety of geodetic data applications: the geoid height value, the title of the organization (abbreviated) which established the station and the one which adjusted the data, the method used to determine the station's elevation, the reference datum for the control point information, and in the description section, the "pack time" and mode of transportation used to reach the station mark. Both Universal Transverse Mercator coordinates and state plane coordinates are given for the point, each expressed in metric units only. In addition, the new data sheet describes both the type of

```
NOAA - NATIONAL OCEAN SERVICE                        HORIZONTAL CONTROL DATA
OFFICE OF CHARTING AND GEODETIC SERVICES             NORTH AMERICAN DATUM 1983
NATIONAL GEODETIC SURVEY                             SEPTEMBER 1984


HORIZONTAL CONTROL STATION:  HILLCREST

   GEODETIC     DEG MIN SEC          SIGMA

   LATITUDE:    29 28 XX.XXXXXN  ±X.XXXXX SEC         QUAD: N29098311  QSN - OO16
   LONGITUDE:   98 34 XX.XXXXXW  ±X.XXXXX SEC         TEXAS - BEXAR COUNTY
                                                      CONTROL DIAGRAM:  NH 14-8, SAN ANTONIO
         CIRCULAR STANDARD ERROR:  ±X.XX METERS       USGS QUAD SHEET:  SAN ANTONIO WEST
         CLASSIFICATION: 1ST ORDER, CLASS I           PROJECT ACCESSION NUMBER: G00270

   AZIMUTH:     223 56 XX.XX       ±X.XX SEC       AZIMUTH MARK:  HILLCREST AZ MK 1943
     (FROM NORTH)
   ELEVATION:   296.2 METERS       ±X.X METERS    ELEVATION DETERMINED BY TRIGONOMETRIC LEVELING AND
                972 FEET           ±X FEET        OBSERVED BY THE COAST AND GEODETIC SURVEY IN 1929.
   GEOID HEIGHT:  XX.X METERS      ±X.X METERS    VERTICAL DATA REFERRED TO NGVD 1929.

   DEFLECTION OF VERTICAL AND SIGMA IN SECONDS:  MERIDIAN +XX.X ±X.X   PRIME VERTICAL +XX.X ±X.X

   THE GEODETIC COORDINATES WERE ESTABLISHED BY CLASSICAL GEODETIC METHODS AND ADJUSTED BY THE
   NATIONAL GEODETIC SURVEY IN 1985.  EARLIEST OBSERVATIONS WERE MADE BY THE COAST AND GEODETIC
   SURVEY IN 1924.

   NAD27 TO NAD83 SHIFT AT THIS STATION IN SECONDS:  LATITUDE +X.XXXXX  LONGITUDE +X.XXXXX

STATE PLANE AND UNIVERSAL TRANSVERSE MERCATOR COORDINATES:

                  X/EASTING    Y/NORTHING   POINT SCALE   CONVERGENCE     GRID AZIMUTH*
    GRID   ZONE    METERS        METERS       FACTOR      DEG MIN SEC      DEG MIN SEC

    TX-L    SC   XXXXXX.XXX   XXXXXXX.XXX   X.XXXXXXXX    +X XX XX.X      XXX XX XX.X
    UTM     14   XXXXXX.XXX   XXXXXXX.XXX   X.XXXXXXXX    +X XX XX.X      XXX XX XX.X
                                        *ARC-TO-CHORD CORRECTION ASSUMED ZERO

STATION MARKS AND REFERENCE OBJECTS INFORMATION:

     CODE          REFERENCE OBJECT        HEADING    DISTANCE          DIRECTION

     DOO      HILLCREST AZ MK                SW      APPROX 0.25 MI    000 00 00.0
     DOO      HILLCREST RM                   NE      26.650 METERS     209 50 10
     D55      HILLCREST RM 2                 S       17.320 METERS     312 37 53

     DOO IS A SURVEY DISK DESCRIBED IN THE DESCRIPTION.
     D55 IS A SURVEY DISK SET INTO A DRILL HOLE IN ROCK OUTCROP.

        THE SURFACE STATION MARK IS A COAST AND GEODETIC SURVEY DISK SET INTO THE TOP
          OF A SQUARE CONCRETE MONUMENT.
        THE UNDERGROUND STATION MARK IS AN IRON NAIL SURROUNDED BY A MASS OF CONCRETE.

STATION RECOVERY HISTORY:

    YEAR RECOVERED   CONDITION OF MARK   RECOVERED BY (CHIEF OF PARTY)

        1924            ORIGINAL        COAST AND GEODETIC SURVEY (HO)
        1934            GOOD            TOBIN SURVEYS INCORPORATED
        1937            GOOD            COAST AND GEODETIC SURVEY (JSB)
        1937            GOOD            TEXAS HIGHWAY DEPARTMENT
        1939            GOOD            COAST AND GEODETIC SURVEY (HAS)
        1943*           GOOD            COAST AND GEODETIC SURVEY (FBQ)
        1953            GOOD            COAST AND GEODETIC SURVEY (FBQ)
        1953            GOOD            COAST AND GEODETIC SURVEY (CTH)
        1953*           GOOD            ARMY MAP SERVICE
    *SEE PUBLISHED TEXT
```

FIGURE 3. Example of horizontal control data sheet. (continued)

```
NOAA - NATIONAL OCEAN SERVICE                    HORIZONTAL CONTROL DATA
OFFICE OF CHARTING AND GEODETIC SERVICES         NORTH AMERICAN DATUM 1983
NATIONAL GEODETIC SURVEY                         SEPTEMBER 1984

CONTINUED FROM PREVIOUS PAGE

HORIZONTAL CONTROL STATION:   HILLCREST
Quad:  N29098311   QSN:  0016

STATION RECOVERY BY THE COAST AND GEODETIC SURVEY IN 1943 (FBQ):

    STATION WAS RECOVERED AS DESCRIBED, WITH THE MARKS IN GOOD CONDITION.  THIS PARTY ADDED A
    REFERENCE MARK 2 AND AN AZIMUTH MARK.  COMPLETE DESCRIPTION FOLLOWS--

    STATION IS ABOUT 8 MILES NW OF THE SAN ANTONIO CITY HALL AND 0.95 MILE NE OF STATE HIGHWAY
    16, ON A VERY PROMINENT HILL ON THE PROPERTY OF M. GOLDSMITH.  IT IS 15 FEET E OF THE FRONT
    ENTRANCE OF A LARGE BROWN CLUBHOUSE.

    THE STATION MARK IS A BRONZE STATION DISK SET IN CONCRETE, STAMPED---HILLCREST 1924---AND
    IS FLUSH WITH THE SURFACE OF THE GROUND.

    REFERENCE MARK 1924, IS NW OF THE STATION.  IT IS A BRONZE REFERENCE DISK SET IN CONCRETE,
    STAMPED---HILLCREST 1924---AND PROJECTS 2 INCHES.

    REFERENCE MARK 2 IS SE OF THE STATION AND 60 FEET SE OF THE SE CORNER OF THE CLUBHOUSE.
    IT IS A BRONZE REFERENCE DISK SET IN A DRILL HOLE IN OUTCROPPING BEDROCK, STAMPED
    ---HILLCREST NO 2 1943---AND IS FLUSH WITH THE SURFACE OF THE GROUND.

    THE AZIMUTH MARK IS A BRONZE AZIMUTH DISK SET IN CONCRETE.  IT IS S OF THE STATION, 5 FEET
    SE OF THE SE END OF AN ORNAMENTAL STONE FENCE, AND 2.5 FEET E OF A WHITE REFERENCE POST.
    IT IS STAMPED---HILLCREST 1943---AND PROJECTS 8 INCHES.

    TO REACH THE STATION FROM THE JUNCTION OF U.S. HIGHWAY 90 AND STATE HIGHWAY 16 AT THE SAN
    ANTONIO CITY HALL, GO N ON HIGHWAY 16 FOR 2.05 MILES AND TAKE THE LEFT FORK, HIGHWAY 16,
    AT THE JUNCTION OF HIGHWAY 16 AND U.S. HIGHWAY 87.  GO NW FOR 4.7 MILES AND THEN TURN
    RIGHT OFF THE HIGHWAY AND ONTO BROADVIEW DRIVE, PASSING BETWEEN TWO STONE PILLARS.  GO 0.7
    MILE TO THE AZIMUTH MARK AT THE STONE FENCE AND KEEP STRAIGHT AHEAD.  GO 0.25 MILE TO THE
    LARGE FRAME CLUBHOUSE ON THE LEFT AND THE STATION.

    STATION CAN BE REACHED BY CAR.

STATION RECOVERY BY THE ARMY MAP SERVICE IN 1953:

    STATION WAS RECOVERED IN GOOD CONDITION.  IT IS IN CLEARED AREA, EAST OF, OR IN FRONT OF
    HOUSE, 29 FEET S-SE OF THE NE CORNER OF HOUSE, AND 30 FEET E-NE OF THE SE CORNER OF
    HOUSE.  IT IS 13 FEET N-NE OF A TREE, AND 26 FEET SOUTH OF A TREE.
```

FIGURE 3. Example of horizontal control data sheet.

monument used to permanently mark the control point and the monument's magnetic property.

EXPECTED COORDINATE SHIFTS FROM NAD 27 TO NAD 83

Now that the procedures for the new adjustment itself and for updating the geodetic network diagrams have been outlined, it is important to examine the effects of the adjustment on related cartographic products. These, in turn, influence a broad spectrum of activities and user groups. Specifically, cartographers need to know how much and in which direction the coordinates of map corners will change after the NAD 83 adjustment. Anticipating this need, NGS describes these expected changes in two publications [19,20]. Succinctly specified are the principal methodology, ellipsoidal parameters, and mathematical formulas used to compute the changes, as well as both the mean and the range of the expected coordinate shifts.

Figures 4, 5, and 6 show the expected coordinate shifts in latitude, longitude, and geoid height from NAD 27 to NAD 83. While the magnitude of these changes may appear significant, especially those for longitude in the western United States, it is important to remember that these values represent average shifts over large geographical areas. For the most part, relative position changes between control points on the local level will be slight. What the coordinate shifts really represent are the restoration of a coherent and highly reliable national network, eliminating the annoyances and major obstacles resulting from the dependence on an outdated reference system.

NGS ASSISTANCE IN TRANSFERRING COORDINATES TO NAD 83 VALUES

To aid the surveying community in transforming current (NAD 27) geodetic coordinates to NAD 83 values, NGS will transform or assist in transforming coordinates based on any surveying system (local rectangular, state plane, Universal Transverse Mercator—UTM, or geodetic). In 1983 NGS issued a policy statement describing the

TABLE 5. Projected Horizontal Control for New Adjustment of the North American Datum 1983.

States*	Stations**	Land area (sq. mi.)	Spacing average (one station per no. sq. mi.)	States*	Stations**	Land area (sq. mi.)	Spacing average (one station per no. sq. mi.)
AL	1,681	51,609	30.7	MT	2,188	147,138	67.2
AK	24,724	586,412	23.7	NE	2,391	77,227	32.3
AZ	3,723	113,909	30.6	NV	2,856	110,540	38.7
AR	2,389	53,104	22.2	NH	563	9,304	16.5
CA	15,417	158,693	10.3	NJ	2,586	7,836	3.0
CO	2,706	104,247	38.5	NM	3,911	121,666	31.1
CT	2,174	5,009	2.3	NY	8,437	49,576	5.9
DE	541	2,057	3.8	NC	14,405	52,586	3.7
DC	634	67	0.1	ND	2,294	70,665	30.8
FL	10,791	58,560	5.4	OH	2,528	41,222	16.3
GA	7,595	58,876	7.8	OK	2,428	69,919	28.8
HI	2,565	6,450	2.5	OR	6,570	96,981	14.8
ID	2,887	83,557	28.9	PA	3,268	45,333	13.9
IL	3,896	56,400	14.5	RI	1,159	1,214	1.0
IN	2,712	36,291	13.4	SC	2,726	31,055	11.4
IA	2,021	56,290	27.9	SD	2,378	77,047	32.4
KS	3,568	82,264	23.1	TN	2,578	42,244	16.4
KY	2,354	40,395	17.2	TX	10,786	267,338	24.8
LA	5,728	48,523	8.5	UT	1,561	84,916	54.4
ME	4,245	33,215	7.8	VT	524	9,609	18.3
MD	4,714	10,577	2.2	VA	6,513	40,817	6.3
MA	5,151	8,257	1.6	WA	9,699	68,192	7.0
MI	3,165	58,216	18.4	WV	983	24,181	24.6
MN	5,314	84,068	15.8	WI	2,994	56,154	18.8
MS	3,096	47,716	15.4	WY	1,349	97,914	72.6
MO	3,272	69,686	21.3	Totals	224,738	3,615,122	16.1**

*U.S. territories and possessions not included.
**Station totals extracted from NGS data base as of January 1980. About 25,000 more stations have been added.

assistance it will provide in converting coordinate values and citing the requirements for submitting data. The policy statement lists three methods for converting coordinate values, described in detail below, and states that the acceptability of the transformation method used will be determined on a case-by-case basis. The most important criterion for a suitable coordinate conversion is the existence of a sufficient number of points in common between the system containing the submitted data and the National Geodetic Reference System (NGRS).

Rigorous Adjustment

The first method is a rigorous adjustment of original field surveying observations. In this procedure a requestor submits the original field observations and station descriptions to NGS for adjustment to NAD 83. The conditions necessary for this rigorous adjustment are that the survey points must be permanently monumented and described; the survey must have been performed to third-order, class I accuracy standards or better; the survey data must be connected by observations to national horizontal network points; and the observations and descriptions must be submitted in a prescribed format [5]. NGS prefers this method because it eliminates distortions that may have been present in the NAD 27 network and ensures that new surveying data are included in NGRS. This method is also advantageous for those submitting data because NGS will perform a least-squares adjustment on the observations and publish the results.

Approximate Method

The second method in the policy statement describes a rigorous coordinate conversion that can be performed by NGS or by the originator. It has been called the "approximate method," in contrast to the above preferred "exact method" of adjusting the original field observations. The

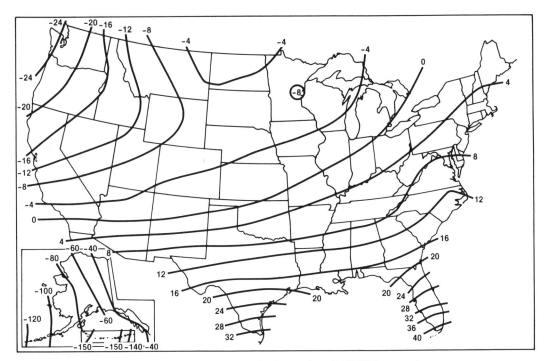

FIGURE 4. Expected latitude change from NAD 27 (in meters).

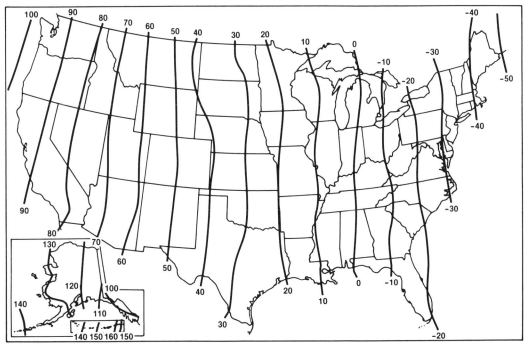

FIGURE 5. Expected longitude change from NAD 27 (in meters).

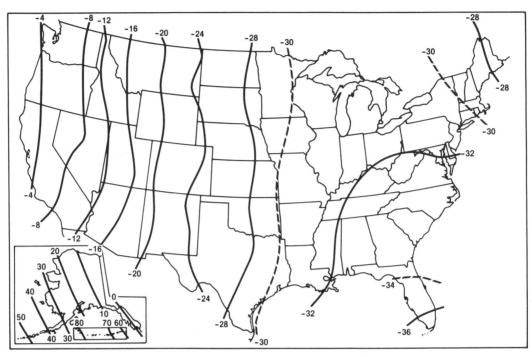

FIGURE 6. Expected geoid height change from NAD 27 (in meters).

procedure for this "approximate" method is as follows: NAD 27 state plane, UTM, or local rectangular coordinates are converted from X and Y values to latitudes and longitudes, the NAD 27 latitudes and longitudes are converted to NAD 83 latitudes and longitudes using a least-squares transformation program and the NAD 83 latitudes and longitudes are converted to UTM or to state plane coordinates using the NAD 83 state plane coordinate system.

The necessary conditions for this method are that (1) the data must satisfy the established minimum requirement for using a least-squares transformation program, that is, a minimum of four common points (the same physical point with coordinates in both reference systems) distributed uniformly throughout the area containing the coordinates, (2) the coordinates must be submitted in a prescribed, computer-readable format, and (3) the coordinates must be given in terms of geodetic (latitude and longitude), UTM, state plane, or local rectangular values. This method does not remove NAD 27 network distortions, and as a result, *NGS will not publish the results.*

Simplified Transformation

The third method, called a "simplified transformation," involves averaging the shifts in latitude and longitude for an area containing stations with both NAD 27 and NAD 83 latitudes and longitudes. In this method the NAD 27 UTM or state plane coordinates are converted from X and Y to latitude and longitude values. Next the average latitude shift and average longitude shift for the area of the survey are used to convert the NAD 27 coordinates to NAD 83 values. The NAD 83 latitudes and longitudes are then converted to UTM or to state plane coordinate values using the NAD 83 state plane coordinate system. Since this third method uses only average coordinate shifts for a project area, the results will likely not be an accurate representation of the actual shifts at each station. As a result, NGS does not advocate using this method, and will provide only technical advice on its usefulness in cartographic applications.

PLANE COORDINATE SYSTEMS AFTER NAD 83 ADJUSTMENT

From the viewpoint of the surveyor, engineer, or other primary user of geodetic information, possibly the single most important product of the new adjustment is the plane coordinate system or systems that will be used to relate local surveying work to NGRS. Even though computers have made it a relatively simple task to adjust and compute surveying data referenced to the ellipsoid, it is likely that most surveying professionals will continue to favor plane coordinates over geodetic positions for some time [4]. The current plane coordinate reference system is derived from NAD 27 geodetic coordinates and applies the mathematics of the Lambert or Mercator map projections to 127 zones that follow county boundary lines [13].

An evaluation of the type of plane coordinate system to best serve the nation resulted in a decision to retain the existing State Plane Coordinate System (SPCS). There are three salient reasons for this decision. First, SPCS had been accepted by the legislatures of 35 states; second, the system has been used for more than 40 years; and third, the procedures used to compute coordinates, and the results obtained, were fundamentally sound [4]. The fact that SPCS includes a large number of separate zones and that it uses several different projections was of little consequence in light of the universal availability of electronic calculators and computers.

As a result, NGS issued a policy statement endorsing the use of both reference systems for surveying and mapping purposes after the new adjustment. This policy was described and published in the *Federal Register* [6]. Excerpts from this notice follow:

Policy on Publication of Plane Coordinates

The National Ocean Service (formerly National Ocean Survey), determined it is in the best interest of the surveying and mapping community that two plane coordinate systems be published and supported beginning in 1983 with the NAD redefinition. These two systems will be identified as the "State Plane Coordinate" (SPC) and the "Universal Transverse Mercator" (UTM) systems.

The UTM system will consist of the transverse Mercator projection as defined in Chapter 1 of the 1958 Department of Army Technical Manual-TM5-241-8, changing only the definition of the datum. The SPC will consist of the same projections and defining parameters as published in USC&GS Special Publication 235 (1974 revision) [15] and legally adopted in 35 states, except for the following changes:

1. The grid will be marked on the ground using the 1983 NAD.
2. Distances from the origin will be expressed in meters and fractions thereof. One additional decimal place should be used for the metric expression of a value previously expressed in feet.
3. The arbitrary numeric constant, presently assigned to the origin, will be unchanged but will be considered as meters instead of feet, except for the following: If a state elects to have a different constant(s) assigned to the origin so that the 1983 NAD plane coordinates will appear significantly different from the 1927 NAD positions, when considering the overall system, then the NGS will consider changing the origin constant. If the state so elects, it must amend its legislation to accommodate this change.
4. Michigan's transverse Mercator system will be eliminated in favor of the legislatively approved Lambert system.
5. Projection equations will be programmed such that the maximum computing error of a coordinate will never exceed 0.1 mm when computing the coordinate of a point within the zone boundaries.

State amendments to the existing projections will be based upon the desires and needs within the states, recommendations of the NGS, and among other things will consider the following items:

1. Refinements to eliminate (for example):
   - Negative Y coordinates for certain islands on the Maine east zone.
   - Negative X coordinates for points on the Dry Tortugas on the Florida east zone.
2. Urbanization that requires either different parameters for existing zones or additional zones such that a metropolitan area would be located in a single zone. For example:
   - New York City
   - Chicago
   - Cincinnati
   - Washington, D.C.
3. A change in the arbitrary origin as discussed above. This can be accomplished in most cases by:
   - Changing the X coordinate constant of 500,000 to 300,000 or 700,000 where the transverse Mercator is used, or changing the X coordinate constant of 2,000,000 to 4,000,000 where the Lambert is used.
   - Changing the Y coordinate constant of zero to 500,000 or 1,000,000.
   - Changing both X and Y.

The NGS will not change projection defining parameters in any state that has legally adopted the State Plane Coordinate System (SPCS) until the state amends its legislation.

VERTICAL NETWORK

Background

The first leveling route in the United States of geodetic quality was performed in 1856–57 under the direction of G. B. Vose, U.S. Coast Survey (now the Office of Charting and Geodetic Services within the National Ocean Service). The leveling survey was required to support currents and tides studies in the New York Bay and Hudson River areas. In 1887 the first leveling line officially designated "geodesic leveling" by the Coast and Geodetic Survey followed an arc of triangulation along the 39th parallel from bench mark A in Hagerstown, Maryland.

By 1900, the vertical control network had grown to 21,095 km of geodetic leveling. These data included work per-

TABLE 6. Elevation of Local Mean Sea Level (MSL) Above National Geodetic Vertical Datum of 1929.

Station number	Station location	MSL*—NGVD 29
8410140	Eastport, ME	0.43 ft. 0.13 m
8413320	Bar Harbor, ME	0.27 ft. 0.08 m
8418150	Portland, ME	0.36 ft. 0.11 m
8419870	Seavey Island, ME	0.40 ft. 0.12 m
8443970	Boston, MA	0.32 ft. 0.10 m
8447930	Woods Hole, MA	0.58 ft. 0.18 m
8452660	Newport, RI	0.37 ft. 0.11 m
8454000	Providence, RI	0.43 ft. 0.13 m
8461490	New London, CT	0.48 ft. 0.15 m
8467150	Bridgeport, CT	0.69 ft. 0.21 m
8510560	Montauk, NY	0.47 ft. 0.14 m
8514560	Port Jefferson, NY	0.53 ft. 0.16 m
8516990	Wilets Point, NY	0.67 ft. 0.20 m
8518750	Battery, NY	0.69 ft. 0.21 m
8531680	Sandy Hook, NJ	0.79 ft. 0.24 m
8534720	Atlantic City, NJ	0.62 ft. 0.19 m
8536110	Cape May, NJ	0.62 ft. 0.19 m
8545530	Philadelphia, PA	1.18 ft. 0.36 m
8557380	Lewes, DE	0.55 ft. 0.17 m
8558690	Indian River Inlet, DE	0.57 ft. 0.17 m
8570280	Ocean City, MD	0.27 ft. 0.08 m
8571890	Cambridge, MD	0.44 ft. 0.13 m
8572770	Matapeake, MD	0.43 ft. 0.13 m
8574680	Baltimore, MD	0.54 ft. 0.16 m
8575512	Annapolis, MD	0.57 ft. 0.17 m
8577330	Solomons Island, MD	0.42 ft. 0.13 m
8594900	Washington, DC	0.71 ft. 0.22 m
8632200	Kiptopeke, VA	0.13 ft. 0.04 m
8638610	Hampton Roads, VA	0.21 ft. 0.06 m
8638660	Portsmouth, VA	0.19 ft. 0.06 m
8635750	Lewisetta, VA	0.51 ft. 0.16 m
8637624	Gloucester Point, VA	0.20 ft. 0.06 m
8651370	Duck Pier, NC	0.22 ft. 0.07 m
8658120	Wilmington, NC	0.81 ft. 0.25 m
8659084	Southport, NC	0.48 ft. 0.15 m
8665530	Charleston, SC	0.52 ft. 0.16 m
8667999	Beaufort, SC	0.43 ft. 0.13 m
8670870	Fort Pulaski, GA	0.42 ft. 0.13 m
8720030	Fernandina Beach, FL	0.41 ft. 0.12 m
8720220	Mayport, FL	0.36 ft. 0.11 m
8721120	Daytona Bch. Shores, FL	0.07 ft. 0.02 m
8773170	Miami Beach, FL	0.43 ft. 0.13 m
8774580	Key West, FL	0.31 ft. 0.09 m
8726520	Saint Petersburg, FL	0.40 ft. 0.12 m
8727520	Cedar Key, FL	0.32 ft. 0.10 m
8725110	Naples, FL	0.54 ft. 0.16 m
8725520	Fort Myers, FL	0.62 ft. 0.19 m
8728690	Apalachicola, FL	0.60 ft. 0.18 m
8729108	Panama City, FL	0.45 ft. 0.14 m
8729840	Pensacola, FL	0.36 ft. 0.11 m
8761720	Grand Isle, LA	1.25 ft. 0.38 m
8770590	Sabine Pass, TX	0.77 ft. 0.23 m

(continued)

TABLE 6. (continued).

Station number	Station location	MSL*—NGVD 29
8771450	Galveston, TX	0.35 ft. 0.11 m
8772440	Freeport, TX	0.31 ft. 0.09 m
8774770	Rockport, TX	0.70 ft. 0.21 m
8779770	Port Isabel, TX	0.19 ft. 0.06 m
9410170	San Diego, CA	0.14 ft. 0.04 m
9410230	La Jolla, CA	0.20 ft. 0.06 m
9410580	Newport Beach, CA	0.08 ft. 0.02 m
9410660	Los Angeles, CA	0.07 ft. 0.02 m
9410840	Santa Monica, CA	0.15 ft. 0.05 m
9412110	Port San Luis, CA	0.10 ft. 0.03 m
9411270	Rincon Island, CA	−0.08 ft. −0.02 m
9413450	Monterey, CA	0.06 ft. 0.02 m
9414290	San Francisco, CA	0.29 ft. 0.09 m
9414458	San Mateo Ridge, CA	0.40 ft. 0.12 m
9414750	Alameda, CA	0.35 ft. 0.11 m
9415144	Port Chicago, CA	0.79 ft. 0.24 m
9415020	Point Reyes, CA	0.29 ft. 0.09 m
9419750	Crescent City, CA	−0.04 ft. −0.01 m
9431647	Port Orford, OR	−0.13 ft. −0.04 m
9432780	Charleston, OR	−0.13 ft. −0.04 m
9435380	South Beach, OR	0.41 ft. 0.12 m
9435827	Depoe Bay, OR	0.34 ft. 0.10 m
9439040	Astoria, OR	1.17 ft. 0.36 m
9440910	Toke Point, WA	0.22 ft. 0.07 m
9443090	Neah Bay, WA	−0.05 ft. −0.02 m
9444900	Port Townsend, WA	0.29 ft. 0.09 m
9447130	Seattle, WA	0.52 ft. 0.16 m
9449880	Friday Harbor, WA	0.12 ft. 0.04 m

*Based on 1960–1978 Tidal Epoch

formed by the Coast and Geodetic Survey, various components of the Corps of Engineers, the U.S. Geological Survey, and the Pennsylvania Railroad. A "mean sea level" reference surface was determined in 1900 by holding elevations fixed at five tide stations. Two other tide stations participated indirectly. Subsequent readjustments of the leveling network were performed by the Coast and Geodetic Survey in 1903 (31,789 km of leveling using eight tide stations); 1907 (38,359 km of leveling using eight tide stations) and 1912 (46,462 km of leveling using nine tide stations) [4].

The next general adjustment of the vertical control network did not occur until 1929. By then the international nature of geodetic networks was well understood, and Canada provided its first order vertical network which was combined with the U.S. network. The U.S. network had grown to 75,159 km of leveling. Canada provided an additional 31,565 km. The two networks were connected at 24 vertical control points (bench marks) that extended from Maine/New Brunswick to Washington/British Columbia. Mean sea level was held fixed at 21 tide stations in the

United States and five in Canada. Although Canada did not adopt the "Sea Level Datum of 1929" determined by the United States, Canadian-U.S. cooperation in the general readjustment strengthened both networks.

Inaccuracies Introduced in the 1929 Datum Results

At the time of the 1929 general adjustment, it was known that local mean sea level at the various tide stations held fixed (at 0.0 elevation) during the adjustment could not, in reality, be considered to be on the same equipotential surface. This was due to the many dynamic forces acting on the oceans. It was thought at that time, however, that the errors introduced by this approach were not significant, being of the same order of magnitude as terrestrial leveling observational errors. However, we now know that errors were introduced into the 1929 general adjustment by considering mean sea level at each tide station to be on the same equipotential surface. The error is estimated to be as much as 1.3 m from Florida to the state of Washington.

Recognition of this distortion and confusion concerning the proper definition of local mean sea level resulted, in 1973, in a change in designation of the official height system from "Sea Level Datum of 1929" to "National Geodetic Vertical Datum of 1929" (NGVD 29) [8]. The change was in name only; the same geodetic height system continues in the United States from 1929 to the present.

Table 6 lists the differences between NGVD 29 and local mean sea level on the 1960–1978 National Tidal Epoch for the National Observation Network Tide stations in the conterminous United States. The tidal epoch is a specific 19-year period adopted by the National Ocean Service as the official time segment over which tide observations are taken and reduced to obtain mean values (e.g., mean sea level) for tidal datums. This standardization is necessary because of the periodic and apparent secular trends in sea level. Since the 1929 vertical adjustment, an apparent secular rise in the sea level has occurred along the coast of the conterminous United States except for a few areas along the west coast. Therefore, mean sea level averaged over the period 1960 to 1978 averages 0.1 m higher than NGVD 29.

Tasks Required by the NAVD 88 Readjustment Project

Approximately 625,000 km of leveling have been added to the National Geodetic Reference System since the 1929 adjustment. In the early 1970s, an extensive inventory of the vertical control network was conducted. The inventory identified thousands of bench marks that had been destroyed, due primarily to post-World War II highway construction as well as other causes. Many existing bench marks had become unusable due to crustal motion associated with earthquake activity, post-glacial rebound (uplift), and subsidence caused by the withdrawal of underground liquids. Other problems were caused by forcing the 625,000 km of

TABLE 7. Source of Distortions in Present National Geodetic Vertical Control Datum of 1929 (NGVD 29).

Source of Distortion	Approximate Amount (m)
"Patching" 625,000 km to Old 75,000 km Net	0.3
Constraining tide gage heights in NGVD 29	0.7
Ignoring true gravity in NGVD 29	1.5
Refraction errors	2.0
Post-Glacial uplift; (Minnesota, Washington, etc.)	0.6
Subsidence caused by withdrawal of underground fluids	9.0
Crustal motions from earthquakes	2.0
Bench mark Frost Heave	0.5

leveling to fit previously determined NAVD 29 height values. Table 7 itemizes these distortions.

CONVERSION OF DATA TO COMPUTER-READABLE FORM

The first major NAVD task to be completed was the conversion to computer-readable form of descriptive data (bench mark descriptions) from paper copy (primarily field records). Today all new bench mark descriptions and recovery notes are automated by NGS field personnel as standard operating procedure for new field projects. This information is merged into existing files with software programmed to adhere to specifications of the Federal Geodetic Control Committee [7]. The descriptions reside on five offline disk packs and can be retrieved by archival cross reference number, state, quadrangle, or county.

The second major task to be completed for the NAVD project was the conversion of archival (historic) observational leveling data to computer-readable form [26]. Because of the large volume of data (1.6 million observed station differences) it was decided that instead of keying individual leveling-rod readings, as is presently done with new NGS surveys, only the stadia intervals, section elevation differences, date, time, "sun code," "wind code," temperature, and number of setups would be converted to computer-readable form.

GEOGRAPHIC POSITIONS FOR BENCH MARKS

To provide automated retrieval capability and apply position-dependent corrections to the observations, a geographic position (latitude, longitude) has been determined for each of the 550,000 bench marks. For those monuments not part of the horizontal control network, the effort involved plotting bench marks on appropriate maps (using the descriptive data mentioned previously) and then determining a "scaled" position using digitizing equipment.

RELEVELING IN SUPPORT OF THE NAVD 88

Basic Framework

An important feature of the NAVD 88 program is the releveling of much of the first-order vertical control network in the United States. The dynamic nature of the vertical control network requires a framework of newly observed elevation differences in order to obtain realistic contemporary height values from the new adjustment. To accomplish this, NGS has identified 83,000 km of the network for releveling.

This basic framework, called Basic Net A, was designed to provide maximum benefit to users, without negatively affecting network geometry. Loop sizes, connections to tide stations, and connections to the Canadian and Mexican geodetic vertical networks were influenced by anticipated future requirements in certain areas. Areas of high priority were coastal areas, major rivers and transportation routes, densely populated areas, and seismically active zones [3].

Other major factors, in order of importance, influenced the selection of lines for releveling:

1. Age. In an area where several first- and second-order lines are available, the line which has not been leveled for the longest period of time is usually selected.

2. Geometry. Guidelines for average loop sizes are 500 km in high-priority areas and 800 km in other areas. A few loops in the Rocky Mountains and in areas of low construction activity, are more than 1,000 km in circumference.

3. Questionable Surveys. Those lines of the net where loop misclosures of previous surveys exceeded tolerance limits are selected.

4. Population Density. Loop sizes are reduced in areas of high population density to provide additional bench marks for users.

5. Tide Station Connections. All leveling lines connecting primary tide stations established by the National Ocean Service are selected if the tide station is less than 1 mile offshore and/or less than 50 km from the nearest Basic Net A line. Secondary and tertiary tide stations are connected to a Basic Net A line if the leveling can be accomplished in ½ day or less.

6. Border Ties to Canada and Mexico. With the cooperation of the other nations of North America, NGS will participate in an adjustment of the geodetic vertical networks in North and Central America. Meetings have been held with representatives of the Geodetic Survey of Canada and the Direccion General de Geografia del Territorio Nacional (DGGTENAL) of Mexico to identify the most beneficial border-junction leveling lines. Twenty-seven connections have been identified along the U.S.-Canadian border, and 13 along the U.S.-Mexican border.

7. Accessibility to the Public: Due to restrictions by railroad companies, as well as for safety reasons, NGS is relocating leveling lines from railroad right-of-ways to highways near the railroads. Interstate highways are avoided if a non-restricted highway is available.

Bench Mark Monumentation

Replacement of disturbed or destroyed monuments precedes the actual leveling. This effort also includes the establishment of highly stable "deep-rod" bench marks, which will provide reference points for future "traditional" or "satellite" leveling systems. Bench marks that are used as reference points for these precise surveys must have high stability and longevity. Prior to the start of any precise survey, the user needs to know the reliability of the bench marks that will be used as reference points. NGS has divided the monuments into the following four classes, (quality codes) based upon their reliability:

A. Monuments of the most reliable nature which are expected to hold their elevations very well.
B. Monuments which probably will hold their elevations well.
C. Monuments which may hold their elevations but are commonly subject to surface ground movements.
D. Monuments of questionable or unknown reliability.

The framework for NAVD 88, includes quality A and B monuments. Although it is desirable to have every monument of quality A, a compromise had to be reached between what is desirable and what is most cost effective. The majority of the monuments established for NAVD 88 are of quality B, with quality A monuments set at 16-km intervals, at junctions of leveling lines and at the "base" of each spur line to water level gages.

Frost heave, certain soil conditions, and local subsidence are some factors which affect bench mark monuments, causing vertical movement. The degree to which monuments maintain vertical stability under the influence of local disturbing effects is a factor in determining the quality of a bench mark. Generally, steel rods or pipes that are driven to a depth sufficient to resist these effects, or disks cemented into large boulders or massive structures, are considered to be of quality B.

NGS currently uses stainless steel rods to establish quality B monuments whenever large boulder or massive structures are not available. The rod sections are made from type-316 stainless steel which is more resistant to corrosion than other affordable alloys. The rod sections are 4 feet long and are connected by stainless steel studs. The rods are assembled and driven to refusal or a minimum depth that is determined from local frost and soil conditions. A type-316 stainless steel cap with a hemispherical end is crimped to the top of the rod as the reference point. A 5-inch diameter PVC pipe, ½ m in length, is placed around the top of the rod for protection. An aluminum logo cap, with a hinged cover for accessing the monument, is stamped with the appropriate bench mark designation and glued to the top of the pipe.

In 1978, NGS developed a method for establishing quality A monuments wherever they are needed. The monument is similar to the quality B stainless steel rod monument. The major difference is that the monument is isolated from local movement caused by frost, soil conditions, and local subsidence by a "sleeve" of PVC pipe. The sleeve is installed by utilizing a drill rig mounted on a large truck. A hole, large enough to accommodate a 1-inch diameter PVC pipe, is drilled to a depth of three times the frost penetration or below the depth of the shrinking and swelling of expansive soils, whichever is greater. A length of 1-inch PVC pipe, approximately 20 cm less than the depth of the hole, is filled with grease and placed into the hole. Stainless steel rods are connected and placed inside the PVC pipe. The rods are then driven to refusal or until a driving rate of 1-foot per minute is obtained. The stainless steel cap, 5-inch PVC pipe, and logo cap are then put in place as for a quality B monument [10].

Field Leveling

Basic A field leveling is being accomplished to first-order, class II specifications, using the "double-simultaneous" method [26]. An increase in leveling progress (while maintaining acceptable accuracy) has been accomplished by equipping NGS field leveling units with a specially modified subcompact truck for rodmen as well as observers. This form of "motorized" leveling has increased production by at least 20 percent as compared to former leveling procedures. Alternate approaches, including high-accuracy trigonometric leveling, are also being evaluated [13,24].

PROCESSING

This procedure converts files prepared according to FGCC input format [7] to the vertical observation library file format, checks the fields for valid entries, calculates and applies corrections to leveling observations, and loads data into the data base.

BLOCK VALIDATION

Block Validation for the vertical adjustment uses methods similar to those described for the horizontal adjustment. All observed elevation differences in a predefined area are combined and analyzed. During this analysis, a first-order primary network consisting of the latest data is selected, analyzed, and documented. Appropriate remaining leveling data are then incorporated into the first-order network. Leveling lines which do not fit (statistically) within the network will not be included in the primary network.

During data analysis, profiles (the differences in elevation along a level line over time), loop closures (primary and secondary), section misclosures, date of data, survey order and class, bench mark stability, previous adjustment reports, and past studies are all utilized in the decision making process.

REFRACTION CORRECTION

Before the data analysis can be conducted in the block validation task, known systematic errors must be removed.

A refraction correction for leveling was developed by Kukkamaki [17] in 1938. However, it was not previously applied to leveling in the United States because its magnitude was considered small, and extra equipment was required to measure temperature variation with height. Most of the experience with measuring these temperatures came from Finland and England where vertical temperature gradients are small. Therefore, it was hoped that by balancing sight lengths, keeping sight lengths reasonably short, and not allowing the sight line to get too close to the ground, refraction error would be insignificant. However, experiments by Holdahl [11] in 1979 and Whalen [27] in 1981 showed that refraction error was significant, especially for long sight lengths. Leveling refraction is proportional to the height difference observed at the instrument station and the vertical temperature gradient. It is also proportional to the square of the sight length, thus accumulating most quickly on long gentle slopes.

Prior to 1957, the maximum sight length permitted by the NGS was 150 m. Since then, the maximum allowable sight length has been held under 50 m for first-order surveys. This reduction was motivated primarily by a change in instrumentation and observing procedures. Theoretically, a 60-m sight has over twice the refraction error of a 40-m sight. Consequently, the observations comprising the network in some regions contain variable amounts of refraction depending on the sight length, slope, and vertical temperature gradient.

The magnitude of the vertical temperature gradient near the ground depends primarily on the intensity of solar radiation. Solar radiation at mid-latitudes is highly variable depending on season and time of day. This causes temperature gradients near the ground to fluctuate similarly. Thus, the amount of refraction error in leveling surveys will generally depend on where and when the measurements were made. Rainfall, cloud cover, and ground reflectivity also have regional and temporal variations which influence vertical temperature gradients.

The remedy of this problem is to correct the measurements for refraction prior to adjustment, and then let the adjustment treat the residual refraction error (remaining because corrections are not perfect) as random. Removing refraction bias from old leveling measurements requires a temperature stratification model, since temperature gradients were not measured. The model, described in detail by Holdahl [12], is based on historical records of solar radiation, sky cover, precipitation, and ground albedo from many locations in the conterminous United States. This model is an important asset to leveling computations because it provides a means of eliminating extreme refraction bias in the absolute heights of historic surveys and makes profiles of relative vertical movements more reliable. Without prior refraction corrections, it may not be possible to distinguish between refraction errors and crustal deformation. An economic advantage is realized by eliminating the need to observe vertical temperature profiles during most modern leveling surveys.

Removal of refraction errors will increase the calculated heights of high mountains in the United States by approximately 3 dm. The impact of refraction removal will be observed in the relative sense even more dramatically. For example, the rapid rise up to the top of the Sierra Nevada Mountains from the west will put approximately 3 dm of correction into existing published height relationships in a distance of only 140 km. This amount is 25 times as great as a normal accumulation of random error over that distance.

The modern surveyor, using sight lengths of less than 40 m, will more easily achieve agreement with the national control network after it has been corrected for refraction. However, the local surveyor who uses long sight lengths to minimize costs may not obtain good checks between published bench marks in the national vertical network. When a discrepancy results, it may be helpful to remember that refraction errors tend to minimize height differences. If the new leveling has been up, or down, long continuous grades and the new height difference is less than the published difference, then refraction is a likely cause of the discrepancy.

MAGNETIC ERROR MODELING

In 1981, Rumpf and Meurish [22] discovered that the "horizontal" line of sight of certain compensator (automatic) leveling instruments was influenced by magnetic fields. Subsequently, NGS began comparing certain lines of leveling to determine empirical magnetic values. These values are determined by comparing repeat leveling over the same line for which at least one of the surveys used instruments that were not significantly influenced by magnetic fields, such as spirit-leveling and certain "non- magnetic" compensator instruments. By comparing these surveys with other surveys that used magnetically influenced instruments, an empirical value of the magnetic effect can be determined [2,29].

These values provide an independent validation of the laboratory calibrations. However, laboratory calibrations could not be performed on the majority of NGS magnetically influenced instruments because the compensators had been repaired or replaced prior to the discovery of the error. For lines surveyed with these instruments, empirically derived correction values proved the only means of reducing the magnetic error contained in the data.

Obviously the best solution, but unfortunately the most expensive, is to relevel affected lines with "nonmagnetic" instruments. NGS intends to keep such releveling to a minimum by judicious application of the laboratory and/or empirical magnetic values.

APPROPRIATE A PRIORI ESTIMATES OF STANDARD ERRORS

When different types of data are combined and adjusted, it is essential to impose a correct relative weighting scheme. This means that a priori standard errors of observations must be estimated for each group of data. This task includes identifying the different groups of leveling observations, and establishing and implementing a procedure to determine the appropriate a priori standard error of observations in each group.

Groups of data have been identified, according to instrumentation and field procedures, but they may be modified after additional analysis [18]. Different methods for estimating variance components in least-square adjustments are being considered. The Interactive Almost Unbiased Estimation (IAUE) technique [19] has been implemented on a HP-1000 minicomputer. Other analyses include: (1) comparing old and new section and loop statistics, (2) reviewing profiles, and (3) studying formal errors in past field techniques. Some problems are anticipated with estimated variance components of leveling data: (1) too small a data sample for a particular group, (2) undetected systematic errors present in the data, and (3) inadequate network design to estimate the internal consistency of groups.

WATER-LEVEL TRANSFERS AND TIDAL INFORMATION

This task includes defining the data formats for water-level transfer and tidal data, loading the data, and estimating their observational accuracies. The 1977–83 water-level transfers and current tidal data (monthly means) in the primary National Tidal Observation Network have been keyed and placed in computer-readable form. The data need to be reformatted and placed in the National Geodetic Vertical Network. Studies have been performed which estimated the accuracies of these data [23,30–32]. Additional analyses will be needed to determine how they should be weighted in the NAVD final adjustment.

INTERPOLATION GRAVITY VALUES FOR BENCH MARKS

The adjustment will be performed using geopotential numbers. This requires estimating actual gravity for all bench marks involved in the project. Phase 1 of this task consists of interpolating gravity value for bench marks in the vertical synoptic file, a file containing primary information associated with the vertical data.

In the second phase, a study will be performed to determine if all gravity values are accurate enough for NAVD 88 purposes. A procedure is being developed in block validation to examine elevation differences, and determine if the estimated gravity values are accurate enough. If additional gravity values are required, then an observation plan for a specific area will be developed and implemented. It is anticipated that few areas of the country will need additional observations, because of the significant coverage of gravity observations in the United States [14].

NAVD 88 CRUSTAL MOVEMENT STUDIES AND PROCEDURES

This task also includes identifying areas of the network influenced by crustal motion, and establishing and implementing a procedure to account for these movements. During the past few years, NGS has been analyzing different numerical techniques in an attempt to model crustal movement. This would allow most data and bench marks to be included in one adjustment. The studies, performed mainly in the Houston-Galveston, Texas area, have been successful in identifying the data required to model movements precisely. But the lack of required geologic and hydrologic information, along with inadequate network design, makes modeling many areas for precise movements very questionable.

Areas of the networks most likely influenced by crustal movement are California, the Texas coast, sections along the east coast, and sections of the U.S. northern border. Other areas will be identified as block validation continues and additional research material is obtained. It may be possible to modify some observations for crustal motion effects, but more likely, most areas will be constrained to the framework network surrounding the area in question after the new adjustment. A plan defining the technique to be implemented in these "moving" areas will be developed and documented for each specific area.

FRAMEWORK ADJUSTMENTS

This task includes designing and analyzing a skeleton primary framework of regional networks. The analysis will be helpful in determining the effects of various datum constraints, magnitudes of height changes from the NGVD 29 datum, influences of systematic errors, deficiencies in network design, and additional releveling requirements.

GLOBAL POSITIONING SYSTEM (GPS) AND NAVD 88

Heights derived from Global Positioning System (GPS) observations cannot be used directly in the NAVD adjustment unless they are combined with precise geoid information. However, height differences and geoid undulation differences will be helpful in detecting and providing an upper limit on systematic errors in leveling data. Studies are being performed on estimating orthometric heights using GPS and gravity [14] and estimating subsidence using GPS-derived heights [24]. The impact of this technology on the vertical datum is yet to be determined.

DATUM DEFINITION

Datum definition is one of the last tasks that will be performed in the NAVD 88 project. Many factors need to be considered before a decision can be made. The solution may be as simple as performing a block shift to minimize differences between NAVD 88 heights and the latest local mean sea level heights. However, there are still some unanswered questions: How do we incorporate and weight tidal heights and water-level transfers? Can we estimate the effects of sea surface topography at tidal stations? How much of a departure is a "world" datum from a "North American" datum? What is the economic impact of each option?

HELMERT BLOCKING

This task consists of partitioning 1.5 million unknowns (most of the 550,000 bench marks have been leveled more than once) and associated observations into manageable blocks and solving a least squares adjustment of the entire data set.

The blocking strategy of NAVD will be easier to define than that of other large adjustments such as the North American Datum (NAD) [20]. After the first-level blocks, it is anticipated that 90 percent of the unknowns will be "interior." A blocking strategy needs to be developed which includes boundaries and the unknowns that are to be carried to the higher levels [5].

PUBLICATION

Publication of the results includes reviewing descriptions on a random basis, and publishing final adjustment heights. However, NGS plans to have NAVD 88 adjusted heights available to the public immediately following the adjustment. They will be loaded into a data base which the public will be able to access directly. In addition, they will also be available in hard copy and microform at a higher cost.

FINAL REPORT

The final report is not scheduled for completion until September 1989. This document will give the history of the network and the NAVD project, and provide information on technical decisions made about error estimation, observation selection, adjustment techniques, and crustal motion modelling. Previously published reports will be the main source of information for the final report.

SUMMARY

The National Geodetic Reference System of the United States has evolved over the last 150 years. Built up yearly by the continuous addition of new surveying data, the horizontal, vertical, and gravity networks have extended into all parts of the nation. The new adjustments, a monumental endeavor which involves all countries of North America, will remove the discrepancies of the existing datums and provide new datums for space-age surveying techniques. The horizontal adjustment, called the North American Datum of 1983, is complete and the vertical adjustment, called the North American Vertical Datum of 1988, will be completed in 1989.

ACKNOWLEDGEMENTS

The authors are grateful for the cooperation received from a number of individuals in the National Ocean Service. J. Ross McKay, Steven A. Vogel, David B. Zilkoski, and Gary M. Young provided valuable input. The review and comments provided by William M. Kaula, Elizabeth B. Wade, Gary M. Young, Emery I. Balazs, Charles W. Charleston, and Thaddeus Vincenty were greatly appreciated. Special thanks are due to Eleanor Z. Andree and Claire A. Wethington for editing and typing services respectively.

REFERENCES (NAD 83)

1. Bossler, J. D., "The New Adjustment of the North American Horizontal Datum," *EOS, 57* (8), 557–562 (1976).
2. Defense Mapping Agency, "Geodesy for the Layman," 5th Edition, reprinted by the National Oceanic and Atmospheric Administration (NOAA), Rockville, Maryland (1983).
3. Dracup, J. F., "New Adjustment of the NAD and the Surveyor," *Proceedings, Second International Symposium on Problems Related to the Redefinition of North American Geodetic Networks,* Arlington, Virginia, pp. 481–486 (National Geodetic Information Center, NOAA, Rockville, Maryland (April 24–28, 1978).
4. Dracup, J. F., "Plane Coordinate Systems," *ACSM Bulletin,* No. 59, 27 (November 1977).
5. Federal Geodetic Control Committee, *Input Formats and Specifications of the National Geodetic Survey Data Base, Vol. 1, Horizontal Control Data,* reprinted 1982, National Geodetic Information Center, NOAA, Rockville, Maryland (1980).
6. *Federal Register, 42* (57), 15913–15914 (1977).
7. Gergen, J. G., "The Observables," *ACSM Bulletin,* No. 51 (1975).
8. Hanson, R. H., "The Network Adjustment," *ACSM Bulletin,* No. 55, 21 (1976).
9. Hothem, L. D., "Doppler Satellite Positioning Program," *ACSM Bulletin,* No. 56, 17 (1977).
10. Isner, J. F., "Another Look at Helmert Blocking," *ACSM Bulletin,* No. 73, 33 (1981).
11. Isner, J. F., "Block Validation and Data Entry," *ACSM Bulletin,* No. 58, 12 (1977).
12. Meiss, P., "Prediction of Roundoff Error," *ACSM Bulletin,* No. 60, 17 (1978).
13. Mitchell, H. and L. G. Simmons, "State Coordinate Systems (Manual for Surveyors)," USC&GS Special Publication 235, revised 1974, reprinted 1981, National Geodetic Information Center, NOAA, Rockville, Maryland.
14. Moritz, H., "Geodetic Reference System 1980," *The Geodesist Handbook, 58* (3), Del Association Internationale De Geodesie, 39ter, Rue Gay-Lussac, 75005 Paris (1984).
15. Mueller, I. I., *Spherical and Practical Astronomy as Applied to Geodesy,* Frederick Ungar Publishing Co. Inc., New York (1969).
16. National Academy of Sciences, *North American Datum,* U.S. Government Printing Office, National Geodetic Information Center, NOAA, Rockville, Maryland (1971).
17. Schwarz, C. R., "Deflections of the Vertical," *ACSM Bulletin,* No. 65, 17 (1979).
18. Strange, W. E. and J. E. Petty, "Geodetic Astronomy," *ACSM Bulletin,* No. 57, 19 (1975).
19. Vincenty, T., "Determination of North American Datum 1983 Coordinates of Map Corners," *NOAA Technical Memorandum,* NOS NGS-6, 8 pp., National Geodetic Information Center, NOAA, Rockville, Maryland (1976).
20. Vincenty, T., "Determination of North American Datum 1983 Coordinates of Map Corners (Second Prediction)," *NOAA Technical Memorandum,* NOS NGS-16, 5 pp. National Geodetic Information Center, NOAA, Rockville, Maryland (1979).
21. Vincenty, T. and B. R. Bowring, "Application of Three-Dimensional Geodesy to Adjustments of Horizontal Networks," *NOAA Technical Memorandum,* NOS NGS-13, 10 pp., National Geodetic Information Center, NOAA, Rockville, Maryland (1978).
22. Vincenty, T., "Height-Controlled Three-Dimensional Adjustment of Networks," *ACSM Bulletin,* No. 72, 33 (1981).

REFERENCES (NAVD 88)

1. Balazs, E. I., "Preliminary Vertical Data Reduction Procedure," *ACSM Bulletin,* 25–26 (1984).
2. Balazs, E. I., "Comparison of Geodetic Levelings, Corrected for Magnetic Errors, to Mean Sea Level between San Francisco and San Pedro, California," presented at the Chapman Conference on Vertical Crustal Motion: Measurement and Modeling, Harpers Ferry, West Virginia, October 22–26, 1984, sponsored by American Geophysical Union, Washington, D.C.
3. Balazs, E. I., "Survey Plans for Basic Net A," *ACSM Bulletin,* 29 (1983).
4. Berry, R. M., "History of Geodetic Leveling in the United States," *Surveying and Mapping, 36* (2), 137–153 (1976).
5. Dillinger, W. H., "Helmert Block Higher Level System," *Proceedings of the Second International Symposium on Problems Related to the Redefinition of North American Geodetic Networks,* Arlington, Virginia, 417–426 (April 24–28, 1978).
6. Engelis, T., R. H. Rapp, and Y. Bock, "Measuring Orthometric Height Differences with GPS and Gravity Data," presented at the Chapman Conference on Vertical Crustal Motion: Measurement and Modeling, Harpers Ferry, West Virginia, October 22–26, 1984, sponsored by American Geophysical Union, Washington, D.C.
7. Federal Geodetic Control Committee, *Input Formats and Specifications of the National Geodetic Survey Data Base Vol. II. Vertical Control Data,* reprinted 1982, National Geodetic Information Center, NOAA, Rockville, Maryland (1980).
8. *Federal Register, 41* (96), 20202 (1976).
9. *Federal Register, 48* (214), 50784 (1983).
10. Floyd, R., "Geodetic Bench Marks," *NOAA Technical Memorandum,* NOS NGS-1, National Geodetic Information Center, NOAA, Rockville, Maryland (1978).

11. Holdahl, S. R., "A Model of Temperature Stratification for Correction of Leveling," *Bulletin Geodesique, 55* (3), 231–249.
12. Holdahl, S. R., "The Correction of Leveling Refraction and Its Impact on Definition of the NAVD," *Surveying and Mapping, 43* (2), 123–140.
13. Hothem, L. D., R. J. Fury, and D. B. Zilkoski, "Determination of Geoid Slopes Using GPS," presented at the Chapman Conference on Vertical Crustal Motion: Measurement and Modeling, Harpers Ferry, West Virginia, October 22–26, 1984, sponsored by American Geophysical Union, Washington, D.C.
14. Hothem, L. D., R. J. Fury, and D. B. Zilkoski, "Orthometric Height Determination with GPS Satellite Surveys," presented at ACSM/ASP Convention, Washington, D.C. (March 10–15, 1985).
15. Huff, L. C., S. F. Clifford, and R. J. Lataitus, "Turbulence Effects on the Rapid Precision Leveling System," *Proceedings of the Third International Symposium on the North American Vertical Datum,* Rockville, Maryland, pp. 461–470 (April 21–26, 1985).
16. Isner, J. F., "Helmert Block Initial Level System," *Proceedings of the Second International Symposium on Problems Related to the Redefinition of North American Geodetic Networks,* Arlington, Virginia, pp. 405–416 (April 24–28, 1978).
17. Kukkamaki, T. J., "Underdie nivellitische refraktion," *Finn. Geod. Inst.,* Helsinki, Finland, Publ. 25, p. 48.
18. Lucas, J. R., "A Variance Component Estimation Method for Sparse Matrix Applications," NOAA Technical Report NOS III NGS-33, 12 pp., National Geodetic Information Center, NOAA, Rockville, Maryland (1985).
19. Lucas, J. R., J. M. Bengston, and D. B. Zilkoski, "Estimation of Variance Components in Leveling Using IAUE," *Proceedings of the Third International Symposium on the North American Vertical Datum,* Rockville, Maryland (April 21–26, 1985).
20. McKay, E. J. and S. A. Vogel, "The North American Datum of 1983: Necessity, Status, and Impact," presented at the *ACSM/ASP Convention,* San Antonio, Texas (September 9–14, 1984).
21. National Academy of Sciences, Geodesy, Trends and Prospects, Committee on Geodesy, National Research Council (1978).
22. Rumpf, W. E. and J. Meurish, "Systematische aunderungen der ziellinie eines orazisions kompensator-nivelliers-insbesondere des Zeiss Ni 1—durch magnetische gleich- und wechselfelder." XVI International FIG Congress, Montreux, Switzerland (1981).
23. Stoughton, H. W., "Investigation of the Accuracy of Water Level Transfer to Determine Geodetic Elevations in Lake Ontario," Ph.D. Dissertation, University of Michigan (published by University Microfilms International, Ann Arbor, Michigan) (1980).
24. Strange, W. E., "Accuracy of GPS for Monitoring Vertical Motions," presented at the Chapman Conference on Vertical Crustal Motion: Measurement and Modeling, Harpers Ferry, West Virginia (October 22–26, 1984).
25. Strange, W. E., "Empirical Determination of Magnetic Corrections for Ni 1 Level Instruments," *Proceedings of the Third International Symposium on the North American Vertical Datum,"* Rockville, Maryland, pp. 363–374 (April 21–26, 1985).
26. Till, J. H., "The Processing of Archival (Historic) Leveling Data," *ACSM Bulletin,* 27–28 (February 1983).
27. Whalen, C. T., "Result of Leveling Refraction Tests by the National Geodetic Survey," NOAA Technical Report, NOS 92 NGS-22, National Geodetic Information Center, NOAA, Rockville, Maryland (1981).
28. Whalen, C. T., "Precise Trigonometric Leveling," presented at the Chapman Conference on Vertical Crustal Motion: Measurement and Modeling, Harpers Ferry, West Virignia (October 22–26, 1984).
29. Whalen, C. T., "Magnetic Field Effects on Leveling Instruments," presented at the Chapman Conference on Vertical Crustal Motion (October 22–26, 1984).
30. Whalen, C. T. and E. I. Balazs, "Test Results of First-Order Class III Leveling," NOAA Technical Report, NOS 64 NGS-4, National Geodetic Information Center, NOAA, Rockville, Maryland.
31. Zilkoski, D. B., "Geodetic Leveling and Mean Sea Level Along the East Coast of the United States," presented at the Chapman Conference on Vertical Crustal Motion: Measurement and Modeling, Harpers Ferry, West Virginia (October 22–26, 1984).
32. Zilkoski, D. B. and V. Kammula, "Comparison of Geodetic Leveling to Mean Sea Level Between Portland, Maine, and Atlantic City, New Jersey," *Proceedings of the Third International Symposium on Problems Related to the Redefinition of North American Vertical Datum,* Rockville, Maryland, pp. 105–119 (April 21–26, 1985).
33. Zilkoski, D. B. and G. Young, "North American Vertical Datum (NAVD) Update," *Proceedings of the U.S. Army Corps of Engineers Surveying Conference,* Jacksonville, Florida (February 4–8, 1985).
34. Zilkoski, D. B. and G. Young, "Status of National Geodetic Surveys North American Vertical Datum (NAVD) Project," *Proceedings of the Third International Symposium on the North American Vertical Datum,* Rockville, Maryland (April 21–26, 1985).

CHAPTER 5

Centimeter Precision Airborne Laser Ranging System for Rapid, Large-Scale Surveying and Land Control

JOHN J. DEGNAN,* WERNER D. KAHN,** AND THOMAS S. ENGLAR, JR.†

ABSTRACT

The Airborne Laser Ranging System (ALRS) is a proposed multi-beam subnanosecond pulse laser ranging system on board an aircraft. It simultaneously measures the distance between the aircraft and six laser retroreflectors (targets) deployed on the Earth's surface with ± 1 cm precision. Depending on the host aircraft and terrain characteristics, the system can interrogate hundreds of small targets distributed over an area as large as 6×10^4 sq kilometers in a matter of hours. Potentially, a total of 1.3 million individual range measurements can be made in one six hour flight. Trilateration techniques are used to derive the intersite vectors between laser ground targets with precisions as high as one part in 10^7. Since all data is initiated, received, collected and processed in the aircraft via totally passive ground targets, there is no need for complex ground instrumentation or skilled personnel in the field. The high precision, speed, and large area coverage of the ALRS make it an attractive and highly cost effective instrument for a variety of large scale surveying, engineering, land management, and geophysics applications, especially when combined with photogrammetric instrumentation.

INTRODUCTION

The Airborne Laser Ranging System (ALRS) is a proposed multibeam laser ranging system operating from an aircraft. Using a single subnanosecond pulse laser, the instrument simultaneously measures the distance between a reference point on the aircraft and up to six retroreflectors deployed on the underlying terrain with a precision of about one centimeter. Depending on the host aircraft and terrain

---

*Advanced Electro-Optical Instrument Section, NASA/Goddard Space Flight Center, Greenbelt, MD
**Geodynamics Branch, NASA/Goddard Space Flight Center, Greenbelt, MD
†The Johns Hopkins Applied Physics Laboratory, Laurel, MD

characteristics, the system can resolve, with centimeter precision, the three-dimensional positions of hundreds of inexpensive, passive targets distributed over an area as large as 6×10^4 square kilometers. At a laser repetition rate of 10 pps, a total of 1.3 million individual range measurements can be made in one six hour flight. Trilateration techniques are used to derive the intersite vectors between ground targets with precisions as high as one part in 10^7. Since all of the data is initiated, received, collected, and processed in the aircraft using passive ground targets, there is no need for complex ground instrumentation or skilled personnel in the field.

Short pulse lasers have been successfully used for over two decades by NASA to precisely measure the distances to earth orbiting artificial satellites and the moon. Modern satellite laser ranging systems have absolute accuracies of one to two centimeters [1]. The data derived from such systems have allowed scientists to measure intercontinental baselines with accuracies of a few centimeters and the slow relative movement of the Earth's tectonic plates which are typically a few centimeters per year. Although the original impetus for the development of the ALRS concept was provided by the geophysics community who wanted an instrument capable of monitoring regional crustal deformation and strain buildup and relief in extended seismic regions [2], ALRS clearly has high potential for large scale surveying and land management applications. For example, the ALRS would allow the rapid verification of existing ground survey networks and permit further densification on a regional scale with target spacings on the order of 5 to 20 Km depending on the maximum altitude capability of the host aircraft. Target retroreflectors placed at Satellite Laser Ranging (SLR) and Very Long Baseline Interferometry (VLBI) sites would serve as fiducial points for tying the ALRS measurements to an appropriate earth-based coordinate system. Local surveyors could, in turn, further densify the grid through more conventional means by referencing their measurements to the target positions established by the ALRS.

Because the aircraft flight path is also resolved with centimeter accuracy along with the target coordinates in the post-flight data reduction process, the system could provide centimeter accuracy navigational data and ground control points for collocated photogrammetric instruments in land management applications. The placement of three or more targets at or near distinctive terrain features within the camera field of view would permit, through high resolution extrapolation, the accurate two dimensional location (latitude and longitude) of all other photographed objects. It is also a relatively simple matter to deflect a portion of the laser energy into a nadir-viewing altimetry channel. Measurement of time-of-flight of the altimetry ground reflection combined with the precise navigational data provided by ALRS would result in highly accurate terrain profiling data.

In addition to large-scale surveying and land management applications, the high precision, speed, and large area coverage of the ALRS make it suitable for a number of scientific measurements including tectonic plate motion, regional crustal deformation, glaciology, and temporal volcanic deformation. Engineering applications include the measurement of ground subsidence due to withdrawal of natural resources (water, gas, oil, etc.), surveying for large scale engineering projects (interstate highways, aquaducts, pipelines, etc.), and the monitoring of surface deformation which can affect the mechanical integrity of dams, pipelines, water canals, and nuclear power plants or waste sites.

In Section 2 of this article, we provide a description of the ALRS, its various subsystems, and a projected mission scenario. In Section 3, we introduce the mathematical models and algorithms used for a covariance error analysis of a postflight estimation process. In Section 4, we discuss the sources and magnitudes of errors in the laser range measurement and present representative satellite laser ranging data to validate the range error model introduced in the previous section. In Section 5, we present the results of covariance error analyses for some representative target grids as an indication of the expected performance of ALRS in actual surveying applications. We conclude with a brief summary in Section 6.

SYSTEM DESCRIPTION

Laser Ranging Subsystem

Figure 1 is a block diagram of the ALRS. The system computer enables the firing of a sub-nsec pulse laser transmitter at a nominal rate of 10 pps. The transmitter is a mode-locked, PTM Q-switched Nd:YAG laser oscillator followed by a double-pass Nd:YAG laser amplifier and a KD*P frequency doubler [3]. On each firing, the transmitter generates a single 150 psec (FWHM) pulse containing several milljoules of energy at the 0.532 µm green wavelength. A beam splitter reflects a very small fraction (<1%) of the outgoing energy into a series of six receiver channels. The remaining energy is divided into approximately six equal parts by beam splitters which direct the energy to six independently controlled pointing systems. The six outgoing pulses pass through the atmosphere to illuminate six ground target retroreflectors. The reflected energy from each target travels back through the atmosphere to the ALRS and is imaged onto the corresponding high speed photomultiplier tube (PMT). Thus, a pair of start and stop pulses, indicating the times at which a given pulse leaves and returns to the instrument, are recorded by each of the six receiver channels. The use of common start/stop receiver components eliminates a potential source of time-dependent range bias which might be introduced by changes in the thermal environment, voltage condition, or other nonstationary processes during flight [1].

A block diagram of an individual range receiver (signal processor) channel is shown in Figure 2. A 5 A° bandpass filter centered at 0.532 µm prevents triggering on extraneous background noise at other wavelengths. The output signal from the high speed photomultiplier (ITT Model F4128) is amplified (ENI Model 601P) and split. One port is input to a low time walk constant fraction discriminator (Tennelec TC454). The discriminator provides a NIM logic pulse to both start, and later stop, a time interval unit (Hewlett Packard 5370B or equivalent) which measures the pulse time of flight. The time interval units (TIU's) in each of the six channels share a common clock input.

Gated charge digitizers (Ortec 227) at the second port of the PMT record the energies of the outgoing (START) and incoming (STOP) pulses thereby enabling the computer to control signal levels into the discriminator via the programmable attenuator (Hewlett Packard 33320G), and to correct, via software, for small (±50 psec) amplitude dependent timing errors in the discriminator.

The range to each target is calculated from time-of-flight, with suitable corrections being made for instrument biases, pulse amplitude effects, and atmospheric refraction delays. Nominal atmospheric refraction corrections are made in-flight for use by the navigation computer. More exact corrections are made during the post-flight data reduction phase. The measurement data is transferred to the system computer for storage and use by the navigational subsystem as described later. Figure 3 provides a sample link calculation for a typical aircraft flying at a maximum altitude of 20,000 ft (6.1 km). With a very modest 1 mJ of transmitted laser energy, a 0.5 degree full beam divergence, and a 5 cm receiver aperture, the expected average signal from the grapefruit sized target is very strong (213 photoelectrons)—even at a worst case elevation angle of 20°. Figure 4 displays the quality of range data at these signal levels. The data, taken over a kilometer path with an attenuated laser beam, shows single shot range uncertainties of 5 mm RMS (one

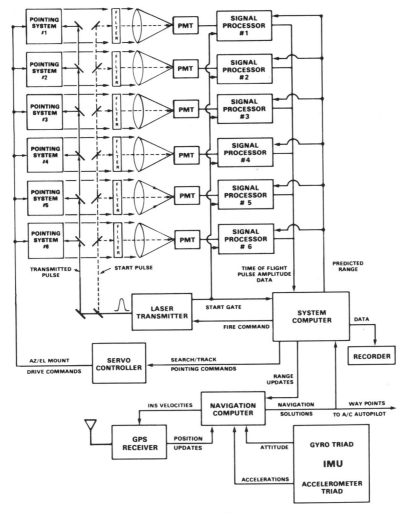

FIGURE 1. Block diagram of the airborne laser ranging system.

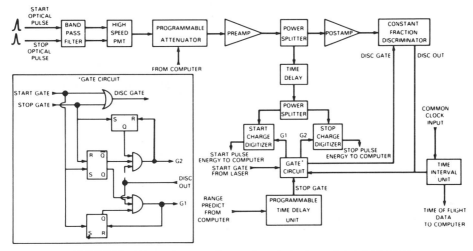

FIGURE 2. Airborne laser ranging receiver single channel signal processor.

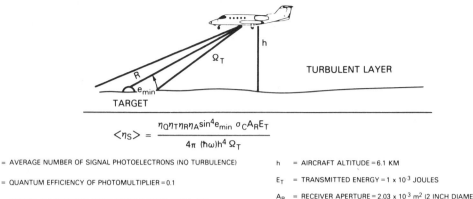

FIGURE 3. Link calculation for low altitude aircraft.

sigma). Recent laser ranging data to the LAGEOS Satellite show 1.5–2.0 cm RMS standard deviations to distances as great as 18,000 km with signal levels of 3–30 photoelectrons [1]. The receiver used in the latter experiments was inferior in performance to that proposed for ALRS.

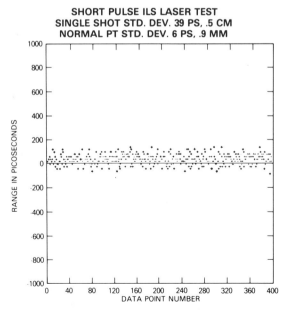

FIGURE 4. Performance of laser ranging subsystem over 1 km horizontal range.

Optomechanical Configuration

The laser transmitter, beam splitting optics, receiver optics, and photomultiplier tubes are mounted on an optical base plate which is isolated vibrationally from the aircraft fuselage. Six azimuth-elevation pointing mounts with 5 cm receiver apertures are rigidly attached to the bottom of the optical bed. Each pointing system consists of a four-mirror coelostat mounted on an azimuthally rotating stage (Figure 5). The laser beam enters the pointing system through a hole in the optical base plate and the rotation stage. The final mirror rotates about an axis parallel to the optical bench to point to a given elevation angle. This particular configuration was chosen because it can be placed very close to the aircraft window to provide near hemispherical viewing capability. A second advantage is that all light beams reflect at a 45° incidence angle independent of the AZ/EL pointing angles. This allows the use of high efficiency dielectric coatings on the mirrors for maximum receiver sensitivity and for potential upgrade of the system to two colors for making direct measurements of the atmospheric refraction correction [4].

Each pointing mount is equipped with two servo systems (azimuth and elevation). In the present concept, these systems are digital controllers that drive precision stepping motors and contain optical encoders to measure angular position. The commands sent to the controller are in the form of angular position, velocity, and acceleration referred to the optical bench. The controller performs the usual loop closure tasks to control the individual servos as directed and to relay information back to the command interface. Command angles and predicted range gates are generated by the system computer using aircraft navigation solutions provided by the navigation and attitude determination subsystem.

With laser beam divergences on the order of 10 mrad, absolute pointing accuracies at the few mrad level are adequate. This performance is a factor 10–20 times less stringent than typically required for ground-based satellite laser ranging systems which have been operational since 1964.

Target Deployment

A number of geodetic monuments will be erected in the region of interest. The "grape-fruit-sized" targets, studded with about six optical cube corners, can be mounted directly on the monuments prior to the survey flight. These can be either permanent installations or the targets can be removed and reused at other locations. In order to tie the target grid accurately to a reference coordinate system, targets would be placed at three or more fiducial points. To simplify the target acquisition sequence, it is desirable that the position of each target be known a priori to approximately 50 m, but this is not a hard requirement. The a priori knowledge of target position does not impact the a posteriori grid resolution achieved by the ALRS. In general, a priori location can be read from a surveyor's map or obtained using radio navigation receivers such as LORAN C, the TRANSIT Satellite System, or the Global Positioning System (GPS).

Target Acquisition and Tracking

The a priori target positions analyzed in the previous section will be stored in the ALRS system computer memory as latitude, longitude, and altitude above sea level. Successful acquisition and tracking of the targets during flight requires an adequate knowledge of the aircraft position, velocity, and attitude. The positional error in a modern inertial navigation system (INS) typically grows at a rate of 0.4–4.0 km/hr. In such a system, vehicle accelerations and attitude are measured by an inertial measurement unit (IMU) consisting of three orthogonal accelerometers and a gyro triad. The navigational computer (NC) performs coordinate transformations and integrates the equations of motion to provide estimates of velocity and position relative to a set of initial conditions. The errors in these estimates have many contributing sources including sensor calibration limitations and inaccuracies, computational errors, and sensor-error-propagation effects.

The performance of an INS can be improved significantly by incorporating additional independent sensors which periodically check and update the navigation solutions. Radio navigation aids (such as OMEGA, LORAN C, and the GPS system) are particularly well suited to the role of auxiliary sensor because of their global or near global coverage. In the ALRS, position information from a GPS receiver is uti-

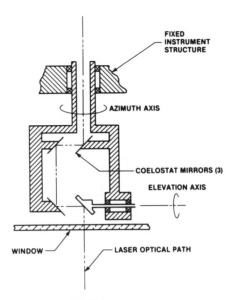

FIGURE 5. Pointing system concept.

lized to update and stabilize the corresponding INS solutions via a Kalman filter algorithm in the NC. The Kalman filter also updates attitude information and provides best estimates of sensor errors such as misalignments, accelerometer biases and scale factors, gyro drift rates, etc. In the event of certain types of failure, the proposed navigational system can be restarted and calibrated in flight.

Flight tests of a LORAN-aided INS [5] performed for the U.S. Air Force have demonstrated a 60 m absolute position accuracy (1 sigma) and angular accuracies of a few tenths of a mrad (R. D. Wierenga, Lear-Siegler Inc., Grand Rapids, Mich. 49508 private communication). LORAN C is significantly more accurate than OMEGA (± 925 m) and, unlike GPS, is already available in most of the northern hemisphere. The Global Positioning System will eventually provide more accurate dynamic fixes in all three coordinates (5 m) and continuous global coverage. The system is expected to become fully operational by 1988.

The navigation data are combined with the stored a priori target positions to compute the estimated range gates and the pointing mount command angles. As the targets come within range, the laser is activated and the presence or absence of range returns is noted. If no returns are detected, a search pattern is executed unit range data are acquired. Selected range data are then passed to the navigation Kalman filter to update the estimate of aircraft position. The ALRS then shifts from the acquisition to the tracking mode.

In the tracking mode, triangulation on the highly accurate laser returns results in a real time aircraft position estimate which is virtually error free (better than a meter standard deviation) so that, except for aircraft attitude estimation

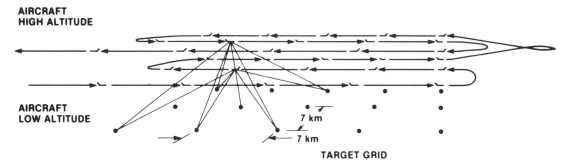

FIGURE 6. Typical ALRS mission scenario perspective showing ranging from two different altitudes.

errors contributed by the gyros, the computed command angles and range gates are essentially correct. The instrument remains in the tracking mode as long as sufficient range data are available to maintain an accurate estimate of aircraft position. In making the transition to a new set of six targets, the system triangulates on laser returns which are common to the new and the previous set. In this way, the aircraft position is known with better than meter accuracy during the transition. The proposed system computer design and a description of the necessary operational software is given elsewhere [6].

Mission Scenario

Simulations [7] have shown that range measurements must be taken at two widely separated altitudes in order to strengthen the geometry sufficiently to recover baselines at the centimeter level. Thus, in a typical mission, the aircraft approaches the target grid at an altitude corresponding to approximately 70% of its maximum altitude (Figure 6). After acquiring the first few targets, the instrument shifts to the tracking mode for the remainder of the mission. After overflying the rows of targets, the aircraft climbs to its maximum cruise altitude for a second set of passes over the target grid. The turning manueuvers between passes can be used to calibrate the onboard attitude sensors.

MATHEMATICAL MODELS AND ALGORITHMS

In this section, the models and equations used for covariance error analysis of a postflight estimation process are presented, together with some remarks concerning the motivation for them. The results of the covariance analyses will be described in detail in a later section of this article.

Range Data Model

Ideally, the ranging system measures the distance (range) between the airborne laser and several retroreflectors deployed on the Earth's surface. Let $Z_{(3\times 1)}(t)$ and $\dot{Z}_{(3\times 1)}(t)$ be the 3-dimensional vectors describing the position and velocity of the aircraft at time t,

$$Z_{(3\times 1)}(t) = \begin{bmatrix} z_1(t) \\ z_2(t) \\ z_3(t) \end{bmatrix}_{(3\times 1)} \text{ and } \dot{Z}_{(3\times 1)}(t) = \begin{bmatrix} \dot{z}_1(t) \\ \dot{z}_2(t) \\ \dot{z}_2(t) \end{bmatrix}_{(3\times 1)} \quad (1)$$

The vector components are expressed in some convenient coordinate system. Also let,

$$U^{(k)}_{(3\times 1)} = \begin{bmatrix} u_1^{(k)} \\ u_2^{(k)} \\ u_3^{(k)} \end{bmatrix}_{(3\times 1)} \quad (2)$$

be the 3-dimensional vector describing the position of the kth retroreflector in the same coordinate system.

The distance between the kth retroreflector and the airborne laser at time t is:

$$d(t, k) = [Z^T(t)Z(t) + U^{(k)T}U^{(k)} - 2Z^T(t)U^{(k)}]^{1/2} \quad (3)$$

The laser range measurement model as developed here is composed of $d(t, k)$ and errors arising from several sources. The principal error source is the refraction increment $r(t, k)$. The model used also includes an explicit measurement bias, b_i, on the ith beam, and a random error $v(t, k, i)$ which is uncorrelated in time or between beams. Thus the ALRS range measurement model is

$$\varrho(t, k, i) = d(t, k) + r(t, k) + b_i + v(t, k, i) \quad (4)$$

The physical basis for each of these error sources coupled with their magnitudes, will be described in detail in a later section. A brief outline of the errors is given here to serve as a foundation for the mathematical model.

The refraction error $r(t, k)$ has been shown by Gardner [8] to depend predominantly on two parameters: p_k, the surface pressure at the kth target site, and q_k, the so-called "gradient of PTK" at the target site. The refraction error can then be expressed as a linear function of p and q,

$$r(t, k) = b_p(t, k)p_k + b_q(t, k)q_k$$

where the functions b_p and b_q depend only upon aircraft/target geometry at time t. The bias b_i represents a constant range error in the ith channel which remains after calibration. The "white" error v is contributed to by timing jitter, atmospheric turbulence, etc. An additional error appears in the position of the aircraft and will be described in the next section.

The Estimation Algorithm

The purpose of the ALRS algorithm is to process the range data $(\varrho_1, \varrho_2, \ldots, \varrho_n)$ and produce an estimate of the retroreflector positions $U_{(3N_r \times 1)}$. It is clear that this must be accompanied, at least implicitly, by estimates of the aircraft positions $Z(t_\ell)$ at the measurement times, t_ℓ. Furthermore, the refraction and the biases must be modelled, and their effects compensated.

Variables which are adjusted in the estimation process are collectively referred to as the estimation state denoted by X, that is

$$X_{N \times 1} = [U_{,}^{(1)} U_{,}^{(2)} \ldots, U_{,}^{(N_r)} Z, \dot{Z}, b_1, \ldots,$$
$$b_{N_B}, p_1, \ldots, p_{N_r}, q, \ldots, q_{N_r}]_{N \times 1}^T \quad (5)$$

where

$$N = 5N_r + N_B + 6$$

and

$p_k, q_k \equiv$ respectively, the pressure and gradient PTK at the kth retroreflector site
$N_B \equiv$ the number of independent laser beams
$N_r \equiv$ the number of retroreflector sites

The inclusion of $Z(t)$ will be explained below.

With an ALRS sampling rate of 10 measurements per second, the aircraft moves only about 10 m between each laser pulse. Because of this, it seems unlikely that the error in aircraft position caused by unmodelled forces acting over that 0.1 seconds could be greater than a few cm. It thus seems reasonable that an estimate of aircraft position and velocity, obtained from the ALRS data, could provide significant improvement in the entire estimation process through having a smaller uncertainty in the aircraft position prior to each measurement. To facilitate this approach, a recursive estimation technique is indicated, and a Kalman filter was thus chosen for this study.

The state elements in [5] have all been assumed to be constants, except for the aircraft position (Z) and velocity ($\dot{Z}$), which were assumed to satisfy a second-order system of the form

$$\ddot{Z} = W \quad (6)$$

where W is white noise. In preliminary simulation studies, it appeared that the estimation of aircraft position did not materially improve overall baseline accuracy [9, p. 13], and therefore the simulations were performed with the conservative assumption that the aircraft position uncertainty was equally large at all measurement points. Except for the numerical differences between recursive and batch computation procedures, this approach yields the same solution statistics as the geometric mode multilateration solution.

The development of the Kalman filter has been treated exhaustively in the literature [10,11] among others. The Extended Kalman Filter (EKF), used in these studies will be presented but not derived.

Let $\hat{X}(t_2, t_1)$ denote the state estimate at t_2 based upon data through t_1. Each measurement at t is represented as a truncated Taylor's series

$$\varrho(t, k, i) = \hat{\varrho}(t, k, i) + \left.\frac{\partial \varrho}{\partial X}\right|_{X = \hat{X}(t|t - \tau)} [X(t) - \hat{X}(t|t - \tau)] + v \quad (7)$$

which, upon forming the six vector $\varrho(t)$ of six individual measurements taken at t, has the form

$$y = H[X(t) - \hat{X}(t|t - \tau) + v] \quad (8)$$

This is now in a suitable form for applying the EKF, in which

$$\hat{X}(t|t) = \hat{X}(t|t - \tau) + K(t)[\varrho(t) - \hat{\varrho}(t)] \quad (9)$$

where

$$K(t) = P(t|t - \tau)H^T [HP(t|t - \tau)H^T + R]^{-1} \quad (10)$$

The matrix $P(t|t - \tau)$ is the error covariance of $\hat{X}(t|t - \tau)$ and is updated by

$$P(t|t) = [I - K(t) H] P(t|t - \tau) \quad (11)$$

where I is a unit matrix of proper dimensionality. The matrix R is defined as follows:

$$R \equiv E(v\,v^T) = \sigma^2 I \quad (12)$$

Because all state components except the aircraft are constant, the propagated matrix $P(t + r|t)$ is the same as $P(t|t)$ except for the aircraft components, in which correlations are set to zero and the variances set back to initial values as described above.

Because the geometry of the (A/C) Aircraft/Grid combination is so important, the covariance matrix, P, does not uniformly approach a condition of asymptotic approach to zero. It is shown in [9] that the second aircraft altitude is required to reduce the variances, and, in fact, not until the last aircraft pass over the target grid is the geometry complete.

Error Variances

The Kalman filter requires that errors be described by covariance matrices. Thus a co-variance $P(0|0)$ of the state X, prior to any measurements, is needed. The initial covariance is usually taken to be diagonal, i.e. errors in target positions, in aircraft position and velocity, in beam biases, and in atmospheric parameters are assumed independent. The values chosen for these variances depend upon the conditions of the survey, but typical values might be target errors (standard deviations) on the order of 5 m., aircraft position errors of 30 m (reinitialized at each time point), beam bias errors of 1 cm, and barometric pressure errors of 1 mbar at instrumented target sites and 100 mbar at non-instrumented sites. The simulations assumed only a few instrumented sites and over regions of 100 km, little gain was obtained from having more atmospheric monitoring (see section on System Performance).

The measurement noise covariance R has been assumed diagonal in most of the studies, with $\sigma^2 = 1$ cm^2, on the basis of field tests of protype instruments.

Coordinates and Constraints

It is well known [9] that the multilateration problem, of which ALRS is an example, is not completely observable; that is, not all of the unknowns of the problem can be determined from the ALRS data. The simplest example of this is the fact that the same set of observations could be obtained if the complete set of aircraft and retroreflector positions were translated and rotated as a rigid body. Thus there is a six-fold degeneracy in the problem. The problem could be written with six station coordinates not appearing. The more flexible approach which was used in this study has all station coordinates appearing in the state. If all coordinates were given the same, large, initial variances, then baseline variances would decrease while a six-space continued to have large variances, possibly inducing numerical problems. To avoid this, six retroreflector position components are chosen to define a local coordinate system and these components have essentially zero error variance. It is important to note that the estimates and variances of the estimates of baseline length are independent of what coordinate system is chosen, and except for the numerical problem previously noted, no particular local coordinate system need be chosen.

These local estimates can be tied back to a larger coordinate system provided independently obtained coordinates in the larger system are available for at least three retroreflectors. These coordinates must be such as to satisfy the basic assumption that internal distances are constant, i.e., close to the time of the ALRS data gathering period, if geodynamical considerations are important.

RANGING COMPONENTS AND ERROR SOURCES

Laser Transmitter

Early satellite laser ranging (SLR) system used Q-switched Ruby lasers. These lasers were capable of generating very high output energies on the order of several joules, in a pulse several tens of nanoseconds wide. Since the ruby laser is very inefficient and requires large power sources to drive the flashlamps, later systems employed more compact and efficient Nd:YAG lasers which also had shorter Q-switched pulsewidths on the order of ten nanoseconds or less. Shorter pulse-widths reduce the random ranging error contribution of the transmitter which varies as $\tau_p/\sqrt{N}$ where τ_p is the FWHM laser pulsewidth and N is the number of received photoelectrons.

It soon became apparent that Q-switched lasers exhibited variable biases which were undesirable in a precision laser ranging system. These biases arise from the multimode nature of Q-switched lasers. In such lasers, individual radiation modes, both "spatial" and "temporal," build up at their own rates and are coupled to the external world by a partially reflecting mirror which forms one end of a two mirror laser cavity. Furthermore, different spatial modes have different far field radiation patterns. This results in a range bias which is dependent on the angular position of the target in the transmitter far field pattern. This effect has become known in the laser ranging community as "wavefront distortion" error. In addition, the magnitude of the bias has been observed to vary slowly with time [12] suggesting that the radiated mode structure is a function of laser temperature.

An early approach to reducing the bias error in Q-switched lasers was to pass the laser pulse through an external "pulse slicer" consisting of an electro-optic Pockels cell situated between two crossed polarizers. Only the central 4 to 5 nanoseconds of the typically 20 nanosecond pulse would be transmitted to the satellite. Although this technique was relatively easy to implement in the field, it was rather inefficient since it rejected a sizable fraction of the laser energy.

A short-lived interim solution to this problem was the development of the Pulse Transmission Mode (PTM) or "cavity-dumped" Nd:YAG laser. This is also a multimode device, but the radiation modes are trapped inside the laser cavity until they are simultaneously "dumped" by an internal electro-optic switch [12]. Thus, since all modes leave the transmitter at the same time, the angularly dependent range bias is greatly reduced. With subnanosecond switching times, cavity-dumped lasers have pulsewidths on the order of 2 to 3 nanoseconds corresponding to the time it takes light to make one round trip in the laser cavity.

The transmitter laser in the most modern field systems consists of a modelocked, Nd:YAG laser oscillator followed by one or more Nd:YAG laser amplifiers. Since modelocked Nd:YAG lasers operate in a single spatial mode (i.e. the fundamental TEM_{oo} mode having a Gaussian spatial profile), they do not exhibit the aforementioned "wavefront distortion" errors. Individual longitudinal ("temporal") modes are "locked" together, by either a passive dye or active acousto-optic or electro-optic modulator internal to the laser resonator, to create pulsewidths as short as 30 picoseconds. A passive nonlinear crystal, such as Potassium Dideuterium Phosphate (KD*P), converts the fundamental infrared radiation of the Nd:YAG laser material at 1.06 micrometers to the 0.53 micrometers green frequency-doubled wavelength, with 40 to 50 percent efficiency, in order to take advantage of more sensitive photodetectors available in the visible wavelength region.

Table 1 lists the ranging performance of various laser types tested at GSFC [13]. The list includes three conventional, stable resonator (denoted by S), Q-switched lasers as well as one unstable resonator device (denoted by U), three PTM Q-switched ("cavity dumped") lasers, and three mode-locked transmitters. Two standard tests were performed for each laser transmitter—a "repeatability test" and a "far field range map" [13].

In the repeatability test, a groundbased target is centered in the laser beam and range measurements are made for approximately 45 minutes. For each set of 100 range measurements, a mean and standard deviation is calculated and plotted versus elapsed time.

The range map test is designed to quantify the range biases resulting from the higher order mode effects ("wavefront distortion") discussed previously as a function of target position in the transmitter far field pattern. In performing the test, the direction of the transmitter beam is varied so that the target lies at different points within the transmitter far field pattern. At each position, two 100 point data sets are taken and a mean range and standard deviation is calculated for each set. The mean range with the target at beam center is then subtracted from the means at the other target positions to determine the dependence of range bias on far field angle.

As can be seen from Table 1, the performance of modelocked transmitters is far superior to that of other laser types. Bias effects introduced by modelocked transmitters, either active or passive, are typically at the subcentimeter level [13].

TABLE 1. Lasers Tested by GSFC.

Type	Manufacturer	Pulsewidth (FWHM)	Repeatability	Rangemap
Q-switched (S)	General Photonics (Moblas)	7 NSEC	Poor	Poor
Q-switched (S)	Modified General Photonics	4 NSEC	Fair (1PPS) Poor (5PPS)	Fair (1PPS) Poor (5PPS)
Q-switched (S)	Westinghouse (Military)	8 NSEC	Poor	Poor
Q-switched (U)	Quanta-Ray	5 NSEC	Very Good	Poor
PTM Q-switched	International Laser Systems (LL102)	4 NSEC	Excellent	Fair
PTM Q-switched	General Photonics	3.6 NSEC	Excellent	Fair
PTM Q-switched	NASA-Modified GP	1.5 NSEC	Very Good	Very Good
Active Mode-Lock	International Laser Systems	225 PSEC	Excellent	Excellent
Passive Mode-Lock	Quantel International YG40 and YG402	60 PSEC, 150 PSEC	Excellent	Excellent

Criteria for Ranging Performance Ratings	
Rating	Peak-To-Peak Variation in 100 Point Mean
Poor	Exceeds 20 cm Peak-To-Peak
Fair	Between 10 and 20 cm Peak-To-Peak
Good	Between 6 and 10 cm Peak-To-Peak
Very Good	Between 2 and 6 cm Peak-To-Peak
Excellent	Less than 2 cm Peak-To-Peak

Photomultiplier

Almost all field SLR systems currently use conventional dynode chain photomultiplier tubes (PMT's). In the latter devices, the incoming light pulse impinges on a photocathode with a quantum efficiency on the order of 10 to 15 percent. The resulting short burst of photoelectrons is then accelerated through a potential difference to an adjacent dynode where electron multiplication, or gain, takes place. The electrons continue to multiply as they propagate from dynode to dynode until they finally arrive at the anode. Gains of 10^6 are typical for PMT's used in SLR systems.

The time it takes the electrons to propagate from the photocathode to the anode via the intervening dynodes is referred to as the "transit time." This represents a fixed temporal delay, or bias, introduced into the time-of-flight measurement. It varies with the PMT bias voltage and can be on the order of tens of nanoseconds (meters). This delay is just one of many optical, electronic or cable delays occurring in a typical ranging system. Fortunately, the overall "system delay," which corresponds to the sum of delays introduced by individual components or propagation lengths, can be accounted for, or possibly eliminated, via calibration procedures or common channel receiver techniques to be discussed later.

Variations in the transit time are referred to as "transit time jitter" and represent a random uncertainty superimposed onto the measurement. Many of the best dynode chain PMT's, such as the "electrostatic" and "static crossed-field" tubes introduced by Varian Inc., attempted to minimize this "jitter" by providing a well-defined path for the electrons to follow through the use of controlling electrostatic or magnetic fields. Unfortunately, this path is only well-defined if the starting point, the focal spot on the photocathode, is also fixed. Pointing errors in a real tracking system, however, will cause the target image to move about within the photocathode. This can cause variations, as large as a nanosecond (15 cms), in the systematic PMT transit time as the target image moves from the center of the photocathode to the edge [14].

Recently, the microchannel plate photomultiplier (MCP/PMT), has been introduced. Photoelectrons generated by the photocathode are accelerated over a short gap of about 0.6 mm ("proximity focusing") by a bias voltage and impinge on the face of a microchannel plate. The latter is made up of a closely packed array of narrow cylindrical channels with diameters on the order of 10 micrometers. Once inside a channel, the photoelectron ricochets repeatedly off the cylindrical inner wall generating more electrons in the process. Typically, three such plates must be stacked together to achieve gains in excess of 10^6.

Since the MCP tubes are characterized by much shorter, better defined electron path lengths, they exhibit much shorter transit times and smaller jitters than conventional dynode chain PMT's. For example, the ITT F4128 MCP PMT has a transit time delay of only a few nanoseconds. More importantly, the path of the electron is tightly controlled by the high bias voltage applied over the narrow acceleration gap and through confinement by the narrow microchannel. This results in an extremely small transit time jitter. Furthermore, since the transit time varies only weakly with the starting position of the photoelectron, the MCP exhibits a greatly reduced sensitivity to image position effects and resulting biases appear to be less than 3 mm [14].

During daylight operations, it is sometimes desirable to "gate" the PMT. Gating of the tube is accomplished by applying the acceleration, or bias, voltage only during a time period when the returning laser pulse is expected. This inhibits the propagation of photoelectrons from the photocathode to the first dynode or microchannel and prevents background optical noise from saturating the electron amplification stages. The ideal "gate" voltage pulse is short in duration (typically a few microseconds or less), has a fast rise and fall time, is very flat in the temporal region where the pulse is expected, and has a highly repeatable amplitude. If the latter requirements are not met, the gate will severely degrade the timing performance of the PMT.

Figure 7 compares the performance of the ITT F4128 MCP PMT with that of an Amperex 2233B dynode chain PMT currently used in the NASA MOBLAS network [15]. The one sigma transit time jitter for an ungated MCP tube is about two centimeters for single photoelectron inputs compared to ten centimeters in the 2233B. For input signal levels of eight photoelectrons or more, the jitter is subcentimeter in the MCP/PMT. The jitter increases between 10 and 25 percent, depending on signal level, for one particular gating configuration developed at Goddard. The MCP/PMT also has an impulse response of 450 picoseconds compared to four nanoseconds for the 2233B. The faster impulse response improves the performance of the timing electronics to be discussed next.

The ITT F4128 contains two internal MCP amplifier stages and has an electron gain of 2×10^5. The F4128 was chosen over the higher gain F4129 model, with three stages and a gain of 3×10^6, because of the former's greater tolerance for the higher background radiation levels associated with daylight operation of the ALRS. The lesser gain can be compensated for by the inclusion of a 1 Ghz bandwidth amplifier available from ENI.

There is some evidence to suggest that the transit time can also be influenced by the amplitude of the input signal or by a higher background of noise photons [16]. Experiments with the Amperex 2233B PMT suggest that the tube transit time decreases with an increase in the photon background rate. Changes in the 2233B transit time on the order of a centimeter have been observed for background count rates of 4×10^6 at typical PMT operating voltages and the transit time decreased rapidly for count rates in excess of 10^7. Similar studies have not yet been performed on microchannel PMT's.

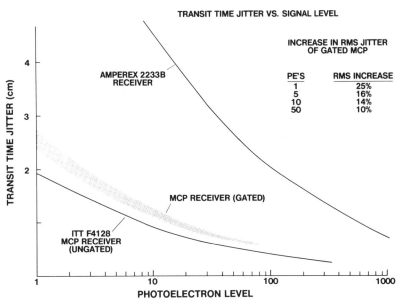

FIGURE 7. A comparison of transit time jitter as a function of signal level for the Amperex 2233B Dynode Chain PMT and the ITT F4128 Microchannel Plate PMT (gated and ungated).

To summarize, factors which can influence the performance of a photomultiplier in a laser ranging system are: (1) impulse response, (2) transit time jitter, (3) the position of the image on the photocathode, (4) the amplitude of the input signal, and (5) background radiation. The microchannel plate photomultiplier offers several important advantages over conventional dynode chain PMTs.

Discriminator

The output from the photomultiplier/amplifier, which is typically a quasi-gaussian waveform with randomly varying amplitude, is input to a constant fraction discriminator (CFD). The discriminator defines the point on the signal waveform on which the timing will be based and generates a rectangular NIM logic pulse having a few nanosecond width which, in turn, starts or stops the time interval unit. Although varying signal amplitudes in other discriminator types (such as fixed threshold, rise time compensated, and hybrids) will cause time biases on the order of half the input pulse width, the bias is highly repeatable and can be taken out with a software correction if the signal amplitude is measured and recorded with each range measurement. The CFD, on the other hand, includes hardwired circuitry which attempts to compensate for a varying signal level. A plot of time bias versus signal amplitude is a measure of the degree to which the aforementioned compensation circuitry has been successfully implemented and is often referred to as the "time walk characteristic" of the discriminator.

Figure 8 displays the time walk characteristics for the older ORTEC 934 CFD, and a relatively new CFD, the Tennelec TC453. Both discriminators were adjusted for minimum time walk with a one nanosecond full width half maximum input pulsewidth. To generate the curve, start and stop pulses separated by a fixed interval are generated electronically. The amplitude of the stop pulse is varied by means of an attenuator. Each point on the time walk characteristic represents the mean of 100 time interval measurements made at a particular stop signal amplitude. Thus, the time walk characteristic is a measure of the expected signal amplitude dependent bias imposed by the discriminator on the range measurement. The standard deviation about the mean value (not shown) is then clearly the random uncertainty associated with the amplitude-dependent bias correction.

The vertical dashed lines in Figure 8 represent the specified dynamic range of the ORTEC unit. As one can easily see from the figure, the TC453 has a much flatter time walk characteristic. The RMS deviation from the nominal zero point is about 0.2 cms over the full dynamic range compared to 1.5 cms for the older ORTEC device.

Tennelec also offers a four channel, gatable version of their basic CFD, the TC454. The discriminator is gated by means of a low voltage, rectangular logic pulse. This per-

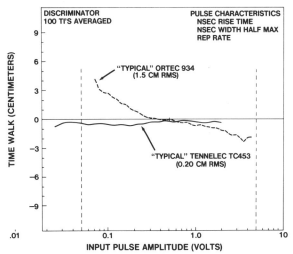

FIGURE 8. A comparison of discriminator time walk as a function of signal level for the ORTEC 934 and the Tennelec TC453.

mits the temporal interval, during which an incoming "stop" signal will be accepted, to be narrowed to as little as 10 nanoseconds (1.5 meters) although other considerations, such as uncertainty in target range, will usually dictate larger gate widths on the order of 100 nanoseconds or more.

Discriminator gating reduces the false alarm rate produced by background radiation through temporal filtering of the incoming counts. Narrowband optical filters in front of the PMT further discriminate against noise through spectral filtering of the incoming light. The discriminator also permits rejection of background noise via the amplitude of the incoming pulse. Amplitude filtering is accomplished by adjusting the discriminator threshold so that it will trigger only when a prescribed minimum voltage level (e.g. 1–3 photoelectrons) is input to it. In practice, the discriminator threshold is usually raised until the frequency of background noise counts are at an acceptable level. Setting the threshold too high will result in a lower probability of detection for the range return and a reduced data yield. Furthermore, since returns that are just above threshold are in the high time walk region of the discriminator characteristic (see Figure 8), it is usually desirable for the mean signal level to lie well above the threshold setting. Thus, the final threshold setting will represent a compromise between these competing effects. At night, the background count is usually so low that one can set a threshold just above the single photoelectron level and even occasionally dispense with the optical filter to increase the optical throughput of the receiver.

Time Interval Unit

The purpose of the time interval unit (TIU) is to measure the time of flight of the optical pulse. The start pulse initiates the time interval measurement and the stop pulse stops it. Almost all field SLR systems use single-stop time interval units. Presently, the Hewlett Packard HP 5370B Time Interval Counter, which has a 20 psec time resolution and 100 psec accuracy, is recommended for field operations.

A noise pulse which enters the receiver prior to the satellite return and exceeds the stop discriminator threshold will disable a single-stop TIU and prevent it from timing the actual range return. For multiple photoelectron systems operating at night, the impact on data yield is negligible. During daylight operation, however, the added background may force the imposition of more severe spectral or amplitude filtering of the incoming radiation with all of the attendant negative influences (e.g., reduced optical throughout, increased discriminator time walk and jitter, etc.) on the quantity and quality of the range data. In these instances, a multistop TIU, capable of detecting and timing one or more stop pulses relative to the start pulse, permits receiver threshold and gain settings to be nearer their optimum values during daylight operation. Unfortunately, most commercial multistop time interval units have inferior time resolutions on the order of a nanosecond. However, developmental multistop TIU's with 78 picosecond resolution exist at the Lawrence Berkeley Laboratories (LBL) and units with 10 psec timing resolution are well within the state of the art [17]. A high quality multistop TIU would also greatly reduce the size of the ALRS range receiver by permitting summing of all six PMT outputs into a single CFD/ TIU combination.

Not all TIU's work in precisely the same manner, but they do have a number of common components and features which permit us to discuss the dominant error sources in a fairly general way. We will use as our model a single stop TIU built for NASA by LBL which has a 9.76 picosecond resolution and a maximum range of 340 milliseconds [17]. The corresponding timing diagram in Figure 9 will be the focus of our discussion.

The heart of the TIU is a very stable clock, or master oscillator, which produces a train of pulses at a "fixed" rate, (typically 50 MHz) and ultimately determines the absolute accuracy and stability of the TIU. The measured time interval T is defined as the temporal separation between the leading edges of a pair of logic pulses applied externally to the start and stop inputs of the digitizer by the corresponding discriminators. The time measurement is split into three parts, i.e. $T = T_1 + T_{12} - T_2$. T_1 is the interval between the leading edge of the start pulse and the second following master clock pulse. T_2 is similarly defined as the interval between the leading edge of the stop pulse and the second following clock pulse. T_{12} is the interval between the two aforementioned clock pulses and is measured by counting the number of intervening clock pulses in a binary scaler, i.e., $N_{12} = T_{12}/T_o$ where T_o is the master clock period.

To accurately measure the smaller components, T_1 and T_2, the latter are "stretched" in two identical time-to-time con-

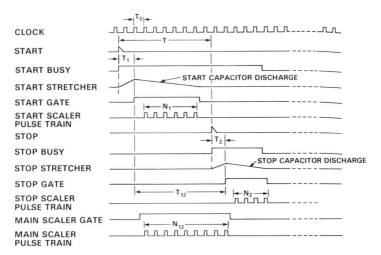

FIGURE 9. Timing diagram for a 9.76 ps resolution time interval unit built for NASA by Lawrence Berkeley Laboratories.

verters called "interpolators." The stretching constant k is an integer and selected to be $k = T_o/T_e$ where $T_o = 20$ nsec is the master clock period and T_e is the desired timing resolution. A value $k = 2048$ in the LBL unit implies a timing resolution $T_e = 9.766$ psec. Competing TIU's generally differ in their approach to the interpolators. In the LBL unit, the arrival of the start pulse causes a capacitor to be charged at constant current over the small time interval T_1. The capacitor is then discharged by a constant current k times smaller than the charging current. A comparator monitors the triangular current waveform across the capacitor and generates a square gate pulse k times longer than the original charging interval. This gate passes a burst of N_1 clock pulses which are in turn counted by a binary start scalar. The number of pulses counted by the binary start scalar is given by $N_1 = kT_1/T_o$. Similarly, an identical interpolator on the stop side counts N_2 clock pulses given by $N_2 = kT_2/T_o$. Thus, the measured time interval is given by

$$T = T_o (N_1/k + N_{12} - N_2/k)$$

Instead of capacitors, the HP5370B TIU uses a second clock, which is slightly offset in frequency from the master oscillator, and coincidence timing to stretch the T_1 and T_2 time intervals. The start pulse initiates the second clock and the start scalar counts pulses from the master clock until coincident pulses from the two clocks are detected to yield the value T_1. The time T_2 is measured similarly. A smaller frequency offset implies a longer period between coincidence pulses and hence a larger stretching constant.

The error introduced by the TIU is now seen to be

$$\Delta T = T \Delta f_o/f_o + T_e (\Delta N_1 - \Delta N_2)$$

where $\Delta f_o/f_o$ is the stability of the external master oscillator and ΔN_1 and ΔN_2 are the interpolator conversion errors.

Considering the first term in the error equation, we see that a fixed offset, Δf_o, in the nominal clock frequency will result in a range dependent bias error. On the other hand, a varying Δf_o, representing clock phase noise, will introduce a random error in the measured range.

To minimize the interpolator conversion errors, ΔN_1 and ΔN_2, the start and stop interpolator components are carefully selected and matched to take advantage of the fact that the two interpolator time intervals are subtracted from each other. Propagation delays of the start and stop input signals through the interpolators are made as short and as equal as possible to eliminate time offset errors and to minimize temperature effects. The measured thermal stability of the LBL-designed unit is about 3.0 psec/°C. High bandwidth components are used to improve the phase synchronization of the interpolators with the master oscillator, and fast rise-time discriminators aid in the precise definition of the beginning and end of the elapsed time interval. Since the start and stop inputs are typically DC-coupled, any noise superimposed on the baseline of the start and/or stop signals can result in a timing error due to a time shift in discriminator firing. Major contributors to timing jitter include component noise, power supply ripple, and clock frequency lock-in noise [17].

Target Effects

We demonstrated in Figure 10 that very high signal levels (and hence high precision and high probability of detection) would be achieved for a ground target cross-section of 10^6 square meters. At normal incidence, the effective optical

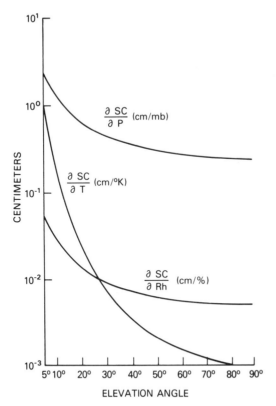

FIGURE 10. Sensitivity of spherical atmospheric refraction correction to measurement errors in surface parameters: pressure P, temperature, T, and relative humidity, Rh.

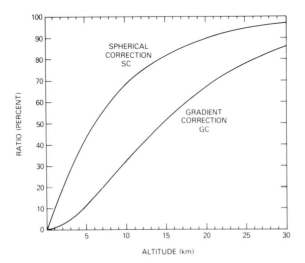

FIGURE 11. Percent of total atmospheric refraction correction applicable to ALRS for different host aircraft altitudes.

cross-section of a single cube corner reflector (CCR) is given by [18]

$$\sigma_T = \frac{4 \pi A^2}{\lambda^2}$$

where $A = \pi r_c^2$ is the area of a single CCR and λ is the wavelength which, for a frequency doubled Nd:YAG laser, is $.532 \times 10^{-6}$ meters. Thus, at normal incidence, a single CCR with a face diameter of approximately one inch (2.54 cms) would easily provide the necessary signal levels. The response of the CCR falls off approximately linearly to zero at an incidence angle of 30°. Thus, to achieve a near uniform target response for all possible viewing angles, the target should consist of a few inch radius hemisphere studded with 10 to 15 CCR's. When one takes into account the physical size and index of refraction for an individual CCR, it can be shown that reflection of the incoming light pulse effectively takes place on the surface of a smaller hemisphere inside the actual target. The radius of this smaller hemisphere must be added to each individual range measurement to compute the actual range from the target center to the ALRS system origin. There are additional small corrections related to precise target orientation relative to ALRS, laser pulsewidth, receiver impulse response, and detection strategy, but even for much larger targets such as the 60 cm diameter LAGEOS spacecraft containing hundreds of CCR's, the peak-to-peak errors are less than 3 mm [1].

Atmospheric Refraction Errors

The atmospheric error, $a(t, k)$, results from two effects. The first, and dominant, component is the effect of the atmospheric refractive index on the group velocity of light. The second component is due to the deviation of the light rays from a straight line path caused by variations in the refractive index with altitude. The group refractivity can, in general, vary in the horizontal as well as in the vertical direction. Therefore, the atmospheric correction can be arbitrarily separated into two parts: a spherical correction (SC) term, and a much smaller gradient correction (GC) term [8]. The SC term utilizes surface measurements of pressure, temperature, and humidity at the target site in a hydrostatic model of the atmosphere to determine the dependence of pressure with altitude. The GC term represents a range correction caused by variations in the meteorological parameters with latitude and longitude near the site. The influence of the atmosphere on laser ranging is the subject of a recent review by Abshire and Gardner [4].

Gardner [8] has demonstrated that the GC terms are relatively small at typical ALRS altitudes (less than a millimeter for aircraft altitudes below 6 km and elevation angles above 20°). Thus, the atmospheric refraction correction can be adequately estimated by the SC term provided the necessary meteorological parameters at the target site are known with sufficient precision.

Figure 10 shows the sensitivity of the spherical correction (SC) term to uncertainties in the three applicable meteorological parameters: pressure, P; temperature, T; and humidity, Rh. The figure assumes that the instrument is flying totally above the atmosphere at spacecraft altitudes. Since the ALRS overflys some fraction of the atmosphere, the sensitivities must be multiplied by an appropriate factor less than unity. This ratio, originally computed by Gardner, is given as a function of aircraft altitude in Figure 11. Thus, an aircraft flying at an extremely high altitude of 18 km (60,000 ft) is subject to almost 90% of the total spherical correction, while an aircraft at 6 km (20,000 ft) is subject to about 50%.

Combining the results of Figures 10-11, we see that, for an elevation angle of 20°, an uncertainty of 1 mbar in surface pressure at the target site results in less than 5 mm uncertainty in the range measurement. The range uncertainties per degree centigrade or per percent humidity are submillimeter for elevation angles greater than 20°. Note, however, that for elevation angles less than 20°, knowledge of surface temperature becomes more critical for centimeter accuracy measurements, and this illustrates a major advantage of the ALRS over ground-based geodimeter measurements which operate over near horizontal paths. In short, for a given lateral distance, the geodimeter beam passes through a much greater effective air mass than a beam from the ALRS and, thus, is subject to more atmospheric refraction error. Furthermore, the range measurements of the ALRS are insensitive to relative humidity unlike VLBI instruments which utilize microwave frequencies. In order to achieve accuracies on the order of 1-2 cm with VLBI techniques, one must supplement the VLBI measurements with highly accurate water vapor radiometer measurements.

To summarize, the atmospheric correction for range measurements to a given retroreflector from the ALRS is adequately accounted for by a model which depends only on the atmospheric parameters at the particular target site. Specifically, the model requires only two parametric inputs: p_k, the surface pressure at the kth target site; and q_k, the so-called "gradient of PTK" at the target site [23]. The quantity "PTK" is simply a product of the surface pressure, P, surface temperature, T, and a coefficient, K, related to the lapse rate. In early simulations [7] it was assumed that surface temperature and pressure in the target region could be modeled by quadratic polynomials in the two surface coordinates, and that the coefficients in the polynomials were determined by ground-based measurements of pressure and temperature at 15 locations (not coinciding with the target locations). It was further assumed that the surface measurements of pressure and temperature were accurate within ± 1 mbar and ± 1.4°C, respectively. The vertical variation in pressure was assumed to be determined by the hydrostatic equation.

More recently, we have demonstrated that the two atmospheric parameters at each target site, p_k and q_k, can be solved for along with the three station coordinates in the reduction of the laser range data, provided we have meteorological data at one or more target locations. This capability eliminates the need for an extensive ground network of meteorological stations and appears to be the result of several factors including the simultaneous multi-beam approach, operating the system from two altitudes, and the variation of pressure with altitude. In the simulations to be described later, the uncertainty in surface pressure is assumed to be ± 1 mbar at monitored sites and ± 100 mbar at unmonitored sites. Simulations have shown that the range accuracy is not strongly dependent on our knowledge of temperature at the target sites and this appears to be in agreement with the sensitivity curves in Figure 10. Furthermore, in keeping with a worst case scenario, the values of the meteorological parameters are assumed to be uncorrelated between sites.

Atmospheric turbulence has a correlation time scale on the order of a millisecond and therefore imposes an additional random error on the range measurement at current system fire rates. For typical atmospheric conditions and elevation angles, the RMS deviations are a few millimeters or less but can grow to a few centimeters under conditions of strong turbulence at low elevation angles (10 degrees) [4]. However, range data is never taken under conditions of strong turbulence or at elevation angles below 20°.

System Delay Calibration

The goal of a laser ranging system is to measure the geometric distance between a fixed reference point within the system (the system "origin") and the external target. A common choice for system origin is the point at which the telescope azimuth and elevation rotation axes intersect. The system actually measures the round trip time of flight of the laser pulse between the two points plus a "system delay." Figure 12 illustrates the origins of system delay in a conventional dual channel laser ranging system. A portion of the outgoing laser pulse is deflected at time $\tau = 0$ by a beam splitter into the start photoodiode at time τ_{p1} while the remainder travels to the target and back to the stop photomultiplier in a time $\tau_{po} + 2\tau_f + \tau_{o2}$. In traveling through the photodetectors, amplifiers, discriminators, and connecting cables to the time interval unit, the start and stop pulses experience additional electronic delays given by τ_1 (T, A_1) and τ_2 (T, A_2) respectively. The latter are functions of the ambient temperature T and the amplitudes of the start and stop signals, A_1 and A_2, as we have seen from our discussion of discriminator time walk. Other operational variables, such as PMT operating voltages, discriminator threshold and electronic attenuator settings, etc., also influence the system delay, but barring operator error, these are kept constant in a given flight.

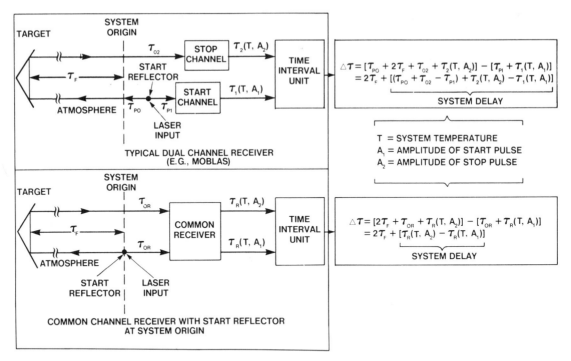

FIGURE 12. Origins and elimination of system delay: (a) conventional dual channel range receiver; (b) "zero delay" common channel receiver.

The method currently employed to estimate system delay in satellite laser ranging systems is to make repeated range measurements (typically a few thousand) to a calibration target a "known" distance away from the system origin immediately before and after each satellite pass ("pre and post calibration"). This "known" time of flight ($2\tau_f$ in Figure 12) is then subtracted from the measured time of flight $\Delta\tau$ to yield the overall system delay. The received signal levels are varied over the dynamic range expected from the satellite by means of an optical attenuator. Range means are then computed for signals falling within each amplitude range. During reduction of the ranging data, the latter information can be used to correct for the amplitude dependence of the system delay.

When Q-switched lasers were used as transmitters, it was not uncommon to see large variations (centimeters) between the pre and post calibration means. The pre and post calibration runs in the most modern field systems which employ modelocked transmitters, however, generally agree to within a few millimeters (see Table 1). Nevertheless, even subcentimeter precision laser ranging systems can have few centimeter absolute errors if the reference standard, i.e. the "known" distance to the calibration target, is in error by this amount. Currently, these standards are obtained by making collocated continuous wave laser geodimeter measurements at the ranging site. Obvious error sources include systematic calibration and random resolution errors associated with the geodimeter reference and variable but systematic atmospheric effects (which can be substantial over several kilometer paths at extremely low elevation angles) [20]. Furthermore, it is usually not possible to position the geodimeter at or very near the system origin since the latter is internal to the ranging system or tracking telescope. Hence, there is an additional error associated with estimating the position of the geodimeter relative to the ranging system origin. Based on repeated site surveys over a period of several years, past experience would suggest that the repeatability in the geodimeter measurements is no better than about two centimeters. As system precisions improved to the centimeter and even subcentimeter levels, the accuracy of the system delay calibration took on increased importance and became a subject for serious investigation.

The basic problem is to define an optical link such that the time of flight, $2\tau_f$, over the calibration range can be truly known with the desired accuracy. To make maximum use of the capabilities of modern hardware, one should strive for an absolute accuracy on the order of a few millimeters. If possible, the calibration should be applied to each and every range measurement so that long term drifts and instabilities in the ranging machine can be taken out of the data.

An obvious improvement to the conventional technique is to move the calibration target sufficiently close to the system reference plane so that the intervening distance can be measured with a measuring tape or rod. This approach would clearly eliminate all of the uncertainties associated with system calibration mentioned previously but is not always feasible for a given optical design in an existing field system. Furthermore, multistop or redundant TIU's would be required to self-calibrate on every outgoing laser pulse. Silverberg [21] has described one such field implementation of a self calibration scheme in the single photoelectron TLRS-1 system. Fiberoptic calibration paths have also been suggested. To be successful, the fibers should be short enough so that uncertainties in the effective index of refraction (group velocity) due to material inhomogeneities, dispersion, fiber bending, and temperature effects, etc. are negligible. Furthermore, common timing electronics must be used to the maximum extent possible.

Figure 12(b) illustrates the common channel approach to system delay calibration introduced previously. In this approach, both the start and stop pulses utilize a common photo-detector and timing channel so that the electronic system delay cancels except for calibratable signal amplitude effects. Thus, any time-dependent variations in the electronic system delay caused by ambient temperature variations, etc. cancel out in this "common channel receiver" approach (which can be implemented with the HP 5370B TIU described previously). If, in addition, the optical propagation paths are made collinear and a reflector is placed at the system origin to deflect the start pulse into the PMT, the start and stop optical delays will also cancel and the system will have an expected bias of zero, i.e. no system delay.

Sample Satellite Laser Ranging Results

In the summer of 1981, a passively modelocked Nd:YAG laser transmitter was installed in the MOBLAS 4 station at the Goddard Optical Research Facility in Greenbelt, Maryland. Satellite tracking tests were initially performed using the standard operational receiver which consisted of the Amperex 2233B PMT, the ORTEC 934 constant fraction discriminator, and the HP 5360 Time Interval Unit. After taking three LAGEOS and one BEC pass with the standard receiver, the 2233B PMT was replaced by a prototype of the ITT 4129 microchannel plate PMT. The latter had only a 5% quantum efficiency compared to the 12 to 15% of current commercial devices. Because of the shorter pulsewidth out of the MCP/PMT, it was necessary to adjust the ORTEC 934 CFD for short pulse operation by means of an external cable delay. The ranging performance of the standard and upgraded systems was evaluated using the software package LASPREP which fits the raw range data to a simple orbit (including geopotential terms up to J3) using best least squares estimates of range and time bias, applies a three sigma filter to the data, and repeats the procedure until there is no further improvement in the RMS of the orbital fit. The software also computes running normal points which are obtained by averaging 50 returns and then dropping the first data point in the subset and adding the subsequent data point to compute the next normal point.

Table 2 summarizes the results of the field experiments [15]. Using the Amperex 2233B PMT with the modelocked laser resulted in single shot RMS precisions to LAGEOS between 2.5 and 4.2 cms, as determined by the LASPREP processor, for three separate passes in September, 1981. The

TABLE 2. Satellite Pass Summary.

Date	Returns	Points Rejected	Single Shot RMS Sys. Cal.	Single Shot RMS Pass	Normal PNT. RMS (50 AV)	Pre/Post	Notes
BEC Passes							
10/8/81	450	16.8%	2.7CM	2.7CM	.5CM	0.0CM	2233PMT
10/9/81	287	10.1%	1.3	2.67	.4	.4	MCPPMT
Lageos Passes							
9/25/81	2420*	3.5%	5.6	3.7	.8	0.0	2233PMT
9/26/81	3717	3.0%	3.6	4.2	1.2	− 8.1*	2233PMT
9/29/81	3510	8.0%	5.5	2.5	1.0	−25.8*	2233PMT
10/9/81	86	17.0%	1.41	1.68	.05	1.97	MCPPMT
10/14/81	2594	6.7%	1.18	1.77	.27	0.0	MCPPMT
10/15/81	5286	4.7%	1.62	2.56	.83	.66	MCPPMT
10/17/81	1877	3.3%	1.15	1.51	.24	.39	MCPPMT
10/20/81	3083	2.6%	1.28	1.68	.30	.88	MCPPMT
10/20/81	2963	1.7%	2.3	2.2/2.0	.4	0.0	MCPPMT
10/21/81	315	24%	2.24	4.1	.79	− .63	MCPPMT
11/3/81	594	4.3%	1.4	2.1	.37	− .58	MCPPMT

*Dual histogram from target pole reflection (see text).

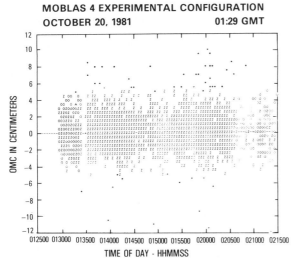

FIGURE 13a. Short arc orbital fit of single shot data OMC range single shot RMS: 1.68cm, number of measurements: 3808; edited: 101 (2.6%).

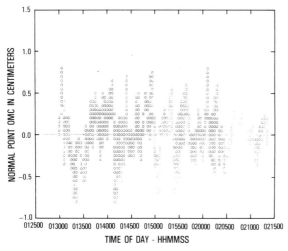

FIGURE 13b. Short arc orbital fit of normal point data (50 pt. average) OMC normal point RMS: 0.30 cms.

RMS precision of the normal points was between 0.8 and 1.2 cms. In these runs, anywhere from 3% to 8% of the raw data was edited out by the iterative processor. Interestingly, the satellite data was better than the ground pre and post calibration data. This apparent inconsistency was later traced to a support pole behind the calibration which reflected spurious pulses into the receiver resulting in a double-peaked range calibration histogram.

After painting the offending pole black and installing the microchannel plate PMT, agreement between the pre and post calibrations was typically subcentimeter with one exception (1.97 cms) for the nine LAGEOS passes. The single shot RMS for the system calibration runs fell between 1.1 and 2.3 cms. The single shot RMS for the LAGEOS data sets was only slightly higher than for the calibration data sets, i.e. typically between 1.5 and 2.5 cms for large data sets. Only 1% to 6% of the raw range data was edited by the processor in obtaining these results. Extensive laboratory tests have suggested that 1.5 cms is about the limit of precision achievable with the older HP 5360 TIU and that the latter was the limiting error source in the field receiver. Similar tests with the newer HP 5370B TIU typically yield a factor of 3 better precision, i.e. a single shot precision of about 0.5 cms. Nevertheless, the satellite normal point RMS was impressive, varying between 0.05 and 0.83 cms over the nine pass LAGEOS data set. Figure 13 displays a LAGEOS data set taken on October 20, 1981 with the micro-channel PMT installed. Figure 13a is a graph of the raw data set totalling 3808 range measurements of which 101 were rejected following ten iterations through the LASPREP processor. The single shot RMS of the edited data was 1.68 cms. Figure 13b is a plot of the corresponding normal points which have an RMS of only 0.3 cms. This is only slightly higher than what would be expected for a totally random error, i.e. $1.68 \text{cms}/\sqrt{50} = 0.25$ cms. The peak-to-peak variation in the normal points was about ± 0.8 cms.

Satellite range tests have begun with the fully upgraded range receiver which includes the Tennelec TC 454 discriminator and HP5370B time interval unit. Preliminary results suggest a single shot RMS of 0.9 cms off the LAGEOS[1] satellite [22], but the data is limited and has not yet been properly reduced and analyzed. Ground data over kilometer to few kilometer horizontal paths using the same or similar receivers have yielded unedited single shot precisions between 5 and 10 mm. The long term repeatability of the mean target range is on the order of a few millimeters.

SYSTEM PERFORMANCE

Simulation studies which have been performed utilized the mission scenario described earlier. For these studies it

---

[1]The *LA*ser *GEO*detic *S*atellite is in a near circular 6000 km orbit.

was assumed that the aircraft initially approaches the target grid at an altitude of 3.9 km. After acquiring the first few targets, the instrument shifts to the tracking mode for the remainder of the mission. After overflying the rows of targets at 3.9 km, the aircraft then climbs to its maximum cruise altitude (say 6 km for the NASA NP3A aircraft) for a second set of passes over the target grid. The turning maneuvers between passes can be used to calibrate the onboard attitude sensors. The spacing between targets will depend on several factors including the scientific objectives of the mission, the aircraft altitude, and terrain limitations. For the NP3A aircraft, the spacing is norminally taken to be 7 km. At typical cruise velocities (i.e. 200 knots) and at most favorable aspect angle, laser range data to a given target is taken for approximately 300 seconds before the pointing system is commanded to acquire a new target. For the laser repetition rate of 10 pps, this corresponds to 3000 range measurements per target. Since a given target is common to a number of six target sets, and data are collected at two altitudes, over 6500 range measurements are typically made to each target. The performance of the ALRS as evaluated from error analysis is illustrated in Figure 14. In this figure, the baseline precision vs. baseline distance from an arbitrary origin is shown. The baseline precision decreases with increased baseline length. For instance, in the absence of atmospheric refraction the baseline precision is 0.65 cm for a 20 km baseline. The simulation was performed for a 15 target grid (3 × 5) under the assumption that the single shot laser range measurement uncertainty was ± 1 cm. The total number of range measurements is 97,959 corresponding to the amount of data collected in approximately 27 minutes of flight time over the grid assuming no data droupouts.

The baseline precision is degraded slightly in the presence of atmospheric refraction. The "with refraction" curve in the figure was determined under the assumption that surface pressure and temperature in the target region could be modelled by quadratic polynomials in the two surface coordinates and that the coefficients in the polynomials were determined by ground-based measurements of pressure and temperature at 15 locations (not coinciding with the target locations). It was further assumed that the surface measurements of pressure and temperature were accurate to ± 1 mbar and ± 1.4°C respectively. The vertical variation in pressure was assumed to be determined by the hydrostatic Equation [23].

Figure 15 shows the importance of refraction errors and also illustrates that an extensive network of meteorological sensors is not required. Atmospheric measurements made at a single site collocated with a laser target within the ALRS grid, will significantly reduce the effects of atmospheric refraction upon ALRS baseline precision. If the information from that single "met sensor" is used and the atmospheric parameters for the other ALRS targets are made part of the set of estimated parameters, a factor of about 7 improvement

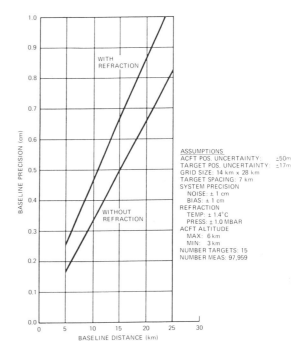

FIGURE 14. Baseline precision vs. baseline distance.

in baseline precision is achieved. The figure also shows that the inclusion of additional meteorological sensors within the ALRS target area does not significantly improve baseline precision. It should be emphasized that Figure 14 assumes an extensive meteorological sensing network in the neighborhood of the target region whereas Figure 15 utilizes meteorological sensors collocated at one or more target sites.

Figure 16 shows the evolution of baseline precision as a function of baseline distance for a region 14km × 112km in which 51 laser retroreflectors are deployed. It can be seen that the baseline precision is degraded at the rate of about 1.7cm/100km. This result compares very closely with that obtained for smaller target grid areas.

In Figure 17, the evolution of baseline precision is given for a series of randomly deployed laser targets. These targets are distributed (as shown in the inset) in a potential ALRS flight test region in the vicinity of Luray, VA. The targets were located at approximately 17 first order survey monuments currently maintained by the U.S. Geological Survey. The simulation indicates that the random target pattern does not significantly affect the baseline precision relative to that presented in Figures 14, 15 and 16. For the simulation, one meteorological sensor was located in the middle of the grid and the meteorological parameters at other sites were determined in the estimation process.

Since the multiple laser measurements from an aircraft

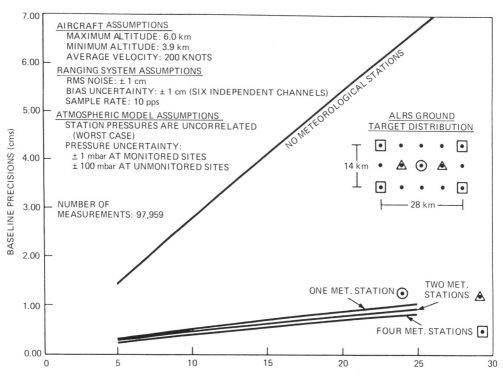

FIGURE 15. Airborne laser ranging system baseline precision as a function of the number of meteorological stations.

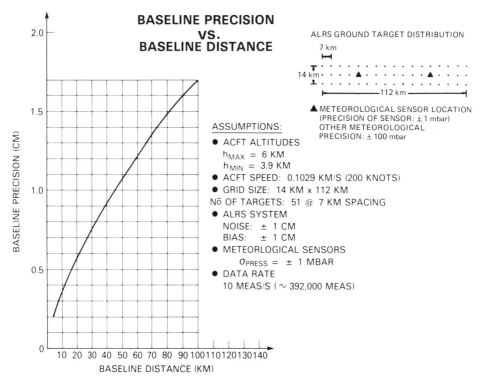

FIGURE 16. Baseline precision vs. baseline distance.

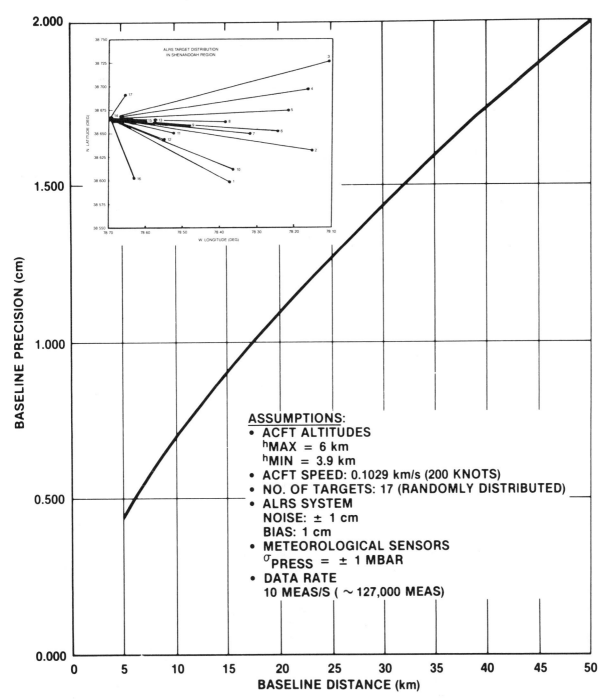

FIGURE 17. Airborne laser ranging system baseline precision for random laser target distribution.

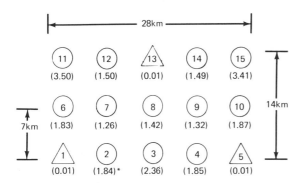

*Numbers in parentheses are errors in normal direction to plane defined by target #'s 1, 5, 13. (cm).
△: Fiducial points of laser target grid
○: Laser targets.

FIGURE 18. Errors in the normal direction to plane defined by fiducial points within the ALRS target area.

(i.e. geometric measurements) are incapable of determining the global location of a grid (Figure 18) certain laser target position coordinates were assumed to be perfectly known, this providing the basis for a well conditioned problem. The position of target #1 is assumed known, thus holding the grid in translation. Then y and z coordinates of target #5 are also fixed. This fixes the direction (though not the length) of the baseline formed by targets 1 and 5. In effect, the x-axis of the coordinate system has been defined. Finally, the z-component of target #13 is assumed known. These mathematical constraints do not affect the interstation baseline lengths, but do affect the vertical distance. For this reason the evolution of the vertical errors is discussed separately. The procedure of defining a local plane facilitates the estimation of target coordinates normal to that plane by defining the vertical coordinate to be normal to that plane. This definition would be effective in detecting subsidence and edifice building. It requires the association of that plane to a larger, more general coordinate system in order to relate the local vertical coordinates to standard surveying coordinate systems.

Because of this definition, however, the errors in the vertical components at target #'s 1, 5, 13 are by definition zero. Examination of the analysis results (i.e. numbers in parenthesis in Figure 18) show that the vertical error tends to increase as the distance from the fiducial points (i.e. target #'s 1, 5, 13) increases. The rate of increase is greater along the boundaries formed by target #'s 1, 6, 11 and target #'s 5, 10, 15 than it is along the boundary formed by target #'s 1, 2, 3, 4, 5 because there is no constraint to bound the error growth. The results presented in Figure 18 correspond to the solution from which the evolution of the baseline error in Figure 14 was obtained.

SUMMARY

The geodetic applications of the Airborne Laser Ranging System (ALRS) include the precise survey and the rapid resurveying of large areas in support of engineering, geodetic, and scientific requirements. The requirements that high density surveys be performed accurately (to a few cm in many cases) and rapidly is one for which the ALRS can be utilized most effectively. For example, the rapid establishment or development of geodetic control networks are required for two classes of geodetic applications:

1. The first of these is the mapping of large scale areas with no existing control or areas in which it is necessary to resurvey an old network. In such areas, maps may be required in such a brief time period that the traditional methods of control cannot be used. Adequate systems with rapid geodetic position capability are desirable and would be used for photogrammetry control of the area.
2. The second application is involved with survey activities in support of a large-scale construction project in an area with no geodetic control or with insufficient control density. For example, a 4000 kilometer long railroad would require establishment of 2000 to 3000 geodetic positions in three dimensions. A typical spacing between points of one to three kilometers with an accuracy of one to two centimeters is required for such an application. Furthermore, vertical control is important for this application. Satisfying these requirements by conventional methods would cost four to six million dollars and take years to complete, whereas ALRS could complete the measurements in a matter of days.

The ALRS has the unique capability of being able to perform the solution for a control network of the type described in a rapid manner.

Under certain situations, land deformation creates environmental hazards. Examples are the deterioration of the Alaskan oil pipeline associated with land deformation which may rupture the pipeline and violate the established environmental protection requirements. Similar problems exist with other oil and gas pipelines, dams, canals, and nuclear waste disposal sites. Due to the length of the pipelines, the required accuracy of measurements, and the frequency with which measurements must be made, such monitoring is expensive. With laser retroreflectors spaced at one or two kilometer intervals on both sides of a pipeline, the measurements could be taken once or twice per year with one to two centimeter accuracy using ALRS.

In addition to large scale geodetic and land management applications, the high precision, speed, and large area coverage of the ALRS make it suitable for a number of scientific measurements including tectonic plate motion, regional crustal deformation, glaciology, and temporal volcanic deformation.

REFERENCES

1. Degnan, J. J., "Satellite Laser Ranging: Current Status and Future Prospects," *IEEE Transactions on Geoscience and Remote Sensing*, Vol. GE-24, No. 4, pp. 398–413 (July 1985).
2. Degnan, J. J., W. D. Kahn and T. S. Englar, Jr., "Centimeter Precision Airborne Laser Ranging System," *Journal of Surveying Engineering*, Vol. 109, pp. 99–115 (August 1983).
3. Harper, L. L., R. H. Williams, K. E. Logan, D. A. Stevens and J. J. Degnan, "Ultrashort Pulse Solid-State Transmitter Development," presented at the Dec. 11–15, 1978, LASERS '78 Conference, held in Orlando, Fla.
4. Abshire, J. B. and C. S. Gardner, "Atmospheric Refractivity Corrections in Satellite Laser Ranging," *IEEE Transactions on Geoscience and Remote Sensing*, Vol. GE-24, No. 4, pp. 413–425 (July 1985).
5. Gaunt, A. E. and Gray, D. L., "The AN/ARN 101 LORAN Receiver," *Proceedings of the Wild Goose Society*, 4 Townsend Rd., Acton, Mass. 01720 (1975).
6. Abshire, J. B., J. L. Friskey and P. L. Fuhr, "Design and Operation of the Airborne Laser Ranging System (ALRS) Computer," NASA X-723-81-30, NASA Goddard Space Flight Center, Greenbelt, MD 20771 (August 1981).
7. Englar, T. S., C. L. Hammond and B. P. Gibbs, "Co variance Analysis of the Airborne Laser Ranging System," Business and Technological Systems, Inc., Seabrook, Md., BTS-FR-81-143 (February 1981).
8. Gardner, C. S., "Atmospheric Refraction Effects in Airborne Laser Ranging," *University of Illinois RRL Publication No. 511* (June 1981).
9. Mueller, I. I., B. H. W. Van Gelder and M. Kumar, "Error Analysis for the Proposed Closed Closed Grid Geodynamics Satellite Measurement System (CLOGEOS)," Department of Geodetic Science Report No. 230, The Ohio State University (September 1975).
10. Jazwinski, A. H., "Stochastic Processes and Filtering Theory," Academic Press (1970).
11. Gelb, A., "Applied Optimal Estimation," M.I.T. Press (1974).
12. Degnan, J. J. and Zagwodzki, T. W., "Characterization of the Q-switched MOBLAS Laser Transmitter and Its Ranging Performance Relative to a PTM Q-switched System," NASA Technical Memorandum 80336, NASA Goddard Space Flight Center, Greenbelt, MD 20771 (October 1979).
13. Degnan, J. J. and T. W. Zagwodzki, "A Comparative Study of Several Transmitter Types for Precise Laser Ranging," in Proc. of 4th Inst. Workshop on Laser Ranging Instrumentation (University of Texas, Austin, Oct. 12–16, 1981), published by Bonn: Geodetic Institute, University of Bonn, pp. 241–250 (1982).
14. Abshire, J. B., NASA Goddard Space Flight Center, Greenbelt, MD, 20771, unpublished results.
15. Degnan, J. J., T. W. Zagwodzki and H. E. Rowe, "Satellite Laser Ranging Experiments with an Upgraded MOBLAS Station," Proceedings of the Fifth International Workshop on Laser Ranging Instrumentation held at the Royal Greenwich Observatory, East Sussex, England, September 10–14, 1985, published by Bonn: Geodetic Institute, University of Bonn, pp. 166–177 (1985).
16. Zagwodzki, T. W., NASA Goddard Space Flight Center, Greenbelt, MD 20771, unpublished data.
17. Turko, B., "A Picosecond Resolution Time Digitizer for Laser Ranging," *IEET Trans. on Nuclear Science*, NS-25, pp. 75–80 (February 1978).
18. Minott, P. O., "Design of Retrodirector Arrays for Laser Ranging of Satellites," NASA Goddard Space Flight Center TX-723-74-122 (March 1974).
19. Fitzmaurice, M. W., P. O. Minott, J. B. Abshire and H. E. Rowe, "Prelaunch Testing of the Laser Geodynamic Satellite (LAGEOS), NASA Technical Paper 1062, NASA Goddard Space Flight Center (October 1977).
20. Abshire, J. B., "Pulsed Multiwavelength Laser Ranging System," NASA Technical Memorandum 83917, NASA Goddard Space Flight Center, Greenbelt, MD 20771 (March 1982).
21. Silverberg, E. C., "The Feedback Calibration of the TLRS Ranging System," *Proceedings of the Fourth International Workshop on Laser Ranging Instrumentation,* pp. 331–337, see Reference [13].
22. Varghese, T., Bendix Field Engineering Corporation, Greenbelt, MD, private communication.
23. Marini, J. J. and Murray, C. W., "Correction of Laser Ranging Tracking Data for Atmospheric Refraction at Elevations Above 10 Degrees," NASA X-591-73-351 (November 1973).

CHAPTER 6

Inertial Surveying

E. F. ROOF*

INTRODUCTION

In 1972 a new survey system based on the use of inertial navigation principles was introduced. The growth in the use of this technology has been phenomenal and should continue to grow. All types of surveys can be accomplished using inertial technology, from the establishment of precise geodetic control—including the measurement of gravity vectors—to the establishment of boundaries, mapping control and other engineering surveys.

In this chapter, the principles and theory of inertial navigation systems used for survey purposes will be stated, inertial survey systems presently in use will be described, operational results will be examined and recommendations for the efficient utilization of inertial survey systems will be made.

BASIC PRINCIPLES OF INERTIAL NAVIGATION

Laws of Inertia

Inertial navigation systems are devices that implement Newton's laws of motion to solve the navigation problem. Newton's laws of motion are stated as follows: a) Every body continues in its state of rest, or uniform motion in a straight line, unless it is compelled to change that state by forces impressed upon it; b) the change of motion is proportional to the motive force impressed, and is made in the direction of the straight line in which that force is impressed; and c) to every action there is always opposed an equal reaction.

The first law indicates the inert character of matter. The measure of this resistance to change in motion or direction is mass. The second law gives the relationship between force, mass and acceleration in the well-known equation $F = ma$. The third law points out that the body being acted upon also exerts a force against the body producing the force. The above laws have equivalents in rotational terms. A body in rotation will tend to continue rotating at the same angular rate and to maintain the same axis of rotation in space until some force is exerted on it.

It should also be noted that Newton's laws of motion are postulated with the assumption that the observations are taken with respect to a frame of reference that is fixed in space. It has been proven that a body can be at rest or in uniform motion in a straight line in one frame of reference and be traveling in a curve and accelerating with respect to another frame of reference. When Newton's laws of motion hold in one frame of reference, they also hold in any other frame of reference that is moving at constant velocity relative to the original frame of reference.

Gyroscopes

A gyroscope is mostly thought of as a rapidly spinning rotor supported on a ball and socket mount, which allows freedom of tilt of the spin axis relative to the base in addition to spin freedom. When such a device is initially set with the spin axis pointed in some direction in space (toward a certain star), the spin axis preserves this direction with a high degree of fixity.

A rapidly spinning rotor imparts three unique properties to a gyroscope: a) It makes the rotor and rotor shaft rigid against angular deflections; b) if a torque is applied about an axis perpendicular to the spin axis, the rotor turns about a third axis orthogonal to the other two axes; and c) when the torque is removed, the rotation of the axis ceases.

A distinguishing feature of any gyroscope is its accuracy either in maintaining its original orientation in inertial space, or in measuring an angular rate correctly. A perfect

---

*USAETL, Fort Belvoir, VA

gyro will maintain its spatial direction forever (supposing no mechanical failure occurs). However, gyroscopes do drift away from their reference direction. Measurements made with respect to the indicated direction are in error by at least the angular drift rate of the gyroscope. A good gyroscope will drift at an extremely low rate so that measurement accuracy will be good over a reasonably long period of time.

Starting with the concept of a rapidly spinning rotor in inertial space, the basic functional design of a gyro may be accomplished in the following five steps:

- Have a rapidly spinning rotor on an axle.
- Mount the axle in a gimbal set having one or two degrees of freedom with respect to the gyroscope housing.
- Provide suitable means for restraining the angular motion of precession of the axle, thereby shaping the transient and steady state response of the gyro to inputs of angular rate or torque.
- Provide a means (pickoff device) for measuring the motion of the external housing relative to the axle.
- Furnish a means (caging device) for initially setting the axle to a desired attitude or position subsequently causing it to move in an ordered precession.

Only two-axes gyroscopes will be described since they are the type primarily used in inertial survey systems. A two-degree-of-freedom gyro can be used to control angular velocity around two orthogonal axes. Figure 1 is a line schematic diagram of a two-degree-of-freedom gyroscope. The inner gimbal axis is 0'-0' and the outer gimbal is 0-0. The two gimbal axes allow the case to be tilted about axis 0'-0' or axis 0-0 without disturbing the space alignment of the rotor spin axis.

Tilts represent an angular disturbance to whatever the gyroscope is attached. They are detected by resolvers that send out error signals to a computer, which sends a message to a servo system to remove the disturbance tilt. The torque generators shown in Figure 1 use data from a computer so that the gyro can be torqued to the desired orientation.

The two basic gyroscopes used in present day inertial survey systems are a floated, two-degree-of-freedom, gas lubricated spin-bearing gyro and a two-degree-of-freedom, electrically suspended gyro.

The floated gyroscope is based on mounting the gyro gimbal in a fluid of the same average density. This means that the fluid, instead of the gimbal bearings, supports the gyro assembly, thereby reducing gimbal friction to extremely low levels. The gyro rotor wheel is driven at a high constant rate of speed (about 24,000 RPM) by a synchronous motor. Usually the gyro motor is inside out with the rotor outside the stator to develop the largest possible angular momentum consistent with requirements for weight and size. Instead of ball bearings, these gyros now use hydrodynamic gas bearings resulting in a significant reduction in gyro drift rate.

The electrostatic gyroscope is based on the rotor being a hollow sphere supported electrostatically. The hollow rotor is the only moving part of the electrostatic gyroscope. The rotor is suspended in an evacuated cavity (pressure about 3×10^{-8} mm Hg) to minimize viscous drag on the spinning rotor. After initial suspension, the rotor is brought up to speed by a rotating magnetic field produced by spin coils located around the equator of the rotor (see Figure 2). When operating speed (about 40,000 RPM) has been reached, a damping coil is energized to damp rotor nutation. There are

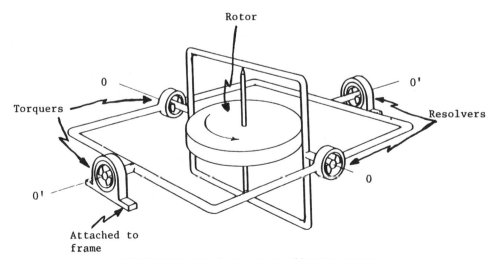

FIGURE 1. Diagram of a two-degree-of-freedom gyroscope.

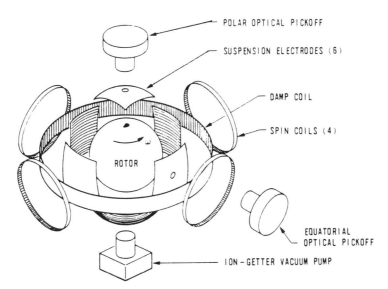

FIGURE 2. Diagram of an electrostatic gyroscope.

optical pickoffs that observe readout patterns scribed on the rotor surface. This readout pattern makes the spin axis visible so that its position relative to the cavity can be determined. Electrostatic gyros have extremely low drift rates.

For a greater in-depth treatment of the theory and use of gyroscopes, refer to the list of references at the end of this chapter.

Accelerometers

Inertial navigation is dependent upon the measurement and double integration of acceleration outputs to successfully accomplish the function of determining three-dimensional positions (three accelerometers mounted orthogonally are used to accomplish this). The output of the accelerometer is first integrated to give velocity. This velocity is then integrated to give displacement. An accelerometer is a precision instrument containing a mass that is coupled to a case through an elastic or an electromagnetic restraint. It must be emphasized that the accelerometer is actually a sensor of a specific force which is the resultant of gravitational force (consisting of both mass-attraction and centrifugal effects) and inertial reaction force.

For an accelerometer to be useful in navigation, the acceleration of gravity must be compensated. The horizontal (north and east) channels are compensated by a feedback system that keeps them level or by the computation of gravity that is used to compensate the accelerometer outputs. The vertical channel (elevation) is more difficult to compensate for the effects of gravity.

There are a variety of precise accelerometers, but only the hinged-pendulum, torque-to-balance type will be described since it is the one presently used in inertial navigation systems for geodetic surveying. Accelerations along the sensitive axis produce torques that induce rotary motion of the pendulum. These rotations are detected by a signal generator that converts them to an electrical signal and transmits this signal to an amplifier or to pulse rebalancing electronics, which drive torquers to maintain the pendulous mass at a null position. The amount of current used by the torquers or the pulses needed to maintain the accelerometer at its null position is a measure of the acceleration being sensed. A diagram of a hinged-pendulum, torque-to-balance accelerometer is shown in Figure 3.

The basic operations of an accelerometer triad can best be explained by the use of a vector diagram (see Figure 4). At $t = 0$, the accelerometer triad is located at point O in Figure 4. Some time later at $t+$ the accelerometer triad is located at point P. Looking at Figure 4 and making the assumption that the accelerometers' sensitive axes are exactly in alignment with N-S, E-W and the plumb line, we can see that the accelerometer with the sensitive axis on line O-Y has measured only the change in latitude along the N-S line. The accelerometer with the sensitive axis on line O-X has measured only the change in longitude along the E-W line and the accelerometer with the sensitive axis on line O-Z has measured only the change in elevation.

There are various error sources that effect the performance of accelerometers. The magnitude of most of these errors is determined by the manufacturer and during a premission calibration routine. They are then compensated for during operational use by software. However, two error sources for the local level, north-oriented inertial systems are best determined by operating the system over a precise

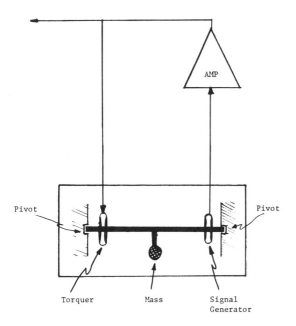

FIGURE 3. Diagram of a hinged-pendulum, torque-to-balance accelerometer.

calibration course. The two error sources are 1) the accelerometer scale factor and 2) accelerometer misalignment. Calibrating for accelerometer scale factor is analogous to calibrating a steel tape or an electronic distance meter and is fairly easy to accomplish. Misalignment errors are caused by procedures in mounting accelerometers on the platform, random changes in gyroscope and platform drift rates, changing gravity vectors during an operational mission and unknown deflections of the vertical at the starting and terminal stations. All of the above listed misalignment errors cause what can be termed as "cross-coupling error." Figure 5 shows two accelerometers. The top diagram shows an accelerometer with its sensitive axis exactly perpendicular to a north–south line, the line along which the accelerometer is moving. The bottom diagram shows an accelerometer with its sensitive axis misaligned 10 arc-seconds from being perpendicular to a north–south line, the line of travel along which the accelerometer is moving. For the perfectly aligned accelerometer, no acceleration has been measured, therefore no change in position is computed from the output. For the accelerometer that is misaligned by 10 arc-seconds, an apparent acceleration is seen. This apparent acceleration, integrated twice, converts into an error in easting or longitude by the amount of distance that the unit traveled

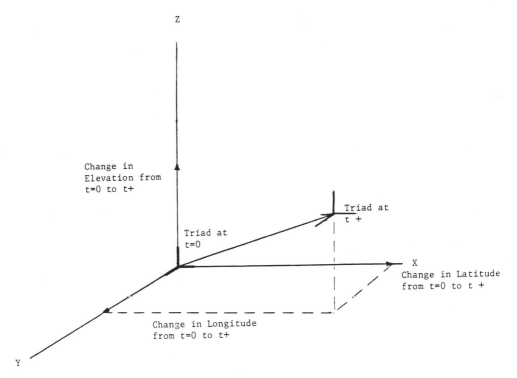

FIGURE 4. Vector diagram of accelerometer triad.

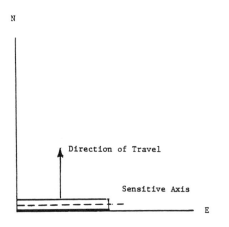 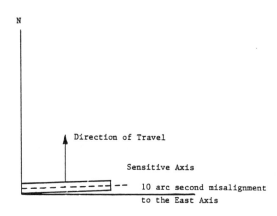

(a) Sensitive axis parallel to the East Direction (b) Sensitive axis misaligned

FIGURE 5. Accelerometer misalignment.

in the northerly direction times the sine of the misalignment angle. If the north distance that the unit traveled was 15,000 feet then the error in the easting or longitude would be (15,000) × (0.00004848), which results in an error of 0.727 feet for longitude.

For a more detailed discussion of accelerometers, refer to the list of references at the end of this chapter.

Inertial Measuring Unit (IMU)

The IMU is the heart of the inertial navigation system. It is defined as a stable platform with its support electronics. The heart of the IMU is a cluster of three accelerometers and two two-degree-of-freedom gyroscopes properly mounted on a stable platform. The stable platform is isolated from the maneuvering of the carrying vehicle by gimbals that allow the case of the IMU full freedom of motion about the stable element.

The gyroscopes are mounted on the stable platform to ensure full platform stabilization. This means that the gyroscopes will be mounted on the platform orthogonally. Since two-degree-of-freedom gyroscopes are used, one gyroscope is used to control the X and Z axes and the other gyroscope is used to control the Y and Z axes. The Z axis of the YZ gyroscope is usually slaved to the Z axis of the XZ gyroscope, but other methods of using the information from this redundant axis are sometimes employed.

The accelerometers are mounted on the stable element so that the sensitive axis of the accelerometers forms an orthogonal system in which acceleration can be measured. Figure 6 shows a four-gimbal stable element.

As stated previously, the stable element is mounted in gimbals to isolate it from its case and the vehicle in which the system is being carried. The first or innermost gimbal contains the stable element. The axis that it rotates about is vertical. This gimbal is called the azimuth gimbal. The axis of the azimuth gimbal is mounted in a second gimbal, which is called the inner roll gimbal. The axis of the inner roll gimbal is mounted in a third gimbal, which is called the pitch gimbal. The axis of the pitch gimbal is mounted in a fourth gimbal called the outer roll gimbal. The outer roll axis is mounted to a frame.

Three gimbals are sufficient to provide isolation for the stable element. However, if the system rolls 90 degrees about the second gimbal axis, the third axis becomes aligned with the first axis and the stable element is no longer free to rotate about the third axis. This condition is called gimbal lock. To prevent gimbal lock, a fourth gimbal called the outer roll gimbal has been added to the system to ensure isolation of the system regardless of vehicle attitude.

The axes of all gimbals are equipped with torquers and resolvers. The torquers are servo motors that can be used to control the attitude of the gimbals. The resolvers are electromechanical components for the purpose of measuring the angle of gimbal rotation. The azimuth resolvers in inertial systems used for surveying are usually resolvers of higher accuracy than those that are used on the other gimbal axes.

IMUs contain considerable electronics. The electronics perform various functions such as controlling the gimbal servos, providing the proper power to various parts of the IMU, translating the information from the gyroscopes and accelerometers for computer use, translating commands from the computer for use in the IMU and providing proper operating temperatures within the IMU. The IMUs used by Litton and Honeywell for inertial surveying are shown in Figures 7 and 8.

For more details concerning IMUs refer to the list of references at the end of the chapter.

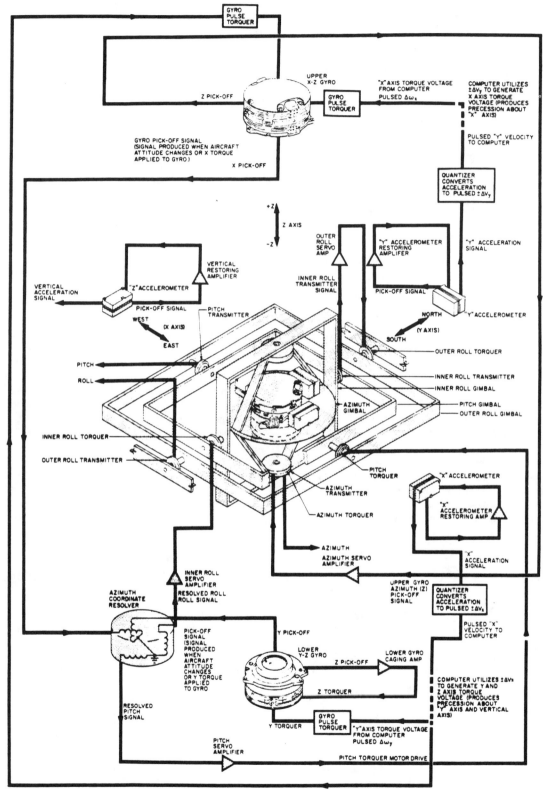

FIGURE 6. Stable element diagram (4 gimballed).

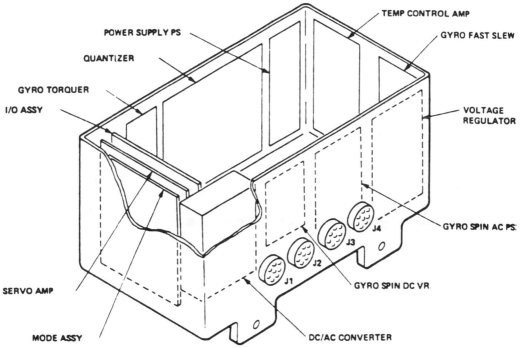

FIGURE 7. Litton IMU.

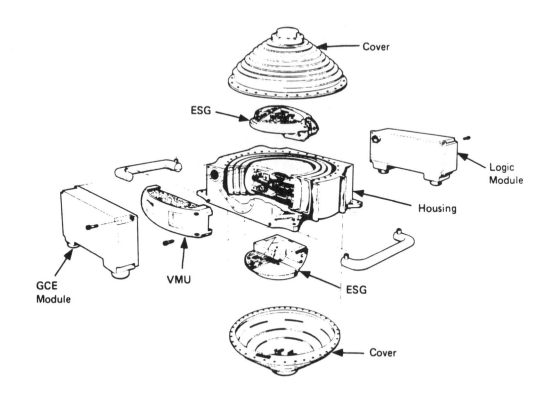
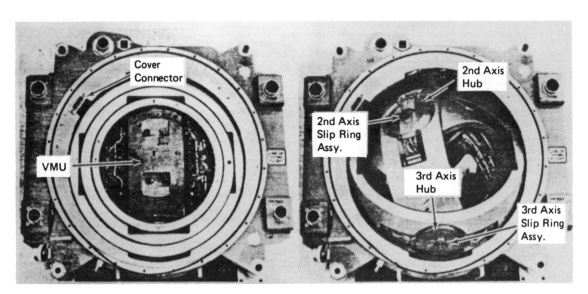

FIGURE 8. Honeywell IMU.

Reference Frames

The reference system for which Newton's laws of inertia are valid is called an inertial reference system and can be defined as a system that maintains a fixed attitude (unaccelerated) with respect to the stars. It is desirable to use a rotating reference frame to navigate or locate a position on the earth. To translate from one reference frame to another, it is necessary to examine how they are used for inertial navigation.

When utilizing inertial navigation systems near the earth's surface, it is preferable to use an earth-centered, non-rotating inertial reference frame with one of the orthogonal axes parallel to the earth's axis of rotation (see Figure 9).

The surface on which geodetic surveying and terrestrial navigation computations are based is an ellipsoid, which approximates the earth with the axis of rotation of the earth being the minor axis of the ellipsoid. This frame is called the geodetic reference frame and has its orthogonal axes aligned with the north, east and elevation on a plane tangent to the ellipsoid at the point of interest. In the interest of

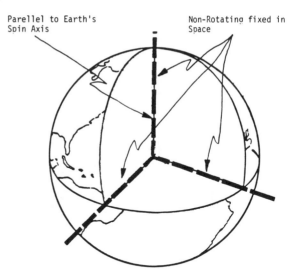

FIGURE 9. Earth-centered, non-rotating inertial reference frame.

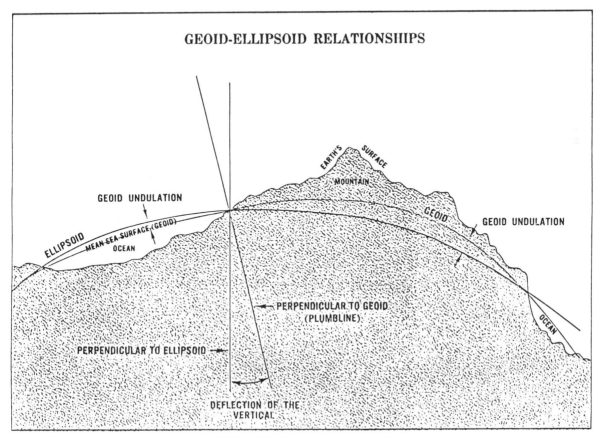

FIGURE 10. Geoid-ellipsoid relationship.

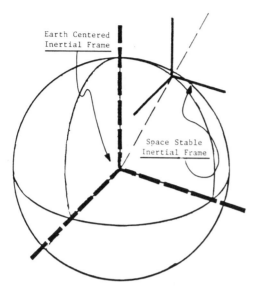

FIGURE 11a. Space-stable inertial reference frame.

practicality and because of the limits of resolution of inertial components presently used, the elevation direction of the geodetic reference frame used by inertial systems is considered normal to the reference ellipsoid. In actuality the elevation direction of the geodetic frame is normal to the geoid, which is the equipotential surface (due to the irregular gravitational field) of the earth's attraction and rotation

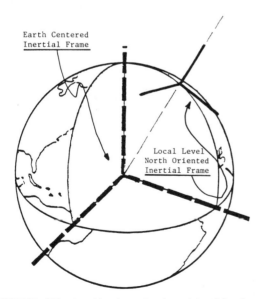

FIGURE 11b. Local-level, north-oriented inertial reference frame.

that on the average coincides with mean sea level in the open sea (Figure 10).

It should be obvious at this point that the geodetic reference frame being tied to a rotating earth requires that the data obtained from an inertial navigation system be mathematically translated and rotated from the inertial reference frame to the geodetic reference frame or that the platform be constantly torqued to keep the level accelerometers properly aligned to the north and east of the reference ellipsoid.

Inertial navigation systems have been designed and fabricated in various configurations. The configuration of the system is dependent on the reference system used to obtain the basic data output. Two types of systems are presently being used for survey purposes.

The first system is the space-stable inertial navigation system (see Figure 11a). In the space-stable system, the three accelerometers form an orthogonal triad that is oriented with the sensitive axis of the Z accelerometer parallel to the earth's axis of rotation and with the X and Y accelerometers parallel to the equatorial plane, the sensitive axis of the Y accelerometer being on the local meridian and the sensitive axis of the X accelerometer pointing east. The accelerometer triad remains in this inertial, non-rotating reference frame with the output of the accelerometers being mathematically rotated and translated from the inertial reference frame to the local geodetic frame.

The second system is the local-level, north-oriented system (see Figure 11b). In this system, the three accelerometers form an orthogonal triad that is oriented to the local geodetic frame. The sensitive axis of the Z accelerometer is oriented to the plumb line of the geoid, with the sensitive axis of the Y accelerometer oriented along the local meridian (north–south) and the sensitive axis of the X accelerometer pointing along the local parallel (east–west). The platform containing the accelerometer triad is continuously torqued to compensate for the rotation of the earth and vehicle movement. This continuous torquing keeps the sensitive axes of the accelerometer triad properly oriented to the local geodetic frame (north–south, east–west and vertical).

THEORY OF INERTIAL SURVEY SYSTEM OPERATION

Effects of System Random Errors

The essential characteristic upon which inertial survey systems are based is the high accuracy of inertial navigation systems over a short time period. Excluding the effects of systematic errors, such as accelerometer scale factor error and misalignment, the navigation error in an inertial system is due to the following:

- initial position error
- uncompensated platform attitude misalignment

- platform drift rate
- accelerometer noise
- uncertainty in the earth's gravity field
- initial velocity error

The effect of initial platform attitude imbalance can be minimized by a well-settled initial alignment of the platform. Initial position error can easily be eliminated by providing "perfect" initial position.

It is well-known that the navigation error induced by platform drift rate bias takes time to accumulate being of the form $(1 - \cos\omega_s t)$ in velocity, and $(\omega_s t - \sin\omega_s t)$ in position, where ω_s is the Schuler frequency. This occurs because the drift rate is first integrated to become platform tilt, which appears as an acceleration error that is subsequently integrated into velocity and position error. In an inertial survey system, the accumulation of such error can be counteracted by allowing the vehicle to "navigate" for only short time periods at the end of which it is brought to a stop and the accumulated tilt and velocity error eliminated by re-levelment.

The effects of accelerometer noise and gravity field uncertainty are indistinguishable, both appearing as an error in the measured vehicle acceleration. Since these acceleration measurement errors integrate directly into velocity and position error, they are a much more serious source of error than platform drift rate. The mechanism of stopping the survey vehicle after appropriately short periods of travel and re-setting velocity error and platform tilt to "null" will reduce the effect of acceleration measurement error on navigation performance. However, the position error accumulated during the vehicle travel period would be excessive if a more sophisticated mechanization for error control were not employed.

The residual system velocity error caused by platform motion due to random disturbances of the "stationary" vehicle when stopped is also a serious source of short-term navigation error when survey accuracies are being attempted. This is because any velocity error integrates directly into position error over the vehicle travel period. Clearly, a simple "nulling" of velocity error at vehicle stops would do nothing to remove the position error accumulated over the previous travel period.

Correlation of Velocity and Position Error in an Open Traverse

The discussion above indicates that random system errors have the characteristic of producing error in system computed velocity over vehicle travel periods, which are subsequently integrated into significant position error. If an accurate model of the manner in which velocity and position error accumulate is known, it is possible to use "observations" of the error in computed velocity at vehicle stops to correct not only the velocity error, but also the position error it has produced. When one deals with random effects, any model of behavior is only approximate. However, for appropriately short vehicle travel periods, it turns out that the stochastic model of velocity and position error accumulation is sufficiently accurate to allow major reductions in the navigation position error accumulated over any one travel period, with a few observations of the velocity error present at the next vehicle stop.

The one sigma growth in velocity error over a travel period is typically of the form depicted in Figure 12a. The one sigma accumulation of position error over a travel period is shown in Figure 12b. At the vehicle stop, the velocity error is nulled to the uncertainty caused by movement of the vehicle. Since the observed velocity error of the vehicle stop is not perfectly correlated with the accumulated position error, a small amount of position error will not be removed at each stop as indicated schematically in Figure 12b. When these increments of position error accumulated over successive travel periods are independent, the total (one sigma) error in position will tend to accumulate as the square root of the number of travel periods, or approx-

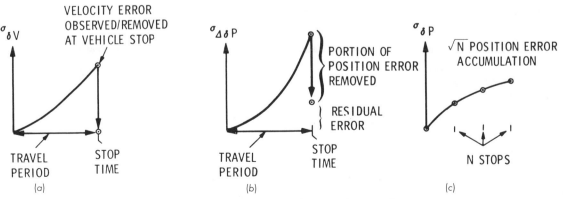

FIGURE 12. Survey position error growth due to random error in open traverse.

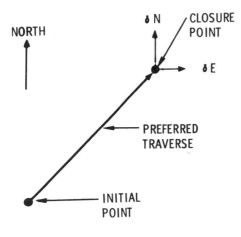

FIGURE 13. Elimination of systematic error on straight, constant rate traverse.

imately as an integral of white noise, as shown in Figure 12c. The magnitude of the power spectral density characterizing this approximating white noise process increases as the following things happen:

- magnitudes of the noise effects cited above increase
- travel time between the vehicle stops increases
- duration of the vehicle stops decreases (resulting in less settled conditions over the next travel period)

Effects of Systematic Errors

In addition to the random errors discussed above, inertial survey system accuracy is affected by certain systematic error when running an open traverse. These systematic errors are a) accelerometer scale factor error and b) accelerometer pointing error (misalignment error); and they induce survey errors in proportion to the distance traveled by the vehicle. For surveys of arbitrary design, certain premission calibration procedures can be employed to substantially reduce the effects of the systematic error sources. However, if the survey is designed to be a) composed of essentially straight-line segments between the starting and closing points or b) performed at a "constant rate," which means that vehicle travel periods and speed between stopping points are essentially constant, then it can be shown that by observing system position error at closure, the systematic errors can be determined and their effect on survey error completely eliminated. This rather powerful result means that any position error at the "surveyed" points between the starting and closing points incurred in a traverse due to systematic error is eliminated by the closure solution. Hence, the strategy of survey can be such that systematic errors induce no survey error.

If the traverse is not performed by keeping the IMU at a constant heading, uncorrectable survey errors due to the mixture of systematic errors can occur. A simple example illustrates this point. Suppose the surveyor proceeds on a generally north heading in an open traverse, except for one east-west excursion to lay a point at some distance from the north-south line. In laying this "off-traverse" point, position error will be incurred due to east accelerometer scale factor error and north accelerometer misalignment. However, when the north-south traverse is closed, no error in the position closure will be charged to these two sources, and, consequently, these systematic errors are not observable. Clearly then, any position error induced by a particular systematic error, or linear combination of systematic errors, that is not observable at traverse closure will also escape correction.

Nature of Residual Survey Error After Traverse Closure

An adequate model of the propagation of inertial survey error in open traverse is of the form:

$$z[t] = H[t]x + \int_0^t \xi[u] \, du \quad 0$$

where:

$z(t)$ is the open traverse position error after system correction at vehicle stop
x is a vector representing systematic errors
$H(t)$ is a matrix expressing the contribution of systematic error to error in survey position in open traverse
$\{\xi[u]\}$ is a white noise process with power spectral density Q, whose integral expresses the accumulation of position error caused by imperfect correction of accumulated position error at vehicle stops due to random system errors

At traverse closure, the terminal position error $z(T)$ is observed and can be used to estimate the linear combination of systematic errors:

$$H\hat{x} = HK_x z[T]$$

where:

$H \triangleq H[t]$ at $t = T$
$K_x = \Sigma H^T [H\Sigma H^T + QT]^{-1}$
Σ^x = the covariance matrix of the systematic error vector, x

Under the assumptions of an essentially straight line survey as depicted schematically in Figure 13, and a "constant rate" of survey as defined above, then:

$$H[t] = H\left[\frac{t}{T}\right]$$

and the estimated effect of systematic error along the survey traverse is:

$$\left[\frac{t}{T}\right] H\hat{x}$$

Since the estimated effect of the component of position error at closure due to the integrated white noise is:

$$\hat{\eta}[T] = K_\xi z[T]$$

where:

$$K_\xi = QT[H\Sigma H^T + QT]^{-1}$$

$$\eta[t] \triangleq \int_0^t \xi[u]\, du$$

and the estimated effect of the integrated white noise along the survey traverse is

$$\hat{\eta}[t] = \left[\frac{t}{T}\right] \hat{\eta}[T]$$

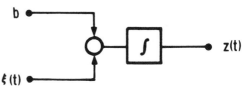

FIGURE 14. Block diagram illustrating integral of sum of bias and white noise.

then the estimated error in the open traverse position survey $z(t)$, is:

$$\hat{z}[t] = \left[\frac{t}{T}\right][H\hat{x} + \hat{\eta}[T]]$$

The error in the correction of the open traverse position survey error is then:

$$\tilde{z}[t] \triangleq [\hat{z}[t] - z[t]]$$

$$= \left[\left[\frac{t}{T}\right] - 1\right]\eta[t] + \left[\frac{t}{T}\right]\eta[T - t]$$

which has the error covariance matrix:

$$\Sigma[t]_{\tilde{z}} \triangleq E[\tilde{z}[t] > <\tilde{z}[t]] = t\left[1 - \left[\frac{t}{T}\right]\right]Q$$

which is totally independent of the systematic error sources, x.

For the elementary scalar system depicted in Figure 14, the results developed for the vector case above yield the solution for the variance of the residual error $z(t)$, as:

$$\sigma^2_{\tilde{z}(t)} = Qt\left[1 - \left[\frac{t}{T}\right]\right]$$

The normalized (one sigma) value of position survey error after correction with the closure solution thus appears as a circular arc when plotted versus the normalized survey time $(t - T)$ as shown in Figure 15, having a peak value at the midpoint of the survey. The maximum (one sigma) value of the position survey error is then:

$$\sigma_{\tilde{z}_{max}} = 0.5\,[QT]^{1/2}$$

and is a function of the survey duration T and the white noise power spectral density Q, characterizing the effect of all sources of system random error.

Since the position survey error after closure is only a function of random system errors, repetition of the traverse will tend to reduce the (one sigma) value of the error by the square-root of the number of traverses. This remains true as long as the gravity disturbance over the traverse is not the major source of survey error.

Simplified Linear Error Model for Theoretical Performance Estimates

The theoretical results presented above are based upon a simplified stochastic linear error model of the inertial survey system, which is depicted in the block diagram of Figure 16. The error equations for this system are expressed as:

$$\frac{d}{dt}[x] = Ax + \xi$$

where the components of the error state vector x are defined as:

$$X \triangleq \begin{bmatrix} \xi E \\ \xi V_E \\ \phi_N \\ b_N \\ V_E \\ \eta \\ \eta_a \\ \Delta_\eta \end{bmatrix} \begin{array}{l} \text{Position error} \\ \text{Velocity error} \\ \text{Platform tilt} \\ \text{Platform drift rate} \\ \text{Accelerometer correlated noise} \\ \text{Deflection of the vertical} \\ \text{Auxiliary state in second-order model of the vertical deflection} \\ \text{Change in the deflection of the vertical from the initial position of the vehicle.} \end{array}$$

The error model dynamics matrix A is defined as:

$$A(t) = \begin{bmatrix} 0 & 1 & 0 & 0 & 0 & 0 & 0 \\ 0 & 0 & -g & 0 & g & g & 0 \\ 0 & R^{-1} & 0 & 1 & 0 & 0 & 0 \\ 0 & 0 & 0 & -Y & 0 & 0 & 0 \\ 0 & 0 & 0 & 0 & -a & 0 & 0 \\ 0 & 0 & 0 & 0 & 0 & -\beta.f & f \\ 0 & 0 & 0 & 0 & 0 & 0 & -\beta.f \\ 0 & 0 & 0 & 0 & 0 & -\beta.f & f \end{bmatrix}$$

where

g = 32.2 fps², is a nominal value of gravity
R = 20.9 × 10⁶ ft, is a nominal value of the earth radius
ω_s^2 = (g/R), Schuler frequency squared
a^{-1} = is the correlation time of the accelerometer noise
β^{-1} = $[2.15V/3600d_o]$, is the correlation time of the vertical deflection (when the vehicle is moving)
V = $[dD/dt]$ is the speed in knots of the vehicle between stops
d_o = the correlation distance of the vertical deflection for the second-order model
f = 1 when the vehicle is moving
  0 when the vehicle is stopped
γ^{-1} = the platform drift rate correlation time

and the components of the white disturbance noise vector are defined as:

$$\xi = \begin{bmatrix} 0 \\ 0 \\ 0 \\ \xi_4 \\ \xi_5 \\ 0 \\ \xi_7 \\ 0 \end{bmatrix}$$

The null velocity observations are processed by a Kalman error control mechanization whose design model is based upon the error model defined above. The initialization of the error covariance matrix, employed for obtaining the theoretical performance estimates as well as the Kalman gains for system correction, is:

$$\Sigma \triangleq E[x_o >< x_o] = [\Sigma_{ij}[0]]$$

$\Sigma_{ij}[0] = 0$, for all $i, j = 1, \cdots, 8$; except:

$$\Sigma_{33}[0] = \sigma_\phi^2$$

$$\Sigma_{44}[0] = \sigma_b^2$$

$$\Sigma_{55}[0] = \sigma_{\tilde{v}}^2$$

$$\Sigma_{77}[0] = 2[\beta\sigma_\eta]^2$$

where

$$E[\xi_4^2] = [2\gamma\sigma_b^2] \cdot \xi[0]$$

$$E[\xi_5^2] = [2a\sigma_{\tilde{v}}^2] \cdot \xi[0]$$

$$E[\xi_7^2] = [4\beta^3\sigma_\eta^2] \cdot \xi[0]$$

σ_ϕ is the initial (one sigma) value of platform tilt.
The following are the (one sigma) values of the indicated variables in steady-state:

σ_b Platform drift rate
$\sigma_{\tilde{v}}$ Correlated accelerometer noise
σ_η Deflection of the vertical

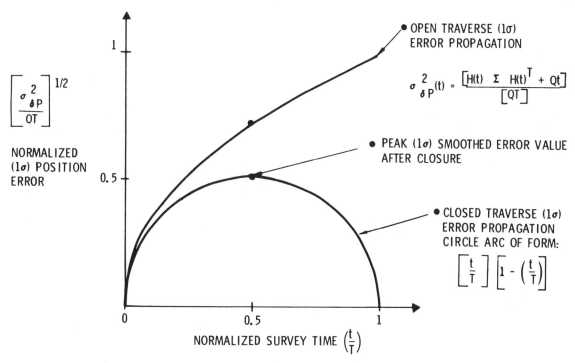

FIGURE 15. Normalized (one sigma) value of error in the raw and smoothed estimate versus time.

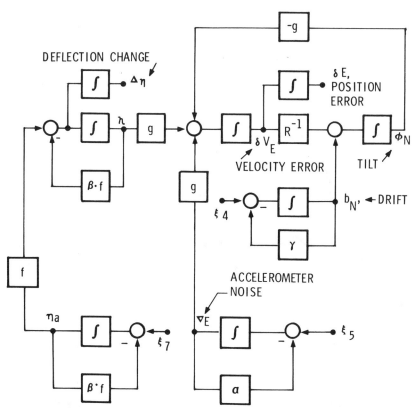

FIGURE 16. Block diagram of simplified error model of the inertial survey system.

The numerical values for the error parameter values are given in the relevant Figures 17 and 18.

At vehicle stops, observations of the residual in-system computed velocity are made. These observations are defined as:

$$\sigma = Hx + \eta_v$$

where the observation matrix H is defined as:

$$H = [0 \ 1 \ 0 \ 0 \ 0 \ 0]$$

and is the random error in the null velocity observation with variance:

$$R_v = E[\eta_v^2]$$

INERTIAL SURVEYING SYSTEMS

Operational Systems

At the present time there are three inertial survey systems in use and one special system being used by the United States Geological Survey for aerial profiling purposes. These systems are the Litton Autosurveyor (LASS); the Ferranti Inertial Land Surveyor (FILS); the Honeywell GEO-SPIN and the Charles Stark Draper Laboratory, Inc., Aerial Profiling of Terrain System (APTS).

The latest version of the LASS, called LASS-2, uses Litton A-200 accelerometers in the horizontal channels and a Litton A-1000 accelerometer in the vertical channel. The orientation of the platform is controlled by two Litton G-300 two-degree-of-freedom gyroscopes. A four gimbal assembly is used to isolate the platform assembly from the rotation of the external case, which is attached to the carrying vehicle. The gimbals have resolvers and torquers for control purposes. Included in the case is an array of electronic assemblies that are used to change analog information from the sensors to digital information for use by the on-board computer, as well as changing digital information from this same computer into analog information for control of the platform. The electronics also control the required voltages, the temperature within the case and the speed of the rotors in the gyroscopes. This complete assembly is called the inertial measuring unit (IMU).

The data processing unit (DPU) of the LASS is a real-time digital computer and is used to perform all the neces-

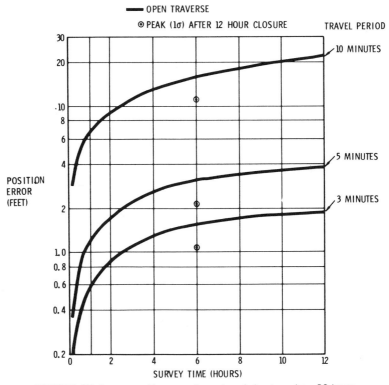

FIGURE 17. Survey position error (one sigma) due to noise—30 knots.

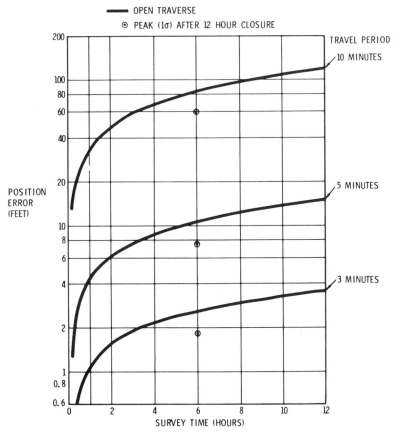

FIGURE 18. Survey position error (one sigma) due to noise—100 knots.

sary computations and give the proper operational commands to other assemblies of the system. It permits limited storage of information gathered during the mission or data to assist in the performance of the mission.

In addition to the IMU and the DPU, other parts of the system include a power supply unit (PSU), which converts 28 volt DC input to the voltage levels required by the system, and a control and display unit (CDU), which permits the operator to interface with the system. For data storage, a data storage unit (DSU) is used. The DSU is a cassette recorder with proper electronic interfacing to the system. This recorder allows direct dumping of data any time during the mission for later processing. The DSU also permits the loading and recording of various programs to and from the DPU. The LASS is shown in Figure 19.

A special alternator for the LASS is mounted on the vehicle and powered by the vehicle's engine. When used in a helicopter, the power system of the aircraft supports the LASS during operation; but an auxiliary power unit must be used during the alignment phase of operation. When the LASS is first energized, it requires about 1700 watts of power. After a few minutes of warmup time, the power requirement drops to 700 watts.

The FILS platform uses Ferranti FA2F accelerometers in all three channels. The platform is controlled by three Ferranti single-degree-of-freedom rate integrating, type 125 gyroscopes. The gimbal assembly is similar to the LASS and serves the same purpose. The platform with its supporting electronics and the DPU are contained in a single case along with the power supply unit. The computer (DPU) serves the same purpose as the DPU in the LASS. A digital recorder is used to transfer mission data from the DPU for post-mission processing. The FILS is shown in Figure 20.

The heart of the Honeywell GEO-SPIN is Honeywell's electrically suspended gyroscopes (ESG), which have very low drift rates. Three Honeywell GG-177 pulse rebalanced accelerometers are used in all three axes of the platform. The space-stable orientation of the platform is controlled by two two-degree-of-freedom ESG gyroscopes. The platform is mounted in a spherical case with some supporting electronics. The other electronics are contained in an interface electronics unit (IEU) and are used for control, power con-

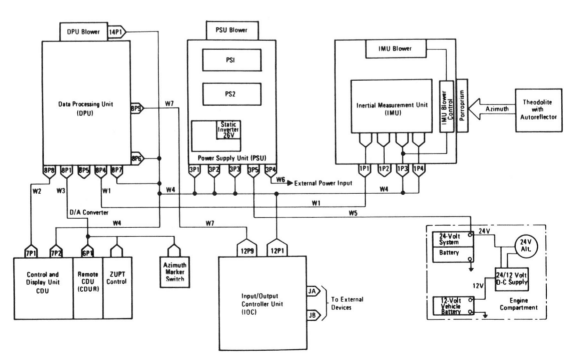

FIGURE 19. Litton Auto-Surveyor System (LASS).

FIGURE 20. Ferranti Inertial Land Surveyor (FILS).

versions and data interfacing to the computer. The DPU is a ROLM 1664 computer. Working with the DPU is a digital storage unit (DSU) used to store data during a mission as well as to load or record programs to or from the DPU. A CDU is used to allow the operator to interface with the system. A Texas Instrument 783 teleprinter is used to transmit data to the system or have data printouts from the system. The GEO-SPIN is shown in Figure 21.

The Aerial Profiling of Terrain System was developed for the United States Geological Survey by the Charles Stark Draper Laboratory, Inc., to provide a precise airborne survey system capable of measuring elevation profiles across various types of terrain from a medium to light aircraft at flight heights up to 1000 meters above the terrain. The system design accuracy goals were a horizontal position accuracy of 60 centimeters and a vertical position accuracy of 15 centimeters with a 90 percent reliability level.

The first three systems mentioned above have had various modifications made to them by the users. These modifications consist of fitting the inertial systems with offset measuring devices to establish stations at points as much as 1 kilometer from the inertial system.

System Operation

The operation of an inertial navigation system for surveying purposes can be divided into three main subgroups. These subgroups are 1) premission calibration, 2) survey mode, and 3) smoothing mode or post mission adjustment.

PREMISSION CALIBRATION

Since there is a difference in the premission alignment of the local-level, north-oriented inertial system and the space-stable inertial system, the premission alignment that is performed automatically under computer control of both systems will be described.

The premission alignment and orientation of a local-level, north-oriented system begins after the power is applied to the system and the position, elevation and time have been entered into the CDU. At the beginning of the alignment the gimbals are caged (aligned) to the attitude of the IMU case. The stable element containing the gyroscopes and accelerometers is leveled to the local gravity field by using the horizontal accelerometers. The gyroscopes are torqued until the

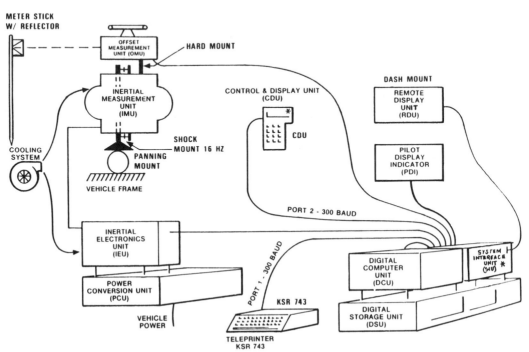

FIGURE 21. Honeywell inertial survey system (GEO-SPIN).

horizontal accelerometers read zero (analogous to leveling a transit or theodolite). This places the Z accelerometer vertical with its sensitive axis along the plumb line. The platform is then oriented to its proper azimuth (sensitive axis of the Y accelerometer pointing north-south and the X accelerometer pointing east-west) by a technique known as gyrocompassing. The gyroscope designated as the X and Z gyroscope is used to perform this function. This gyroscope, because of its alignment, senses earth rate (rotation) if the input axis (east-west axis) is not exactly perpendicular to the north-south line. When the gyroscope senses a component of earth rate, a command is given by the computer for the platform to turn in azimuth until the component of earth rate being sensed is reduced to zero. While the platform is being oriented in azimuth and leveled to the local gravity field, an internal calibration is also being performed. During this calibration, the gyroscope and accelerometer biases, as well as the gyroscope drift rates, are estimated. These parameters are held fixed until the next premission calibration is performed. The calibration parameters are used during the mission for predicting platform performance during the survey. When the premission calibration and orientation is completed, the system automatically goes into the survey mode.

The premission alignment and orientation of the Honeywell space-stable inertial system begins after the power has been applied to the system and the position, elevation and time have been entered into the CDU. At this time the alignment phase starts. During alignment the gimbal resolver errors, the accelerometer scale factor, misalignment angles and biases are computed by torquing the platform to each of 21 prestored test positions where the output of each accelerometer and the average gimbal angle for each of the four gimbles is determined. In addition to the above parameters, the platform and gyroscope drift rates are also determined. Upon completion of the alignment, the platform then aligns itself so that the sensitive axes of the accelerometer triad are parallel to the earth-centered inertial reference frame. This is accomplished by first going through a coarse alignment where the sensitive axis of the Z accelerometer is placed parallel to the earth's polar axis. Next, the X accelerometer is placed with its sensitive axis pointing from the earth's axis of rotation to the local meridian and parallel to the local latitude plane. The Y accelerometer is parallel to the local latitude plane, orthogonal to the other two accelerometers and pointing east. When the coarse alignment is finished, the system goes through a fine alignment. When the fine alignment is completed, the system automatically goes into the survey mode.

SURVEY MODE

The survey mode for both the local-level, north-oriented and the space-stable inertial systems is performed in the same manner. When the premission alignment is completed and the inertial system enters the survey mode, traversing to establish survey control can be performed. The traverse is started by initializing the system over a control station with known values. To start the traverse, a station ID number, coordinate values (position, elevation and offset values) are entered into the CDU.

Following completion of the initial update, the system is driven to other points where survey parameters are required. While in transit from one point to another, the system is continuously determining its change in position. When performing a survey mission, the vehicle must be stopped every 2 to 5 minutes to perform zero velocity updates (ZUPTS) for the purpose of controlling the error growth of the system. The greater the accuracy required in determining the survey parameters, the closer together the ZUPTS must be performed.

ZUPTS are used by the system to provide external information to the error control system as explained in the section on theory. At the time the ZUPT is being performed, the output of the accelerometer should be zero—(null) but very seldom is it zero at the start of the ZUPT. These errors are due to position errors, various gyroscope and accelerometer errors and changes in the gravity field.

When the vehicle arrives at a survey point at which geodetic parameters are required, the index mark of the system is placed over the point. The system, on command of the operator, does a mark. At the end of the mark period the positional and other data as required by the program is stored in the DSU. This process is repeated at each survey point where data is required.

SMOOTHING MODE

After data has been gathered at each of the required points, the mission must be terminated over a known control point (a different point than the one from which the system started the mission). When the terminal station of the traverse is reached and the update performed (position and elevation entered into the CDU), there is a difference between the computed values, as determined by the system, and the real values despite the use of an error controlling system during the traverse (such as the use of a Kalman Filter mechanization). These residual errors increase as a function of time, distance traveled, the route and type of terrain over which the traverse is being run and the interval between ZUPTS. The principal error sources are accelerometer scale factor, misalignment angles and biases, gyro biases, changes in the gyro drift rates and changes in the gravity field in which the system travels. Most inertial systems have an onboard post-mission smoother used in an attempt to correct the effects of these residual errors on the computed positions at each of the stations at which coordinates are determined. All of the system onboard smoothers are barely adequate. Better adjusting techniques have been developed by various users of inertial systems.

System Accuracies

The accuracies of inertial survey systems are dependent upon many factors. Among these factors are system component errors, systems mechanization errors, operator induced errors, environmental induced errors and the errors in the basic horizontal and vertical stations.

There are many system component error sources. The effect of most of these error sources can be overcome by proper mathematical modeling. Some of these error sources are predictable gyro drift rate, scale factor, misalignment and g-sensitivity, as well as accelerometer drift rate, scale factor, misalignment and null uncertainty. System component error sources that cannot be compensated for by mathematical modeling can be grouped into what is generally termed as component sensitivity errors (an example is the smallest acceleration that an accelerometer is consistently capable of measuring accurately).

System mechanization errors are present only because the companies making inertial survey systems have not optimized the navigation equations in the original aircraft inertial navigation systems to a precise inertial survey system.

Like any other survey system, the results obtained with an inertial survey system can be effected by the person operating the system. However, the chances of errors caused by the operator of an inertial survey system are reduced to his action at the starting and ending stations of a traverse and his care of operating the vehicle. Gross errors are only possible at the starting and ending stations. Operator errors are easily traced in an inertial system and can usually be eliminated when doing a post-mission adjustment.

The two largest sources of environmental induced errors are temperature changes and the earth's gravity field. Gyroscopes and accelerometers are extremely sensitive to temperature changes. Very small changes in temperature cause a change in the gyro drift rate and the rate of acceleration being measured by the accelerometers. The effects of internal temperature gradients (caused by heat generated by electronic components), as well as changes in electromagnetic fields generated by various electrical components, are the main causes for the environmental induced errors in the gyro and accelerometer outputs. These changes are difficult to eliminate using mathematical models but can be partially compensated with proper modeling and further controlled by careful operation of the survey system.

The errors caused by the various system components (primarily the accelerometers) that sense the earth's changing gravity field as changes in acceleration of the system can be partially overcome by proper mechanization within the software and proper operation of the system. Further improvement can be made by utilizing information obtained from the inertial survey system during the mission in a post-mission adjustment.

All basic horizontal and vertical control stations, regardless of how they were established, are not error free in reference to their absolute position to the starting or closing stations or any other control point. This positional uncertainty can be from one or two centimeters to as much as one meter, depending on how the points were established. Since inertial systems must start and end each mission on different control stations, consideration must be given to the accuracy relationship these stations have to each other. This is why it is recommended that only first order control stations be used to initiate or terminate inertial survey missions.

A considerable amount of horizontal and vertical control data obtained with various inertial systems has been analyzed. This data was obtained during the testing of inertial systems over courses containing high order control stations. Comparisons of the values obtained with inertial survey systems against the published values of the control stations gave the following results: for closely controlled single run traverses, a difference for horizontal values of 20 to 25 centimeters and a difference for vertical values of 8 to 12 centimeters; for multi-run traverses (2 forward and reverse runs), a difference for horizontal values of 10 to 15 centimeters and a difference for vertical values of 3 to 6 centimeters. A few special tests were run using 1 to 1.5 minutes between zero velocity updates. These tests showed a slight improvement in horizontal values, but vertical positioning was greatly improved showing a difference of 1 to 2 centimeters for multi-run traverses.

Litton's LASS also has the capability of recovering the gravity vector. In average-type runs of about 40 kilometers, this system has measured gravity anomalies to an accuracy of 1.4 mgals and deflections of the vertical to an accuracy of 0.6 arc-seconds. In closely controlled runs of about 15 kilometers, this system has measured gravity anomalies to an accuracy of 0.8 mgals and deflections of the vertical to an accuracy of 0.15 arc-seconds.

Work has recently been completed with a system that has better accelerometers in the horizontal channels. Preliminary results with this system show an improvement in the horizontal channels. Accuracies of 4 to 6 centimeters against known control are being obtained.

REGIONAL AREA ADJUSTMENT PROGRAM FOR INERTIAL SURVEY DATA

Introduction

Since most existing on-line, real-time error control mechanizations can only control explicitly a subset of inertial survey system errors, off-line error adjustment programs have been developed to contend with the remainder of the existing system errors. In this section, one off-line regional adjustment program (RAP) software package is presented that determines, via a minimum variance algorithm, the adjustment of three (3) axes of position and gravity data from an inertial survey system using data from different traverses

INERTIAL SURVEYING **117**

throughout a region. The control point data consists of known or partially known position and/or gravity components as well as points of intersection or of repetition. The system is required to stop for null velocity updates as it travels through the network. At each stop point, the estimates of the position and gravity components are recorded for input to the regional adjustment program. At each stop point, the RAP software yields minimum variance position and gravity estimates relative to the presumed model of residual system errors.

For years, most surveys utilizing inertial navigation were done on a single traverse basis where the survey system is driven/flown between control points and then back again, stopping periodically for ZUPTS. More recently, network surveys have taken advantage of the crossing of traverses. This has resulted in denser surveys of larger areas. The RAP software discussed herein is unique in that it belongs to the latter set, and requires only a minimal change to the on-line software of existing systems and a minimal amount of recorded data.

Smoother Module Description

The inertial survey off-line adjustment method operates best when it reflects the operating characteristics of the hardware with which it is associated. Once the underlying models of the 3-axis position and gravity vector estimates have been established, it is possible to conceive of appropriate modularization of the software to implement the overall adjustment algorithm. Figure 22 shows the overall basic structure of the filter/smoother configuration showing the modularization of the smoother. In addition, this figure shows the results of the on-line filter step; these are the best (filtered) estimates of the 3-position components (ϕ, λ, η) and the 3-gravity components (ξ, η, sg). These inputs are compared against the commensurate closure values in the smoother.

The smoother software consists of seven (7) modules, which alternate between smoothing and adjusting. These modules and the variables modeled therein are listed in Table 1.

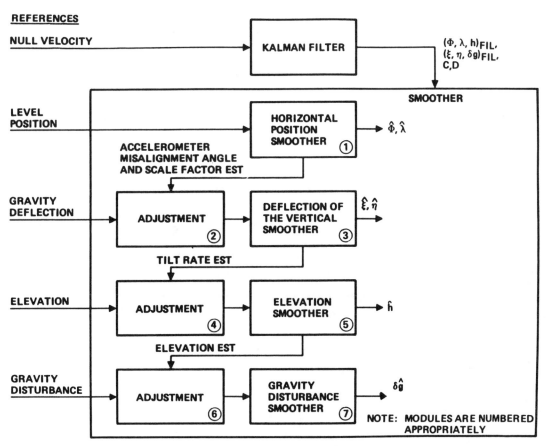

FIGURE 22. Modules for regional area adjustment program.

TABLE 1. Characteristics of Smoother Software by Module.

Module Number	Module Name	Variable Estimated or Adjusted
1	Horizontal Position Smoother	Azimuth error due to initial error and misalignment between level accelerometers
		East, North accelerometer scale factors
		Azimuth gyro drift rate bias
2	Gravity Deflection Adjustment	East, North accelerometer scale factors
		Misalignment between level accelerometers
3	Gravity Deflection Smoother	East, North, Azimuth gyro biases
4	Elevation Adjustment	Platform tilt rates
5	Elevation Smoother	Misalignment of vertical accelerometer about the level axes
6	Gravity Disturbance Adjustment	Elevation
7	Gravity Disturbance Smoother	Vertical accelerometer trend

The horizontal position smoother obtains smoothed estimates of the level position components using an overdetermined set of equations. Within the horizontal position smoother (module 1) are modeled systematic errors that are estimated in the process using either level position closure data and/or level position repetition data from intersection points.

Since some of these systematic errors (accelerometer misalignment angles and scale factors) are common to horizontal position and gravity deflection, it is possible to remove their effects from the gravity deflection input by using the estimates obtained from the previous module. Therefore, the adjustment for gravity deflection (module 2) is used. The adjusted gravity deflection values are then used in the deflection of the vertical smoother (module 3) to obtain smoothed deflection estimates. In a similar manner the elevation input is adjusted (module 4) via a common error from the deflection smoother (tilt rates) and the adjusted value is used in the elevation smoother (module 5). The elevation estimate from the elevation smoother is then used to adjust the gravity disturbance input (module 6) and the adjusted disturbance input obtains the smoothed disturbance value (in module 7).

Smoothing Algorithm

The model for each of the smoother modules can be fitted into the form:

$$Y = HX + V$$

where

Y = Measurement error vector resulting from closure and/or intersection points
X = Systematic error vector to be estimated
H = Measurement coefficient matrix, which depends on the hardware used and the error characteristics
V = Measurement noise vector, which depends on the hardware as well as the on-line filter

Using the smoother model above, the systematic error X can be estimated via least squares as

$$\hat{X} = \Sigma H^T R^{-1} Y; \quad \Sigma = [H^T R^{-1} H + \Sigma_0^{-1}]^{-1}$$

R = Measurement noise covariance matrix
Σ_o = Initial matrix of estimate error covariances
Σ = Current matrix of estimate error covariances

The measurement error estimate is then given by

$$\hat{Y} = H\hat{X} + \hat{V}$$

where V is the estimate of the noise resulting from linear interpolation. It can be shown that the optimal smoothed estimate of the integral of white noise V at each mark point is obtained by linear interpolation (see reference Huddle 1976) from the residuals of the known closures and/or corrected intersection points.

The measurement equation consists of the closure and intersection point errors. Since an intersection point is a stop point that is encountered more than once and no reference value is given, a unique estimate of the intersection data is desired. This is accomplished by introducing a new state for each intersection point coordinate. The adjustment algorithm can be summarized as follows using the horizontal position module as an example:

- Obtain $\hat{X}$ by a least square technique, which yields estimates—
  1. Systematic errors
  2. Intersection point position errors
- Correct the mark and intersection point computed positions for systematic errors and intersection point position errors
- At each closure point obtain residual (r_c) by comparing corrected position to reference position

$$r_c = Y - H\hat{X}$$

1. Obtain the best estimate of noise (V_c) by weighting the closure residual with the inertial system position error variance ($\sigma^2_{\delta P_c}$) and the reference error variance, which is the sum of the error variance in the response at the alignment point and the error variance at the closure point (defined as $\sigma^2\text{Ref}_c$).

$$\hat{V}_c = \frac{\sigma^2_{\delta P}}{\sigma^2_{\delta P} + + \sigma^2\text{Ref}_c} r_c$$

2. For each measurement at an intersection point, the residual (r) is obtained by comparing the corrected position to the real-time computed value of the intersection position for each encounter.

$$r_{Ii} = P^c_i - \hat{P} - H_i X$$

where

- P^c_i is the computed value for encounter i $\hat{P}$ is the corrected position (smoothed value)
- subscript i indicates the ith encounter.

3. Obtain the best estimate of the noise ($\hat{V}_I$) by weighting the intersection residual error r_I, using the intersection system position variance ($\sigma^2_{\delta P})_{Ii}$ and the reference error variance, which is the error variance at the alignment point (defined as $\sigma^2 \text{Ref}_I$).

$$\hat{V}_I = \frac{\sigma^2_{\delta P}}{\sigma^2_{\delta P_I} + \sigma^2\text{Ref}^2_{cI}} r_{Ii}$$

4. By linear interpolation $\hat{V}$ at each mark point between the known $\hat{V}_{Ii}$, $\hat{V}_c$ along the traverse to correct mark point positions for noise.
5. Obtain the measurement error estimate using

$$\hat{Y} = H\hat{X} + \hat{V}$$

6. Adjust the position data based on $\hat{Y}$.

Description of Software Logic

The software consists of a main program and seventeen subroutines, which implement modules 1–7. The function of the main program is to read the required information, store the information internally, control the execution of the subroutines, and print out the final results (for modules 1–7). For explanation purposes, the level position smoothing module (first five subroutines) will be used to explain the overall logic of the program.

There are five steps involved in the level position module. In step 1, the main program accepts filtered (on-line) position, reference position and azimuth gyro coefficients for modeling purposes (not shown in Figure 1). It then calls a subroutine in step 2 to compute the observable difference and the preliminary estimates of intersection points. In step 3, another subroutine is called to set up matrices Y, H, R, and Σ_o for the least square solution. Then a least square subroutine is called in step 4 to solve the over-determined system of equations. The systematic error and intersection state estimates are thus obtained. Finally, the adjustment subroutine is then used in step 5 to adjust the intersection position, obtain systematic error correction, linearly interpolate the noise error estimates, and, finally, adjust the mark point position.

Adjustment Results

The RAP was checked using a simulator that had been used on previous programs. The results obtained with the simulator show accuracies for an area survey using 3 minutes between ZUPTS, as 6 cm for the horizontal, 5 cm for the vertical, 1 mgal for gravity disturbance and 0.4 arc-seconds for gravity deflections. Using 1 minute between ZUPTS, the results obtained were 2 cm for horizontal, 1.3 cm for vertical, 0.9 mgal for gravity disturbance and 0.21 arc-seconds for gravity deflections.

Application of RAP on Microcomputers

The original version of RAP was coded in standard FORTRAN and installed on a main frame computer. To increase the mobility of the RAP software package, a microcomputer version of RAP has been developed and tested on the HP 9836. Results indicate that with two megabytes of memory, 100 survey points can be adjusted in the field.

APPLICATIONS AND PLANNING FOR INERTIAL SURVEY SYSTEMS

Applications

Inertial survey systems have proven to be very cost effective on projects that are large in number of points to be surveyed or that cover large areas. Other projects where inertial survey systems have been very effective are surveys conducted in remote areas where logistic support for survey personnel is difficult; where weather, terrain, environmental restrictions or traffic make work difficult for conventional surveyors; and where limited time exists to complete the survey because of time or contract restrictions.

There are many applications for inertial survey systems. Among these applications are the following:

1. Control surveys—densification of existing control as well as the recovery and checking of existing control.

2. Topographic surveys—establishment of control for all types of photogrammetric mapping.
3. Engineering surveys—route surveying for pipelines, highways, railroads and transmission lines; hydroelectric complexes (construction and reservoir profiling); new city planning.
4. Mine surveying—all types of surveys for both surface and subsurface mining.
5. Boundary surveying—public land and other types of property surveys. A very fast way to survey boundaries when the lines do not have to be cleared.
6. Hydrographic surveying—primarily along river channels due to the requirements for ZUPTS.
7. Geophysical surveying—can determine free air gravity anomalies to an accuracy of 1 mgal. Can be used to determine elevations for use with gravity meters. Can also be used to measure deflections of the vertical between two known deflection stations.
8. Vertical profiling—when used in conjunction with a laser can profile to accuracies of better than one half a foot at the rate of XYZ positioning of eight times a second. To maintain this accuracy, ZUPTS must be performed every three minutes or less.

Planning

In order to ensure a well-executed, cost-effective survey project using inertial technology, it is essential that efficient planning be performed. Support planning, such as fueling of all types of vehicles and monumenting of control stations, and justifying demand greater attention than conventional type surveys because of the speed with which inertial systems accomplish surveys and because the systems are time dependent if accuracy is to be maintained.

Following are some guidelines for accomplishing inertial surveys. These guidelines will affect both the accuracy and cost of the surveys.

- Basic control—the accuracy of any inertial survey cannot be any better than the control to which it is tied. Inertial surveys should always use the best available control. If good control is not available, consideration should be given to establishing it with GPS using translocation or the differential mode. Proper coordinates as well as positive identification of all control to be used on a project must be obtained.
- System calibration—the inertial system must be calibrated accurately to the local control over lines between five to ten miles in length. For the local-level, north-oriented system, these lines must be as close as possible to a north-south line and an east-west line. Closure (raw values) should be better than 15 centimeters for both scale factor and misalignment for the accelerometers. Gyro drift should be less than six seconds per hour.
- Linear traverses—the best results are obtained when the rate of change for height, latitude and longitude is linear with respect to time. Since this is often not practical for field work, the best compromise is to keep all the work within a diamond-shaped figure, from which no point to be established is more than ⅓ the distance from the nearest control point from a straight line between the two control points for that traverse. One of the most important restrictions is that there must not be any doubling back on any portion of the traverse.
- System alignment—during the premission alignment, the vehicle in which the system is mounted should be as motionless as possible. Make sure that the vehicle does not move during the alignment. Keep sheltered from wind gust. Do not do an alignment near heavy traffic and always park the vehicle parallel to the direction of the wind where possible. Prior to alignment, the system should be warmed up for a minimum of 30 minutes.
- Traverse time and distance—all traverses must be carefully planned so that the survey can be run with the minimum time and distance between known control stations. Known control must be located prior to the traverse and not during the traverse. It must be remembered that accuracy is dependent on the distance and time between control stations.
- ZUPTS—accuracies obtained with an inertial survey system are highly dependent on the time between ZUPTS. When high accuracy is wanted, the time between ZUPTS must be reduced. Horizontal accuracy appears to be best when using time between ZUPTS of two to three minutes. Vertical accuracies of a couple of centimeters have been obtained with time between ZUPTS of one minute.
- Gravity anomalies—large changes in gravity between ZUPTS must be avoided. This can best be done by reducing the time between ZUPTS.
- Uniformity—to increase the accuracy obtained with an inertial system, it is essential that uniformity be used for the following items: 1) speed of the vehicle between ZUPTS and 2) time between ZUPTS.

REFERENCES

1. Britting, K. R. *Inertial Navigation Systems Analysis.* John Wiley & Sons, Inc. (1971).
2. Broxmeyer, C., *Inertial Navigation Systems.* McGraw-Hill, New York, NY (1964).
3. Fernandez, F. and G. R. Macomber, *Inertial Guidance Engineering.* Prentice Hall, Inc., Englewood Cliffs, NJ (1964).
4. Gregerson, L. F., "Inertial Instrumentation at the Geodetic Survey of Canada," Commonwealth Survey Officers Conference at Cambridge, England (August 1975), and the General Assembly of the IUGG-IAG in Grenoble (August and September 1975).
5. Gregerson, L. F., "Inertial Geodesy in Canada," American Geophysical Union Conference, San Francisco, CA (December 1975).

6. Hannah, J., "The Development of Comprehensive Error Models and Network Adjustment Techniques for Inertial Surveys," Dept. of Geodetic Science Rep. 305, Ohio State Univ., Columbus, Ohio (1982).
7. Hannah, J., "Inertial Rapid Geodetic Survey System (RGSS) Error Models and Network Adjustment," Dept. of Geodetic Science Rep. 332, Ohio State Univ., Columbus, Ohio (1982).
8. Huddle, J. R., "Navigation to Surveying Accuracy with an Inertial System," *Proc. of the Bicentennial National Aerospace Symposium*. Warminster, PA, available from Litton Guidance and Control Systems Division, Woodland Hills, CA (1976).
9. Huddle, J. R., "The Measurement of the Change in the Deflection of the Vertical with a Schuler-Tuned North Slaved Inertial System," Litton Guidance and Control Systems Division, Woodland Hills, CA (1977).
10. "Inertial Technology for Surveying and Geodesy," *Proc. of the 1st International Symposium on Inertial Technology for Surveying and Geodesy.* Ottawa, Canadian Institute of Surveying, Box 5378, Station "F" Ottawa, Canada, K2C 3J1 (Oct. 12–14, 1977).
11. "Inertial Technology for Surveying and Geodesy," *Proc. of the 2nd International Symposium on Inertial Technology for Surveying and Geodesy.* Banff, Canadian Institute of Surveying, Box 5378, Station "F" Ottawa, Canada, K2C 3J1 (June 1–5, 1981).
12. Leondes, C. T., "Theory and Applications of Kalman Filtering," Advisory Group for Aerospace Research and Development (NATO), Clearing House for Federal Scientific and Technical Information, U.S. Dept of Commerce (1970).
13. O'Donnell, C. F., *Inertial Navigation.* McGraw-Hill Book Co., New York, NY (1964).
14. Pitman, G. R., *Inertial Guidance.* John Wiley & Sons, New York, NY (1962).
15. Roof, E. F., "Results of Tests Using an Inertial Rapid Geodetic Survey System," U.S. Army Engineer Topographic Laboratories, Ft. Belvoir, VA.
16. Roof, E. F., "Inertial Survey Applications to Civil Works," U.S. Army Corps of Engineers, Engineer Topographic Laboratories, Fort Belvoir, VA (1983).
17. Savet, P. H., *Gyroscopes: Theory and Design.* McGraw-Hill Book Co., New York, NY (1961).
18. Schwarz, K. P., "Inertial Surveying Systems-Experience and Prognosis," FIG Symposium on Modern Technology for Cadastre and Land Information Systems, Ottawa, Canada (1979).
19. Schwarz, K. P., "Error Characteristics of Inertial Survey Systems," *Proc. FIG XVI International Congress,* Montreux, Switzerland (Aug–Sept. 1981).
20. Slayter, J. M., *Inertial Guidance Sensors.* Reinhold Publishing Corp., Chapman & Hall, Ltd., London.
21. Wei, S. Y., "Offline Optimal Adjustment of Inertial Survey Data," Litton Guidance and Control Systems Division, Woodland Hills, CA (1985).
22. Wrigley, W., W. M. Hollister, and W. G. Denhard, *Gyroscopic Theory, Design and Instrumentation.* MIT Press, Cambridge, MA (1969).

SECTION TWO
Construction

CHAPTER 7	An Organizational Model for Building Construction Firms in the Computer Age	125
CHAPTER 8	CPM in Construction	189
CHAPTER 9	Environmental Concerns During Construction	197
CHAPTER 10	Construction: Traffic Control at Freeway Work Sites	211
CHAPTER 11	Determination of Power Spectral Density Functions from Design Spectra	231
CHAPTER 12	Chance Constrained Aggregate Blending	241
CHAPTER 13	Constitutive Modelling of Ballast	257

CHAPTER 7

An Organizational Model for Building Construction Firms in the Computer Age

GIANFRANCO DIOGUARDI*

INTRODUCTION

I will discuss in this chapter a model of building construction firm having an organizational structure most able to make the best possible use of computerized services.

The computer today offers an aid which can prove essential to any type of organization, especially to production firms. It seems nevertheless important to stress how the *use* of computer systems often becomes an *abuse*. This happens when the machine (hardware) and its programs (software) become part of a firm's organizational structure without there having been the necessary preparation, which must spring from a renewed meditation of the operational context through which the firm strives day after day to reach its goals. This is why it becomes especially important to discipline all information procedures, including the sorting out of the type of information that has to be made available and the forms necessary to carry it, thus highlighting the items—which must always be sufficient and never superabundant—necessary to the rational and correct management of the firm.

Once the information process and its supporting structure have been set up, it becomes fairly easy to transfer them to software so that the entire process can be managed by using a computer. There is of course the possibility for each organization to create its own totally new computer programs, but it is probably best to take advantage of the rich and fast-changing production offered by the market. The various programs can easily be adapted to the needs of each individual activity within the firm.

The most important facet of my approach is precisely the fact that each department or activity within the single firm must be viewed as an autonomous organization with an individuality of its own.

It is illusory to think it possible to apply immediately and automatically to these company departments or activities, standard structures, procedures and software. The organizational process must have the support of the experience gathered by the firm, and all information instruments must keep on adapting themselves to the everyday realities which characterize each organization.

This holds true for any organization process, and therefore also for the best possible use of computers.

This chapter, by discussing the organization in the firm, is intended to promote a meditation about the problems which are typical of the building construction firm. I begin by analyzing some specific problems having to do with the production process and particularly with managing construction sites, in order to obtain a model for a general organization structure complete with possible strategies for reaching the desired results. I then propose a series of operational standard forms that can solve the problems pertaining to an integrated M.I.S. (Management Information System). These forms have been designed so that they can be readily transferred on software packages commonly available, or easily obtainable through the services of an expert in this field. In any case, the examples in this chapter must always be considered as guidelines, and modified in order to meet the specific needs that any entrepreneur recognizes when tackling the construction site's problems of his or her company.

The chapter ends with some of the important aspects that the building construction industry presents in its training, innovation, research and development processes.

The various subjects discussed, and especially the standard forms illustrating this chapter, result from experiences had in a medium-sized construction firm active in the public works sector in Italy.

I have sought to give general meaning to basic problems, so that the examples originating in a particular firm can ac-

---

*Universita degli Studi di Bari, Facolta di Ingegneria, "Economia Industriale è Organizzazione Aziendale," Bari, Italy

quire wider connotations and be useful in sketching a theory. That is why I have emphasized those general organizational characteristics which, although typifying the construction firm, are present also in other productive structures, and are therefore rich in useful suggestions. In some cases, the entrepreneurial model born within the building industry can represent a novel organization example for other sectors as well.

This explains why I have included a sub-chapter discussing organization "theories" as they result from the analysis of company models existing in some production areas, for example in electronics, in various manufacturing sectors and also in some services. I am convinced that any paper dealing with organization should stimulate in the reader the desire to formulate his or her own organization theory.

The ancient Romans—who certainly dealt successfully for a long time with the problems of organizing and managing a worldwide empire—were fond of saying, *"Faber quisque fortunae suae."* ("Each man is the maker of his own fortune.") And they were right.

ENVIRONMENT

Building construction is a sector strongly characterized by peculiarities which inevitably condition the firms that operate in it.

Unsteady Growth

The building industry appears economically important both for companies directly active in it and for a number of collateral sectors. This fact often induces to consider it as politically instrumental in attempts to relaunch a depressed national economy. When this happens, the industry's growth becomes unsteady, being stimulated by periods of fast development followed by periods of more or less pronounced recession, both limited in time. Survival, then, becomes difficult for any firm with rigid organizational structures.

Peculiarities of Contracts

The activity of construction firms is based on contracts usually spanning several years, always involving large amounts of money, so that success in obtaining these contracts has considerable consequences for the very survival of each firm. There is always the danger of having to face production peaks with organizational structures either too large or too small. This too is a problem endangering firms characterized by rigid organizational structures.

Problems of Construction Sites

Each construction site can be compared to a factory having very special characteristics: a limited life-span (seldom exceeding five years) and activities that appear constantly influenced by the surrounding environment, by the type of soil on which works have to be executed and by frequently erratic weather. All of these conditions make it difficult to follow work schedules closely.

Why Outside Help

Each job must be completed through a series of totally heterogeneous operations. This means employing, in each site, work crews differing in specialization and technical know-how. The problems that each single firm has in forecasting future contracts make it impossible to plan ahead the work of specialized personnel, in order to guarantee to work crews a continuity of employment. This is why it becomes necessary to resort to external firms, often quite small.

Some Negative Effects of Competition

In the building industry and in the various sectors that operate alongside it, firms differ greatly in size and degree of specialization. Building construction itself is characterized by the existence of a few large companies and a great many small artisan firms. In connected sectors the situation is similar. This seems to sharpen competition, and a well-balanced development becomes difficult. This problem is rendered more acute by a tendency to favor sales promotion to the point where some building firms become real estate companies and sales are seen as more important than production.

The Question of Cash Flows

In work contracts obtained by construction companies, cash outflows connected to production costs run faster than any inflow. This is easily explained when considering that construction materials (raw, intermediate goods and components) must be paid at least partially when placing the order, and the balance within a period of 30 to 90 days (when credit is tight this period tends to become shorter and shorter). On the other hand, the cost of human resources employed has to be dealt with every week. After having obtained the technical approvals and after the completion of various bureaucratic procedures, invoicing becomes possible: it is the final stage before actually receiving the agreed payment. Settlement, however, will seldom be prompt because clients stand to gain from making payments at the latest possible date, thanks to interests earned. This means a negative cash flow so that each firm ends up financing the client for the entire duration of the contract, although this is not one of its institutional tasks.

This situation, which applies to each individual contract, synergistically accumulates its negative effects over the entire sphere of the firm's operations, especially when the financial climate becomes turbulent and inflation pushes the interest rates for credit up to extremely burdensome levels. This causes proper (when money is obtained through bank credit) or improper (when money is obtained through internal financing) financial charges that in either case impact the job's economic balance.

Interest paid becomes an increasingly important item of the balance sheet, to the point where it often endangers – or definitely upsets – the financial viability of the single job and, generally speaking, of the firm. Furthermore, the traditional designation of interest paid among overhead expenses does not facilitate an immediate correlation between causes (delayed cash inflows, immediate outflows) and effects (financial charges), and therefore hampers costs control. What can happen is that construction jobs shown as profitable by cost accounting can in fact lose money due to the financial charges being larger than the profits.

This is why it becomes necessary to extend cost accounting controls to the "potential" single cash flows of each job, since in their totality these cash flows, when balances are negative, represent for the firm debts and their charges.

This extension of controls cannot be limited to cost accounting; a concomitant investigation must be carried out beginning with a contract acquisition study. It is the only way to design specific financial strategies that will allow a rational handling of the phenomenon.

The various points I have discussed cause instability in economic planning and bring confusion to entrepreneurial roles. This instability and confusion create the typical turbulent character of the construction industry's environment. From it the difficulty arises in forecasting and executing rigid plans both within companies and in single construction sites. We must live with – and manage – this uncertainty.

PRODUCTION

The purpose of building construction is the on site production of works as provided in the contract. The construction site can be viewed as a factory with a limited life-span. Work operations stem from the approved project and from the soil's geological nature. A number of work operations can be closely scheduled on the basis of a detailed project design: these we will call "programmable works." For operations having to do with site characteristics (surveys, technical explorations, earth moving works, foundations, general arrangements), a detailed program becomes difficult since these are operations that depend on the type of soil; we will term them "non-programmable works." The difference between these two groups of works is very important and has to be kept well in mind when deciding the construction site's general work organization.

Programmable Works and Their Problems

"Programmable works" present specific problems resulting from the peculiarities of building construction itself.
For example:
1. Operations are carried out in geographically fixed areas, often far apart from each other, usually with long time-period requirements (one, two or more years).
2. Work operations in single sites are varied in type and technology employed.
3. Since tasks are not of a repetitive nature, nor subject to standardization, there is often a substantial lack of professional specialization among workers.
4. It becomes difficult to plan permanent work for variously skilled labor, since it is practically impossible to have, at the same time, a series of contracts geographically close to each other and suitably staggered in time so as to enable each category of specialized operatives to find continuous employment. This fact leads to the use of specialized outside firms.
5. There is a tendency to prevent the specialization of human resources in order to employ them at random in different work operations. It is a tendency also favored by the lack of comprehensive training. Their experience is often casual or biased toward certain artisan skills.
6. Scant professional qualification and the absence of organized training bring about inadequate compliance with safety regulations. This is also due to companies' tendency to ignore safety regulations laws.
7. It is almost impossible to apply on single work operations detailed and analyzed planning, organization and control procedures. In other words, there are limited possibilities for activating a control system similar to the one employed in "time and methods" organizations when work is not repetitive. It is therefore necessary to organize the general execution of the project so that overall goals can be controlled at the level of homogeneous completed tasks.

These factors generate some suggestions for better organizing building construction and, in particular, for allowing a planning and control procedure.

Planning Construction Work

This type of production originates from two characteristic moments which can be more or less important but which coexist always. Some products are premanufactured in factories or plants, i.e., they are prepared before the opening of a construction site. This can cause problems in coordinating orders and in controlling contract compliance on the part of subsuppliers.

Next we have the site phase, with work on soil and subsoil and the construction of the product. This phase consists of the assembling of premanufactured components and of on-the-site production.

Therefore, we can say that construction work is characterized by heterogeneous tasks covering long periods of time, usually years. Planning it would mean coordinating operations which are frequently subcontracted, and hence autonomous from the firm's productive capacity; even when considering direct work operations, it is extremely difficult to rigidly program elementary operations as Taylor has theorized for similar problems to be solved in the factory. In building construction, each work operation always presents an "artisan" aspect rooted in the single worker's habits and in the fact that each task has to be carried out in constantly differing environments. Quite often the single employee's work habits prevail over the ergonomics principles and any attempt to optimize his performance on the basis of these principles will not only obtain poor results but could easily give way to demotivating frustrations.

This has been proven true also in the manufacturing industry where repetitiveness of single elementary operations has permitted their optimization as theorized by Frederick Taylor, the father of "scientific management," but, on the other hand, has caused serious forms of mental alienation. This is why there is a general tendency to modify the organization concept of the production line.

Building construction is characteristic also because while the various work operations can be at least in part identified and planned—with all the unknowns due to weather conditions—still other operations concerning soil and subsoil (excavations, foundations, etc.) escape a true planning possibility. This is because the scope and quality of these operations can be defined only during execution on site.

Heterogenity of operations, premanufactured components and assembly on site, difficulty in estimating in advance quality and quantity of the work to be done: all of these are limiting factors that conditions, in building construction, both planning and control.

Segmentation of the Production Cycle

It becomes necessary, then, to envision planning methods based on the concept of the sequential coordination of finished actions, and not on the analysis of single elementary components of those same actions. These methods must also originate an information feedback procedure, i.e., an active and retroactive control process that allows interventions on elementary actions and not merely to ascertain, at the end of the entire process, the failure to reach the intended goals.

The basic concept is that of a "segmentation" of the entire production cycle, obtained by identifying elementary operations of the finished type as simple as possible to control, in order to organize around them an optimizing sequential process, time and cost-wise. This can be obtained through a detailed study of the preliminary bill of quantities which should analyze the complete project in its various elementary operations. But this document, of top importance for site management, is often missing and, on the other hand, the technical time for preparing it is always extremely limited.

A clear advantage derives from the development of a two-level procedure: a first, synthetic one for finding a mechanism that permits organizing non-programmable work operations on soil also. From this, a second and more thorough process should derive, in which it will prove extremely advantageous to make use of CPM network-based planning schedules accompanying budget planning.

The first level of a contract's general planning can be carried out in four standard phases: an organization phase, a planning phase, a construction phase and a completion phase. Each one of these can be subdivided into stages that can be coordinated into specific block graphs pointing out sequences between the various operations and the operational formalities to be analyzed through precise operational procedures.

A brief description of the operations that make up each phase can be useful at this point.

The Organization Phase

This phase (Figure 1) is represented by a general analysis of the work to be done and by a market research concerning local prices for materials, labor, subcontracts, etc. Contract documents are studied together with the project design in its various aspects (architectural, structural, mechanical systems) and the general construction site organization is set up. Afterwards, while the planning phase itself is being started, all non-programmable works are envisioned, grouping those preliminary tasks which escape rigid planning.

To start with, the general organization of the construction site has to be implemented (plant, provisional works, etc.). In addition, all tasks concerning site investigation, surveys and technical exploration will have to be carried out, together with those tied to local characteristics and therefore difficult to program (demolitions, earth movements, excavations, curtain walls, foundations, etc.). Non-programmable works are ususally the first to be carried out, and during their execution it is necessary to complete the organizational set-up so as to obtain a feasible schedule.

Construction work itself will start after the finalization of a general project time schedule and cost analysis. It is not difficult to prepare this since the number of tasks at this stage is usually limited, and forecasts should be kept fairly flexible.

Work must be preceded also by the preliminary study of the construction site plan, by the lay-out for machines and transportation means and all other mechanical systems which presumably will be employed.

This preliminary study will be verified beginning with the non-programmable works, and it will be possible to modify it on the basis of the first stage of works. Afterwards, the program will become final and will be translated into operational procedures. At the end of this first stage, the preliminary financial program will have to be updated and the results will be included in the general plan which will then result complete in all of its parts.

Thanks to this procedure—i.e., the execution of works according to a variable time table allowing all necessary analysis for the completion of a general plan—it also becomes

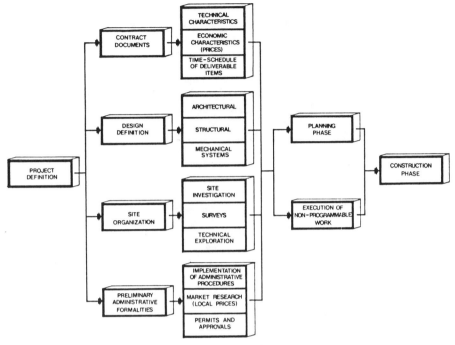

FIGURE 1. Organization phase.

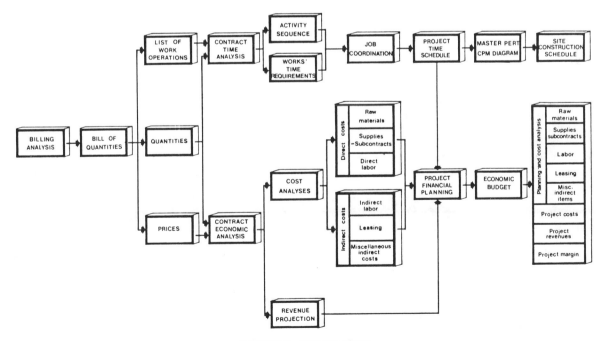

FIGURE 2. Planning phase.

possible to attempt to logically organize the casual episodes tied to the specific characteristics of each construction site.

The Planning Phase

This phase (Figure 2) is based on the analysis of the project's estimated bill of quantities, which examines in detail the various work operations to be carried out in the future construction site. The bill of quantities becomes the very heart of the entire site organization. Therefore, it must be based on data gathered during the organizational phase and helps in identifying analytically and with precision the different necessary tasks, their quantity and the cost of each single one.

The items which will make up the bill of quantities will prove useful both for the site technical organization (list of tasks and their number) and for the project financial planning (quantity and cost).

The bill of quantities is also useful for planning the complete site production cycle, since it allows forecasting and checking its economic life in terms of direct and indirect costs and revenues.

One of the basic purposes of the bill of quantities is to point out already in the planning phase the gain or loss balance from each work operation; this will allow a study, beforehand, of the necessary methods for better organizing work, thus correcting and improving these situations. By so doing, in the execution phase the construction site will be able to show satisfying revenues and direct costs—capable of absorbing overhead expenses with a resulting net profit.

When an estimated bill of quantities exists (either as an official or informal document), it will have to be completed, after a more thorough analysis, with all those tasks, even limited in scope, that are usually unforeseeable when the contract is being stipulated. While preparing the bill of quantities, the project must ideally be subdivided in all the basic operations that characterize the job itself.

Forgetfulness in this stage will result in a more or less pronounced time loss in the future. Since, as we have seen, the bill of quantities represents a basis for the entire work organization, it is necessary to give the utmost care to its preparation.

This takes place through subsequent approximations: it must be readied at the beginning of the organization phase on the basis of the project design; then, as the project itself is modified and completed so that it can become executive, the bill will be updated until it can be considered final. It is precisely at this point that it also becomes the basic document for the execution of works, and all ensuing modifications in the project will have to be transferred immediately in accounting terms on the bill itself.

The analysis of the bill of quantities permits the contract's time and economic analysis and therefore also the second level technical and economic planning, using instruments like the CPM diagram and the economic budget, which are the basis for a rational control of the execution phase.

The Uses of PERT and CPM

Program Evaluation and Review Technique (PERT) studies the development of a complex project through statistics, taking into account only the time factor (PERT TIME), while some of its variations also allow analysis and optimizing cost behavior from a time point of view (PERT COST, CPM, etc.).

Particularly with CPM (Critical Path Method) we want to optimize both factors through a more deterministic reasoning which permits identifying a way to reach the intended final result with the least expenditure of time and cost.

These techniques are especially interesting because the segmentation of a complex event into single basic complementary tasks allows immediate awareness of the general effects that a partial episode can have on the entire network and therefore also on the final result.

The single stages into which the total project is fractioned are characterized by elementary events and activities. The events, usually associated with a deadline, are moments in time when an activity starts and finishes while the activity itself (any type of action which permits the advancement of the project) is characterized by duration in time and in some cases by a cost factor. Cost and duration in some instances can even result nil (for example, for suspended activities) and their function is purely indicative.

The project, sectioned and analyzed, can have a dual graphic projection thanks to particular networks based on *activity*, when events are identified by circles or similar means (squares or rectangles), connected to each other by arrows representing activities; or, dually, on *events*, when circles, representing activities, are connected to each other by incoming or outgoing arrows which indicate events starting and finishing the activities themselves (Figure 3). Of these two representations, the first clearly shows the logical dependencies and connections between the various activities. This constitutes a factor of the uppermost interest.

In PERT-type planning, the first approach to the problem evolves through a *planning* phase in which, after having listed all foreseeable elementary actions, these will be logically connected. Next, there will be the identification of those finish events that must necessarily precede the start of each examined activity, those that cannot be started before the finish of the preceding one and, lastly, those activities which can be carried out at the same time.

The planned duration of each activity is usually established on the basis of probabilities, almost always by using the so-called "beta" distribution function, where duration is expressed in simplified terms by the formula $T = a + 4n + b/6$: a and b represent the best and the worst foreseeable duration according to conditions encountered while work-

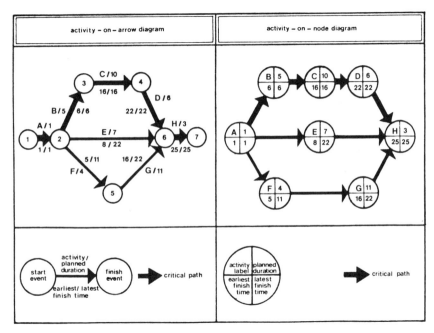

FIGURE 3. Example of network-based planning schedule.

ing, while *n* represents duration in perfectly normal working conditions.

When the duration of each activity and their linkage become known, start and finish times can be established.

Linkage will usually determine parallel activity paths which, beginning from an event given as start, will reach the finish event. The longest path between these two events is called "critical," because any delay in single activities along such a path will immediately influence the completion of the project. *Activities* which have shorter paths can register delays or postponements to be absorbed by the difference between the critical execution time and the earliest finish time connected with the same path. The difference constitutes the maximum delay allowable. When preparing a PERT or CPM network program, three pairs of singular time can be identified, the first pertaining to events, and the other two to activities, earliest time, latest time, earliest finish, latest finish, earliest start and latest start.

For events and activities, delays can be identified using the difference between minimum and maximum duration. These are generally called event slack, free float, total float.

The identification of these temporal entities—and the resulting network-based planning of the production activity—allows, while planning, to point out activities—and therefore also work operations—of crucial importance (*critical path*). Also it allows simulation of the effects of each possible delay or anomaly on the entire production cycle; the analysis of the degree of employment of equipment and of work crews; the optimization of the distribution of resources. (Usually we seek a layout of resources that renders equally critical all paths and reduces as much as possible the total duration.) It also becomes possible during controls to concentrate on *critical activities*, to evaluate the acceptable delays and to identify the best means for coping with anomalies in the production process.

This type of planning is usually done by computer; it also supplies a Gantt-type barchart for single work operations, very simple to use for controls during site operations (Figure 4).

In this chart the scheduled contract package shows with lines (by days or months) each work operation. There is the possibility of indicating the program of subcontracts with the total duration and scheduled deadlines for each order.

To be effective these charts must be continuously updated, and it is this task that often generates problems. Here too the use of a computer is advisable, since it allows automatic updating through an input of control data.

By using PERT-type programs it is possible to better organize and control the economic behavior of the construction site by making it part of the general budget.

PERT-type programs also allow a continued control of the construction phase.

The planning process represents the necessary instrument for starting the construction phase, during which the control stage concerning the true program performance will be verified, as will the technical and production terms of the various work operations.

The Construction Phase

This phase (Figure 5) is characterized by the execution of work programs and by the recording of data necessary for controlling the execution of the various works time-wise, economically and financially, and for suggesting possible corrective actions when data are different from what had been anticipated.

Resources will be procured and scheduled; work operations will be started while developing a programmed series of controls.

The construction site controls will be technical, temporal, and economic, as well as administrative and general, in order to have a direct feedback process allowing immediate corrective reactions.

Quality control must take in both products which enter the construction site (raw materials and intermediate goods) and work operations (both direct and indirect, i.e., subcontracted). Control of work operations is tied to the management of human resources and to their employment in specific work operations, assistance, and general site operations. Other specific controls can deal with safety measures as applied on the site.

Generally speaking, these direct controls are carried out by the site manager, who can be compared to a plant manager and must therefore have a strong personality, since it is up to him or her to coordinate a series of very heterogeneous events. This can also have negative effects since his or her personality could hamper the application of organization processes capable of rationalizing production. This is because the site manager is the only true and perhaps irreplaceable contract manager, and has therefore a tendency to make day-to-day, casual decisions that give him or her a full knowledge of information thus emphasizing in him or her a pre-established authority, a fact strengthened by the existence of objective difficulties. All of these factors can create in the site manager a tendency to complete autonomy, to a mentality which in the building construction industry is quite common but which hampers an organization concept that takes in planning and control procedures according to an industrial one that wants to rationalize operative processes.

The economic course is usually controlled by means of monthly records which analyze the site's direct and indirect costs and revenues with the necessary recovery supplement for different lots. These records, obtained on site by technicians, have an ex-post-type accounting control.

Controls implemented by the administration re-propose the site's economic course on the basis of information gathered after the fact by the financial period analysis. The balance between the technical site results and the administrative ones allows an interesting control by exception that can point out possible economic situations that deserve to be further analyzed technically or from an accounting point of view.

The accounting or administrative department will further check financial episodes tied to the construction site, especially those concerning cash income which becomes regular only thanks to a careful accounting of works done. In this way the client has the means and the time to execute his or her own controls.

The Completion Phase

The completion phase (Figure 6) puts an end to the construction site cycle. It should be characterized by a thoughtful reconsideration of all that has happened so that data can be logically and rationally registered after having been discussed and analyzed in a way that transforms it into a motivating and rich reservoir of practical experiences, a useful know-how for those future offers through which new contracts will be obtained.

In summing up we can say that production planning should allow, thanks to the analysis of a list of work operations, the bill of quantities, and the costs and revenues, an optimizing of the progress of single operations both eco-

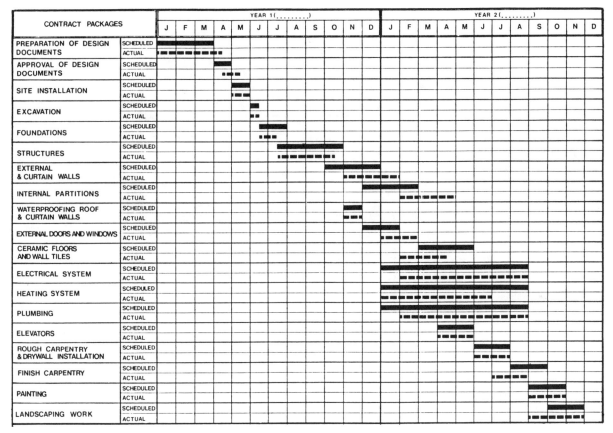

FIGURE 4. Gantt-type barchart for building works analysis.

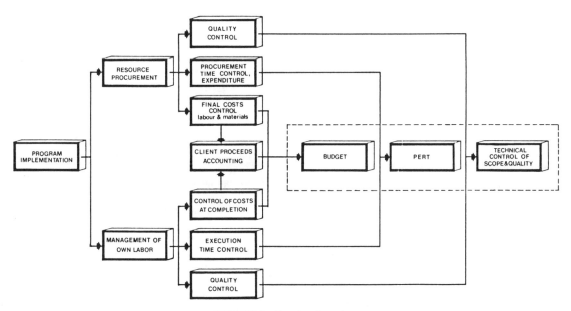

FIGURE 5. Construction phase.

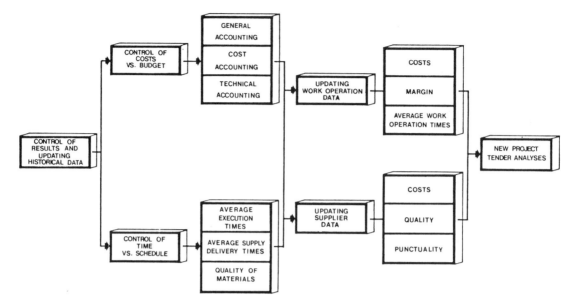

FIGURE 6. Completion phase.

nomically and time-wise, with a resulting improved coordination of the various tasks.

The completion phase is also a phase for site technical controls which take in work programs, costs and revenues and also allow keeping under control the financial course pertinent to the site's revenues and expenditures. This can be facilitated by a final reckoning at completed sites, with an analysis of past mistakes and of internal and external labor employed in the production cycle.

THEORIES

Today both the concept of business enterprise and the organization models that characterize it are changing.

What Is an Organization?

To the question, "What is an organization?" W. G. Ouchi (1980, p. 140) answers: "An organization, in our sense, is *any stable pattern of transactions* [emphasis added] between individuals or aggregations of individuals. Our framework can thus be applied to the analysis of relationships between individuals or between subunits within a corporation, or to transactions between firms in an economy".

This definition is sufficient to suggest a novel approach to the various problems pertaining to the firm. Specific definitions have also been attempted in order to better describe these ongoing changes which could mark the firm in the near future. As we have seen, Ouchi defines the firm as "a stable pattern of transactions." Others call it a "transactional" or a "relational" firm, or describe it as characterized by "transactional costs" (Williamson, 1975) or a "dynamic network organization" (Miles-Snow, 1984); still others term it a "quasi-firm" (Eccles, 1981; Kay, 1983) or polydivisional or multipolar or "macrofirm" (Dioguardi, 1983).

The "Make" and "Buy" Logic

At the onset of industrial development, the firm coincided with the production site, i.e., with the factory or plant. The factory, in other words, identified the firm.

It became natural enough to privilege the "make" rather than the "buy" logic, with its relative economies of scale. Investments for plants and production equipment acquired essential significance since the purpose was to concentrate the largest possible amount of processing within the factory.

This meant concentrating strategic decisions when designing plant projects; the result was a rigidity in subsequent operational management, with ensuing medium to long-term programs very often difficult to carry through due to the unstable situations both of the final market and of the intermediate one pertaining to property plant and equipment. Specific functions within the firm, such as marketing, were structured so as to bend the consumer's will to the firm's production requirements. In other words, it was the consumer at the service of the firm, and not vice-versa. Regardless of efforts, however, positive results were always difficult to get since general conditions changed constantly and rapidly while the individual consumer became increasingly his or her own decision-maker, owing also to improved average living conditions.

The Nature of Modern Business Enterprises

Later, there appears a new entrepreneurial level. Its task is the management of a number of factories and the search for the right solution to the organizational problem of coordinating interactivities between various production poles.

Alfred D. Chandler, Jr., the topmost historian of business organizations, in his book *The Visible Hand—The Managerial Revolution in American Business*, gives us a detailed analysis of this problem: "The theme propounded here is that modern business enterprise took the place of market mechanisms in coordinating the activities of the economy and allocating its resources. In many sectors of the economy the visible hand of management replaced what Adam Smith referred to as the invisible hand of market forces. The market remained the generator of demand for goods and services, but modern business enterprise took over the functions of coordinating flows of goods through existing processes of production and distribution, and of allocating funds and personnel for future production and distribution" (1977, p. I).

The reasons at the root of these phenomena are condensed by Chandler into eight propositions (1977, pp. 6–11) which give some indications as to the nature of the modern business enterprise, its outset, development, and consolidation.

The first of these propositions appears especially meaningful when it states that modern multi-unit enterprises took the place of small traditional enterprises when administrative coordination allowed greater productivity, lower costs, and higher profits than allowed by coordination of market mechanisms. It is a proposition resulting directly from the definition of a modern business enterprise which came into being and continued to grow by establishing or acquiring business units that in theory were able to operate as independent enterprises, by internalizing the activities that had been or could be carried on by several business units and the transactions between them.

This analysis deserves an afterthought, since it refers to an organizational set-up substantially similar to the one we find in the building construction industry, where each construction site can be viewed as a sufficiently autonomous production unit, a firm in itself.

In the building construction industry the firm's growth through the acquisition of a number of construction sites leads to certain managerial methods within the organizational structure and to an increasing resort to subcontracts. Thus, we have an internalization process—which becomes first of all the coordinating process—in coexistence with an externalization process which occurs through transactional ties with affiliated companies. This phenomenon can be noted at the very outset of building construction enterprises, as Chandler himself indicates (1977, p. 53) by pointing out that building and construction firms expanded in order to meet the rising demand by employing and training younger craftworkers. As the cities grew, master carpenters and builders often contracted to build a series of houses at one time, and so kept a number of journeymen and apprentices at work under their direction. Those who took over the task of laying down and paving city streets were small local contractors using local labor. Normally an engineer or city official supervised their work, and their workers continued to use traditional tools and skills.

According to Chandler, the internalization of the firm's activities led to the establishment of a managerial hierarchy (his second proposition) and to the expansion of the general volume of activities. Because of this, "administrative coordination [became] more efficient and more profitable than market conditions (III). Managerial hierarchy and administrative coordination became a source of power and of continued growth (IV), of professionalism (V), [which became] increasingly separated from ownership (VI) and [preferred] politics that favored long-term growth of enterprises instead of maximized current profits (VII)."

Chandler concludes with his eighth and final proposition ". . . large enterprises grew and dominated major sectors of the economy, they altered the basic structure of these sectors and of the economy as a whole. . . . What the new enterprises did was to take over from the market the coordination and integration of the flow of goods and services from the production of the raw materials through the several processes of production to the sale to the ultimate consumer."

Coordinating Interactivities

Today an enterprise not only controls a number of production plants, but resorts to specialized outside firms which in the long run almost become part of the enterprise's organization. Thus, coordinating interactivities becomes the most important issue. Transactions with outside firms were carried out according to contracts which to all effects set down rules for behavior and activities.

The question of internal or external transactions, of resulting costs, and of the identification of the firm with these specific problems has been noted in its technical aspects in 1937 by Ronald H. Coase in an essay that today we consider a classic, "The Nature of the Firm." Among other things, Coase regards the firm as a "combination" and "integration" of transactions: "Outside the firm, price movements direct production, which is coordinated through a series of exchange transactions on the market. Within a firm, these market transactions are eliminated and in place of the complicated market structure with exchange transactions is substituted the entrepreneur co-ordinator, who directs production" (Coase, 1937, p. 388).

Referring to Coase's 1937 classic paper, Oliver E. Williamson writes: ". . . He, like others, observed that the production of final goods and services involved a succession of early stage processing and assembly activities. But whereas others took the boundary of the firm as a parameter and examined the efficacy with which markets mediated exchange in intermediate and final goods markets, Coase held that the boundary of the firm was a decision variable for which economic assessment was needed" (1981, p. 550).

According to another author, N. M. Kay in *The Evolving Firm—Strategy and Structure in Industrial Organization* (1982, p. 34), Coase identifies costs as the reason for the existence of firms.

At the time Coase expressed these thoughts, they represented enlightening intuitions which do not seem to have been completely understood, regardless of progress registered in this area of thought. Williamson is well aware of this fact when he points out that transaction cost analysis is an interdisciplinary approach to the study of organizations that puts together economics, organization theory and aspects of contract law, providing a unified interpretation for a disparate set of organizational phenomena (1981). Although additional applications have been made, the limits of transaction cost analysis have yet to be reached, according to Williamson, and there is reason to believe that the surface has merely been scratched.

An even more dramatic statement is expressed by Coase himself 35 years after his 1937 paper which, he says, was much cited and little used. Williamson and Ouchi explain this with the failure to make the issues operational over the interval (Williamson-Ouchi, 1981, p. 349). Furthermore, according to Williamson, "firms and markets are treated separately rather than in active juxtaposition with one another. The propositions that (1) firms and markets are properly regarded as alternative governance structures to which (2) transactions are to be assigned in discriminating (mainly transaction cost economizing) ways are unfamiliar to most organization theory specialists and alien to some" (1981).

The Need For Specialization

Today, many axioms of the classical theory of organization seem to have lost much of their usefulness (Van de Ven-Joyce, 1981). Some authors, for instance, J. C. Tarondeau (1982), divide the evolution of company strategy into three stages, the first dominated by production, the second by sales (therefore conditioned, in a consumer society, by the market), the third characterized by innovation and by a reconciliation between production, flexibility, and quality. In this last stage a need is felt for increasing specialization within production units thanks to a flexible organization.

Today we are experiencing the revolution brought about by electronic automation and by computerized information systems, which allow the more stressing and least interesting jobs to be executed by human-controlled robots (Butera-Thurman, 1984). As T. Reve and R. E. Levitt point out in their paper "Organization and Governance in Construction" (published in *Project Management*, February, 1984): ". . . Information and communication turn out to be key factors in project coordination, but the quality of the information again depends on trust between the integrator and the other contractors." And in "Organizational Adaptation to Small Computers" (which has appeared in October 1984 in *Applications of Small Computers in Constructions*), H. G. Irwig and R. E. Levitt report how a case study ". . . has demonstrated the strong inter-relationship between the environment of a firm, its organizational strategy and structure, and the way in which it utilizes computer technology. . . . A transition to a more decentralized structure, which has been long and arduous for many firms using older more rigid mainframe technology, was relatively painless for this firm whose crisis of control coincided with the availability of well designed mini-computer hardware and software."

As a revolution, this one which is being determined by computers is often more complex than the industrial one, owing also to the new role of technological progress whose evolution must be guided by specific global company strategies (Gaudin, 1982).

Why a Flexible Structure

It therefore becomes necessary to continuously generate innovations, to promote personnel training thus raising cultural levels, to manage financial phenomena with a new approach. Particularly important is the need to widen the context of strategic decisions from the moment when investments are decided to the entire period of specific competence of their operational management. This also means radical changes in financial strategy, consistently with the general environment: rate of inflation, the usually steep cost of money that can substantially condition firms' profit and loss accounts, and insufficient credit availability.

In other words, the concept of the firm is changing for a whole series of reasons.

The macroeconomic environment, more and more turbulent and discontinuous, forces the firm to acquire a more flexible structure in order to adjust more easily to changes in the environment itself. The ever increasing and indispensable integration between the firm and its economic interlocutors imposes the study of new coordination, planning, and control processes.

As with all socioeconomic episodes, there are differences within this reality. Changes are sharper in some production sectors, for example in electronics and manufacturing in some services. In other fields (particularly in chemical processing) these changes have not yet acquired a clear configuration.

The Building Construction Model

In discussing the nature of construction contracts as ways of governing construction transactions, T. Reve and R. E. Levitt (February 1984) state that ". . . Contracts can take on more or less of the characteristics of an employment relationship or of a market relationship." And referring to the turbulent environment of the building sector, P. R. Lawrence and D. Dyer in *Renewing American Industry* write: "We have argued that three factors shaped the structure of the residential construction firm: the characteristics of the product, the site-bound technology it requires, and the cyclical fluctuations of the housing market. The same factors militate against the predominance of large-scale housing firms and an oligopoly structure in the industry. The quasi-firm (with the exceptions noted below) appears to be the standard structure of construction firms regardless of type or size" (p. 153).

This model is then typical of the building construction industry, a sector where awareness to changes and organizational imagination have never been especially evident. But practical experimentation confers indisputable validity to this model. In fact, its concept is not merely theoretical: it arises casually in answer to the peculiar demands of an environment (building construction) which has always been turbulent and unstable, characteristics which today can be ascribed to any economic sector.

The answer chanced upon, which in time became indispensable practice, carries the riches of a practical experience that, although limited to a sector, could certainly prove useful to organizations active in different production and service areas.

This explains why the building construction model is being cited by some authors as among those which will prove significant for tomorrow's organizations. For example, R. E. Miles and C. C. Snow, in an article published in 1984, write that ". . . managers are considering a new organization form and are experimenting with its major components and processes. The reality of this new form simply awaits articulation and understanding. In this century there have been three major breakthroughs in the way organizations have been designed and managed. The first occurred at the turn of the century in the form of functional organization. . . . Next came the divisional form, which facilitated even more organizational growth and, more importantly, facilitated diversification in both products and markets. The third breakthrough was the matrix structure in which elements of the functional and divisional forms were combined into a single system able to accommodate both standard and innovative products or projects."

In a thesis submitted in 1984 to Stanford University by Lloyd McCara Waugh I find this interesting analysis of building construction:

"Construction is a unique business because:

(a) Raising financial capital is not a problem—for two reasons
   (i) very little financial capital is required, construction firms need minimal plant and equipment relative to other (especially manufacturing) firms, thus many construction firms are essentially self-financed out of retained earnings or the proprietor's personal assets; and
   (ii) in the case of many contractors the need for financial capital is passed on to lower levels—subcontractors, speciality contractors, and material suppliers;
(b) Construction firms generally sell a service which is expertise rather than capital-intensive. This is particularly true of general contractors and design firms."

Organization for the New Century

Waugh then asks, "What resource is the firm most dependent on?" He writes: "For a manufacturing plant the answer may be financing (investors); for a construction firm it may be expertise (managers); and for a volunteer group it may be labor (workers). If expertise is the resource on which a construction firm is most dependent, then in some situations such a firm may be acting perfectly logically by defying certain financial principles. Our contention is that in construction, either consciously or subconsciously, firms attempt to maximize 'return on expertise' rather than 'return on equity.' "

According to the authors I have cited, then, a promising new organization form is emerging, one that appears to fit the fast-approaching conditions of the next century. As has happened to previous forms, its elements are sprouting in several companies and industries simultaneously. The construction industry has long been known for its use of subcontracting to accomplish large, complex tasks. Today, the size and complexity of construction projects can be immense. Under such circumstances, companies must be able to form a network of reliable subcontractors which have never before worked together; some companies, therefore, have found it advantageous to focus only on the overall design and management of a project, leaving the actual constructions to their affiliates. Miles and Snow, in their 1984 article, explain how certain firms in the electronics and computer industries are already dealing with conditions that in the future will be widespread: rapid change, fragmentation, high technology, abundance of information, and so on. In these companies, product life-cycles are often short and all firms live under the constant threat of technological innovations that can change the structure of the industry. Individual firms must keep on redesigning their processes around new products. Across the industry, spin-off firms are continually emerging. From these examples, state the two authors, some key characteristics of the new organization form are clearly visible. Organizations of the future are likely to be *vertically disaggregated*: functions typically encompassed within a single organization will instead be performed in independent organizations. That is, the functions of product design and development, manufacturing, and distribution, ordinarily integrated by a plan and controlled directly by managers, will instead be brought together by *brokers* and held in temporary alignment by a variety of *market mechanisms*.

Given these characteristics, Miles and Snow have found it useful to refer to this emerging form as the *dynamic network organization*: "The full realization of this new type of organization awaits the development of a core activating and control mechanism comparable to those that energized the previous organization forms (e.g., the profit center in the divisional form). Our prediction is that this mechanism essentially will be a broad-access computerized information system. . . . Properly constructed, the dynamic network organization will display the technical expertise of the functional form, the market focus of the divisional form, and the efficient use of resources characteristic of the matrix. And, especially important, it will be able to quickly reshape itself whenever necessary. . . . We have argued that minimal fit is necessary for survival, tight fit is associated with corporate excellence, and early fit provides a competitive advantage that can lead to the organization Hall of Fame. Tomorrow's Hall of Fame companies are working on new organization forms today."

THE FIRM

The building construction firm which operates in the environment described above must be able to organize and coordinate the production taking place in a number of construction sites, while developing the necessary services and all activities indispensable to its survival.

In order to accomplish all this, we can envision a firm model with three organizational levels (Figure 7).

The Corporate Level

On the first level, which can be called the "corporate level," mainly managerial and service functions must be carried out: management, administration, personnel, general affairs, secretarial services, and so on. Also, production coordination through information, planning and control procedures must be carried out on this level. People responsible for the corporate level must be able to function as leaders who know how to stimulate others toward successful accomplishment of the common company goals.

This is why the corporate level can be described as an aggregation of people capable of formulating a program and of keeping it constantly up to date on the basis of information gathered from the external scene and of the results obtained by the firm.

No income is produced on the corporate level, but only costs independent from production, i.e., fixed costs that accumulate in time. These must be absorbed by profit margins obtained by the production poles which in exchange receive various services.

The Operations Level

One of the corporate level's principal tasks is the coordination of the operational poles which represent a second organizational level which I term "operations level." It is on

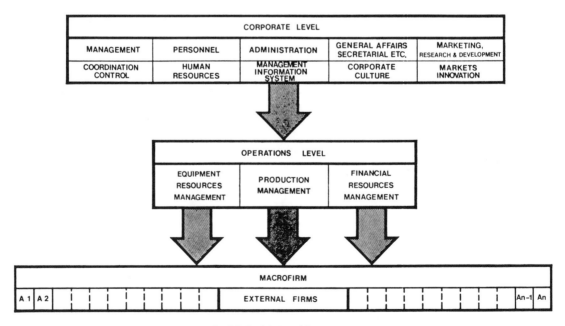

FIGURE 7. Model of firm structure.

this level that I locate the management of those processes that must generate profit margins capable, among other things, of absorbing fixed costs. These processes concern real flows which interest the firm's administration in regards mainly to acquired resources that will be transformed into finished products, the acquisition of equipment resources and the consequent management of services supplied by them, and, finally, the financial flow which intervenes principally in phenomena of economic nature.

For each of these three main flows (which also become processes for generating possible profits), it is necessary to identify a person in charge, as if each process is an independent operative pole, with the possibility of creating an independent external market while continuing to be coordinated by the corporate level.

The three processes concerning production, equipment resources, and financial resources evolve through three flows of resources (production, production equipment, money) acquired according to different rules and time periods (continuous or point input) and are transformed or used according to specific needs and in more or less extended time periods (point or continuous output).

Why a Strategic Equipment Resources Management

Of these processes the most important one is certainly production, but the turbulent climate calls special attention also to the other two. A phenomenon which should be pointed out here is the need to render strategic equipment resources management. This means identifying a leader who will be responsible for the acquisition of such resources, giving him or her the task of managing them with the maximum autonomy in order to obtain a satisfying economic result during their life-cycle inside the firm.

The goal can be reached by considering the acquisition and usage of such equipment as duties of an independent leasing company that offers its services wherever there is a demand—in the firm's production pole or even outside the firm.

The person in charge of production will likewise be free to acquire the equipment necessary to his or her production activity either inside the firm's pole responsible for the acquisition of such resources or outside the firm whenever this proves technically, economically, and financially advantageous.

The manager of equipment resources must be in a condition to carry out all the necessary operations for optimizing their use, including the sales of such resources that are judged technically obsolete and deserving of being substituted with more advanced and efficient ones. There is in this the attempt to modify the investment concept from strategic but "punctual" in time and binding future operating periods for the equipment's full life-cycle, to strategically deferred in time thus rendering the person in charge free to optimize the management of that same equipment in order to suit it to the internal demand or to the competitive one which generates in the external market.

Formally speaking, these responsibilities can be turned over directly to the purchasing manager who then departs from his or her traditional staff position tasks by assuming an additional direct responsibility: that of generating profit.

The Financial Resources Management

The person in charge of financial resources must manage for best results the monetary resources. He or she will do so by trying to postpone outflows and quicken inflows, and by procuring money at the most favorable credit terms. This also becomes a strategic management which must generate profits or in any case curtail financial charges which can be burdensome to building construction due to this sector's characteristics, especially when money is particularly costly. The total management of monetary phenomena and of the financial ones (management of debts and credits) tied to them can therefore become the subject of a specific operative pole having the responsibility for obtaining the described goals.

The Macrofirm Level

All poles of the multipolar firm (production, equipment resources management, financial management) constantly interact with each other and with external firms. With these, stable transactions come about which represent true *programs* disciplining contacts between the poles of the multipolar firm and the external firms tied to it. These interactions represent real stable links between subjects not necessarily always the same. The firm organization in its fullest meaning evolves precisely through these stable interactive links between its own poles and the outside firms. This creates a relationship network that in its totality can be compared to a small economic system whose engine is activated by the corporate level through the operations or multipolar level. This network of stable relationships which opens the firm towards the outside world represents the third organizational level which I will call "macrofirm" (Dioguardi, 1984).

In summing up, we can say that a firm, in general, and a building construction firm, in particular, could be represented by a series of operations evolving in a model having three organizational levels: corporate level (services and coordination control), operations level (production, equipment, money), and macrofirm (relations with external firms).

The model can also be activated through forms of temporary associations, joint-ventures, or consortiums with external firms both on the operations level (associations with financial companies giving credit or with firms active in industrial leasing), and, more often, on the macrofirm level.

From the traditional, internalized firm model where the

"make" logic prevails, we move on to a firm model where, through various organizational levels, strategic decisions are reached which concern what should be made and what should be bought. In this set-up it often happens that the people in charge of internal production tend to create new specialized independent firms that will become suppliers of the main company. Thus, we can say that the model is able to foster new entrepreneurship. The operations structure which we have described can be brought to life through subcontracting or temporary associations with specialized sub-suppliers by means of consortium or joint-venture forms.

Creating linkage systems between the various operating firms will determine more individual responsibility concerning the common goals to be reached thanks to the specialized activity of each firm (Figure 8).

New Meanings for Traditional Duties

This being the picture, new meanings can be acquired by functions like the financial and purchasing services. Especially for the financial function, in addition to the usual duties a special role will have to be assigned to the person responsible for credit recovery so as to accelerate as much as possible the moment of monetary inflow tied to credits which stem from production revenues.

The purchasing department will have to manage internal interactivities between the operations level and outside firms that constitute the macrofirm. In strategical terms, it must also take care of investments in production equipment.

The production pole proper remains the most important one, both because it has the principal responsibility for generating profit and because of its organization problems. The activity of this pole can be carried out partly internally ("make" within the firm logic) and partly by opening to outside firms ("buy" logic which determines the more important transactional relationships pertaining to the macrofirm) (Figure 9).

The first area requires the direct employment of human resources, while the second must acquire subcontractors through the logic of market competition. In this case, there is the problem of reconciling the economic and quality aspects of services to be acquired.

The Need for a Task Force

In any case, useful are specialists capable of controlling subcontractors and also of creating and managing independent work crews which may have to take the place of subcontractors when emergency situations arise.

What has been stated concerning production has shown

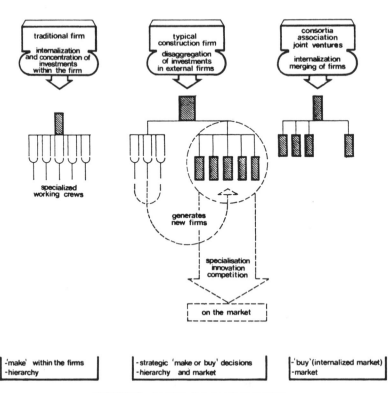

FIGURE 8. Possible firm structure models.

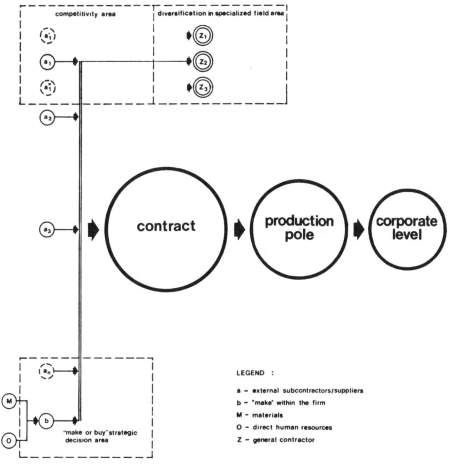

FIGURE 9. Strategic decision areas within the production pole sphere.

that the building construction firm can cope with its turbulent market by arriving at a production organizational model having these main specific characteristics:

a. Existence of a firm whose main task is coordinating and organizing.
b. Ample use of specialized subcontractors so as to curtail the firm's fixed costs. Experience shows that relationships between the main firm (general contractor) and the operational firms (subcontractors) become in time almost stable; this favors labor employment, if we observe the phenomenon from a general point of view but also almost monopolistic situations among more habitual subcontractors.
c. The need within the leading firm of a highly qualified task force capable of controlling the activity of subcontractors and of coping with possible emergencies concerning subcontractors for works already started or for those to be started in a general climate favorable to full employment of resources.

The Quasifirm Theory

The transactional aspect which emerges from the analysis of the macrofirm in its strictly productive aspect has been discussed by experts both in general terms and as applied to the building industry. For example, Robert Eccles, in one of his articles, presents his theory of the "quasifirm" and discusses it in the context of the building industry: "Technical characteristics of construction result in a preferred contracting mode intermediate between bilateral and unified governance structures when certain environmental conditions exist. Using Woodward . . . classification of production technologies into unit and small batch, large batch and mass, and process, Thompson . . . classified construction as an intensive (unit) or job shop technology. . . , I have argued

... that there are strong reasons for the general contractor to make extensive use of subcontractors, based on the theory of subcontracting using concepts borrowed from Stigler ..., Thompson ..., and Williamson ...." (1981).

A specific field analysis made on 38 building construction firms of the San Francisco area (the results can easily be extended to other building construction markets) have highlighted the persisting ties between general contractors and subcontractors.

This phenomenon is explained by the fact that general contractors usually do not participate in formal tenders, but tend instead towards market research—and a new, ampler marketing role should be found; this enables them to assign works to firms which remain tied to them also by the expectancy of continuing work. In the already mentioned research, only 19.6% of general contractors used the lowest bid-type tender for selecting subcontractors.

Eccles (1981) states that when construction projects are not subject to institutional regulations requiring competitive bidding, relations between the general contractor and subcontractors seem stable over fairly long periods of time. According to Eccles, this type of quasi-integration results in what he calls the "quasifirm."

But also in the case of formal tenders there appears to be a certain stability in the selection of subcontractors. One of the reasons is the experience gathered by the subcontractors after a first association, which betters his position in subsequent tenders (R. E. Levitt-S. Gunnarson, 1982).

Williamson, for example, explains (1979) the phenomenon by stating that the distinction between ex ante and ex post competition is essential. What commonly is an effective large-numbers bidding situation at the beginning, at times is transformed into a bilateral sort of trading relation. The winning subcontractor obtains advantages over other bidders also for future works "because non-trivial investments in durable specific assets are put in place (or otherwise accrue, say in a learning-by-doing fashion) during contract execution." Elsewhere Williamson (1979) writes: "The efficient governance of recurring transactions will vary as follows: classical market contracting will be efficacious wherever assets are nonspecific to the trading parties; bilateral or obligational market contracting will appear as assets become semispecific; and internal organization will displace markets as assets take on a highly specific character."

Other motives can be identified in the tendency within the corporate level to seek the cooperation of firms with whom the problems of starting a relationship have already been solved. And so we have the description of the "quasifirm" as a structure sufficiently similar to the one described by Williamson as capable of activating an "internal contract system." Eccles specifies, as follows, the problem within the building industry: "The general contractor and special trade subcontractors can form a stable organizational unit when conditions permit . . . called here 'quasifirm' " (1981).

As for the motives that push entrepreneurs to adopt the described model, the research conducted by Eccles revealed that 26.6% of the general contractors interviewed use subcontracting for the purpose of curtailing overhead expenses, while instability due to seasonal or economic cycles was of scant importance in decisions to subcontract.

The Importance of a Flexible Structure

As I see it, the main motives in institutionalizing this model can be listed as follows:

(a) The need to keep the firm structure flexible (therefore curtailing fixed costs with the aim of absorbing cycles' ups and downs, whose effects intervene when deciding to use subcontractors).
(b) The opportunity to have a better specialization in various works.
(c) The opportunity to foster process and product innovation.
(d) The convenience to be found in easier organizational management based mainly on the coordination of various work stages, with beneficial effects also for the financial planning.

The first of these points seems the most important one, as Levitt-Gunnarson (1982) have rightly pointed out, while other authors often do not emphasize the problem, probably because they are influenced by the easy mobility of human resources which allows fast structural modifications according to the existing production conditions inside the firm. But this phenomenon causes considerable problems in institutionalizing a stable process of professional training in the building industry.

Problems of the Macrofirm

The "transactional" firm or "quasifirm" or "macrofirm model" does not escape some criticism.

A first problem derives from the fact that subcontractors do not easily develop and maintain a really specialistic vocation, capable of fostering, as much as possible, training and innovation processes within their activity. These firms tend instead to become general contractors, distorting the meaning of the term "general" which ends up by describing the approximation characterizing the non-qualified and least dependable entrepreneurship. This on the market can result in imperfect competition between firms, based on the different quality of work being supplied. This in turn deteriorates, quality-wise, processes and products, and spoils the image of the entire building construction sector.

It would be necessary, and also useful, that this situation become the subject of controlling actions on the part of general contractors and of clients when assigning work.

Another problem is the direct negotiation and the continued relationship between the general contractor and subcontractors. This, as we have seen, can lead to monopolistic situations with negative effects on price-costs competition, with resulting distortions also on the favorable prices that the general contractor can offer to the client. This is even more important when acting in a climate of full employment of resources because the lack in the market of free subcontractors can determine problems in finding transactional associations.

Corrective measures are easier with subcontractors represented by those firms which have been started by former personnel of the general contractor's productive poles. The measure would consist in a direct intervention in the ownership of such firms through equity issues allowing a control, even if only partial, on operational decisions. The general contractors thus acquire holding company characteristics. A more direct corrective measure is the creation and existence of a small "task force" which in emergency situations can independently organize specialized crews able to take the place of subcontractors.

Still another corrective measure could be represented by motivating subcontractors on the basis of a formal promise of continued relationship based on stable economic conditions even when the market is less generous.

The inconveniences tied to a quasifirm-type structure have been thoroughly explored by Williamson (1975, pp. 96-97). He lists five of them and adds three defective incentives within the system. To list the most important:

- the already mentioned danger of a bilateral monopoly between parties, controllable only in part
- the difficult regulation of components flow
- biased process innovations in favor of labor saving as against materials saving and innovations
- insufficient incentives for product innovation

These problems have already been discussed, at least in part. I especially disagree about the dangers regarding innovations. The macrofirm can instead prove beneficial insofar as each specialized firm is called upon to modify and renew its production according to the needs of a number of general contractors: this on the market promotes innovation.

Eccles transfers this concept to the project context: ". . . subcontractors work according to the plans and specifications of architects, who identify the materials to be used; hence, materials innovation is out of the hands of both the general contractor and the subcontractor and all that the subcontractor *can* do is find labor-saving process innovations; and again, product innovation is the responsibility of the architect so there are no advantages of integration over subcontracting" (1981).

These thoughts are acceptable in part. It must be kept in mind that one of the specific duties of the construction firm is the quality and critical control on the field of product performances, so as to transmit, in a continuous feedback process, useful suggestions and innovation ideas to the people responsible for designing the product.

STRUCTURE

A formal organization structure is made up by the totality of roles, functions, and procedures that can be clearly and formally described in a company's organization chart (Figure 10). But it is also necessary to take into account all those relations that come about between responsible people within the system and that usually fall outside formally established rules, and that in their totality are defined as "informal organization."

People and Organization

Informal relations must be understood and disciplined; whenever possible they should even be utilized for improving operational efficiency.

Any type of organization is formed either by people inclined towards a bureaucratic approach (similar in some ways to the subjects of McGregor's X theory) or by people with entrepreneurial-like motivations (McGregor's Y theory). It is this second category that generates real results within the firm, by constructively reacting to problems and obstacles and by assuming leadership roles in the drive towards success. The other category includes people capable of recording events; they often call attention to negative aspects but do not suggest constructive countermeasures. Their role is vitally important since it is only thanks to their alertings, even when pessimistic, that operators with entrepreneurship are stimulated into action against negative events.

On the other hand, it is important, on an informal basis, to identify entrepreneur-like people so that they can become part of a "task force" directly connected to top management. Members of this task force should be placed in key positions within the firm in order to activate a stimulating and leading action descending directly from the managing director or the general manager. To emphasize the importance given to this really informal structure, it would be best to name it, in the organization chart, "management task force," highlighting its stimulating functions of liaison between the operations centers and top management.

Committees

The committees are formal organizational structures whose efficiency depends on the atmosphere in which the meetings are held. These meetings should be considered the natural places in which to exchange information, both in terms of strategic decisions and of controls. It is here that the various facets of a situation filter through a pluralism of opinions and reach a constructive dialectic synthesis in which the attitude of each committee member regarding a particular problem is summarized. Committees should be formed by members of top management and by the more important representatives of the various tasks dealt with by

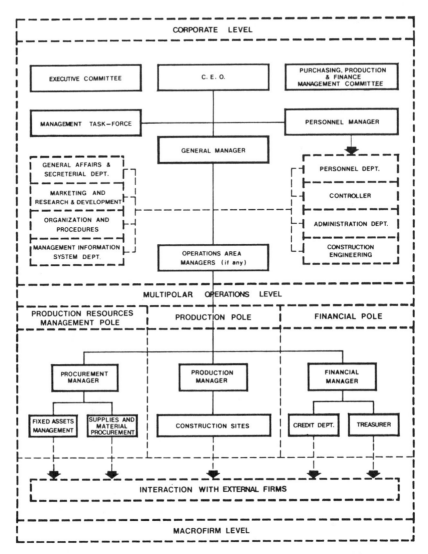

FIGURE 10. Example of formal organization chart for a building firm.

each committee; in some cases, also by external consultants, whose role becomes especially constructive when, during a committee's work, they face problems with more objectivity than the one expressed by those directly involved in it. In any case, consultants, even when only occasionally tied to the firm, must receive clearly defined agendas before the scheduled meeting.

A committee derives its importance from its structure which is similar but at the same time profoundly different from the board of directors. The difference exists also in a superior operational capability which receives an impetus from the prevalent attendance of people belonging to the firm.

The consultants—who can also be members of the board of directors—represent a useful idea-producing external mechanism, useful also because it will not limit the passage of experience exclusively to the firm's culture.

A committee's efficiency depends on the informality given to each meeting. A phenomenon which must be especially avoided and which involves people having responsibilities in the firm, could be defined as the "confession effect." It happens for example when a manager who has not been capable of obtaining positive results in his or her work, feels "absolved" of any responsibility by the very fact of having reported his or her failure to the committee, whose members—it must be pointed out— have a prevailing staff role. When the committee comes up with an operational decision, it will always have to be passed on to the person in charge of the activity tied to that particular decision. This is why the committee cannot take on responsibilities that do not fall under its competence. The committee is essentially a consultive and not an operational structure. The "confession effect" often emerges also in the delegating process, when a person to whom a responsibility has been delegated makes a mistake; by reporting this to the delegating party, he or she might convince himself or herself of having done all of his or her duty regardless of the results. And this of course is not right.

The Executive Committee

It is necessary to provide for an executive committee whose task will be to analyze medium and long-term strategic problems; short-term ones will be dealt with when discussing the year's problems and results; this last analysis will be thoroughly handled by other specific committees (purchasing, production, finance).

These various committees will be presided over by the people most directly interested in the specific problems that each meeting will be called to discuss; it should also be possible to invite, from time to time, firm or outside people who are in a position to give suggestions or to express opinions about specific problems (Figure 11).

The executive committee is undoubtedly one of the more important, owing to its general characteristics which allow during its sessions the opportunity to handle all of the firm's main problems. Its operational procedures, which in general should also discipline the other committees, must call attention to the precise goals to be reached, to the meeting's subjects and organization. These points can be summarized as follows.

Goals: analysis and discussion of the more relevant management problems, so as to rationalize the decision-making process concerning the short, medium, and long-term period. Problems submitted to the committee must then be examined in order to point out:

- internal and external conditioning situations
- their distribution and priority in the medium and long-term plans of the company
- proposals for alternative solutions
- proposals for steps to be undertaken
- existence of clear-cut priorities in the execution phase

Subjects: generally speaking, these are the matters to be discussed by the committee:

- definition of the firm's short and long-term goals
- definition of strategies necessary for obtaining these goals
- analysis of practical development strategies

DIOGUARDI		POSSIBLE COMPOSITION OF THE VARIOUS COMMITTEES			YEAR	DEPT.: FORM:
ITEMS		EXECUTIVE COMMITTEE	PRODUCTION COMMITTEE	PURCHASING COMMITTEE	FINANCE COMMITTEE	
1	C	C.E.O.	GENERAL MANAGER	PROCUREMENT MANAGER	FINANCIAL MANAGER	
2	P	GENERAL MANAGER	PRODUCTION MANAGER	GENERAL MANAGER	GENERAL MANAGER	
3	P	PERSONNEL MANAGER	PERSONNEL MANAGER	PRODUCTION MANAGER	ADMINISTRATION MANAGER	
4	P	ADMINISTRATION MANAGER	(AREA MANAGER)	(AREA MANAGER)	CREDIT MANAGER	
5	P	MARKETING MANAGER	(SITE MANAGER)	(AREA MANAGER)	(PRODUCTION MANAGER)	
6	P	(CONSULTANTS)			(PROCUREMENT MANAGER)	
7	P	(CONSULTANTS)				

LEGEND: C = Committee chairman
P = Permanent member
(P) = Occasional member

FIGURE 11. Possible composition of the various committees.

To be checked in particular:

- consistency with existing plans and their periodical updating
- existence of an interest in their execution
- feasibility and convenience from a strategic and economic point of view
- problems concerning the organizational structure that could provoke significant changes
- problems concerning personnel originating in the external scenery in which the firm is active
- problems connected to sales and promotion, to public relations, to tender participation
- new contracts acquisition course
- main contracts course
- year's economic and financial course

Organizing Meetings

Here is a suggested guideline for organizing committee meetings:

- Before each meeting the committee's secretary office, after having consulted the various participants, will prepare an agenda to be approved by top management and then distributed to participants so that each one will have sufficient time to prepare for discussion.
- During the meeting, the various problems on the agenda will be debated, possibly following the guideline already indicated and aiming for conclusions and suggested actions. If a general agreement is not reached, a final decision will be made by the person chairing the meeting.
- The committee's secretary office will prepare minutes of the meeting to be distributed to all participants.

Here it should be pointed out that an essential factor in the successful outcome of meetings is a fixed periodicity and a well-planned time budget; a climate favorable to cooperation but not to passive attitudes is also called for. The necessary factor is not a general agreement, but a general discussion so that the best possible results and decisions can be gathered from the variety of opinions expressed.

A basic function of the executive committee is to analyze and control the various aspects pertaining to the organizational structure and, consequently, the various problems concerning personnel management.

A typical matter to be debated in this committee is the acquisition of new contracts. In general, each new contract poses the problem of selecting the right strategies in choosing personnel.

If the firm has people already available, then it is necessary to make a very competitive offer in order to try to obtain at all costs the work which will enable the firm to employ all of its personnel. In case the firm is already employing its personnel to the fullest, it must consider that the acquisition of a new contract will mean hiring new people. This means not only selecting efficient workers, but also anticipating the effects of a natural increase of fixed costs on the firm's future.

Strategically speaking, each possible new contract must be examined also from a technical point of view. For example, in the case of jobs requiring new construction techniques or a particularly brief execution time, all economic, technological, and organizational risks must be taken into account together with the effects, positive or negative, that they might provide for the firm's general image.

The Production Committee

Committees pertaining to operational activities will deal specifically with technical, economic, and organization problems that are tied to the management of poles which produce income.

The production committee must discuss all construction site problems by thoroughly analyzing each site's general performance.

One of the problems that could be brought up in this committee concerns the possible global negative results relative to all or some projects being executed. For example, in the first case the problem could originate from lack of sufficient management control on people responsible for the various work sites. It will therefore be necessary to steer the actions of top management so as to balance the energy spent on the external environment and that which is being spent inside the firm. In other words, it will be important that the firm's more vital forces concentrate on production's economic problems in order to bring about new processes capable of limiting costs and of widening the income area.

But a situation could arise where only some jobs are giving negative results. In this case, it is necessary to concentrate on how to increase productivity. In particular, cost-reducing processes could be activated by better organizing work, thus containing production costs. Or, client strategy can be implemented by proposing technical variants capable of improving the general economic course of that particular job.

In some cases both strategies could become necessary and the production committee is precisely the place to decide this.

The Purchasing Committee

One of this committee's important duties is to buy fixed production equipment so that it produces profits. This means that the use of such equipment must be paid by the production pole in profitable terms without losing competitivity in the external market.

Some discussions in this committee will deal with the company's policies concerning expansions of subcontracts and of subsupplies, or the acquisition of more workers in order to carry out single work operations quickly.

Another significant problem is the need to mediate between the purchasing department's thrifty tendencies, the financial department's desires to delay as much as possible payments to suppliers, and the attitude of production people, who must always be the only ones responsible for the entire job and therefore also for the quality and punctuality in delivering on the part of subcontractors and subsuppliers.

This mediation between the financial department and the production poles must have the main purpose of maintaining production people as the only line-responsible party. It is precisely this mediating function of the purchasing committee that must be regarded as one of the more complex tasks within the firm. This committee is also the right place in which to discuss possible delays in payments to suppliers. These are delays that can have negative effects on the economy of the cost of supplies. On the other hand, if and when quicker payments are possible thanks to cash liquidity, they should be negotiated with the supplier in order to obtain further discounts.

The Finance Committee

The finance committee must cope with problems concerning finances, treasury and cash in terms of programs and of the operating account.

One of the most recurrent problems in this committee concerns investigating the causes of a possible excessive accumulation of client's credits. The ways for cashing in quickly on these credits should also be debated. In circumstances like these it will be necessary to involve job costs accountants so that all figures can be submitted to clients at a faster pace. This could mean increasing the number of technical accountants in the various construction sites and so it really becomes a question of choices. Should we call in external consultants or use the firm's technicians by assigning them temporarily to these accounting duties? A third solution would consist in hiring new employees, thus expanding the organizational structure. In this case the matter would have to be submitted to the executive committee where questions of structure and personnel are discussed.

Also relations with clients should be examined in order to obtain more prompt payments at more favorable conditions.

The financial committee should also examine financial problems inevitably connected to each new job. It is important to check the firm's supply of capital so as to insure that the economic increase due to new jobs will not result in a general undercapitalization and in the need to resort to costly bank credit. This situation would curtail the firm's income and consequently the firm's self-financing capacity, thus priming a dangerous financial spiral.

A New Model for the Organization Chart

Formally, the organization chart can be subdivided in three separate areas corresponding to the organization's corporate, operations, and macrofirm levels.

In the first of these levels we have top management and the various committees. In line with top management there is the already discussed management task force and the personnel manager which characterizes this level. From the personnel manager depend the personnel department, in a staff position together with the other services (general affairs and secretarial department, marketing and research and development, organization and procedures, management information system department, controller, administration department, and construction engineering).

In line with the general manager are the operations area managers, active when the firm, having to operate in different geographical areas, decides it is necessary to have a person in charge of each one of them.

We have then, in line position, the three managers that represent the operations or multipolar level. The first of these poles handles not only normal operations purchases, but is also responsible for buying all that is necessary for the acquisition of production equipment. The procurement manager will be in a position to offer this equipment both to the manufacturing pole or to the existing market demand, and will therefore sell it wherever the best results are obtainable.

The manufacturing pole must organize, coordinate and control the various construction sites. In so doing, according to our model, the responsible parties are free to acquire fixed assets for production wherever it is more convenient for them: in the procurement pole that also handles investments, or in the external free market by selecting the more convenient work offers made by subcontractors. Thanks to this procedure, the production pole is free to use internal or external equipment services, in order to obtain the best economic advantages.

It is the financial pole responsibility to administer with positive economic results the firm's finances. It does so trying to obtain extension of payment terms, to hasten as much as possible credit collection, coping with lack of cash either with its own means or through bank credit at the best terms that the market is offering. In this picture, the easiest variable is the one concerning client credits, with the firm contacting clients in order to speed up payments. When the cost of money is high, each and every delay causes interest changes which can negatively affect even a possible positive economic course of the firm. This explains the importance of being sure the person in charge is someone with credit collection experience and who knows how to speed up cash inflows as much as possible.

The three operations poles interlock and interact continuously with each other and with external firms in relations that produce inflows (acquisition of resources, of fixed assets, of money) and outflows (production, fixed assets performance, payments of debts, etc.) When these sufficiently stable interactions concern the external environment they give life to the third organizational level of the firm, the one I have called "macrofirm."

Site Management

In this analysis of the organization chart for a building firm, I would like to call special attention to the organizational structure concerning the management of each construction site (Figure 12).

At the very top of the site's organizational structure there is the project manager upon whom depend the site superintendent and the foremen for specialized direct work. In a staff position under the project manager there is the construction engineer (who formulates the estimated bills of quantities for clients), while in a staff position under the superintendent we find the administrative accountant (who is in control of each lot of resources that enter or exit the construction site).

The project manager must personally handle relations both with the client and with the site, since it is his or her duty to accomplish the technical and economic results needed by the project whose various aspects are managed by him or her.

Among other things the project manager has to check on the economic estimate of the offer and elaborate the eventual technical and economic variations that could improve the estimated results. His or her responsibility is to handle formal relations with the client and with public bodies for permits and authorizations. He also handles relations with possible free-lance collaborators and verifies the assignment and organization of crews for the various site tasks. He prepares general work schedules including the employment program of the crews (which must be submitted to personnel), the program for the procurement of materials (to be submitted to the procurement department), and the program of works that must be carried out monthly complete with revenues and means (to be submitted to the financial management). He maintains coordinating relations with the various departments of the "corporate level" and supervises both the hiring of new workers and relations with labor unions. In addition, the project manager cooperates with purchasing for market research concerning suppliers and subcontractors.

One of the most important tasks of the project manager consists of checking periodically the general work schedule. In doing so he must highlight all departures from the original program and suggest the necessary corrective measures (general program updating), while supervising works ac-

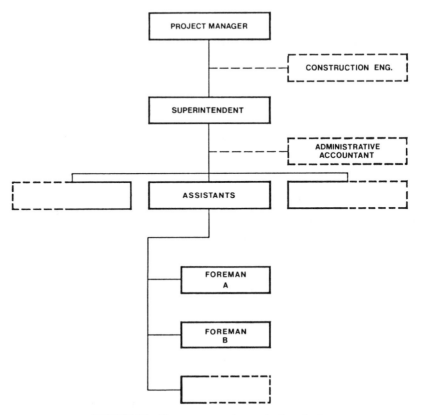

FIGURE 12. Site management organization chart.

counting. He must also compile every month the project's economic summary necessary for updating the construction site monthly budget.

The site superintendant works directly under the project manager. This is the person responsible for organizing the site and for overseeing the various work operations. He must therefore cooperate with the project manager in the preparation and in the execution stages by organizing, among other things, the workers assigned to him for that particular job; he also designs the general site layout and when doing so he lists the necessary equipment. He is responsible for managing works according to the technical and economic program and to the firm's policies regarding direct and indirect site labor (hiring, dismissals, selection of the various site assistants, etc.). One of his special duties is to see that safety regulations are duly applied and in this sense he must have complete freedom in deciding upon the installation of safety equipment, shelters, covers, etc. This is because he will have to answer for all accidents that might happen on the site.

He must cooperate with the site construction engineer for a more efficient and prompt technical accountancy, while the administrative accountant—who is in a staff position—must handle all quality and quantity control of materials entering and exiting the construction site, checking that orders correspond to approved purchasing programs and to planned delivery deadlines. He coordinates and controls all work assigned to subcontractors and issues payment approvals for completed works.

Site assistants work directly under the superintendent and it is their duty to check and coordinate the execution of the site's various work stages, forwarding all necessary information to the foremen. They must compile the daily reports on usage of direct and indirect labor, of materials, of equipment, and coordinate them with works executed by subcontractors. It is their task to check the daily application of programs and to see that safety measures are applied; they must report to the procurement department all maintenance necessities concerning the site's equipment.

Site Controls

There is a whole series of controls that must be set up on the construction site, especially those concerning labor employed in the various work operations, and materials in arrival.

Each single work can be subdivided as follows:

1. Work operations pertaining to the site's general organization
2. Direct work operations
3. Subcontracts and supplies
4. Services

Work operations concerning groups 1 and 4 do not take into account usage of materials, and therefore quality control will consider only labor; for the other groups, there must also be controls on materials used. All direct work operations must be expressly checked to determine costs. These operations must be preceded by subtasks which transform raw materials into intermediate goods to be used in direct work operations which are those that result in finished works. Production takes place by employing work crews that use raw materials obtained directly from the site's stores, and products that have undergone a first transformation process in the operations. Raw materials subjected to these processes are, for example, aggregates, binders, iron for reinforced concrete, while some subtasks' products are structural concrete, coatings, wrought iron.

In direct works are included those transformation and repair works that do not include the use of raw materials and which can be executed mainly by the sole use of labor, or with an added limited usage of intermediate goods.

Under "subcontracts and supplies" we list works for which subcontractors and suppliers are fully responsible. In some cases the firm—or general contractor—limits its contribution to labor for assistance work. Materials necessary for these operations arrive on site as intermediate goods ready to be laid out. Usually their arrival coincides with installations, and subcontractors active in the project assume responsibility for storage operations. The firm or general contractor checks on these materials from an organizational more than from an economic point of view since usually the buying order calls for the delivery of finished and already installed products.

Finally, "service" represents a group of works for which no materials are necessary, and so controls are limited to labor.

Materials and labor are primary components of the various work operations, and are subject to different types of controls.

In order to execute a series of controls on materials it is necessary to have in the construction site a storage deposit area through which all raw materials pertaining to the construction work will transit. This deposit can be managed by using a personal computer connected to the firm's main one. This will allow automatic checking of the daily usage of raw materials, and recording of quantities employed in the various work operations and subtasks.

A check-in takes place when materials arrive; these will be divided into intermediate goods to be used by subcontractors, and into raw materials to be stored for future use in direct work operations. The intermediate products are forwarded to the spot where they will be installed and where they will be further checked in by people in charge. These raw materials to be stored will be also checked so as to identify their place of departure, quantity, and quality. A new control will take place at the moment of usage.

Should the final destination be the work center itself, a last control will be carried out there, and also a daily check of quantities used and of those left on spot. This information will be needed to calculate the production capacity of work crews assigned to the various operations.

When the request for materials originates in subtask operations, it becomes necessary to control both raw materials actually employed and the intermediate goods production. Their final destination should always be recorded; it then becomes possible to know the quantities used in single work operations in order to establish the costs for each one.

As for workers employed in the construction site, a preliminary control must be carried out by the site superintendent when he or she decides the composition and assignment of the various crews on the basis of the approved general plans.

These work crews can be subdivided as follows:

1. Crews for subtasks
2. Crews for work operations proper
3. Crews for brickwork

For each work operation there will be a subsequent control of the actual employment of labor, which must also determine possible shiftings of workers from one operation to the other; these can happen during any workday and cannot be foreseen when scheduling (for example, a motive can be the anticipated arrival on site of materials that have to be unloaded). The superintendent must of course see to it that these shiftings be as limited as possible and in any case that they take place within each homogeneous work operation group and not between different groups.

The totality of information resulting from these controls, gathered by the person in charge of administrative accounting, will help to furnish all necessary data for checking the project's economic course on the basis of the approved programs.

Delegation as a Problem

A problem concerning the firm's entire organizational structure is the specific and formal assignment of responsibilities, which means giving authority to various people. The ensuing responsibilities must be subdivided between these same people and will concern relations with employees, purchasing, lock-ups, payments, signature of contracts and revenues certifications, bank operations, specific and general mail, general and site insurance. It is important that each responsible party be clearly aware of the area within which he or she must be perfectly free to act; this, of course, means that he or she will also be called to answer for his or her actions.

My comments about the organizational structure of the firm have begun with an explanation of the limits pertaining to "formal" organization, and of the problems of the "informal" one. I think it useful to close these comments with a statement by Theodore Levitt: "There is only one way to manage anything and that is to keep it simple" (1978). Impressed by this wisdom, I have jotted down a list of suggestions for all work associates, to help them become enterprising people who know how to overcome obstacles that always seem to show up. By so doing, they will avoid becoming passive bureaucrats.

APPRECIATE YOUR INDIVIDUALITY

This means being able to invent, to seek, to create, to keep provoking new conditions for mastering your work in order to fight mediocrity and to obtain results which will make you proud and envied by others.

ACT WITH ENTHUSIASM

This means recapturing your creative powers and enjoying the novelty of each day that must see you a protagonist. This means injecting excitement in your activity, enjoying it so that you will come to consider it an attractive new challenge to be conquered time and time again.

ACT SIMPLY

In all activities keep your common sense, the good old wisdom, the steady mind of the family head. Especially in production areas you must never forget that activities must reap positive results (income superior to costs) without quality losses.

ESTABLISH PRIORITIES

This means trying your best to understand the important goals of each action. These goals must receive priority since all things cannot be of equal importance.

ACT CONCRETELY

This means producing results and not papers, facts and not words, solutions and not problems, creative actions and not excuses.

USE DIRECT COMMUNICATIONS

This means living as much as possible on the field, *cooperating* with other people and explaining your problems to those who can give you useful suggestions. This means trying always to make yourself richer by communicating with others and by recapturing the pleasure of pointing out the successful results you have obtained. This means communicating directly with all interested parties, especially when you must ignore formal rules, which is always possible if it means getting better results.

FEEL LIKE A "LEADER," NOT LIKE A "BUREAUCRAT"

Each one of us in his or her own activity area must be in a position to make independent decisions. Each one of us must feel engaged in an endless struggle to obtain results. Each one of us must be able to pass on this approach to all those that we work with, so that they also feel like leaders and not like bureaucrats. They should make the following concept their own: I am and must be always at war with outside events. I must reach my goals by continuously "reinventing" my work-day. I must be obsessed by results, I must become a fanatic in wanting even to surpass the goals I had initially set for myself.

STRATEGY

"Strategy and structure" are the initial words of a famous book by Alfred D. Chandler. The title goes on to say, "Chapters in the History of the American Industrial Enterprise." The analysis of these two most important factors in organizational behavior and firm development helps to understand the evolution of America's major productive organizations.

Organizing, Managing, and Developing

Chandler's study suggests that organizational complexity is a variable which increases as firms, having accumulated resources, plan to rationalize management in order to optimize their usage for the purpose of fostering growth, thanks to the added accumulation of resources to be rationalized in the expansion stage.

Chandler's book is not only a basic text for studying the history of firms organization, but it also proves to be most useful for those who wish to understand how to organize, manage, and develop any type of company in its production development.

By defining strategy and structure it is possible to suggest behaviors that are scientifically exact and certainly useful for those that have to cope daily with these functions.

First of all, Chandler proposes a distinction between the formulation of policies and procedures and their implementation. This formulation can be defined as either strategic or tactical: "*Strategic* decisions are concerned with the long-term health of the enterprise. *Tactical* decisions deal more with day-to-day activities necessary for efficient and smooth operations" (1962, p. II).

Then he proceeds to define specifically the concepts of structure and strategy. *Structure*, he writes, is the organization design through which a firm is administered. It includes the lines of authority and communication between the various administrative offices and people and all information and data that flow through these lines. Both these lines and information are essential for coordinating, appraising, and planning in order to reach the basic goals and to knit together the resources of the company which include finances, equipment such as plants, offices, warehouses, marketing and purchasing facilities, sources of raw materials, laboratories, and the global know-how of personnel.

Chandler gives the maximum emphasis possible to human resources which he considers the element that most characterize the firm's structure and therefore the firm itself. It is the human resources that establish strategies and then try to implement them. According to Chandler, *strategy* is the determination of goals of an enterprise, and the adoption of actions and the allocation of resources needed for carrying out such goals.

It follows, according to this author, that structure stems from strategy and that even the most complex structure results from several basic strategies. "*Expansion of volume* led to the creation of an administrative office . . . *geographical dispersion* brought the need for a departmental structure and headquarters to administer several local field units . . . while the developing of new lines of products or continued growth on a national or international scale brought . . . the multidivisional structure with a general office to administer the different divisions" (1962, p. 14).

Nevertheless, the starting point for entrepreneurial development is always the pre-existing structure, through which have been accumulated resources that are partially destined to the firm's self-financing. This is why I have chosen to deal first with *structure*, pointing out some basic principles which should be kept in mind when studying it. As Chandler indicates, when a firm has accumulated considerable resources, the need of a steady employment of its workers, money, and materials provides a continuous stimulus to move in new areas, thus finding new markets, to develop new products, etc.

This often happens casually and not in any rational form, so that production growth is not accompanied by a prompt structure adjustment from the point of view of organization, finances, economy. Thus the growth process could lead to a crisis in the entire entrepreneurial system. This in single firms happens in a way fairly similar to what emerges from the history of large American companies as discussed by Chandler: "The failure to develop a new internal structure . . . was a consequence of overconcentration on operational activities . . . or from . . . inability . . . to develop an entrepreneurial outlook. . . . They may also have resisted administratively desirable changes because they felt structural reorganization threatened their own personal position, their power, or most important of all, their psychological security" (1962, p. 15–16).

Centralization and Decentralization

Specifically in construction firms, the organizational structure suggested in this chapter is in many ways similar to the one described by A. P. Sloan, Jr. in his *My Years in General Motors*, in which he himself explains: "[a] good management rests on a reconciliation between centralization and decentralization, or 'decentralization with co-ordinated control' " (1972, p. 505). The operational decentralization implemented by delegating formal responsibilities to project managers and site superintendents is coordinated by general management; when the model allows sufficient centralized control, there are positive effects on the production phenomena decentralized in elementary operations. But in this model correcting the structure on the basis of the actual needs of operations and the strategy for an optimal destination of all accumulated resources fall within the sphere of what I have defined as *corporate level*, where the major strategic action will have to be evolved in order to insure the firm's survival through operations capable of optimizing all operational processes with positive results in regards to the goals that the firm has programmed.

In the specific building sector the turbulent factor already discussed makes it difficult to manage the firm, and this has an influence on the resulting entrepreneurial strategies.

The firm, on its two organizational levels (*corporate* and *operations*) offers a sufficiently flexible structure for managing a variable number of projects. This flexibility is facilitated and increased by interactive relations with external firms within the organizational level I have called "macro-firm."

But regardless of this flexibility, the firm can enter a crisis because of insufficient production (lack of contracts) or because of excessive work acquisitions. In the first instance, the problem has a specific economic character. The stable organization of the firm, especially in its *corporate level*, determines costs which accumulate in time, independently from production, and which must therefore be absorbed by those profit margins that it will be able to determine. When production falls below the safety level, it cannot obtain profit margins capable of absorbing such costs and so the overall economic results become negative. In the long run this fact can lead to unhappy and irreversible consequences for the firm.

When this happens, the entrepreneur responsible for the firm tries his or her best to find new contracts, lowering prices as much as possible to become more competitive on the market.

Acquisition Strategy

This is true for all industrial activities, but in the building sector it becomes especially important because of the economic, financial, and organizational weight of each contract. This explains why particularly in the building industry the new contract acquisition strategy has to be carefully thought out, especially from the economic and financial point of view. In order to establish strategic management of the firm, it is then necessary to have an updated, day-to-day situation showing new contracts, contracts still being discussed and possible contracts (Figure 13) so as to prepare a medium-term economic plan based on certain, probable, and uncertain data to be carried over to the short period pertaining to the year being examined (Figure 14)

Economically, the offered price must result from a careful and thorough study of costs of each work operation and of global operations, in order to avoid in the executive implementation phase a negative balance between revenues and direct work operations costs, which would cause not only the impossibility of absorbing general overhead, but also a net loss already at the site level.

To reach this goal, after having inspected the places where works will be done, all available documents will be studied and a tabulation will summarize all general data (Figure 15). The bill of quantities will be prepared or, if it is already available, it will be checked against project designs and job descriptions. As for the economic aspect, it is generally difficult to have a detailed cost analysis for each work operation. Therefore, it is useful to proceed by means of sample analysis, studying direct site costs for the economically more important work operations, thus covering about 70-80% of the total work operations costs. Those of the remaining 20-30% will be obtained statistically from normal price lists, reducing them by a percentage obtained analogically from what has been established for operations already analyzed. It will then be possible to compile a specific technical card that summarizes the economic analysis of the offer (Figure 16). This form shows in real figures the site direct costs (divided into analyzed and non-analyzed costs) and the site general overhead. Shown are the contribution, in percentage, for the various public and private bodies, insurance costs and, finally, still in percentage form, it becomes possible to indicate the profit margin with which we want to acquire the job.

It is precisely on this item that management strategy develops, so as to acquire the contract by lowering or increasing the percentage of such margin.

Then, through a proportional calculation between total costs, divided by total profit and costs percentage, we can determine the final price on which, by backwards percentage calculations, we will be able to identify all necessary parameters so as to issue the offer in global price form, or as discount or increase of the prices proposed for each work operation.

The financial aspect must also be thoroughly analyzed. It is essential to remember that each work operation creates new cash imbalances to be financed. Cost outputs are generally more rigid and run faster than inputs deriving from contract revenues (usually more flexible and less certain). All of this generates two kinds of problems. The first, of a direct nature, causes improper charges of a financial nature which add to site costs with a resulting negative contract balance even when the economic revenues seem superior to costs.

Secondly and more generally, global indebtedness of the firm can exceed its borrowing capacity, thus blocking the possibilities for further credit.

The effects are dangerous for normal administration; there is the risk of not being able to cope with possible increased financial demands, even if only temporary. Furthermore, the accumulation of interests payable could in some cases determine a real loss for the firm. Thus, the strategy which initially had an economic and market characteristic (search for new contracts) can be modified by strategic con-

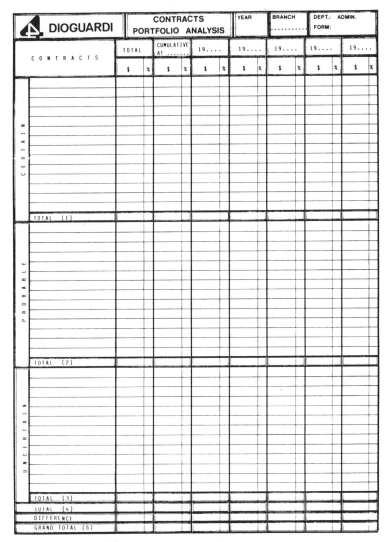

FIGURE 13. Form for contracts portfolio analysis.

siderations of a financial nature, suggesting that economic expansion be contained.

Economic and Financial Strategies

Usually economic strategies seem to prevail in smaller firms, while financial ones are more typical of larger ones.

In small firms there is a pronounced need for success in the market. Strategies therefore privilege immediate economic growth, emphasizing some specific functions, marketing, for example. The financial elements represent variables which, being easy to check, facilitate subsequent controls of the firm's performance. Control on cash flow phenomena often does not use the right instruments for checking the qualitative composition of outflows (determined by debts caused by costs and payments for investments and other assets (for example, acquisitions of shares, etc.) and from inputs (credits or inflows from debts or liabilities). This can cause misinterpretations, especially in times of easy credit, when it is easy to obtain money by short- or long-term debt. The consequent increase in financial charges liability has a negative influence on the profit and loss account and consequently on the quality of the firm's growth tendency, which may find itself undercapitalized.

An optimal situation could be determined by trying to combine as best as possible the two strategies: an interest

DIOGUARDI	PRE-AWARD FORECAST OF CONTRACT	YEAR	DEPT.: ADMIN.: FORM.:

N.	CLIENT	CONTRACT	CONTR. SIGN. FORESN	WORKS START FORESN	DELIVERY OF WORK		PROJECTED CONTRACT REVENUES			REVENUES FOR THE YEAR	COMMENTS	OUTCOME
					CONTR.	PROJECTED	BASE CONTR.	EXTRA	TOTAL			

FIGURE 14. Model for pre-award contract analysis.

DIOGUARDI	TENDER ANALYSIS GENERAL CHARACTERISTICS	Project	Date	Manager	Dept.: Technical Form: Page.........

Site inspection	Technical personnel
Site location and soil-subsoil characteristics	
Area and surrounding condition	Transportation access
Site installation and particular conditions	Quarries and disposable areas

UTILITIES CONNECTIONS		AVAILABILITY OF ITEMS REQUIRED	
	Water		Labour
	Electricity		Subcontractors
	Telephone		Transportation
	Water mains		Sand, gravel, etc.
	Sewers		Other materials
	Miscellaneous		Miscellaneous

FIGURE 15. Tender general characteristics analysis.

DIOGUARDI	TENDER ANALYSIS COSTS AND RETURNS	Project	Date	Manager	DEPT.: FORM.:

		AMOUNTS SUBTOTALS	AMOUNTS TOTALS
DIRECT COSTS	PURE COST OF JOBS ANALYSED (1)	....	
	PRESUMED PURE COST OF JOBS NOT ANALYSED (2)	....	
	CONTINGENCIES	....	
	CONTINGENT	....	
	TOTAL FORESEEN PURE COSTS (3)		....
MISCELLANEOUS SITE EXPENSES	SITE INSTALLATIONS	....	
	SITE DEMOBILIZATION	....	
	SCAFFOLDING AND SAFETY EQUIPMENT	....	
	SITE MANAGEMENT	....	
	MAINTENANCE AND TESTING	....	
	INDIRECT WAGES & SALARIES	....	
	TRAVEL EXPENSES	....	
	LEASING OF EQUIPMENT	....	
	BONDS	....	
	INTEREST COSTS	....	
	TOTAL MISCELLANEOUS SITE EXPENSES	(%..........)	....
SUNDRIES	SUNDRIES	....	
	....	....	
	TOTAL SUNDRIES		....
	POSSIBLE EXTRA PROCEEDS FROM CHANGE ORDERS		....
CONTRIBUTIONS	PERMIT AND MISCELLANEOUS FEES	...........%	
	ALL RISKS INSURANCE	...........%	
	....	...........%	
	MARGIN FOR HEAD OFFICE	...........%	
	TOTAL (4)	%	
	CARRY FORWARD (5)		

BROUGHT FORWARD (5)	....
100 − (4) = (6)	
.... (5) x 100 / (6) .... (7)	
CONTRIBUTIONS......... % = ... (7)=	....
ALL RISKS INSURANCE... % = ... (7)=	....
.... % = ... (7)=	....
PROFIT & MARGIN HEAD OFFICE % = ... (7)=	....
TOTAL BASIC AMOUNT (7)	

INCIDENCE OF CHARGES & EXPENSES ON PURE COST

TOTAL BASIC AMOUNT (7) / FORESEEN PURE COST (3) .... = ...

PERCENTAGE INCREASE TO BE APPLIED ON ANALYSIS

TOTAL BASIC AMOUNT (7) / FORESEEN PURE COST (1)+(2) .... = ...

DECREASE OR INCREASE TO BE APPLIED TO TENDER STARTING FIGURE

TENDER FIGURE (8) ....

TOTAL BASIC AMOUNT (7) ....

DIFFERENCE (9) ....

INCREASE / DECREASE RESULTING

(9) .... / (8) .... x 100 = ... %

NOTES

DATE ________________ SIGNATURE: ________________

FIGURE 16. Model for tender economic analysis.

concentrated exclusively on economic facts might not highlight the financial problems which would then emerge only at a later date, with negative consequences on the operating account; an attention dedicated only to financial strategies could hide economic problems which if noticed only at a later time could become extremely difficult to solve. These phenomena are sharpened proportionally to the existing difference between the financial and the economic cycles, which at times reflects the difference between the investment cycle and a product life-cycle.

This is not easy to arrive at because, to acquire a "normal" contract portfolio capable of activating within the limits of operational normality the organizational structure of the firm, it is necessary to convey towards the market a number of offers larger than the essential minimum. The risk is that the number of contracts acquired will remain limited regardless of competitive prices, because of excessive competition; there is also the danger of acquiring more contracts than are necessary, weighing on financial phenomena and on the firm's structure. This being the case, the macrofirm mechanism can prove of great help since it allows increased relations with outside firms and—in border cases—the recourse to total subcontracting, leaving to the general firm financial functions and the time and quality control of work produced by other firms.

How to Improve Financial Performance

We have already seen that in any case the financial management becomes important for the firm's general stability. Operations for improving the management financial performance can take place mainly in three areas. It is possible to act on the external financial system, especially banks, when there is no self-financing, in order to optimize relations qualitatively (type of financing obtainable), and on the proposed cost of money. Other actions for reducing financial charges can be targeted on assets (revenue) or liabilities (costs) of the profit and loss account in order to expedite the cash inflow while slowing up as much as possible the outflow.

In regards to this, there are some situations that must be pointed out. First of all and generally speaking, any type of financial behavior in the firm must be accompanied by a constant knowledge of the limits within which negotiations can develop without problems, i.e., without causing direct or indirect damages to the firm, even if it is only of a psychological nature. It is a typical case of application in the management context of the physics "threshold" concept, which is the maximum intensity degree before effects become dangerous. In no field as in the financial one crossing the threshold in order to obtain some goals can cause dangerous consequences with psychological overtones that can very rapidly become highly dangerous.

The threshold problem, for example, must always be taken into account in relations with banks, when discussing the quality of a requested credit and the economic terms that the system wants to apply. Not to bargain (i.e., not to operate below a certain threshold) can have a negative effect, since banks in that case tend to feel they can impose terms which are convenient only to them. In that case, the firm's image can be spoiled by signs of carelessness in controlling the different factors that make up the operating income.

On the other hand, to insist above a certain limit in order to obtain especially favorable terms could imperil relations with banks. This phenomenon proves damaging particularly when it is necessary to ask for transitory cash availability, i.e., exceeding the credit lines already granted. Another negative effect can surface in the long-term period, when credit already granted has to be routinely renewed.

It is necessary to intervene not only on the cost of money, but on the quality of credit as well. A useful financial strategy consists in trying to transform bank credit stemming from current account, typical of short-term cash elasticity, into more stable medium to long-term forms (similar to borrowed capital) but connected to specific administrative operations (for example, advances for certain contracts or against invoices envisioning deferred payments).

Threshold problems must be considered also when dealing with clients in order to cash in on payments. Too much pressure on clients can deteriorate the firm's image, especially in the case of a relationship based on trust. The same is true in the case of subsuppliers, when attention must be given to their specific and real financial needs: relations too burdensome for suppliers can endanger the quality of work. Nor should the "bad publicity" originating from these same suppliers be overlooked.

Purchasing Strategy

An important problem pertaining to the purchasing office and which should be envisioned as mainly strategic concerns the decision to subcontract specialized works to a single firm or to assign each specialized work to a wider number of suppliers. From the point of view of negotiations and of the consequent financial planning, a limited number of subsuppliers and subcontractors simplifies things, both when negotiating and in the control phase. But this increases the risks of insufficient specialization in a given sector. There are some dangers deriving from this, for example the monopolistic competitive one, or the danger stemming from the economic and financial trustworthiness of the supplier; or, again, the danger of marring the firm's image, since the firm in this case tends to underemphasize its vocation to coordinate different production firms. This, especially, can annul the very function of the macrofirm as a complex system of productive and specialized firms coordinated so as to reach a common goal. It appears therefore necessary to add an additional segmentation of work and consequently an increase of the number of subsuppliers. It is also useful to have efficacious planning instruments, for example those supplied by the PERT, CPM, etc. network systems.

Tender Strategy

Strategically speaking, the process leading to the acquisition of new contracts can result from chance relations. More often, however, the firm has to hunt for work by taking part in tenders which it will attempt to win thanks to more convenient offers (Figure 17).

But it also becomes necessary to try to entice the client by suggesting the possibility of offering additional services. For example, there are usually two proposals that increase an offer's specific value. During work operations a continuous and rational quality control can be applied independently from the client's requests. This has positive effects and can even be the subject of a separate offer. The added economic burden increases construction costs, but also lengthens the

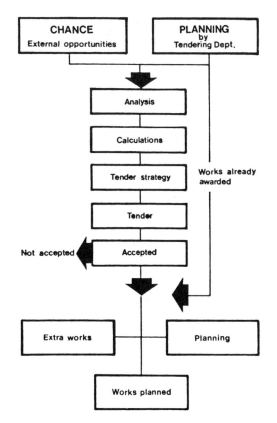

FIGURE 17. Market strategy for contract acquisition.

life-cycle of the product, thus lowering future global costs, i.e., construction plus maintenance costs during the entire life-cycle of the construction facility. All of this suggests that a planned maintenance activity can also be offered, and can earn an extra payment from the client. But the very fact of having pointed it out by means of a separate offer in the context of a tender strategy will prove interesting for those clients that think in terms of future protection for their investments.

MANAGEMENT

The operations management of the firm must coordinate the work schedule of each site, so that they be compatible with each other.

The Budget

This coordinating process is implemented by means of the budget, i.e., a general program of the firm's total activity, based on an integrated system of forecasts. Each forecast is formulated on the basis of specific goals, and the progress towards each one of them is checked on short-term basis. Whenever results depart from the approved plans, it becomes necessary to adopt immediate corrective actions. Thus, the budget must not simply forecast the firm's administration in its economic and financial aspects, but must also deal with the physical organization of resources that have to do with production. Operationally speaking, the budget must represent an integrated system of forecasts, and differs in this from an estimate.

The estimate generally deals with future facts and is used in trying to anticipate the event in its totality; the budget, instead, is articulated in a series of estimates which, closely connected, cover the entire company system in its future evolution, and so becomes an integrated system not only of forecasts, but above all of goals.

The life of the firm is therefore analyzed in all its possible manifestations, by connecting single events to general situations, and by identifying short-term phenomena in the wider context of medium- to long-term plans.

The budgetary forecast, furthermore, must evolve essentially within the firm's area, where single operators, while forecasting, are also determining the results they must attempt to obtain. All ensuing data is sent to top management, who has the function of coordinating and summarizing information.

The operations for getting these results must not be forced top-down but have to be identified—through personal and rational choices—by each operator when preparing the budget.

The result will be a higher degree of peripheral delegation, both in the execution phase and when results are checked against programs. This explains why an instrument like the budget is not only a method of management, but also a philosophy which fosters an integrated management through set goals and in this way facilitates the delegation of authority and responsibility.

The main subject of production is the contract. The resulting work usually takes up two years or more and so the budget must cover the same time-span and the different forms must be designed accordingly.

Operational budgets must in all cases stem from long-term forecasts based on possible alternatives in which general company policies are projected and then transferred, quantitatively, in short- and medium-term budgets. These will carry economic and financial analysis that will become a "money" budget following a series of mainly technical budgets that will carry the study of quantitative future needs and the planned distribution of resources based on the scheduled production.

Financial and Economic Management

The financial and economic analysis I have just described will be of help in preparing the study of the operating account in terms of costs and revenues, which will then be transformed in debts and credit situations before appearing in actual income and outflow form. A comparative analysis will determine possible unbalances which, if positive, can signify cash availability or, if negative, the need for finances obtainable from sources not directly connected to the production cycle. These availabilities or needs are tied to flows in the balance sheet, whose modifications, within the administration, represent typical financial phenomena. At the beginning of the financial period the system presents itself as a consolidated balance sheet, made up of assets and liabilities and with a capacity determinated by the difference resulting from these two groups of items. During the firm's activity in the year there can be increases in assets (investments) and a decrease in liabilities which will cause cash outputs. Decreases in assets (retirement of fixed assets) or increases in debts resulting from added credits will instead determine cash inflows (Figure 18).

In the firm these financial manifestations continuously entwine with those pertaining to the economic administration; this way, the availabilities resulting from the economic account represent autonomous forms of financing and therefore increase available financial sources, while needs covered in the economic administration by possible negative financial unbalances will have to be met by means of borrowing. The financial phenomena resulting from the economic administration and from assets (investments) and liabilities (debts) will join in forming the cash in and outflow. Forecasting on these items will determine cash budget behavior and the foreseeable balance at the end of the accounting period. This balance will make it possible to determine the future capital situation of an operation which will conclude the cycle preceding the formulation of the op-

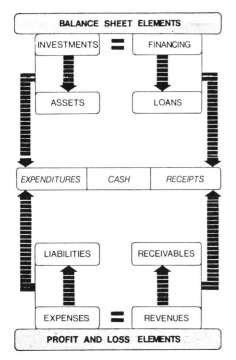

FIGURE 18. Connection between cash and profit and loss and balance sheet elements.

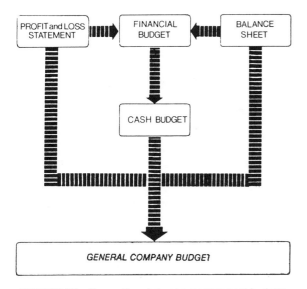

FIGURE 19. Connections between management budgets.

erational budgets (Figure 19). As for the specific financial budget, a constructive approach is to subdivide this type of phenomena in two main categories: those rigidly tied to predetermined deadlines, and those that can be viewed as more or less flexible. This allows an immediate control on the "sure" imbalance between revenues and outputs; this imbalance will emerge unmistakably since at the same time events with differing degrees of certitude will have to be identified.

It then becomes possible during each periodic control (weekly, monthly, quarterly) to study how to act on flexible phenomena in order to speed up those events that allow the firm to cope with cash imbalances ascertained as rigid and on which any expediting action or request of extension would in fact prove useless.

In other words, for each standard accounting period it is necessary, when facing a rigid cash budget, to consider creating an emergency fund to be established around the three operational components (clients, suppliers, banks) so that it can be used in the least maneuverable situations.

Coping with Credits

It should be emphasized how, from a structural point of view, two functions become especially significant, due to their interactions with the parties responsible for the firm's finances: credit collection and purchasing. The first of these functions should in fact include expediting actions on credits: each time that a payment due arrives before the foreseen deadline, it means for the firm's administration an economic improvement which increases as the cost of money goes up. A special "issued bills book" will simplify operations for the people in charge who, on the other hand, must show considerable stamina and a very special sensitivity in order not to cross the threshold discussed above when contacting clients.

Coping with Debts

The problem for the purchasing department is complicated by overt or latent tensions between the need to buy the best quality with the shortest delivery time, thus satisfying in full the supplier, and the necessity of obtaining the most favorable financial payment terms.

The conflict can be solved by drawing a boundary line between the duties of those who are responsible for purchasing and production, and the duties of those who are responsible for finances. Business negotiations concerning economics and quality must be carried out by purchasing and production after having defined standard payment terms (for example, 30–60 days after delivery of goods). When this type of negotiation has been concluded, the person responsible for the firm's finances can step in and negotiate with the supplier the improvement or worsening of the agreed terms in the presence of modified standard financial terms. By anticipating or postponing agreed payment terms on the basis of the existing firm's liquidity, it might become possible to modify positively or negatively (for supplier and with his or her approval) the economic terms that had been agreed upon. This negotiation must be totally independent of the preceding one and must evolve through a special financial formula indicating, on the basis of the current cost of money, the economic advantages or disadvantages resulting from speedier or slower payments.

Budget-Making Procedures

In the final stage the procedure will be similar. The cash flow will represent the first and more immediate control instrument, allowing a tracing back to the financial manifestations of events on which, at the same time, the final economic analysis must be developed.

The formulation of budgets in the preliminary phase and their control in the execution phase must obey predetermined procedures according to a calendar of deadlines in which the various events and data are allocated and recorded. In other words, it becomes necessary to introduce in the firm's system a rigid time schedule for office work; the various tasks, consequently, have deadlines that make the various operators answerable to each other when carrying out their duties.

Budgets are formulated through a preordained series of standard forms, on which forecasted and final figures are written in, together with the balance between these two groups of elements. This facilitates a control by exception. The form itself must be designed so that along the route from the firm's periphery to its management area, information becomes increasingly concise.

Allocating Human Resources

Especially in building construction firms, an important problem has to do with the allocation to the various sites of the people who will manage them. It is necessary to control the complete program concerning their employment and the time each one will spend on the assigned project or projects.

The use of a simple bar-chart can be of help (Figure 20). In it, for each subject, the managing personnel is indicated on the basis of the already described organization chart (project manager, superintendents, accountants, foremen) together with the time-period forecasted for completing the project. The fact that each site usually remains active for more than one year suggests that the form be designed for at

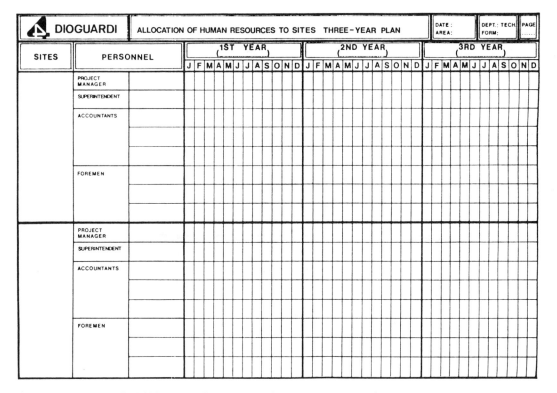

FIGURE 20. Form for allocation of human resources to sites programme.

least a three-year period. Quite often some of the site's top personnel will be assigned to more than one site at a time. In this case it is necessary to calculate the time percentage and therefore the cost allocated to each site so as to obtain, in an analytical bookkeeping, the real cost of each site.

Controlling Suppliers and Subcontractors

Also important for each project is the control of the main suppliers and subcontractors, representing the external terminals of the transactional organization I have defined as a macrofirm. It is necessary to be able to compare a series of offers for each needed supply. This allows choosing the more advantageous one, quality- and quantity-wise, and also to contact other firms which could substitute the selected one if and when problems arise. A standard form (Figure 21) simplifies the analysis of the different offers by comparing them to forecasts in the bill of quantities. The different items will thus result more homogeneous, while negative or positive differences for each item can be noticed immediately.

It might be useful, here, to give a bit more thought to purchasing. The planning phase for production and supplies is handled by special offices, and if often happens that they completely ignore financial problems; furthermore, they do not usually like to see their work coordinated with that of the financial operators.

Here too it is important to evolve a correct financial planning procedure, capable of delaying as much as possible the issuing of orders and consequently the financial outflows. This means positive returns on the year's profit and loss account.

It is an approach not without risks, since it can cause problems for production times. In fact, when attempting to delay as much as possible orders and consequent payments, it is necessary to do so without negatively conditioning the general production schedule. The task can be rendered easier by acting first of all on those activities that in a PERT-type program do not fall along the critical path. Subcritical activities should be carefully examined to see which could have a significant financial weight. They must be compared to those on the critical path, in a cost/profit type analysis, in order to verify the possible advantages in delaying these activities to the point of rendering them critical, with resulting financial gains.

In any case, it would be quite useful if the offices in charge of production's economic and time planning also furnish a document showing, for each supply or work operation, the maximum time limits for orders pertaining to that particular supply or operation.

Once again, as it is always the case in a firm, the necessity arises to mediate between the needs to optimize the economic and temporal course of production (needs that, in conformance to the "everything now" logic and financial

DIOGUARDI	COMPARISON OF OFFERS FOR THE SUPPLY OF..	SITE DESTINATION	Date N.	DEPT.: PROCUREMENT MOD.:

SUPPLIES		OWN ESTIMATED				ACTUAL	SUPPLIERS							
ORDER N.	DESCRIPTION	UNIT OF MEASUREMENT	QUANTITY	UNIT COST	TOTAL	QUANTITY	UNIT PRICE	TOTAL	UNIT PRICE	TOTAL	UNIT PRICE	TOTAL	UNIT PRICE	TOTAL

REMARKS	THE MATERIAL MUST BE ORDERED FROM THE FIRM:
	SIGNED FOR APPROVAL:

FIGURE 21. Model for price comparison of subcontractors offers.

DIOGUARDI | SUBCONTRACT WORK PAYMENT CERTIFICATE | CODES: SITE, SUBCONTRACT

STATEMENT N° .... provisional / final as at .... subcontract works executed / supplies delivered

for the work site ....

by the firm .... in accordance with order / contract n° .... dated ....

brief description of works / materials supplied ....

....

SUMMARY

VALUE OF WORKS EXECUTED

- .... $ ....
- .... $ ....
- .... $ ....
- .... $ ....

Value of works executed as at .... $ ....

Contractual retention .... % .... $ ....

.... % .... $ ....

DIFFERENCE $ .... $ ....

DEDUCTION

- for partial payments already effected $ ....
- .... $ ....
- .... $ ....
- .... $ ....

TOTAL DEDUCTIONS $ .... $ ....

NET TOTAL .... $ ....

NET SUM TO BE PAID .... $ ....

date ...., ...

COMPILED BY: PROJECT MANAGER

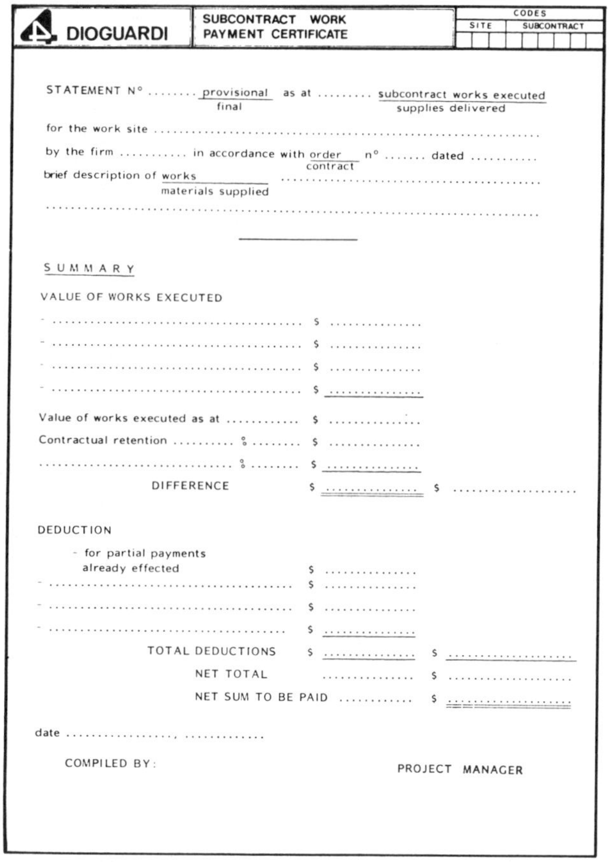

FIGURE 22. Form for subcontract work payment certificate.

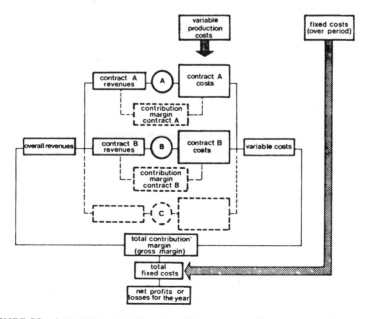

FIGURE 23. Accounting model for determining year's profits or losses by direct costing.

strategies that want to extend as much as possible that same waiting time).

Management and Control

The budget concept is accompanied by that of the control that must continuously evolve for the entire duration of works.

Special attention will be given to the costs formulation in order to check, at the end of the work cycle, what has been planned in the preliminary phase.

One of the most important items concerns payments for subcontract works. The authorizations will have to be issued by using a form (Figure 22) compiled by the site's administrative accountant and signed by the project manager. The firm allows the administration department to issue payments and the budget service to update the final figures that have to be compared to plans. Costs control must take in all construction sites' supplies and update all figures concerning stored materials. Furthermore, labor employment figures must be recorded, together with costs that will be charged to single work operations.

Both contract costs (which vary with production) and structural costs (fixed as compared to production since they are independent from it) will be checked. The formulation of costs presents some problems, the solution of which can condition the period's results. The most important of these problems is undoubtedly the one concerning the way to handle those structure (overhead) expenses which cannot be directly charged to production proper.

In building construction it is best to operate with direct costing structured so as to eliminate all arbitrary criteria when allocating (Figure 23). Thanks to this type of accounting system, expenses are listed under production fixed costs and production variable costs, viewing the first group as real period costs and allocating the others to the specific item being produced, thus handling them as production costs.

The year's profits or losses will emerge from the balance between profits and pure direct costs for each contract, thus determining the contribution margins whose total (gross margin) will be used to cover variable costs over a period and, therefore, to produce the eventual net profits for the year. This procedure avoids polluting the accounting system with ambiguous criteria in allocating overhead expenses, which will instead be globally absorbed by the pertinent period.

Cost control has to be accompanied by a control of profits and consequently by the timely issue of pertinent payment

DIOGUARDI | SUMMARY OF CERTIFIED WORK | YEAR | N.º | DEPT.: ADMIN. FORM:

PROGR. N.	DATE	CLIENT	CONTRACT PROGR. N.	WORKS EXECUTED			PAYMENT REQUESTED				AMOUNT PAID (INVOICE)			COMMENTS
				NET OF ADVANCED DEDUCTION	PREVIOUS SUM	WORKS EXECUTED	RETENTIONS	NET AMOUNT	ADVANCE	AMOUNT REQUESTED	AMOUNT	DATE	INV. REF.	
TOTAL VALUE OF WORKS EXECUTED							TOTAL AMOUNT REQUESTED IN PAYMENT					TOTAL AMOUNT PAID		

FIGURE 24. Form for summary of certified works.

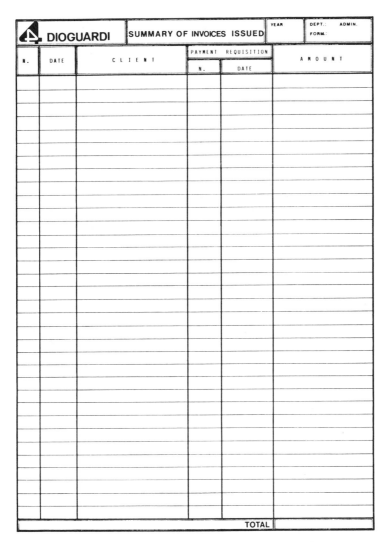

FIGURE 25. Form for summary of invoices issued.

documents which must be constantly overseen by the credit collection manager so as to obtain payments as quickly as possible.

The situation of finished works and of pertinent documents issued to clients, including those for payments and invoices that determine the firm's credits (i.e., the debts that the client will have to pay) must be summarized in forms that have the function of pointing out these phenomena and their deadlines so as to allow pressure for payment.

For example, one of the forms (Figure 24) will show, for each client, the progress of executed works and the payment requested. Another form (Figure 25) will list, date by date, invoices issued thus showing the general credit picture with foreseeable payment deadlines and possible returns to banks that have granted credits against advance payments on contracts or invoices issued for progressive stages of works (Figure 26).

Budget Economic Control

The budget economic control takes place by checking the time progress of single work operations, and by monthly updating both the Gantt-type barcharts programs used in construction sites, and the more complex and global network-type charts (PERT or CPM). Especially for this second type of chart, updating is rendered easier by the use of a computer complete with a starting program. The computer will automatically update programs of each work operation

DIOGUARDI			TRADE RECEIVABLE SITUATION			DATE:	DEPT.: ADMIN. FORM:	
CLIENT SITE	DOCUMENT REF.	REFERENCE WORKS CERTIFICATE PRICE UPDATE	AMOUNTS			FORECASTED MONTH OF PAYMENT	ACTUAL COLLECTED AMOUNT	DATE
			INVOICES	RETENTION	TOTAL			
		TOTAL						

FIGURE 26. Form for trade receivables situation.

highlighting differences which can be absorbed in the future and those that tend to become critical for final delivery of works.

Controls must always be constructive and never repressive, in order to generate actions capable of keeping the system constantly in line with the approved goals. From an organizational point of view, two kinds of control become necessary. First of all, a form of peripheral self-control must be activated by parties directly involved, so that the feedback can be prompt thus expediting eventual corrective measures. On the other hand, this type of analytical and limited control, involving short time-periods, often fails as an efficient instrument for the concise analysis of the problem. The evaluations of the operator result quite correct if viewed at the time, but in their totality they can cause significant variances, even when working on predetermined guidelines which insert single operations in a more general picture. It is therefore absolutely important that the controlling activity carried out by operators be supported by a top management critical revision in which problems are evaluated with a detachment not always possible in people involved daily in these problems. The decentralized control activity must, in other words, be accompanied by a centralized control by exception concerning only those events whose weight and whose unusual persistency bring them under the jurisdiction of a superior operational level.

The information logic I am proposing is therefore of the cyclic type, since single items come up from the bottom to the top of the firm and then return to the bottom. This logic is typical of the planning process. Even the periodicity of control must reflect this methodological approach, and all scheduled controls must be supported by occasional and non-scheduled controlling activities. This will allow to see the situation from points of view differing from the usual ones that can be negatively influenced by routine.

INFORMATION

Managing a firm means being informed every day of what is happening and of what is being done. The process of handling information is of the cyclic bottom-up/top-down type.

The Management Information System

Information is gathered from the periphery and from the external environment and sent towards the center, where abundant and individually not very significant data is elaborated into important and specific information.

Top decisions are sent towards the operational periphery which must apply them in order to obtain the intended results. The control of these results takes place during the work stage, so that top management is promptly informed of what is happening and corrective moves can be suggested, enabling the firm to better reach its goals (Figure 27).

When the information process takes place by means of a computer (EDP) it is possible to handle in a very short time a great volume of information, even of high quality. But to obtain fast performance it is indispensable to create beforehand a flow of data capable of arriving at the EDP center continuously, regularly, and without delays. It is also necessary to create, afterwards, a selection of information to be transmitted to interested parties. This way, the people engaged in operational tasks will be put in the condition to receive the quantity of necessary information without being forced to undertake tiresome selection work. It is possible to obtain this through rigidly planned office work, so that each person in charge is faced with precise deadlines as regards the transmission of data concerning his functions. In other words, it is necessary to organize a network of procedures for operations having a repetitive character.

Forwarding Information

One of the more important problems is the correct transmission of information, both when data leave and arrive. Even in a small firm where human contacts are much more immediate, it is always hard to eliminate those distortions that are caused by swinging personal moods. Of course, time itself has a corrective effect: as we go along, we learn to recognize the various temperamental tendencies of ourselves and others and this fact works as a corrective filter. But it is quite different when the firm's expansion determines a more complex organizational structure and consequently less frequent human exchanges.

The information system has two different action areas, depending on whether it deals with strategic decisions concerning what to do, or the control of what has been done, i.e., results already obtained. The procedure, although cyclic, can be variously influenced (Figure 28).

Controlling Information

The information data necessary to strategic decision-making are transmitted from the operational base (results obtained) to top management, where information arriving from the external environment are also gathered. There, together with the decisional process there is an automatic control action aiming to correct distortions present in the information. This action, often subconscious, must be rendered fully conscious so as to increase its operational potential. It is thanks to it that top management succeeds in more or less correctly expressing its role and therefore the strategic lines of action. When these are communicated to the operative people, they might arrive marred by distortions very difficult to detect and correct. When the information has been transmitted in hard-to-understand form, the distorting phenomena happens independently of the will of the operators. But there can also exist a distortion which we can define as "forced," caused by the operational base for its own ends.

In such cases the effects of distortion are indeed serious, since they can result in a lack of corrective actions from top management. When this happens control is seldom prompt and needs to verify the initial results, causing excessive delays in possible repairing actions.

These difficulties can be limited, although not entirely eliminated, by trying to emphasize purely quantitative information (i.e., expressed by objective numerical data) instead of qualitative ones (implying behavioral data) which are easier to manipulate.

A typical example of bottom-up distortion is the case of economic estimates of work operations compiled by site management, necessary for the year's profit and loss account. It is always difficult to verify if these estimates, stemming as they do directly from the professionality of each party involved, are optimistic or pessimistic. True, by increasing the number of construction sites, there is on the whole a certain and almost automatic compensation, but in any case the problem continues to exist in single sites. There have been attempts to solve it by routinely asking each operator to issue a double estimate (optimistic and pessimistic), thus forcing him or her to be more rational. It is an approach that has shown some good results.

On the other hand, we can try for something more substantial through the planning process which allows dealing with foreseeable data and which is directly controlled by the operators involved.

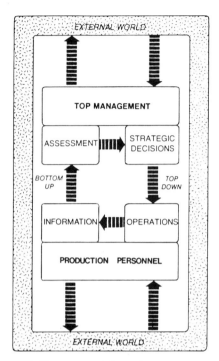

FIGURE 27. Management information system as a cyclic process.

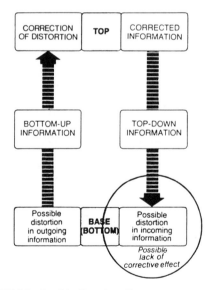

FIGURE 28. Possible distorting effects on management information system.

Linking Financial and Economic Phenomena

The main objective of an efficient information system is to positively link, through planning, fast financial inputs to the slower economic ones, so that when studying effects, it becomes possible to immediately recognize the real causes and to act on them and not only on final events.

The financial area always presents an intrinsic character of immediacy—financial information travels much faster than economic information. So it is necessary to try to have the financial control information input coincide as much as possible with the economic and financial causes that have produced them. This is because it is precisely the economic phenomena that determines the main financial effects; consequently, it is necessary to envision those instruments that allow to immediately reach their slower input. Financial information, because of a whole series of occasional and not very important causes, can distort judgement on the overall management of the firm. The instruments that prove useful in formulating the financial analysis coincide with the financial phenomena and have to do with the planning process. If and when the year's activity is disciplined by an analytical and detailed initial program, i.e., by a budget that analyzes investments and the economic administration, and from these extracts the monetary terms of the problem, it becomes easy to rapidly establish a control on the financial course. It will then be possible to proceed to economic and financial phenomena, thus arriving at a more correct and general appraisal of the operating account (Figure 29).

Standard Forms

Another indispensable support for the information system is a standard set of forms that supplies data in a repetitive manner. This will allow routine analysis of these same data without having to invest too much time in understanding their general design which must be, once and for all, standardized.

Typical of the building construction industry are contracts taking one or more years. This explains why all information linked to the estimates for each contract must necessarily consider time periods longer than one year.

Generally speaking, in order to be useful to the management of a construction firm, an information system must begin by analyzing the tendencies that have surfaced in the preceding years. This means examining contracts both acquired and probable so as to weight the effects of their introduction in the budget. This data is needed in order to formulate a medium-term economic plan based on sufficiently reliable information concerning the corporate and operational levels fixed costs and the economic behavior of acquired contracts. The evaluation will be less reliable for contracts being negotiated and which are therefore in need of more or less accentuated competitive strategies in order to accumulate the volume of work necessary for obtaining an economic balance (Figure 30).

Useful for this type of study are the forms designed for analyzing specific trends pertaining to owner's equity, profit and loss accounts and some significant trend ratios (Figures 31 and 32).

From the program of certain and probable contracts it is possible to formulate, for each single contract, an overall economic project budget (Figure 33) which is then summarized in a medium term profit program (Figure 34).

Control will be carried out in single construction sites and for each project, through analysis of revenues, direct and indirect site costs, provisional sums for finished works not yet certified, or certified and not yet invoiced, or invoiced and not yet paid.

The economic data arriving from single sites will be checked by the administration and listed, for each site, on a form which analyzes budget estimates and the final figures for each item as time progresses (Figure 35).

Afterwards, for management use, a contract profit analysis form (Figure 36) will summarize the most significant information for each contract (total revenues and direct production and site margin, i.e., minus general overheads); the same information can be transferred in a form for contract economic budget and control (Figure 37) which allows immediate and global tracking of the progress of various projects. The totality of events concerning each work will finally be summarized in a consolidated annual budget (Figure 38).

At the same time a financial analysis of each contract will take place, with special attention to receivables and credits due while, in total, revenues and expenditures will have to be indicated, both with respect to construction sites and banks should there have been initial advances (Figure 39). All economic, financial and organizational data will be summarized on forms (Figures 40 and 41) which will indicate on a monthly basis all that is of significance for each work in progress. The financial items pertaining to all contracts will be recapitulated in a final financial budget for the year (Figure 42).

From these documents, which show the firm's general trend and are therefore useful for the definition of the general strategies, it will be necessary to extract data specifically tied to the year's administration and to credit collection.

Finally, a form recapitulating the various figures will make it possible to analyze the total assets and liabilities from the point of view of what is actually happening as compared to what had been predicted (Figure 43).

ABC Analysis

At this point I would like to call attention to an analysis procedure seldom used but which has proven most helpful in pointing out the degree and quality of the firm's de-

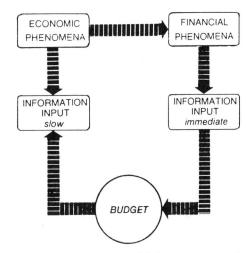

FIGURE 29. The budget as a link between economic and financial phenomena.

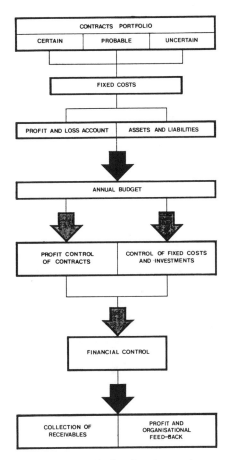

FIGURE 30. Operating budget functions and management.

DIOGUARDI | TREND ANALYSIS | DEPT.: ADMIN. FORM:

	DESCRIPTION / YEAR					
		$	$	$	$	$
ASSETS	CURRENT ASSETS					
	FIXED ASSETS					
	TOTAL ASSETS					
LIABILITIES	CURRENT LIABILITIES					
	LOANS					
	TOTAL LIABILITIES					
	RESERVE FUNDS					
	TOTAL LIABILITIES AND FUNDS					
EQUITY	CAPITAL AND RESERVED					
	NET PROFIT (LOSS)					
	NET EQUITY					
	REVENUE					
	VARIABLE COSTS					
	GROSS MARGIN					
	FIXED COSTS					
	NET INTEREST CHARGES					
	GROSS PROFIT					
	NET PROFIT AFTER TAX					
SOURCES	OPERATING CASH FLOW					
	ASSETS DECREASES					
	DEBT INCREASES					
	NET CAPITAL INCREASE					
	TOTAL					
UTILIZATION	OPERATION CASH FLOW					
	ASSET INCREASES					
	DEBT REDUCTIONS					
	NET CAPITAL REDUCTION					

FIGURE 31. Form for firm's trends analysis.

pendence on the external world. This will allow me to better explain the macrofirm.

The "ABC Analysis" (Pareto's Rule) is a classic approach generally applied to purchases, but I would like to suggest it for all firm's relations with the external environment.

At the corporate level, this type of analysis can be used for purchases pertaining to work in progress, for production equipment investments, and for financial relations. But the ABC analysis is also applicable to relations with buyers of the firm's products and services. The purpose is always to indicate the number of relations with outside parties and consequently the degree of the firm's dependency on the external world and its macrofirm characteristics, highlighting those relations that deserve added attention.

Considering for example the year's normal purchases, this analysis groups the various relations pertaining to these purchases into three homogeneous categories. Then, by associating these relations to their financial amounts, and by arranging them in decreasing order on a diagram carrying in the abscissa the number (percentage) of relations to be carried out and in the ordinates the values pertaining to them (percentage of total amount). The result will be a series of points forming a characteristically decreasing curve (Figure 44). This curve shows that the factors being examined can generally be grouped in three categories: class A has fewer factors, something like 10 to 15 percent of the entire volume, but it covers the greater part of the total yearly value for this type of relations (about 70 percent). It is immediately obvious that the analysis and control of these purchases should be of top interest for the firm.

DIOGUARDI	TREND RATIO ANALYSIS	☐ ESTIMATE ☐ ACTUAL	Dept. : Admin. Form :

	ITEMS	PERIOD %	PERIOD %	PERIOD %
COMPANY ASSETS	EQUITY / LIABILITIES			
	CURRENT ASSETS / TOTAL ASSETS			
	CURRENT LIABILITIES / TOTAL LIABILITIES			
	EQUITY / NET FIXED ASSETS			
	DEPRECIATION / FIXED ASSETS			
	TOTAL LIABILITIES / INVESTED CAPITAL			
FINANCIAL	CURRENT ASSETS / CURRENT LIABILITIES			
	CURRENT ASSETS / REVENUES			
	CURR.ASSETS-CURR.LIABILITIES / REVENUES			
	INVESTED CAPITAL / REVENUES			
	OPERATING CASH FLOW / REVENUES			
	TRADE RECEIVABLES / REVENUES			
EFFICIENCY / PROFITABILITY	INCOME BEFORE TAXES / REVENUES			
	INCOME AFTER TAXES / REVENUES			
	INCOME AFTER TAXES / EQUITY			
	INCOME BEFORE INTEREST EXPENSE / INVESTED CAPITAL			
	INCOME AFTER TAXES BEFORE INTEREST EXPENSE / REVENUES			
	REVENUES / INVESTED CAPITAL			

FIGURE 32. Form for trend analysis ratios.

DIOGUARDI	PROJECT BUDGET	CONTRACT	YEAR	DEPT ADMIN FORM

	ITEMS	TOTAL CONTRACT	1ST JEAR	2ND YEAR	3RD YEAR	4TH YEAR	5TH YEAR	6TH YEAR
1	REVENUES FROM WORKS CERTS.OR INVOICES							
2	CHANGE ORDERS							
3	TOTAL REVENUES (1+2)							
4	WORKS IN PROGRESS-FINAL							
5	WORKS IN PROGRESS-INITIAL							
6	WORKS IN PROGRESS BALANCE (4-5)							
7	TOTAL REVENUES (3+6)							
8	MATERIALS							
9	LABOR							
10	SUBCONTRACTS							
11	FREIGHT AND TRANSPORT							
12	SUNDRY SITE DIRECT COSTS							
13	TOTAL DIRECT COSTS							
14	SITE MARGIN (7-13)							
15	SITE OVERHEADS							
16	SITE CONTRIB. MARGIN (14-15)							
17	TECHNICAL ADJUSTMENTS							
18	ADJUSTED MARGIN							

FIGURE 33. Model for economic overall budget.

DIOGUARDI | MEDIUM-TERM PROFIT PROJECTION | YEAR | DEPT.: ADMIN. FORM:

	ITEMS		19		19		19		19		19		19	
			$	%	$	%	$	%	$	%	$	%	$	%
1	REVENUES FROM WORKS CERTIFICATES OR INVOICES													
2	CHANGE ORDERS													
3	OTHER REVENUES													
4	TOTAL REVENUES (1+2+3)													
5	WORK IN PROGRESS - FINAL													
6	WORK IN PROGRESS - INITIAL													
7	WORK IN PROGRESS-BALANCE (5-6)													
8	ACCRUED REVENUES (4+7)													
9	MATERIALS													
10	LABOR													
11	SUBCONTRACTS													
12	FREIGHT AND TRANSPORTATION													
13	MISC. DIRECT SITE COSTS													
14	TOTAL DIRECT COSTS													
15	SITE MARGIN (8-14)													
16	TECHNICAL STAFF'S SALARIES													
17	DEPRECIATION & LEASING													
18	MISCELLANEOUS													
19	SITE OVERHEADS													
20	SITE CONTRIB. MARGIN (15-19)													
21	SALARIES AND FRINGE BENEFITS													
22	FEES TO 3RD PARTIES													
23	TRAVEL & ENTERTAINMENT													
24	GENERAL EXPENSES													
25	MISCELLANEOUS													
26	FIXED COSTS													
27	MISC. REVENUES & CONTING. ASSETS													
28	GROSS OPERATIVE PROFIT (20-26+27)													
29	INTEREST RECEIVED	BANKS												
		OTHERS												
30	INTEREST EXPENSES	BANKS												
		OTHERS												
31	TOTAL INTERESTS (29+30)													
32	NET PROFIT BEFORE TAX													
33	INCOME TAX													
34	NET PROFIT AFTER TAX (32-34)													

FIGURE 34. Forms for medium term profit program.

DIOGUARDI | CONTRACT PROFIT ANALYSIS | CONTRACT: | YEAR | DEPT.: ADMIN. FORM:

	ITEMS	CONTRACT TOTAL		CUMULATIVE TO....		CUMULATIVE TO....		Jan. - Feb.				Jan. - Apr.				Jan. - June				Jan. - Aug.				Jan. - Oct.				YEAR'S TOTAL			
								BUDGET		ACTUAL		BUDGET		ACTUAL		BUDGET		ACTUAL		BUDGET		ACTUAL		BUDGET		ACTUAL		BUDGET		ACTUAL	
		$	%	$	%	$	%	$	%	$	%	$	%	$	%	$	%	$	%	$	%	$	%	$	%	$	%	$	%	$	%
1	REVENUES FROM WORK CERTS. OR INVOICES																														
2	CHANGE ORDERS																														
3	TOTAL REVENUES (1+2)																														
4	WORK IN PROGRESS FINAL																														
5	WORK IN PROGRESS INITIAL																														
6	WORK IN PROGRESS BALANCE (4-5)																														
7	TOTAL REVENUES (3+6)																														
8	MATERIALS																														
9	LABOR																														
10	SUBCONTRACTORS																														
11	FREIGHT AND TRANSPORT																														
12	MISCELLANEOUS DIRECT SITE COSTS																														
13	TOTAL DIRECT COSTS																														
14	SITE MARGIN (7-13)																														
15	SITE OVERHEADS																														
16	SITE CONTRIB. MARGIN (14-15)																														
17	ADJUSTMENTS																														
18	ADJUSTED MARGIN (16-17)																														

FIGURE 35. Form for contract profit analysis.

FIGURE 36. Form for contracts profit analysis summary.

FIGURE 37. Summary for contract budget and control.

DIOGUARDI

BUDGETARY CONTROL

YEAR

DEPT. ADMIN. FORM

	ITEMS		TOTAL AS AT 31/12/	JANUARY FEBRUARY		JANUARY APRIL		JANUARY JUNE		JANUARY AUGUST		JANUARY OCTOBER		TOTAL FOR YEAR			
				BUDGET	ACTUAL	BUDGET	ACTUAL	BUDGET	ACTUAL	BUDGET	ACTUAL	BUDGET	ACTUAL	BUDGET	ACTUAL		
1	REVENUES FROM WORKS																
2	CHANGE ORDERS																
3	TOTAL REVENUES																
4	WORK IN PROGRESS FINAL																
5	WORK IN PROGRESS INITIAL																
6	WORK IN PROGRESS BALANCE (5-4)																
7	OVERALL REVENUES (3-6)																
8	MATERIALS																
9	LABOR																
10	SUBCONTRACTS																
11	FREIGHT AND TRANSPORT																
12	MISC. DIRECT SITE COSTS																
13	3RD PARTIES' FEES																
14	TOTAL DIRECT COSTS																
15	SITE MARGIN																
16	TECHNICIANS' SALARIES																
17	DEPRECIATION & LEASING																
18	MISC. INDIRECT COSTS																
19	CONTRACT'S OVERHEADS																
20	CONTRACT'S CONTRIB. MARGIN																
21	SALARIES																
22	FEES TO THIRD PARTIES																
23	TRAVEL & ENTERTAINMENT																
24	GENERAL EXPENSES																
25	OTHERS																
26	FIXED COSTS																
27	GAIN ON SALES OF ASSETS																
28	GROSS OPERATING PROFIT																
29	INTEREST RECEIVED	BANKS															
		OTHERS															
30	INTEREST EXPENSES	BANKS															
		OTHERS															
31	TOTAL FINANCIAL RECEIPTS & EXP.																
32	NET PROFIT BEFORE TAX																
33	INCOME TAX																
34	NET PROFIT AFTER TAX																

FIGURE 38. Consolidated annual budget form.

DIOGUARDI

CONTRACT FINANCIAL BUDGET AND CONTROL

YEAR

DEPT. ADMIN. FORM

ITEMS			31/12/ 19	AT 28TH FEBRUARY		AT 30TH APRIL		AT 30TH JUNE		AT 31ST AUGUST		AT 31ST OCTOBER		AT 31ST DECEMBER	
				BUDGET	ACTUAL	BUDGET	ACTUAL	BUDGET	ACTUAL	BUDGET	ACTUAL	BUDGET	ACTUAL	BUDGET	ACTUAL
WORK IN PROGRESS	DUE	CONTRACT ITEMS													
		CHANGE ORDERS													
		NEW PRICES													
		TOTAL DUE (a)													
	REQUESTED	CONTRACT ITEMS													
		CHANGE ORDERS													
		TOTAL REQUESTED (b)													
	TOTAL WORK IN PROGRESS (c)=(a+b)														
INVOICES ISSUED & TO BE COLLECTED (d)															
GUARANTEE RETENTION															
		TOTAL RETENTION (e)													
TOTAL RECEIVABLES (f)=(c+d+e)															
TOTAL PAYMENTS COLLECTED (g)															
INITIAL ADVANCE															
RESIDUAL BALANCE OF ADVANCE															

FIGURE 39. Form for contract financial budget and control.

DIOGUARDI	REPORT OF EXCEPTION	CLIENT:	CONTRACT:	YEAR: MONTH:

INTERACTIONS WITH CLIENT

SITE WORKS TECHNICAL PROGRESS

PERSONNEL UTILIZATION AND PLANNING

– Technicians

– Labor force

SUBCONTRACTS (Macrofirm)

– Schedule of contracts to be entered into

– Interactions with suppliers

PHYSICAL AND FINANCIAL PROGRESS

MISCELLANEOUS

FIGURE 40. Narrative report of exception.

DIOGUARDI	FINANCIAL AND PROFIT ANALYSIS CONTRACT:	YEAR	DEPT.: ADMIN. FORM:

PROFIT ANALYSIS	CONTRACT TOTAL BUDGET		MONTH: BUDGET		MONTH: ACTUAL		YEAR TO DATE: BUDGET		YEAR TO DATE: ACTUAL		CUMULATIVE FROM CONTRACT'S BEGINNING	
1 REVENUES FROM WORKS												
2 CHANGE ORDERS												
3 TOTAL REVENUES												
4 WORK IN PROGRESS - FINAL												
5 WORK IN PROGRESS - INITIAL												
6 WORK IN PROGRESS - BALANCE												
7 OVERALL REVENUES		100		100		100		100		100		100
8 MATERIALS												
9 LABOR												
10 SUBCONTRACT AND INSTALLED MATERIALS												
11 FREIGHT AND TRANSPORT												
12 SUNDRY DIRECT SITE COSTS												
13 FEES TO 3RD PARTIES												
14 TOTAL DIRECT COSTS												
15 SITE MARGIN												
16 TECHNICAL STAFF'S SALARIES												
17 DEPRECIATION & LEASING												
18 INTEREST EXPENSES												
19 MISCELLANEOUS INDIRECT COSTS												
20 SITE OVERHEAD												
21 CONTRACT CONTRIB. MARGIN												
22 ADJUSTMENTS												
23 ADJUSTED MARGIN												

FINANCIAL SITUATION AS AT ______

WORK IN PROGRESS	DUE	CONTRACT ITEMS	
		PRICE UPDATES	
		TOTAL DUE (A)	
	REQUESTED	WORKS	
		PRICE UPDATES	
		TOTAL REQUESTED (B)	
		TOTAL WORKS IN PROGRESS (C)=(A+B)	
		INVOICED AMOUNTS TO BE COLLECTED (D)	
GUARANTEE RETENTION		WORK IN PROGRESS	
		PRICE UPDATES	
		TOTAL RETENTION (E)	
		TOTAL RECEIVABLES (F)=C+D+E	
		TOTAL PAYMENTS COLLECTED (G)	
		TOTAL REVENUES PAYMENTS (H)=F+G=7	
		INITIAL ADVANCE	
		RESIDUAL BALANCE OF ADVANCE	

INFORMATION ON CONTRACT

DESIGNER

SUPERVISING ENGINEER

TEST AND INSPECTION ENGINEER

PERSONNEL	NAME	TOTAL COST	FOR THIS CONTRACT %TIME	FOR THIS CONTRACT COST
PROJECT MGR.				
SUPERINTENDENT				
WORKS ACCOUNT				
ADMINISTRATIVE CLERK				
ASSISTANT				
TOTAL COST OF PERSONNEL		TO		

CRAFTSMEN (DIRECT & INDIRECT)

SKILLED No.	SEMI-SKILLED No.	LABORERS No.	TOTAL No.

COMMENTS

FIGURE 41. Form for financial and profit analysis.

DIOGUARDI	FINANCIAL BUDGET	YEAR	DEPT.: ADMIN FORM:

DESCRIPTION	JANUARY	FEBRUARY	MARCH	APRIL	MAY	JUNE	JULY	AUGUST	SEPTEMBER	OCTOBER	NOVEMBER	DECEMBER	TOTAL	
GROSS INVOICE RECEIPTS														
BANK LOAN REPAYMENTS														
NET INVOICE RECEIPTS														
ADVANCES FROM CLIENTS														
INTEREST RECEIVED														
BANK ADVANCES														
LOANS														
TOTAL RECEIPTS														
OPERATING EXPENDITURES														
TAXES														
LOANS PROMISSORY NOTES														
INTERESTS PAYED														
TOTAL EXPENDITURES														
BALANCE FOR PERIOD														
BANKS C/A BALANCE														
BANKS ADVANCES C/A BALANCE														
PROMISSORY NOTES														
NET BANK BALANCE														
TOTAL DEBTS WITH BANKS														
CREDIT FACILITIES														
CLIENTS' ADVANCES BALANCE														

FIGURE 42. Form for financial budget.

DIOGUARDI — ASSETS AND LIABILITIES BUDGET AND CONTROL — YEAR — DEPT. ADMIN. FORM:

ITEMS	SITUATION AS AT 31/12	SITUATION AS AT 28/2/		SITUATION AS AT 30/4/		SITUATION AS AT 30/6/		SITUATION AS AT 31/8/		SITUATION AS AT 31/10/		SITUATION AS AT 31/12			
		BUDGET	ACTUAL	BUDGET	ACTUAL	BUDGET	ACTUAL	BUDGET	ACTUAL	BUDGET	ACTUAL	BUDGET	ACTUAL		
CASH															
BANKS: CURRENT ACCOUNTS															
BANKS: DEPOSIT ACCOUNTS															
TRADE RECEIVABLES															
OTHER RECEIVABLES															
SUPPLIERS ADVANCES															
ACCRUED ASSETS															
DEFERRED ASSETS															
WORKS IN PROGRESS															
INVENTORY															
CURRENT ASSETS															
FIXED ASSETS															
SHAREHOLDINGS															
TOTAL ASSETS															
BANKS: CURRENT A/CS															
BANKS: ADVANCES															
BANKS: NOTES ACCEPTED															
ACCOUNTS PAYABLE, TRADE															
OTHER LIABILITIES															
ACCRUED LIABILITIES															
CLIENTS: ADVANCES															
CURRENT LIABILITIES															
LOANS AND OTHER LONG-TERM DEBTS															
RESERVES															
TOTAL LIABILITIES															
COMPANY CAPITAL STOCK															
RESERVES															
NET EARNINGS OR LOSSES FROM PREVIOUS YEAR															
NET EARNINGS OR LOSSES OF CURRENT YEAR															
TOTAL EQUITY															
TOTAL LIABILITIES AND EQUITY															

FIGURE 43. Form for assets and liabilities budget and control.

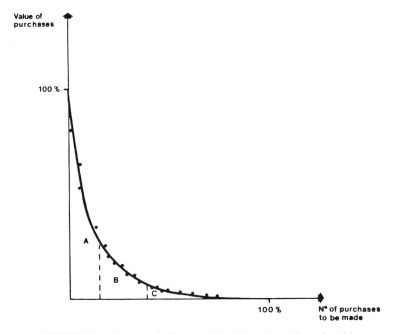

FIGURE 44. Curve A, B, C for analysis of purchases (Pareto's rule).

Class B is represented by a high number of factors, about 20 to 30 percent, which accounts for a significant part of the total yearly purchases (about 25 percent). Also in these cases the purchasing department should implement a rationalizing and controlling action.

Class C is represented by a high number of factors (about 60 to 70 percent) which however account for a minimum part of the total value of the year's purchases (only about 5 to 10 percent). Obviously, extensive action in this class would prove costly, due to the high number of relations, and without significant effects or results.

By analyzing and carefully disciplining class A purchases and by adopting simplified procedures for class B and C, it becomes possible to obtain an adequate control on about 70 percent of operating costs. This implies a thorough intervention on only 10 to 15 percent of total expense items.

This approach changes structurally when dealing with other type of relations, for example with banks or clients. But as far as method is concerned, this type of analysis continues to be significant because it allows identification of the areas deserving of more attention and, consequently, the acquiring of firmer and more complete information.

The control of the various situations described, and the continuous updating of the operational forms I have illustrated, represent for the firm the so-called "Management Information System."

The use of a central computer with terminals in the various construction sites allows automatic updating of all the information. It suffices to feed in all basic data gathered in each site. The market offers a number of programs capable of developing this type of analysis. Each firm has its own individual life, and differs always from all similar organizations. This is why the programs sold on the market must be modified and personalized so as to faithfully represent the characteristics, the style of management, in other words the "corporate culture" of each firm.

TRAINING

When compared with other industries, building construction lacks organized training programs for its top managers, its intermediate groups, and, especially, for its labor force.

It is a situation that gives rise to a whole series of problems. In particular, management training in the construction sector can be a good opportunity for stimulating the implementation of productivity, cost reduction, and innovation.

Training For Stimulating Innovation

Construction firms tend to hold on to their best people, even when production is slow. This lowers personnel availability on the market, and finding really experienced and qualified managers is anything but easy. A training process of sorts is indeed present in practically every firm, but it is generally informal, causing time and energy expenditures not supported by sufficient motivating stimuli both in those that "teach" the trade and in those that should learn it. If in the case of medium- to high-level groups the problem is serious, it becomes dramatic where it concerns the labor force. It is among these people that inexperience and a lack of specific professional culture often become the main causes of quite frequent and serious accidents.

This is why in building construction firms it is necessary to promote an attitude favorable to training at all levels. A training process must be developed and planned within each single firm because with few exceptions, there is a lack of consultants actually able to evolve externally really effective training courses. One of the reasons, of course, is that this type of training has to be above all practical, based on the experience of those that operate daily in the field.

The training of groups at all levels, seen as an entrepreneurial function to be developed on a permanent basis within the firm, can be viewed as an innovation process inside the traditional structure of the building firm; people so trained will represent the main factor for future research and development, and not—as it frequently happens—a hindrance for introducing innovation in the firm. Therefore, in this sense, training itself can be considered a chapter of research and development not yet thoroughly defined. Training is a research and innovation process to be applied not generally, but individually in each firm, with the assistance and supervision of top management or of the ownership when it coincides with top management.

Permanent Training

Human resources represent without any doubt the main chance factor within organizational systems. On the other hand, it is interesting to note how precisely this factor often becomes one of the major obstacles for innovation.

Resistance to change can be noticed at almost all levels in the organization, since it naturally tends—through the work specialization process—to repetition of learned processes which generates habit. And habit prefers to avoid future unknowns, to favor not innovation but present and hard-fought-for certainty. Overcoming this situation should stay high in the organization's agenda of challenges, both formally and informally. The right solution, in any case, always depends on higher professionals, obtained thanks to a well-planned professional training program.

Human resources must be trained to consider possible changes, and also be made aware of problems connected to changes thanks to a training program that has to be planned within the firm so as to be permanent. First of all, this program should involve intermediate groups, and through them it should be extended to operational people.

Permanent training means organizing the entire program so as to automatically identify the necessary instructors for the various levels, both outside and, more important still, inside the firm. It is a process that must be carried out on the basis of careful selections favoring career planning. Thus, the permanent training process develops alongside the routine operational task duties, which is like saying that training must not involve an employee or a worker only when they are just hired or when new duties are bestowed on them. Training must be a process which prepares each operator *now* for the future and for the duties he or she will encounter along his or her career as it has been planned. This way, in training human resources we will also have a core of generalized administration of the future, allowing the firm to positively anticipate future needs.

The firm's requirements for human resources can arise first of all from natural causes tied to the retirement of existing personnel, from age limits (a phenomenon that in normal conditions can usually be predicted), and by administrative reasons (dismissals, resignations, etc.). When production and investments are steady, the number of people to be hired should be equal to that of people leaving the firm, if and when the lack of experience in new arrivals can be considered as balanced by an increase in proficiency, and therefore in efficiency, in the firm's stable personnel. These phenomena can keep the organization's metabolism at the minimum level necessary for the system's survival. But it is difficult to manage them, especially in the building sector where the unknowns connected with the arrival of new personnel usually increase the degree of inefficiency more or less existing in any organizational system.

It is important to understand this concept, and to know that increased professionalism in company groups improves the organization's efficiency and therefore also production and profit. It also allows a better delegating process, fostering a positive future company continuity.

Instructors

In general, the permanent training model must involve the best performing people in the different sectors in a program addressed to new personnel or to the younger workers. The usually informal process must instead be planned, so that both the trainer and the trainee become aware of their roles and of the goals that have been set. This also means finding various ways for motivating instructors and trainees alike if and when the objectives of the training programs are reached. The purpose is to obtain a field training process where the trainee learns directly from older and more experienced people according to a philosophy not very different from the one applied on the Quality Control Cycles (QCC).

The characterizing concept of the QCC is that the person normally doing a certain task makes the best teacher because he or she knows more than anyone else about that particular work. This makes it possible to arrive at a sort of "job enrichment" process by grouping crews of workers who, together with their foremen, pass on to each other experiences and together search for the best ways of improving the efficiency and quality of their performances. This concept is especially applicable in site works where proficient workers can operate with so much more efficiency, solving problems before they materialize and become hard to solve, especially when safety is concerned.

It is particularly important that personnel functioning as instructors be made aware of the leadership role they have when carrying out their training duties.

The Role of the Macrofirm

Programs for practical training will have to be implemented through a technical explanation of what has been learned and through on-the-spot controls, especially for what concerns labor. For coping with the first of these problems it might prove useful also to call upon outside instructors, but with the purpose of methodologically organizing each subject matter, pin-pointing what has been learned on the field. Controlling actions, especially those concerning safety, must evolve in such a way as not to seem repressive, the purpose being to teach through controls the most recurrent problems.

The organizational system proper to the building construction firm which expresses itself through a series of relations typical of the macrofirm can easily activate a training process to be developed both inside the firm and among the principal companies that cooperate with the general contractor. This system will act as the coordinating and responsible party for the efficiency of work performances.

It is precisely in this peripheral context that we find the informal training typical to the artisan tasks. It is also possible to organize in each firm a formal training process whose implementation should be disciplined by specific rules. In such a case, the macrofirm could propose itself for coordinating and controlling the training program and, in some cases, it should at least partially supply technological and financial aids. The investment necessary for this program could be charged in part to the client, especially if it is a public body interested in this type of professional development and safety improvements, and in part to the subcontractors, who can find in these learning tools a way of expanding.

Safety and Subcontractors

Training and the resulting improved professionalism of workers should result in a better understanding of safety problems. Each worker should be stimulated to evolve in his or her own professionalism the capacity to independently assess the risks present in different work operations. Consequently, he or she should be able to convince himself or herself to adopt rigid safety measures, and to ask that everything be done so as to minimize risks for himself or herself and others. When adequate safety measures fail to be introduced, the general contractor should also be held responsible, not only the subcontractors. This, in fact, should be formalized in written agreements and can justify the training process as described above.

Such measures are absolutely necessary to insure real safety on work sites. They should start at the top and then spread toward the base, where working conditions are directly experienced. Needless to say, all decisions in this sense must be constantly supported by the top.

Stability, professional training, social awareness, improved work conditions, and salary increases tied to production performance should put the building construction industry in a position to attract those few human resources which are available and which would react more positively to better staying conditions offered by the single firm and therefore by the entire sector.

A must for obtaining these goals is discarding the concept that building firms can directly hire all the labor force needed to execute the entire series of tasks connected to each specific activity.

In other words, it is a mistake to think it is possible to gather "good-for-everything" people. Experience shows that these same people may be lacking specialization and professionalism.

Some Operational Suggestions

Training programs could, operationally speaking, be organized around four basic factors:

1. A theoretical text introducing the general picture of the specific area being studied. The text could be prepared by internal instructors (at least one) with the possible aid of external consultants.
2. A practical experience period directly in the field, under the responsibility of experienced mentors pre-selected from within the firm.
3. A diversified experience period, involving work operations to be executed in differing but similar situations, under the responsibility of the person managing the sector involved.
4. A series of technical seminars helpful to organize methodologically what has been assimilated in the field. These seminars will be conducted by internal or external instructors. Here all trainees' criticisms concerning the firm's existing structures will be recorded. These comments will be summarized in special forms to be forwarded to top management. The organization service together with the personnel department will prepare reports capable of innovating existing procedures and structures on the basis of comments recognized as valid.

These four factors of the training program can evolve along a number of basic routes. First, a *general area* will have the purpose of training people to be in charge of general secretarial work, of organizational procedure services, and general affairs. It will also deal with personnel problems and eventually with management of peripheral offices.

A *technical area* will train people especially in charge of the production sector and of areas tied to it (site purchasing office, storage and maintenance service, project managers and site superintendents, site assistants and foremen, account engineers, site administrative accountants, people in charge of safety).

The *administrative area* training will concern accounting offices, budget, control, information and problems tied to financial services.

For each one of these areas here are some suggestions for specific programs.

Some subjects common to all these areas will concern organization and organizational structures, the analysis of existing procedures and of the information system including its organizational supports (standard forms).

In the general area, problems to be discussed concern organizational procedures, filing, procurement, and personnel management.

The technical area will further investigate problems concerning its own standard forms, site supplies, the technical and economic analysis of bills with consequent site planning (PERT, CPM).

Also to be thoroughly examined are the general legal aspects concerning the contracts, specifications, price lists, accounting books, accounting, tests and reserves.

Other matters include bills of quantities, technical costs analysis, suppliers and subcontractors analysis and consequently direct and general costs analysis. Also to be discussed are problems concerning industrial relations and those with labor unions, with site management, with clients.

New technologies should also be considered, together with management of equipment and its maintenance, and organization and layout of sites.

Special attention should be given to safety problems and regulations in the construction sites, by better defining prevention and control and by promoting safety engineering as far as specific investments are concerned.

The question should also be examined from a theoretical point of view, referring to new studies that are being conducted in various universities.

For example, R. E. Levitt, of Stanford University Department of Civil Engineering, in a foreword to James Edward Koch's "Liability and the Injured Worker in the Construction Process' states:

> For the last 15 years our department has been involved in a series of studies of construction safety which indicate the strategies and procedures used by successful foremen, project managers, and company top managers in achieving reduced accidents on their construction projects. Our most recent report suggested that construction owners could also

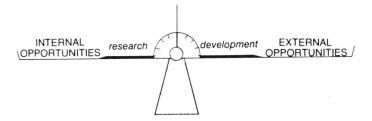

FIGURE 45. Opportunities to stimulate research and development in the firm.

have a major impact on construction safety through selection of safer contractors and through various techniques for orienting and monitoring the contractors working of their projects. One of the issues which arose in suggesting that owners could achieve reduced accidents by becoming more involved was the question of liability. . . .

These studies should be passed on to all operators so that, through the training process, they can acquire a better understanding of one of the main problems of organization, particularly building construction firms, both in regard to work sites and to advanced technologies that can promote automation especially in office work.

INNOVATION

In general, the traditional construction firm has not been known to promote important research and development processes in its organizational structures. The innovations in the production cycle of this industry are few and not very significant.

It is a function that must be strengthened.

Research and Development

Innovations can interest products and processes, thanks to opportunities arising in the firm or to stimuli arriving from the external environment (Figure 45). The research and development function must be able to recognize these opportunities and to transform them into a cyclic process generating continuous improvements (Figure 46). As a sector, building construction usually innovates processes more than products.

Therefore, innovation can be developed in the organizational processes of production, but they can also concern the life-cycle of finished products. All progresses which tend to modify the single product, viewed as an aggregate of various components, must be studied and applied in the designing stage.

Research and development can be stimulated by pre-existing situations and by a specific structure which every firm should organize for its own strategic needs. Each new contract represents a test bench. The client's needs, the variability of work sites, and the time necessary to complete each job are always new elements on which to apply a research

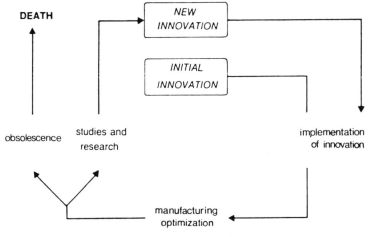

FIGURE 46. Cyclical process of innovation.

for specific solutions that afterwards can be widely implemented. Thus, they become for the firm a reservoir of technological experience.

The organizational system that I call a macrofirm represents a natural enough mechanism for fostering, in the smaller firms, innovations that are then passed on to general contractors and to the entire sector. This happens because specialized firms work for a large number of general contractors. They have to do field research to satisfy specific needs that each general firm expresses in its orders, for implementing the quality necessary for its own prestige and for satisfying the standards set by the final client. This way, every specialized firm has to continuously adjust its technology and by so doing favors continuous improvements. The results of this research for improving production on the basis of the general contractor's specific needs should be automatically transferred to their clients—the general contractors—and therefore to the market, thus spreading innovation. But it is important that general contractors learn to profit from this process by absorbing and rationalizing all innovative suggestions and indications as they arise from single sites. It means taking these new suggestions into account when working on new projects and specifications, thus stimulating the specialized firms. They should keep innovation active in their special work. The training process can prove fundamental, since existing human resources always have the primary ideas for research. By stimulating research through training it becomes possible to confer new impetus to innovation.

Logically, it is within the general contractor's organization that the function of research and development should be coordinated, in order to profit as much as possible from all suggestions arriving from work sites or from the various practical situations that management encounters daily. It is here that conditions should be set so as to execute in the best possible ways these activities. But not only these. Any process of research and therefore of innovation has to be connected to a control on the finished product's quality. This control must be continuously implemented either directly or through the specialized subcontractors, and must take place during work operations. But it can also continue after the product has been handed over to the client through a process of planned preventive maintenance.

The Question of Maintenance

Maintenance in building construction means the totality of operations necessary for insuring the continuous efficiency of the specific product and of all its components.

Maintenance, then, means a continuity of intervention based more on prevention work than on restoration or rebuilding.

Contingent or extraordinary maintenance executed through programs of emergency or periodical interventions (including, by extension, the recovery of pre-existing building and infrastructures) is accompanied by preventive or ordinary maintenance on new buildings, which can be systematically programmed.

Controls for establishing the need of maintenance work should be disciplined by precise regulations and can be requested by those using the construction product.

At this point it becomes necessary to differentiate between building structures continuously used (apartment houses, office buildings) and those irregularly used (sports and cultural centers, etc.). For the first group it is necessary to establish an information structure capable of immediately recording the need for intervention on wear reported by dwellers, thus implementing an educational action for the best use of the product.

Maintenance in building construction becomes, therefore, socially and not only technologically significant.

When building structures are irregularly used, the firm responsible for maintenance must strengthen its program of direct controls which can be described as follows:

a. permanent, i.e., continuing in time
b. periodical, i.e., with rigid deadlines
c. contingent, i.e., expressly requested by possible users
d. occasional, i.e., programmed but without rigid deadlines

The preventive maintenance process can also add interest to the offer made to the client, and in this sense it can prove useful for the acquisition of the contract itself. There are other important functions tied to the maintenance process.

Product Assistance

Especially for apartment buildings, it is useful to undertake also a social function, so as to check that the building project and its execution correspond to the real needs as expressed by actual dwellers. The "Evaluation Survey" or P.O.E. (Post Occupancy Evaluation) are methods which allow adjustment in the field of a building project on the basis of the dwellers' usage and behavior.

In this sense, analytical models have been studied by a number of associations, for example by the EDRA (Environmental Design Research Association), that in the last 15 years during its congresses has organized special sessions dedicated to P.O.E.

An increasing number of scholars have investigated this problem in the U.S. and in other countries, for example, in Italy where Giandomenico Amendola, professor at the University of Bari, has made interesting theoretical and practical studies within the urban laboratory context.

And so we have a new concept for firms active in building construction: product assistance after delivery to the client. This means also finding specific instruments that can support new needs, instruments for which the use of computerized systems seem especially advisable. The purpose, of course, is to check that the products can satisfy the needs for which it has been built.

At the same time, it is necessary to teach those who utilize it how to do so the right way.

Homes and the city itself must be viewed as a sociotechnical system in symbiosis with people that live in them, and which are at the same time their "users" and their "managers." To accomplish all this, the building firm must acquire new instruments which can be simply technological but organized on socio-technical basis.

The Urban Laboratory

The concept of "urban laboratory" stems from these realities and should be viewed as an independent area within the firm's organizational structure. Its purpose is to "produce" maintenance and to reconstruct buildings already existing in the urban context. This "urban laboratory" represents an auxiliary unit of the building construction firm, a socio-technical consulting laboratory, which must also provide for research and development of new technologies and production procedures.

The laboratory itself can be divided into a number of specialized sections. A technical assistance section can be set up for maintenance proper, while at the same time it is possible to promote studies for the improved usage of energy in the home. An urban sociology division will work for the improvement of project work through the feedback of information gathered directly in the field. The functions of the laboratory should also include diagnostic advice and experimentation, in addition to strict maintenance work.

To accomplish all this it is necessary to equip the laboratory with a wide and complex range of highly technological instruments, especially for diagnosis (instruments for thermography, for fumes and humidity analysis, for energy leakage control, ultrasonic apparatus). The use of a personal or mini-computer can facilitate handling data for a variety of purposes: statistical processing of gathered data for diagnosis pertaining to inputs, recording of data pertaining to executed works and solutions adopted, listing of technical and legal regulations, and estimating and planning maintenance and reconstruction work. A computer in the "urban laboratory" can improve operational efficiency and optimize its specific character of socio-technical instruments operating in real time inside the complex metabolism typical of any urban system. These different urban laboratory functions give rise to the need to gather and record data which must be processed and distributed to the operational bodies. The economic advantages, the fast pace and the efficiency of these operations can be very much improved by the use of a computer when viewed as an instrument of an efficient and well distributed information system.

The urban laboratory, therefore, must become an intelligent terminal of the firm's operational structure, so as to reach goals such as the best management of equipment, organize maintenance work, social assistance to dwellers, and to owners for maintenance and requalification of buildings. It can also function as an entity which controls directly the field project results and gathers all suggestions that can contribute in improving both the existing situation and future similar projects.

Within the firm all these undertakings must favor applied research with the aid of instruments that can formalize it. This means that maintenance activity concerning finished products can be developed through the organization of a real firm for city maintenance.

Maintenance for the City

The "city maintenance firm" is an organization that can insure continuous maintenance processes to any diversified private and public development (Dioguardi, 1984). All cities can be viewed as machines for communicating and exchanging information. Each city has its own history but also its daily life, its own metabolism through which it experiences the phenomenon of aging (De Rosnay, 1975).

Its components (apartment houses, public buildings, infrastructures) experience wear and obsolescence due to the passage of time and too often incorrect use.

Here the word "wear" is being used to indicate the loss of performance of single components of the construction system, while "obsolescence" means the loss of functionality.

Therefore, the maintenance process must develop a prevailing preventive action focused on the wear in structures, or, when wear is already noticeable, a recovery action if there are sufficient economic or other motives. This is especially true when a sole owner, public or private (insurance companies, pension funds, banks, etc.), has to provide maintenance on buildings geographically far apart and differing in quality (residential housing projects, public buildings, sports and cultural centers, infrastructures, etc.).

In these cases, there are problems in planning maintenance work. This means that the owner probably will be very interested in dealing with a single organization set up to manage the complete program.

A city maintenance firm must be flexibly structured so as to cope both with planned and emergency work. A polycentric structure should be envisioned; each homogeneous urban area (city, neighborhood, more or less large housing developments) will be presided over by a sufficiently independent operational unit, which will relate to a central nucleus responsible for general instructions. Each nucleus, which has functions similar to those described for the urban laboratory, can use a personal or mini-computer for initial processing of information received from control and therefore for disciplining in real time the necessary emergency actions.

The various poles can be electronically connected with each other and with a central computer which will process basic data and all programs transmitted to it.

For planned and emergency interventions the maintenance firm can look for and obtain the cooperation of smaller specialized external firms, especially of the "artisan" type. Relations with these should be disciplined by annual contracts. The coordinating and connecting activities with the operational firms can take place directly through the central pole or by delegating the peripheral poles while the central pole implements controls. The choice of one of these solutions usually depends on territorial extension, which will also suggest whether the central pole and each peripheral one should have deposits for spare parts and standard material to be used for urgent operations, and also various instruments for diagnostic operations.

This organizational structure can very well become another operational function of the macrofirm and have its operational connections with the multipolar firm and with the corporate level. Thanks to the central computer it will be possible to check field works executed by each specialized firm.

An operation sequence for a maintenance firm can be expressed as follows:

1. *Intervention program* pre-established on the basis of the project or derived from controls and the consequent diagnosis elaborated by central or peripheral computers.
2. *Educational contacts with users*, based on a "maintenance booklet," taking into account the components used during construction.
3. *Control program* and also a plan for informative contacts with the users, concerning the utilization and deterioration of the single components of the residential system.
4. *Standardization of results* and of controls, with retroactive updating of computer programs.
5. *Maintenance actions* planned in advance, and recording of same in the programs so as to update them continuously.
6. *Gathering of information* for emergency works and their scheduling (immediate or with pre-established deadlines).
7. *Execution of emergency works* and recording of same in existing programs with subsequent updating.

On the basis of pre-established programs it is necessary to identify the specialized firms which can execute these works, to be sure of the nature, time, and cost of each intervention. It will also be necessary to organize a spare parts warehouse allowing quick interventions in case of emergency.

Furthermore, a crew of specialized workers can be assigned to each peripheral pole, in order to execute both planned and extraordinary operations, and the principal controls.

From all of this, it emerges quite clearly that the urban laboratory, and more generally speaking the city maintenance firm, have their "brain" in the computer which interconnects peripheral terminals that have been rendered "intelligent" by specially designed software programs.

The building construction firm that dedicates itself to maintenance and therefore to the conservation and development of the city can, through the expert use of the computer, create new frontiers for social actions which in the future could find interesting and widespread application in underdeveloped areas, by combining building construction, industrial development, and social growth.

Some Current Examples

In the United States and in Europe we find some interesting cases, for example the American experience of the "City Venture Corporation," in which public institutions, religious organizations, and private firms (among them one of the major computer industries) cooperated toward "the urban revitalization and the development of new communities through the application of diverse resources available." As the pamphlet presenting this undertaking explains: ". . . City Venture works with residents, neighborhood organizations, local government institutions, private contractors and suppliers to establish businesses in depressed inner-city areas, create jobs, employ disadvantaged residents of the immediate area, and develop a stable work force with a range of job options. City Venture assembles the management skills, focuses public and private funds, and employs the full range of physical and social science necessary to implement holistic solutions to urban problems."

Another example in the making refers to an Italian company—Bonifica S.p.A. of the IRI group—and possible future for South Bronx in New York. The hypothesis is to upgrade a shabby and run-down neighborhood with a revitalization project which envisions the creation of a whole series of plans for Italian firms active in the U.S. market. It would mean increasing opportunities for certain industries, certainly, but also this local area's social and economic situation would improve.

These are but two examples of the role of the building construction firm can expand and acquire new meanings, to the point where job enrichment becomes not only applicable to workers' activities, but to the entrepreneurial undertaking as a whole. This new role involves particularly the organizational level I have defined as *corporate* and which is capable of coordinating not only the specific activities of the multipolar firm, but any other action having to do with the innovative potentialities of the art of entrepreneurship. This can get extraordinary help from computerized automated processes.

AUTOMATION

Automation and the computer are more and more conditioning the operational management of firms, and, in particular, their information systems and automatic data processing.

The use of these instruments, however, must not become abusive.

Computers and Organization

The ease with which data are multiplied through computer use can produce information which is exuberant and not only useless but also damaging because in the end they hide news essential for the rational management of the firm. There is another danger that must be pointed out. Computers, even when expressing in the best possible way their characteristics of "artificial intelligence," are and remain machines (hardware). To function, these machines must be equipped with the right programs (software) created to satisfy the specific needs of single firms. But this is not all. It is also necessary that computers, even when fed sophisticated programs, be placed at the center of an efficient organizational procedures system able to supply on schedule the input data for processing and the processed output data to be distributed to interested operators.

In general, we have to accept as absolutely true the words of Federico Butera, when he writes: "The first organizational elements affected by automation are human tasks which can be considered as the building blocks of organization. In this subsection we assess the effects of elimination of tasks, or of certain components, the creation of new tasks and the transformation of tasks, as well as tasks of designing technology and organization, and the structure of tasks relationships" (1984, p. 70).

Without an organizational network tuned to these tasks, computers remain beautiful toys which can entertain and even astonish, but not improve efficiency in managing a firm. Their enormous quick processing potentialities will not even be tapped without the support of an organization tuned to computers' extremely fast-working rhythms. In this sense I could propose a general rule: before introducing a computer system in the operational structures of the firm, be sure to test the necessary organizational procedures to control its income and outputs. A thorough training action in the use of computers must be implemented, as is well stated by Friel and Trella in a foreword for an analysis pertaining to the construction sector: ". . . the importance of education and training as key factors common to [all] companies leading to a successful utilization of computer technology. Education can be seen allowing a large number of people in each organization to efficiently use microcomputers" (1984, p. 19).

The Importance of Computer Language

It could be useful to begin with a computer service external to the firm. This can allow the general organization to be prepared before acquiring a computer (hardware) for direct use.

The organization must become a cyclic process for the continuous reformulation of procedures and programs (software), disciplined by the computer on the basis of experiences gathered along the way, of known mistakes, of new requirements, thus establishing a process of continual optimization which must characterize the whole sociotechnical system represented by the organizational structure and by the computer.

Computer usage creates the necessity of a specific language for software. All organizational procedures for sending data to the computer or for receiving information from it must be designed starting from such a language.

These data and information are often generated by or distributed to people or groups outside the firm. Irwig and Levitt, on this topic, state: "This case study . . . has demonstrated the strong interrelationships between the environment of the firm, its organizational strategy and structure, and the way in which it utilizes computer technology" (1984, p. 56). This problem, therefore, becomes especially true for data destined for firms which are interacting with the multipolar firm or general contractor. The computer represents a vital element for the management of the macrofirm and for interrelations that it determines. That is why it is essential to find a common language by which the general contractor or multipolar firm relates with external firms in regards to the various operations in terms of schedules, punctuality, quantity and quality, reciprocal costs and profits.

The same problem arises with clients, but in more rigid terms since in this case it is the firm that has to adjust its language to the client's requirements, although the high number of relations facilitates the task as compared to normal relations with subcontractors and suppliers. With these parties it becomes necessary to establish a real training process enabling them to unify procedures, thus creating a common language. To aid these functions it is possible to supply specific data processing services and programs prepared especially for external firms which interact with the macrofirm. These are typical tasks of the coordinating and controlling functions of general contractors. These tasks are destined to proliferate in the future, since through these functions a macrofirm can better assert itself thanks to stable interdependencies and the best possible management of stable or semi-stable interactions.

Custom-made Automation

The industrialization process in building construction requires necessarily that firms learn to manage information through data processing. When considering the building sector we must always keep in mind the great amount of different work operations that characterize it.

Works executed directly on site by subcontractors must be rationalized through a prevailingly organizational process, which must include the identification of sequential methods capable of optimizing the interdependencies existing between the various operations through the use of planning instruments like PERT and CPM.

Friel and Trella have commented that: "It is a matter of

getting people to utilize the technology correctly [and] convince them that it is there to assist them, not to replace them or watch over them" (1984, p. 6). In fact, in order to acquire real efficiency these methods must be supported by a change in attitude and understanding within the firm so as to allow using the most complete and comprehensive information (cost estimates, bill of quantities, standard costs analysis, etc.).

This problem is clearly explained by J. A. Fadem. Towards the end of his work *Automation and Work Design in the United States*, he points out the need that each firm has to find its own way to automation. It is a mistake to try the rigid optimization of system with the intent of obtaining the "one best way." There is no single "cookbook" applicable in all instances. Fadem recalls how Taylor himself insisted that scientific management was ". . . something that varied as it was adapted to particular cases, but always involved a mental revolution of employer and employee toward their work and toward each other . . . [it] fundamentally consists of a certain philosophy which can be applied in many ways . . . recognizing as essential the substitution of exact scientific investigation and knowledge for the old individual judgement or opinion in all matters relating to the work done in the establishment."

The general contractor must become the main promoter of system themselves (hardware and software), and should even consider the possibility of leasing them to the smaller firms with which it cooperates. It is only with a mechanism of this kind (top to bottom) that information systems can also spread successfully in smaller companies. Any action having a purely consultive character would prove unrealistic with the only result that firms would continue to operate according to traditional methods, without trying for a true rationalization of their procedures, regardless of the fact that it is a must for the general evolution of the system's productivity. Irwig and Levitt have explained this concept as follows: "A transition to more decentralized structure, which has been long and arduous for many firms using older, more rigid mainframe technology, was relatively painless for those firms whose crisis of control coincided with the availability of well designed mini-computers hardware and software" (1984, p. 56).

Computers and the Building Construction Industry

Within the context of these general suggestions and of this type of organization, the use of a computer can be applied to specific areas and company functions, with results which prove especially significant in the construction sector, where the computer can greatly aid the decisional process. The introduction of automation in this field is indeed very complex but also very promising. Management decisions concern planning and technical, economic, and financial control as well as project designs and cost assessments. And it is precisely in these areas that, thanks to data processing, it becomes possible to apply the fundamental instruments (PERT/CPM, budget, cost accounting, cost analysis for each organizational unit, etc.) for an efficient management information system.

In the building construction sector automation can be applied more on information and administration systems of single firms and sites than within single technologies which, even in the most sophisticated cases, are tied to specific work in the field and continue to have a craft style of organization.

The information system utilized by building firms must be of the "distributed type," so that it can include all the firms operational poles. The ideal solution would be to install one or more terminals in each construction site.

EDP (Electronic Data Processing) can then be developed so as to administer data acquired ex post and planning-wise, general administration aspects.

In the first case EDP can help with managerial functions like general accounting and salaries calculations, with all their traditional administrative tasks (invoicing, storage, etc.). In these cases, computer data processing substitutes for the manual execution of operations already institutionalized, compulsory or necessary.

The second area of application within the firm's administration is more difficult to cope with, but it is also the one that can give the most interesting results. Here it would be possible to act in financial and economic planning, thus substantially influencing project-making during the implementation of works.

Therefore, it becomes possible to program organizational instruments for automated information (PERT, budget, analytical accounting of elementary costs, etc.) which represent the very base of rational and shrewd management of the firm.

Choosing the Right Computer System

When a firm decides to set up a management information system, some typical problems usually arise. For example, choice of the computer in terms of size, terminals potential for peripheral office or construction sites, future memory improvements. Another problem concerns the acquisition of software packages already available on the market (for general accounting procedures, banks, suppliers, clients and personnel administration and so on). Furthermore, it is necessary to understand problems tied to the formulation of specific programs answering the particular needs of the firm, and to the information that we want to administer (for example, cost accounting, planning and control of site works, control of purchases and stocks).

These different factors and the machines that make up hardware must be examined from different points of view. For example:

(a) The system's dependability (easy maintenance and, therefore, quick and continuous service insured for the entire life-cycle of the acquired system).

(b) System performance and the possibility of expansion for memory and processing capabilities of the central unit, in terms of added peripheral units.
(c) Software packages already available on the market and the ease with which they can be applied to the firm's requirements. The importance of software development, time and cost-wise, should be stressed, since software is often more expensive than hardware.
(d) Possibility of using the system with a limited presence of specialized personnel within the premises, and few training problems for peripheral users.
(e) Easy adoption of software packages and easy creation of data-banks.

The offers of various specialized suppliers must be examined by taking into account some specific standard parameters, among which are:

- validity of the system in relation to the firm's own needs
- total costs including service for at least three years or more
- amount of available information
- expansion of the system's possible applications
- ease of operation
- service being offered

Computer Applications

The main purpose of an EDP should be the realization and the implementation of an efficient process of planning and controlling important information pertaining to the firm.

It can prove useful in a number of areas.

In the secretarial center it can be used not only as a work processor for letters and reports, but also for managing archives of parties with whom the firm is dealing, thus forming a base for the development of public relations services.

In the technical area the aid of the computer is useful for standard project work with the control of static computation, the planning of projects and controls as works progress. It is possible to prepare offers to clients complete with bills of quantities and consequent price and cost analysis through special pre-established software programs for price lists.

The administration department can process all general accounting operations, cost accounting, and administration procedures for clients and suppliers. Also possible are control procedures for incoming and out-going invoices and subsequent payments.

Personnel will be able to administer salaries and automatically control attendance, to plan the best allocation of human resources, and to establish career projections.

In the purchasing department, the computer allows a centralized management of resources and equipment procurement, and also coordinates the various subcontractors and subsuppliers. Through the EDP system equipment and stocks can be checked, together with equipment movements, thus disciplining maintenance in order to avoid premature obsolescence.

Typical applications in the production area are those pertaining to time planning and to the control of works in progress with the consequent planned coordination of subcontractors' operations. Technical accounting will be possible both for the client and for an internal control of costs as compared to budget estimates. Other important activities will be the control of human resources and of stocks in deposit in each construction site, concentrating, then, the information in the main offices.

The financial department will be able to discipline relations with banks, credit collection, and payments due to suppliers, and to merge central and peripheral cash pertaining to the various territorial areas and work sites.

Without any doubt an efficient automation process in its totality (hardware/software/organizational procedures before and after the processing activity) represents the necessary condition for optimizing the administration of the building construction firm in its three levels: corporate level, operations level, and macrofirm level.

REFERENCES

1. Amendola, G., *Uomini e case—I presupposti sociologici della progettazione architettonica*, Edizioni Dedalo S.p.A., Bari, Italy (1984).
2. Amendola, G., et al., *Segni & Evidenze—Atlante sociale di Bari*, Edizioni Dedalo S.p.A., Bari, Italy (1985).
3. Argyris, C., *Management and Organizational Development. The Path from XA to YB*, McGraw-Hill, New York (1971).
4. Arrow, K. J., *The Limits of Organization*, W. W. Norton & Company, New York, London (1974).
5. Butera, F., "Environmental Factors in Job Design," *The Quality of Working Life*, L. E. Davis, A. B. Cherns, eds., New York, Free Press (1976).
6. Butera, F., J. E. Thurman, eds., *Automation and Work Design*, North Holland, Amsterdam, New York, Oxford (1984).
7. Chandler, A. D., Jr., *Strategy and Structure, Chapters in the History of the American Industrial Enterprise*, MIT Press, Cambridge, Mass. (1962).
8. Chandler, A. D., Jr., *The Visible Hand: the Managerial Revolution in American Business*, Harvard University Press, Cambridge, Mass. (1977).
9. Chandler, A. D., Jr., "Evolution of the Large Industrial Corporation: an Evaluation of the Transaction Cost Approach, *Report*, Harvard University Press, Cambridge, Mass. (1982).
10. Chandler, A. D., Jr., H. Daems, *Managerial Hierachies: Corporation Perspectives in the Rise of the Modern Industrial Enterprise*, Harvard University Press, Cambridge, Mass. (1980).
11. Ciborra, C., "Markets, Bureaucracies and Groups in the Information Society," *Information Economics and Policy*, 1, 2, pp. 145–160 (1983).

12. Ciborra, C., P. Migliarese, A. Romano, "A Methological Inquiry of Organizational Noise in Socio-Technical Systems," *Human Relations* (1984).
13. Coase, R. H., "The Nature of the Firm," *Economica N.S.* (n. 4), pp. 386-405 (1937).
14. Coase, R. H., "The Problem of Social Cost," *Journal of Law and Economics* (Oct. 1960).
15. Coase, R. H., "Industrial Organization: a Proposal for Research," *Policy Issues and Research Opportunities*, V. R. Fuchs, ed., Industrial Organization, National Bureau of Economic Research, New York (1972).
16. Coase, R. H., "Adam Smith's View of Man," *Journal of Law and Economics* (No. 19) (Oct., 1976).
17. De Rosnay, J., *Le Macroscope: vers un vision globale*, Editions du Seuil, Paris, France (1975).
18. Dioguardi, G. F., "Il laboratorio di quartiere: aspetti organizzativi e imprenditoriali," *Impresa e Societa*, Cedis, Roma, Italy (No. 12), pp. 11-20 (June 30, 1981).
19. Dioguardi, G. F., *Organizzazione come strategia*, Isedi-Mondadori, Milano, Italy (1982).
20. Dioguardi, G. F., *Nuovi modelli organizzativi per l'impresa —il caso dell'edilizia*, Etas, Milano, Italy (1983).
21. Dioguardi, G. F., "Macrofirm: Construction Firms for the Computer Age," *Journal of Construction Engineering and Management, ASCE, vol. 109* (No. 1), pp. 13-24 (March, 1983).
22. Dioguardi, G. F., "L'impresa di manutenzione della città," *Impresa e Società*, Cedis, Roma, Italy (No. 6), pp. 6-10 (March 31, 1984).
23. Drucker, P. F., *Managing for Results*, W. Heinemann, London, England (1964).
24. Drucker, P. F., *Managing in Turbulent Times*, W. Heinemann, Ltd., London, England (1980).
25. Eccles, R. G., "The Quasifirm in the Construction Industry," in *Journal of Economic Behavior and Organization* (No. 2) (1981), from Reprint Series, Division of Research, Graduate School of Business Administration, Harvard University (1981).
26. Eccles, R. G., "Control with Fairness in Transfer Pricing," *Harvard Business Review* (No. 6), pp. 149-161 (November-December, 1983).
27. Fabris, A., *L'organizzazione di una societa in cambiamento*, Etas, Milano, Italy (1975).
28. Friel, M., G. P. Trella, "Microcomputers in the Construction Industry Implementation and Problems," *Report* (December 10, 1984).
29. Galbraith, J., *Designing Complex Organizations*, Addison-Wesley Publishing Co., Inc., Reading, Mass. (1973).
30. Gans, J., *People and Plans*, Basic Book, New York (1968).
31. Gutman, R., *People and Buildings*, Basic Books, New York (1972).
32. Hanika, F. de P., *New Thinking in Management*, Hutchinson, London, England (1965).
33. Kay, N. M., in *The Evolving Firm—Strategy and Structure in Industrial Organization*, MacMillan, London, England (1982).
34. Katz, G. L., R. E. Levitt, "The Construction Executive Program—Stanford University's Aid to Managers," *The Military Engineer*, pp. 364-367 (September, 1984)
35. Irwig, H. G., R. E. Levitt, "Organizational Adaptation To Small Computers," *Applications of Small Computers in Construction*, Proceedings of a Session sponsored by the Construction Division of the American Society of Civil Engineers in conjunction with the ASCE National Convention, San Francisco, California, W. C. Moore, ed., pp. 41- 59 (October, 1984).
36. Lawrence, P. R., D. Dyer, "Residential Construction: A Hidden Resource," *Renewing American Industry*, The Free Press, New York, Collier MacMillan Publishers, London, pp. 146-165 (1983).
37. Learned, E. P., A. T. Sproat, *Organization Theory and Policy: Notes for Analysis*, R. D. Irwin, Homewood, Illinois (1966).
38. Levitt, R. E., "The Effect of Top Management on Safety in Construction," *Technical Report No. 196*, Stanford University, Department of Civil Engineering, Stanford, California (July, 1975).
39. Levitt, R. E., "The Organization of Work in Construction," *Human Factors—Ergonomics for Civil Engineers and Construction Managers*, Wiley & Sons, New York (1981).
40. Levitt, R. E., "Problems in the Organization of Very Large Projects," *Transactions*, Gruppo di Ricerca sul Management, Bari, Italy (Issue No. 21), pp. 8-35 (1982).
41. Levitt, R. E., "Foreword" to "Liability and the Injured Worker in the Construction Process" by James Edward Koch, *Technical Report No. 283*, Stanford University, Department of Civil Engineering, Stanford, California, p. i (June, 1984).
42. Levitt, R. E., "Superprojects and Superheadaches: Balancing Technical Economies of Scale Against Management Diseconomies of Size and Complexity," *Project Management Journal, Vol. XV* (No. 4), pp. 82-89 (December, 1984).
43. Levitt, R. E., C. C. Bourdon, *Union and Open Shop Construction*, Lexington Books, Lexington (1980).
44. Levitt, R. E., S. Gunnarson, "Is a Building Construction Project a Hierarchy or a Market?" presented at the 7th Internet Congress, Copenhagen, Denmark (1982).
45. Levitt, R. E., R. D. Logcher, "The Human Element in Project Control Systems," presented at the October, 1976, Project Management Institute Annual Symposium held at Montreal, Canada.
46. Levitt, R. E., R. D. Logcher, N. H. Qaddumi, "Impact of Owner-Engineer Risk Sharing on Design Conservatorism," *Journal of Professional Issues in Engineering, ASCE, Vol. 110* (No. 4), pp. 157-168 (October, 1984).
47. Levitt, T., "A Heretical View of Management 'Science'," *Fortune*, pp. 50-52 (December 18, 1978).
48. Likert, R., *The Human Organization: Its Management and Value*, McGraw-Hill, New York (1967).
49. Logcher, R. D., R. E. Levitt, "Human Information- Handling Capacity as a Factor in the Design of Project Control Systems," presented at the May 1976, International Council for Building Research, Studies and Documentation Working Symposium held at Montreal, Canada.

50. Logcher, R. D., R. E. Levitt, "Organization and Control of Engineering Design Firms," *Issues in Engineering—Journal of Professional Activities, ASCE, Vol. 105* (No. E11), Proc. Paper 14302, pp. 7–14 (January, 1979) and *L'Industria delle Costruzioni*, ANCE, Rome, Italy (No. 122), pp. 32–38 (December, 1981).
51. Lombardini, S., "L'entreprise motrice et la distribution spatial de l'activité économique," *Economie Appliquée* (1976).
52. Lombardini, S., *Il metodo della scienza economica: passato e futuro*, UTET, Torino, Italy (1983).
53. Lorenzoni, G., *Una politica innovativa nelle piccole e medie imprese*, Etas Libri, Milano, Italy (1979).
54. Lorenzoni, G., "Le strategie di impresa fondate su sinergie esterne," *L'Impresa—Rivista Italiana di Management*, Torino, Italy (No. 1) (1980).
55. Maisel, S. J., *Housebuilding in Transition*, University of California Press, Berkeley (1953).
56. McGregor, D., *The Professional Manager*, McGraw-Hill, New York (1967).
57. Miles, R. E., C. C. Snow, "Fit, Failure and the Hall of Fame," *California Management Review, Vol. 26* (No. 3) (Spring, 1984).
58. Moss, S. J., *An Economic Theory of Business Strategy*, Martin Roberson and Co., Ltd., Oxford, England (1981).
59. Ouchi, W. G., "The Relationship between Organizational Structure and Organizational Control," *Administrative Science Quarterly* (No. 22) (1977).
60. Ouchi, W. G., "The Transmission of Control through Organizational Hierarchy," *Academy of Management Journal* (No. 21) (1978).
61. Ouchi, W. G., "Markets, Bureaucracies and Clans," *Administrative Science Quarter* (No. 25), pp. 129–141 (March, 1980).
62. Ouchi, W. G., *Theory Z—How American Business Can Meet the Japanese Challenge*, Addison-Wesley Publishing Co., Reading, Mass. (1981).
63. Pfiffner, J. M., F. P. Sherwood, *Administrative Organization*. Prentice Hall, New York (1960).
64. Piano, R., et al., *Antico è bello—il recupero della città*, Laterza, Bari, Italy (1980).
65. Reve, T., R. E. Levitt, "Organization and Governance in Construction," *Project Management, Vol. 2* (No. 1), pp. 17–25 (February, 1984).
66. Schumacher, E. F., *Small is Beautiful: A Study of Economics as People Mattered*, London Blond and Briggs, Ltd., London, England (1973).
67. Sheldon, O., "Taylor the Creative Leader: An Analysis of Taylor's Contribution to the Problem of Human Welfare," in *Bulletin of the Taylor Society: Critical Essays on Scientific Management, Vol. 10* (No. 1), pp. 76–77 (Feb., 1925).
68. Scott, R. W., *Organizations, Rational Natural and Open Systems*, Prentice-Hall, Inc., New Jersey (1982).
69. Simon, H. A., *The Sciences of the Artificial*, MIT Press, Cambridge, Mass. (1969).
70. Simon, H. A., *Administrative Behavior*, The Free Press, New York (1976).
71. Simon, H. A., *The Shape of Automation*, Harper-Row, New York (1965).
72. Sloan, A. P., Jr., *My Years with General Motors*, Doubleday & Co., Inc., Garden City, N.Y., 1963 (Anchor Books Edition, 1972).
73. Stigler, G. J., "The Division of Labor is Limited by the Extent of the Market," *Journal of Political Economy* (No. 31) (1951).
74. Sylos-Labini, P., *Oligopoly and Technical Progress*, Harvard University Press, Cambridge, Mass. (1962).
75. Tanner, R., A. G. Athos, *The Art of Japanese Management—Application for the American Executives*, Simon and Schuster, New York, N.Y. (1981).
76. Tarondeau, J. C., "Les trois âges de la stratégie industrielle," *Revue Française de Gestion*, Paris, France (No. 36), pp. 29–38 (Juin–Juillet–Août, 1982).
77. Teece, D. J., "Technology Transfer by Multinational Firms: The Resource Cost of Transferring Technological Knowhow," *Economic Journal* (No. 87) (June, 1977).
78. Teece, D. J., "Economies of Scope and the Scope of the Enterprise," *Journal of Economic Behavior and Organization* (No. 1) (1980).
79. Teece, D. J., "Internal Organization and Economic Performance: an Empirical Analysis of the Profitability of Principal Firms," *The Journal of Industrial Economics* (Dec. 1981).
80. Teece, D. J., *Some Efficiency Properties of the Modern Corporation: Theory and Evidence*, Graduate School of Business, Stanford University (1982).
81. Thompson, J. D., *Organization in Action*, McGraw-Hill Book Co., Inc., New York, N.Y. (1972).
82. Van de Ven, A. H., W. F. Joyce, eds., *Perspectives on Organization Design and Behavior*, Wiley & Sons, New York (1981).
83. Waugh, L. M., "Decision Making in Construction: Returns on Equity vs. Returns on Expertise," Thesis submitted to the Department of Civil Engineering and the Committee on Graduate Studies of Stanford University in partial fulfillment of the requirements for the Degree of Engineer (June, 1984).
84. Welsch, A. G., *Budgeting: Profit-Planning and Control*, 3rd Ed., Prentice-Hall, New York (1970).
85. Williamson, O. E., *The Economics of Discretionary Behavior*, Englewood Cliffs, N.J., Prentice-Hall (1964).
86. Williamson, O. E., *Corporate Control and Business Behavior*, Englewood Cliffs, N.J., Prentice-Hall (1970).
87. Williamson, O. E., "The Vertical Integration of Production: Market Failure Considerations," *American Economic Review* (No. 61) (May, 1971).
88. Williamson, O. E., *Markets and Hierarchies*, Free Press, New York (1975).
89. Williamson, O. E., "Transaction Cost Economics: The Governance of Contractual Relations," *Journal of Law and Economics* (No. 22), pp. 233–261 (October, 1979).
90. Williamson, O. E., "On the Governance of the Modern Corporation," *Hofstra Law Review* (Fall, 1979).
91. Williamson, O. E., "Organizational Innovation: The Transac-

tion Cost Approach," *Discussion Paper No. 83*, Center for the Study of Organizational Innovation, University of Pennsylvania (Sept., 1980).

92. Williamson, O. E., "The Organization of Work," *Journal of Economic Behavior and Organization* (No. 1), pp. 5–38 (1980).

93. Williamson, O. E., "The Economics of Organization: The Transaction Cost Approach, *American Journal of Sociology* (No. 87), pp. 548–577 (Nov. 1981).

94. Williamson, O. E., "The Modern Corporation: Origins, Evolution, Attributes," *Journal of Economic Literature* (No. 19) (Dec., 1981).

95. Williamson, O. E., W. G. Ouchi, "The Markets and Hierarchies and Visible Hand Perspectives," *Perspectives on Organizational Design and Behavior*, A. H. Van de Ven, W. F. Joyce, eds., Wiley & Sons, New York (1981).

96. Williamson, O. E., M. L. Wachter, J. E. Harris, "Understanding the Employment Relation: The Analysis of Idiosyncratic Exchange, *Bell Journal of Economics* (No. 6), pp. 250–278 (Spring, 1975).

CHAPTER 8

CPM in Construction

EDWARD G. DAUENHEIMER* AND ROBERT K. BLOCK**

INTRODUCTION

The Critical Path Method (CPM) is a tool for planning, scheduling and ultimately controlling construction projects. It uses a network or diagram approach as opposed to the instinctive method of "running a job."

The contractor's instincts are not totally disregarded in the CPM planning. These instincts, which are a result of knowledge gained through experience, are essential in developing a good plan and schedule.

The main objective of having a detailed plan before a project begins is to shorten the construction time. This can be accomplished by identifying potential problem areas and then by either minimizing or eliminating the problem before it becomes a costly surprise. A good working CPM schedule will shorten construction time, resulting in cost savings to contractors which will be passed along to owners.

HISTORY

CPM was first developed between 1956 and 1958 by the E.I. du Pont Nemours Company. In 1956 they formed a group to study applications of new management methods in their engineering duties. The company was involved in construction of chemical plants and the planning and scheduling of these projects became one of their first priorities.

The group had a computer and decided to evaluate its use in the scheduling of construction. Mathematicians used an approach where they theorized that if the sequence and duration of each activity which makes up a total project are fed into a computer, then a schedule of work could be generated. In early 1957, computer experts from the Remington Rand Company joined the research group. The original work was revised and their results became the basic CPM, which to date has had no fundamental changes.

The design and testing of CPM continued up to 1960. In the next 10 years, practical usage of CPM developed. CPM activity was greatest in construction. A 1965 survey showed that only 3 percent of the nation's contractors actively used CPM, but they were large contractors and accounted for 20 percent of the nation's major construction. Of the contractors using CPM, 90 percent were satisfied with the investment of time and effort required. It was difficult to put an actual dollar value on savings in scheduling time and costs but they estimated that their savings exceeded 10 percent in many cases.

From 1970 to 1980, the acceptance and use of CPM increased for many reasons. First, engineering schools added CPM techniques with computer applications to their undergraduate curricula.

Second, the decade was an important one for the development of construction management. Its major selling point was the promotion of management control. This could be achieved through CPM planning and scheduling.

Third, the evolution in computer capability not only made basic network systems more available but also provided for correlation of schedules to costs and resources.

Finally, the increase in construction litigation, especially of the type citing delay as a cause of damages, had caused schedules and their utilization to become important to both plaintiff and defendant. The proper use of a valid CPM plan can be important in supporting a contractor's claim.

The same CPM schedule, under different circumstances, can also be important in defending the owner or owner's representative on a project. The CPM schedule, therefore, must be taken seriously by all parties in order to be a useful tool in reducing construction time and costs.

---

*Department of Civil & Environmental Engineering, New Jersey Institute of Technology, Newark, NJ
**Conti Construction Company, South Plainfield, NJ

MECHANICS OF CPM

Diagrams—Networks

The Critical Path Method is founded on a graphical model of a project. Each activity in a project is represented by a symbol. The symbols are arranged to show the logical sequence in which the activities of the project are to be performed. The flow chart created is called either an arrow diagram or a network diagram. Without a good network diagram, any further effort at CPM scheduling is a waste of time.

Two approaches to network diagramming are found in construction practice: Critical path arrow diagram method (ADM) and critical path precedence diagram method (PDM). The ADM is most popular with contractors. The essential difference between these two critical path methods is in the network diagramming technique. This difference is illustrated below:

Except for the network technique, the two methods are essentially the same. The PDM approach is more flexible when lags and other types of logical relationships (besides the finish-to-start type shown on the above diagram) are introduced.

Please note that another type of network scheduling is the Program Evaluation and Review Technique (PERT). PERT has not been accepted by the construction industry. The ADM shall be explained in detail below.

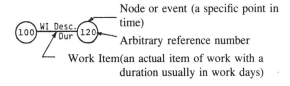

Node or event (a specific point in time)
Arbitrary reference number
Work Item (an actual item of work with a duration usually in work days)

CPM-ADM
(Activity-on-the-Arrow Network)

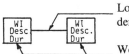

Logical connector (defines precedence between work items)
Work Items (actual items of work with durations)

CPM-PDM
(Activity-on-the-Node Network)

Figure 1 is an example of a network. It is a simplification of the portion of a $20 million project which involved installation of a force main.

Preparing a network diagram forces decision making. The network, by itself, does not make decisions. The preparation helps the planner to understand a project by identifying the activities required to complete a project.

CPM is a logical and organized planning method. Therefore, the physical layout of the network should be logically organized. If it is, then presentation to people unfamiliar with the network will be easily understood. Good organization could also bring to light activities that were not originally planned for but are essential to the completion of a project and, if overlooked early, could lead to disaster.

The intersection of two or more arrows is an event and has a zero duration. All activities leading into an event must be complete before any activities leading out of the event can be started. This is a basic rule of the network logic. Important events are called milestones. In the example shown in Figure 1, event 14 is a milestone. At that point pumping through the new pipeline to a new sewerage treatment plant will begin and demolition of the old plant can begin.

The events in Figure 1 are numbered. This numbering is not performed until after the initial network is completed and then rearranged in a more orderly form with important chains of activities usually forming the center or backbone. The numbering should be done carefully in order to make the network easier to use. It will also help to identify logic loops which will not work in practice, pointing out a "which comes first, the chicken or the egg" type problem whose early solution could save time and money.

The number given to the beginning of an activity is referred to as the i and the number given to the completion is referred to as the j. The i-j number for an activity can be used to describe an activity.

There are a couple of basic rules to be followed when assigning event numbers to a network. One is that each activity must have a unique i-j. In the example in Figure 1, a restraint or dummy had to be added to avoid two activities having the same i-j description, 1-5. This restraint's i-j description is 2-5.

Another rule is that the j number for an activity should be greater than its i number. By following this rule loops are eliminated.

Event and Activity Time Computations

Once the plan is completed in the form of an arrow diagram, durations must be associated with each activity. This is usually shown on diagram below each activity description. Whatever unit is used must be consistent throughout the network. Days or weeks are usually used. The example project uses days. Accurate estimates of the duration of each activity should be used. Average daily production rates should be used, not best or worst rates. Figure 2 has durations associated with each activity.

Early event times are calculated by adding the durations as you work from left to right in the network. When there is a choice between two or more values, the larger value is used. Early event times are shown on Figure 2 in the square

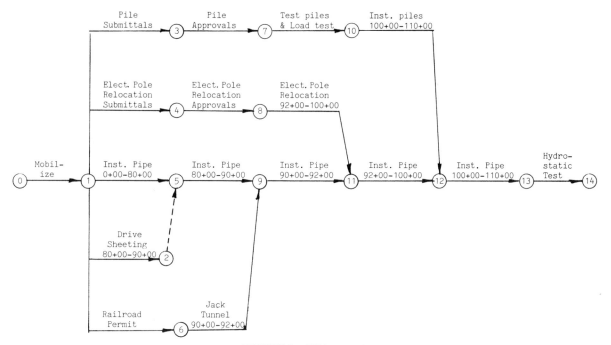

FIGURE 1. CPM network.

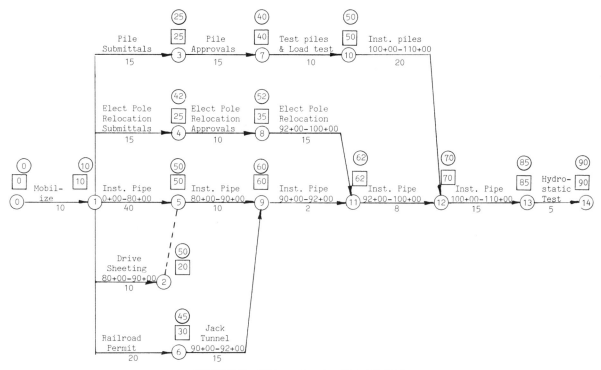

FIGURE 2. CPM newtwork with event times.

box above the event number. The shortest time in which the work can be completed is 90 days or 18 weeks.

Late event times are calculated by subtracting the durations as you work from right to left in the network. The late event time at event 14 is equal to the early event time. This is true for all terminal events. Where there is a choice between two or more values, the earlier value is used. Late event times are shown in the circles above the early event times in Figure 2.

The early event time is the latest that the activities converging at an event will finish. The late event time is the earliest that the activities starting at that event will commence. And the shortest project time is the result of the longest path, which is known as the critical path. This all sounds contradictory, but it works.

Early and late event times are not as useful to us as activity times, which are derived from the event times. The four activity times are early start, early finish, late finish and late start.

Early start = ES = early event time (i)
Early Finish = EF = ES + duration = ES + D
Late finish = LF = late event time (j)
Late start = LS = LF − duration = LF − D

Table 1 gives the activity times for our example CPM network. It also shows which activities are on the critical path and which are not. An activity is critical if it meets these three conditions: 1) The early and late event times at the beginning of the activity must be equal; 2) The early and late event times at the completion of the activity must be equal; and 3) The difference between the early start and late finish must equal the duration of the activity. It is important to test for the third rule because a non-critical activity in parallel with two critical activities which are in series will have equal early and late event times as its nodes. If its duration is less than the sum of the durations of the critical activities it will not be critical.

Float time is also shown in Table 1. This is the amount of time that the start of a non-critical activity can be delayed without delaying the planned completion date of the project. In the example project, the start of activity 1–2, driving of sheeting, can be delayed 30 days from the early start date and not delay activity 5–9, pipe installation inside the sheeting. If the sheeting takes more than 10 days, this activity will become critical and delay the job as many days as it overruns.

This is one reason why it is not a good practice to start items with float on their late start day. It might also put succeeding activities on a critical path, which is undesirable. In the example project, if activity 1–4 started on its late start date, then activities 4–8 and 8–11 would lose their float time and become critical.

The final task is to convert project days to calendar days. Table 2 lists early and late start and finish dates by calendar day. Float is also listed and if zero float is indicated the activity is considered critical.

Computer Versus Manual Computation

Computers are invaluable for networks which contain 50 or more activities when dozens of revisions and updates will

TABLE 1. Activity Times in Project Days.

Activity (i-j)	Duration (days)	Description	ES	EF	LS	LF	Float (days)	Critical
0–1	10	Mobilize	0	10	0	10	0	Yes
1–2	10	Drive sheeting 80 + 00 to 90 + 00	10	20	40	50	30	No
1–3	15	Pile submittals	10	25	10	25	0	Yes
1–4	15	Elect. Pole relocation submittal	10	25	27	42	17	No
1–5	40	Inst. pipe 0 + 00 to 80 + 00	10	50	10	50	0	Yes
1–6	20	Railroad permit	10	30	25	45	15	No
2–5	0	Restraint	—	—	—	—	—	—
3–7	15	Pile approvals	25	40	25	40	0	Yes
4–8	10	Elect. relocation approvals	25	35	42	52	17	No
5–9	10	Inst. pipe 80 + 00 to 90 + 00	50	60	50	60	0	Yes
6–9	15	Jack tunnel 90 + 00 to 92 + 00	30	45	45	60	15	No
7–10	10	Test piles and load tests	40	50	40	50	0	Yes
8–11	10	Elect. relocation 92 + 00 to 100 + 00	35	45	52	62	17	No
9–11	2	Inst. pipe 90 + 00 to 92 + 00	60	62	60	62	0	Yes
10–12	20	Inst. piles 100 + 00 to 110 + 00	50	70	50	70	0	Yes
11–12	8	Inst. pipe 92 + 00 to 100 + 00	62	70	62	70	0	Yes
12–13	15	Inst. pipe 100 + 00 to 110 + 00	70	85	70	85	0	Yes
13–14	5	Hydrostatic testing	85	90	85	90	0	Yes

be needed. No matter how carefully thought out, actual activity durations will vary from those originally estimated. Updates are therefore necessary to find out how the overall schedule is affected. If the completion date is delayed substantially, then revisions will be necessary. Also, some specifications require that the contractor furnish monthly updates.

It is recommended that early event times should be computed manually even if a computer is used. This would provide an overall time frame and locate mistakes in the network. After the manual check, the computer would be given the necessary data and would make the repetitive network calculations.

Another advantage of the computer is that costs and manpower levels can be associated with the schedule. If this feature is used properly, it could be an excellent means of keeping track of costs. The program then becomes an excellent overall managerial tool.

CPM NETWORK PREPARATION/PLANNING

Information Collection

Collection of information is the first step in preparing a network. Our approaches used for information collection are: 1) conference, 2) executive, 3) consultant, and 4) staff planning.

In the conference approach, key project personnel take part. This includes people from the contractor's office and field groups, as well as representatives from the owner or engineer or both. The group should not be so large that it cannot function properly and must have the authority to make commitments on the sequence of work.

The group discusses the project from start to finish while drawing an arrow diagram. Due to many changes, a blackboard is helpful for the rough draft. After agreement on work sequence is reached, the information is put on paper.

This approach could create an excellent plan if all of the key people can be brought together at the same time. This can be a problem due to these people having responsibilities on projects that can be thousands of miles apart. If the contractor is not committed to a good CPM schedule, these people will never get together. There will be numerous excuses.

Also, having representatives for the owner and engineer can be useful in locating possible problems that the contractor can be unaware of or overlook. A situation that the owner might know about would be other construction starting in the area and possibly affecting the schedule. If it is pointed out early enough, a possible extra or claim might be avoided. An engineer might point out a difficult item in the project that he will insist on being performed in a way that is more time consuming to the contractor and could change the critical path. An engineer or owner might also have important information on the length of time it takes to obtain certain permits. This could be a critical item that might require early expediting.

In cases where there are too many key people for a manageable conference or too few, an executive approach could be used. This group is usually made up of a project man-

TABLE 2. Activity Times in Calendar Days.

Activity (i-j)	Duration (days)	Description	ES	EF	LS	LF	Float (days)
0–1	10	Mobilize	11–7	11–18	11–7	11–18	0
1–2	10	Drive sheeting 80 + 00 to 90 + 00	11–18	12–5	1–4	1–18	30
1–3	15	Pile submittals	11–18	12–12	11–18	12–12	0
1–4	15	Elect. pole relocation submittal	11–18	12–12	12–14	1–6	17
1–5	40	Inst. pipe 0 + 00 to 80 + 00	11–18	1–18	11–18	1–18	0
1–6	20	Railroad permit	11–18	12–19	12–12	1–11	15
2–5	0	Restraint	—	—	—	—	—
3–7	15	Pile approvals	12–12	1–4	12–12	1–4	0
4–8	10	Elect. relocation approvals	12–12	12–27	1–6	1–20	17
5–9	10	Inst. pipe 80 + 00 to 90 + 00	1–18	2–1	1–18	2–1	0
6–9	15	Jack tunnel 90 + 00 to 92 + 00	12–19	1–11	1–11	2–1	15
7–10	10	Test piles and load tests	1–4	1–18	1–4	1–18	0
8–11	10	Elect. relocation 92 + 00 to 100 + 00	12–27	1–11	1–20	2–3	17
9–11	2	Inst. pipe 90 + 00 to 92 + 00	2–1	2–3	2–1	2–3	0
10–12	20	Inst. piles 100 + 00 to 110 + 00	1–18	2–15	1–18	2–15	0
11–12	8	Inst. pipe 92 + 00 to 100 + 00	2–3	2–15	2–3	2–15	0
12–13	15	Inst. pipe 100 + 00 to 110 + 00	2–15	3–7	2–15	3–7	0
13–14	5	Hydrostatic testing	3–7	3–14	3–7	3–14	0

ager, superintendent, and staff CPM engineer or CPM consultant. This type of group can work more quickly, but there is a loss in communication among key personnel.

This is the most common approach used. Most of the input from key field personnel will not be given at a conference table, but in the field or via telephone. It seems like there are never enough knowledgeable key people available at the same time and sometimes their effort is not strong enough to be useful.

The consultant approach is a modified version of the executive method. A staff engineer or CPM consultant talks the project through with the contractor's superintendent and key people and then prepares a diagram. The advantage of this approach is that the time demands on the key people are minimized. The disadvantage is that they might not accept the diagram as readily as one they helped prepare.

The fourth approach is staff planning. Large contractors who use CPM often set up staff planning groups. The major disadvantage of this method is that the group tends to lose touch with the needs of the field portion of the project group. It often results in the field group servicing the planning group instead of the planning group servicing the field group like it should.

Subcontractors

Subcontractors' work is sometimes on the critical path and sometimes it would be if planned for realistically. The reason is that the general contractor has to estimate the duration of the subcontractors' activities and decide if it can run concurrently with other activities. This is usually done without any input from the subcontractor. They are left out of the planning due to the fact that they are usually not subcontracted at that point and sometimes because the general contractor thinks he knows everything. And if you ask a prospective subcontractor for a schedule, he might give an overly optimistic one, hoping it might get him the subcontract.

It is therefore important when a subcontract is given out that the time estimates and sequence of work be reviewed and revised accordingly. Another consideration is that the subcontractor should be fully aware of the specifications he must meet.

Materials

Shop drawing or submittal approval should be incorporated in the CPM plan for materials with long delivery times. These deliveries of certain types of equipment for a project often turn up on the critical path. When this is clearly shown on the CPM it becomes obvious that submittals should be timely and the engineer should be made aware of the urgency of his review.

A good CPM schedule will also help the contractor order non-critical materials as they are needed. This is especially important when storage space is limited and when early deliveries only cause extra handling costs, which are already a large part of construction costs.

THE CPM SCHEDULE

Analysis/Manipulation

The critical path should be the first area examined when the end fate of the initial plan is later than desired. Check the path for series activities that could be put in parallel. If this is not sufficient or possible, then critical activities with long durations should be investigated.

Possible alternatives for shortening critical activities would be: 1) increase number of crews; 2) increase manpower or equipment; 3) work 10–12 hours per day; 4) work 6–7 days per week; 5) work double or triple shifts, etc. The cost associated with any of these possibilities should be carefully considered before any changes are made, especially if in the bidding stage, no overtime was estimated. It could be possible that the cost of liquidated damages would be more economical.

Always figure on the actual completion date to exceed the first CPM end date. This is due to weather, changed site conditions, labor problems, equipment breakdowns, change orders, etc. A contingency can be added, but usually is not. Therefore, updates are necessary with revisions if the end date for the schedule must be met. It is usually best to have the end date of the schedule set sufficiently behind the required completion date of the project. This will allow for the contingency factors.

Progress/Updates

Project progress should be marked on the CPM network. This will give a current status report for the project. It will also keep the field people familiar with the network logic.

Updates of the network should be at regular intervals, usually monthly on long-term projects. These updates should include a report describing the following: critical path, activities completed, activities in progress, and problem areas. The updates and reports should be coordinated with the jobsite meetings.

Cost Control

To use CPM for cost control, a cost must be assigned to each activity. This cost breakdown should be realistic and within the framework of the bid. An important use of the activity cost breakdown could be for making progress payments on lump sum projects or for breaking down lump sum items on unit price projects. If there are many activities associated with an item or category, then the estimated cost for that item should be broken down for the activities.

The cash requirements for a project can be forecast by computer with the use of the CPM cost estimates. These forecasts can help owners in investing the construction funds to earn the highest yields. Contractors will be able to determine their financial needs and methods better.

A cost system based on CPM can help estimate the cost of expediting a project. It is even possible that a project can be completed early at a lower cost due to this type of expediting.

All of the positive possibilities for the CPM-based cost system are not being fully realized currently. This is because the existing cost collection and accounting systems are not based on construction activities. It takes time, money, and the willingness to change accounting systems. Just like CPM scheduling, savings are not easily calculated and the transition will be slow.

Resource Planning

In the planning and scheduling of projects, it is generally assumed that labor and equipment are available as needed. This is not the case in the real world. These resources are kept level by performing the critical path and low-float activities first and those with more float are used as fill-in work. It is possible in some cases that due to limited resources, the end date can be extended beyond the date arrived at from the original network.

Heavy equipment and key men seem to be the major resources that have to be juggled. This type of leveling usually involves several projects. In the example project there is only one pipe laying crew due partly to this resource leveling which was intuitively incorporated in the original CPM network. This created two critical paths instead of one: the pile chain of activities and the pipe installation chain of activities. It would have been uneconomical to rent more equipment and hire another pipe foreman in order to expedite some pipe installation activities and create float in them.

There are numerous computer programs for resource leveling. They can be very useful for large, complex projects. The quantity of resources available is inputted and a new schedule needed to meet these levels is generated. When the resources are limited, the project duration will either be greater than or equal to the normal time duration.

SUMMARY OF THE ADVANTAGES OF CPM

The basic advantages of CPM are being able to predict the time required to complete a project and being able to identify the critical activities in a project.

Prediction of time requirements will help to determine if the planned progress is acceptable. If it is not, then it will be easier and less costly to expedite critical items early in a project. Care must be taken, though, not to overlook non-critical activities.

CPM, if fully utilized will make it possible to economically schedule components of a construction project. Scheduling of equipment, materials, labor, subcontractors, and money can all be optimized.

It can also help to quickly answer "what if" questions. If a project has to be shortened, various alternatives can be readily evaluated. These evaluation qualities are also good for determining the time effect of a delay or addition of a change order.

The Critical Path Method is based on logic, or common sense. The interrelationships between the activities in a project must be considered. Conventional means of scheduling, such as bar graphs, do not force this type of pre-planning. This pre-planning is extremely important. If the preliminary network is the limit of how far CPM is implemented, much benefit will have been realized.

CONCLUSIONS – IMPLEMENTATION OF CPM

Although use of CPM seems to be the cure-all for planning and scheduling of the "organized confusion" known as construction, it has not been accepted as such. Reasons for increased usage, such as management control, correlation of costs and resources to schedule, and construction litigation have been outweighed by lack of acceptance, due in part to lack of education on the benefits and mechanics of the method. However, the major problem appears to be the natural desire of construction people on all levels to maintain the status quo. The mechanics of CPM are there, but the people are not.

The various levels of management view the problem differently. In many cases top-level management is unaware of any scheduling problems. As far as they know, their quality of planning is excellent. CPM would be perfect for this level in that they would only become involved when needed for the problem areas, e.g., "management by exception."

Middle management, engineers, supervisors, office managers, and others between the top level and the field, usually develop the plans, schedules, and bar graphs. They are receptive to CPM, but are restricted by the top level. Many of the younger engineers have already been trained in CPM and courses are readily available for educating the others. As the engineers with exposure to CPM move into higher positions, they will increase its usage.

Field superintendents are a difficult group to convince. Many learned their jobs the hard way and are usually against learning to use a new tool. They are often afraid of computers and do not want to give their knowledge out to a young engineer. They are worried about job preservation. These superintendents must be carefully introduced to CPM as a possible money saver and a tool to prevent slowdowns and stoppages of production work that they are responsible for. They should be trained in the common sense approach of CPM and some should attend courses on CPM.

Pushing projects in the field is important, but the office support is the key to profitability. Good scheduling with CPM, along with its benefits of coordinating materials, equipment, labor, subcontractors, cash requirements, etc., will enable general contractors to have better control over more projects and greater efficiency. If time and cost savings are estimated at 5 to 10 percent with a cost of only 0.5 percent of a project, then use of CPM for multiple projects should increase these savings.

A force that is increasing CPM use is the requirement for a CPM schedule in many construction contracts, although a good CPM reference standard would help the cause. This contractual obligation to use CPM and young engineers moving into top management positions will be two of the major impetuses behind increased utilization in the future.

REFERENCES

Antil, J. M. and R. W. Woodhead, *Critical Path Methods in Construction Practice,* New York: Wiley-Interscience (1982).

O'Brien, J. J. and R. G. Zilly (ed.), *Contractor's Management Handbook,* New York: McGraw-Hill, pp. 7-1–7-30.

O'Brien, J. J., *CPM in Construction Management,* New York: McGraw-Hill (1984).

CHAPTER 9

Environmental Concerns During Construction

WALTER KONON*

The increased sensitivity of regulatory agencies and the general public regarding the environmental impacts of construction activities have resulted in an effort by contractors to minimize adverse construction impacts such as noise, vibration and soil erosion and sedimentation. These efforts are often mandated by federal or local regulations or specified in the contract documents.

Environmental specifications may prohibit certain construction practices, delineate specific control measures, and identify critical areas where special limits are applicable. When job specific environmental regulations are not specified, contractors must adhere to applicable regulations and to what is regarded as good industry practice. This chapter discusses environmental impacts and generally accepted industry practices as they apply to soil erosion and sedimentation, noise, and ground vibrations at construction sites.

SOIL EROSION AND SEDIMENTATION

During construction that involves land disturbance such as highways, pipelines or housing construction, the removal of natural vegetation and topsoil cover requires the initiation of protective measures to reduce erosion and possible future dredging. Construction projects should be planned, designed and constructed with sediment control in mind. While construction sites are not the sole source of eroded sediments, construction excavation operations can produce significantly increased rates of soil erosion and sedimentation. This form of pollution can be relatively clean, or at times can be quite toxic, when it becomes a carrier and then a storage area for sewerage and chemical pollutants that are introduced from non-construction sources.

It is estimated that the streams and rivers of the United States carry approximately one-billion tons (907.2 billion Kg) of sediment to the oceans each year, making this material by its sheer weight and volume the greatest pollutant of our waters. The National Commission on Water Quality estimated the sediment from uncontrolled construction sites at 200 million tons (181.4 billion Kg) per year. These figures indicate that uncontrolled construction operations contribute a significant percentage of the total sediments carried by our nation's waters.

Today, environmental impact statements require the evaluation of construction related impacts on all major federally funded projects [3]. The submission of environmental protection plans by builders prior to the start of construction, especially with regard to erosion and sedimentation control, has become a prerequisite for obtaining a building permit in many areas. Most of these environmental protection regulations are performance-oriented and they encourage contractors to be innovative in their environmental control measures [5]. Some construction contracts such as the Bureau of Reclamation dam and tunnel contracts in Colorado provide specific owner designed sedimentation and turbidity control measures or allow suitable contractor designed alternates [4]. These controls are specific bid items of the contract with provisions for payment for both installation and maintenance of the control measures.

The federal government and a large number of states have placed emphasis on controlling soil erosion, sedimentation, and turbidity originating at construction sites through legislation and the adoption of water quality standards. This includes federal legislation such as:

- The National Environmental Policy Act (1969)
- The Water Pollution Control Act amendments (of 1970, 1972, 1977)
- The Surface Mining Control and Reclamation Act (1977)

and state legislation such as:

---

*Department of Civil & Environmental Engineering, New Jersey Institute of Technology, Newark, NJ

- The Soil Erosion and Sedimentation Control Acts of Maryland, North Carolina and New Jersey
- Environmental Quality Review Acts of California and New York
- Water Quality Standards of Colorado, Arizona, Alaska and New Mexico

A definition of terms may be useful at this point:

Erosion is the loosening and removal of earth materials from the earth's surface. Rain is a major cause of erosion. It can wear away a thin layer of soil surface by a process called sheet erosion or the concentration of rainfall runoff in channels can remove soil by gully erosion.

Sedimentation is the process of erosion, entrainment, transportation and deposition of soil materials.

Turbidity is the reduction of transparency of water due to the presence of suspended materials such as clay, silt, or fine sand, and organic and inorganic matter. It is an expression of the optical property of a sample of water to transmit light.

There are distinct problems that are associated with erosion, sedimentation and turbidity. Streams, rivers, and reservoirs can silt up, causing reduced floodwater capacity which results in more frequent and higher floods and deposition of sediment on highways, streets, lawns, and other adjacent areas. Sediment deposits in drainage facilities, irrigation canals and navigation channels create loss of service and increased cleanout costs. Sediments can injure the gills and breathing structures of certain types of fish such as trout, and turbidity can affect the size, population and species of fish in streams. Turbidity produces muddy water which inhibits water sports and impairs municipal water supplies.

In general, many problems arise from erosion, sedimentation, and turbidity. However, completely stopping sedimentation is neither physically nor economically possible since even seeded lawns and mature forest lands produce some sedimentation. Representative rates of erosion for a variety of land use activities are presented in Table 1.

Construction sites are not the sole source of erosion but construction excavation work does produce increased rates of erosion when compared to other land use activities. What we must strive for is a rate of erosion, sedimentation, and turbidity produced by construction excavation work that, when added to the normal natural rate, will not cause damage to our environment.

To achieve this goal, it is necessary for us to be able to:

- predict the amount of erosion and subsequent sedimentation and turbidity produced by construction excavation
- determine sedimentation and turbidity in adjacent waters
- estimate the environmental consequences of additional increased sediment and turbidity levels
- design and implement economically feasible erosion, sedimentation and turbidity control measures at construction sites that will provide the required level of environmental control

Let us review these factors in terms of a hypothetical construction site.

Predicting Erosion

Sheet erosion is produced by the kinetic energy of raindrops impacting on the ground surface and loosening soil particles. These particles are then moved by the rainfall runoff down adjoining slopes where additional erosion occurs. The amount of sheet erosion produced by a construction site is a probabilistic value which is a function of rainfall intensity and duration, ground cover, ground disturbance, soil erodability, slope, area of exposure and time of exposure. The Universal Soil Loss Equation [1], which was originally developed to predict sheet erosion from agricultural lands, is most often used to predict erosion from construction sites.

UNIVERSAL SOIL LOSS EQUATION

$$E = (LS) \, R \, K \, C \, P$$

E = soil erosion – tons/acre/year
(LS) = slope length factor
R = rainfall factor

TABLE 1. Representative Erosion Rates for Different Land Use Activities.

Land Use	Erosion	
	Tons/Acre/Year	(Kilograms/Square Miles/Year)
Forests	0.5	(0.11)
Developed Urban Areas	0.5	(0.11)
Grasslands	5	(1.12)
Croplands	40	(8.97)
Harvested Forests	70	(15.69)
Active Surface Mines	100	(22.42)
Uncontrolled Construction	100	(22.42)

Use of fiber mesh and hay mulch to control erosion on severe slopes.

Construction site silt basin with evidence of sediment buildup.

Soil erosion at cleared construction site.

Erosion at highway embankment under construction.

Highway embankment with topsoil cover and hay mulch.

K = soil erodability factor
C = ground cover index factor
P = surface condition practice factor

A determination of the erosion from a hypothetical eight-acre (32,376 m²) construction excavation site located in Ridgebury, New Jersey, shown in Figure 1, is presented. For the construction site, the slope is 4 percent with a slope length of 600 feet (183 m). This produces an (LS) factor of 1.01. The rainfall intensity factor (R) is obtained from a U.S. Conservation Service rainfall factor chart for the State of New Jersey. Ridgebury, New Jersey, is located in the $R = 175$ region of the state. A soil investigation of the construction site shows a fine sandy loam soil (SM) in the horizon to be disturbed by the excavation work. This yields a soil erodability factor (K) of 0.3. The cover index factor (C) will vary from a maximum value of 1.0 for no vegetation cover to a value of 0.01 for a sod cover. Wood chips, temporary seeding, hay mulch or fiber matting can all be used to protect the soil from erosion, and reduce the cover index factor. If the site is exposed for the full year without any temporary cover, the cover index factor (C) would be 1.0. The surface condition practice factor for construction sites (P) will vary with the type of surface earthwork being done. Compact and smooth earth as scraped with a bulldozer or scraper up and down a slope will produce a factor of $P = 1.3$. The same slope raked with a bulldozer root rake across the slope will create small horizontal grooves across the slope and will reduce the factor to 0.9. Leaving sections of our hypothetical site with the proper surface condition at the end of the day will reduce erosion. An estimated composite value of $P = 1.1$ is made for our site. Substituting these factors into the equation yields an estimated erosion rate of 58 tons/acre/year (13 Kg/m²/year) from our site.

The amount of sediments that will reach the Recreational Pond adjacent to our hypothetical construction site will depend on the percentage of eroded material that will reach the Playground Brook. Since the brook is in close proximity to the point of erosion, most of the material will reach the pond. If an estimated 95 percent of the eroded material reaches the pond, the sediment load will be $(.95)(58)(8) = 441$ tons/year (400,075 Kg/year).

Measuring Sediment Loads

Sediment measuring techniques are well developed and the A.S.C.E. special publication No. 54 is an excellent reference. The U.S. Soil Conservation Service and the U.S. Geological Survey maintain sediment load records for many rivers and streams, and are good sources of information.

Allowable Sediment Loads

The additional amount of sediments that can be introduced into adjacent waters without environmental damage depends on the present quality of the water, the type of aquatic life these waters support, the way the water is used downstream of the construction site, and where the eroded sediments will finally be deposited. Obviously, a construction site that will produce additional sediment and turbidity in a clear mountain trout stream that is used as a municipal water supply downstream of the construction site will require very different control measures than a site that is located at the mouth of the Mississippi River where the river is already carrying a high sediment load, and is about to enter the Gulf of Mexico.

For our construction site, the sediments can cause increased siltation and turbidity in the waters of the Recreational Pond, impairing the pond's use as a recreational resource. A specific determination of environmentally acceptable increased levels of sedimentation and turbidity in the stream and pond would require a multidisciplinary examination of the problem.

Erosion, Sedimentation, and Turbidity Control

The technology for controlling soil erosion, sedimentation and turbidity is fairly well established and it is the responsibility of those engaged in construction work to learn to utilize the proper level of available control procedures. While basic soil erosion practices should be implemented on all construction sites as a normal part of good excavation procedure, the utilization of extensive mulching, large sedimentation basins and turbidity filtration plants should be used only where these are really necessary. Instituting the most stringent control measures presently possible at every construction site would be just as great a waste of our national resources as was our past policy of no controls at all.

Typical construction site erosion and sedimentation control measures used in New Jersey [7] include:

1. The diversion of off-site rainfall runoff away from construction sites so that this water does not run over bare earth areas.
2. The control of the amount of area exposed at the site at any one time.
3. Limitations on the time of bare earth exposure.
4. Protection of completed excavation areas by temporary mulching or seeding when immediate landscaping is not feasible.
5. Channelling of on-site runoff water to silt basins, silt fences, or silt barriers.
6. Holding or filtering of water pumped from dewatering operations to remove suspended solids.
7. Protection of slopes by the use of diversion ditches, diversion terraces, temporary slope drains and jute or man-made mesh products.

For our construction site, several of these measures might be implemented. The slope length factor (LS) could be reduced by installing diversion ditches at 200 foot (61 m) intervals if this is compatible with the construction work. This may require the protection of the diversion ditches with jute mesh so that the decrease in sheet erosion is not offset by an increase in gully erosion in the diversion ditches. Temporary mulching or seeding could simultaneously be used to protect the surface after excavation work in a portion of the site has been completed. Reducing the bare earth exposure to a four-month period by using a two ton per acre (0.45 Kg/m^2) hay mulch application to cover the site for the remaining eight months would further reduce erosion. The intermediate and bottom diversion ditches (shown in Figure 2) will carry eroded material to a silt basin. The effectiveness of the silt basin would depend on the type of soil to be trapped and the size of the basin. Our hypothetical silt basin could be expected to catch 50 percent of the eroded sediments. Instituting all the abovementioned measures would reduce our erosion to 11 tons/acre/year (2.47 Kg/m^2/year), and reduce our sediment load to 42 tons/year (38,102 Kg/year).

The implementation of an erosion and sedimentation control plan in New Jersey is not only a legal and environmental necessity, but is also good public relations and can often be cost beneficial. The cost of bringing in new fill material and regrading of eroded slope areas can be avoided when proper control measures are instituted. The frequency of lawsuits against contractors for damaging trees and lawns on adjacent properties and for silting up adjacent lakes and streams can also be reduced.

Dredging

Those construction and non-construction sediments that eventually do settle out may have to be removed by dredging. Dredging is used to reopen navigable channels, for laying underwater pipelines, for mining fill materials, and for restoring recreational areas.

The problems of dredging are usually twofold: (a) the dredging operation itself which can produce severe siltation and turbidity, and (b) the disposal of the dredged spoil material. Legislation such as the Water Pollution Control Act of 1972, with Section 404 regulating spoil disposal in inland waters, and the Marine Protection Research and Sanctuary Act of 1972, with Section 103 (the Ocean Dumping Act) regulating ocean disposal, have focused attention on the spoil disposal problem.

Spoil disposal in suitable dry upland areas is expensive due to the transportation costs involved in the shipment of the spoil. Disposal in adjacent wetland areas is no longer an environmentally acceptable solution. Dumping of spoil into deep water ocean areas such as the New York Bight area is being more strictly regulated. The Ocean Dumping Act now requires bio-assay testing of potentially dangerous spoil materials to determine their toxicity to marine organisms. These bio-assay tests consist of over one hundred individual tests that usually require one to three months to complete and can cost upward of $10,000.

Some navigable waterways such as the Houston Ship Channel in Texas and Newtown Creek in Long Island City, New York, are believed to contain so many harmful chemicals mixed in with the bottom sediments that spoil material from these areas may not pass the bio-assay tests. Not only

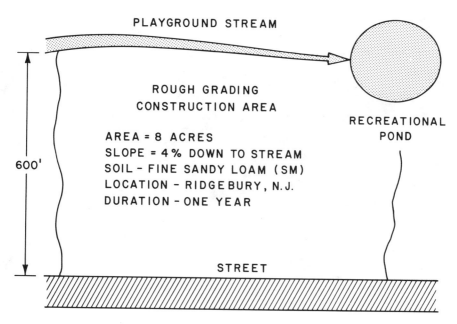

FIGURE 1.

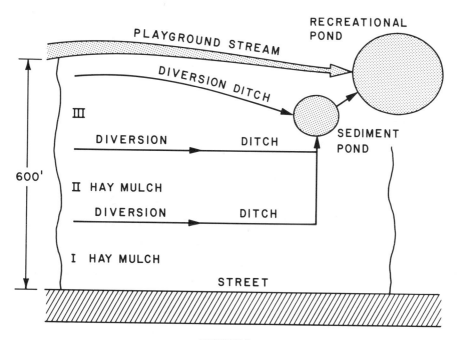

FIGURE 2.

Settling tank for subway excavation wellpoint dewatering effluent.

will it be difficult to find a suitable spoil disposal site for this type of sedimentation, but the release of toxic pollutants during removal may prohibit dredging operations in the first place.

SUMMARY

The control of soil erosion, sedimentation and turbidity originating as a result of construction excavation operations has received increased legislative attention.

The federal government and many states are seeking to minimize construction related soil erosion through legislation and the enactment of water quality standards. While construction sites are certainly not the sole source of soil sediments, uncontrolled construction work can produce significantly increased levels of soil erosion.

Fortunately, the technology for minimizing soil erosion from construction sites is well developed. What is now needed is that engineers, planners and builders become more familiar with basic good practice soil erosion control techniques and adopt them as an integral part of construction excavation work.

GROUND VIBRATIONS

Ground vibrations caused by construction operations such as blasting, pile driving, demolition, and compaction have a potential to cause damage to adjacent structures and can be a major source of public complaint.

Vibration producing construction operations cause waves to propogate through the surrounding soil and rock. The waves are transmitted outward from the source and gradually are attenuated with distance due to geometric and frictional dumping. The parameter that is most often used to describe and measure ground motion is particle velocity. This is the velocity of displacement of an individual particle as the vibration wave passes that particle. Particle velocity can be measured by an engineering seismograph and is expressed in inches per second.

Many years of collective experience in monitoring construction vibrations have established that particle velocity is useful in predicting potential damage to adjacent structures. Ground vibrations are complex sinusoidal-type wave forms and wave characteristics other than particle velocity such as frequency and duration also affect damage potential.

The frequency of ground vibrations is an important factor when it approaches the natural frequency of an affected structure. This can produce a resonance condition where the motion at the foundation level is magnified in the structure. Most building structures and their components have natural frequencies between 5 and 40 hertz. In some instances, it is possible to reduce the damage potential to a building by changing the vibratory frequency of a piece of construction equipment away from the natural frequency of an adjacent structure.

Total duration of the vibration is another key element in predicting damage potential. For equivalent wave characteristics, long lasting steady state vibrations such as those produced by vibrocompaction and vibratory pile driving tend to cause more damage to structures than impulse or transient

Foundation blasting work in urban areas requires ground vibration and noise monitoring.

Pile driving operations close to occupied buildings can produce excessive noise levels.

vibrations such as those produced by blasting. The safe level of intensity for steady state vibrations should be one-half the safe level for transient vibrations.

In addition to the direct damage potential effects of vibrations on structures, there is also the effect of the vibrations on soils through which the vibration wave passes. It is well documented that vibratory energy can be used to compact fill materials for roads and dams. The densification of the soil due to vibrations is undesirable, however, when such soils are supporting a structure, as differential settlements and structural damage to the building are possible. Dry, loose to medium dense sands, and saturated loose to medium dense sands are most susceptible to densification [19]. Partially saturated or moist sands are less susceptible to settlement. The settlements in sands occur during the construction operation, as opposed to occurrence at some future time. Cohesive soils such as clays are not particularly sensitive to vibrations.

Vibration Criteria

Construction contractors and the general public both need adequate safeguards based on factual data to protect their specific interests. Contractors need a reliable vibration damage criteria on the basis of which they can plan and conduct their construction activities without causing structural or environmental damage. The public benefits by the absence of conditions which would produce damage.

During the 1970s, the most widely accepted damage criterion for safe ground vibration levels had been that the peak resultant particle velocity at the affected structure should not exceed 2 inches per second. This criterion was based on statistical studies of blast data near wood frame single family dwellings by Nicholls, et al. [31] for the U.S. Bureau of Mines. The 2 in/sec. safe vibration limit was a probabilistic type of criterion with a 95% reliability level. This meant that a small percentage of affected structures may experience some minor damage when vibration levels were less than 2 in/sec., while other structures could tolerate vibration levels significantly in excess of 2 in/sec. with no observable damage.

With the publication of the Bureau of Mines Report No. 8507 in 1980 by Siskind, et al. [24], recommended vibration criteria for residential structures were modified. Safe levels of peak particle velocity were now a function of the frequency of vibration and ranged from 0.5 to 2.0 in/sec. as shown in Figure 3. Historic and sensitive older buildings require more conservative vibration criteria ranging from 0.25 to 0.5 in/sec for transient vibration levels [32].

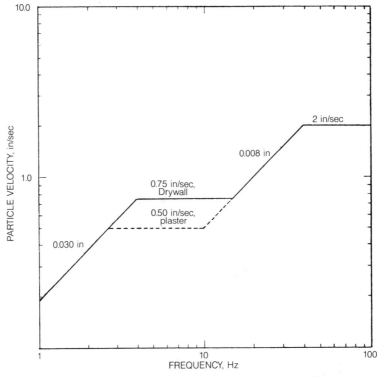

FIGURE 3. Safe levels of blasting vibration for houses using a combination of velocity and displacement. Courtesy: Siskind, et al., USBM—RI 8507 Ref. [24].

Air Blast

During blasting operations, an air blast is produced as a result of venting of gases to the atmosphere. Air blast effects are measured in pounds per square inch of over-pressure produced at the most critical location. The USBM recommends a safe airblast over-pressure level of 0.5 psi [31]. Confinement of the blast in the boreholes by the proper use of stemming and mats reduces the air blast by allowing a more gradual release of gases. Except in extreme cases (no stemming or mats) control of blasting procedures to limit ground vibrations to levels below 2.0 in/sec automatically limits over-pressure to safe levels. Air blast, therefore, is an insignificant factor in causing damage to buildings, but may be a major source of complaint from residents especially during periods of temperature inversion.

Effects of Vibrations on Humans

Vibration levels which are safe for structures can be annoying and uncomfortable when viewed subjectively by human beings. Goldman analyzed human response to steady-state vibrations [33]. These results were converted to

TABLE 2. Distance Versus Weight of Explosives Method.[1]

Distance to a Building		Weight of Explosive per Delay pounds	Distance to a Building		Weight of Explosive per Delay pounds
feet over	feet not over		feet over	feet not over	
0	to 5	¼	250	to 260	45
5	to 10	½	260	to 280	49
10	to 15	¾	280	to 300	55
15	to 60	[2]	300	to 325	61
60	to 70	6	325	to 350	69
70	to 80	7¼	350	to 375	79
80	to 90	9	375	to 400	85
90	to 100	10½	400	to 450	98
100	to 110	12	450	to 500	115
110	to 120	13¾	500	to 550	135
120	to 130	15½	550	to 600	155
130	to 140	17½	600	to 650	175
140	to 150	19½	650	to 700	195
150	to 160	21½	700	to 750	220
160	to 170	23¼	750	to 800	240
170	to 180	25	800	to 850	263
180	to 190	28	850	to 900	288
190	to 200	30½	900	to 950	313
200	to 220	34	950	to 1000	340
220	to 240	39	1000	to 1100	375
240	to 250	42	1100	to 1200	435
			1200	to 1300	493

[1]This table over 60 feet is based upon the formula $W = D^{1.5}/90$.
[2]One tenth of a pound of explosive per foot of distance to a building.
Courtesy: Explosives, N.J.A.C. 12:190, Ref. [29].

particle velocities and presented in USBM Bulletin 656 [31] (see Figure 4). Since frequencies generated by blasting are usually in the range of 10 to 40 hertz, people in a building subjected to a particle velocity of 0.5 in/sec produced by construction blasting typically view this level as unpleasant.

The technical problem of quantifying human response to vibrations is complicated by window and dish rattling and by the presence of air blast during blasting. Siskind, et al. [24] observed that "5 to 10 percent of the neighbors will judge peak particle velocity levels of 0.5 to 0.75 in/sec as less than acceptable based on direct reaction to vibration. Even lower levels cause pychological response problems, and thus social, economic, and public relations factors become critical."

Due to the subjective nature of ground vibrations, it is important that a contractor institute a public education program to convince nearby residents that rattling is to be expected and is not damaging. When blasting, a siren warning system can be used to inform the public of each blast prior to detonation. Blast warning signals are usually: two long signals—first warning; one long signal—final warning one minute to blasting; two short signals—all clear. In the

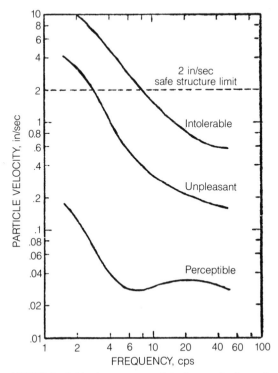

FIGURE 4. Subjective response of the human body to vibratory motion (after Goldman). Courtesy: Nicholls, et al., USBM Bul. 656, Ref. [31].

TABLE 3. Comparison of Construction Equipment Noise Ordinances.

Type of equipment (1)	Los Angeles ordinance effective date (2)	Los Angeles ordinance noise limits, in decibels at 50 ft (3)	City of Chicago ordinance effective date of manufacture (4)	City of Chicago noise ± limits, in decibels at 50 ft (5)	City of New York effective date (6)	City of New York noise limits in decibels at 1 m (7)
Construction and industrial machinery, such as crawler-tractors, dozers, rotary drills and augers, loaders, power shovels, cranes, derricks, motor graders, paving[1] machines, off-highway trucks, ditchers, trenchers, compactors, scrapers, wagons, pavement breakers, compressors, and pneumatic powered equipment, etc.	July 7, 1973	90	January 1, 1973	88		
	July 1, 1975	83	January 1, 1975	86		
	July 7, 1977	75	January 1, 1980	80		
					December 31, 1973	94[1]
					December 31, 1975	90[1]

[1]Paving machines or paving breaker.
Courtesy: Review of Construction Noise Legislation, ASCE, Ref. [34].

absence of a public relations program, a ground vibration level of 0.5 in/sec in a community can be expected to produce a 15 to 30 percent annoyance level [24].

Seismic Monitoring

Peak particle velocities should be monitored with an engineering seismograph equipped with an air wave detector to monitor air blast. When blasting, prudent construction practice will dictate the use of delay firing and also that initial blast charges be kept below that expected to produce the maximum allowable vibration levels. Successive blasts can then be "worked up" gradually to develop a calibrated relationship at a particular site between charge and expected vibration level.

Some building codes permit the use of the scaled distance relationship for blast vibration control in lieu of seismic instrumentation. The scaled distance relationship $D/\sqrt{W}$, where D = distance in feet and W = charge weight per delay in pounds, can be related to partical velocity. Typically a scaled distance of 20 ft/$\sqrt{\text{lbs}}$ can be used for a peak partial velocity criteria of 2.0 in/sec. For example, for a scaled distance factor of 20 ft/$\sqrt{\text{lbs}}$, if a potential damage point is 100 ft. away from the blast source, then 25 pounds of explosive per delay can be detonated without exceeding the safe vibration criterion of 2.0 in/sec.

The state of New Jersey code regulating the use of explosives, N.J.A.C. 12:190-7.26, presents the scaled distance relationship by means of a distance versus weight of explosive method shown in Table 2.

Along with seismic monitoring, a comprehensive construction control program consisting of a preconstruction survey of nearby structures, horizontal and vertical survey control, and strain tell-tales at existing distress points should be undertaken.

Construction Noise

Construction activities and the operation of construction equipment will produce noise. Many cities in the U.S. have enacted noise legislation dealing specifically with construction noise. In 1974, the EPA proposed noise emission standards for portable air compressors. These measures and a growing public awareness of noise served to alert contractors to include noise control as a factor in planning, cost estimating, and performing construction work in populated areas.

SOUND

The human ear is an extremely sensitive organ for detecting vibrations in air called sound waves. These waves can be detected by the human ear if the frequencies lie between 20 and 20,000 hz and if their intensities are within a certain range. The human ear is most sensitive to sounds of about 3,000 hz. Loudness is the subjective response of the human ear to changing sound wave intensities and frequencies. Noise is simply unwanted sound.

Sound levels are measured in decibels (db). The decibel scale starts at 0 from a chosen reference value and compares sound intensities to that reference value. The chosen reference value is the threshold of hearing of a young person. It must be noted that the decibel scale is logarithmic in nature. Sound intensity levels in decibels from independent sources cannot be added directly. For example, if two sound sources measure 60 dB each, they would measure 63 dB when operating together. A 3 dB change which is actually a doubling of the sound intensity level is typically just perceptible to humans. A 10 dB increase is needed to cause a subjective doubling of loudness.

In measuring sound levels, it is important that the sound descriptor be capable of correlating well with human reaction to sound. To this end, the "A" weighted sound pressure level (dBA) was developed. It is designed to approximate the response of the human ear to sound and correlates well with human subjective assessment of the loudness of sound. A second important noise description is the equivalent sound level (Leq) which is the average sound level integrated over some specified period of time, for example 60 seconds (60 sec Leq). The (Leq) measurement provides a numerical value of a fluctuating sound level which is equivalent to a steady-state sound with the same sound intensity level. The level of sound measured for Leq is usually taken to be A-weighted.

NOISE REGULATIONS

The lack of uniformity of existing community noise regulations is a source of considerable difficulty to the construction industry. Equipment that is acceptable at one location may be in violation of noise ordinances at another location. A comparison of construction equipment noise ordinances for three U.S. cities is presented in Table 3 [34]. While these are typical, a check of local codes must be made before starting a project.

Construction machinery manufacturers are also aware of the noise problem and strive to produce equipment that meets customer and regulatory demands. In 1974, the EPA in 39 CFR 7954-95 established noise emission standards for portable air compressors. This required new portable air compressors to have a maximum average sound level of 76 dBA when measured at a distance of 23 ft. from the surface of the compressor enclosure. It is up to the contractors to maintain the compressor's ability to meet noise regulations. A listing of construction equipment sound levels is presented in Table 4 [34].

When construction equipment is inherently too loud, contractors can use sound barriers. Depending on the type of equipment and its noise level, noise reduction barriers can be plywood panels, walls, curtains, or partial enclosures. In New York City, a contractor can apply for a permit to exceed code levels if it can be demonstrated that there is no reasonable way to prevent the levels from being exceeded.

The Occupational Safety and Health Administration (OSHA) construction regulations, 1926.52, dealing with occupational noise exposure, require that employees not be exposed to daily noise doses exceeding 90 dBA for an 8-hour day, or greater noise levels for shorter periods. When employees are subjected to sound levels exceeding OSHA standards, feasible administrative or engineering controls shall be utilized. If such controls fail to reduce sound levels, personal protective equipment such as ear plugs or ear muffs shall be provided and used to reduce sound levels to acceptable OSHA norms.

It is the responsibility of the contractor not to generate excessive noise at the construction site. While community noise reactions are subjective in nature, violation of local noise codes can result in fines and work stoppage. To avoid this, the contractor should be fully aware of noise regulations, maintain equipment in good operating condition, monitor sound levels, and include noise control as a factor in planning the work.

TABLE 4. Sound Pressure Levels for Construction Equipment.

Type of equipment (1)	Levels, in decibels, (A-weighted) fast (2)
Scrapers	89–95
Scrapers, elevating	88
Graders	77–87
Dozers	87–89
with squeaky tracks	90–93
Dozers, sheepsfoot	82–88
Rollers	72–80
Rollers, vibrating	80–85
Loaders, bucket	80–81
Loaders, terex	96
Backhoe	79–91
Gradall	87–88
Crane	80–85
Trucks, off highway	81–96
Trucks, asphalt	69–82
Trucks, concrete	71–82
Trucks, cement	91
Trucks, 14-wheel	88
Tractors, with water pump	73–80
Pavers	82–92
Autograder	81
Compressors	71–87
Rock drills (handheld, pneumatic)	88
(track mounted)	91
Concrete saws	87
Concrete saws, chain	88–93
Water pumps	79
Concrete pumps	76
Generators	69–75
Concrete plant	93
Asphalt plant	91
Pile driver (Vulcan No. 1)	90

Courtesy: Review of Construction Noise Legislation, ASCE, Ref. [34].

REFERENCES

1. *Agricultural Handbook No. 282*, Washington DC, Department of Agriculture (1965).
2. Billings, L.G., "The Evolution of 208 Water Quality Planning," *Civil Engineering*, Vol. 46, No. 11, pp. 54-55 (Nov. 1976).
3. Cheremisinoff, P. N. and A. C. Morresi, *Environmental Assessment and Impact Statement Handbook*, 1st Ed. Ann Arbor Science, Ann Arbor, Michigan (1977).
4. *Control of Turbidity at Construction Sites*, U.S. Department of the Interior, Bureau of Reclamation (Dec. 1977).
5. Dallaire, G., "Controlling Erosion and Sedimentation at Construction Sites," *Civil Engineering*, Vol. 46, No. 13, pp. 73-77 (Oct. 1976).
6. Pisano, M., "208: A Process of Water Quality Management," *Civil Engineering*, Vol. 46, No. 11, pp. 73-77 (Nov. 1976).
7. *Standards for Soil Erosion and Sediment Control in New Jersey*, New Jersey State Soil Conservation Committee (1974).
8. *Valleys and Hills, Erosion and Sedimentation*, U.S. Department of Agriculture, 1955 Yearbook of Agriculture, pp. 135-143.
9. Koehn, E. and J. A. Rispoli, "Protecting the Environment During Construction, *Journal of the Construction Division, ASCE, Vol. 108*, pp. 233-246 (June 1982).
10. "Erosion Control During Highway Construction—Manual on Principles and Practices," Transportation Research Board, National Research Council, Washington, DC (Apr. 1980).
11. Borg, R.F., "Social and Environmental Concerns in Construction," *Journal of the Construction Division, ASCE*, Vol. 102, pp. 1-20 (Mar. 1976).
12. Ashley, C., "Blasting in Urban Areas," *Tunnels & Tunneling*, Vol. 8, No. 6, pp. 60-67 (Sept. 1976).
13. Bata, M., "Effects on Buildings of Vibrations Caused by Traffic," *Building Science*, Vol. 6, pp. 221-246 (1971).
14. Chae, Y. S., "Design of Excavation Blasts to Prevent Damage," *Civil Engineering*, ASCE, Vol. 48, No. 4, pp. 77-79 (Apr. 1978).
15. Edwards, A. J. and Northwood, T. D., "Experimental Studies of the Effects of Blasting and Structures," *The Engineer*, Vol. 210, pp. 538-546 (Sept. 30, 1960).
16. Esrig, M. I. and A. J. Ciancia, "The Avoidance of Damage to Historic Structures Resulting from Adjacent Construction," Reprint 81-052, ASCE (1981).
17. Esteves, J. M., "Control of Vibrations Caused by Blasting," Memoria 498, Laboratorio National De Engenharia Civil, Lisboa, Portugal (1978).
18. German Institute of Standards, "Vibrations of Buildings; Effects on Structures," Pastfach 1107, D-1000 Berlin 30, DIN 4150, Part 3 (Mar. 1983).
19. Heckman, W. S. and Hagerty, D. J., "Vibrations Associated with Pile Driving," *Journal of the Construction Division*, ASCE, Vol. 104, No. CO4 (1978).
20. Hendron, A. J., Jr. and L. Oriard, "Specifications for Controlled Blasting and Civil Engineering Projects," *Proceedings, AIME*, Vol. 2, pp. 1585-1609 (1972).
21. Medearis, K., "Rational Damage Criteria for Low-Rise Structures Subjected to Blasting Vibrations," Part 2, Vol. 65, pp. 611-621 (Sept. 1978).
22. Richart, F. E., J. R. Hall, and R. D. Woods, *Vibrations of Soils and Foundations*, Prentice-Hall, pp. 93-139 (1970).
23. Rudder, F. F., Jr., "Engineering Guidelines for the Analysis of Traffic-Induced Vibrations," *Federal Highway Administration Report No. FHWA-RD-78-166* (Feb. 1978).
24. Siskind, D. D., et al., "Structure Response and Damage Produced by Ground Vibration from Surface Mine Blasting," *United States Bureau of Mines Report of Investigation 8507* (1980).
25. Swiss Consultants for Road Construction Association, "Effects of Vibration on Construction," *VSS-SN640-312* Seefeldstrasse 9, CH 8008, Zurich, Switzerland (Nov. 1978).
26. Whiffin, A. C. and D. R. Leonard, "A Survey of Traffic-Induced Vibrations," *RRL Report LR418*, Road Research Laboratory, Department of the Environment, Great Britain (1971).
27. Wiss, J. F., "Construction Vibrations: State-of-the-Art," *Journal of the Geotechnical Engineering Division*, ASCE, Vol. 107, No. GT2 (1981).
28. Wiss, J. F., "Damage Effects of Pile Driving Vibrations," *Highway Research Board Record 155*, pp. 14-20 (1967).
29. "Explosives," N.J.A.C. 12:190, New Jersey Department of Labor, Division of Workplace Standards, Office of Safety Compliance, C 386, Trenton, NJ 08625 (Oct. 15, 1982).
30. Gaziogler, S. M. and R. E. Langston, "How to Control Blast Vibrations in an Urban Area," *Journal of Explosive Engineering*, Vol. 2, No. 2 (July/August 1984).
31. Nicholls, M. R., et al., "Blasting Vibrations and Their Effects on Structures," USBM, Bulletin 656 (1971).
32. Konon, W. and J. R. Schuring, "Vibration Criteria for Historic and Sensitive Older Buildings," ASCE Reprint 83-501 (Oct. 19, 1983).
33. Goldman, D. E., "A Review of Subjective Responses of Vibrating Motion of the Human Body in the Frequency Range—1 to 70 Cycles per Second," Naval Medical Res. Inst. (Mar. 1948).
34. Task Committee on Noise and Vibration of the Committee on Social and Environmental Concerns in Construction of the Construction Division, "Review of Construction Noise Legislation," *Journal of the Construction Division, ASCE*, Vol. 103, pp. 123-137 (Mar. 1977).
35. *Community Noise*, Pub. No. NTID 300.3, EPA, Washington, DC (Dec. 1971).
36. Hersh, A. S., "Construction Noise Its Origins and Effects," *Journal of the Construction Division, ASCE*, Vol. 100, pp. 433-448 (Sept. 1974).
37. von Gierke, H. E., "Noise—How Much is Too Much?," *Noise Control Engineering*, Vol. 5, No. 1 (July/August 1975).
38. Romano, F. M., "Prediction and Analysis of Highway Construction Noise," ASCE Annual Convention (Oct. 21, 1983).

CHAPTER 10

Construction: Traffic Control at Freeway Work Sites

ZOLTAN A. NEMETH,* AJAY K. RATHI,** AND NAGUI M. ROUPHAIL[†]

INTRODUCTION

Freeway driving is characterized by usually free flowing traffic, by the absence of interruption by traffic control, and by generous design standards. There are no sudden changes in the alignment nor in the width of the traveled way.

Reconstruction, rehabilitation, and maintenance operations, however, disrupt these normal conditions. The combination of high speed traffic, constricted road conditions, and distracting construction activities near the traveled way produce potentially unsafe conditions. This has been recognized and documented by several accident studies [1,2,3]. The function of traffic control at work sites is to move traffic safely and predictably past the work area and also to provide protection for the construction or maintenance crew.

The Manual of Uniform Traffic Control Devices [4] is published by the Federal Highway Administration and presents the basic standards of the design and application of traffic control devices. It is usually referred to in transportation engineering as the *MUTCD*. Part IV of the *MUTCD*, "Traffic Controls for Street and Highway Construction and Maintenance Operations" provides the minimum standards for traffic control at the most common work site situations on public streets and highways.

The *MUTCD* is augmented by the *Traffic Control Devices Handbook* [5], also published by the Federal Highway Administration. It links the *MUTCD* standards with the relevant traffic engineering practices and traffic control devices.

The Federal-Aid Highway Program Manual [6] also includes sections on work site traffic control standards and procedures required on Federal-aid projects.

The area within which traffic is affected by work site traffic control is called *traffic control zone*. A typical traffic control zone is divided into five segments as shown in Figure 1, reproduced from *MUTCD* [4]. The purpose of the *advance warning area* is to alert drivers to the approaching construction site and to provide sufficient time to respond to the condition. Many freeway work sites require the closure of one or more lanes. The transition area provides the last opportunity for drivers to merge into the open lane. The gradual closing of the lane by channelizing devices is referred to as the establishment of a *taper*. If construction work requires the closing of both lanes, then traffic is channelized through a crossover into the other side of the freeway which is to be operated as a bidirectional or two-lane, two-way roadway. The *buffer space* between the taper and the work area provides a safety margin or escape area for drivers crashing through the taper. It should be kept free of construction equipment and material to allow such drivers to recover and stop before reaching the work site. The *work area* is the site occupied by construction equipment, material and construction workers. The work area is often shielded by barriers to protect workers. The *termination area* permits drivers to return to normal operation.

The material presented in this chapter is intended to serve as a comprehensive reference on traffic control in the special case of freeway work sites. For background information and for more details the reader is referred to the list of references at the end of this chapter. An effort was made to include only readily available materials without compromising the comprehensiveness of the list.

Basic Principles of Traffic Control at Freeway Work Sites

There are some basic principles which will tend to increase safety when applied in the planning and operation of

---

*Department of Civil Engineering, The Ohio State University, Columbus, OH
**KLD Associates, Huntington Station, NY
[†]The University of Illinois at Chicago, Chicago, IL

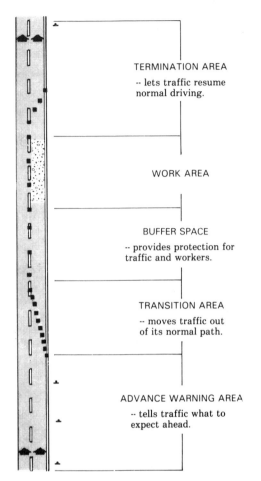

FIGURE 1. Subareas of traffic control zone. Source: Reference [4].

traffic control at freeway work site. Most of these principles apply to non-freeway type of work sites as well and are included in the *MUTCD* [4]. The basic principles include the following:

1. All construction projects should be planned in a manner that treats traffic *safety* an *integral* and important element of the plan.

   Formal Traffic Control Plans should be completed by qualified personnel. All key members on the construction force need to understand the Traffic Control Plan and appreciate the significance of it. This Traffic Control Plan may play a significant role in a potential liability case resulting from an accident.

2. Traffic flow should be kept *as free of restrictions* as possible. Unless some unusual situation clearly requires speed reduction, speed zoning is not advisable. Advisory speed signs are notoriously ineffective means of reducing approach speeds. If speed reduction is deemed necessary, then additional measures have to be taken to induce motorist compliance (e.g., have highway patrol present). Sudden and frequent variations in geometrics should be avoided as much as possible. In anticipation of run-off-the-road incidents, a roadside recovery area should be kept as free of construction equipment and material as practical.

3. Motorists should be given adequate information on the conditions prevailing at the work site.

   Advance warning should be given as far in advance as needed to prepare approaching motorists for possible congestion upstream of the work site. Additional advance warning should be given on adjacent facilities when entrance ramps are located between work sites and advance warning signs.

4. Exposure to the motorists to potential hazards should be minimized.

   Construction time should be kept as short as possible. Construction at night or during low travel seasons are means by which exposure can also be reduced. Work vehicles should not operate on the traveled way without extra caution.

5. When lane closure is required, a merging zone of sufficient length must be provided to enable traffic in the closed lane to merge with the open lane traffic.

   In general, geometrics and traffic control should be as comparable as possible to normal conditions. Crossovers and bidirectional zones should be avoided when possible. When bidirectional zones are introduced, physical separation of opposing flow by barriers should be considered. If barriers are not justified, motorists need to be constantly reminded of the two-way separation.

6. Routine inspection of traffic control devices is an absolute necessity.

   The inspection is to be performed by individuals familiar with the principles of effective traffic control. The individual responsible for safety at the work site must have sufficient authority to modify traffic control or working conditions as needed. Traffic control devices which are no longer needed should be removed as soon as possible.

7. The provision of safety requires uninterrupted care throughout the duration of the project.

   Such signs of unsafe conditions as traffic control devices damaged by vehicles or a number of skid marks in the area should be evaluated. If accidents occur at the work site, accident records should be examined. Modifications in traffic control should be made in response to safety problems thus identified.

Driver Information Needs

Drivers need adequate information on conditions to be encountered at freeway work sites in order to be able to perform the proper driving tasks safely and efficiently. Interdisciplinary teams of researchers have studied the relationship between driving tasks and information needs. Guidelines have been developed for effective transmission of the proper information for drivers in many different situations [7], including at work zones [8,9]. Human factor principles were applied to develop an effective approach to evaluate driver information needs in potentially hazardous situations. The approach is based on the Positive Guidance concept [10,11]. It is founded on the premise that drivers can handle a variety of potentially hazardous situations if sufficient information is given on the nature of the hazard and on the action needed by the driver to safely perform the driving tasks. Hazards are defined as objects, conditions, or situations [9]. *Object hazards* may be fixed or moving and will produce severe damage when struck by a vehicle. Some type of barriers used at work sites fall into this category. Construction equipment and construction material hauling trucks are examples of moving object hazards. The recognition of fixed object hazards is a simple task, but it is slightly more difficult to evaluate moving object hazards. The latter involves the proper assessment of the relative speed and path of construction vehicles or other moving object hazards.

The recognition of highway *condition hazards* is more demanding. The often poor condition of pavement within the traffic control zone and reduced lane width are typical examples of condition hazards.

The most difficult to detect danger is the *situation hazard*. It involves a combination of elements which are not necessarily hazardous by themselves. These elements often exist for a short period of time or over a short segment of roadway only. An example might involve the combination of construction related dirt on the roadway, rain, and an inattentive driver overtaking slower traffic in a single lane zone. These situation hazards are responsible for many run-off-the-road type accidents at work sites.

The Positive Guidance approach is based on a conceptual model of the driving task [11]. It identifies three distinct levels of driver performance, each relying on different information sources. The three levels of driver performances are navigational, control, and guidance levels. The *navigational* level of driver performance involves trip planning and maintaining the proper direction to reach the desired destination. The needed information is provided by maps and guide signs.

The *control* level of driver performance involves speed control and steering. In their efforts to maintain a steady speed and to stay on course, drivers rely on visual cues, such as the position of lane and edgemarkings on the pavement. In normal freeway driving, characterized by gentle geometrics and free flowing traffic, this level of driving is performed without conscious attention.

When conditions change, for example at freeway work sites involving lane closures, drivers appraise the situation, make decisions, and follow up with action. Drivers now perform at the *guidance* level. Driving at this level is a decision making process. The basic decision involves judgment regarding speed selection and lane placement. This judgment is a continuous process in response to changing roadside conditions. Decisions related to avoiding hazards at the work sites are integrated into the lane placement and speed selection task. The speed selection is the basic decision involved in a car following task. In order to maintain a safe gap drivers continue to adjust their speed in response to the changes in the speed of vehicles ahead of them. A somewhat more complex decision is involved in overtaking and passing since both speed control and adjustment in lane placement are involved. Other typical guidance level performances are gap acceptance, merging, and responding to traffic control information. Traffic control devices providing information include pavement markings and delineators. Temporary channelizing devices and construction signs are the major carriers of information at work sites, often supplemented by attention getting devices. Some basic understanding of how drivers receive information and respond by speed or lane placement decisions is necessary. Two concepts that are especially important are *primacy* and *expectancy*. The concept of primacy recognizes that at any one instant some information is relatively more important to the driver than other information. The main factor in judging the primacy of different types of information is the potential severity of consequences of failing to receive an information. Consider the different levels of driver performance: information missed in the *control* or *guidance* level performance is more likely to result in an accident than the missing of a *guide* sign. One implication is that guide signs (e.g., interchange signs) located in the transition zone at a lane closure site are likely to be missed by drivers involved in scanning the open lane traffic in preparation to merge. From the traffic engineer's point of view it also means that more effort should be made to successfully transmit information to drivers on approaching crossovers and bidirectional zones than on less demanding and thus less hazardous types of traffic control zones.

Another criterion used in assessing the primacy of information is the distance factor. Thus, information on a nearby hazard has higher priority than that related to a more distant hazard of similar importance. The concept of *expectancy* asserts that past experience influences drivers in their interpretation of the given information. Prior experience is important in recognizing hazards in traffic control zones at freeway work sites. The credibility of signs at freeway work sites has been damaged by the fairly common practice of not removing or not covering up signs which are no longer relevant or not relevant at a given time. A survey of drivers revealed [12] that a significant proportion of drivers regularly disregard signs at construction sites and take their cues from visible construction activities or from other drivers. Consistently reliable use of warning information induces credibility and relieves drivers of the responsibility of recognizing the hazard themselves. Traffic engineers have at their disposal a variety of tested traffic control devices to help the driver safely through the traffic control zones. When drivers fail to select the appropriate speed or path, it can often be related to inadequacies in the transmission of the proper information [11]. The problem may be too little information or conflicting information. Confusion also may be caused by too much information given relative to the driver's ability to absorb the information during the time available.

In conclusion, the provision of driver information in traffic control zones should satisfy the following needs: adequate advance information regarding the nature and location of the traffic control zone to be encountered and driver actions needed or prohibited in order to safely negotiate the traffic control zone; clear delineation of the path to be followed, using only devices and markings consistent with driver expectancy based on prior driving experiences; proper maintenance of the information system to ensure that the information provided to the drivers corresponds to the actual current conditions.

FUNCTIONAL REQUIREMENTS OF TRAFFIC CONTROL

This section discusses the major functions that are to be served by traffic control devices at freeway work sites.

National standards of design and application of traffic control devices are given in the *MUTCD* [4]. Section G of Part VI of the *MUTCD* deals exclusively with the standards applicable to freeways.

In its broadest definition, traffic control at freeway work sites encompasses requirements, all of which are aimed at moving traffic safely, and efficiently past freeway work sites. These requirements can be summarized as follows:

1. To provide *advance warning* for traffic approaching the work site; in the event of a total roadway closure, advance warning information must be supplemented with information regarding the availability of alternate routes.
2. To provide a clear and unambiguous view of the physical hazard to motorists entering the work site; this is accomplished by proper *delineation* of the work site during the day and at night.
3. To provide clear and unambiguous information of the proper travel path, by means of effective *guidance* of motorists around the physical hazard.
4. To provide, when necessary, a *positive separation* between opposing traffic movements or between traffic and the work force, as a means of preventing head-on collisions and providing added protection to the working crew.
5. To provide an *acceptable level of service* to moving traffic, characterized by short delays, moderate speeds and few stops.

These objectives are accomplished through the careful planning, selection, placement, and maintenance of the traffic control devices at the construction site. In this regard, the term "device" will be used to refer to all on-site measures that are applied at freeway construction sites for the purpose of traffic control. These include:

1. Advance warning construction signs
2. Variable message signs
3. Regulatory and guide signs
4. Channelizing devices
5. Warning lights
6. Arrow boards
7. Temporary barriers and crash cushions
8. Pavement markings and rumble strips
9. Flagmen and highway patrolmen
10. Traffic control or shadow vehicles at moving operations
11. Highway advisory radio

Traffic control procedures at freeway construction sites, such as lane or shoulder closures, median crossovers, detours, are essentially generated by *combining several of the above devices into a distinct pattern*, such that the five basic requirements of traffic control are met. The mechanism for developing an optimum combination and location of devices is the Traffic Control Plan. This plan is communicated to several entities including the highway agency, the contractor, enforcement agencies, utility companies, and the general public.

Advance Warning Function of Traffic Control

Freeway construction zones produce disruptions to normal traffic flow patterns and often require drivers to perform unexpected maneuvers, especially in rural-type environments.

Thus a primary function of an advance warning system is to provide drivers with *timely* information about road conditions ahead, and of the anticipated maneuvers required to negotiate these conditions.

Typically, the warning function is imparted by a set of advance warning signs. These signs are often supplemented on freeways with high-visibility devices such as flashing arrow boards and changeable message signs. In general, design standards for traffic control devices are proportional to the design speed on the facility. As such, freeway construction zones require the highest standards regarding device size (48″ × 48″ signs), lettering, visibility (one to two miles upstream of work area) and frequency (warning signs on both sides of the road, also known as gating).

Studies by Humphreys [13] and Hostetter [8] indicated that a significant portion of work zone traffic control problems may be alleviated by simply adhering to current design standards and guidelines and avoiding contradictory, misleading, or non-specific information.

FUNCTIONAL REQUIREMENTS OF ADVANCE WARNING DEVICES

At a minimum, all traffic control devices utilized at freeway construction zones for the purpose of advance warning must provide the following information [8].

Feature Warning

This class of information tells the driver what type of traffic control to expect (e.g., lane closure, two-lane two-way operation).

Maneuver Warning

This class of information tells the driver what he is expected to do (e.g., lane change or stop) at the work site.

Feature Location

The driver must be given the specific distance to the feature (e.g., lane closure) and the distance available to complete the required maneuver (i.e., merging).

Prohibition and Restriction

At times certain actions which are common in freeway driving need to be prohibited at freeway work sites (e.g., passing or lane changing). The distance must also be given to the beginning and ending of restrictions.

Speed Zone

When speed reduction is considered desirable, the use of advisory speed signs (black on orange) is the common method of informing drivers about the probable maximum safe speed. Some agencies prefer to establish speed zones by posting regulatory speed signs (black on white). In the latter case the use of speed limit change warning signs is also recommended.

Route Guidance

When the construction work requires the closing of all lanes of a freeway, information needs to be given on detours to alternate roadways. Once drivers are informed that all traffic must leave the freeway, the location of the follow-up signs depends on the location of intersections or interchanges where alternate route choices are available.

Confirmation

The frequency of confirming information should be correlated with the severity of the hazard involved with failing to respond to the information. For example, where opposing traffic is not positively separated in a bidirectional zone, drivers continuously need to be reminded of the need to stay in lane. Soft shoulders or shoulder dropoffs are other examples of hazards that drivers must be warned of at close intervals. If engineering judgment warrants it, signs should be placed close enough that one sign be legible to drivers at all times.

SPECIFICATION OF WARNING DISTANCES

The successful handling of a potentially hazardous situation by drivers is dependent on their performance of a sequence of tasks: *detect* the presence of the object, condition or situation hazard; *recognize* it as a hazard; *decide* and *prepare* to respond accordingly [11]. These four pre-maneuver steps require time which is called reaction time. The total hazard avoidance process time also includes the time required to complete the maneuver. A conceptual model of the hazard avoidance process involved in a lane closure situation is presented in Figure 2.

Hazard detection is dependent upon the interaction of many factors, including the hazard's conspicuity, its primacy, and the number of information sources competing for the driver's attention.

The driver's ability to recognize a hazard as such is influenced very much by prior experience. This prior experience does not necessarily need to be related to driving when the hazard is an object in the path of the driver. The recognition of highway condition hazards or situation hazards is more dependent on expectancy. The same is true for the driver's ability to select the appropriate choice of action.

Many factors influence the driver's ability to select the appropriate path and speed in a given condition, such as the predictability of the behavior of nearby vehicles and the delineation of the selected new path. Experience also plays an important role in this process. Information handling task involved in the selection and performance of the appropriate action is quite complex. The time required to complete each step, as illustrated in Figure 2, can be translated into distance if a certain travel speed is assumed. This distance is called Decision Sight Distance [14]. The basic principle guiding all advance warning at freeway work sites asserts that this Decision Sight Distance must be considered in the location of the advance warning traffic control devices. Table 1 is a presentation of recommended Decision Sight Distance at a range of design speeds.

SPECIFICATION OF WARNING DEVICES

This section provides an overview of the advance warning devices which are typically used in freeway construction zones. For detailed description of the devices and their performance, the reader is advised to consult the *MUTCD* [4] and the list of references at the end of this chapter.

Signs

The selection of an advance warning sign sequence is made from typical *MUTCD* layouts. In some cases, these are supplemented with high-level warning devices. A systematic procedure for identifying driver requirements in terms of sign information content (e.g., feature warning, etc.) and sign information location had been developed by Hostetter [8]. It serves the dual purposes of selecting and adjusting

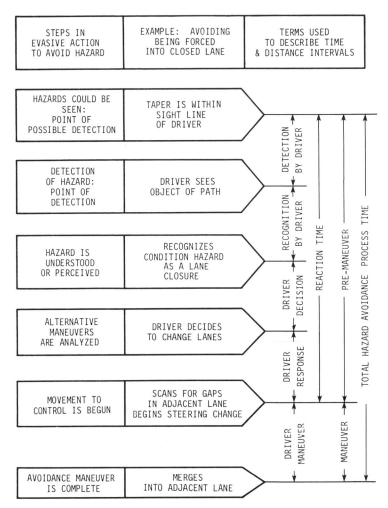

FIGURE 2. Hazard avoidance process of freeway lance closure. Based on Reference [14].

MUTCD sign sequences to meet sign information type and content requirements. An example of the procedure applied to a lane closure on a multilane highway is shown in Table 2. References [9] and [15] are examples of research which investigated the relationship between driver response and the choice of such factors as sign sequence, color, and mode of message transmission (e.g., word or symbol).

Flashing or Sequencing Arrow Boards

An arrow board (or arrow panel) is a high-visibility device designed to supplement the advance warning signs at freeway construction zones. Guidelines for the application of arrow boards have been developed in a study by Graham and Migletz [16], and are summarized below:

1. Arrow boards are more effective in lane closure situations than in traffic diversions, splits or shoulder closures.
2. Arrow boards are most effective when placed on the shoulder of the roadway near the beginning of the lane closure taper.
3. Arrow board needs are determined based on prevailing work activity and highway type. Urban freeways and rural multilane highways are indicative of conditions where arrow board use is a high priority.

While common practice is to place one arrow board near the transition taper, studies in Texas indicated that conditions exist where a second arrow board is warranted [17]. The study shows that when the available sight distance to the lane closure drops below 1,500 feet, a significant proportion of drivers remained stranded in the closed lane near the

TABLE 1. Recommended Decision Sight Distance.

Design speed (mph)	Times (Seconds)				Decision sight distance (Feet)	
	Pre-Maneuver					
	Detection & recognition	Decision & response initiation	Maneuver	Summation	Computed	Rounded for design
30	1.5–3.0	4.2–6.5	4.5	10.2–14	449–616	450–625
40	1.5–3.0	4.2–6.5	4.5	10.2–14	598–821	600–825
50	1.5–3.0	4.2–6.5	4.5	10.2–14	748–1027	750–1025
60	2.0–3.0	4.7–7.0	4.5	11.2–14.5	986–1276	1000–1275
70	2.0–3.0	4.7–7.0	4.0	10.7–14	1099–1437	1100–1450
80	2.0–3.0	4.7–7.0	4.0	10.7–14	1255–1643	1250–1650

Source: Reference [8].

taper. This number declined sharply when an advance arrow board was added at the site. It is cautioned, however, not to place the advance board too far upstream (over 4,000 feet) so that drivers may not move back to the closed lane of traffic.

Arrow boards have also been employed to control traffic in moving maintenance operations [18]. In this case the board is best mounted on the back of a shadow vehicle which follows the maintenance crew.

Section 6E-7 of the *MUTCD* [4] specifies minimum legibility for flashing or sequencing arrow boards.

Changeable Message Signs

Variable or changeable message signs are often included as advance warning devices at highway work zones [19]. Changeable message signs have the capacity of warning motorists of unusual conditions as they occur. These conditions may involve congestion, accidents or reduced visibility. They have been tested in token freeway traffic monitoring and control systems [20]. In many instances, changeable message signs were installed on frontage roads and entrance ramps to divert traffic away from congested freeway segments.

A study examined the feasibility and effectiveness of changeable message signs at highway work zones [21], by means of extensive field observations of traffic flow with and without the device in place. In addition, motorists were surveyed to determine their preference and understanding of the various messages generated by the devices. The following guidelines were recommended by the author:

1. The format for changeable message devices should permit maximum amount of information to be displayed at a glance. Specifically, a three line presentation format is recommended, with a maximum of two messages sequences per cycle.

TABLE 2. Selection of Warning Devices at Lane Closures.

Information type	Suggested information content	Applicable *MUTCD* devices
Feature Warning	Specify lane(s) closed Specify distance to closure	(W20-5) Right lane closed ½ mile Left lane closed ½ mile and (W4-2), Symbolic Lane closure sign
Maneuver Warning	Specify merge maneuver Specify merge direction	(W9-2) Lane ends, Merge Right Lane ends, Merge Left
Feature Location	Identify taper beginning	(W1-6) Directional Arrow Sign High Visibility Arrow Board

Source: Reference [8].

2. Changeable message devices should be located 3/4 of a mile in advance of freeway lane closure.
3. Changeable message devices are supplemental devices at work zones, and should not be considered as an alternative to flashing arrow boards.

Guidance Function of Traffic Control

When construction activities require the closing of part of the traveled way, traffic must be safely guided through and around the work site. The primary devices used are channelizing devices (cones, barricades, etc.). At times flagmen and highway patrolmen are also needed. The purpose of a flagman is to direct traffic away from the work area and, when applicable, to slow down or stop traffic altogether, using a STOP/SLOW paddle. Temporary barriers which are installed to separate traffic and work crew or opposing traffic streams are utilized primarily to provide positive separation (prevent vehicle penetration) [22]. A common traffic control type zone where temporary barriers are often used is the two-lane two-way operation. This situation is created when one side of the freeway must be fully closed for construction and the traffic which normally uses it is crossed over the median to the opposing direction, where it shares the roadway with the opposing traffic.

For construction zones operating beyond daylight hours, special consideration must be given to provide adequate nighttime visibility for all devices. This may be accomplished by using reflectorized devices (reflect vehicle lights upon contact), warning lights (in a steady or flashing mode) or both. The concept of Decision Sight Distance has been applied in the development of visibility requirements for channelizing devices [23].

SPECIFICATION OF GUIDANCE DEVICES

When traffic must divert from its normal path, it is first guided away from the work area by means of a transition taper. A taper is comprised of several channelizing devices.

The *Traffic Control Devices Handbook* [5] incorporates specific guidelines for the placement of channelizing devices in the work zone.

Channelizing Devices

There is a variety of channelizing devices which can be used at freeway traffic control zones. Primary devices include cones, tubes, vertical panels, drums, and barricades. These intermittently spaced devices are distinguished from continuous positive barriers utilized at some construction sites for the purpose of providing positive separation of traffic movements.

The design, use and effectiveness of channelizing devices was the subject of a comprehensive study [25]. The study included laboratory experiments of device configurations, controlled field studies of the most promising designs and a field study of three devices with design and layout variations at two work zone types, namely lane closure and diversion. The principal findings of the study are summarized below:

1. All primary channelizing devices, when designed and used properly, adequately perform the function of channelization, both in daytime and a night. No individual device elicited unique driver responses which may be considered hazardous.
2. Drivers do not respond so much to the presence of individual devices, but rather to a general pattern displayed by an array of devices. Therefore, it is important that care be taken in the layout and maintenance of these devices.
3. For all the devices tested, there was an average loss of 1,000 feet in device detection (i.e., first point at which the driver actually sees the device) between daytime and night conditions.

Other aspects of channelizing devices including driver preference and understanding of specific design characteristics, such as stripe configuration, stripe width, color ratio (orange vs. white), barricade width to height ratio; these have been studied in laboratory-type settings [26,27].

Concern over the safety of motorists at bidirectional or two-lane two-way zones has prompted interest in developing low-cost channelizing devices to be used on the two-way segment. It is necessary that the channelizing devices meet higher standards with respect to visibility, structural integrity and cost effectiveness [28]. Only when traffic exposure becomes significant (generally for over 10,000 average daily traffic) or when the construction activity is highly concentrated (two-way section length is less than one mile) must a positive barrier (portable concrete, or steel beam on barrels) be considered.

Performance criteria were developed for channelizing devices utilized at bidirectional zones, so that manufacturers of the potential devices develop products which meet the performance requirements [29]. A ranking of potentially promising devices has emerged from this study, based on ratings given by a panel of experts. The list of the proposed devices is shown in Table 3.

Lighting Devices

Urban freeways must service heavy traffic volumes for a considerable portion of the day. Therefore, it is often necessary to carry out construction activities in off-peak and nighttime hours [30,31]. During night operations, supplemental lighting devices must be affixed onto the barricades to indicate hazards and delineate the safe path of travel.

Guidelines for the use of flashing and steady-burn barricade warning lights have been set forth by the Institute of Transportation Engineers [32]. These pertain to such aspects as optical requirements, flash rate requirements, lens size, lens illumination and painting.

TABLE 3. Ranking of Channelizing Devices for Bidirectional Zones.

Rank[a]	Description of channelizing device
1.	Continuous Raised Median with Supplemental Devices
2.	Intermittent Array of High Visibility and High Durability Devices
3.	Intermittent Array of Flexible Self-Restoring Devices
4.	Continuous Raised Median with Supplemental Devices
5.	Car Deflector Designed to Deflect Cars at Small Angle
6.	Large Raised Pavement Markers
7.	Longitudinal Rumble Strip
8.	Intermittent Array of Vertical Panels

[a]Based on a composite, multifactor rating procedure by a panel of potential users, researchers, consultants, manufacturers and national organizations.
Source: Reference [28].

It is advisable that, in addition to warning lights, floodlighting be provided at the work area so that (a) the work crew is able to work efficiently and (b) drivers are warned of the crew's presence. Care must be taken, however, in locating floodlights at sufficient heights to prevent glare effects.

Finally, all advance warning devices must be either reflectorized or illuminated if they are to be used at night.

Flagmen

Flagging is used at freeway construction zones to reduce traffic speeds, to separate moving traffic from work activity occurring near the travel lanes and to stop or slow traffic in advance of a temporary road blockage (e.g., hauling equipment across the road). Flagging should only be employed when required to control traffic or when all other methods of traffic control are inadequate to warn and direct drivers [4]. When flagmen are located at freeway construction zones it is imperative that their presence and location be visible to approaching traffic at all times. Flagmen visibility is enhanced by the following measures:

1. Advance flagger signs (*MUTCD* sign W20-7a).
2. Separation of flagman from work crew.
3. Proper attire including orange vest and hat.
4. Outside garments must be reflectorized for nighttime flagging.
5. Effective signalling techniques [33].

Requirements for reflectorized garments have been described in the literature [34]. A study aimed at developing a design of work zone flagger's vest has indicated that the vest should be constructed of fluorescent orange material; for night use, white or silver retroreflective tape trimming which outlines the flagger's figure recommended [35].

It is important that flagmen be adequately trained prior to their field assignment. Several states have developed training courses and manuals for flagmen [36].

Finally, flagging may also be performed by trained highway patrolmen. Studies have indicated that the presence of patrolmen decreased vehicle speeds, but that the speed reduction was only effective for a short length of the work zone. Moreover, some patrolmen have expressed concern over their own safety and doubts about the effectiveness of flagging on high-speed facilities [37].

Delineation Function of Traffic Control

Traffic control zones at freeway work sites involve a divergence in the normal driving path of traffic, generally in the form of lane drops, lane shifts or both. Since drivers utilize lane edge and centerline markings to obtain tracking cues for lateral vehicle control, they must be made aware of the impending changes in the roadway alignment. This is especially true in periods of limited ambient visibility, such as during rain, fog, and at night.

Delineation objectives are satisfied, in part, by the channelizing devices at transition tapers. However, if the revised lane patterns are to be in effect for a considerable period of time, then temporary pavement markings must be affixed onto the pavement to delineate the new path. This will require proper removal and obliteration of old markings to prevent driver confusion between two conflicting markings.

Temporary pavement markings consist of paints, reflectorized marking tapes and removable raised pavement markers. The major drawback of marking paints is that they are difficult to remove. The practice of over painting (with black paint or asphalt) is a common yet ineffective way of obliterating old markers; the end result is a glossy contrasting surface which increases its nighttime visibility [38].

Other acceptable methods of marking removal include sand blasting, high pressure water or air jets, application of chemical solvents, burning, grinding, high temperature burning with excess oxygen and scoring-jetting [38,39,40, 43,44].

A study which compared the performance of paint markings to removable reflectorized traffic tape [41], found that the tape provided delineation comparable to the paint. In addition, tape installation time (3 min for a 360 feet section) and removal time (15 min for the same section) were minimal. Also noted was the fact that the tapes did not exhibit any tear, splits and cracks during the 19 days they were in place.

Raised pavement markers have been typically installed at high hazard locations, such as sharp horizontal and vertical curves to supplement existing pavement markings. They provide a visual and audible delineation of the work zone, and have proven to be very effective during inclement weather. Most problems associated with raised markers pertain to their installation and removal. The process of cementing individual markers to the pavement can be rather time consuming [41]. In addition, markers installed on concrete pavement experience high rates of damage and loss.

Separation Function of Traffic Control

Intermittent channelizing devices, such as traffic cones and tubes provide only nominal separation between vehicular traffic and work crew, or between opposing traffic movements at bidirectional zones. Errant vehicles are not prevented from penetrating the work area, or from encroaching onto the opposing lane of traffic. This creates a potential hazard to both motorists and workers. It is evident that the frequency and severity of such incidents will depend on: (a) vehicle operating speeds and (b) amount of exposure of the work space to motorists. The latter is a function of traffic volumes and the spatial extent of the construction activity.

Temporary barriers provide positive separation between traffic and the work area. They are designed to prevent vehicles from accessing some portion of the highway and to redirect errant vehicles with minimum damage to the vehicle and its occupants. Guidelines for the use of temporary barriers have been largely adopted from those used for permanent barriers [42]. The lack of specific guidelines has resulted in untested designs to be used in construction areas. Furthermore some researchers believe that, because of their temporary nature, these barriers should not be required to meet the more rigid standards for permanent barriers [43].

A systematic process for determining the need for temporary barriers in highway work zones has been formulated in a U.S. Department of Transportation technology sharing reprint report [44]. It considers the following parameters:

1. Consequences of vehicle encroachments
2. Roadway and roadside features (slopes, side clearances)
3. Expected frequency of encroachments

It is recommended that barrier costs and accident experience also be incorporated in the decision to install temporary barriers at freeway construction zones.

Portable safety shaped concrete barriers have been extensively used at freeway work zones [45]. These barriers (also known as New Jersey barriers) consist of a number of precast, 10 feet concrete sections that are fastened together to form the continuous barrier. The cross section is designed to effectively restrain and redirect vehicles upon impact, as depicted in Figure 3.

A Federal Highway Administration report [43] on temporary barriers at construction zones is a good source of further information. An interesting barrier concept which utilizes discarded automobiles and can be erected at a work site in a very short time is described in a Texas Transportation Institute report [46].

Regardless of the type of barrier selected, it is imperative that all safety features associated with the design of permanent barriers, such as end treatment and joint treatment, be adapted to temporary barrier design as well.

Level of Service Impact of Traffic Control

Freeway construction zones represent disruptions to the normal flow of traffic. The amount of delay experienced by the motorist is a function of several factors which are not directly related to the design of the individual traffic control zone. Key elements in determining the level of service at a freeway construction zone include:

1. The number and width of freeway lanes and shoulders open to traffic in the work zone [47,48].
2. The type and complexity of preparatory maneuvers (e.g., lane changing and speed change) to be performed by traffic entering the work zone.
3. The type, intensity, and schedule of the construction activity.
4. The presence of speed reduction devices, such as advisory or regulatory signs, flagmen or patrolmen, rumble strips, and complex transition tapers [37,49].
5. The imposition of demand control measures upstream of the work zone, including selective ramp closures, advance warning of impending congestion and alternate routes, highway advisory radio, etc. [50].
6. The attractiveness of alternate routes and modes.

To maintain an adequate level of service over the freeway facility during construction, maximum utilization of

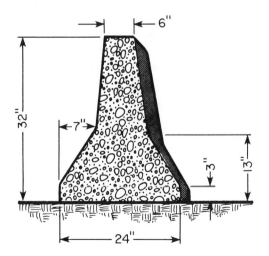

FIGURE 3. Portable safety shape barrier. Source: Reference [5].

available roadway capacity is essential. Techniques such as lane splitting, traffic shifting over to shoulder lanes were found to be effective in increasing overall capacity [51]. Demand-oriented measures such as night scheduling of construction activities and detours may also be considered in conjunction with capacity improvement techniques. For an in-depth analysis of the operational impacts of traffic control, the reader is referred to a study by Wang and Abrams [47].

PLANNING FOR TRAFFIC CONTROL

Systematic development and implementation of a traffic control scheme in a work zone requires an integrated project planning and management process. The major activities which define the work zone project planning and management process include the following:

1. Project Scheduling
2. Preparation of the Traffic Control Plan
3. Public Information
4. Training

For construction projects on freeways, the basic responsibility for the safety of both motorists and work crews rests with the unit of government which has operational control over the freeway (state or turnpike authorities).

Project Scheduling

When the preliminary schedules of construction projects are selected, consideration should be given toward minimizing traffic conflicts and congestion. The schedule should avoid, if possible, work during peak hours and holidays, several consecutive projects or excessively long work zones.

In developing the schedule, consideration to a controlled staging of the construction is desirable. These considerations are indicated in the Traffic Control Devices Handbook [5], and include:

1. The location of work (on roadway or shoulders), the number of lanes required for the work activity, and the length of the work area.
2. Hours of a day during which a lane may be closed and whether work may progress simultaneously in both directions of traffic.
3. Time involved, such as curing of pavement or bridge decks.
4. Hazards created by the work activity within the recovery area, such as boulders, drainage basins, pipe, headwalls, blunt ends of guardrail, and sign supports. Minimize time of exposure to hazards such as dropoffs.
5. Delays during traffic control set-up and take-down time (preferably done during low traffic volume periods).

Traffic Control Plan

Although all construction projects need to have a traffic control plan, the degree of detail depends on the needs. In some cases reference to standard plans, such as found in the *MUTCD* [4], in the *Traffic Control Devices Handbook* [5] or in state manuals will suffice. Most freeway construction projects will need detailed traffic control plans.

A Federal Highway Administration publication presents one fairly detailed 20-step procedure for the development of traffic control zone plans [8].

The first steps involve the identification of the traffic control features, such as lane closures, crossovers, bidirectional zones, or closed shoulders. These features are located on a plan view with tentative start and termination points. Some of these features, such as the starting point of the transition zone, can be relocated if the need arises; others might be fixed by such factors as exit or entrance ramps. This plan view will reveal the necessary driver maneuvers, such as lane change or turn, associated with each feature. Approach speeds to each feature also need to be estimated.

The second sequence of steps is aimed at the determination of the reception points where the drivers should receive the information on the features and/or necessary maneuvers. These reception points are the location where warning signs should become legible to the approaching drivers. The determination of these points is based on the Decision Sight Distance Concept.

The next step involves the elimination of possible conflicts produced by this tentative assignment of the information reception points. There are two major types of situations that lead to conflicts: warning signs corresponding to different features may fall in the same area; or there is insufficient distance between two adjacent features to provide warning information at the desirable locations. The potential for confusion, however, needs to be evaluated using engineering judgment. The solution may require the relocation of a feature or the application of the concept of primacy in the determination of the final warning distances. High-level warning devices may also be used to reduce the possibility of missed information.

The finalization of the signing plan involves: the actual location of the warning signs; the signing of the start and termination points of the traffic control zone, the integration of route guidance signs; and possibly the introduction of speed zones or other restrictive warnings [8].

In addition to the plans of the traffic control zones showing the location of traffic devices, the traffic control plan should include, when applicable: special manpower needs, such as flaggers or enforcement personnel; special activity schedules, such as times when certain lanes must be kept open to traffic; and phone numbers of highway agency officials to be contacted in emergency.

Especially in urban areas, coordination with several agencies is essential. These agencies may include police and fire department, ambulance services, transit authorities and utility companies.

Public Information

A public information program can provide valuable payoffs in terms of eliminating many potential problems created by freeway traffic control zones, especially in urban areas. A good public information program provides reliable information before the start of a project and throughout the duration of the project to the affected motorists. The means of dissemination of information on the start and progress of the work may include:

1. Public hearings, in public buildings and on local interest programs on television or radio.
2. Frequent press releases to newspapers, radio and television stations.
3. Special mailings, agency newsletters, civic group publications, notices included in utility billings.
4. Signs giving information as to coming traffic restrictions and detours.

Judgment needs to be used as to the method and degree to which these techniques are used. Some of the factors which need to be considered are the severity of the problem to be caused by the traffic control zone and the duration of the problem.

For more details in the area of public information needs and techniques the reader is referred to Reference [52].

Training

Each person whose actions affect safety at construction work sites should receive training appropriate to the job they perform and the decisions they make. This includes upper level management personnel as well as construction field personnel.

Only those individuals who are qualified by sufficient training in safe traffic control practices and have basic understanding of the traffic safety principles and standards, such as the *MUTCD*, should supervise the selection, placement and maintenance of traffic control device on construction and maintenance work zones [5]. For information on training courses available for personnel involved in traffic control at freeway work sites the reader needs to contact the state department of transportation. The Federal Highway Administration has sponsored the development of such training courses.

TYPICAL WORK ZONE TYPES

This section illustrates typical traffic control zones at the two most critical types of freeway work sites: those that require lane closures and those that require crossovers and two-lane two-way (bidirectional) zones.

The typical applications presented here are based on the *MUTCD* [4] and *Traffic Control Devices Handbook* [5]. These layouts represent minimum desirable requirements for normal situations.

Unusual conditions, such as reduced sight distance, heavy truck traffic and traffic backups, may require higher-type treatment at a given freeway work site. This increased level of treatment can be achieved in a variety of ways, including the following:

1. Provide additional devices, such as more signs, more channelizing devices, and flashing arrow boards.
2. Provide upgraded devices, such as larger signs or changeable message signs, and use barriers in place of channelizing devices.
3. Increase the time and space available for motorist action be providing longer advance warning area and longer tapers.
4. Improve geometrics at crossovers.

Lane Closure

In this application, the work zone is protected by the closing of one or more traffic lanes of a freeway, leaving one or more lanes open to traffic.

Single lane closures represent a frequently used control strategy at freeway construction and maintenance operations. Compared to other more extensive control strategies (e.g., detours, crossover), this approach has considerably less negative effects on freeway traffic operation. This type of control can be adopted for both short term and long term operations. A typical right lane closure is shown in Figure 4.

The *MUTCD* sets the standard for the taper length L as follows: the minimum length of the taper in feet is the product of the width of the closed lane in feet and the posted speed limit in miles per hour.

The minimum spacing between channelizing devices in a taper should be approximately equal in feet to the speed limit [4]. The truck shown in Figure 4 in the buffer zone is typically used on short duration or moving maintenance operation. The arrow board shown at the start of the taper might be placed on the truck in this situation. In more permanent installation the termination area should also have a taper as shown in Figure 1. While these closing tapers are optional [4] they may help delineate the path back to normal operations. They should be 50 to 100 feet long, with the channelizing devices spaced at maximum of 10 to 20 feet apart, respectively.

If two lanes are to be closed on a six or eight lane freeway, traffic should be channelized laterally one lane at a time, separating the two tapers by a distance of $2L$. The beginning of the traffic control zone should be marked by high-level warning signs (i.e., equipped with flashing lights). The closing of two lanes requires a careful analysis of the capacity of the open lane or lanes. It might be necessary to shift traffic over to the shoulder to avoid severe congestion. For more details consult References [48,51].

On Figure 5 a relatively difficult to handle work site is shown. The work area is in the center lane and the other two lanes are both to remain open. To close the center lane, the traffic from the left lane is channelized to the center lane. The traffic in the center lane can then be guided past the work area by a second transition taper. The use of flashing arrow boards is required in both transition areas. The taper length is the same as in Figure 4. Old pavement markings should be removed in both transition areas, as shown in Figure 5.

If the construction work is near an interchange, access to the ramps should be maintained. The path to the exit ramp should be clearly delineated by channelizing devices as shown in Figure 6. The minimum taper length L should be the same as on Figure 4 and as described in the text. If the maintenance of access is not feasible, the ramp may be closed by using signs and barricades. Sufficient advance warnings of ramp closure should be given to the motorists and alternative routes suggested. Prior cooperation is needed with the agency that has jurisdiction over the roadway to which access is to be closed.

Two-Lane, Two-Way Operation

When both lanes of a four lane freeway are closed in one direction, then the other two lanes have to be operated as a two-lane two-way roadway. Crossovers need to be established to channel traffic from the closed lanes through the median to the other side of the freeway where a bidirectional zone is established. Crossovers and bidirectional zones have both been found to experience safety problems in the past [3]. Careful consideration should be given, therefore, to other options before this particular type of traffic control zone is selected. The *Traffic Control Devices Handbook* [5] recommends that the following questions be considered in the decision making process:

1. Can the construction project be completed by closing only one lane at a time without creating an unduly hazardous condition to construction personnel?
2. Is an acceptable detour available?
3. Can temporary lanes be constructed in the median or can traffic be shifted over to the median? Are there lateral or vertical clearance restrictions that need to be considered? Is the shoulder sound structurally to support traffic?
4. If the two-lane two-way operation is selected, will it result in shorter construction time? Will it enable the contractor to reduce construction costs by more efficient performance?

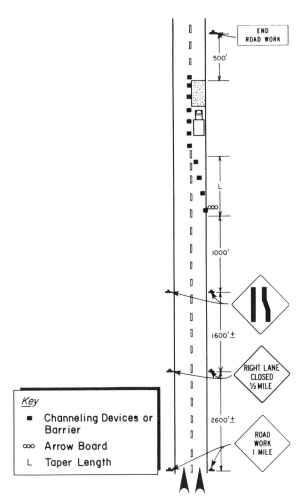

FIGURE 4. Traffic control zone at freeway lance closures. Based on Reference [5].

A typical traffic control zone is shown on Figure 7. Signs are shown for only one direction of travel. The signs and channelizing devices shown are considered minimum by the *MUTCD* [4]. Warning lights should be used at night on channelizing devices as needed or required by current standards. Other traffic control devices that may be required by the traffic control plan include: variable message signs; directional arrows painted on the pavement; signs advising motorists of the length of the bidirectional zone remaining; and high-level warning devices at the beginning of the traffic control zone.

The *MUTCD* needs to be consulted regarding the details of current standard barricades and channelizing devices. The treatment of the centerline in the bidirectional zone is a critical element of traffic control. On federal-aid projects

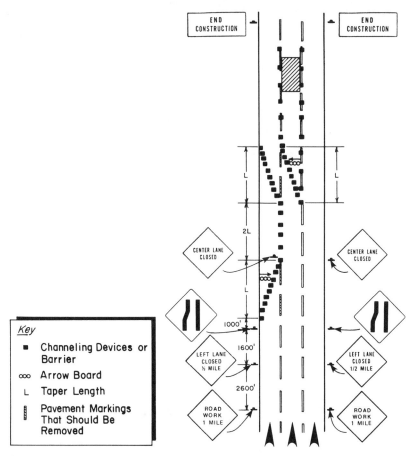

FIGURE 5. Traffic control zone at center land closures. Based on Reference [5].

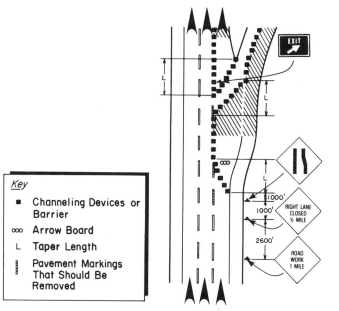

FIGURE 6. Access to exit ramp within traffic control zone. Based on Reference [5].

CONSTRUCTION: TRAFFIC CONTROL AT FREEWAY WORK SITES 225

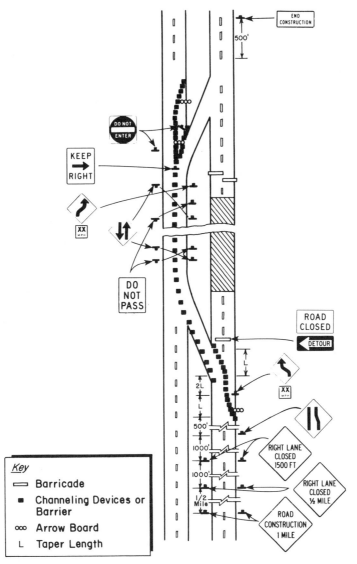

FIGURE 7. Crossover and two-lane two-way operation. Based on Reference [5].

Reference [6] requires positive separation in many cases. Reference [23] provides information on past experiences with a variety of centerline treatments. That reference recommends portable concrete barriers for short length projects, such as bridge reconstruction or rehabilitation projects. It also recommends that asphalt curb with tubular markers be considered where the lane width reduction caused by concrete barriers is unacceptable. The asphalt curbs tested were 4 inches high and 18 inches wide. The installation cost was found to be about 80 percent lower than that of the portable concrete barriers. If portable concrete barriers are used, the *Traffic Control Devices Handbook* [5] warns that the ends of these barriers present a hazard to the motorists. To eliminate this hazard, it is recommended that these temporary barriers be flared away from the traveled way or connected to existing permanent barriers when applicable. The most effective treatment, however, is the use of crash cushioning devices, such as sand-filled plastic barrel systems.

The geometric design of the crossover is a critical element in the safety of this type of traffic control zone. The diagonal design shown in Figure 7 is preferable to the design which employs two reverse curves. Trucks are most likely to encounter difficulties at crossovers, indicated by accident studies [9,23]. Load shifts were found to contribute to the loss of control in some cases. Both studies found that accidents were occurring at both entering and exiting crossovers. Advance warning for the exiting crossover is needed, such as the arrow board at the beginning of the closing crossover shown on Figure 7. Concrete barriers do not compensate for inadequate geometrics at crossovers. Generous shoulders and an area clear of object hazards adjacent to the crossover will significantly increase safety at crossovers.

The length of the two-lane two-way or bidirectional zone is usually dictated by the nature of the construction activity. Cost effectiveness from the contractor's point of view may require shorter segments in some cases while other type of activities favor long sections. From the point of view of safety, however, shorter segments are preferable [5]. On longer sections emergency vehicles will have more difficulty in getting through. Also, longer queues will form behind slower vehicles. This increases the probability of rear-end accidents. Where the centerline treatment involves flexible devices such as self-restoring tubular markers, the incidence of illegal passing will increase on longer sections due to growing impatience. Rough terrain often encourages freeway designers to plan long sections of independent roadway alignment. This may force the establishment of undesirably long bidirectional zones. Intermittent passing sections may provide a solution to the frustration created by slower moving vehicles [5].

INSTALLATION AND EVALUATION OF TRAFFIC CONTROL

The person responsible for traffic control should inspect all traffic control materials that are to be installed. Damaged and non-standard material should be replaced. Reflectorized material is especially prone to damage. Sufficient inventory of traffic control devices must be maintained to be able to replace those damaged by traffic during construction.

Installation and Removal of Traffic Control Devices

It is recommended by the *Handbook on Traffic Control Devices* [3] that traffic control devices should be placed in the same order in which drivers encounter them. If traffic is subject to control in both directions, installation should begin at the same time at each end of the work site. At two-lane two-way zones and crossovers it is essential that the opposing traffic be channelized out of its lane before permitting traffic from the other direction into that lane. Existing signs that are not applicable during construction work should be covered or removed. The same rule applies to pavement markings as well.

Construction traffic control signs that are no longer applicable, should also be removed promptly and stored out of sight. If traffic control devices are to be removed and installed again several times during the construction, a procedure must be worked out so that the installation can be completed quickly and properly. The presence of the workers involved in this procedure is itself a hazard and high-level warning devices should be used to alert motorists.

Inspection and Evaluation of Traffic Control

The traffic control installation needs to be inspected routinely by the person responsible for traffic control. On a larger project this person is usually an employee of the highway agency. The first inspection should be conducted shortly after the traffic control installation is completed. The purpose of this inspection is to establish that the traffic control installation functions as intended. The inspector should observe traffic control from the motorists' point of view while driving through the work site.

The object is to assess the elements of traffic control system installed in terms of their effectiveness of communicating to the driver the required information. Hostetter suggests [8] the application of three criteria which were derived by the application of human factors principles in an assessment of construction site information requirements.

THE DRIVER RESPONSE CRITERION

This criterion alludes to the most likely response of the motorists to the information transmitted by traffic control device. The response that a given device is intended to elicit can involve a vehicle maneuver, a driver decision, or simply a state of cognizance.

Traffic control devices that instruct drivers to perform a specific vehicle maneuver need to be reviewed with the most care. An example of such a device is the flashing arrow board placed at the beginning of the transition area that instructs all vehicles still in the closed lane to merge into the open lane. The other important type of work zone signs require a specific driver response other than vehicle maneuvers. The TURN OFF 2-WAY RADIOS sign is a good example of this category of signs. An example of the second type of sign, which requires drivers to make a decision without any immediate overt action, might be a sign informing the motorists of an alternative route or optional detour. The sign SOFT SHOULDERS is an example of the third class of signs. It informs drivers of a condition they should be aware of and remain to be alerted to, but no overt action is needed to be performed. When a traffic control device is critically viewed from the driver's response point of view, the questions that need to be considered are:

1. Does the sign or any other traffic control device clearly communicate to the driver what the required response is?
2. Can the device communicate the information effectively

by itself or is there a need for supplementary information?
3. If supplementary information is needed, what are the consequences of not receiving the supplementary information?
4. The last question deals with the potential ambiguity. Can the instruction be easily misinterpreted because of the message content? Will it be understood by drivers with average reading level? In general, potential causes for confusion or uncertainty must be evaluated.

THE REFERENCE LOCATION CRITERION

The second criterion which should be applied in the evaluation refers to the location where the driver must respond to the information presented to him relative to the location of the source of information. Some signs refer to responses that need to be completed prior to reaching the traffic control device; a WRONG WAY sign used at the end of a two-lane two-way traffic control zone is an example (see Figure 7). Other devices require that the driver respond immediately at the location of the sign. The arrow sign indicating crossovers (see Figure 7) is intended to be installed this way. The most critical evaluation is needed in the third case where the device may not require the motorist to respond until some point downstream. Advance warning signs typically fall into this category.

The main question that needs to be considered by the inspector is: "Does the sign clearly state where the driver needs to respond?" A second question that is relevant in many cases asks: "When specific distance information is provided, is it desirable to update the information as drivers progress down the roadway?" One such situation mentioned earlier is a long two-lane two-way operation where information on the length of the remaining bidirectional zone should be given.

THE ROAD CONDITION CRITERION

The third information criterion that needs to be considered by the inspector concerns information given about specific road conditions that provide the rationale for a specific instruction. A BLASTING ZONE 1000 FT sign for example will explain the reason for responding to the TURN OFF 2-WAY RADIO sign.

The inspector must try to answer questions such as: "Does the sign clearly describe the road condition?" or "How specific is the information about the road condition?" For example, the symbolic lane closure sign may indicate to the driver that a lane changing maneuver will be required but a RIGHT LANE CLOSED 1500 FT sign will also tell the driver the distance available to complete merging. For more details the reader is referred to Reference [8].

Continuing inspection is needed to assure that traffic control continues to perform properly. The inspector should have the authority to modify traffic control as conditions demand. If safety problems develop, prompt action is essential. In the event of accidents the likelihood of lawsuits is real [53]. A careful documentation of efforts made to protect the safety of motorists and crew alike is the best advanced preparation for such an event. These records should indicate the time and location of the work and describe the traffic control in effect. A daily diary is to be kept and work orders or change orders must be filed. Special notes on the traffic control plan sheet also serve as a reference.

The frequency of inspection depends on the complexity of the traffic control which depends on the nature and duration of the construction. The frequency of incidents involving damaged devices and the deficiencies recorded during past inspections also should be considered.

The inspection involves the following steps: note signs of potential problem; identify and evaluate the seriousness of the problem; select appropriate solution to the problem; and inplement solution [38].

Three common problems and the usual signs suggesting the presence of these problems are listed below. Potential solutions to these problems are also presented.

SAFETY PROBLEM CHARACTERIZED BY INAPPROPRIATE SPEEDS

This problem can be suggested by one or more of the following signs: conflicts and near misses; control devices damaged by traffic; crew member injured by motorists; and rear end accidents or skid marks indicating action to avoid rear end accidents.

If the speeds are too high for conditions the most effective solution is to arrange for marked law enforcement vehicles to be present [37]. Some speed reduction can also be expected when temporary rumble strips are installed. Efforts to reduce speeds by changing alignment to encourage motorists to slow down or placing channelizing devices closer to the traveled way have been less successful.

If speeds are too low, the following measures may provide some increase: lengthen channelizing taper; smooth out alignment; or screen work area from motorists view.

SAFETY PROBLEM CHARACTERIZED BY INADEQUATE TRAVEL PATH

The existence of this problem is usually indicated by the following signs: damage to channelizing devices; run-off-the-road type accidents; tire mark on shoulder or median; and slow speeds and erractic vehicle movements.

The solution to this problem may call for changes in the channelization or it may be necessary to provide more travel lanes by utilizing the shoulder or constructing by-pass lanes. It may also be necessary to increase the level of protection provided to the work force.

CAPACITY PROBLEM

There are several readily detectable signs of capacity problems at freeway work sites. The most conspicuous ones are slow speeds and traffic queues. Merging problems and accidents involving merging vehicles also indicate capacity problems. There are two approaches to the solution and both may need to be applied. First, capacity can be increased by increasing lateral clearance, increasing width of lanes, or actually increasing the number of traffic lanes. The second method involves the reduction of the flow of traffic through the work sites. A well organized public information problem may encourage drivers to take alternate routes if suitable detours are available. It is imperative that the information given be specific, current, and accurate.

REFERENCES

1. Nemeth, Z. A., and D. J. Migletz, "Accident Characteristics Before, During and After Safety Upgrading Projects on Ohio's Rural Interstate System," *Transportation Research Record, 672*, TRB (1978).
2. Hargroves, B. and M. Martin, *Vehicle Accidents in Highway Work Zones, Report No. FHWA/RD-80/063*, Washington, D.C. (December 1980).
3. Nemeth, Z. A. and A. K. Rathi, "Accident Characteristics at Work Zones on a Turnpike Facility," *Traffic Quarterly*, Eno Foundation for Transportation (January 1983).
4. *Manual of Uniform Traffic Control Devices for Streets and Highways*, U.S. Department of Transportation, Federal Highway Administration (1978).
5. *Traffic Control Devices Handbook*, U.S. Department of Transportation, Federal Highway Administration (1983).
6. Federal Highway Administration, *Federal Aid Highway Program Manual Volume 6*, Chapter 4, Section 2 Subsection 12, Washington, D.C., U.S. Department of Transportation.
7. King, G. F., and H. Lunenfeld, "Development of Information Requirements and Transmission Techniques for Highway Users," *NCHRP Report 123*, HRB, Washington, D.C. (1971).
8. Hostetter, et al., *Determination of Driver Needs in Work Zones, Report No. FHWA/RD-82/117*, Washington, D.C. (September 1982).
9. Pain, R. and B. G. Knapp, "Motorists Information Needs in Work Zones," *ITE Journal, 49:*4 (April 1979).
10. Alexander, G. J. and L. Harold, "Satisfying Motorists Need for Information," *Traffic Engineering, 43:*1 (October 1972).
11. Post, T. J., G. J. Alexander, and H. Lunenfeld, *A User's Guide to Positive Guidance, Report No. FHWA-TO-81-1*, Washington, D.C. (December 1981).
12. Rockwell, T. and Z. Nemeth, *Development of Driver Based Method for Evaluating Traffic Control Systems at Construction and Maintenance Zones*, The Ohio State University, *Final Report, Volume II, Engineering Experiment Station, Report 581*, Columbus, Ohio (October 1981).
13. Humphreys, J., H. Mauldin, and T. Sullivan, *Identification of Traffic Management Problems in Work Zones, Report No. FHWA-RD-79-4*, Washington, D.C. (December 1979).
14. Alexander, G. J. and H. Lunenfetz, *Positive Guidance in Traffic Control*, U.S. Department of Transportation, Federal Highway Administration, Washington, D.C. (1975).
15. Rockwell, T. and V. Bhise, *Development for a Methodology for Evaluating Road Signs*, The Ohio State University, *Engineering Experiment Station Report 315B*, Columbus, Ohio (June 1970).
16. Graham, J. and D. J. Migletz, *Guidelines for the Application of Arrow Boards*, Midwest Research Institute, *Report No. FHWA-RD-79-58* (December 1978).
17. Faulder, M. and C. Dudek, *Field Evaluation of Flashing Arrow Boards as an Advance Warning Traffic Control Device at Freeway Work Zones*, Texas Transportation Institute, *Research Report 228-5* (April 1981).
18. Bryden, J., *Effectiveness of Flashing Arrow Boards During Moving Maintenance Operations*, New York Dept. of Transportation Research, *Report FHWA/NYIRR-79/73* (October 1979).
19. Changeable Message Signs, *NCHRP Synthesis of Highway Practice 61*, TRB (July 1979).
20. McDermott, J., "Freeway Surveillance and Control in Chicago Area, *Transportation Engineering Journal of ASCE, Vol. 106*, No. 1 (May 1980).
21. Hanscom, F., "Effectiveness of Changeable Message Displays in Advance of High-Speed Freeway Lane Closures," *NCHRP Rept. 235*, TRB (September 1981).
22. McGee, H. and B. Knapp, "Visibility Requirements for Traffic Control Devices in Work Zones," *Transportation Research Record 703*, Washington, D.C. (1979).
23. Graham, J. and D. J. Migletz, *Design Considerations for Two-Lane Two-Way Work Zone Operations*, Midwest Research Institute, *Report No. FHWA-RD-83/112*, Washington, D.C. (October 1983).
24. Graham, J., D. Harwood, and M. Sharp, "Effects of Taper Length on Traffic Operations in Construction Zones," *Transportation Research Record 703*, TRB, Washington, D.C. (1979).
25. Pain, R. F., H. McGee, and B. Knapp, "Evaluation of Traffic Control for Highway Work Zones," *NCHRP Report 236*, TRB, Washington, D.C. (1979).
26. Cottrell, B., *Evaluation of Chevron Patterns for Use on Traffic Control Devices in Street and Highway Work Zones*, Virginia Highway and Transportation Research Council, *Report No. FHWA/VA-80/32* (February 1980).
27. Pain, R. F. and B. Knapp, "Experimental Evaluation of Markings for Barricades and Channelizing Devices," *Transportation Research Report 703*, TRB, Washington, D.C. (1979).
28. Lewis, R. M., *Review of Channelizing Devices for Two-Lane, Two-Way Operations, Report No. FHWA/RD-83/056*, Washington, D.C. (June 1983).
29. Lewis, R. M., *Performance Criteria for Channelizing Devices Used for Two-Lane, Two-Way Operations, Report No. FHWA/RD-08/057*, Washington, D.C. (August 1983).
30. Lee, D., "Night-Time Construction of Urban Freeways," *Traffic Engineering, 39:*6 (March 1969).

31. Gillis, L., "The Feasibility of Night-Maintenance and Construction," *Public Works* (April 1969).
32. ITE Standards for Flashing and Steady-Burn Barricade Warning Lights, *Traffic Engineering, 41:*10 (August 1971).
33. Richards, S., N. Huddleston, and J. Bowman, *Driver Understanding of Work Zone Flagger Signals and Signaling Devices*, Texas Transportation Institute, *Research Report 228-3* (January 1981).
34. Smith, G. M. and M. H. Janson, *Reflectorized Flagman Vests*, Michigan Dept. of Highways and Transportation, *Research Report No. R-1021* (September 1976).
35. Gordon, D. A., *Design of Work Zone Flagger's Vest, Report No. FHWA/RD-83/0898,* Washington, D.C. (June 1983).
36. Stevenson, J., "How Utah Trains its Flagmen," *Traffic Safety* (July 1974).
37. Richards, S., R. Wunderlich, and C. Dudek, *Controlling Speeds in Highway Work Zones*, Texas Transportation Institute, *Research Report 292-2* (February 1984).
38. Byrd, Tallamy, MacDonald and Lewis, *Traffic Control for Street and Highway Construction and Maintenance Operations*, Training Course Prepared for the U.S. Department of Transportation, Federal Highway Administration, Third Edition (1978).
39. Stripe Removal by High Temperature Burning with Excess Oxygen, *Implementation Package 77-16*, FHWA, (June 1977).
40. Removal of Pavement Markings by Scoring-Jelting, *Technology Sharing Report 77-123*, FHWA (June 1977).
41. Davis, T., *Construction Zone Safety and Delineation Study*, New Jersey Dept. of Transportation *Research Report FHWA-NJ-83-005* (February 1983).
42. American Association of State Highway and Transportation Officials, *Guide for Selecting, Locating, and Designing Traffic Barriers*, Washington, D.C. (1977).
43. Bronstad, M. and C. Kimball, *Temporary Barriers Used in Construction Zones, Report No. FHWA/RD-80/095*, Washington, D.C. (December 1980).
44. Hargroves, B., "Warrants for Temporary Positive Barriers in Highway Work Areas," *Technology Sharing Reprint Report*, U.S. Dept. of Transportation (May 1982).
45. "Temporary Concrete Barriers Guide Traffic Through a 10-Mile Work Site," *Rural Roads* (January 1974).
46. Sicbing, D., et al., *A Portable Traffic Barrier for Work Zones*, Texas Transportation Institute, *Research Report No. 262-3* (October 1982).
47. Abrams, C. and J. Wang, *Planning and Scheduling Work Zone Traffic Control, Report No. FHWA/RD-87/049*, Washington, D.C. (August 1981).
48. Dudek, C. and S. Richards, "Traffic Capacity Through Urban Freeway Work Zones in Texas," *Transportation Research Record 869*, TRB, Washington, D.C. (1982).
49. Graham, J., R. Paulsen, and J. Glennon, *Accident and Speed Studies in Construction Zones*, Federal Highway Administration *Report No. FHWA-Rd-77-80* (June 1977).
50. Dudek, C., S. Richards, and J. Faulkner, *Feasibility of Changeable Message Signs and Highway Advisory Radio for Freeway Maintenance*, Texas Transportation Institute, *Research Reprint 228-9/263-2* (July 1981).
51. Dudek, C., S. Richards, and M. Faulkner, *Traffic Management During Urban Freeway Maintenance Operations*, Texas Transportation Institute, *Research Report No. 228-10F* (January 1982).
52. Franklin, J. D., et al., *Traffic Controls for Construction and Maintenance Work Zones*, for U.S. D.O.T. FHWA by the American Public Works Association, Washington, D.C., *Technology Sharing Report No. FHWA-TS-77-204* (1977).
53. Humphreys, J. B., "Highway Liability—A Review of Work Zone Accident Cases," *ITE Journal, 49:*4 (April 1979).

CHAPTER 11

Determination of Power Spectral Density Functions from Design Spectra

DIETER D. PFAFFINGER[*]

INTRODUCTION

In dynamic structural analysis the response spectrum method is widely used to estimate the maximum values of structural response quantities under the dynamic loads. The loads are characterized by their response spectra and the structural quantities are calculated by combining the corresponding maximum modal contributions. The response spectrum of a time-dependent load is obtained from the maximum displacements of a simple oscillator under that load at different values of its natural frequency and its damping ratio. Usually response spectra show very irregular behaviour especially for small damping ratios. This is inconvenient for design purposes. Therefore, design spectra have been developed for several dynamic loads such as earthquake excitations from single response spectra by averaging and smoothing. The corresponding structural response quantities then represent average or expected maximum values.

Assuming the excitations as stationary random processes with normal distributions, a random analysis of the structure also furnishes the expected maximum values of the structural response quantities, their variances and confidence intervals. In the probabilistic approach, the loads are characterized by their power spectral and cross spectral density functions. If the random analysis is performed in modal coordinates, the steps involved to determine the structural response are similar to the steps in a response spectrum analysis. There are many cases, however, where the response spectra of the loads are known rather than their power spectral density functions or their time histories. Thus, methods are required which permit the determination of power spectra from given response spectra.

In the following, it is assumed that the given response spectra are average spectra obtained from several excitations. The underlying dynamic loads are taken as a stationary Gaussian process. The relationship between response spectra and power spectra is established from the probability density function of the extreme values. Describing the power spectral density function by piecewise polynomials containing free parameters, the parameters can be determined iteratively by a least square fit. The method can also be used to calculate a least square fit value of the duration of the stationary output process.

THEORETICAL BACKGROUND

Stationary Gaussian Processes

A random process $x(t)$ consists of n functions $x_i(t)$, $i = 1, \ldots n, n \to \infty$, which are related to the same physical phenomenon. The process is said to be stationary if all statistical averages are independent of time. For a stationary process the reference system can be always shifted such that the mean or expected value

$$\mu = E[x(t)] = \lim_{n \to \infty} \frac{1}{n} \sum_{i=1}^{n} x_i(t) \qquad (1)$$

will be zero where t denotes an arbitrary but fixed time. The variance of such a process then becomes

$$\sigma_x^2 = E[(x(t) - \mu)^2] = E[x(t)^2] \qquad (2)$$

In the following a zero mean value is always assumed.

A stationary Gaussian process is completely charac-

---

[*]P + W Engineering, Zurich, Switzerland

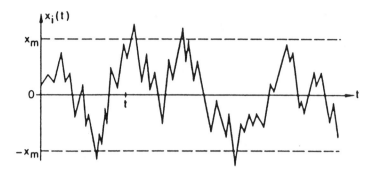

FIGURE 1. Member $x_i(t)$ of the process $x(t)$.

terized in the probabilistic sense by its power spectral density function (p.s.d.f.)

$$S_x(\Omega) = \lim_{\substack{T \to \infty \\ n \to \infty}} \frac{1}{n} \sum_{i=1}^{n} \frac{2|X_i(\Omega)|^2}{T} \quad (3)$$

where $X_i(\Omega)$ denotes the Fourier transform of $x_i(t)$ and T is the duration. It is seen from Equation (3) that $S_x(\Omega)$ is an even and non-negative function of Ω. The variance of the process is obtained from

$$\sigma_x^2 = \frac{1}{2\pi} \int_0^\infty S_x(\Omega) d\Omega \quad (4)$$

Hence, the p.s.d.f. furnishes the contribution at frequency Ω to the variance of the process.

Of particular interest is the distribution of the extreme values of the process $x(t)$. Considering the number of crossings of the barrier $\pm x_m$ (Figure 1) during a duration T to be a rare event, the Poisson distribution furnishes the cumulative probability $P(x_m)$ that the extreme values will not exceed $\pm x_m$ as

$$P(x_m) = \exp(-2fTe^{-1/2(x_m/\sigma_x)^2}) \quad (5)$$

where

$$f = \frac{1}{2\pi} \left[\frac{\int_0^\infty \Omega^2 S_x(\Omega) d\Omega}{\int_0^\infty S_x(\Omega) d\Omega} \right]^{1/2} \quad (6)$$

denotes the expected frequency of the process. The expected frequency is the average number of crossings of $x = 0$ per unit time by the random process. The probability density function $p(x_m)$ of the maxima is obtained by differentiating the cumulative probability function, Equation (5). According to Davenport [4], the expected maximum value $|x|_{max}$ and the corresponding variance σ_{max}^2 can be estimated from

$$|x|_{max} = E[x_m] = \int_0^\infty x_m p(x_m) dx_m$$

$$\cong \sigma_x \left(\sqrt{2\ln 2fT} + \frac{\gamma}{\sqrt{2\ln 2fT}} \right) \quad (7)$$

$$\sigma_{max}^2 = E[(x_m - |x|_{max})^2]$$

$$= \int_0^\infty (x_m - |x|_{max})^2 p(x_m) dx_m \cong \sigma_x^2 \frac{\pi^2}{12\ln 2fT} \quad (8)$$

where $\gamma = 0.577216$ denotes Euler's constant. The Poisson distribution assumes statistical independence of the crossings of $\pm x_m$. If the statistical dependence between the crossings is taken into account, different cumulative probability functions $P(x_m)$ are obtained [12,13]. Usually the descriptions of the dynamic loads contain considerable inaccuracies, which also lead to inaccuracies in the expected extreme values. It therefore seems justified to use only the simple expression Equation (7) in the following.

Random Response of Simple Oscillator

Figure 2 shows a simple oscillator of mass m, stiffness k and viscous damping constant c. The oscillator is excited by a load $p(t)$ or by a support motion $x(t)$. In reality, structures are always damped. Hence $c > 0$ will be assumed.

The equation of motion for the relative displacement $q(t)$ is

$$m\ddot{q} + c\dot{q} + kq = p \qquad (9)$$

In the case of support motion, p is replaced by

$$p = -m\ddot{x} \qquad (10)$$

Assuming $p(t)$ and $q(t)$ as stationary random processes with the corresponding power spectral density functions $S_p(\Omega)$ and $S_q(\Omega)$, the transfer function theorem states that

$$S_q(\Omega) = |H(\Omega)|^2 S_p(\Omega) \qquad (11)$$

with

$$|H(\Omega)|^2 = \frac{1}{m^2} \frac{1}{(\omega_0^2 - \Omega^2)^2 + 4\zeta^2\omega_0^2\Omega^2} \qquad (12)$$

where

$$\omega_0 = \sqrt{\frac{k}{m}} \qquad (13)$$

denotes the circular frequency of the undamped oscillator and

$$\zeta = \frac{c}{2\sqrt{km}} \qquad (14)$$

is the damping ratio. In the following $m = 1$ will always be assumed. Denoting the p.s.d.f. of the load $p(t)$ or of the support accelerations $\ddot{x}(t)$ by $S_x(\Omega)$ thus

$$S_q(\Omega) = \frac{S_x(\Omega)}{(\omega_0^2 - \Omega^2)^2 + 4\zeta^2\omega_0^2\Omega^2} \qquad (15)$$

follows. It should be noted, that the transfer function theorem relates a stationary input process with the stationary output process. If the oscillator is excited, however, the output process will become stationary only after a certain time which depends on the damping. Denoting the duration of the stationary part of the output process $q(t)$ by T_q, Equation (7) can be used to determine the expected maximum relative displacement

$$|q|_{max} = \sigma_q \left(\sqrt{2\ln 2 f_q T_q} + \frac{\gamma}{\sqrt{2\ln 2 f_q T_q}} \right) \qquad (16)$$

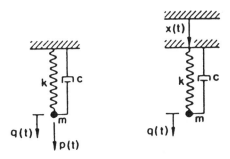

FIGURE 2. Simple oscillator.

where

$$\sigma_q^2 = \frac{1}{2\pi} \int_0^\infty S_q(\Omega) d\Omega \qquad (17)$$

$$f_q = \frac{1}{2\pi} \left[\frac{\int_0^\infty \Omega^2 S_q(\Omega) d\Omega}{\int_0^\infty S_q(\Omega) d\Omega} \right]^{1/2} \qquad (18)$$

The spring force is directly proportional to the displacement. The above results thus also furnish the expected maximum spring force.

Design Spectra

The displacement response spectrum $S_d(\omega_0, \zeta)$ of a load constituting a random process will be defined as the expected maximum relative displacement $|q|_{max}$ of a simple oscillator with circular frequency $\omega_0 = \sqrt{k/m}$ and damping ratio ζ under this load. By varying ω_0 and ζ, a set of curves for $S_d(\omega_0, \zeta)$ is obtained. The design spectrum is then determined by averaging the maximum relative displacement due to several members of the random process and subsequent smoothing of the resulting curves. It is assumed that $S_d(\omega_0, \zeta)$ constitutes such smooth functions of ω_0 and ζ. In the applications, the dynamic excitation is frequently characterized by its pseudo-velocity spectrum

$$S_v(\omega_0, \zeta) = \omega_0 \sqrt{1 - \zeta^2} \, S_d(\omega_0, \zeta) \qquad (19)$$

rather than by $S_d(\omega_0, \zeta)$.

The displacement response spectrum $S_d(\omega_0, \zeta)$ as well as Equation (16) furnish the expected value of the maximum relative displacement. Hence

$$\sigma_q \left(\sqrt{2\ln 2 f_q T_q} + \frac{\gamma}{2\ln 2 f_q T_q} \right) = S_d(\omega_0, \zeta) \quad (20)$$

holds. Squaring both sides and using Equation (19)

$$\sigma_q^2(\omega_0, \zeta) \left[g(\omega_0, \zeta, T_q) + \frac{\gamma^2}{g(\omega_0, \zeta, T_q)} + 2\gamma \right]$$
$$= \frac{S_x^2(\omega_0, \zeta)}{\omega_0^2(1 - \zeta^2)} \quad (21)$$

follows with

$$g(\omega_0, \zeta, T_q) = 2\ln 2 f_q T_q \quad (22)$$

If the p.s.d.f. $S_x(\Omega)$ of the exciting process is known, Equation (21) together with Equations (15), (17), and (18) furnishes immediately $S_d(\omega_0, \zeta)$. If, on the other hand, $S_d(\omega_0, \zeta)$ is given, Equation (20) or (21) constitutes a nonlinear integral equation for $S_x(\Omega)$ which in general cannot be solved in closed form. In the following an approximate solution of Equation (21) is obtained which is based on a least square fit procedure for $S_x(\Omega)$ represented in discretized form.

DETERMINATION OF POWER SPECTRA

Discretisation

POLYNOMIAL APPROXIMATION

$S_x(\Omega)$ is described in the interval $0 \leq \Omega \leq \Omega_{max}$ by its functional values S_i at discrete frequencies Ω_i as shown in Figure 3. In addition the values of derivatives may be introduced as parameters. Between the frequencies Ω_i $S_x(\Omega)$ is represented by piecewise polynomial interpolation. Thus,

$S_x(\Omega)$ can be written in the form

$$S_x(\Omega) = \sum_{i=1}^{n} p_i P_i(\Omega) \quad (23)$$

where the p_i denote the n parameters of the discretisation and the $P_i(\Omega)$ are interpolation polynomials

$$P_i(\Omega) = \sum_{k=0}^{k_i} a_{ik} \Omega^k \quad \Omega_{\ell_i} \leq \Omega \leq \Omega_{u_i} \quad (24)$$

defined between the corresponding lower and upper bounds Ω_{ℓ_i} and Ω_{u_i}. As $S_x(\Omega)$ is an even and non-negative function of Ω, the conditions

$$\frac{d}{d\Omega} S_x(\Omega) \big|_{\Omega=0} = 0 \quad (25)$$

and

$$S_x(\Omega) \geq 0 \qquad 0 \leq \Omega \leq \Omega_{max} \quad (26)$$

have to be satisfied by Equation (23).

The interpolation can be done in many different ways. In the simplest case, the parameters p_i represent functional values only. $S_x(\Omega)$ can be interpolated by piecewise linear functions between successive functional values or by higher order polynomials comprising several functional values. Due to the oscillations of high order polynomials, the maximum order should not exceed five. By including the values of the first derivatives of $S_x(\Omega)$ in the parameters p_i at the beginning and end of each piecewise polynomial, continuity of $S_x(\Omega)$ also with respect to the first derivative can be ensured. If the first derivatives are included at each discrete frequency Ω_i, the interpolation can be accomplished by 3rd order polynomials, i.e. splines, between Ω_i and Ω_{i+1}. In order to satisfy Equation (25), the first derivative at $\Omega = 0$

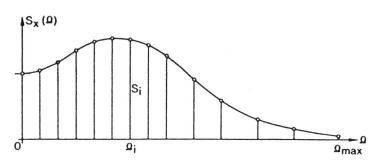

FIGURE 3. Discretized power spectral density function $S_x(\omega)$.

can be included for the polynomial representation and will then be constrained to zero. In general, also the values of higher derivatives can be included in the parameters. By including, for instance, second derivatives at specific frequencies Ω_k and constraining them to zero, points of inflection can be prescribed. In most cases, however, the representation of $S_x(\Omega)$ by its functional values and the values of first derivatives at discrete frequencies is sufficient.

INTEGRATION

Substituting Equation (23) into Equation (15) furnishes

$$S_q(\Omega) = \sum_{i=1}^{n} p_i \frac{P_i(\Omega)}{(\omega_0^2 - \Omega^2)^2 + 4\zeta^2\omega_0^2\Omega^2} \quad (27)$$

To determine σ_q^2 and f_q according to Equations (17) and (18) the integrals

$$\int_0^\infty \Omega^{2j} S_q(\Omega) d\Omega = \sum_{i=1}^{n} p_i I_{2j,\,i}(\omega_0, \zeta) \quad j = 0, 1 \quad (28)$$

with

$$I_{2j,\,i}(\omega_0, \zeta) = \int_{\Omega_{\ell_i}}^{\Omega_{u_i}} \frac{\Omega^{2j} P_i(\Omega)}{(\omega_0^2 - \Omega^2)^2 + 4\zeta^2\omega_0^2\Omega^2} d\Omega \quad (29)$$

$$j = 0, 1$$

have to be evaluated. It is seen from Equation (24) that $I_{2j,\,i}$ can be built up by linear combination of the contributions

$$J_m(\omega_0, \zeta) = \int_{\Omega_\ell}^{\Omega_u} \frac{\Omega^m}{(\omega_0^2 - \Omega^2)^2 + 4\zeta^2\omega_0^2\Omega^2} d\Omega \quad (30)$$

$$m = 0, 1, \ldots$$

where Ω_ℓ and Ω_u denote the appropriate bounds of the piecewise polynomial under consideration. For $m \leq 3$ the integrals Equation (30) are readily evaluated analytically by partial fractions [10] furnishing

$$J_m(\omega_0, \zeta) = \frac{x_1}{2} \ell n \frac{(\Omega_u - a)^2 + b^2}{(\Omega_\ell - a)^2 + b^2}$$

$$+ \frac{x_2 + ax_1}{b} \arctan \frac{b(\Omega_u - \Omega_\ell)}{(\Omega_u - a)(\Omega_\ell - a) + b^2}$$

(31)

$$+ \frac{x_3}{2} \ell n \frac{(\Omega_u + a)^2 + b^2}{(\Omega_\ell + a)^2 + b^2}$$

$$+ \frac{x_4 - ax_3}{b} \arctan \frac{b(\Omega_u - \Omega_\ell)}{(\Omega_u + a)(\Omega_\ell + a) + b^2}$$

with

$$\left. \begin{array}{l} a = \omega_0 \sqrt{1 - \zeta^2} \\ b = \omega_0 \zeta \end{array} \right\} \quad (32)$$

and

$$\begin{Bmatrix} x_1 \\ x_2 \\ x_3 \\ x_4 \end{Bmatrix} = \begin{bmatrix} \dfrac{-1}{4\omega_0^3\sqrt{1-\zeta^2}} & 0 & \dfrac{1}{4\omega_0\sqrt{1-\zeta^2}} & \dfrac{1}{2} \\ \dfrac{1}{2\omega_0^2} & \dfrac{1}{4\omega_0\sqrt{1-\zeta^2}} & 0 & \dfrac{-\omega_0}{4\sqrt{1-\zeta^2}} \\ \dfrac{1}{4\omega_0^3\sqrt{1-\zeta^2}} & 0 & \dfrac{-1}{4\omega_0\sqrt{1-\zeta^2}} & \dfrac{1}{2} \\ \dfrac{1}{2\omega_0^2} & \dfrac{-1}{4\omega_0\sqrt{1-\zeta^2}} & 0 & \dfrac{\omega_0}{4\sqrt{1-\zeta^2}} \end{bmatrix} \begin{Bmatrix} r_0 \\ r_1 \\ r_2 \\ r_3 \end{Bmatrix} \quad (33)$$

where $r_m = 1$ and $r_i = 0$, $i \neq m$. In the case of $m > 3$, i.e.,

$$m = 3 + n \quad n = 1, 2 \ldots \quad (34)$$

the integrand can always be broken down into its polynomial and its rational part using the recursive relationship

$$\frac{\Omega^{3+n}}{(\omega_0^2 - \Omega^2)^2 + 4\zeta^2\omega_0^2\Omega^2} = \Omega^{n-1}$$

(35)

$$+ \frac{2\omega_0^2(1 - 2\zeta^2)\Omega^{n+1} - \omega_0^4\Omega^{n-1}}{(\omega_0^2 - \Omega^2)^2 + 4\zeta^2\omega_0^2\Omega^2}$$

The rational part is again integrated analytically as outlined above, whereas the integration of the polynomial part is elementary.

Having built up the integrals $I_{2j,i}(\omega_0, \zeta)$ for specific values ω_0 and ζ, σ_q^2 and f_q are obtained from Equations (17), (18) and (24) as

$$\sigma_q^2 = \frac{1}{2\pi} \sum_{i=1}^{n} p_i I_{0,i}(\omega_0, \zeta) \qquad (36)$$

and

$$f_q = \frac{1}{2\pi} \left[\frac{\sum_{i=1}^{n} p_i I_{2,i}(\omega_0, \zeta)}{\sum_{i=1}^{n} p_i I_{0,i}(\omega_0, \zeta)} \right]^{1/2} \qquad (37)$$

Least Square Fit

SOLUTION OF THE NONLINEAR EQUATIONS

The parameters p_i are either free parameters or have assigned fixed values. By setting, for instance, the parameter corresponding to the first derivative at $\Omega = 0$ to zero, Equation (25) can be satisfied. As the p.s.d.f. approaches zero for high frequencies, also the functional value at Ω_{max} may be constrained to zero. Substitution of Equations (36) and (37) into Equations (21) and (22) furnishes

$$F(p_i, \bar{p}_i, T_q, \omega_0, \zeta) = \frac{S_v^2(\omega_0, \zeta)}{\omega_0^2(1 - \zeta^2)} \qquad (38)$$

with

$$F(p_i, \bar{p}_i, T_q, \omega_0, \zeta) = \sigma_q^2 \left(g + \frac{\gamma^2}{g} + 2\gamma \right) \qquad (39)$$

where the $\bar{p}_i$ and p_i denote the free and the fixed parameters, respectively. The free parameters can be determined by a least square fit. By linear expansion of F around values p_i^ℓ of the free parameters

$$F(p_i, \bar{p}_i, T_q, \omega_0, \zeta) = F(p_i^\ell, \bar{p}_i, T_q, \omega_0, \zeta) + \sum_i \frac{\partial F}{\partial p_i} \Delta p_i \qquad (40)$$

is obtained, where the summation is extended over all free parameters. The Δp_i represent corrections to the values p_i. Substituting Equation (40) into Equation (38) furnishes the linear equation

$$\sum_i \frac{\partial F}{\partial p_i} \Delta p_i = \frac{S_v^2(\omega_0, \zeta)}{\omega_0^2(1 - \zeta^2)} - F(p_i^\ell, \bar{p}_i, T_q, \omega_0, \zeta) \qquad (41)$$

From Equation (39) together with Equations (36) and (37)

$$\frac{\partial F}{\partial p_i} = \frac{1}{2\pi} \left[I_{0,i} \left(g + \frac{\gamma^2}{g^2} + 2\gamma \right) + \left(I_{0,i} - \frac{I_{2,i}}{4\pi^2 f_q^2} \right) \left(\frac{\gamma^2}{g^2} - 1 \right) \right] \qquad (42)$$

is obtained.

The pseudo-velocity spectrum $S_v(\Omega_0, \zeta)$ is known at N pairs of values Ω_k, ζ_k, $k = 1, \ldots N$. With a given duration T_q hence N linear equations

$$\sum_i a_{ki} \Delta p_i = r_k \qquad k = 1, \ldots N \qquad (43)$$

with

$$a_{ki} = \frac{\partial}{\partial p_i} F(p_i, \bar{p}_i, T_q, \omega_k, \zeta_k) \bigg|_{p_i = p_i^\ell} \qquad (44)$$

and

$$r_k = \frac{S_v^2(\omega_k, \zeta_k)}{\omega_k^2(1 - \zeta_k^2)} - F(p_i^\ell, \bar{p}_i, T_q, \omega_k, \zeta_k) \qquad (45)$$

are established. These equations can be written as

$$[A]\{\Delta p\} = \{r\} \qquad (46)$$

where $[A]$ contains the derivatives of F with respect to p_i according to Equation (44), $\{\Delta p\}$ are the corrections and $\{r\}$ contains the residua r_k as defined in Equation (45). It is assumed that the number of equations is larger than the number of free parameters. The minimum condition

$$\{r\}^T\{r\} = \min. \qquad (47)$$

of a least square fit then furnishes

$$[B]\{\Delta p\} = \{R\} \qquad (48)$$

with the symmetric and positive definite matrix

$$[B] = [A]^T[A] \qquad (49)$$

and

$$\{R\} = [A]^T\{r\} \qquad (50)$$

After the solution of Equation (48) new improved values of

the free parameters are obtained from

$$p_i^{\ell+1} = p_i^\ell + \Delta p_i \quad (51)$$

The fitting process is repeated until convergence has been achieved.

DETERMINATION OF THE DURATION

In the iterations, so far a fixed value has been assigned to the duration T_q. It is possible, however, to consider T_q to be a free parameter as well. In this case, the left hand side of Equation (41) has to be augmented by the term

$$\frac{\partial F}{\partial T_q} \Delta T_q = \frac{2\sigma_q^2}{T_q}\left(1 - \frac{\gamma^2}{g^2}\right)\Delta T_q \quad (52)$$

where ΔT_q denotes a duration increment and T_q is replaced by a known value T_q^ℓ. From the accordingly augmented Equation (48) thus also a least square fit value of

$$T_q^{\ell+1} = T_q^\ell + \Delta T_q \quad (53)$$

is obtained. It is seen from Equation (16), that T_q essentially plays the role of a scaling constant. On the other hand, the free parameters of $S_x(\Omega)$ contain also implicitly a scaling constant. Hence, the system of equations for the least square fit will be better conditioned for a fixed value of T_q than for T_q being a free parameter and more iterations are usually required in the latter case. T_q is the duration of the stationary output process $q(t)$ of the simple oscillator under the stationary input process $x(t)$ which may be a load or a support acceleration. In a random analysis T_q is needed to calculate the expected maximum response quantities of the structure. As already mentioned, $q(t)$ becomes stationary only after some time T_0, which depends on the period and the damping ratio of the oscillator. By adding an average value of T_0 to T_q, also the duration of the input process $x(t)$ may be estimated. To start the iterations, initial values have to be assigned to the free parameters. These can readily be obtained by assuming $S_x(\Omega)$ to be a constant S_0 over the interval $0 \leq \Omega \leq \Omega_{\max}$. The value of S_0 is determined in one step from a least square fit as outlined above.

NUMERICAL EXAMPLES

Verification Example

For earthquake excitation, the p.s.d.f. of the ground acceleration can be taken in the form

where ω_g and ζ_g denote the dominant circular frequency and the corresponding damping ratio of the ground, respectively, and S_0 is a scaling factor. $S_x(\Omega)$ is zero for $|\Omega| > 10\,\omega_g$. Thus, no energy is transmitted in frequencies above $10\omega_g$. An insight into the accuracy of the presented method can be obtained by determining the response spectrum from Equation (54) and then using this spectrum to calculate the underlying p.s.d.f. The resulting discretized representation of $S_x(\Omega)$ can directly be compared with Equation (54). With the numerical values

$$\omega_g = 15.5 \text{ rad } s^{-1};\ \zeta_g = 0.642;\ S_0 = 1.0 \quad (55)$$

the pseudo-velocity spectrum for $\zeta = 0.02$ was calculated from Equation (54) at 75 discrete frequencies

$$\omega_i = \frac{2\pi}{T_i} \quad (56)$$

with

$$\begin{array}{ll} T_i = 0.06 + i \cdot 0.005 & i = 1,2,\ldots 18 \\ T_i = 0.15 + i \cdot 0.05 & i = 1,2,\ldots 57 \end{array} \quad (57)$$

using Equations (15), (16), (17), (18) and (19). The integrations were carried out numerically. The duration of the output process was taken as $T_q = 10\,s$. A coarse and a fine discretized representation of $S_x(\Omega)$ was investigated. In the coarse representation the functional values at $\Omega = 0, 7.5, 15, 30, 60, 100$ rad s^{-1} and the first derivatives at all frequencies except at $\Omega = 100$ rad s^{-1} were chosen as parameters. The fine representation contained the functional values at $\Omega = 0, 5, 10, 15, 20, 30, 40, 60, 80, 100$ rad s^{-1} and the first derivatives at all frequencies except at $\Omega = 80$ and $\Omega = 100$ rad s^{-1}. In order to satisfy Equation (25), the first derivative at $\Omega = 0$ was constrained to zero in both cases. For the least square fit, the duration T_q was considered a free parameter as well. Thus, the coarse representation contained a total of 11 and the fine representation a total of 18 free parameters. The interpolation was done by third order polynomials. The only exception occurred in the coarse representation between $\Omega = 60$ and $\Omega = 100$ rad s^{-1} where a second order polynomial was used. Figure 4 shows the resulting discretized power spectral density functions as well as the analytic expression according to Equation (54) for $0 < \Omega < 100$ rad s^{-1}. It is seen that the agreement is very good. Table 1 gives the number of free parameters NF

$$S_x(\Omega) = S_0 \frac{\left[1 + 4\zeta_g^2\left(\dfrac{\Omega}{\omega_g}\right)^2\right]\left[1 - 0.03\left(\dfrac{\Omega}{\omega_g}\right)^2 + 0.002\left(\dfrac{|\Omega|}{\omega_g}\right)^3\right]}{\left[1 - \left(\dfrac{\Omega}{\omega_g}\right)^2\right]^2 + 4\zeta_g^2\left(\dfrac{\Omega}{\omega_g}\right)^2}\quad 0 \leq |\Omega| \leq 10\,\omega_g \quad (54)$$

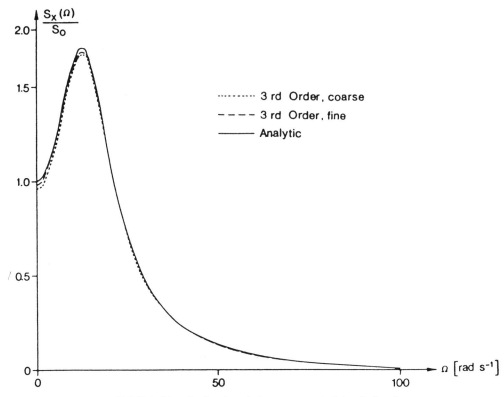

FIGURE 4. Discretized and analytic power spectral density functions.

of the discretized representations, the number of iterations NI for convergence, the variances σ^2, the expected frequencies f and the durations T_q. For Equation (54) σ^2 and f were evaluated by numerical integration. Table 1 also contains the value of

$$D = [\Sigma (S_v - \bar{S}_v)^2]^{1/2} \qquad (58)$$

after the last iteration where S_v and $\bar{S}_v$ denote the given and the calculated pseudo-velocity spectrum, respectively, and the summation has to be extended over all values used in the least square fit. Convergence is said to have been achieved, when the difference between the values of D at two subsequent iterations is smaller than 10^{-10}. As expected D is smaller for the fine representation than for the coarse representation. It is also seen from Table 1 that the calculated optimum values of T_q agree reasonably well with the value of $T_q = 10$ s assumed in the calculation of the response spectrum.

Housner's Design Spectra

Figure 5 shows the pseudo-velocity spectrum $S_v(T,\zeta)$ after Housner [6] for $\zeta = 0.02$ as a function of period T.

For the least square fit 85 discrete values of $S_v(T,\zeta)$ were used. The p.s.d.f. $S_x(\Omega)$ was characterized by third and fifth order continuous polynomials using the functional values at $\Omega = 0, 5, 10, 15, 20, 30, 40, 60, 80, 110$ rad s^{-1}. In addition first derivatives were introduced namely at $\Omega = 0, 5, 10, 15, 20, 30, 40, 60$ rad s^{-1} for the third order polynomials and at $\Omega = 0, 15, 40$ rad s^{-1} for the fifth order polynomials. By this, the last polynomial in the latter case was only of order 4. Again Equation (26) was satisfied by constraining the first derivative at $\Omega = 0$ to zero in both cases. The least square fit with T_q as a free parameter furnished an optimum value of $T_q = 2.84$ s for the third order approximation and of $T_q = 4.97$ s for the fifth order approximation. A recalcula-

TABLE 1. Results from Least Square Fit.

	NF	NI	D	σ^2	f	T_q
Coarse	11	6	0.0184	6.63	4.01	10.97
Fine	18	6	0.0025	6.71	4.00	10.42
Equ. (54)				6.80	4.14	10.00

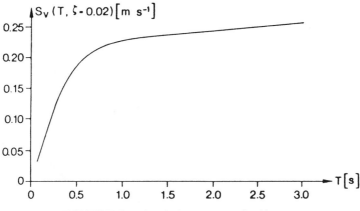

FIGURE 5. Pseudo-velocity spectrum after Housner.

tion was then done with a fixed value of $T_q = 4\ s$, which is close to the average value. The resulting functions $S_x(\Omega)$ are shown in Figure 6. It is seen that both representations agree very closely. The response spectra calculated from $S_x(\Omega)$ show excellent agreement with Figure 5 with deviations which are below drawing accuracy. In Table 2 again some results from the least square fit are given.

CONCLUSIONS

The above described method to determine the power spectral density function $S_x(\Omega)$ of a stationary Gaussian process from given smooth response spectra is based on the probability distribution of the extreme values of such a process. For the expected extreme value the approximate expression of Davenport is used. $S_x(\Omega)$ is described in discretized form by parameters, which represent functional values and values of first derivatives at discrete frequencies, and by piecewise polynomials. The parameters are determined from a least square fit. Due to the polynomial representation of $S_x(\Omega)$, all integrations required to establish the system of normal equations can be carried out analytically. Thus, numerically accurate integral values are obtained in a computationally efficient way. The fitting procedure can be extended to determine also a least square fit value of the duration of the output process.

The accuracy of the method depends on the number of parameters used in the description of $S_x(\Omega)$, on the order of the interpolation polynomials and on the number of equations considered in the least square fit. These quantities can be chosen in a flexible way according to the accuracy require-

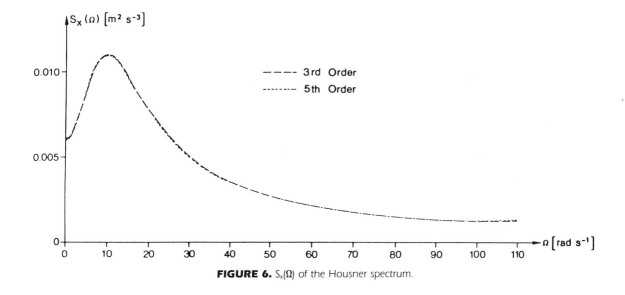

FIGURE 6. $S_x(\Omega)$ of the Housner spectrum.

TABLE 2. Least Square Fit Results for Housner's Design Spectra.

		NF	NI	D	σ^2	f
Housner	3rd order	17	4	0.0040	0.0664	6.82
	5th order	12	4	0.0039	0.0664	6.83

ments in a particular case. The numerical examples indicate that already rather simple representations of $S_x(\Omega)$ furnish results with sufficient accuracy for engineering purposes.

Some open questions remain. One of them is the determination of the duration of the stationary input process from a least square fit value of the duration of the output process. The time T_0 after which the response of a simple oscillator reaches stationarity depends on the period and the damping ratio of the oscillator. As several periods and eventually also several damping ratios are considered in the least square fit, some average value of T_0 should be chosen. Another open question concerns the influence of the use of more refined expressions for the expected extreme value of a stationary random process on $S_x(\Omega)$. Finally it should be mentioned that the above described calculations have to be done on a computer. A FORTRAN program in which all calculations have been implemented has been written for this purpose.

ACKNOWLEDGEMENT

This chapter is partly based on [11]. The American Society of Civil Engineers has kindly granted permission to reprint the appropriate material which is hereby gratefully acknowledged.

REFERENCES

1. Ang, A., "Probability Concepts in Earthquake Engineering," *Applied Mechanics in Earthquake Engineering*, W. D. Iwan, ed., AMD-Vol. 8, American Society of Mechanical Engineers, New York, N.Y., pp. 225–259 (1974).
2. Clough, R. W., "Earthquake Analysis by Response Spectrum Superposition," *Bulletin of the Seismological Society of America*, Vol. 52(3), pp. 647–660 (July, 1962).
3. Clough, R. W. and Penzien, J., *Dynamics of Structures*, McGraw-Hill, Inc., New York (1975).
4. Davenport, A. G., "Note on the Distribution of the Largest Value of a Random Function with Application to Gust Loading," *Proceedings of the Institution of Civil Engineers*, Vol. 28, paper No. 6739, pp. 187–196 (1964).
5. Der Kiureghian, A., "On Response of Structures to Stationary Excitation," *Report No. EERC 79-32*, Earthquake Engineering Research Center, University of California, Berkeley (Dec., 1979).
6. Housner, G. W., "Behaviour of Structures during Earthquakes," *Journal of the Engineering Mechanics Division*, ASCE, Vol. 85, EM4, pp. 109–129 (Oct., 1959).
7. Newmark, N. M., Blume, J. A., Kapur, K. K., "Seismic Design Spectra for Nuclear Power Plants," *Journal of the Power Division*, ASCE, Vol. 99, PO2, pp. 287–303 (Nov., 1973).
8. Pfaffinger, D. D., "Comparative Seismic Analysis," *Proceedings of the European NASTRAN User's Conference*, The MacNeal-Schwendler Corp., Munich (1976).
9. Pfaffinger, D. D., "Die Methode der Antwortspektren aus der Sicht der probabilistischen Tragwerksdynamik," *Report Nr. 90*, Institute of Structural Engineering, Swiss Federal Institute of Technology, Zuerich (1979).
10. Pfaffinger, D. D., "Analytical Evaluation of Modal Covariance Matrices," *Computer Methods in Applied Mechanics and Engineering*, Vol. 24(3), pp. 269–286 (Dec., 1980).
11. Pfaffinger, D. D., "Calculation of Power Spectra from Response Spectra," *Journal of the Engineering Mechanics Division*, American Society of Civil Engineers (January, 1981).
12. Ruiz, P. and Penzien, J., "Probabilistic Study of the Behaviour of Structures during Earthquakes," *Report No. EERC 69-3*, Earthquake Engineering Research Center, University of California, Berkeley (1969).
13. Vanmarcke, E. H., "On the Distribution of the First-Passage Time for Normal Stationary Random Processes," *Journal of Applied Mechanics*, Vol. 42, pp. 215–220 (March, 1975).

CHAPTER 12

Chance Constrained Aggregate Blending

SANG M. LEE* AND DAVID L. OLSON**

ABSTRACT

Problems of material size variance and the existence of multiple, conflicting objectives in aggregate blending for construction operations are considered. Linear programming, chance constrained programming, and goal programming models for the aggregate blending problem are presented for a simple case. Computer code availability is discussed. A linear approximation for chance constraints is presented to allow use of linear programming codes. The inaccuracy of this approach is discussed.

INTRODUCTION

Construction operations often involve the mixture of materials into an aggregate blend satisfying specified size gradations. These aggregate blends are used in surfacing mixtures of asphalt, concrete, or gravel, as well as for concrete production and many other applications. It is often possible to select materials for blending from a variety of sources. Design of material mixtures is a crucial element of many civil engineering design and operation activities.

Alternative materials have differing characteristics. The intent is often to use larger material as the basis of a blend and smaller materials to fill voids. A major characteristic is material size as measured by sieve analysis. Description of these characteristics can be expressed in percent retained or passing by sieve size. However, most materials are not totally consistent, and a more complete description of material characteristics would include the variance of these readings. Another characteristic of importance is material cost. In general, material that is consistently graded has added quality and may be expected to cost more, all other factors being equal. However, cost is also determined by local availability. The design of aggregate involves the need to consider physical characteristics as well as cost. The purpose of this chapter is to consider the tradeoff in cost and risk in the design of aggregate mixtures, with emphasis upon the variance of materials and its impact upon aggregate design.

The problem involved in designing mixtures when component materials vary in size is that intended product properties are often not met. First, final product that is designed with average properties may yield samples that fail to meet specifications. This can result in rejection of particular batches of product, which can have enormous cost impact. Secondly, if product that varies from design specifications is used, the quality of the final product may be less than desired. In addition, it is often necessary for field personnel to make significant adjustments to proportions of material blended in order to readjust average output to specified ranges. Whenever severe adjustments are made, the blending process can be disrupted, resulting in hidden costs due to a widely varying flow of material, as well as additional operational time and material waste.

Present practice often uses a trial and error process, trying out various combinations of materials available. While estimators can become extremely proficient in designing economical and operationally efficient blends, a trial and error procedure would often miss solutions which can improve the quality of the blend at the same time that cost could be reduced. The aggregate blending problem can be formulated as linear programming model, providing the ability to obtain the solution that would satisfy design specifications on average at minimum cost. However, linear programming requires a number of dangerous assumptions for this application, and would likely result in very dangerous designs from the aspect of specification satisfaction. And linear programming assumes no variance in material,

---

*Department of Management, University of Nebraska, Lincoln, NE
**Department of Business Analysis, Texas A&M University, College Station, TX

which is unrealistic. Chance constrained programming provides a means of incorporating the variance properties of input materials in order to obtain design blends with prescribed probabilities of satisfying each particular gradation specification. A new problem develops, in that judgement concerning the relative degree of risk versus cost must be addressed. And chance constrained programming algorithms are not as readily available as linear programming algorithms.

This chapter presents, through a small aggregate blending example based upon a real asphaltic concrete aggregate design for a contractor with a Nebraska State Highway project, how a number of alternative techniques can be applied. Emphasis will be upon the limitations to each technique. The main intent is to provide tools to support the decision-making judgement of civil engineers. Goal programming is presented as a means to directly address the judgemental aspects often required in real design problems. Finally a linear goal programming approximation to the problem allowing use of readily available computer codes and allowing consideration of variance is presented.

SAMPLING INFORMATION

The design problem begins by identifying size characteristics of available input materials. Statistical sampling of materials is a well developed process. Traditionally, the mean percentage (retained or passing) is used as the basis for design. However, the mean conveys no information concerning the consistency of materials, one of the major problems in aggregate blending. The variance of materials can be obtained from the same information used to identify the mean.

The problem used to demonstrate alternative design techniques is to blend a coarse material with sand and fines to satisfy the properties given in Table 1.

Three materials are assumed available. Field testing will yield measures of each of these materials size characteristics. For this problem, Table 2 presents 40 observations of the sieve samples for the coarse material.

Whenever samples are taken and results are not identical, the mean does not provide all available information about material size. Variance of materials can also be calculated.

TABLE 1. Design Specifications.

Sieve Size	Minimum Percent Retained	Maximum Percent Retained
#10	56	
#50	80	84
#200	90	92

A number of statistical distributions of true material properties are possible. Very often, naturally occurring materials are described by the normal distribution. The distribution that applies can be identified by plotting observed readings. The normal distribution is identified by a symmetric graph, with the greatest number of observations occurring at the mean. Figure 1 presents a graph of the percent by dry weight retained on the #10 sieve for the coarse material. This appears to be generally normally distributed, given the fairly wide variance in material and that there are only 40 samples to base the judgement upon. The measures for the coarse material on the #50 sieve and #200 sieve involve much less variance, with about the same number of observations above the mean as below the mean. Therefore, the coarse material appears to be at least roughly normal in distribution.

While the mean percentage retained (or passing) for materials is often used, the variance is not. Calculation of the variance based upon sampling is well developed, however, and requires no additional testing data. An accurate measure of variance will likely require more observations. Table 3 presents the formula for sampling variance and the calculations for the #10 sieve for the coarse material. Note that the sum of squared differences between each observation and the mean is divided by the number of observations minus one, to adjust for sampling bias.

Table 4 presents the means and variances by sieve size for all three available materials, as well as cost per cubic yard. All of this information is typically available, and with the exception of the variance, is normally used as the basis for aggregate design. Each construction problem is quite different, involving the need to blend combinations of materials to meet specified requirements.

This decision can be viewed as constrained optimization, with the intent being to identify the proportion of the three materials in such a way as to minimize cost. This section will review how this problem can be formulated as a linear programming model. Note, however, that there are a number of assumptions required for application of linear programming, all of which cause some difficulty. Later sections will present alternative modeling procedures which minimize these difficulties, as well as chance constrained programming, which utilizes available variance information to obtain a solution satisfying specified probability requirements. Finally, goal programming modeling will be presented as a means of reconciling the inherent tradeoff between the probability of satisfying the required specifications and cost.

The aggregate blending problem involves the decision of the proportion of materials to combine to satisfy specifications. The decision variables in the example problem are:

F—fines as a percent of total aggregate
S—sand as a percent of total aggregate
C—coarse as a percent of total aggregate.

TABLE 2. Sieve Data for Coarse Material.

Sample	Sieve- #10	#50	#200
	Total percent retained on sieve—dry weight		
1	64	96	98
2	67	96	98
3	72	97	99
4	72	96	99
5	67	96	99
6	63	95	98
7	79	97	100
8	81	97	100
9	78	96	99
10	74	96	99
11	69	95	98
12	70	96	99
13	71	96	99
14	70	96	99
15	68	95	98
16	75	96	99
17	71	96	99
18	74	96	99
19	70	96	99
20	70	96	99
21	68	95	99
22	73	96	99
23	74	96	99
24	81	97	100
25	73	96	99
26	75	96	99
27	73	96	99
28	78	97	99
29	78	96	99
30	82	97	100
31	73	96	99
32	74	96	99
33	73	96	99
34	74	96	99
35	73	96	99
36	74	96	99
37	74	96	99
38	73	96	99
39	72	96	99
40	73	96	99
average	72.825	96.050	98.975
variance	17.943	.254	.230

(sample data to nearest percent)

TABLE 3. Variance Calculation for Coarse Material on #10 Sieve.

Sample	X_j	$[X_j - \bar{X}]$	$[X_j - \bar{X}]^2$
1	64	8.825	77.8806
2	67	5.825	33.9306
3	72	.825	.6806
4	72	.825	.6806
5	67	5.825	33.9306
6	63	9.825	96.5306
7	79	6.175	38.1306
8	81	8.175	66.8306
9	78	5.175	26.7806
10	74	1.175	1.3806
11	69	3.825	14.6306
12	70	2.825	7.9806
13	71	1.825	3.3306
14	70	2.825	7.9806
15	68	4.825	23.2806
16	75	2.175	4.7306
17	71	1.825	3.3306
18	74	1.175	1.3806
19	70	2.825	7.9806
20	70	2.825	7.9806
21	68	4.825	23.2806
22	73	.175	.0306
23	74	1.175	1.3806
24	81	8.175	66.8306
25	73	.175	.0306
26	75	2.175	4.7306
27	73	.175	.0306
28	78	5.175	26.7806
29	78	5.175	26.7806
30	82	9.175	84.1806
31	73	.175	.0306
32	74	1.175	1.3806
33	73	.175	.0306
34	74	1.175	1.3806
35	73	.175	.0306
36	74	1.175	1.3806
37	74	1.175	1.3806
38	73	.175	.0306
39	72	.825	.6806
40	73	.175	.0306
Sum	2913		699.7740

$$\text{Variance} = \frac{\Sigma(X_j - \bar{X})^2}{n-1} = \frac{699.7740}{39} = 17.9429$$

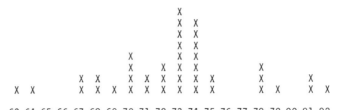

FIGURE 1. Percent coarse material retained on #10 sieve.

The objective to be optimized is cost. The function measuring cost is

$$\text{Minimize } \$2.44\,F + \$3.30\,S + \$4.50\,C$$

The first constraint required in this model is to force the component percentages to add to 100%.

$$F + S + C = 100$$

At this stage, it is apparent that a mixture consisting totally of fines would be the least expensive. However, that solution would not satisfy the required size specifications. Sieve limits can be included by the following constraints:

#10 sieve: $\quad 0\,F + 48.35\,S + 72.825\,C \geq 56$
#50 sieve: $\quad 1\,F + 80.80\,S + 96.050\,C \geq 80$
$\quad\quad\quad\quad\quad 1\,F + 80.80\,S + 96.050\,C \leq 84$
#200 sieve: $11\,F + 96.10\,S + 98.975\,C \geq 90$
$\quad\quad\quad\quad 11\,F + 96.10\,S + 98.975\,C \leq 92$

Solution of this model by a linear programming computer package yields the following solution:

$F = 6.294\%$
$S = 50.016\%$
$C = 43.690\%$
Cost $= \$3.770$/cubic yard

This solution appears attractive, in that it is the solution satisfying the required constraints with the least possible cost.

TABLE 4. Available Materials.

Sieve Size	FINES Mean	FINES Variance	SANDS Mean	SANDS Variance	COARSE Mean	COARSE Variance
#10	0	—	48.35	4.079	72.825	17.943
#50	1.00	.103	80.80	2.636	96.050	.254
#200	11.00	.513	96.10	.708	98.975	.230
Cost/Cubic Yard	$2.44		$3.30		$4.50	

There are two features of this solution, however, that are not attractive. First, linear programming solutions assume continuous possible decision variable values. Practical application requires rounding the solution, either to the nearest percent or possibly to the nearest tenth of a percent. Rounding the continuous solution can lead to two unattractive outcomes. The rounded solution is no longer guaranteed to be the absolutely least expensive rounded solution. The other unattractive feature is that the rounded solution may not strictly satisfy the required constraints. The solution obtained results in the following expected performance:

	Continuous Solution	Rounded Solution
F	6.294%	6%
S	50.016%	50%
C	43.690%	44%
Cost	$3.770	$3.776
Expected % Retained		
#10	56.000*	56.218
#50	82.440	82.722
#200	92.000*	92.259**

It can be seen that the continuous linear programming solution includes two constraints that are exactly met (the lower limit on the #10 sieve and the upper limit on the #200 sieve). The rounded solution turned out to be safer on the #10 sieve, but would be expected to fail on average on the #200 sieve. This practical limitation of linear programming could be overcome through the use of integer programming packages, which would identify the optimal solution to models including the restriction that solutions be limited to integer values. However, integer programming is a good deal more costly than linear programming. In addition, the problem of varying input material characteristics is not addressed. The design should include a buffer to lower the risk of failing to meet specification limits. Linear programming solutions by their nature are extreme and squeeze all available cost by taking advantage of constraint limits. The real blending decision requires a tradeoff between cost and risk.

CHANCE CONSTRAINED PROGRAMMING

Chance constrained programming modifies linear programming by treating model coefficients as variables. In the model we have been using, while cost coefficients and target limits are constant, we know that the coefficients for percentage of material retained on each sieve will vary. In addition, we have an estimate of the variance of these coefficients, and that the distribution of percentage retained appears to be normal. Chance constrained programming provides a means of incorporating this information into our model. The form of the sieve limit constraints now will be probabilistic, requiring some specification of the probability of required constraint satisfaction. These probabilistic constraints are nonlinear, meaning we no longer can use linear programming algorithms for solution.

For a normally distributed variance, the #10 limit will now be: $0 F + 48.35 S + 72.825 C - z$ (sum of coefficient variance)$^{1/2} \geq 56$. This form of constraint will replace the five sieve limit constraints. The z value is determined from normal distribution tables. Some common z values and associated probabilities are given in Table 5.

For a normally distributed variable, the z value multiplied by the standard deviation (the square root of the variance) will give the prescribed penalty to satisfy a target. A z value of 0 would be equivalent to a .50 probability of satisfying a target, which would be the same as the original linear constraint.

Given our model, and incorporating the sampling information we have, the chance constraints would be:

#10 lower limit
$$0 F + 48.35 S + 72.825 C +$$
$$- z(0 F^2 + 4.079 S^2 + 17.943 C^2)^{1/2} \geq 56$$

#50 lower limit
$$1 F + 80.80 S + 96.050 C +$$
$$- z(.103 F^2 + 2.636 S^2 + .254 C^2)^{1/2} \geq 80$$

#50 upper limit
$$1 F + 80.80 S + 96.050 C +$$
$$+ z(.103 F^2 + 2.636 S^2 + .254 C^2)^{1/2} \leq 84$$

#200 lower limit
$$11 F + 96.10 S + 98.975 C +$$
$$- z(.513 F^2 + .708 S^2 + .230 C^2)^{1/2} \geq 90$$

#200 upper limit
$$11 F + 96.10 S + 98.975 C +$$
$$+ z(.513 F^2 + .708 S^2 + .230 C^2)^{1/2} \leq 92$$

TABLE 5. Z Values and Probabilities.

Z	Probability
.253	.60
.524	.70
.841	.80
1.282	.90
1.645	.95
2.054	.98
2.327	.99

The variance term serves as a buffer or penalty. If we choose a highly variable material, the penalty is greater. The sign of the z value indicates the direction of penalty. For a greater than or equal constraint (lower limit), the penalty forces overachievement of the expected value of the function to at least equal the penalty. Conversely, for less than or equal constraints (upper limits), the penalty is positive.

The use of chance constraints calls for additional judgement on the part of the modeler. The tradeoff in cost and risk must be addressed by setting risk levels in the chance constraints. Goal programming provides one means of reconciling these conflicting objectives. At this stage, we can demonstrate chance constrained programming by examining the impact of specifying various risk levels in the form of the probability specified for constraints.

There are comments in order at this stage. First, chance constrained programming used here implies a normal distribution of coefficient values. That distribution often is the true distribution. However, the variance used is based upon sampling information, introducing error. Probability accuracy for an application would require testing. The probabilities used would be accurate if the coefficients actually were normally distributed and if the estimates of mean value and variance were precise. The concept of probability is still very useful, in that a higher probability level assigned to a constraint will result in reduced risk. Second, we will be able to assign a probability level of satisfaction required for each individual constraint. The algorithm will seek to do this. This is not the same as specifying a particular probability of satisfying all such constraints simultaneously, which may be the real objective. That is a much more intractable mathematical problem. References for this area of research are cited in the reference section. We will use the joint probability of satisfying all constraints simultaneously as a measure of evaluation, but we can only model requirements for each constraint to be met independently. Again, the desired general impact is obtained.

By adding the penalty for risk to each constraint, we increase the risk that there will be no feasible solution to our model (we may make the constraint set so restrictive that there is no combination of available materials that could satisfy our specifications). As can be seen by the z values given

TABLE 6. Chance Constrained Solutions.

Probability Specified	LP(50%)	60%	70%	80%	90%
Coarse	43.690%	46.395%	49.483%	53.408%	
Sand	50.016%	47.084%	43.756%	39.552%	infeasible
Fines	6.294%	6.521%	6.761%	7.040%	
Expected % Retained					
#10 sieve	56.000*	56.552	57.192	58.018	
#50 sieve	82.440	82.671	82.951	83.327	
#200 sieve	92.000*	91.884	91.769	91.644	
Probability of Satisfying Limit					
#10 lower limit	.500*	.600*	.700*	.800*	
#50 lower limit	.998	.9996	.9999	.9999	
#50 upper limit	.968	.952	.918	.833	
#200 lower limit	.9999	.9999	.9999	.9999	
#200 upper limit	.500*	.600*	.700*	.800*	
Joint Probability	.242	.343	.450	.533	
Cost/cubic yard	$3.770	$3.801	$3.836	$3.880	

*at prescribed limit

in Table 5, there is a great deal more penalty associated with seeking high levels of probability. Increasing the probability from 90 to 95 percent involves more penalty than increasing probability from 60 to 70 percent. In a later section, we will present a linear approximation of these nonlinear constraints and examine the cost of that approximation.

Table 6 presents solutions for chance constrained models set at various levels of probability for the satisfaction of gradation limits. The 50 percent column represents the earlier linear programming solution. The solutions can be evaluated in terms of cost as well as expected probability of satisfying all gradation limits simultaneously. As before, it can be seen that the lower #10 limit and the upper #200 limit are at the prescribed limits. For the 50 percent solution, if there is any variance in material, there is a 50 percent chance that one of these limits will fall out of specification. The expected percent retained on the #50 sieve is near the middle of the specified range. Therefore, there is a strong probability that both the upper and lower limits will be satisfied. Assuming normality and disregarding sampling error, the probability of satisfying the lower #50 limit with the 50 percent solution can be calculated by the formula:

$$z = |\text{target value} - \text{expected value}|/\text{standard deviation}$$

or

$$z = \frac{(82.440 - 80)}{[(.103 \times (.06294)^2) + (2.636 \times (.50016)^2) + (.254 \times (.43690)^2)]^{1/2}}$$

$$z = 2.440/.8416 = 2.899$$

Once the z score is obtained, the probability of satisfying a limit is obtained from a table of normal curve areas. The area of satisfying the lower #50 limit will be .998, found by identifying the normal curve area included by a z value of 2.899. The probabilities for each of the five gradation limits for the 50 percent solution are:

$z_{l\#10} = (56 - 56)/2.10843 = 0$ Probability .500
$z_{l\#50} = (82.440 - 80)/.8416 = 2.899$ .998
$z_{u\#50} = (84 - 82.440)/.8416 = 1.854$ .968
$z_{l\#200} = (92.0 - 90)/.47220 = 4.2354$ .9999
$z_{u\#200} = (92 - 92.0)/.47220 = 0$ .500

The joint probability of satisfying all five limits simultaneously is the product of the five probabilities, or .242.

Review of Table 6 demonstrates that increasing the specified probability of constraint satisfaction has the expected impact of increasing use of less variable materials at the expense of the highly variable sand. Accordingly, cost tends to rise as greater prescribed probabilities are applied. And as the risk level is increased, the model becomes infeasible. In this case, it is not possible to combine the available materials in such a way as to obtain .90 confidence of satisfying all contraints simultaneously.

TABLE 7. Rounded Chance Constrained Solutions.

Probability Specified	LP(50%)	60%	70%	80%
Coarse	44%	46%	49%	53%
Sand	50%	47%	44%	40%
Fines	6%	7%	7%	7%
Expected % Retained				
#10 sieve	56.218	56.224	56.958	57.937
#50 sieve	82.722	82.229	82.687	83.297
#200 sieve	92.259	91.466	91.552	91.667
Probability of Satisfying Limit				
#10 lower limit	.541	.541**	.665**	.791**
#50 lower limit	.9994	.997	.9998	.9999
#50 upper limit	.936	.987	.959	.842
#200 lower limit	.9999	.9994	.9998	.9999
#200 upper limit	.292**	.880	.845	.784**
Joint Probability	.148	.468	.539	.522
Cost/cubic yard	$3.776	$3.792	$3.828	$3.876

**failed prescribed limit

Table 7 presents expected performance of solutions rounded to the nearest percent. As can be seen, rounding can lead to significant divergence from designed performance, especially at the lower limit on the #10 sieve. If production equipment is sufficiently precise to allow inputs specified to tenths of a percent, much of this distortion can be avoided.

GOAL PROGRAMMING

Goal programming provides a means of considering multiple, possibly conflicting, objectives. Goal programming operates as a process, requiring the decision maker to identify objectives of importance, place quantitative targets of attainment on these objectives, and then rank order the relative importance of these objective targets. A goal programming solution is obtained much the same as in linear (or nonlinear) programming, with the exception that infeasibility is avoided by seeking to get as close to a specified target as possible without sacrificing targets assigned a higher priority.

For the aggregate blending problem we have been considering, there are two clear objectives: minimize cost and maximize the confidence of satisfying gradation requirements. Other real objectives are possible as well. For operational reasons, limits on use of specific materials could exist. Fines have relatively low variance in our model and at high levels of specified confidence, solutions include increasing proportions of fines. There may be reasons for not including more than a particular level of fines, as they may clog machinery, or provide poor product characteristics other than size. Goal programming models allow the opportunity to include any number of additional constraints to the model.

Once objectives have been identified, target levels of attainment for relative comparison with other objectives need to be determined. One set of possible target levels could be:

Cost	Confidence level	Fines limit
$4.00/cubic yard	.50	≤8%
$3.90	.70	≤7%
$3.85	.80	
minimize	.90	
	.95	

The grid of objective targets could be made as fine as possible. For planning purposes, a rough grid could be used until the tradeoffs involved are identified. Once a solution is obtained, a finer grid in the vicinity of the obtained solution could be used.

When target levels are identified, it is necessary that relative importance be identified. Preemptive goal programming operates in the same manner as linear programming, with the exception that those objective targets assigned higher priorities are required to be satisfied before consideration of targets assigned less priority. A possible set of ranking for this problem could be:

Priority	Objective	Target Level
1	Expected gradation limits met	≥.50 probability
2	Cost	≤$4.00/cubic yard
3	Gradation limit confidence	≥.70 probability
4	Fines	≤8 percent or less
5	Cost	≤3.85/cubic yard
6	Gradation limit confidence	≥.80 probability
7	Fines	≤7 percent or less
8	Gradation limit confidence	≥.90 probability
9	Cost minimized	$0.00
10	Gradation limit confidence maximized	≥.99 probability

Goal programming models require a modification of linear programming constraints and a new type of objective function. Constraints are all converted into equality relationships with the addition of a negative and positive deviational variable for each goal constraint. The negative deviation variable for constraint i (d_i^-) represents the functional value below the specified target. The positive deviational variable for constraint i (d_i^+) represents functional value for a solution above the target value. If the desired target is a minimum level of specified attainment, the negative deviational variable is to be minimized. Conversely, if the target is a maximum level, the positive deviational variable is to be minimized. The preemptive priority system assures that a more important target level is not sacrificed for attainment of a less important target level.

Required constraints can be included in a goal programming formulation. In our aggregate blending model, no deviation from the requirement that the sum of the three materials add to 100 percent is allowed, and that constraint would not be modified.

First Priority

The first priority objective is to design the mix to satisfy the gradation limits on average. The variance of materials would have no impact at this stage, so in effect, the z value is implicitly 0.

$$0\,F + 48.35\,S + 72.825\,C + d_1^- - d_1^+ = 56$$
$$1\,F + 80.80\,S + 96.050\,C + d_2^- - d_2^+ = 80$$
$$1\,F + 80.80\,S + 96.050\,C + d_3^- - d_3^+ = 84$$
$$11\,F + 96.10\,S + 98.975\,C + d_4^- - d_4^+ = 90$$
$$11\,F + 96.10\,S + 98.975\,C + d_5^- - d_5^+ = 92$$

The first priority would be met if a solution were obtained with d_1^-, d_2^-, d_3^+, d_4^-, and d_5^+ equal to zero. The complementary deviational variables (d_1^+, d_2^+, d_3^-, d_4^+, and d_5^-) could take on any positive value without affecting the solution.

Second Priority

The second priority would be to minimize d_6^+ where

$$2.44\,F + 3.30\,S + 4.50\,C + d_6^- - d_6^+ = 4.00$$

Third Priority

The third priority includes chance constraints. However, there is no real difference in form. The z value would be set to .524, representing a confidence level of .70.

$$0\,F + 48.35\,S + 72.825\,C - .524$$
$$\times\,(0\,F^2 + 4.079\,S^2 + 17.943\,C^2)^{1/2} + d_7^- - d_7^+ = 56$$

$$1\,F + 80.80\,S + 96.050\,C - .524$$
$$\times\,(.103\,F^2 + 2.636\,S^2 + .254\,C^2)^{1/2} + d_8^- - d_8^+ = 80$$

$$1\,F + 80.80\,S + 96.050\,C + .524$$
$$\times\,(.103\,F^2 + 2.636\,S^2 + .254\,C^2)^{1/2} + d_9^- - d_9^+ = 84$$

$$11\,F + 96.10\,S + 98.975\,C - .524$$
$$\times\,(.513\,F^2 + .708\,S^2 + .230\,C^2)^{1/2} + d_{10}^- - d_{10}^+ = 90$$

$$11\,F + 96.10\,S + 98.975\,C + .524$$
$$\times\,(.513\,F^2 + .708\,S^2 + .230\,C^2)^{1/2} + d_{11}^- - d_{11}^+ = 92$$

The third priority would be satisfied if d_7^-, d_8^-, d_9^+, d_{10}^-, and d_{11}^+ were zero.

Fourth Priority

The fourth priority would be satisfied if d_{12}^+ were 0 where

$$F + d_{12}^- - d_{12}^+ = 8$$

Fifth Priority

The fifth priority was assigned to keeping cost at or below $3.85 per cubic yard. The cost function, identical to goal constraint six, but with deviational variables d_{13}^- and d_{13}^+, would be set to the target of 3.85. This goal would be satisfied if d_{13}^+ were zero. If that was not possible, given satisfaction of priorities one through six, the minimum cost given satisfaction of prior goals would be the goal programming solution.

Sixth Priority

The sixth priority would require five more constraints, identical to goal constraints 7 through 11, with the exception that the z value used would be .841, representing a .8 confidence level, and these constraints would include deviational variables d_{14} through d_{18}. The goal would be to make d_{14}^-, d_{15}^-, d_{16}^+, d_{17}^-, and d_{18}^+ as close to zero as possible while maintaining satisfaction of higher goals. If feasible, these five deviational variables would be zero.

Seventh Priority

The goal constraint would be: $F + d_{19}^- - d_{19}^+ = 7$. This goal would be satisfied if d_{19}^+ were zero.

Eighth Priority

The five gradation limit constraints would be used, with deviational variables d_{20}^-, d_{21}^-, d_{22}^+, d_{23}^-, and d_{24}^+ minimized. The z value used would be 1.282.

Ninth Priority

The objective is to minimize cost. This can be attained by setting the cost function equal to zero and minimizing the positive deviational variable (d_{25}^+).

Tenth Priority

The five gradation limit constraints would be used with deviational variables d_{26}^-, d_{27}^-, d_{28}^+, d_{29}^-, and d_{30}^+ minimized. The z value used would be 2.327.

Goal programming models can be as complex as the designer desires. The list of goals given here is broad, covering a wide range of cost and probability levels for the initial analysis. Once a general idea of the tradeoffs involved are identified, a much smaller list of goals reflecting fine tuning would be used. A chance constrained goal programming algorithm would proceed by seeking to minimize the deviational variables assigned first priority subject to the required constraints (five in this case). If it is feasible to minimize all deviational variables assigned priority 1 to zero, the algorithm would consider priority 2, with the added requirement that the five deviational variables satisfied at priority 1 not be sacrificed. Once a goal is encountered that cannot be totally satisfied, the objective is to minimize the value of the deviational variable or variables assigned that priority, subject to satisfaction of all more important goals and required constraints. If this solution is identified, those goals given less important priority would be considered only to break ties between solutions minimizing the deviational variables representing the unsatisfied priority goal.

The first goal programming solution has as its objective the minimization of the following deviational variables in the priority order listed:

Priority	Deviational variables
1	$d_1^-\ d_2^-\ d_3^+\ d_4^-\ d_5^+$
2	d_6^+
3	$d_7^-\ d_8^-\ d_9^+\ d_{10}^-\ d_{11}^+$
4	d_{12}^+
5	d_{13}^+
6	$d_{14}^-\ d_{15}^-\ d_{16}^+\ d_{17}^-\ d_{18}^+$
7	d_{19}^+
8	$d_{20}^-\ d_{21}^-\ d_{22}^+\ d_{23}^-\ d_{24}^+$
9	d_{25}^+
10	$d_{26}^-\ d_{27}^-\ d_{28}^+\ d_{29}^-\ d_{30}^+$

Given this set of goals, the solution would be as given in Table 8.

The first unsatisfied goal of this model is priority 6. The cost of this solution is exactly $3.85/cubic yard, the goal assigned priority 5. All probabilities of satisfaction are greater than .7, the priority 3 goal. Fines are below 8 percent, the priority 4 goal. The lower #10 limit has a probability less than the desired .8 assigned priority 6. In this case, the goals assigned priorities 7 through 10 would not impact the solution. Priority 7 turned out to be satisfied, although it played no part in selecting the solution.

The sensitivity of the solution can be identified by rearranging priority assignments. Goal programming solution 2 reverses the priority assigned to the cost target of $3.85/cubic yard ($d_{13}^+$) and the target probability of .80 (d_{14}^-, d_{15}^-, d_{16}^+, d_{17}^-, and d_{18}^+). Table 9 presents the resulting solution.

This solution sacrifices some cost in order to attain the desired probability level for each gradation constraint. Two of the gradation limits are exactly met (lower #10 and upper #200). The goal to keep fines at or below 7 percent is not met by this solution. Note that the joint probability of satisfying all five gradation constraints simultaneously has dropped slightly, because the probability of satisfying the upper #50 limit was overachieved more in goal programming solution 1 (see Table 1). As stated before, the goal programming model can only specify satisfaction of individual constraints, and cannot address the objective of joint probability. Therefore, joint probabilities for models with lower

TABLE 8. Goal Programming Solution 1.

C = 50.824%   Cost = $3.85
S = 42.213%   Expected values:
F = 6.963%   #10—57.423%
                  #50—82.994%
                  #200—91.636%

Probability of satisfaction:

lower #10—.731
lower #50—.9999
upper #50—.915
lower #200—.9999
upper #200—.800

Joint probability—.535

Priority	
1	satisfied (all expected values within specifications)
2	satisfied (cost below $4.00/cubic yard)
3	satisfied (probability of satisfaction all greater than .7)
4	satisfied (fines below 8%)
5	satisfied exactly (cost $3.85/cubic yard)
6	unsatisfied (lower #10 probability below .8)
7	satisfied (fines below 7%)
8	unsatisfied (two limits have probability below .9)
9	unsatisfied (cost greater than 0)
10	unsatisfied (three limits below .99 probability)

TABLE 9. Goal Programming Solution 2.

C = 53.408%   Cost = $3.880
S = 39.552%   Expected values:
F = 7.040%   #10—58.018%
                  #50—83.327%
                  #200—91.644%

Probability of satisfaction:

lower #10—.800
lower #50—.9999
upper #50—.833
lower #200—.9999
upper #200—.800

Joint probability—.533

Priority	
1	satisfied (all expected values within specifications)
2	satisfied (cost below $4.00/cubic yard)
3	satisfied (probability of satisfaction all greater than .7)
4	satisfied (fines below 8%)
5*	satisfied (cost of satisfaction all .8 or more)
6*	unsatisfied (cost greater than $3.85/cubic yard)
7	unsatisfied (fines greater than 7%)
8	unsatisfied (three limits less than .9 probability)
9	unsatisfied (cost greater than 0)
10	unsatisfied (three limits below .99 probability)

*priority levels switched from Goal Programming Solution 1.

cost and probabilities specified can be higher than models with higher probability levels. However, test sampling is normally done in batches. The use of individual constraints makes each limit independently safer. Average groups of samples will have much higher probability of passing required limits than specific observations. This probability should be greater if all probabilistic constraints are considered with higher lower limits. Therefore, there is rational reason to consider each limit independently, as is modeled.

Thus far, we have considered the aggregate blending problem involving the need to mix materials to meet specifications. Without consideration of material variance, this problem could be formulated as a linear programming model with the objective of minimizing cost. However, the presence of variance in material size leads to consideration of chance constraints, and the balance of risk with cost as well as any other decision objectives that might exist. The problem of balancing risk and cost requires judgement on the part of the designer. One way to support this judgement is through goal programming.

We have discussed models of the problem thus far, but have not discussed solution technology. Linear programming codes are widely available, both for main frame and microcomputer systems. Linear goal programming codes are not as commercially developed, but are available. Both authors have codes for both main frame and micro systems. Chance constrained codes are more theoretically appropriate, but very few codes exist. The authors have a code suitable for main frame computers. To date, this code is not suitable for microcomputer use.

It would, therefore, be useful to obtain a linear programming model that penalized gradation constraints for material variance. Such a linear approximation could be used in both linear programming as well as goal programming models.

LINEAR APPROXIMATION

The source of nonlinearity in the chance constrained models is the compensating variance of the independent materials. There is greater penalty involved with mixtures including high concentrations of a particular material than

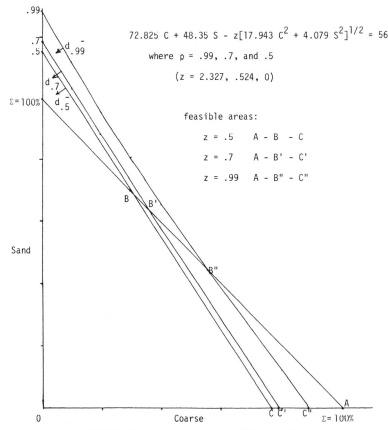

FIGURE 2. Chance constraint—#10 lower limit.

when balanced mixtures are used. Because only two variables contribute to the #10 sieve (fines are expected to completely pass through), this chance constraint can be graphed. Figure 2 presents the proportion of C (coarse material) and S (sand) that would be required to satisfy the required limit at .50 probability (a linear constraint), .70 probability, and .99 probability. Because the variable S averages 48.35% retention on the #10 sieve, a blend consisting only of S would not satisfy the limit of 56%. That would mean the required percentage of sand alone would be 115.8%. That, of course, would fail the required constraint that the sum of the variables be 100%. In this case, because fines are not considered for the #10 limit, we can graph all feasible combinations as that area where $S + C \leq 100$. Using expected values only, it would be feasible to select a mixture anywhere in the triangle defined by points A, B, and C.

The impact of chance constraints is apparent from Figure 2. As the specified probability is increased, the chance constraint curves away from the linear expected value (.50 probability). Combinations of S and C, however, have less penalty than the extremes (only S or only C), because if S happens to have an extreme occurrence, it may be compensated by an opposite occurrence for C and vice versa. Satisfaction at each of these three probability levels is indicated by $d^-_{.5}, d^-_{.7}, d^-_{.99}$. For satisfaction of a particular chance constraint, the deviational variable must be 0. The impact of increasing the level of probability is to shrink the feasible triangular area. It must be remembered that the other four limits must be satisfied as well. It can be seen why high levels of probability are infeasible.

Note that the extreme case is when only one of the variables is used. A linear approximation of chance constraints can therefore be developed, which will always be stricter than the real chance constraint. Due to the limited feasible area, further restricting chance constraints has the undesirable feature of increasing the likelihood of infeasibility. However, variance of material can be considered, while utilizing a more readily available linear programming code. Linear approximations to chance constraints can be obtained by taking the square root of each variable's variance, multiplying that value by z, and adjusting the linear coefficient for that variable.

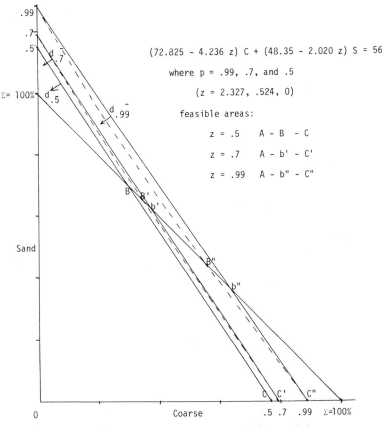

$(72.825 - 4.236 z) C + (48.35 - 2.020 z) S = 56$

where p = .99, .7, and .5

(z = 2.327, .524, 0)

feasible areas:

z = .5   A - B - C

z = .7   A - b' - C'

z = .99   A - b" - C"

FIGURE 3. Linear approximation—#10 lower limit.

TABLE 10. Linear Approximation Solutions.

Probability Specified	LP(50%)	60%	70%	80%	90%
Coarse	43.690%	47.358%	51.460%	57.732%	
Sand	50.016	46.028	41.586	34.326	infeasible
Fines	6.294	6.614	6.954	7.942	
Expected % Retained					
#10 sieve	56.000*	56.743	57.582	58.640	
#50 sieve	82.440	82.744	83.098	83.440	
#200 sieve	92.000*	91.833	91.661	91.001	
Probability of Satisfying Limit					
#10 lower limit	.500*	.631	.751	.850	
#50 lower limit	.998	.9998	.9999	.9999	
#50 upper limit	.968	.946	.894	.813	
#200 lower limit	.9999	.9999	.9999	.993	
#200 upper limit	.500*	.645	.784	.993	
Joint Probability	.242	.385	.526	.681	
Cost/cubic yard	$3.770	$3.811	$3.858	$3.924	

*at prescribed limit

$$72.825\ C + 48.35S - z(17.943C^2 + 4.079S^2)^{1/2} \leq$$
$$\leq [72.825 - z(4.236)]C + [48.35 - z(2.020)]S$$

if $z = 0$ (probability .5) $72.825\ C + 48.35\ S \geq 56$; $z = .524$ (probability .7) $70.605\ C + 47.292\ S \geq 56$; $z = 2.327$ (probability .99) $62.968\ C + 43.649\ S \geq 56$. Figure 3 presents the approximations for the last two constraints. It can be seen that the linear approximation is nothing more than connecting the extreme solutions.

In Figure 3, the feasible areas for the required constraint and the #10 limit is identical for the .5 level of confidence, because the linear approximation is identical to the actual constraint (z is 0). For probability levels greater than .5, however, the feasible area for the linear approximation is restricted. There is very slight modification at the .7 level ($A-B'-C'$ for the actual chance constraint, $A-b'-C'$ for the linear approximation). Note that the coarse intercepts are the same and this combination of constraints would include a feasible area at the coarse intercept. At the .99 level of confidence, there is a more distinct difference in the chance constraint and its approximation. The error of approximation is greater at larger confidence levels, and b'' is further from B'' than b' and B'.

Table 10 presents the solutions to the chance constrained models given in Table 6 when the linear approximations are used. In all cases, the penalty for variance in Table 10 is greater than the penalty in Table 6. In Table 6, the #10 lower limit and the #200 upper limit were always binding or exactly met. In the linear approximation, a buffer to these limits is observed. Overachieved limits are present at target probability levels greater than .5. For binding constraints, the direction of error introduced by the approximation is always on the safe side. This, of course, increases the likelihood of higher cost.

Linear approximations provide a means to solve the aggregate blending model with a linear code. The direction of error for linear approximations also provides a consistent error which may compensate for the random error due to rounding. Table 11 provides comparable solutions for rounded solutions which can be compared with those in Table 7. Of course, less precision in the solution is observed than would be obtained by using the actual chance constraints.

LINEAR GOAL PROGRAMMING

The approximations can be used for the goal programming model as well. This would result in modification of the chance constraints (priority levels three, six, eight, and ten). The z value would vary for each of these. The new constraints would be:

Third Priority

The z value would be .524, representing .70 confidence. $[0 - .524(0)]\ F + [48.35 - .524(4.079)^{1/2}]S + [72.825 - .524(17.943)^{1/2}]C + d_7^- - d_7^+ = 56$ would equal $0\ F + 47.292\ S + 70.605\ C + d_7^- - d_7^+ = 56$. In like manner,

$$.832\ F + 79.949\ S + 95.786\ C + d_8^- - d_8^+ = 80$$
$$1.168\ F + 81.651\ S + 96.314\ C + d_9^- - d_9^+ = 84$$

TABLE 11. Rounded Shear Linear Approximation Solutions.

Probability Specified	LP(50%)	60%	70%	80%	90%
Coarse	44%	47%	51%	58%	
Sand	50%	46%	42%	34%	infeasible
Fines	6%	7%	7%	8%	
Expected % Retained					
#10 sieve	56.218	56.469	57.448	58.678	
#50 sieve	82.722	82.382	82.992	83.261	
#200 sieve	92.259*	91.494	91.609	90.960	
Probability of Satisfying Limit					
#10 lower limit	.541	.585*	.734	.853	
#50 lower limit	.9994	.999	.9999	.9999	
#50 upper limit	.936	.981	.917	.881	
#200 lower limit	.9999	.9996	.9999	.991	
#200 upper limit	.292*	.869	.817	.995	
Joint Probability	.148	.498	.550	.741	
Cost/cubic yard	$3.776	$3.804	$3.852	$3.927	

*failed prescribed limit

TABLE 12. Approximation Goal Programming Solution 1.

C = 51.460%
S = 41.586%
F = 6.954%

Cost = $3.86
Expected values:
  #10—57.582%
  #50—83.098%
  #200—91.661%

Probability of satisfaction:

  lower #10—.751
  lower #50—.9999
  upper #60—.894
  lower #200—.9999
  upper #200—.784

Joint probability—.526

Priority
1. satisfied (all expected values within specifications)
2. satisfied (cost below $4.00/cubic yard)
3. satisfied (probability of satisfaction all greater than .7)
4. satisfied (fines below 8%)
5. unsatisfied (cost over $3.85/cubic yard)
6. unsatisfied (lower #10 probabilities below .8)
7. satisfied (fines below 7%)
8. unsatisfied (three limits below .9 probability)
9. unsatisfied (cost greater than 0)
10. unsatisfied (three limits below .99 probability)

TABLE 13. Approximation Goal Programming Solution 2.

C = 57.732%
S = 34.326%
F = 7.942%

Cost = $3.92
Expected values:
  #10—58.640%
  #50—83.266%
  #200—91.001%

Probability of satisfaction:

  lower #10—.850
  lower #50—.9999
  upper #60—.878
  lower #200—.993
  upper #200—.993

Joint probability—.841

Priority
1. satisfied (all expected values within specifications)
2. satisfied (cost below $4.00/cubic yard)
3. satisfied (probability of satisfaction all greater than .7)
4. satisfied (fines below 8%)
5.* satisfied (probability of satisfaction all greater than .8)
6.* unsatisfied (cost greater than $3.85/cubic yard)
7. unsatisfied (fines greater than 7%)
8. unsatisfied (two limits below .9 probability)
9. unsatisfied (cost greater than 0)
10. unsatisfied (two limits below .99 probability)

*priority levels switched from Goal Programming Solution 1.

$$10.625\ F + 95.659\ S + 98.724\ C + d_{10}^- - d_{10}^+ = 90$$
$$11.375\ F + 96.541\ S + 99.226\ C + d_{11}^- - d_{11}^+ = 92$$

Sixth Priority

The z value would be .841, representing .80 confidence.

$$0\ F + 46.652\ S + 69.263\ C + d_{14}^- - d_{14}^+ = 56$$
$$.730\ F + 79.435\ S + 95.626\ C + d_{15}^- - d_{15}^+ = 80$$
$$1.270\ F + 82.165\ S + 96.474\ C + d_{16}^- - d_{16}^+ = 84$$
$$10.399\ F + 95.392\ S + 98.572\ C + d_{17}^- - d_{17}^+ = 90$$
$$11.601\ F + 96.808\ S + 99.378\ C + d_{18}^- - d_{18}^+ = 92$$

Eighth Priority

With a z value of 1.282 (.90 probability), the constraint set would be:

$$0\ F + 45.761\ S + 67.395\ C + d_{20}^- - d_{20}^+ = 56$$
$$.589\ F + 78.719\ S + 95.404\ C + d_{21}^- - d_{21}^+ = 80$$
$$1.411\ F + 82.881\ S + 96.696\ C + d_{22}^- - d_{22}^+ = 84$$
$$10.082\ F + 95.182\ S + 98.360\ C + d_{23}^- - d_{23}^+ = 90$$
$$11.918\ F + 97.018\ S + 99.590\ C + d_{24}^- - d_{24}^+ = 92$$

Tenth Priority

$$0\ F + 43.650\ S + 62.968\ C + d_{26}^- - d_{26}^+ = 56$$
$$.253\ F + 77.022\ S + 94.877\ C + d_{27}^- - d_{27}^+ = 80$$
$$1.747\ F + 84.578\ S + 97.223\ C + d_{28}^- - d_{28}^+ = 84$$
$$9.333\ F + 94.433\ S + 97.859\ C + d_{29}^- - d_{29}^+ = 90$$
$$12.667\ F + 97.767\ S + 100.091\ C + d_{30}^- - d_{30}^+ = 92$$

Using exactly the same priority structure as in goal programming solution 1 (Table 8), the linear approximations to the chance constraints yield the solution given in Table 12. Note that it was no longer possible to obtain a cost at the $3.85 specified in priority 5 while satisfying the linear approximations to the chance constraints. Therefore, the goal programming algorithm effect minimized cost subject to a probability of satisfaction (as linearly approximated) of .70. The actual probability of expected satisfaction is overachieved in reality, although the linear approximations for the lower #10 limit and the upper #200 limit are binding.

Reversing priorities 5 and 6 as in goal programming solution 2 results in the solution given in Table 13. This places more importance upon a probability of attainment of .80 (as linearly approximated) than upon the cost target of $3.85 per cubic yard. The linear approximations are all satisfied, lower #10 and upper #50 limits binding. This results in a cost of $3.92, as opposed to $3.88 in the chance constrained goal programming model, while the lowest level of expected satisfaction is .850 for the lower #10 limit. This added cost is the amount of inaccuracy incurred by the linear approximation.

CONCLUSION

The aggregate blending problem requires a decision which can be supported by optimization procedures, yielding solutions that are cost efficient. However, linear programming solutions are distorted due to the need to assume linearity in gradation constraints. Linear programming models do provide the flexibility of further constraining the model to include a variety of constraints. A simple aggregate blending model was presented to convey the principle concepts. Much more involved models can be built and rapidly solved by existing computer technology. All methods presented in this chapter could be further constrained to include limits on proportions of material with specified structural or chemical characteristics, in addition to size constraints.

Chance constrained programming provides a means to more appropriately model the aggregate blending decision. This in turn leads to the need to reconcile cost minimization with risk minimization. Chance constrained goal programming provides a technique capable of obtaining solutions that match specified decision maker objectives. Use of linear approximations sacrifice some cost efficiency, but allow use of more readily available computer codes.

Two additional problems were addressed. Use of statistical sampling information and chance constrained programming will result in solutions with efficient cost given increased probability of satisfying gradation constraints, or other constraints representing product quality. However, in practice, solutions must be rounded to some degree. The more precise the solution, the less impact rounding will cause. Secondly, while concepts of probability were used, the actual probabilities of satisfying gradation constraints is affected by sampling error and rounding. Relative probabilities for solutions should be correct. Actual probabilities could be verified by observation of actual operations. The primary benefit of considering material variance is that product quality should improve.

Computer codes would be required for solution of the optimization models given in the chapter. Linear programming codes are widely available. Linear goal programming codes, both for main frame and microcomputer systems, are available from either author. Chance constrained goal programming code is available as well, although microcomputer code has not been developed to date. Solution times for aggregate blending problems are not limiting in any of the cases.

Because of the restricted availability of chance constrained goal programming codes, a linear approximation of chance constrained gradation limits was presented.

The chapter presented the concepts necessary for application of optimization to the aggregate blending problem. The following reference section provides sources of information on these concepts.

REFERENCE INFORMATION

Statistical sampling theory is well developed in many disciplines. The theory that applies to other fields applies to gradation sampling as well.

Linear programming is also highly developed. Any number of texts, in engineering or business, should be readily available.

Chance constrained programming was originally developed by Charnes and Cooper [1,2,3]. Most references are research articles, applied to business. However, Kelle [4] has provided a more recent application to asphalt mixture.

Goal programming was also originally developed by Charnes and Cooper, along with Ferguson [5]. Good reference books include Lee [6] and Ignizio [7,8].

Chance constrained goal programming references are more limited. Lee and Olson [9] presented the concept using a more comprehensive mixture example. Theoretical aspects appear in another Lee and Olson article [10]. The concept of linear approximation has been submitted for publication. The impact of joint chance constraints was addressed by Miller and Wagner [11] and Jagannathan and Rao [12].

Lewis [13] has presented a recent means of formulating aggregate blending problems using linear programming. Torrent, Alvaredo, and Poyard [14] have presented an alternative method to obtain a combined gradation matching desired and actual gradation as closely as possible without cost consideration.

An evaluation of main frame goal programming codes is given in Olson [15]. A number of presentations for microcomputer goal programming codes have been given by a variety of authors, none published to date.

REFERENCES

1. Charnes, A. and W. W. Cooper, "Chance Constrained Programming," *Management Science, 6* (1), 73–79 (1959).
2. Charnes, A. and W. W. Cooper, "Chance Constraints and Normal Deviates," *Journal of the American Statistical Association, 57*, 134–148 (1962).
3. Charnes, A. and W. W. Cooper, "Deterministic Equivalents for Optimizing and Satisficing under Chance-Constraints," *Operations Research, 11*, 18–39 (1963).
4. Kelle, P., "Chance Constrained Inventory Model for an Asphalt Mixing Problem," in *Recent Results in Stochastic Programming*, P. Kall and A. Prekopa, eds., Springer-Verlag, NY, 179–189 (1979).
5. Charnes, A., W. W. Cooper, and R. O. Ferguson, "Optimal Estimation of Executive Compensation by Linear Programming," *Management Science, 1* (2), 138–151 (1955).
6. Lee, S. M., *Goal Programming for Decision Analysis*, Auerbach Publishers, Philadelphia (1972).
7. Ignizio, J. P., *Goal Programming and Extensions*, D.C. Heath, Lexington, MA (1976).
8. Ignizio, J. P., *Linear Programming in Single and Multiple Objective Systems*, Prentice Hall, Englewood Cliffs, NJ (1982).
9. Lee, S. M. and D. L. Olson, "Chance Constrained Aggregate Blending," *Journal of Construction Engineering and Management (ASCE), 109* (1), 39–47 (March 1983).
10. Lee, S. M. and D. L. Olson, "A Gradient Algorithm for Chance Constrained Nonlinear Goal Programming," *European Journal of Operational Research*, to appear in 1985.
11. Miller, B. L. and H. M. Wagner, "Chance Constrained Programming with Joint Constraints," *Operations Research, 13*, 930–945 (1965).
12. Jagannathan, R. and M. R. Rao, "A Class of Nonlinear Chance-Constrained Programming Models with Joint Constraints," *Operations Research, 21*, 360–364 (1973).
13. Lewis, P. A., "Estimating Aggregate Proportions by a Single Computer Method," *Proceedings, Institute of Civil Engineers, Design & Construction: Part 1, 74*, 515–518 (August 1983).
14. Torrent, R. J., A. Alvaredo, and E. Royard, "Combined Aggregates: A Computer Based Method to Fit a Desired Grading," *Materials & Structures*, Rilem No. 98, 139–144 (March-April 1984).
15. Olson, D. L., "Comparison of Four Goal Programming Algorithms," *Journal of the Operational Research Society, 35* (4), 347–354 (April 1984).

CHAPTER 13

Constitutive Modelling of Ballast

R. JANARDHANAM*

INTRODUCTION

Ballast is a common and important material used in the support of railroad track beds. Ballast also provides a free draining medium to railroads. Under vehicle loading, the materials in the track bed are subjected to general three dimensional states of stress. The loads are repetitive arising from the wheel loads during travel of a single train. Thus, the track bed is subjected to a series of loading, unloading and reloading cycles. Two of the problems related to the performance of ballast materials are the excessive elastic deformations caused by the rapid application and removal of wheel loads and the accumulation of large plastic deformations resulting from many repetitions of individual wheel loads.

Excessive elastic deformations in the ballast can cause shortening of the rail-tie life because of fatigue resulting from increased bending stresses. In addition, the ride quality is reduced if the elastic deformations are excessive. The plastic deformations necessitate continual realignment of the rail-tie system by addition of ballast. Continued maintenance is not a satisfactory solution. Modern analytical methods can be used to improve the present experience oriented design of rail-tie support systems. Numerical methods need adequate input in the form of material response parameters for reliable results. To accurately predict the response of material, the test method used to evaluate the properties of ballast should simulate the in service stress conditions.

Most laboratory tests in the past for ballasts have used cylindrical triaxial test devices employing both static and cyclic loading. The results from these tests have proved to be useful in defining the behavior of ballast. Ties are generally 8 inches wide and spaced at every 20 inches. Thus, compacted ballast forms rectangular columns under the ties surrounded by relatively less compact ballast. Under such conditions, among the available testing devices in the past, conventional triaxial testing is believed to be more representative of field conditions than other devices such as plane strain testing. Of course, three dimensional stress state would be more representative and desirable. The ballast can be tested in three dimensional state of stress using a truly triaxial test device. Then the stress-strain response can be modelled from a series of loading-unloading-reloading cycles under slow repeatedly applied static loads.

REVIEW

The behavior of ballast under repetitive load has been studied by many investigators. Here, the conventional (cylindrical) triaxial equipment has been used to study graded aggregates. Repeated deviator stress, and either constant or pulsed confining pressures have been used. However, most of the research has been directed toward studies of nonlinear elastic properties of the materials and relevant equations have been developed. Two of the more widely used equations follows:

$$E_r = KJ_1^n \qquad (1)$$

and

$$E_r = K^1 \bar{\sigma}_3^m \qquad (2)$$

where E_r = resilient modulus, K, K^1, n, and m = material constants determined from the laboratory data, J_1 = the first stress invariant and $\bar{\sigma}_3$ = effective confining pressure.

A number of factors such as stress history, frequency and duration of load application, geometric characteristics of aggregates, gradation, degree of compaction, degree of saturation and stress levels can influence the magnitude of this

---

*Department of Civil Engineering, University of North Carolina at Charlotte, Charlotte, NC

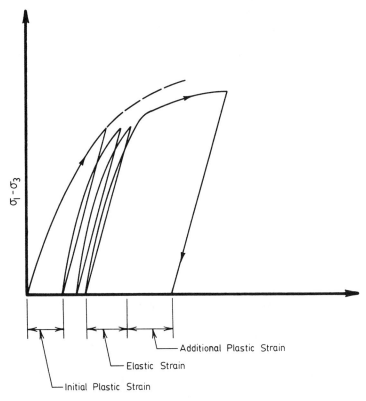

FIGURE 1. Schematic of repeated loading response for triaxial compression.

resilient modulus and, in general, the constitutive behavior of ballast. Resilient modulus increases with the number of load applications, but reaches a state thereafter the sequence of stress applications has no significant effect on the results. Frequency and duration of load applications have virtually insignificant influence on E_r. Geometric characteristics, namely the shape, angularity and surface texture of ballast differ according to the material type (limestone, basalt, granite, slag). Two of the methods for measuring shape, angularity, and surface texture are the particle index test and the pouring test. As particle index increases, both maximum stress difference and angle of sharing resistance increases.

The effect of degree of compaction on resilient modulus is not well understood. For dense graded granular materials, increased levels of saturation generally resulted in decreased modulus values. Significant factor affecting the resilient modulus is observed to be the stress state. To obtain good laboratory test results, the specimen size (diameter) should be 4 to 20 times the maximum particle size. High proportions of the large size particles require the ratio to be closer to 20.

Most of the factors affecting resilient behavior discussed probably influence the repeated load plastic strain response. However, in the past, the plastic behavior of the ballast has not received appropriate attention. Accumulated plastic strains are important because they can be principally the cause for differential settlements. The two equations for predicting plastic strain are given below as Equations (3) and (4).

$$\epsilon^p = (\sigma_1 - \sigma_3)/E_i \left[1 - \frac{(\sigma_1 - \sigma_3)(1 - \sin\bar{\phi}) R_f}{2(\bar{c} \cos \bar{\phi} + \bar{\sigma}_3 \sin\bar{\phi})} \right] \quad (3)$$

Where ϵ^p = permanent strain, E_i = initial tangent modulus, $\bar{c}$ = effective cohesion, $\bar{\phi}$ = effective angle of internal friction, R_F = the ratio of the stress difference, $(\sigma_1 - \sigma_3)$, at failure to the stress difference at infinite strain.

Office of Research and Experiments of the International Union of Railway has given an alternative equation for defining the plastic strain:

$$\epsilon^p = 0.082(100n - 38.2)(\sigma_1 - \sigma_3)^2(1 + 0.2 \log N) \quad (4)$$

where N = number of repeated loading cycles, n = initial porosity.

Results from cyclic cylindrical triaxial testing are used in the foregoing predictive equations. It has been observed that a substantial portion of permanent deformation occurs in the first few cycles and the deformation results at a decreasing rate for subsequent cycles. In Equation (4), the first 100 cycles of loading are extremely important. However, in the actual case some additional plastic strain accumulates with each loading cycle. If after several repeated loading cycles the specimen is subjected to a deviator stress greater than that previously experienced, the stress strain curve will continue in the direction of the original curve as shown in Figure 1. To replicate the true in-service conditions and to test the ballast in all possible stress paths, a comprehensive series of tests have been conducted on scaled down ballast in a truly triaxial test device. Main attention here is given to the factors such as state of stress, stress path, plastic strains and particle size. The tests are performed on dry samples, hence, the stresses are considered to be effective stresses. Details of truly triaxial device, laboratory testing program, analysis of test results and constitutive models developed are given in the following sections.

DESCRIPTION OF MATERIALS

Several types (limestone, granite, slag, etc.) and gradations of materials are used for ballast. The standard gradations recommended by the American Railway Engineering Association (AREA) are shown in Table 1; the No. 4 and No. 5 gradations are used more frequently than are the others. Limestone and granite are used more often than others for ballast. The ballast tested in the truly triaxial device is from the Urban Mass Transportation Administration (UMTA) Test Section at the Transportation Test Center (TTC), Pueblo, Colorado. The grain size distribution of the materials labelled as Ballast I, which is granite gneiss, obtained from the field site, (UMTA) is shown in Figure 2. It is a uniformly graded material with an uniformity coefficient C_u of about 1.20. The results of the characterization tests are: Specific Gravity = 2.68, Los Angeles Abrasion Loss % = 32.4, Particle Index = 13.48, Gradation Parameter = 1.84, Soundness Loss % = 0.25, and Crushing Value = 25.6.

TRULY TRIAXIAL DEVICE

The truly triaxial device used herein is capable of applying homogeneous and independently controlled three-dimensional stress states on a 4 × 4 × 4 inch cubical specimen through flexible membranes. The triaxial cell and an exploded view with details of one cell wall are shown in Figure 3. It is a stress controlled test device. A cubical specimen made out of the ballast of desired density is inserted in the cubical cavity of the frame. It is loaded on each face through an assembly of flexible pads and a hydraulically pressurized membrane. The pressure is contained by walls which also serve as basis for deformation detecting proximity transducer probes. The load and deformation readings are recorded electronically and are processed by a data acquisition system to yield stresses and strains.

DETAILS OF BALLAST TESTED

With the 4 × 4 × 4 inch size specimen permitted in the device, it is difficult to test the field ballast with the mean particle size of 30 mm. Hence, the size of the field material is scaled down to two smaller sizes, called Ballast II and Ballast III. The technique followed is that proposed by Lowe which involves development of the grain size distribution of the test material parallel to that of the field material. The grain size distribution curves of the two scaled down sizes are shown in Figure 2. The mean grain size of the parent material, Ballast I, is 30.00 mm and that of the scaled down

TABLE 1. Area Recommended Ballast Gradations (Reference [33]).

Size No.	Nominal Size Square Opening	Amounts Finer Than Each Sieve (Square Opening) Percent by Weight									
		3"	2-1/2"	2"	1-1/2"	1"	3/4"	1/2"	3/8"	No. 4	No. 8
24	2-1/2"–3/4"	100	90–100		25–60		0–10	0–5			
3	2"–1"		100	95–100	35–70	0–15		0–5			
4	1-1/2"–3/4"			100	90–100	20–55	0–15		0–5		
5	1"–3/8"				100	90–100	40–75	15–35	0–15	0–5	
57	1"–No. 4				100	95–100		25–60		0–10	0–5

(1 inch = 2.54 cm)

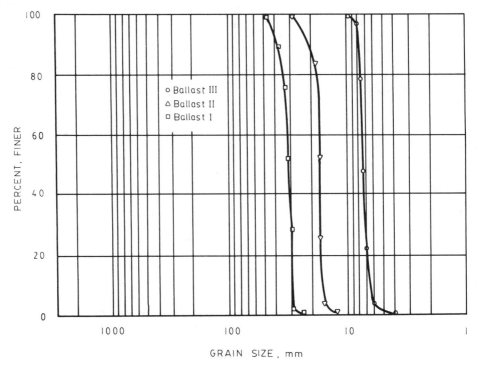

FIGURE 2. Grain size distribution curves for ballast.

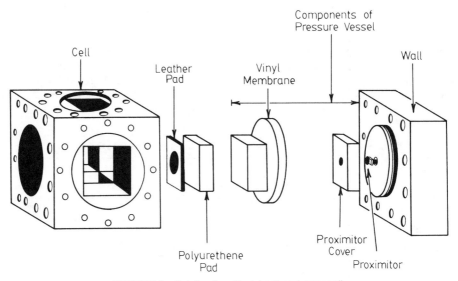

FIGURE 3. Details of multiaxial cell and one wall.

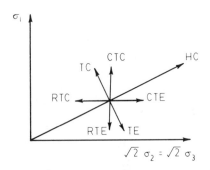 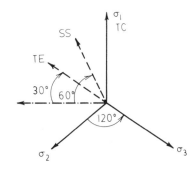

FIGURE 4. Stress paths used in truly triaxial tests. (a) Stress paths on triaxial plane; (b) Stress paths on octahedral plane.

ballasts, Ballast II and Ballast III, is 16.00 mm and 7.00 mm respectively.

SAMPLE PREPARATION

A special sample preparation mold made of four thin aluminum plates 4 × 0.25 inches (10 × 0.64 cm) is used for this purpose. Pure latex dental dam hoop is used as the membrane. The mold serves to contain the sample compacted to the field density. Because of the open graded nature of ballast, vibratory compaction is selected. The material is spooned into the mold which sits on a base plate, and then compacted by vibration. The shape and size of the specimen are maintained by applying vacuum through a thin teflon tube, one end of it being buried in the test ballast sample. The mold is carefully removed and the sample is inserted in the apparatus through the side of cell frame by using a thin lubricated plexiglass sheet as a ramp. After fixing all the walls to the frame, the vacuum pressure is disconnected.

TESTING PROGRAM

Unlike conventional triaxial device, where only biaxial testing can be done on sample, here, in truly triaxial device, the sample can be loaded in any desired stress path. Tests are conducted by following number of significant stress paths. In addition to investigating the influence of gradation on strength and deformation properties, stress path dependency is also studied. As the loads (stresses) on three mutually perpendicular directions can be controlled independently in the device, any desired stress path can be followed as shown in Figure 4. Loads (stresses) are applied on the six faces of the sample; the stresses on the opposite faces being the same. The resulting deformations are recorded after each increment of load is applied as the particular stress path is followed. Details of various stress paths are given below, together with test results for some typical stress paths for Ballast II. Table 2 shows the tests performed on Ballast II with details of the initial density of samples, confining pressure applied and stress paths followed. Similar tests were performed for Ballast III; however, the typical test results only for Ballast II are included herein. Typical results for Ballast III are used for comparison wherever necessary. The various stress paths followed are Hydrostatic Compression (HC), Conventional Triaxial Compression (CTC), Conventional Triaxial Extension (CTE), Reduced Triaxial Compression (RTC), Reduced Triaxial Extension (RTE), Simple Shear (SS), Triaxial Compression (TC), and Triaxial Extension (TE). Brief descriptions of some stress paths are given below.

TABLE 2. Details of Various Tests Run on Ballast II.

Types of Stress Path Followed	Confining Pressure psi	Initial Density of Sample pcf
Hydrostatic Compression (HC)	—	93.00
Conventional Triaxial Compression (CTC)	5.00 10.00 15.00 20.00	
Reduced Triaxial Compression (RTC)	30.00 40.00 50.00	as above
Triaxial Compression (TC)	15.00 25.00 35.00	
Simple Shear (SS)	15.00 25.00 35.00	

1 psi = 6.89 kPa
1 pcf = 15.97 kg/m³

Hydrostatic Compression (HC) Test

Stresses of the same magnitude are applied on all sized faces of the specimen. From the initial hydrostatic stress, loading is continued with the stress increments of same magnitude applied on all six faces of the specimen, up to a desired stress level, then the sample is unloaded to the initial hydrostatic stress state. The sample is then reloaded incrementally to a higher stress level and then the load is reversed. The sample is subjected to a number of such loading cycles. Typical results in terms of axial strain and normal stress are shown in Figure 5. These and subsequent results exhibit the influence of anisotropy in the specimens.

Conventional Triaxial Compression (CTC) Test

The sample is loaded hydrostatically up to the predetermined confining pressure and thereafter deviator stress is applied in one direction. The confining pressure is kept constant and deviator stress is increased to a predetermined level and reversed. The sample is subjected to a number of load-unload-reload cycles. The strains are measured in all the three directions. Number of CTC tests are performed on samples with different confining pressures. Typical stress-strain response in terms of axial strain and lateral strains vs. the deviatoric stress is shown in Figure 6.

Reduced Triaxial Compression (RTC) Test

The Ballast specimen is loaded to a desired hydrostatic stress level by increasing the pressure in steps, each increment applied after the stabilization of load at the previous level. Strains are measured in all the three directions. Then σ_1 is held constant while σ_2 and σ_3 are reduced in stages with decrement $\Delta\sigma_2$ equal the decrement $\Delta\sigma_3$. Here σ_1, σ_2, and σ_3 are the three principal stresses acting on the specimen along the three principal directions. As σ_2 and σ_3 are decreased, the confining pressure drops and the specimen is sheared. The confining pressure is varied cyclically to have number of load.unload.reload cycles and typical test results are shown in Figure 7.

Simple Shear (SS) Test

The sample is subjected to a predetermined hydrostatic stress level and then deviated in simple shear stress path. Intermediate principal stress σ_2 is kept constant. σ_1 and σ_3 are varied in such a way that increment $\Delta\sigma_1$ is equal to decrement $\Delta\sigma_3$ so that σ_{oct} ($= 1/3(\sigma_1 + \sigma_2 + \sigma_3)$) remains constant. Strains are measured in all three directions. Typical results of simple shear test are shown in Figure 8.

Tests following other stress paths like triaxial compression, triaxial extension are also conducted and the tests

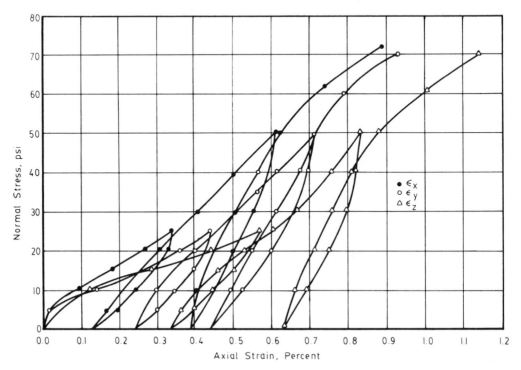

FIGURE 5. Stress-strain response curves for hydrostatic compression test (1 psi = 6.89 kPa).

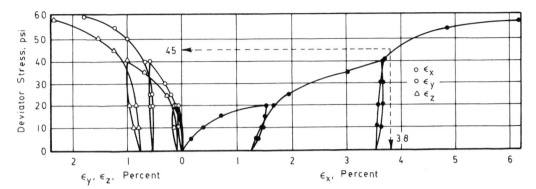

FIGURE 6. Stress-strain response curves for conventional triaxial compression test. $\sigma_2 = 15.00$ psi (104 kpa); (1 psi = 6.89 kPa).

results (not reported here) are used for the development of constitutive law(s) for the material.

ANALYSIS OF TEST RESULTS

The test results indicate that the ballast exhibits inelastic or plastic behavior involving both elastic and plastic deformations. It is also observed that it experiences plastic deformations from the very start of the load-deformation response. It can be concluded, based on test results, that the important factors influencing the repeated load plastic strain behavior of ballast are the degree of compaction and the stress level. Also, unlike resilient response, permanent deformation behavior of ballast is found to depend on loading history. The increase in plastic strain in general is seen to be inversely proportional to the number of loading cycles, for a constant stress ratio equal to repeated deviator stress divided by confining pressure.

Under heavily loaded trains, considerable ballast resistance will be caused by high frictional resistance. Proper compaction of ballast bed during laying and maintenance will avoid the possibility of deformations occurring in the ballast due to a large number of repetitive loads. Most of the tests have been performed at low pressures only, since the height of ballast in track is small and the confining pressures are small. At low cell pressures, the density of placement has an immense effect on the ballast failure resistance.

It is possible to use the tests results for developing plasticity models based on the concepts of yield and flow rules and hardening behavior. However, from a practical viewpoint, simplified models based on the resilient modulus

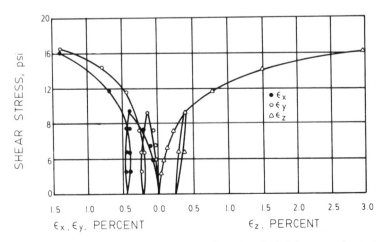

FIGURE 7. Stress-strain response curves for reduced triaxial compression test. $\sigma_2 = 40.00$ psi (275.60 kpa); (1 psi = 6.89 kPa).

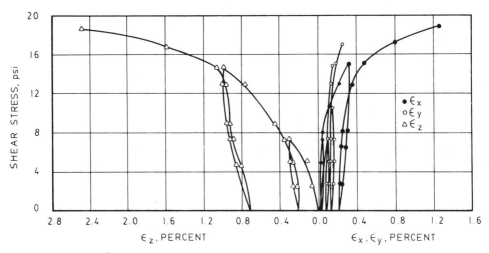

FIGURE 8. Stress-strain response curves for simple shear test. $\sigma_{oct} = 35.00$ psi (241 kpa); (1 psi = 6.89 kPa).

concept and variable moduli idea are developed. These models can provide simple and useful representation of the behavior of ballast.

RESILIENT MODULUS

The resilient modulus E_r is defined as the repeated deviator stress divided by the recoverable axial strain. The resilient modulus for a granular material is not constant but varies with the state of stress. Analysis of the E_r values at various stress levels is used to show the stress dependent behavior. In this study, the unloading modulus in the CTC tests can be regarded as the resilient modulus, E_r. It is computed as an average of the values of E_r for various unloading-reloading loops. The resilient moduli for Ballast II and Ballast III as determined from various test results with different confining pressures are plotted in Figure 9. From this figure, E_r for Ballast II is found to be

$$E_r = 1300 \, \sigma_3^{0.87} \quad (5)$$

Similarly the resilient modulus for Ballast III is found to be

$$E_r = 2600 \, \sigma_3^{0.27} \quad (6)$$

From the available data obtained by Knutson for a ballast composed of granite gneiss following American Railway Engineering Association (AREA) Grade 4 material, tested in a conventional cylindrical triaxial device, an expression for the resilient modulus is obtained as

$$E_r = 4800 \, \sigma_3^{0.6} \quad (7)$$

Effect of Particle Size

The variations of the resilient modulus with the mean grain size of the material for different confining pressures computed from Equations (4) to (6) are shown in Figure 10. The resilient modulus increases with the mean (effective) grain size. The resilient modulus appears to have almost a linear relationship with the mean grain size at low confining pressures. This trend is similar to that obtained from field measurement for ballast at the Facility for Accelerated Service Testing (FAST) Section, TTC, Pueblo, Colorado, by Bosserman in which the vertical track modulus was found to increase with mean particle diameter.

Table 3 shows comparison of strains at different stress levels that Ballast II and III have experienced under hydrostatic compression. The particle size does not appear to influence significantly the strains at various stress levels. However, as the pressure increases, the strains for Ballast III becomes slightly greater than those for Ballast II.

Table 4 shows values of strains at typical levels of deviator stress for three different confining pressures for CTC tests. It can be seen that in general, the smaller sized ballast experiences greater strains with the deviator stress for a given confining pressure. From the foregoing results, it is seen that (1) the resilient modulus is dependent on particle size, (2) the volumetric behavior is not affected significantly by particle size and (3) the shear behavior is dependent on particle size.

ACCUMULATED STRAIN

The results under slow cyclic test obtained herein are compared with those from fast cyclic tests with cylindrical

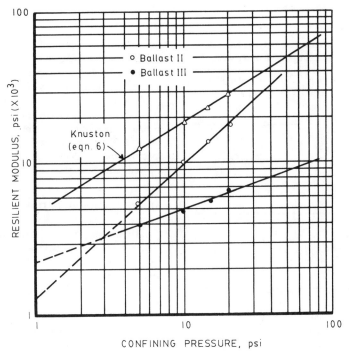

FIGURE 9. Variation of resilient modulus with confining pressure (1 psi = 6.89 kPa).

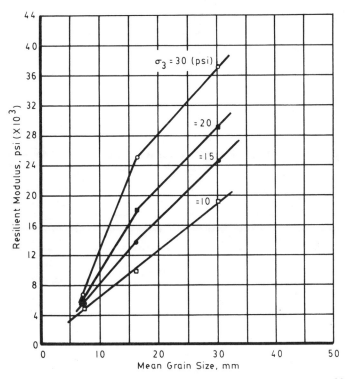

FIGURE 10. Variation of resilient modulus with grain size (1 psi = 6.89 kPa).

TABLE 3. Comparison of Strains at Different Stress Levels for Hydrostatic Compression Test.

Mean Pressure (psi)	Ballast II Mean Strain, percent	Ballast III Mean Strain, percent
15.0	0.26	0.19
25.0	0.44	0.41
35.0	0.56	0.56
45.0	0.67	0.70
55.0	0.75	0.82
65.0	0.88	0.95

(1 psi = 6.89 kpa)
Mean pressure = $(\overset{*}{\sigma}_1 + \overset{*}{\sigma}_2 + \overset{*}{\sigma}_3)/3$
Mean strain = $(\epsilon_1 + \epsilon_2 + \epsilon_3)/3$

triaxial test reported by Knutson. Because the grain sizes used are different in the two investigations, the comparisons are essentially presented as a qualitative evaluation. In both cases, however, the samples were composed of granite gneiss with initial compacted density of about 93.0 pcf (1485 kg/m³).

Figure 11 shows the axial strains with number of cycles of load applications in the fast cyclic tests for amplitude of deviator stress equal to 45.0 psi (310.50 kPa) and confining pressure equal to 15.0 psi (103.50 kPa) with ballast similar to Ballast I. The maximum strain occurs after about 5000 cycles and is equal to 3.4 percent. The axial strain for the deviator stress equal to 45.0 psi in Figure 5 is about 3.8 percent for Ballast II. In another test with confining pressure equal to 5.0 psi (34.45 kPa) and deviator stress amplitude of 20.0 psi (127.8 kPa) the maximum strain with fast cyclic loading was observed to be 0.20 percent. The corresponding strain from the slow cyclic tests in this investigation was found to be 0.24 percent.

The foregoing results pertain to only two tests for the material with one density. In general, behavior of ballast would be dependent on the rate of loading. It is appropriate to perform (rapid) cyclic tests by varying factors such as deviator stress, confining stress, (initial) density and particle size. These results are presented simply to indicate a specific aspect of the behavior of the ballast in question with a given field density.

ADVANCED CONSTITUTIVE MODELS

Early attempts to mathematically model the behavior of ballast under both static and/or dynamic loadings were based on the assumption that the ballast could be approximated as a linear elastic material. Of course such a model can be of extremely limited validity. They were subsequently replaced by simple elasto-plastic models of von-Mises or Drucker-Prager type, in which a yield function was used to describe the material failure under specific combinations of the shear stresses and pressure. Recently, more advanced elasto-plastic models have been used to model the action of the ballast under a wide variation of applied pressure. In these models the yield condition depended upon the pressure in a general way. Different pressure-volume relations were used for initial loading and for subsequent unloading and reloading.

Although these models reproduce actual ballast behavior quite adequately in both static and cyclic uniaxial strain tests, they do not do so in triaxial compression tests. But this mathematical material model, namely, variable moduli model in which the basic constitutive law is an isotropic relation between the increments of stress and strain repro-

TABLE 4. Comparison of Strains at Different Levels for Conventional Triaxial Compression Test.

Confining Pressure, psi	Deviator Stress, psi	Ballast II Axial Strain, percent	Ballast III Axial Strain, percent
10.0	15.0	0.32	0.44
	25.0	0.92	1.52
	35.0	2.58	2.96
15.0	15.0	0.70	1.05
	25.0	1.93	2.25
	35.0	3.02	3.10
20.0	15.0	0.45	0.50
	25.0	0.90	1.05
	35.0	1.87	2.12

(1 psi = 6.89 kpa).

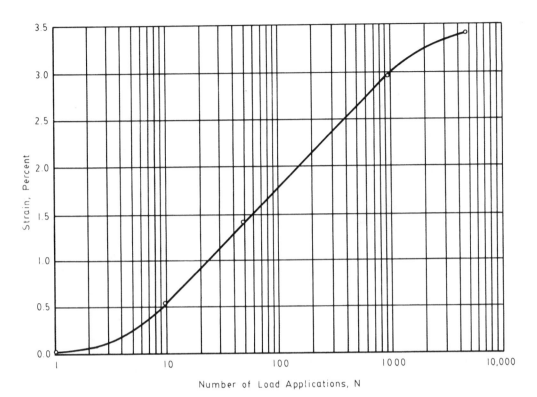

FIGURE 11. Ballast behavior for fast cyclic tests.

duces adequately, the behavior both in static and dynamic uniaxial and triaxial tests. The model has no explicit yield condition. The bulk and shear moduli, however, are functions of the stress and/or strain invariants.

It is possible, from the test data, to develop advanced plasticity models. However, in view of the past experience and the test results herein, it is felt that a variable moduli model based on two parameters bulk modulus K, and shear modulus G, may be appropriate.

Details of Variable Moduli Model

The incremental form of the stress-strain relation in variable moduli model can be expressed as

$$\Delta S_{ij} = 2G \, \Delta \epsilon_{ij} \qquad (8)$$

$$\Delta p = 3K \, \Delta \epsilon_{ij} \qquad (9)$$

where ΔS_{ij} and $\Delta \epsilon_{ij}$ are the deviatoric stress and strain increments respectively, and Δp and $\Delta \epsilon_{ij}$ are the mean stress and strain increments. In writing Equations (8) and (9) an implicit assumption is made that the material is isotropic, because the separation of the constitutive relation into deviatoric and volumetric parts precludes any coupling between them as is observed in granular media.

There have been three major models in vogue. They are:

1. Constant Poisson's Ratio Model,
2. Variable Moduli Model based on the Invariants of the strain tensor,
3. Combined stress-strain variable moduli model.

Constant Poisson Ratio Model

The ratio of the bulk modulus to shear modulus is assumed to be constant. The two moduli model can be function of mean pressure or volumetric strain or both. The model can be expressed as

$$\frac{K}{G} = \frac{2(1 + \nu)}{3(1 - 2\nu)} = \text{constant} \qquad (10)$$

This model has been found to be satisfactory only for the uniaxial states of strain and it gives contradicting predictions in the triaxial conditions.

Variable Moduli Model With Invariants of Strain Tensor

The bulk modulus K and the shear modulus G are assumed to be functions of the first invariant of the strain tensor, I_1, and the second invariant of the deviatoric strain tensor, I_{2D} and can be expressed as

$$K = K_0 + K_1 I_1 + K_2 I_1^2 \tag{11}$$

$$G = G_0 + G_1 I_{2D} + G_2 I_1 \tag{12}$$

K_0 and G_0 are initial bulk and shear moduli respectively. K_1, K_2, G_1, and G_2 are material parameters.

This model fails to take into account unloading part. Another limitation is that, in many existing finite element procedures, stresses are stored rather than strains and hence the implementation of this model in an existing program can require additional difficulties.

Combined Stress-Strain Variable Moduli Model

Both the shear modulus and the bulk modulus are assumed to depend upon the stress and strain invariants. Different functions G and K apply in initial loading and subsequent unloading and reloading. For initial loading, the bulk modulus K which is expressed as a function of the mean strain and a shear modulus G which is a function of first two stress invariants or more specifically the pressure p and square root of the second invariant of the deviatoric stress tensor. They can be written as

$$K = K_0 + K_1 I_1 + K_2 I_1^2 \tag{13}$$

The bulk modulus is chosen to be a quadratic in I_1 where I_1 is the first invariant of strain tensor. The shear modulus, when expressed in terms of J_1 and J_{2D} where J_1 is the first invariant of stress tensor and J_{2D} is the second invariant of deviatoric stress tensor is

$$G = G_0 + \gamma_1 J_1 + \gamma_2 \sqrt{J_{2D}} \tag{14}$$

All these models are based on experimental curves. Equations (13) and (14) are the first terms in the series expansions of more general analytic functions K and G of the stress and strain invariants. The quantity $\sqrt{J_{2D}}$ is preferred rather than J_{2D} itself, since it is of the same order as J_1 and the components of the stress tensor. At zero stress and strain the bulk and shear moduli reduce respectively to K_0 and G_0, the "linear elastic" values, which are related in terms of the "elastic" Poisson's ratio ν_0.

$$\frac{K_0}{G_0} = \frac{2(1 + \nu_0)}{3(1 - 2\nu_0)} \tag{15}$$

with γ_1 positive and γ_2 negative, the material hardens in shear with increasing pressure and softens with increasing shear stress.

By specializing the above relation for triaxial state of stress as

$$\sqrt{J_{2D}} = \frac{1}{\sqrt{3}} (\sigma_1 - \sigma_3) \tag{16}$$

and

$$J_1 = p = \frac{\sigma_1 + 2\sigma_3}{3} \tag{17}$$

G can be expressed as

$$G = G_0 + \frac{\gamma_1}{3}(\sigma_1 + 2\sigma_3) + \frac{\gamma_2}{3}(\sigma_1 - \sigma_3) \tag{18}$$

So a necessary condition for G to decrease as σ_1 increases is that

$$\gamma_1 + 3\gamma_2 < 0 \tag{19}$$

In other words $\gamma_1 > 0$ and $\gamma_2 < 0$.

The five parameters to fully describe the model are

$$K_0, K_1, K_2, G_0, \gamma_1 \text{ and } \gamma_2 \tag{20}$$

If the stress quantities are non-dimensionalized, they can be written as

$$\frac{G_0}{K_0}, \frac{K_1}{K_0}, \gamma_1, \gamma_2, \frac{K_2}{K_0} \tag{21}$$

The ratio G_0/K_0 is positive and is related by the Equation (4.13). The higher order terms K_1/K_0 and K_2/K_0 may be positive and negative. The values are restricted by the condition that $K > 0$.

The material behavior under loading is quite different from unloading conditions. In case of uniaxial state of stress, loading and unloading can easily be seen as the state is only one-dimensional. Under three-dimensional state of stresses it is not easily seen. Different relationships for G and K are to be assumed under unloading conditions, thus accounting for inelastic behavior in load-unload cycles. In this investigation, it is assumed that the response under unloading and reloading up to the maximum past state of stress is essentially elastic.

In fact, this model differs with plasticity model by the way unloading is defined. In plasticity models, unloading is defined by a yield criterion which represent both deviatoric and hydrostatic states of stresses, whereas, in variable

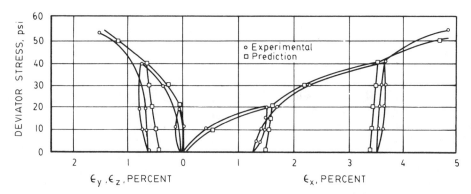

FIGURE 12. Comparison of predictions from variable moduli model with CTC test, $\sigma_2 = 15$ psi (104 kPa).

moduli model, behavior under deviatoric and hydrostatic states of stresses are decomposed and described independently.

Both G and K vary continuously with the states of stress in this model. Therefore solution of boundary value problems have to be done using incremental procedures only. The incremental stress-strain relationship in terms of G and K can be expressed as

$$d\sigma_{ij} = Kd\ \epsilon_{KK}\sigma_{ij} + 2G\left(d\ \epsilon_{ij} - \frac{d\epsilon_{KK}}{3}\delta_{ij}\right) \quad (22)$$

Under plane strain and axisymmetric idealizations, the incremental stress-strain relationship is

$$\begin{Bmatrix} d\sigma_{11} \\ d\sigma_{22} \\ d\sigma_{12} \\ d\sigma_{33} \end{Bmatrix} = \begin{bmatrix} K+\frac{4G}{3} & K-\frac{2G}{3} & 0 & K-\frac{2G}{3} \\ K-\frac{2G}{3} & K+\frac{4G}{3} & 0 & K-\frac{2G}{3} \\ 0 & 0 & 2G & 0 \\ K-\frac{2G}{3} & K-\frac{2G}{3} & 0 & K+\frac{4G}{3} \end{bmatrix} \begin{Bmatrix} d\epsilon_{11} \\ d\epsilon_{22} \\ d\epsilon_{12} \\ d\epsilon_{33} \end{Bmatrix}$$

(23)

The fourth row and column are meaningful for axisymmetrical idealization; for the plane strain conditions, the constitutive matrix is (3 × 3).

Determination of Material Parameters

The problem of choosing material parameters to fit the data is greatly simplified when there are tests available in which as few of the independent variable as possible are varied simultaneously. The hydrostatic test, in which the pressure and volumetric strain are measured is the best test to offer data for computing the values of K which appear in Equation (13). Shear tests like simple shear test or conventional triaxial compression test where shear stress and shear strain are measured would be desirable tests from which the values of σ in the shear moduli (Equation (14)) can be determined.

Material Parameters for Ballast II

In the combined stress-strain variable moduli model, K, the bulk modulus is expressed in terms of the first invariant of the strain tensor and the shear modulus, G, is expressed in terms of the first invariant of the stress tensor and the second invariant of the deviator stress tensor. It is possible to represent the unloading and reloading moduli as functions of the stress and strain invariants like the loading moduli, Equation (13) and (14). As a simplication, the unloading and reloading moduli can be adopted as the initial modulus (at zero stress level) or a multiple of the initial modulus. They can be also adopted as an average value of the (mean) slope of the unloading and reloading portions of the stress-strain curves. The parameters of the model are determined by using a process of curve fitting and are given by

$K_0 = 4.00 \times 10^3$ psi (27.56 × 10^3 kpa)
$G_0 = 1.71 \times 10^3$ psi (11.78 × 10^3 kpa)
$\gamma_1 = 2.59$
$\gamma_2 = -60.17$
$K_1 = -5.49 \times 10^5$ psi (37.82 × 10^5 kpa)
$K_2 = 2.54 \times 10^7$ psi (17.50 × 10^7 kpa)

The material model can now be expressed as

$$K = 4 \times 10^3 - (5.49 \times 10^5)\ I_1 + (2.54 \times 10^7)\ I_1^2 \quad (24)$$

and

$$G = 1.71 \times 10^3 + 2.59 \frac{J_1}{3} - 60.17 \, J_{2D}^{1/2} \quad (25)$$

In order to verify the above model for the loading part, the CTC test in Figure 5 was used. An envelope of loading curve corresponding to ϵ_x was used, and an average curve for ϵ_y and ϵ_z was used by assuming them to be equal. Here the cubical test sample was discretized in a three-dimensional nonlinear finite element analysis which included the variable moduli model. The CTC stress path (Figure 6) was simulated in the finite element analysis; the predictions and observations are compared in Figure 12. The correlation is very good.

As an approximation, the unloading-reloading behavior was assumed to be linear elastic with moduli to be equal to K and G. With these values, the unload-reload response in Figure 6 was computed, and is shown in Figure 12. The unloading was started from the computed points on the primary loading curves. The computed response is linear elastic but compares well with the observations. As stated earlier, if required, it is possible to define unloading and reloading moduli as functions of stress and strain invariants.

It may be noted that a nonlinear elastic model such as the variable moduli is limited in its capacity to predict plastic response. Although such models can provide satisfactory predictions for some problems, for more accurate predictions, it may be necessary to develop models based on the theory of plasticity.

CONCLUSIONS

Since ballast in (railroad) track support is subjected to three-dimensional states of stress, it is appropriate to perform truly triaxial tests in order to develop appropriate constitutive models. A series of truly triaxial tests involving slow cyclic loading were performed under various stress paths on a ballast with two different gradations. The stress-strain response exhibits both elastic and plastic behavior. Here the commonly used resilient modulus model and the variable moduli model have been developed as simplified characterizations. The results indicated that the moduli are dependent on state of stress, confining or mean pressure, stress paths and particle size. The results from the tests can be used to develop nonlinear elastic and plasticity models which can be incorporated in nonlinear analysis for field or boundary value problems.

REFERENCES

1. American Railway Engineering Association, *Manual of Recommended Practice*, 59 East Van Buren St., Chicago (1969).
2. Barksdale, R. D., "Repeated Load Test Evaluation of Base Course Materials," Ph.D. Thesis, Georgia Institute of Technology, Atlanta, Ga. (1972).
3. Bosserman, B. N., "Ballast Experiments at FAST," *Proceedings, Facility for Accelerated Testing (FAST) Engineering Conference,* Denver, Colorado, pp. 45–54 (November 1981).
4. Desai, C. S., and H. J. Siriwardane, *Constitutive Laws Engineering Media*, Prentice-Hall, Englewood Cliffs, N. J. (1983).
5. Desai, C. S., R. Janardhanam, and S. Sture, "A High Capacity Truly Triaxial Device and Applications," *J. of Geotech. Testing,* ASTM (March 1982).
6. Desai, C. S., H. J. Siriwardane, and R. Janardhanam, "Interaction and Load Transfer Through Track Guideway Systems," Report, Contract No. DOT-US-80013, Department of Transportation, Office of University Research, Washington, D.C. (1980).
7. Janardhanam, R., "Constitutive Modelling of Materials in Track Support Structures," Ph.D. Dissertation, Virginia Polytechnic Institute and State University, Blacksburg, Va. (1981).
8. Knutson, R. M., "Factors Influencing the Repeated Load Behavior of Railway Ballast," Ph.D. Thesis, University of Illinois, Urbana, Il. (1976).
9. Lowe, J., "Shear Strength of Coarse Embankment Dam Materials," *Proceedings 8th Congress of Large Dams*, pp. 745–761 (1964).
10. Nelson, I. and M. L. Baron, "Application of Variable Moduli Models to Soil Behavior," *Int. J. Solids Structures*, Vol. 7, pp. 399–417 (1971).
11. Office for Research and Experiments of the International Union of Railways (O.R.E.), "Stress and Strains in Rails, Ballast and in the Formulation Resulting from Traffic Loads, Question D71," Report No. 10, Vols. 1 and 2, Utrecht, Netherlands (1970).
12. Office of Research and Experiments of the International Union of Railway Ballast Under Repeated Loading, Report D, 117/RP5, Utrecht, Netherlands (1974).
13. Raymond, G. P., "A Study of Stress and Deformation Under Dynamic and Static Load Systems in Track Structure and Support," Report No. 75–10, Queens's Univ., Kingston, Canada (1975).
14. Raymond, G. P., et al., "Railroad Ballast Load Ranking Classification," *J. of Geotechnical Engineering Div.*, ASCE, Vol. 105, No. GT2, pp. 305–322 (Feb. 1979).
15. Raymond, G. P., et al., "Repeated Load Triaxial Tests on a Dolomite Ballast," *J. of Geotechnical Eng. Div.*, ASCE, Vol. 104, No. GT 7, pp. 1013–1029 (July 1978).
16. Siriwardane, H. J. and C. S. Desai, "Computational Procedures for Nonlinear Three-Dimensional Analysis with Some Advanced Constitutive Laws," *Int. J. Num. and Analyt. Methods in Geomechanics*, Vol. 7, No. 1 (1983).
17. Selig, E. T., et al., "Ballast and Subgrade Response to Train Loads," Report to Transportation Research Board, Washington, D.C. (Jan. 1978).
18. Thompson, M. R. and R. M. Knutson, "Permanent Deformation Behavior of Railway Ballasts," Presented at the 57th Annual Meeting of the Transportation Research Board, Washington, D.C. (Jan. 1978).

SECTION THREE
Transportation

CHAPTER 14	Public Transportation in Urban America	273
CHAPTER 15	Mass Transportation: Carpool Assignment Technique Application	335
CHAPTER 16	Transit Performance Evaluation	345
CHAPTER 17	Matching of Transportation Capacities and Passenger Demands in Air Transportation	365
CHAPTER 18	Pedestrian Flow Characteristics	393
CHAPTER 19	Simulation Model Applied to Japanese Expressway	407
CHAPTER 20	The Role of Driver Expectancy in Highway Design and Traffic Control	429
CHAPTER 21	Technique for Identifying Problem Downgrades	457
CHAPTER 22	Transportation Planning and Network Geometry: Optimal Layout for Branching Distribution Networks	473
CHAPTER 23	Factors Affecting Driver Route Decisions	499

CHAPTER 14

Public Transportation in Urban America

WILBUR S. SMITH*

INTRODUCTION

Urban densities, and residential densities in particular, are a result of many factors. These include city age, history, setting, culture, and economy. Consistently, transport technology has been a factor which has influenced travel patterns, with a resulting worldwide diversity of city patterns of movement.

Mass transportation is an integral part of the total transportation system, especially in urban areas, but its popularity has declined in recent decades as auto ownership and usage increased.

While the impact of the motorbus has been undeniable, the freedom of movement, or ease of route changes cannot be correlated with many basic urban characteristics. Because of its great flexibility, bus transportation is adaptable to all types of residential development. It is obvious that most private developers do not usually make investments based on the presence of bus service, if indeed mass transportation is a factor in their decisions.

Types of Transit Service

A variety of specialized transit systems have been devised to meet the transportation needs of urban areas. They are primarily concerned with moving people, and for which a fare is usually charged.

Transit service may be provided by motor buses, electric trolley buses, rail cars and trains, or by specialty systems such as monorails and moving walkways. These may be operated on public streets, along with other traffic, or in private rights-of-way at ground level, in subways, or on elevated structures, either completely or partially separated from other traffic. Transit vehicles vary widely in size and capacity, although vehicles seating less than nine persons are generally classified as taxicabs, or limousines. Rapid transit relates to rail, bus, or other transit forms which operate over exclusive rights-of-way separate from other traffic.

The general classes of transit are:

1. *Bus transit* moves passengers in vehicles supported by rubber tires and operating on normal roads and streets. In most cases the vehicles are mixed with other traffic, but may also have use of exclusive lanes. Buses normally seat 12 to 60 passengers, plus standees. They operate on fixed routes and schedules.
2. *Rail rapid transit,* sometimes known as "heavy rail," is a system that moves passengers in large numbers in trains operating on exclusive rights-of-way. They may be in a subway, elevated, or at grade. Different degrees of automation are applied. Heavy rail transit is generally considered to be duo-rail electric traction service; however, a number of recent designs include rubber-tired systems and monorails as well as non-traction propulsion, such as linear electric motors. Vehicles are usually greater than 45 ft. in length, and 8–10 ft. wide, with a weight range of 750–1,200 lbs./ft.-length [1].
3. *Light rail transit* moves passengers in intermediate size groups on short trains, or in single cars over a variety of rights-of-way: grade-separated, reserved (as in a street median), or shared with other street traffic, the latter operation being known by tradition as streetcars or trams. Light rail transit is generally considered to be duo-rail electric traction service with lightweight construction of rolling stock, approximately 750–950 lbs./ft.-length [1].

The information herein relates to the United States and Canada, except in a few cases where comparisons are made with overseas cities.

*Columbia, SC

273

4. *Specialty systems* are adapted to special applications or places, such as airports, include ferry operations, peoplemovers, paratransit, and perhaps other modes.

Evolution and Development

HISTORY OF TRANSIT SYSTEMS

Table 1 shows the beginning dates of the transportation systems discussed, the dates of peak mileage and ridership, and the end of service, where applicable. The dates when mileage and ridership of a system bottoms out followed by renewal growth are also shown for appropriate systems.

Horse-Drawn Public Transport

Public urban transportation essentially began with horse-drawn vehicles designed for roadway use. Transit services were initially provided by adapting existing vehicles, such as the stagecoach, which could only carry about five or six passengers. Larger horse-drawn coaches (also called omnibuses) were soon being built to carry up to 18 passengers with systems operating in Paris in 1829 and in New York by 1831.

Horse-drawn vehicle service was widely accepted on city streets with Philadelphia, Boston, and Baltimore adopting service in 1831, 1835, and 1844, respectively. For a quarter of a century the horse-drawn coach was the only regular means of public urban travel with systems such as the London General Omnibus Company expanding to a size of 17,000 horses and 1,400 omnibuses by 1905 [1].

Operationally, the horse-drawn coach was much the same as the present-day urban motor bus. An average vehicle velocity of 6 miles per hour and headways of 15 seconds were not uncommon.

With the advent of street railway systems, electric streetcars, and the motor bus, the horse-drawn coach, with its passenger capacity limitations, began to lose ground. The introduction of the motor bus, especially, resulted in the disappearance of the horse-drawn coach in New York by 1908.

The horse-drawn streetcar, operated on rails, first began operations in New York City in 1832. New Orleans authorized a similar project shortly thereafter; however, it was not until the 1850s that other cities began using horsecar rail systems.

By putting horse-drawn vehicles on rails, it was possible to get more output from a team of horses. The average empty weight was about 4,000 pounds and passenger capacity varied from 20 to 25 persons. The rail horsecars were faster than horse-drawn coaches and soon became serious competition, although they still had the problems of street congestion and resultant delays.

The 1880s were the heyday years for horsecars. By 1882 there were more than 100,000 horses and mules pulling

TABLE 1. Chronological Benchmarks in the Development of Urban Transit in the United States.

Transit System	Prototype operation	Beginning of reliable service, with subsequent growth	Peak system mileage with subsequent growth	Peak ridership with subsequent decline	End of service, if any	System mileage bottoms out, with subsequent growth expected	Ridership bottoms out, with subsequent growth expected
Horse-drawn coach	1827	1831	1860	1860	1908	—	—
Horsecar	1832	1855	1890	1890	1923	—	—
Rapid transit (elevated and subways)	1867	1870	1937	1946 (1929)[a]	—	1954	1977
Cable car	1873	1873	1894	1894	(1955)[b]	—	—
Electric streetcar (incl. light rail)	1884	1888	1917	1923	—	1975	1977[d]
Trolley bus	1910	1923	1952	1949	—	1978	1978
Motor bus	1907	1920	[c]	1948	—	—	1972

[a]Peak year excluding World War II.
[b]Since 1955, remaining cable car lines in San Francisco have been preserved for historical and tourist reasons.
[c]Motor bus route-mles of service have been continually expanding and have not reached a distinct peak.
[d]Some light rail systems are being completed in the eighties, while experimental tests were conducted in 1985 with railbus (motor bus on rails) service in Cleveland and Buffalo.

Source: B. S. Pushkarev and J. M. Zupan, *Urban Rail in America: An Exploration of Criteria for Fixed-Guideway Transit*; Report UMTA-NY-06-0061-80-1, Urban Mass Transportation Administration, Washington, DC, Nov. 1980.

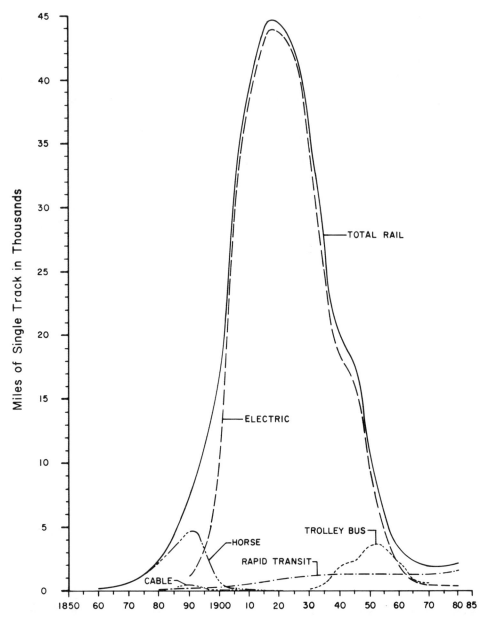

FIGURE 1. System mileage of urban rail in the United States, 1855–1985. Source: B. S. Pushkarev and J. M. Zupan, *Urban Rail in America: An Exploration of Criteria for Fixed-Guideway Transit*; Report UMTA-NY-06-0061-80-1, Urban Mass Transportation Administration, Washington, DC, Nov. 1980.

18,000 streetcars over 3,000 miles of track in cities of the United States [1]. Within the next eight years, horsecar trackage had more than doubled with a peak trackage in 1890 of over 5,600 miles, as shown in Figure 1. With the advent of the cable railways and the electric streetcars the horsecar era began to end and the 1890s showed a steady decline in animal powered vehicles.

The Cable Car

San Francisco became the first city in the world to have a cable railway transit system in 1873. Chicago had an operational system by 1882 and in 1883 cable car systems were also installed in Philadelphia, New York, St. Louis, Oakland, Denver, Cleveland, Seattle, Washington, Baltimore, and Providence. By the 1890s there existed in the

United States approximately 500 miles of cable tracks over which 5,000 cars carried about 400 million passengers per year [1].

The cable car was a subsystem innovation which basically replaced the horses with stationary steam plant powered cables as the streetcar propulsion source. Capital costs were higher, but operational costs were much lower, edging the cable cars over horsecars after a period of time. With the development of the electric motor, another subsystem innovation, the electric motor driven streetcar became a practical reality which had significant advantages over the cable powered car it displaced. Philadephia and Baltimore converted from cable cars to electric streetcars in 1897. By 1902 the total extent of cable car trackage in the United States was half of the 1890 peak mileage. In 1940 San Francisco remained the only city with a cable railway which today is a historical monument and is still operable [1].

Electric Streetcars

Although much of the early development work was done in Europe, the first practical application was in Montgomery, Alabama in 1886 with the adaption of horsecars to be driven by electric motors and chain drives [1]. The system installed in Richmond, Virginia, in 1882 is widely

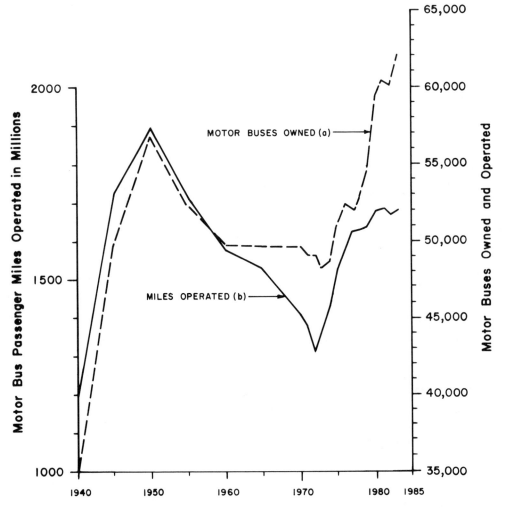

FIGURE 2. Trends in motor bus usage. Notes: Data for years 1979 and later includes all types of buses and vans operated by transit systems. Transit ridership increased dramatically in World War II due to gasoline rationing and other factors. Source: American Public Transit Association, *Transit Fact Book*, Washington, DC, 1985.

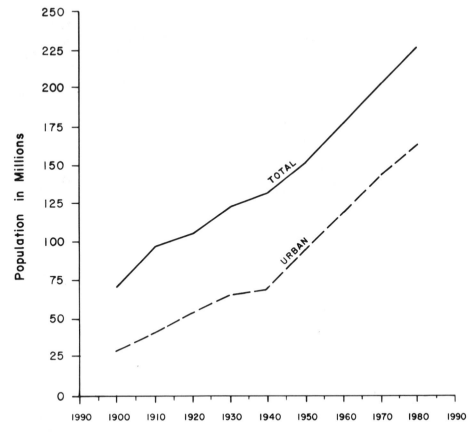

FIGURE 3. U.S. population trends. Source: U.S. Bureau of the Census, *Historical Statistics of the United States, Colonial Times to 1970*, Series A57, Washington, DC, 1975; U.S. Bureau of the Census, *Statistical Abstract of the United States*, Table No. 23, Washington, DC, 1985.

credited with having perfected the technology for reliable commercial service and starting an expansion of electric streetcar lines throughout the world. The electric streetcar was in its heyday in the 1890s and early 1900s with peak trackage of 44,000 miles in 1917, as Figure 1 shows [2].

Electric railway executives in the U.S.A. formed the Electric Railway President's Conference Committee (ERPCC) in 1929. This group developed an advanced streetcar design, the PCC car. The PCC cars were manufactured in the United States in 1936–1952 and were also widely copied in Eastern Europe and other parts of the world. The new car design was superior in speed, braking systems, and passenger comfort.

Elevated and Subway Systems

Heavy rail transit service started in London about 1863, using steam locomotives in underground subways. Elevated service started in New York in 1870 using cable propulsion, but was replaced within a year with steam locomotion [2]. Elevateds remained mainly a U.S. institution with Chicago's South Side Elevated Railway beginning operation in 1892 [1].

New York electrified their elevated in 1898–1903. Objections to underground transit because of the hazards of smoke and steam became moot; but, subway construction in New York finally began in 1900 with subsequent operations starting in 1904.

By this time electric subways were already operating in London, Budapest, Glasgow, and Paris. Rapid transit track miles in the United States quadrupled between 1902 and 1937. Over 90 percent of the mileage was in New York and Chicago with Philadelphia and Boston accounting for the remainder [2].

Bus Transit Systems

With the advent of the automobile the motor bus also came into existence in New York in 1907. By 1914 the horse-drawn coaches (omnibuses) of London were entirely replaced by more than 3,000 motor buses. Trends in motor bus usage in the United States are shown in Figure 2.

Streetcar companies changed to motor buses in the 1930s because buses were generally cheaper and provided much more flexible services than streetcars. Likewise, as it became increasingly more expensive for passenger revenues to replace and maintain street trackage, more and more roads were being paved. Motor bus routes could be extended or changed to meet changing demands without heavy financial requirements. Between 1904 and 1940, paved mileage in the United States increased from 9 to 47 percent of the total street and road mileage [2].

The jitney, a five and six passenger automobile, offered more flexible and faster service than streetcars, and diverted as much as 50 percent of the streetcar passengers during peak hours in some travel corridors. Streetcar companies fought the jitney and by 1919 had effectively caused them to be regulated; but by then, the motor bus had already proved its worth and patronage was on the increase.

In an effort to avoid the capital cost of replacing tracks but still make use of electric traction, the trolley bus came into public service in the United States around 1911. The biggest build up for trolley buses was in the early 1920s with a peak in 1952 of 3,700 miles of dual overhead wire, as shown in Figure 1.

POPULATION TRENDS

Population in United States cities of over 10,000 people grew from 11 million to nearly 45 million, or almost one-half of the national total, between 1880 and 1920, as the

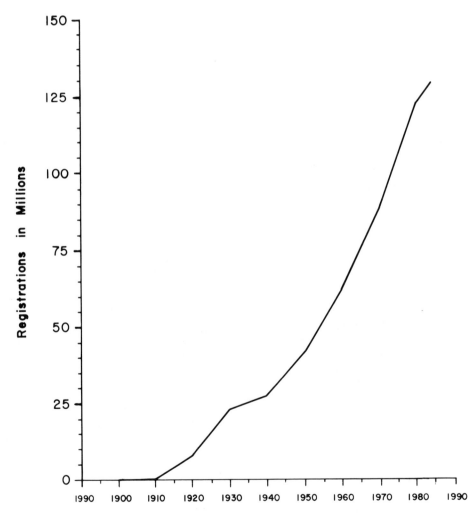

FIGURE 4. Privately owned passenger car registrations. Source: U.S. Federal Highway Administration, *Highway Statistics*, Washington, DC, Annual.

streetcar lines reached out from city centers. Urban rail ridership increased from 0.6 billion to 15.5 billion trips annually. The development pattern of urban cores and of the nearby suburbs was shaped during this period. It was a pattern of moderately high densities compactly arranged within walking distance of streetcar lines. The population of these areas grew and then declined in subsequent years as the costs of providing infrastructure services increased and as governmental organizations and management became more difficult. In the 1920s, patronage of urban transit began to slow and it no longer kept pace with urban population growth.

Between 1926 and 1980, the nation's population grew from 113 million people to 226 million while the urban population during this same period grew from 61 million to 167 million as shown in Figure 3. In 1920, urban population was 51 percent of the total population while in 1950 and 1960 the percentage had increased to 64 and 70 percent, respectively. The 1980 percentage was 74 pecent and by the year 2000, the urban to total population ratio will be 85 percent if trends continue as expected [3,4].

Another growth pattern trend that effects mass transit occurs within the central city area. Traditionally transit has best served residential areas of high population density relatively close to the central business district (CBD). Since the 1950–1960 decade, the bulk of population increase has taken place within urbanized metropolitan areas with the majority of that increase taking place outside the CBD. The CBD area population has actually declined in most areas.

AUTOMOBILE TRENDS

As the automobile, the bus and other motorized vehicles became more reliable and popular in the 1920s, areas that had not been reached before by transit became easily accessible. This "filling in" of outward expansion between transit corridors continued and accentuated the decentralization process already started by street railway, rapid transit, and commuter rail lines. Although these changes began about 1925, the trend toward decentralization was constrained by depression and war for almost a quarter century.

As shown in Figure 4, the estimated number of privately owned passenger car registrations has grown over four fold since 1945. Estimates are that privately owned passenger cars topped 120 million for the first time in 1980 and represent over 99 percent of the privately and publicly owned passenger cars in the U.S.

EFFECTS OF URBAN CHARACTERISTICS ON TRANSIT

The social, demographic, and land use characteristics of the urban area of American cities influences both the travel behavior of the residents and the demands for transportation systems. However, the freedom of movement permitted by the automobile has transformed the patterns and densities of urban growth along lines which have often proven difficult to serve with mass carriers.

Each transport mode (streetcars or trams, rail transit lines, buses, and automobiles) expanded the area of urban development, and each controlled in its own way the population concentration at the city center and dispersion at the periphery. Residences, shops, and work places are clustered along the radial transit lines, and near rapid transit and commuter rail stations.

Public Transport and Downtown Areas

There is a strong interdependence between public transport and the center city. Transit permits compact, consolidated, mutually reinforcing downtown developments and provides high radial transport capacities into the center city.

Many downtown areas are becoming centers of commerce, government, and finance, with a corresponding growth in office employment. In general, downtown retailing, manufacturing, and entertainment functions will, at best, remain at present levels. Despite downtown's lesser relative growth, it will remain the city's largest, most complex traffic generator. But, there appears to be a limit to the number of motor vehicles that can enter or leave downtown in peak periods as shown in Figure 5.

Trends in Transit Patronage

Transit ridership responds to short-term conditions, like the Great Depression and World War II, as the historic pattern in Figure 6 shows. Between 1935 and 1940, overall transit use increased by a modest amount as the national economy emerged from the depression era. Transit use virtually doubled during World War II due to the halt in new car production and the rationing of gasoline. After the War, use of public transport declined as new cars became available.

By 1955, transit riding had dropped below prewar levels. It continued to decline throughout the 1950s, then nearly leveled off in the 1960s. Since only a few miles of urban freeway were in use prior to 1960, transit's loss of patronage was almost entirely due to the rapid rise in personal car ownership rather than to the urban freeway networks which began to materialize during the 1960s. Ridership growth in the seven decades prior to 1927 were related to expanded economic activity in compact urban environments. From 1927 to 1972 transit ridership declined, while a three-fold increase in auto ownership per capita occurred. The renewed growth after 1972, despite continuing urban dispersal and rising auto ownership, was started mainly by improved bus service. As bus patronage increased, rapid transit use decreased, mainly in areas like Manhattan, Philadelphia, and Cleveland. Figure 6 shows a trend of increased rapid transit ridership after 1977 which was due to improved

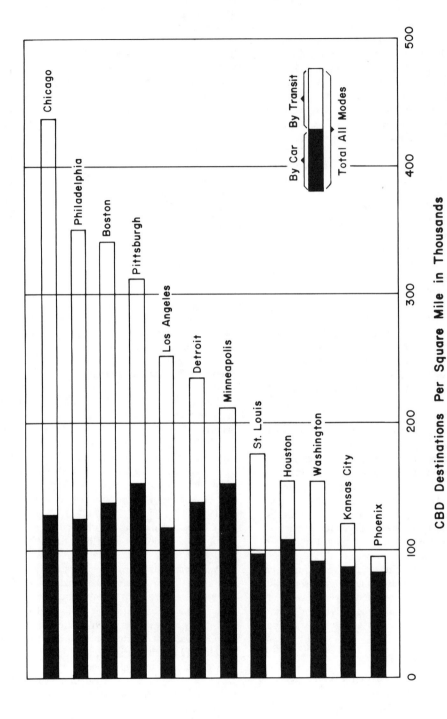

FIGURE 5. Public transpotation contribution to intensity of several downtown areas. Source: H. S. Levinson, "Modal Choice and Public Policy in Transportation," Article in *Urban Transportation—Perspectives and Prospects*, Eno Foundation for Transportation, Inc., Westport, CT, 1982.

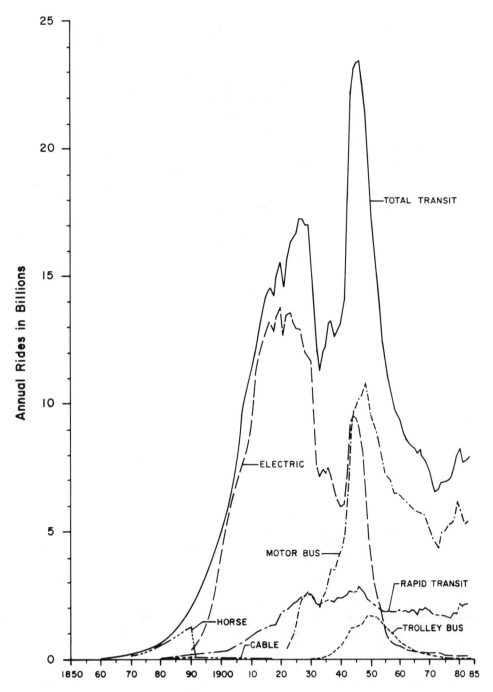

FIGURE 6. Passenger trips on urban transit in the United States by mode, 1855–1983. Notes: Procedures for collecting data were changed in 1980 in accordance with Urban Mass Transportation Act, Section 15. Source: B. S. Pushkarev and J. M. Zupan, *Urban Rail in America: An Exploration of Criteria for Fixed-Guideway Transit*; Report UMTA-NY-06-0061-80-1, Urban Mass Transportation Administration, Washington, DC, Nov. 1980.

economic conditions in Manhattan and new systems in Washington, San Francisco and Atlanta. Fuel shortages in 1974 and 1979 followed by rising fuel prices contributed to the upswing of both bus and rapid transit use.

Factors Affecting Transit Use

The effectiveness of various transport investment decisions depends on the extent that transit as well as highways can attract travelers. The use of autos and public transit for travel to the center city depends on many interrelated factors, including [5]:

- center city employment
- trip purpose
- parking availability and costs
- car ownership and availability
- barriers to travel that serve to restrict or concentrate movements
- capacities of major approach highways
- public attitudes towards highways, transit and parking
- transit and auto travel times

Modal choice varies with the level of car ownership or availability with travelers making modal decisions based on time and out-of-pocket cost differences. Fast transit services coupled with high driving and parking costs are conducive to transit use.

POPULATION DENSITY

The effects of density on urban travel patterns and transportation demands are well documented. Many origin-destination and travel demand surveys throughout the world have shown how residential and employment densities influence the number and types of trips, and the modes of travel used.

As population density rises, there is an increase in the total number of person trips, including pedestrian trips, and a corresponding decrease in the number of trips in vehicles. This is because many shopping, social, and school trips and

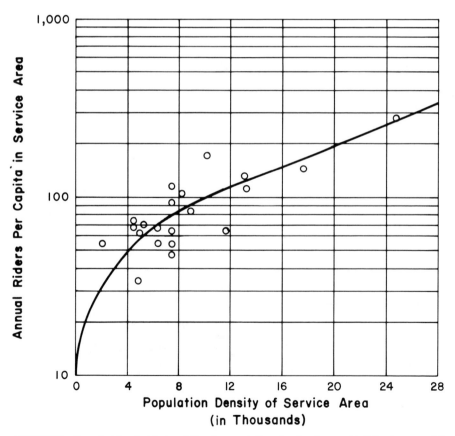

FIGURE 7. Transit riding habits in relation to population density. Source: H. S. Levinson and F. H. Wynn, "Effects of Density on Urban Transportation Requirements, Community Values as Affected by Transportation" *Highway Res. Rec. No. 2*, Highway Research Board, Washington, DC, 1963.

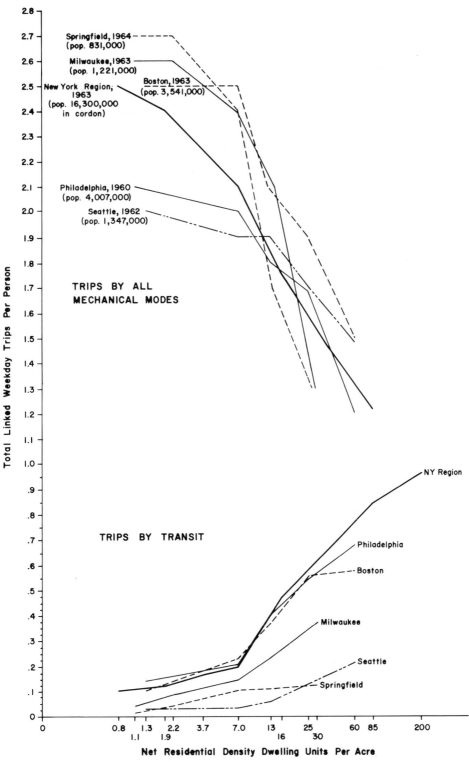

FIGURE 8. Total trips per person by mode related to residential density in six urban areas. Source: B. S. Pushkarev and J. M. Zupan, *Public Transportation and Land Use Policies*, Indiana University Press, Bloomington, Indiana, 1977.

some work trips are made on foot in these areas. A greater proportion of the non-walking trips also are made by public transport in high-density environments. Moreover, in high-density areas, income and car ownership are normally less—these are the important explanatory factors that influence trip rates.

Figure 7 shows how transit ridership relates to population density. In almost every case high density (in combination with low car ownership) tends to increase transit use. But the relationships are also influenced by factors such as service and fare differentials, concentrations of activities within the central area, and clustering of population along specific corridors that influence ridership.

In almost every city, overall trip rates decrease as cities or neighborhoods become more dense. This is apparent from the relationships between person trips and residential densities shown in Figure 8 for six United States urban areas [6]. It is seen that an individual living on a one-acre lot in one of the six urban areas shown—ranging in size from Springfield, Massachusetts (population 531,000), to the New York Region (population 16,300,000 within the survey area)—made anywhere from 2.0 to 2.6 trips by all vehicular modes on an average weekday. The propensity for making trips did not change much as the density increased to three dwellings per acre. However, as the density increased from 3 to 30 dwellings per acre, the number of trips per person

TABLE 2. Radial Population Distribution in 29 Urban Areas, 1970.

Urban Area	1970 Populaton ($\times 10^3$)		SMAS Populaton at 20 Miles ($\times 10^3$)	Percent of SMSA Population Within Given Mileage from Center									
	Urbanized Area	SMSA		2%	4%	6%	8%	10%	12%	14%	16%	18%	20%
Urbanized Area Population over 5,000,000													
New York[a]	16,207	11,572	12,129	3	12	28	43	56	71	83	91	98	104
Los Angeles	8,351	7,041	6,164	2	8	17	27	37	50	61	71	79	88
Chicago[a]	6,715	6,978	5,075	1	8	19	31	43	52	59	65	68	73
Urbanized Area Population 2,000,000 to 5,000,000													
Philadelphia[a]	4,021	4,822	4,038	6	20	32	45	56	63	69	76	80	84
Detroit	3,971	4,204	3,592	2	9	19	31	43	55	67	75	81	86
San Francisco[a]	2,988	3,108	2,282	14	25	32	37	48	63	74	83	91	98
Boston[a]	2,653	2,754	2,850	5	22	39	52	65	74	81	89	95	103
Washington[a]	2,481	2,862	2,676	5	18	34	53	67	76	84	88	91	93
Dallas-Ft. Worth	2,015	2,318	1,478	2	12	21	33	40	48	56	57	61	64
Urbnized Area Population 1,500,000 to 2,000,000													
Cleveland[a]	1,960	2,064	1,930	3	15	29	49	65	74	83	87	90	93
St. Louis	1,833	2,363	2,066	2	15	27	41	52	63	73	78	82	87
Pittsburgh[a]	1,846	2,401	1,887	5	18	32	44	52	59	64	69	74	78
Houston	1,678	1,985	1,671	4	18	38	46	72	78	81	82	82	84
Baltimore[a]	1,580	2,071	1,857	12	27	47	61	71	77	80	84	87	90
Urbanized Area Populaton 1,000,000 to 1,500,000													
Milwaukee	1,252	1,404	1,310	8	30	51	66	76	80	83	87	91	93
Seattle	1,238	1,425	1,307	4	15	29	42	55	63	74	82	88	92
Miami[a]	1,220	1,268	1,427	7	24	38	56	71	83	91	101	109	112
San Diego[a]	1,198	1,353	1,107	4	18	36	50	65	73	77	79	81	82
Atlanta[a]	1,173	1,390	1,342	7	21	36	50	61	71	80	87	93	97
Buffalo[a]	1,087	1,349	1,221	10	25	44	59	67	76	77	82	89	91
Denver	1,047	1,230	1,105	9	27	48	69	81	85	87	88	88	90
Urbanized Area Population Under 1,000,000													
New Orleans	962	1,046	1,000	18	46	68	78	88	92	92	94	95	96
Tampa-St. Petersburg	864	1,013	733	7	18	28	36	39	43	43	48	57	72
Portland	825	1,007	949	11	30	46	63	74	83	86	89	92	94
Indianapolis	820	1,111	941	7	27	46	63	69	74	78	82	83	85
Providence	795	914	1,157	13	34	51	61	70	81	91	105	121	126
Columbus	790	916	894	8	35	58	74	83	89	91	94	97	98
San Antonio	773	864	834	11	40	65	81	90	91	93	96	97	97
Dayton	686	850	878	12	33	49	61	72	77	84	91	96	103

[a]Urban areas with existing or committed fixed guideway systems.
Source: U.S. Depatment of Transportation, *Urban Data Book*.

TABLE 3. Changes in Transit Usage, Private Car Ownership, and Population, 1935–1980.

Year	Transit Passenger Trips			Private Passenger Cars Registered			Total Resident Population			Persons Per Private Passenger Car	Transit Rides Per Capita
	Total Trips/Rider ($\times 10^3$)	5-year Change (%)	Per Cent of 1935 Level	Passenger Cars Registered ($\times 10^3$)	5-Year Change (%)	Per Cent of 1935 Level	Resident Population ($\times 10^3$)	5-Year Change (%)	Per Cent of 1935 Level		
1935	12,226	—	100.0	22.5	—	100.0	127.2	—	100.0	5.7	96
1940	13,098	+ 7.1	107.1	27.4	+21.7	121.7	132.5	+ 4.2	104.2	4.8	108
1945	23,254	+77.5	190.2	25.7	− 6.2	114.2	133.4	+ 0.7	104.9	5.2	174
1950	17,246	−25.8	141.1	40.2	+56.4	178.7	151.9	+13.9	119.4	3.8	114
1955	11,529	−33.1	94.3	52.0	+29.4	231.1	165.1	+ 8.7	129.8	3.2	70
1960	9,395	−18.5	76.8	61.4	+18.1	272.9	179.9	+ 9.0	141.4	2.9	52
1965	8,253	−12.2	67.5	74.9	+22.0	332.9	193.6	+ 7.6	152.2	2.6	43
1970	7,332	−11.2	60.0	88.8	+18.6	394.7	203.2	+ 5.0	159.7	2.3	36
1975	6,972	− 4.9	57.0	106.1	+19.5	471.6	213.6	+ 5.1	167.9	2.0	33
1980	8,235[a]	+18.1	67.4	122.6	+15.6	544.9	226.5	+ 6.0	178.1	1.8	36

[a]Unlinked Transit Passenger Trips beginning in 1980 based on data collection procedures defined by Urban Mass Transportation Act, Section 15. Series not continuous between 1975 and 1980.
Source: *Statistical Abstract of the United States*, 1966 (population); Federal Highway Administration, Bureau of Public Raods, *State Motor Vehicle Registrations* (various tables); and American Transit Association, *Transit Fact Book*, 1968 (transit riders).

declined anywhere from 50 percent in Springfield and Milwaukee to 16 percent in Seattle, with the New York Region about in the middle with a 30 percent reduction in total trips. This reduction in total travel demand is due to a reduction of trips made by auto. In contrast, transit trips increased dramatically with the rising density. This demand did not increase much up to a density of seven dwellings per acre. However, with a density increase from 7 to 30 dwellings per acre, transit demand roughly tripled in the New York Region, in Philadelphia, and in Boston to around 0.6 trips per person per day [6].

Public transport is an activity whose locus is primarily the nation's older cities. Two such metropolitan areas, New York and Chicago, account for 7 percent of the nation's total population and 9 percent of its urbanized population. But the two cities generate over 40 percent of the nation's mass transit patronage. While the national index of annual per capita transit rides is a mere 48, in New York City the ratio is 168 and in Chicago it is 93. The nation's five largest urbanized areas account for 14 pecent of the country's population, and generate over 50 percent of the transit rides [7].

In the major urban areas of the United States, exclusive of New York City, surface transit vehicles accommodate the majority of all transit patronage. There is a strong relationship between the age of the city and transit riding. In general, those cities which experienced their major growth after the private car came into general use have the lowest transit usage. This reflects the lower population densities in these cities. The radial population distribution shown in Table 2 depicts the accumulated percent of total population up to 20 miles from the city center for 29 large urban areas in the United States. Cities with major rail transit systems in existence or under construction in 1980 are footnoted accordingly. Generally, metropolitan areas of 2 to 10 million average some 90 percent of their population within a 20 mile radius while cities on the order of 1 million average 90 percent of the population within 17 miles of city centers but have fewer total people within a given distance of the city center. This shows that a rail line of certain length in a smaller city will have fewer potential riders than a line of similar length in a larger city.

AUTOMOBILE OWNERSHIP

The greatest impact on transit has been from the increase in private car ownership and use, as shown in Table 3. In the 45-year period, 1935–1980, the nation's population increased by 78 percent, while private car registrations increased 445 percent and transit riding decreased 31 percent [8].

The ratio of population to cars owned in the United States dropped from 5.7 in 1935 to 1.8 in 1980, while the average number of transit rides per capita decreased from 96 to 36 annually (disregarding the wartime disruption of trends) [8].

Typical relationships among car ownership, trip generation, and travel mode for selected cities are summarized in Table 4. The first car generally has the most reductive effect on transit ridership. In smaller cities the overall rates of person-trip-making are approximately the same as for larger cities within a given level of car ownership. However, the proportion of trips by public transport is much less, even among zero-car households.

Table 5 shows person-trips by trip purpose and mode. It shows that 83 percent of all trips are by car, station wagon, van or pickup. While at the same time, public transportation

TABLE 4. Trip Generation by Car Ownership in Selected Urban Areas.

Item	Cars/Dwelling Unit			
	0	1	2	3+
Total trips/person				
London	0.93	2.03	2.56	3.06
Chicago	1.08	2.61	3.46	4.01
San Francisco	0.99	2.17	2.63	3.03
Detroit	0.70	2.08	3.20[a]	
Minneapolis	1.60	2.83	3.51	
Pittsburgh	1.12	1.70	2.40[a]	
Seattle	0.90	2.60	3.20	3.40
Indianapolis	0.93	2.13	2.88	2.25
San Juan	0.99	2.00	3.28[a]	
Springfield, MA	0.71	2.27	3.13[a]	
Knoxville	0.85	2.49	3.36	3.87
Lexington	0.76	2.06	2.91	3.26
Transit trips/person				
London	0.78	0.60	0.55	
Chicago	0.62	0.36	0.31[a]	
San Francisco	0.61	0.20	0.14	0.12
Detroit	0.42	0.25	0.24[a]	
Minneapolis	0.42	0.09	0.04	
Pittsburgh	0.74	0.38	0.26[a]	
Seattle	0.37	0.12	0.06	0.06
Indianapolis	0.32	0.09	0.04	0.10
San Juan	0.73	0.38	0.24[a]	
Springfield, MA	0.20	0.07	0.04[a]	
Knoxville	0.32	0.09	0.04	0.03
Lexington	0.28	0.10	0.07	0.05

[a]Two or more cars per dwelling unit.
Source: Computed from origin-destination studies in each urban area conducted by H. S. Levinson.

was only used for 3 percent of the trips. Figure 9 shows the person-trips by mode for all purposes based on data in Table 5.

The combined effects of household income and car ownership on trip rates are shown in Table 6 and Figures 10 and 11. Figure 10 gives average daily person-trips per household by number of autos per household versus income for various population levels while Figure 11 provides a guide to estimating trips by purpose for various levels of income and car ownership. Table 6 and Figures 10 and 11 all tend to suggest that trips per household increase as income increases within given population limits and for various trip purposes. The only exception is that work trips tend to remain constant for given levels of car ownership.

WORK TRAVEL MODE

There is a high degree of correlation between downtown employment density (i.e., employees per acre) and public transport use (Figure 12). More than 90 percent of the travelers into Manhattan, where employment approximates 800 persons per acre, arrive by transit, as compared to Denver, where 20 percent arrive by transit, and employment approximates 150 persons per acre. It is clear that as employment density increases there is greater reliance on public transport.

As Table 7 shows, household income is a significant determinant of how a person travels to and from work. More than 9 of every 10 persons travel to work by private motor vehicle whether they live in small communities or large central cities, as shown in Table 8. Tables 9 and 10 show that the use of private motor vehicles for travel to work increased between 1909 and 1977 while the length of virtually all such trips declines during that period.

URBAN LAND USE

Each type of urban land generates person and vehicle trips in accordance with the nature and intensity of its use. Commercial land may generate more than 150 person-destinations per acre; public space, fewer than 5. The strong relationship between the amount of downtown floorspace, employment and transit ridership is shown in Figure 13. The high ridership levels in New York, Chicago, Philadelphia and Toronto relates closely to office floor space in each CBD.

The important correlation between employment density and transit ridership bears out the fact that high residential density by itself will do little for transit if there is no dominant place to go.

CBD PARKING AND TRANSIT

Formulation of parking policies in any given city should be consistent with the downtown and surrounding area transportation goals, and should either encourage or restrain automobile usage in accordance with that city's urban activity patterns. By restraining automobile usage in large cities with high downtown development intensities and extensive transit services, the downtown commuter is more apt to utilize public transit instead of the automobile. However, this approach has limited applicability in automobile oriented cities where transit services are neither adequate or competitive with the automobile.

Commuter parking is the main parking problem in large cities with suitable transit options. Work trips account for about a third of all parkers in cities under 250,000 as compared with half or more in larger populated urban areas. In transit-oriented cities, automobile usage is best regulated by controlling parking supplies and pricing. Limiting commuter parking in the CBD, while at the same time increasing fringe parking on the CBD perimeter and at outlying express transit stations, will help divert commuters to public transit. Likewise, parking rates should complement rather than compete with transit. All day parking rates in congested downtown areas should be high enough to discourage commuter parking.

Central city parking policy should not discourage automobile travel to and from downtown for shopping and other purposes. Shoppers put a lot of emphasis on the flexibility of the automobile. If reasonably priced parking is not available in the CBD within close proximity of shopping areas, shoppers will patronize regional shops instead of traveling to the CBD by public transit. Two major reasons for avoiding public transit relate to safety and the handling of packages in congested transit systems.

Parking policies and programs should be designed not only to improve access, but also to reduce air and noise pollution, conserve energy, and improve public safety. They should continue to improve the accessibility, amenity, and vitality of cities. They should reflect each community's specific goals and needs [9].

1. Parking for the CBD should be viewed as part of a total development concept and transportation system. Curb parking should be reassessed in terms of overall downtown street use priorities relating to:
   - pedestrian circulation
   - service requirements
   - transit vehicles
   - automobile movement, and
   - curb access

   Free standing garages in central areas should be discouraged; the emphasis should be on multi-use development and continuity of block frontage. Where street and land-use conditions permit, "auto-free zones" should be established.

2. Short-term parking should be encouraged in the CBD wherever adequate transit is available to serve long-term parkers. Rates and hours of opening should be adjusted to favor short-term parkers, provided they do not create serious inequities to garage operators or to persons who must drive.

3. Zoning for parking in the city center should be restructured to stipulate *maximum* as well as *minimum* amounts of parking.

4. Outlying parking should be located along express transit lines at locations where:
   - land is relatively inexpensive
   - environmental impacts are minimal, and
   - highway access is good

   Costs to use fringe parking and express transit should be *less* than the costs to drive and park downtown. Facilities should generally be four-to-ten miles out and provide 1,000 to 2,000 spaces; and,

5. Improved municipal and metropolitan management of *parking* will be necessary to provide proper coordination of transit, highway, and parking improvements to institute proper controls over public and private facilities and to assure high levels of parking enforcement.

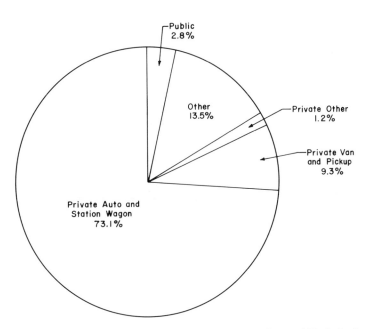

FIGURE 9. Person trips by mode of transportation. Source: U.S. Federal Highway Administration, *National Personal Transportation Study*, Washington, DC, 1977.

TABLE 5. Person-Trips, by Trip Purpose, and Mode of Transportation.

	Mode of Transportation									
	Private				Public					
Trip Purpose	Auto and Station Wagon	Vans and Pickups	Other[a]	Total	Bus and Street Car	Subway, Elevated Rail	Other[c]	Total	Other[b]	All
Earning A Living										
Home-to-Work	75.3%	12.0	2.0	89.3	3.1	0.8	0.7	4.7	6.0	100.0%
Work Related	66.9%	19.5	1.4	87.8	0.8	0.3	1.1	2.2	10.0	100.0%
Subtotal	74.0%	13.2	1.9	89.0	2.8	0.7	0.9	4.3	6.6	100.0%
Family and Personal Business										
Shopping	81.0%	7.9	0.9	89.8	1.0	0.1	0.3	1.4	8.9	100.0%
Doctor or Dentist	87.9%	4.7	0.4	93.0	2.9	0.4	1.0	4.3	2.6	100.0%
Other	72.1%	12.3	1.5	86.0	0.9	0.1	0.2	1.3	12.7	100.0%
Subtotal	77.6%	9.6	1.1	88.4	1.0	0.1	0.3	1.5	10.2	100.0%
Civic, Educational and Religious	49.8%	4.0	0.3	54.2	4.6	0.3	0.1	5.0	40.8	100.0%
Social and Recreational										
Visiting Friends	75.5%	9.0	1.4	85.8	1.1	0.1	0.2	1.4	12.8	100.0%
Pleasure Driving	75.3%	15.3	5.6	96.2	0.3	0.2	0.0	0.5	3.3	100.0%
Vacations	75.0%	8.6	0.7	84.3	2.9	0.0	2.8	5.7	10.0	100.0%
Other	76.3%	8.0	1.1	85.5	0.8	0.2	0.3	1.3	13.2	100.0%
Subtotal	67.3%	7.1	1.0	75.3	2.2	0.2	0.2	2.6	22.1	100.0%
Other	78.7%	7.9	0.9	87.4	2.4	0.7	1.4	4.4	8.1	100.0%
All Purposes	73.1%	9.3	1.2	83.7	2.0	0.3	0.5	2.8	13.5	100.0%

[a]Includes other trucks, motorcycle, self-contained recreational vehicle and taxi (personal use).
[a]Includes bicycle, walk, school bus, moped, and other modes not elsewhere classified.
[c]Includes train, airplane and taxi.
Source: U.S. Federal Highway Administration, *National Personal Transportation Study*, 1977.

TABLE 6. Average Annual Vehicle Trips By Household Income and Vehicle Ownership, 1977.

Number of Household Vehicles	Annual Household Income						
	Under $5,000	$5,000–9,999	$10,000–14,999	$15,000–24,999	$25,000–34,999	$35,000–49,999	$50,000 and over
None	36	[a]	[a]	[a]	[a]	[a]	[a]
One	777	1012	1185	1259	1338	1413	[a]
Two	1228	1661	1841	2037	2092	2194	2198
Three	2263	2131	2138	2586	2961	3130	2426
Four or more	[a]	2278	2414	3212	3337	3662	3077
Vehicle-owning Households	965	1324	1616	2028	2315	2555	2254
All Households	535	1099	1514	1972	2278	2474	2201

[a]Data insufficient for presentation, but these households are included in the rates for all households.
Source: Federal Highway Administration, *Nationwide Personal Transportation Study*, Report 9, "Household Travel," Washington, DC, 1982.

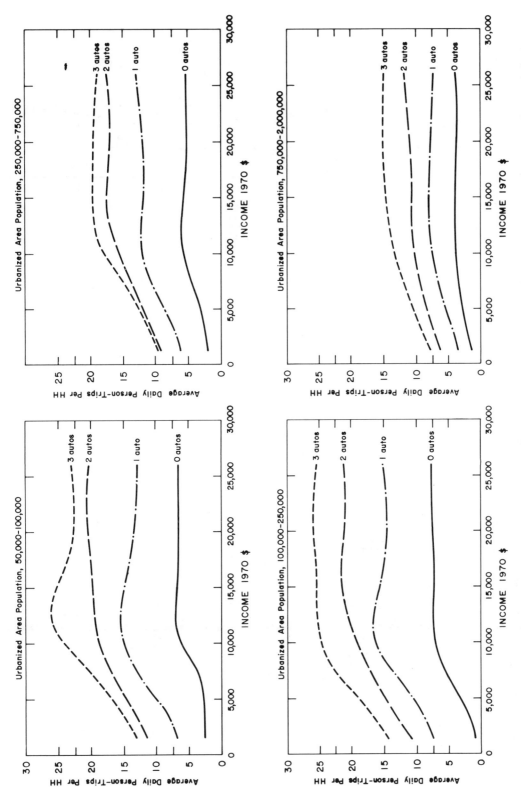

FIGURE 10. Average daily person trips per household by number of autos per household vs. income, for various urban area populations. Source: A. Sosslau, et al., *Quick Response Techniques and Transferable Parameters-User's Guide*, National Cooperative Highway Res. Prog. Report 187, Transportation Research Council, Washington, DC, 1978.

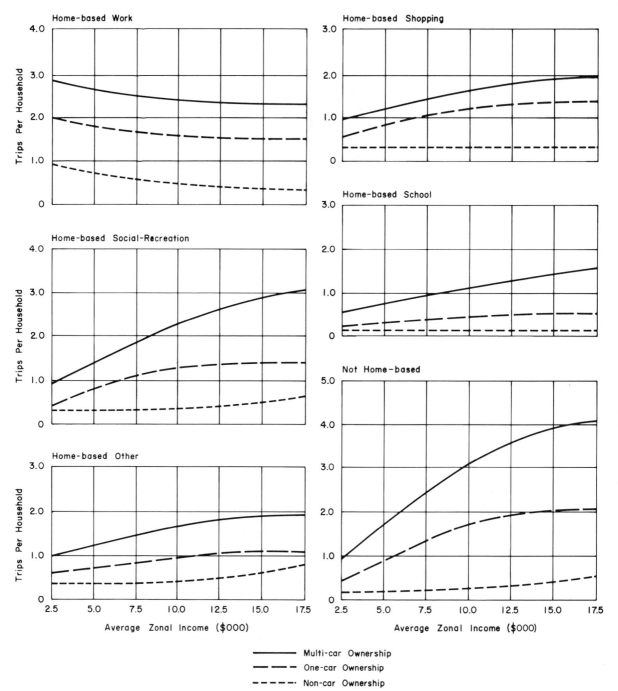

FIGURE 11. Trips per household by car ownership and income. Source: Wilbur Smith and Associates, *Ohio-Kentucky-Indiana Regional-Transportation and Development Plan*.

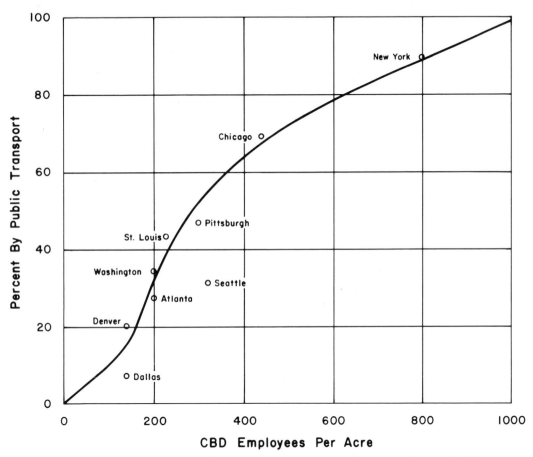

FIGURE 12. CBD trips by public transport. Source: H. S. Levinson, "Modal Choice and Public Policy in Transportation," Article in *Urban Transportation—Perspectives and Prospects*, Eno Foundation for Transportation, Inc., Westport, CT, 1982.

TABLE 7. Home-to-Work Trips by Mode and Income.

Mode	Household Income							
	Less than $5,000	$5,000–9,999	$10,000–14,999	$15,000–24,999	$25,000–34,999	$35,000–49,999	$50,000 or more	All
Private Vehicle								
Driver	55.0%	65.6%	72.2%	77.3%	76.7%	76.5%	78.6%	72.7%
Passenger	24.6	20.4	18.8	15.9	15.3	15.9	15.4	17.8
Total (Private Vehicle)	79.6	86.0	91.0	93.2	92.0	92.4	94.0	90.5
Public Transportation	7.2	7.2	4.1	3.3	3.9	4.5	3.4	4.5
Other	13.2	6.8	4.9	3.5	4.1	3.1	2.6	5.0
Total (All modes)	100.0%	100.0%	100.0%	100.0%	100.0%	100.0%	100.0%	100.0%
Average Trip Length (miles)	6.4	8.5	8.9	9.8	10.1	9.9	8.6	9.2

Source: U.S. Federal Highway Administration, *Nationwide Personal Transportation Study*, 1977.

TABLE 8. Home-to-Work Trips by Mode and Residence.

Mode	Residence						
	Outside SMSA			Inside SMSA			All
	Under 5,000	Over 5,000	Average	Outside Central City	Within Central City	Average	
Private Vehicle (PV)							
Auto	71.7%	77.0%	74.3%	79.9%	75.8%	78.2%	76.8%
Truck/Rv/Van	20.9	14.9	17.9	10.8	7.4	9.1	11.9
Motorcycle/Moped	1.3	1.4	1.3	0.9	1.3	1.1	1.2
Bicycle	0.6	0.7	0.7	0.5	0.6	0.5	0.6
Total (PV)	94.5%	94.0%	94.2%	92.1%	85.1%	88.9%	90.5%
Public Transportation (PT)							
Bus	0.1%	0.5%	0.3%	2.2%	6.2%	4.1%	2.9%
Train	0.1	0.0	0.1	0.9	0.6	0.7	0.5
Streetcar	0.0	0.0	0.0	0.1	0.2	0.1	0.1
Subway	0.0	0.0	0.0	0.3	2.0	1.1	0.8
Taxi (commercial use)	0.0	0.1	0.1	0.0	0.5	0.2	0.2
Total (PT)	0.2%	0.6%	0.5%	3.5%	9.5%	6.2%	4.5%
Other							
Walk	4.8%	5.0%	4.9%	3.8%	5.2%	4.5%	4.6%
Other	0.5	0.4	0.4	0.6	0.2	0.4	0.4
Total (other)	5.3%	5.4%	5.3%	4.4%	5.4%	4.9%	5.0%
Total (All modes)	100.0%	100.0%	100.0%	100.0%	100.0%	100.0%	100.0%

Source: U.S. Federal Highway Administration, *Nationwide Personal Transportation Study*, 1977.

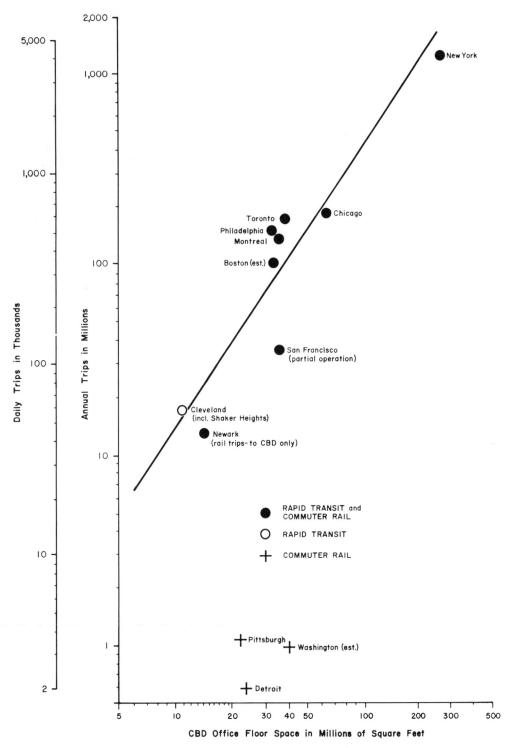

FIGURE 13. Trips by rapid transit and commuter rail related to downtown office floorspace. Source: Regional Plan Association, New York, NY.

TABLE 9. Workers, Mode of Travel to Work, 1969 & 1977.

Mode	1969	1977
Private Vehicles		
Automobile	82.7%	80.5%
Truck/RV/Van	8.1	12.5
Total (Private Vehicles)	90.8%	93.0%
Public Transportation	8.4	4.7
All Other	0.8	2.3
Total (All Modes)	100.0%	100.0%

Source: U.S. Federal Highway Administration, *Nationwide Personal Transportation Study*, 1977.

TABLE 10. Average Home-to-Work Trip Length by Mode of Travel.

	Trip Length (Miles)	
Mode	1969	1977
Private Vehicles		
Automobile	9.4	9.2
Truck	14.2	10.6
Motorcycle	9.2	10.3
Public Transportation		
Bus and Streetcar	8.7	7.2
Subway	12.1	9.7
Train	31.1	24.2
Taxi	4.5	3.3
All Modes	9.9	9.2

Source: U.S. Federal Highway Administration, *Nationwide Personal Transportation Study*, 1977.

TABLE 11. Transit Industry Equipment and Passengers.

Year	Heavy Rail Cars	Light Rail Cars	Trolley Coaches	Motor Buses[b]
	Passenger Vehicles Owned			
1940	11,032	26,630	2,802	35,000
1945	10,217	26,160	3,711	49,670
1950	9,758	13,228	6,504	56,820
1955	9,232	5,300	6,157	52,400
1960	9,010	2,856	3,826	49,600
1965	9,115	1,549	1,453	49,600
1970	9,338	1,262	1,050	49,700
1975	9,608	1,061	703	50,811
1976	9,714	963	685	52,382
1977	9,639	992	645	51,968
1978	9,567	944	593	52,866
1979	9,522	959	725	54,490
1980	9,693	1,013	823	59,411
1981	9,801	1,075	751	60,393
1982	9,867	1,016	763	62,114
1983[a]	9,943	1,013	686	62,093
	Passenger rides/trips (millions)[c]			
1940	2,382	5,943	534	4,239
1945	2,698	9,426	1,244	9,886
1950	2,264	3,904	1,658	9,420
1955	1,870	1,207	1,202	7,250
1960	1,850	463	657	6,425
1965	1,585	276	305	5,814
1970	1,881	235	182	5,034
1975	1,673	124	78	5,084
1976	1,632	112	75	5,247
1977	1,610	103	70	5,488
1978	1,706	104	70	5,721
1979	1,777	107	75	6,156
1980	2,108	133	142	5,837
1981	2,094	123	138	5,594
1982	2,115	136	151	5,324
1983[a]	2,167	137	160	5,422

[a]Preliminary
[b]Includes vans owned and leased by transit systems beginning in 1979.
[c]Total Passenger Rides from 1940 through 1979 based on individual transit data collection procedures. Unlinked Transit Passenger Trips beginning in 1980 based on data collection procedures defined by Urban Transportation Act, Section 15. Series not continuous between 1979 and 1980.
Source: American Transit Association, *Transit Fact Book*, 1985.

It must be concluded that by controlling the supply of CBD parking spaces and their cost, travelers should be encouraged to switch from the automobile to public transit in most cities. Studies conducted in places like Baltimore on transferring parking from the CBD to the fringe areas have not been conclusive, but several studies have shown that various types of fringe parking have worked well as a means of reducing automobile usage in favor of public transit.

BUS AND RAIL TRANSIT SYSTEMS

In any given city, each transit mode is best suited to particular conditions of population density, land use, employment, environmental characteristics and other related constraints. Each transit mode is best utilized when coordinated with other modes. Rail transit service, for example, is most efficient when coordinated with a feeder bus network. Each of the nine largest urban areas, as well as an additional 15 smaller urban areas, in the United States is served by at least two transit vehicle modes. Population centers like New York City and San Francisco–Oakland have a greater variety of transit vehicle types to choose from [10].

The most common mode of transit service in the United States is the motor bus which carries over 68 percent of all transit riders on some 2,486 transit systems operating motor buses [10]. As can be seen in Table 11, the number of buses grew from 59,411 in 1980 to 62,093 in 1983. At the same time, the number of passenger trips on buses grew to 5,422,000,000. The number of new transit vehicle deliveries for the years 1940–1983 is given in Table 12. Noteworthy of this table is the large number of motor bus deliveries as compared with other transit modes shown.

Vehicle Design and Performance

Dimensions and performance characteristics for four major modes of transit are summarized in Table 13.

DIMENSIONS AND CAPACITIES

Typical ranges for key dimensional and capacity characteristics of various transit vehicles are shown in Table 14. Passenger capacities of vehicles in the above table are determined by apportioning the net floor area of a vehicle between standing and seated passengers by selected ratios and areal allowances and levels of service and comfort based on management policy. Net floor area does not include operator and/or conductor space, on-board fare collection equipment, if any, and no standing passenger areas required for safe vehicle operation. Typical passenger space design range limits include [11]:

1. Widths of seats per passenger typically vary from 16 to

TABLE 12. New Transit Passenger Vehicles Delivered.

| Year | Light Rail Cars | Heavy Rail Cars | Trolley Coaches | Motor Buses[a] | | | |
				29 seats or fewer	30–39 seats	40 seats or more	Total buses
1940–44[b]	1,525	189	1,377	[c]	[c]	[c]	21,842
1945–49[b]	2,130	665	3,492	6,369	10,817	16,114	33,300
1950–54[b]	79	599	1,003	441	3,879	9,120	13,440
1955–59[b]	0	1,771	43	19	854	9,165	10,038
1960–64[b]	0	2,588	0	22	620	12,279	12,921
1965–69[b]	0	1,878	0	202	1,131	11,725	13,058
1970–74[b]	0	1,248	3	823	910	13,127	14,860
1975	0	127	1	419	128	4,714	5,261
1976	4	472	260	395	251	4,099	4,745
1977	62	506	198	549	308	1,580	2,437
1978	35	172	0	610	222	2,973	3,805
1979	70	94	141	408	130	2,902	3,440
1980	32	130	98	287	143	4,142	4,572
1981	188	276	0	153	171	3,735	4,059
1982	10	126	0	67	138	2,757	2,962
1983[d]	30	88	0	151	74	3,856	4,081

[a]Buses or bus-type vehicles only, excludes vans and passenger automobiles.
[b]Five-year totals
[c]Data not available
[d]Preliminary
Source: American Transit Association, *Transit Facts Book*, 1985.

TABLE 13. Characteristics of the Urban Transit Fleet as of December 31, 1980.

Charactersitic	Motor Bus	Heavy Rail	Light Rail	Trolley Coach
Number of Vehicles	59,411	9,693	1,013	823
Number of Vehicles Equipped with Air Conditioning	42,891	4,690	132	271
Number of Vehicles Equipped with Two-Way Radios	38,469	7,198	315	594
Number of Vehicles Equipped with Wheelchair Lifts or Ramps	6,133	0	0	110
Average Age, Years	8.8	18.0	28.4	9.1
Average Length	38'3"	58'4"	52'8"	39'6"
Average Number of	45.6	53.6	50.1	47.4
Propulsion Power	Diesel: 96.1% Gasoline: 3.3% Propane: 0.6%	Electricity	Electricity	Electricity
Length/Gross Weight of a Typical Vehicle	40' 34,000 lbs.	60' 81,000 lbs.	47' 56,000 lbs.	40' 33,500 lbs.
Average Operating Speed in Revenue Service	11.8 mph	19.8 mph	9.6 mph	8.3 mph

Source: American Transit Action, *Transit Facts Book*, 1981.

TABLE 14. Ranges of Geometric Dimensions and Passenger Capacities of Transit Vehicles.

Transit Vehicle Type	Length[a] (ft)	Width[a] (ft)	Height[b] (ft)	Design Capacity of Single Unit			Design Capacity of Maximum Train	
				Seats	Standees	Total Passengers	Cars	Total Passengers
Van	15–18	5.5–7.2	7–9	10–16	—	10–16	—	—
Minibus	18–25	6.5–8.0	7.5–10	15–25	0–15	15–40	—	—
Transit bus								
Single Unit	25–40	7.5–8.5	9–11	30–55	10–75	40–115	—	—
Articulated	54–60	8.0–8.5	9.5–10.5	35–75	30–125	95–185	—	—
Double-deck	30–40	7.5–8.5	13–14.5	50–85	15–50	90–130	—	—
Streetcar								
Single Unit	40–55	6.5–9.0	10–11	20–60	40–80	75–130	3	225–440
Articulated	60–90	7.5–9.5	10–11	30–85	120–200	100–275	3	300–825
Rail Transit Car								
Steel wheel	45–75	8.5–11.0	10–13	40–85	50–250	100–330	8–10	1000–2700
Rubber-tired	48–60	8.0–9.5	11–12	35–55	70–130	110–170	9	1000–1500

[a] Lengths and widths shown are external body dimensions.
[b] Heights are from pavement or top of rail to roof.

Source: *Lea Transit Compendium*, Vol. 11 (1975), Nos. 5, 6, 9; Vol. III (1976–77), Nos. 5, 6, 9; M. S. Sulkin and D. R. Miller, "Some State-of-the-Art Characteristics of Rubber-Tired Rapid Transit," 27th California Transportation and Public Works Conference, Pasadena, California, 1975; various databooks and specifications.

24 inches with 17 to 20 inches typical of vehicles used in local and high-density service.
2. The area per seated passenger typically varies from 2.9 to 5.7 square feet with 3.2 to 4.2 square feet appropriate for local service vehicles.
3. The minimum area per standing passenger under easy standing conditions usually varies from around 2.5 to 4.0 square feet with the area under crush peak period conditions being as little as 1.6 square feet.
4. Usually the minimum aisle widths between transversely positioned seats range from 21 to 31 inches.
5. On most transit vehicles, doorway width per passenger lane for boarding and alighting vary from 22 to 30 inches.

PERFORMANCE

Maximum rates of acceleration, deceleration, and rate of change of acceleration and deceleration (jerk) in normal transit service must be related to the tolerance of a standee who is not able to hold on to any kind of hand grip. The upper limits of acceleration and deceleration rates are usually from 3.0 to 3.5 mph/s^2 with 20 mph/s^2 being the preferred maximum jerk rate [11]. Figure 14 provides speed-time-distance curves for a typical urban transit bus, indicating maximum acceleration, cruising, and deceleration rates. Figure 15 provides the same data for a rail rapid transit train.

Since transit vehicles start and stop frequently, the traction and braking efforts are an important design consideration. The adhesion between vehicle wheels and the roadway or guideway surface determine the extent to which the tractive effort or braking force can be utilized. The maximum force that can be transferred between wheels and guideway surface is expressed by the formula [11]:

$$F = \mu \times W_n$$

where

F = friction (adhesion) force; same units as W_n
μ = dimensionless friction or adhesion coefficient
W_n = force normal to the guideway surface

Design values for rubber-tired vehicles on dry roadway surfaces or guideways usually range from 0.73 to 0.79 with vehicle tires having little or no wear. Values for rail cars normally range from 0.25 to 0.30 for steel wheels on dry steel rails. When road or rail surfaces are wet, lower design friction coefficients have to be assumed to allow for slippery conditions. Values can drop as low as 0.27 and 0.18 for rubber-tired vehicles and rail cars, respectively. In addition to the types of materials used in the travel-way surface and wheels, the textures of the contact surfaces of these materials, the condition of the roadway or guideway (including slipperiness when damp or wet) and of the wheels, loaded vehicle weight, grades, and other factors influence adhesion characteristics.

The speed terms associated with transit operations are as follows [11]:

- *Maximum speed*—the maximum vehicle performance capability with seated load on a level alignment.
- *Platform speed*—the service speed over a route, including allowance for passenger stops and traffic delays.
- *Schedule speed*—the average speed on a route when time spent at the terminals for layover is also included.

Table 15 shows some typical speed ranges for various types of transit vehicles. Table 16 presents weight ranges for various types of transit vehicles as well as the ranges of weight/unit design capacity ratios.

Urban Transit Routes and Service

Transit planning should reflect the specific needs and operating requirements of each urban area. The best possible service should be provided to the greatest number of people within the governing economic constraints. Planning must balance the amount and type of services provided with the net costs of providing the service. This trade-off underlies all service planning decisions. Services should be carefully related to existing and potential markets and concentrated in heavy travel corridors with the greatest service frequency and route coverage on the approaches to the city center. Route structure should be clear and understandable, and service duplication should be minimized.

KINDS OF ROUTES

Geographically, several kinds of routes can be identified [11]:

1. *Radial routes* radiate from the CBD and usually carry the largest numbers of passengers per unit length of route. Most rail routes can be classified as radial.
2. *Circumferential routes* provide service between different outlying areas without requiring travel through the CBD. By intercepting radial routes, they also serve as feeder/distributors for the latter.
3. *Crosstown routes* are usually short routes with fairly straight alignments that run tangential to the CBD and perpendicular to radial routes.
4. *Feeder routes* connect areas outside the transit corridors to radial routes.
5. *Shuttle routes* provide service between two major traffic generators, such as between the CBD and an outlying parking facility or railroad terminal.

Transit route networks vary widely in complexity and in their interrelationships with other transportation networks. Network concepts offer the user a wide variety of destinations, usually via a single line that may incorporate a bal-

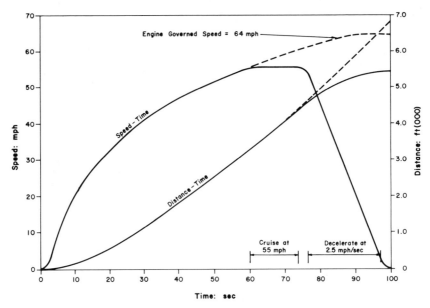

FIGURE 14. Speed-time-distance performance curves for typical urban transit bus. Note: Data based on full seated load on level, tangent road, with all accessories, including air conditioning, in operation. Source: GMC Truck & Coach Division, 1978.

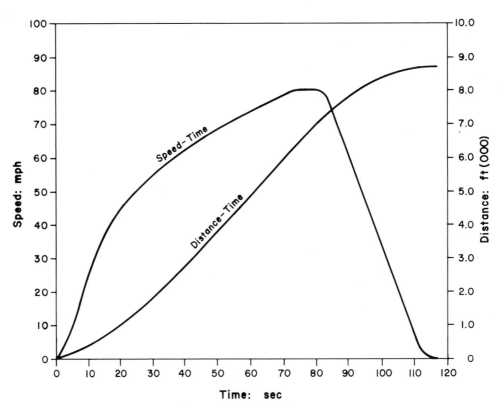

FIGURE 15. Speed-time-distance performance curves for rail rapid transit. Note: Data based on four-car San Francisco Bay Area Rapid Transit (BART) District train with full seated load on level, tangent track. Source: Parsons, Brinckerhoff-Tudor-Beuhtel, general engineering consultants to BART.

TABLE 15. Typical Vehicle Velocities and Stop Spacings.

Transit Vehicle and Service Type	Maximum Performance Speeds (mph)	Platform Speeds (mph)	Linear Stop Spacing		
			CBDs (ft)	Non-CBD	
				Traditional Practice (ft)	Some Modern Systems With Longer Stop Spacings (ft)
Urban bus					
Local	50–65	8–14	500–1000	500–800	1000–1500
Limited stop	50–65	12–18	500–1000	1200–3000	2000–5000
Express	50–65	16–32	a	4000–30,000	5000–15,000
Streetcar, local	40–60	8–15	500–1000	500–800	1000–1500
Light rail transit	50–65	15–35	1000–2000	—	2000–5000
Heavy rail transit	50–70	15–35	1000–2500	1700–3500	3500–8000
Regional rapid transit	70–85	35–55	2000–3000	—	6000–30,000

^aUsually stops at only one or two terminals in or adjacent to CBD.
Source: Institute of Transportation Engineers, *Transportation and Traffic Engineering Handbook*, (W. S. Homburger, ed.), Prentice-Hall, Inc., Englewood Cliffs, NJ, 1982.

anced approach that utilizes a combination of heavy rail, light rail, buses, and other types of transit systems as appropriate. Planning for new transit networks should be influenced heavily by service philosophies and policy decisions with criteria for planning and design evolving from policy determinations regarding the type, extent, frequency, duration, and nature of service to be provided. Consideration has to be given to service area coverage, demand pattern, transit stop requirements, vehicle and system performance capabilities, right-of-way and facilities site availability, and interface with other transit systems.

TYPES OF SERVICE AND COVERAGE AREA

Service along most urban transit routes usually takes the form of local service which serves all stops along a route. Service which does not serve all stops along a route is called either limited-stop or express service and becomes feasible when radial routes extend more than 3 miles from the CBD, and when patronage volumes are large enough to economically sustain such service. Often, limited-stop and express service is warranted only during peak periods, although special service, such as to airports, will be scheduled all day. Limited-stop service operates along city streets with stops mainly at major transfer points beyond the CBD. Express service involves even faster operation, often on freeways and other major highways parallel to the local route that it supplements.

To reach a transit stop or station, patrons will typically travel up to maximum distances shown in Table 17. A few patrons may travel considerably farther. Excluding elderly and handicapped persons, acceptable walking distances from origin to boarding stop and from alighting stop to destination is usually between 0.3 and 0.4 miles.

Area coverage, or accessibililty, depends upon the spacing of routes. Standards, such as those shown in Table 18, may be used in the design of adequate transit networks.

TABLE 16. Ranges of Weight of Transit Vehicles.

Transit Vehicle Type	Empty Weight (lb × 1000)	Weight/Unit Design Capacity (lb/psgr)
Van	5–7.5	200–650
Minibus	7–17	200–700
Transit bus		
Single unit	14–26	175–340
Articulated	28–36	160–360
Double-deck	15–28	200–230
Streetcar		
Single unit	36–52	320–575
Articulated	45–110	250–600
Rail transit		
Steel wheel	35–90	200–500
Rubber tired	44–75	275–450

Source: *Lea Transit Compendium*, Vol. II (1975), Nos. 5, 6, 9; Vol. III (1976–1977), Nos. 5, 6, 9; various other databooks and specifications.

TABLE 17. Typical Maximum Distance Traveled to Reach Urban Transit Stops and Stations.

Access Mode	Most Patrons (mi)	Some Patrons (mi)
Walk	0.4–0.6	0.6–1.0
Bicycle	1.0–2.0	2.0–3.0
Feeder transit; motorcycle	2.0–4.0	4.0–8.0
Auto		
Kiss-ride; taxi	3.0–4.0	4.0–6.0
Park-ride	4.0–6.0	6.0–10.0

Source: Henry D. Quinby, "Coordinated Highway-Transit Interchange Stations," *Origin and Destination: Methods and Evaluation*, Highway Res. Rec. 114, Highway Research Board, Washington, DC, 1966, pp. 99–121; various other reference data.

These standards are usually modified to provide better accessibility under existing conditions, i.e., closer spacing of routes where terrain hampers walking. They do not apply to special services provided for the elderly or handicapped.

The following service planning guidelines reflect current practices in the United States and Canada [12]:

- *Service Criteria*—Bus service should be provided where population density exceeds 2000 per square mile and ridership exceeds 20 to 25 passengers per bus hour on weekdays, 15 on Saturdays, and 10 on Sundays. Route continuity and transfer requirements may lower these values.
- *Service Area*—Overall service areas are usually defined by legislation. Where population density within this area exceeds 4000 per square mile, or three dwelling units per acre, 90 percent of the residences should be within 0.25 mile of a bus line. Where population density ranges from 2000 to 4000 per square mile, 50 to 75 percent of the population should be within 0.5 mile of a bus line.
- *Route Structure and Spacing*—Bus routes should fit major street patterns. Basic grid systems are appropriate where streets are in a grid pattern; radial or radial-circumferential systems are applicable where radial or irregular street patterns exist. Bus routes should be spaced at approximately 0.5 mile in urban areas and 1 mile in low-density suburban areas. Closer spacing should be provided where terrain inhibits walking.
- *Route Directness and Length*—Circuitous routings should be avoided. A route not more than 20 percent longer than the comparable trip by car is desirable. Route deviation should not exceed 8 minutes per round trip (based on at least 10 customers per round trip). There should usually not be more than two branches per trunk-line route, and never more than four. Routes should be as short as possible to serve their markets, and generally should not exceed 25 miles or 2 hours per round trip. Where long routes are required, travel times should be increased because of the difficulty in maintaining schedules. Through routes, which can save costs and reduce congestion in the CBD, should be provided wherever possible.
- *Route Duplication*—There should be one route per arterial except on approaches to the CBD or a major transit terminal. A maximum of two routes per street (or two branches per route) is desirable. Express service should utilize freeways or expressways to the maximum extent possible. Express and local services should be provided on separate roadways, except where frequent local service is provided.
- *Service Period*—Regular service should be provided on weekdays from about 6 A.M. until 11 P.M. Suburban feeder service should be provided on weekdays from 6 A.M. until 7 P.M., or only during the morning and afternoon rush periods. In larger cities, 24 hour service should be provided on selected routes. Saturday and Sunday service should be provided on principal routes; however, Sunday service may be optional in smaller communities.

Bus Priority Systems

The most common types of bus priority in use, both freeway and arterial, are as follows:

1. With-flow bus lanes
2. Contra-flow bus lanes
3. Reserved bus lanes on freeways
4. Priority access to freeways
5. Bus-only streets
6. Busways
7. Priority at traffic signals

There are other less spectacular methods of giving buses priority but only those listed above are discussed.

TABLE 18. Accessibility Standards for Urban Transit.

Population Density (thousands/mi²)	Average Route Spacing		Route Density (route-mi/mi²)
	Radial Routes (mi)	Circumferential Routes (mi)	
Over 12	0.40	0.60	4.00
10–12	0.50	0.75	3.33
8–10	0.60	0.90	2.67
6–8	0.80	1.20	2.00
4–6	1.00	1.50	2.67
2–4	1.00	—	1.00
Under 2	2.00	—	0.50

Source: Adapted from Massachusetts Bay Transportation Authority, *Service Policy for Surface Public Transportation*, Boston, 1975.

WITH-FLOW LANES

With-flow bus lanes are the most common type of bus priority in use. As the name implies, they are reserved for buses traveling in the same direction as that of the normal traffic. They are usually located along curbs; however, in a few cases, they may be in the street medians. This type of priority is applied mostly to important roads in town centers and to main radial roads. In general, curb lane installations should be considered when there are at least 30 to 40 buses and 1,200 to 1,600 people one way in the peak hour. Median lanes require wide streets, passengers are required to cross active traffic lanes to reach bus stops, and left turns must be prohibited, or controlled, to minimize interference with buses and bus passengers. In general, they should not be installed unless there are a minimum of 60 to 90 peak-hour buses serving 2,400 to 3,600 people using each median lane.

CONTRA-FLOW LANES

Contra-flow lanes enable buses to operate opposite to the normal flow of traffic and are usually, but not always, in a one-way street system. Buses using contra-flow lanes are separated from other traffic flows. Thus they are removed from the conflicts with other vehicles and so are unaffected by peak-hour congestion (or backups) at signalized intersections. The lanes are relatively self enforcing, have high visibility, and can provide more direct bus routings, thereby reducing bus travel times and distances. They likewise allow two-way bus service on one-way downtown and radial streets, especially where it is desirable to provide direct service to major passenger generators. Contra-flow lanes should be considered for installation when there are at least 40 to 60 buses carrying 1,600 to 2,400 people one way in the peak hour. Where lanes are in effect all day, at least 400 buses should use the lanes.

RESERVED BUS LANES ON FREEWAYS

Reserved freeway lanes for buses provide a cost-effective approach to bus priorities in radial highway corridors with peak-hour congestion and heavy bus volumes. They apply freeway traffic operations and control techniques to reserved lanes for buses and/or other designated vehicles (such as emergency vehicles, trucks, or high-occupancy cars). They involve minimum physical construction; and, they can speed bus service where stations or intermediate access are not required.

Reserved bus lanes should be provided only where the total number of bus passengers in the heavy direction of flow is equal to or greater than the "typical" lane-carrying capacity of automobile passengers.

Contra-flow or "wrong way" bus lanes using a portion of the roadway that serves relatively light opposing freeway traffic flow will not reduce peak-directional highway capacity or efficiency. Contra-flow bus lanes should be applied only on roadways with more than four lanes where peak-hour traffic is highly unbalanced.

The minimum number of peak-hour buses to warrant a contra-flow lane can be derived from the following approximate relationship [13]:

$$B = V_o(t_1/t_2)(o_1/o_2)$$

in which

B = minimum number of buses per hour
V_o = total traffic in peak hour in the off-peak direction
t_1 = time lost per vehicle in the off-peak direction
t_2 = time savings per bus using the bus lane
o_1 = persons per vehicle in the off-peak direction
o_2 = persons per bus using the bus lane

This equation states that the minimum number of buses needed to warrant a contra-flow bus lane must be equal to or greater than the number of automobiles in the off-peak direction weighted by the ratio of car-to-bus passenger occupancies for the off-peak and peak directions, respectively, and the ratios of expected car and bus travel times. Results of the equation, based on the speed-volume relationships identified in the *1965 Highway Capacity Manual* [14] are given in Table 19 and shown in Figure 16. On eight-lane freeways where car volumes in the off-peak direction can be accommodated in two lanes, a buffer lane should separate bus and car travel and provide bus passing break-down opportunities (Figure 17).

PRIORITY ACCESS TO FREEWAYS AND OTHER FACILITIES

Preferential bus entry to freeways can be provided at locations that are controlled by ramp metering. Special traffic signals on entrance ramps allow only those vehicles to enter the freeway that can be accommodated without reducing mainline speeds. Ramp metering with bus bypass lanes is especially applicable in corridors with relatively low peak-hour bus passenger demands and frequent peak-hour congestion. It enables buses to enter and leave freeways for passenger loading with minimum delay.

Design-year peak-hour bus volumes ideally should exceed 10 buses, with an average time savings of 1 minute per bus. However, in view of the low installation costs (usually pavement markings and signs), bus priorities could be installed with lower volumes. The bus bypasses could also be used by car pools and trucks at selected locations.

Exclusive bus ramps can be provided by constructing an exclusive bus ramp, or by converting an existing auto ramp to exclusive bus use. The choice will depend on balancing the costs of new ramps against the impacts of auto ramp closures on freeway and arterial street traffic operations.

Bus bypass of queues should be considered wherever there are more than 60 design-year peak-hour buses, and where each bus is expected to save at least 5 minutes. Adequate sight distances at the beginning and end points are essential. Conditions should allow providing bus priority facilities

TABLE 19. Approximate Minimum Bus Volumes Required for Contra-Flow Bus Lane.[a]

Total Peak Direction Demand (veh/hr)	Off-Peak Direction Volume (veh/hr)												
	900	1200	1500	1800	2100	2400	2700	3000	3300	3600	3900	4200	4500
3600	34	41	90	135	205	288	365	495	693	1701	3130	5355	7594
3900	14	22	36	54	82	115	146	198	277	680	1252	2142	3038
4200	10	17	28	42	63	89	112	152	213	524	963	1648	2337
4500	8	13	21	32	48	68	86	116	163	401	736	1260	1787
4800	5	9	15	22	33	46	58	79	111	272	501	857	1215
5100	4	6	10	15	23	32	41	55	77	189	342	595	844
5400	2	3	5	8	12	17	22	30	42	102	187	320	454
6300	1	1	2	3	4	6	8	11	15	37	68	117	166
7200	1	1	1	2	3	4	5	6	9	23	42	72	103
8100	—	—	—	1	1	2	2	4	5	13	23	40	57

[a]Assumes an occupancy factor of 1.5 and 50 for automobiles and buses, respectively. The domain of practical application involves hourly bus volumes ranging from about 40 to 200. Volumes of under 40 buses per hour do not usually warrant contra-flow lanes. Volumes larger than 200 buses per hour exceed most urban bus fleets and fall outside the domain of practical application.
Source: H. S. Levinson, et al., *Bus Use of Highways: Planning and Design Guidelines*, National Cooperative Highway Research Program, Report No. 155, Transportation Research Board, Washington, DC, 1975.

without reducing highway capacity through the bottleneck. Where bus bypasses at toll stations are provided, there should be at least 10 buses per hour, each saving 5 minutes or 15 buses per hour, each saving 2 to 3 minutes [15]. Adequate approach reservoir capacity is essential to minimize car queues across entry to the bus lanes.

BUS-ONLY STREETS

As a rule, bus streets should provide priority access to the CBD and major activity centers in the city center. They should also serve major bus flow concentration that results when individual lines converge onto a single street. Bus streets should be conducive to both pedestrian and bus mall development in the CBD. They should penetrate major retail and office centers and provide reasonable access to shops and offices. There should be parallel routes for displaced automobile traffic with no automobile parking garages located on bus streets to help discourage automobile traffic. Access should be allowed on bus streets only to essential deliveries and emergency services.

The various aims behind the creation of a bus-only street are: (1) to speed up buses by removing other traffic from the street; (2) to create sufficient carriageway space so that buses can stop and wait, if necessary, without hindrance to other buses (bus terminals are often located on bus-only streets, so that the street becomes a mini-bus station); (3) to assist pedestrians to cross the road more easily and safely; and (4) to improve the environment. Bus streets may be warranted where high bus volumes traverse narrow streets as part of downtown redevelopment proposals. They may include the last block of an arterial street, a dead-end street at the end of several bus routes, a "bus loop" necessary to change directions at a major bus terminal, a downtown bus mall, and bus circulation through an auto-free bus zone. Typical applications include:

- terminal approach
- bus loop
- short connector links
- bus-pedestrian mall
- auto-free zone

At least 200 buses should use the bus street daily in each direction of travel and a minimum of 20 to 30 buses and/or 800 to 1,200 people should use the bus street in each direction during the peak hour [15].

Reserving streets for public transport can improve bus service. However, care must be taken to select streets that provide maximum advantage to public transport without excessively hindering other traffic and access to adjacent premises.

BUSWAYS

Busways are separated roadways, or freeways, designed for exclusive or predominant use by buses. Busways may provide (1) line-haul express transit service to the city center; (2) feeder service to rail transit lines; and, (3) short bypasses of major congestion points. They should segregate buses from other types of traffic and should include ancillary passenger-bus interchange and parking facilities. They may be constructed at, above, or below grade, either in separate rights-of-way, or within freeway corridors. They may be designed as "open" systems in which buses can enter or leave at intermediate points, or as "closed" systems in which buses operate over the entire length of busway. Consideration should be given to developing line-haul busways wherever the following basic conditions are met [15]:

1. Design-year urban population should exceed 750,000; CBD employment, 50,000; and peak-hour CBD cordon person-volumes, 35,000 (symmetrical city).
2. There should be at least 40 buses and 1,600 passengers potential to the busway in the design year. (Where railroad rights-of-way are utilized, these volumes could be reduced under certain circumstances.)
3. Buses utilizing the busway should save at least 5 minutes over alternative bus routings.
4. Current highway demands in the corridor exceed capacity; and environmental, social cost, and/or traffic conditions preclude providing additional road capacity.

Downtown busway development should be considered where peak-hour bus speeds are less than 5 to 6 mph, where the congested area extends for more than a mile, where surface-street priority options cannot be effectively developed, and where potential bus loadings are high.

With cutbacks in both operating subsidies and capital funding, busways are becoming more popular with transit agencies as a solution to a variety of problems. They can be converted from existing city streets or roadways, from abandoned railroad tracks or streetcar routes, or may be constructed especially for this purpose. Regardless, they require less right-of-way and are cheaper to construct.

The alternatives to busways are expensive conventional transit facilities. Rail transit facilities require large investments in right-of-way acquisition and construction, as well as high operation and maintenance costs.

Two busway systems recently went into operation in the

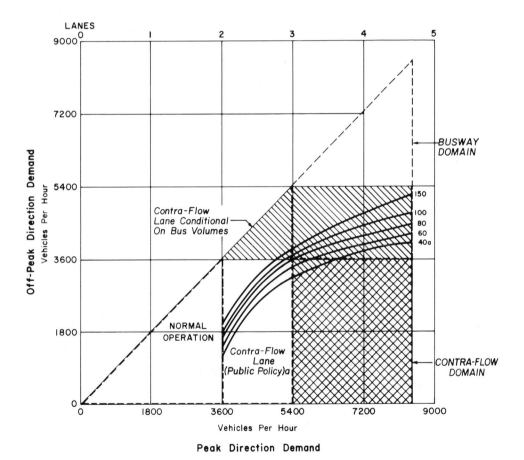

FIGURE 16. Contra-flow bus lane concept, 6-lane freeway. Note: Curved lines show approximate minimum bus volumes for contra-flow lane to minimize opposing traffic. a. Bus volumes under 40/hr meter ramps. Source: H. S. Levinson, et al., *Bus Use of Highways: Planning and Design Guidelines*, National Cooperative Highway Research Program Report 155, Transportation Research Board, Washington, DC, 1975.

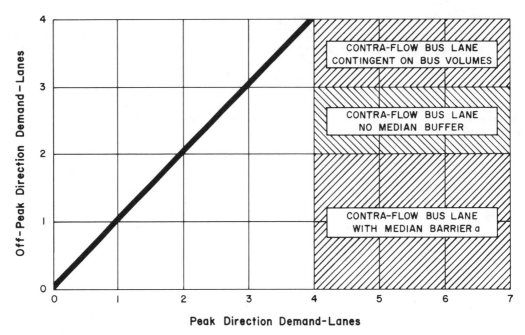

FIGURE 17. Contra-flow bus lane concept, 8-lane freeway. Note: a. Reversible car lanes may apply when bus volumes are less than 40. Source: H.S. Levinson, et al., *Buse Use of Highways: Planning and Design Guidelines*, National Cooperative Highway Research Program Report 155, Transportation Research Board, Washington, DC, 1975.

Pittsburgh area while others are being planned or are under construction in Houston, Seattle, Washington, and Baltimore [16]. The 11 mile San Bernadino Freeway Busway between Los Angeles and suburban El Monte was built in the 1970s and has had a steady increase in patronage ever since. A one mile extension is presently under construction. When the 11 mile busway was completed in mid 1974, it carried over 10,000 bus passengers daily. In 1976, car pools with three or more occupants were allowed and bus ridership had increased to 20,000 persons per day. Today, bus ridership is 22,000 per day. In addition, 16,000 people riding in more than 5,000 cars use the busway. Bus ridership is expected to go to 60,000 people per day by the year 2000 [16].

PRIORITY AT TRAFFIC SIGNALS

Where bus flows do not justify such priority treatments as previously discussed, priority can be given to buses at some signalized intersections by adapting the signal timing and phasing to decrease the delay to approaching buses. Adjustments of traffic signal timing at intersections to facilitate bus flow can substantially reduce average bus waiting times and can improve operating economy; bus delays at traffic signals usually represent 10 to 20 percent of overall bus trip times and nearly one-half of all delays.

Transit Stops and Terminals

Terminals and interchange facilities are best suited to places where large volumes of passengers board or transfer daily and are generally provided in conjunction with rail rapid transit, busways, and contra-flow freeway bus lanes. They are most applicable in large urban areas where the population exceeds 750,000, downtown employment is greater than 50,000, and other travel conditions help justify terminal feasibility.

BUS STOPS

There are many factors that affect choice of location. The main objectives that any location should include are: maximum passenger convenience, maximum safety for passengers boarding and alighting, and, minimim interference with traffic by buses entering and leaving stop zones. Frequency of stops should generally not exceed 8 to 10 per mile. Superfluous stops should be avoided since they add delays.

TERMINAL REQUIREMENTS AND CONSIDERATIONS

Although the requirements of the passenger, operator, and community all must be considered in a well-designed terminal, the needs of the passenger are the most important.

Basic requirements to be satisfied when considering a terminal are minimum transfer time and walking distance as well as convenience. Not only should a facility be efficient for the operator, but it should also be attractive to both the passenger and the community.

To be functional and financially feasible, a terminal facility must achieve maximum integration of various transport modes as well as handle large volumes of passenger boardings and transfers daily. Facilities should generally be provided in conjunction with rail rapid transit lines, busways, and contra-flow freeway bus lanes which make them more applicable in urban areas with populations in excess of 750,000 and downtown employment of 50,000 or higher. Other planning considerations for potential facilities are [15]:

1. Express bus or rail transit routes with high passenger concentrations and service frequency are desirable.
2. Route structure and land-use patterns make it feasible to break bus routes and/or provide crosstown and circumferential services.
3. Patrons rely on cars and have access to outlying express transit stations.
4. Downtown access and parking are inhibited by high cost, peak-hour congestion, and inadequate approach road capacity.

CAPACITY CONSIDERATIONS

The capacity of a stop or station is the maximum number of vehicles per hour that can use the facility without extensive waiting. The maximum capacity of a bus stop (curbside or in a station) is expressed by [11]:

$$F_{max} = \frac{1800}{T_d} \text{ and } H_{min} = 2T_d$$

where

$T_d = A \times T_a + B \times T_b + T_c$ two-way flow at busiest door

$T = \max \begin{Bmatrix} A \times T_a \\ B \times T_b \end{Bmatrix} + T$ one-way flow through doors

F_{max} = max frequency (capacity), buses per hour
H_{min} = min average headway, in seconds
T_d = average dwell time per bus, in seconds

TABLE 20. Average Boarding and Alighting Intervals for Transit Vehicles.

Operation	Physical Conditions	Operational Conditions	Seconds Passenger/Lane[a]
Boarding	High-level platform (rapid transit)	Fares paid at fare gates	1.0
		Fares paid off vehicle (at fare gates or by passes)	2.0
	Low-level or no platform (buses and streetcars)	Single-coin or toaken fare paid on vehicle	3.0[b]
		Multiple-coin fare paid on vehicle	4.0[b]
		Zone fares prepaid; tickets registered on vehicle	4.0–6.0
		Zone fares paid on vehicle	6.0–8.0[b]
Alighting	High-level platform	No ticket checking at vehicle doors	1.0
	Low-level or no platform	No ticket checking at vehicle doors	1.0
		Ticket checking or issue of transfers at vehicle doors	2.5–4.0

[a] A lane represents one file of persons, 22–44 in. wide. Assumes that all lanes are used equally; however, allowance is usually made for the fact that whereas some lanes are used to capacity, others operate below that flow rate.
[b] Where "exact fares" are required and drivers do not make change, times may be somewhat less.
Source: H. S. Levinson, et al., *Bus Use of Highway: Planning and Design Guidelines*, National Cooperative Highway Res. Program, Report 155, Transportation Research Board, Washington, DC, 1975.

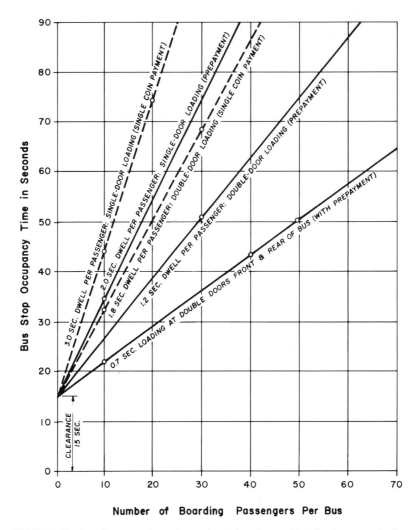

FIGURE 18. Boarding capacity and stop dwell time. Source: H. S. Levinson, et al., *Bus Use of Highways: Planning and Design Guidelines*, National Cooperative Highway Research Program Report 155, Transportation Research Board, Washington, DC, 1975.

A, B = no. of alighting/boarding passengers
T_a, T_b = average alighting/boarding time, seconds/passenger
T_c = sum of other delays at the stop

The capacity depends more on the loading positions, since interference from other buses can reduce the effectiveness of multiple berth stops. The number of persons per hour per bus per channel, rather than the number of buses per hour, is the key variable to maximize. When dealing with trains in rail transit stations, capacity also becomes a function of train length, acceleration, deceleration, and train control. Average boarding and alighting times for transit vehicles are given in Table 20.

Dwell time, as used in the capacity formula, is the standing time of transit vehicles at a passenger stop. Components that make up dwell time are boarding and alighting times and the time required for doors to open and close (usually 1 to 4 seconds for each). When a delay occurs, it also includes time waiting to enter stop when the position is not clear for entering and/or time required to move from stop area into moving bus lane. Typical effects of vehicle door capacities and passenger loading intensities on bus dwell times at stations are shown in Figure 18.

TABLE 21. Observed Peak Hour Passenger Volumes on Urban Transit Buses, United States.

City	Location	Buses/Hour	Headway (s)	Passengers/Hour in Peak Direction	Passengers/Bus
City Streets					
San Francisco	Market Street	155[a]	23	9,900	58
New York	Hillside Avenue	170[a]	21	8,500	50
Philadelphia	Market Street	143[a]	25	8,300	58
Washington	K Street NW	130	28	6,500	50
Chicago	State Street	151[a]	24	6,100	40
Freeways					
New York[b]	I-495	490	8	21,600	44
Washington[c]	Shirley Highway	110	33	5,500	50
Chicago	North Lake Shore Drive	80	45	4,000	50
New York	Long Island Expressway	89	40	3,560	40
Philadelphia	Schuylkill Expressway	78	46	2,800	36
Tunnel, bridges					
New York	Lincoln Tunnel	735	5	32,500	44
San Francisco	Oakland Bay Bridge	360	10	14,900	41
Philadelphia	Ben Franklin Bridge	137	26	5,065	37
New York	George Washington Bridge	108	33	4,245	39
Washington	Memorial Bridge	100	36	4,020	40

[a]Buses use more than one lane.
[b]Location is on New Jersey side of Hudson River.
[c]Location is on Virginia side of Potomac River.
Source: H. S. Levinson, et al., *Bus Use of Highways: State of the Art*, National Cooperative Highway Research Program, Report 143, Transportation Research Board, Washington, DC, 1973; Data from University of California, Institute of Transportation Studies.

TABLE 22. Observed Peak-Hour Passenger Volumes on Rapid Transit Systems, 1978–1980.

City	Location	Trains/Hour	Headway (s)	Length of Trains (ft.)	Passengers/Hour in Peak Direction	Passengers/Train
New York (IND)	53rd St./East River	29	124	600	48,070[a]	1660
Paris (Metro)	Gare de l'Est	37	97	294	40,160	1085
New York (IRT)	59th St./Lexington Ave.	24	150	510	35,120[a]	1460
Toronto	Bloor/Wellesley	30	120	450	31,784	1060
Montreal	Mount Royal Station	25	144	492	24,000	960
London (Tube)	Liverpool St. Station	30	120	413	23,000	770
Hong Kong	Nathan Road	20	180	300	20,000	1000
London (Underground)	Baker Street Station	28	129	423	17,000	610
Chicago	35th/Ryan	21	171	384	16,500	785
Stockholm	Skanstull	30	120	463	14,765	490
Boston	Andrew Station	17	212	280	13,045	770

[a]That these figures do not represent capacity is illustrated by the fact that higher volumes of passengers were carried before 1960. At the 53rd/East River location, 61,400 passengers were counted in the peak hour, on Lexington Avenue Line 44,500.
Source: Institute of Transportation Engineers, *Transportation and Traffic Engineering Handbook*, (W. S. Homburger, ed.), Prentice-Hall, Inc., Englewood Cliffs, NJ, 1982.

System Performance

HEADWAY

Limiting conditions both upstream and downstream of a particular route section can determine headway along that particular route section. Tables 21 and 22 give headways for urban buses and rail rapid transit, respectively. Maximum headway is determined by policy. Headways are usually divisible into 10, such as 10 or 20 minutes, so patrons can easily remember them with minimum reference to timetables. Headway is the schedule interval of time between successive transit buses, or trains, measured from the front end of the leading bus or train to the front end of the following bus or train as they pass a point.

Minimum achievable headways depend on such factors as the vehicle speed, braking rates, degree of safety, system response time, train length (if applicable), station dwell time, and random influences where right-of-way is not exclusive or controlled. Station stops are often the major physical factor in limiting minimum headway on rail-transit systems.

PASSENGER CAPACITY

The passenger carrying capacity of a bus or rail transit line is often the operational parameter of most interest to planners. Capacity is the number of people that can be moved past a fixed point per unit of time and may be expressed as follows [17]:

$$C = \frac{3600\ p}{h}$$

where

C = capacity in passengers per hour
p = number of passengers per train
h = headway in seconds

In simple terms, the capacity of a bus or rail line depends on bus or train size and frequency, but is actually a complex function involving:

- vehicle size
- number of passengers (seated and standing) per vehicle
- number of vehicles per train (rail only)
- vehicle speed
- track curvature and grade limitations (rail only)
- roadway grade limitations (bus only)
- acceleration and deceleration rates
- braking distance
- frequency of station stops
- dwell time at stations
- right-of-way restrictions
- system reaction time
- practical allowance for operational tolerances

Tables 21 and 22 give peak-hour, peak-direction passenger volumes for buses and rail transit, respectively. Critical speed of a transit vehicle is the one that permits minimum headway and maximum capacity. As speed increases beyond the critical value, headway increases and capacity decreases. In the case of rail vehicles, the more cars per train, the higher the critical speed.

SPECIALTY MODES AND SYSTEMS

Transit modes discussed in this section are not as familiar to the average transit rider, mainly because there are far fewer vehicles and systems in operation than the familiar ones mentioned previously and they carry only a fraction of the total urban transit ridership. Table 23 compares various statistics of several different transit modes from the popular motor bus to the less familar types of service such as Automated Guideway Transit.

Specialty systems can be fixed route, flexible route, or demand responsive and are usually tailored to meet specialized local requirements. Examples of modes used would include the urban ferry boat, cable car, inclined plane, aerial tramway, and others not listed in Table 23 such as paratransit vehicles. Services offered to riders of speciality systems encompass a broad range from automated people movers at major activity centers to special buses and tour services that handle community and neighborhood transportation needs such as car pools and dial-a-bus service.

Urban Ferry Boats

As can be seen in Table 23, ferry boats are the next most common type of transit after the bus, heavy rail and light rail. In 1983 there were some 60 ferry boats operated by 13 operators which is a decrease from 1980 when there were 68 ferry boats operated by some 16 operators.

Public ferries operate scheduled service between fixed terminal points and utilize vehicles that, in some cases, can carry up to 2,500 passengers, making them the largest transit vehicles in the nation. The state of Louisiana operates six ferry boats in New Orleans that offer free Mississippi River crossings for vehicles and passengers. These boats make around 370 crossings per day.

Cable Cars

The evolution and development of the cable car was discussed in the Introduction. It became the second generation of light rail when it replaced the horse car in San Francisco. Although it served some 12 U.S. cities at its peak around 1890, the electric streetcar replaced it everywhere except San Francisco which today has the only remaining operating system in the country. Electric propulsion was superior to

TABLE 23. United States Transit Industry Modes in 1983.

Mode	Number of Systems	Vehicles Owned and Leased	Vehicle Miles Operated (Millions)	Unlinked Passenger Trips (Millions)	Estimated Passenger Miles (Millions)	Average Unlinked Passenger Trip Length (Miles)
Motor Bus	1,031	62,093	1,677.8	5,422	20,047	3.7
Heavy Rail[a]	12	9,943	407.5	2,167	10,350	4.8
Light Rail	11	1,013	16.0	137	391	2.9
Trolley Coach	5	686	15.0	160	325	2.0
Cable Car	1	39	b	b	b	b
Inclined Plane	5	10	0.3	b	b	b
Urban Ferry Boat	13	60	1.0	52	234	4.5
Aerial Tramway	1	2	b	b	b	b
Automated Guideway Transit	1	45	b	b	b	b
Total	1,080	73,891	2,117.6	7,938	31,347	

[a]Includes one Monorail System.
[b]Data not available.
Source: American Transit Association, *Transit Facts Book*, 1985.

the cable car in both cost and performance, as well as onboard amenities that could be offered with electric power in each car. The San Francisco Cable Car System is a historical monument which is only in operation today because the city passed an ordinance in 1955 to assure "in perpetuity" operation of a 10.5 mile section of track. The cable car was initially an adaption of street railways in San Francisco because the horse cars could not negotiate the steep hills where 20 percent grades were not uncommon.

Peoplemovers

The term peoplemovers covers a variety of innovative systems that are used to carry people on short hauls in congested downtown areas and other major activity centers. They usually operate at slow speeds along fixed guideways or routes. Generally, they operate in automatic mode without the use of an attendant or driver. The more notable systems in use today include light guideway transit, inclined planes, aerial tramways, and moving way transit systems like moving walkways.

LIGHT GUIDEWAY TRANSIT

Automated peoplemovers or light guideway transit (LGT) moves small groups of people in vehicles operated singly or in short trains over an exclusive guideway. Trains are usually operated under automatic control with no attendants on board and utilize high level station platforms. In most applications, the systems are short and operate in either a loop or shuttle configuration. Most vehicles tend to be bottom supported on rubber tires although there are some exceptions in practice and other designs are possible and being considered.

Table 24 presents data on peoplemover systems that represent most of the technology developments as of 1974. The technological development represented in this table are what are being planned for most future installations. Tampa International Airport was one of the few places that had a rubber tired, bottom supported system before 1972. Monorail systems were more prevalent prior to 1972 with installations at places like Disney World, Disneyland, and Tokyo International Airport. These systems are well suited for use at airports, in shopping centers, and in congested central business district areas and other types of activity centers. They are often the most feasible approach for ferrying people to and from distant parking areas at airports and shopping centers, and from the central terminal area to satellite arrival and departure terminals at airports.

INCLINED PLANES AND AERIAL TRAMWAYS

Inclined planes and aerial trams are peculiar to mountainous or steep hill applications and, therefore, have only limited potential. On level terrain these systems do not compare with LGT and other systems from a feasibility standpoint. As of 1980 there were only five cities in the country with inclined planes and the only use of an aerial tramway is a facility connecting Roosevelt Island with midtown Manhattan in New York City [10]. Tramways similar to the New York system carry tourists to and from major amusement or recreation attractions in Georgia, Tennessee, Arizona and Wyoming.

MOVING WAY TRANSIT

This transit class incoudes systems with and without cabs. The systems without cabs are more prevalent and can be found in airports, shopping centers, public buildings and

TABLE 24. Representative Peoplemover Systems.

System[a]	Vehicle Size Length	Vehicle Size Width	Vehicle Size Height	Seats	Maximum Passenger Load	Maximum Speed (mph)	Energy Source	Propulsion	Controls	Switching	Minimum Curves (ft)	Stations Number	Stations Type	Service	Length of Route (miles)	Number Cars	Completion Date	Total Cost ($millions)
Seattle-Tacoma International Airport	37'	9'4"	11'3"	14	106	25	electrical, from between wheels 600V, 30, AC	dc series 100 hp	fully automatic, unattended service, fixed block	Guideway	96	6	on-line	scheduled	1.69	9	1973	5.3[b]
West Virginia University Morgantown	15'6"	6'0"	8'9"	8	21	30	electrical, from wayside 575V, 30, AC	dc series 60 hp	fully automatic, unattended service, central & multicentral controls, point-follower moving block	On board	30	3	off-line	scheduled (demand off-peak)	2.2	15	1972	53
Dallas-Ft. Worth Regional Airport	21'3"	7'4"	10'2"	16	40	17	electrical, from wayside 480V, 30, AC	dc series 75 hp	fully automatic, unattended service, central & wayside controls, fixed blocks	Guideway	45	6+	off-line	scheduled	13	68	1973	31
San Diego Zoo Park	39'	8'0"	12'9"	62	62	30	electrical, at guidebeam 500V, 10, AC	dc series 50 hp	manual, with automatic separation, fixed blocks	Guideway	100	1	on-line	scheduled	5	18	1972	—
Pittsburgh	35'	9'2"	10'8.5"	28	110	60	electrical, between wheels 600V, DC	—	fully automatic, unattended service, central & wayside controls, fixed blocks	Guideway	90	11	on-line	scheduled	10.5	150	1975	—

[a]Note: All systems are bottom supported on pneumatic rubber tires and utilize horizontal wheels on a center guidebeam for guidance
[b]Does not include tunnel costs.
Source: F. L. Schell, *Peoplemovers: Yesterday, Today, and Tomorrow*, Article in *Urban Transportation Prospectives and Prospects*, Eno Foundation for Transportation, Inc., Westport, CT. 1982.

TABLE 25. Demand-Responsive Guidelines for Provider Selection.

	Local Public Agency	Public Transit Authority	Existing Private Operator	Experienced Management Firms	Non-Profit Organizations
Cost (Labor)	More costly than private, less costly than public transit authority	More costly (union labor); may resist competitive service.	Least costly; non-union bus usually more costly than taxi but 1/2–3/4 less than public transit authority.	Costs vary, but are typically similar to non-union private bus operators.	Varies; potentially least costly due to charitable contributions & volunteers; least reliable.
Implementation Ease	Longer and more difficult implementation as personnel need to be located, hired & trained. New capital investment also required for vehicles and equipment.	Quickest and easiest implementation using established authority whose personnel and capital commitment exist; if no existing authority, has longest implementation period.	Fairly rapid and easy implementation using existing entity, where skilled staff & capital commitment exist.	Moderately easy implementation using experienced firm with skilled staff.	Faster than public agency implementation, yet probably most difficult.
Funds	Directly eligible for a variety of Federal and state funds.	Directly eligible for many Federal and state funds.	Ineligible for many Federal & state funds; subsidies must be funneled through public body.	Ineligible for many Federal & state funds; subsidies must be funneled through public body.	Ineligible for many Federal & state funds unless publicly-owned.
Control	Greatest degree of public control.	Moderate to high amount of public control.	Least degree of public control.	Moderate to small degree of public control.	Moderate degree of public control.
Market	Can operate for limited mobility and/or general public service; probably best suited for the latter.	Best suited for general public service; difficult to tailor service for limited mobility users.	Provides economical travel to the limited mobility user; expensive to subsidize for the general public. Difficult for severely disabled to use taxis.	Can operate service to meet the limited mobility user and/or general public markets; constrained only by public sponsor.	Best suited for the limited mobility user segments of population; not suited for the general public.

Source: J. W. Billheimer, et al., *Paratransit Handbook, A Guide to Paratransit Implementation*, 2 Volumes, Systan, Inc., Prepared for U.S. Department of Transportation, Transportation Systems Center, Cambridge, Massachusetts, January 1979.

other areas with high density pedestrian traffic. The most noted examples are the moving walkway and the escalator. These systems satisfy a transit demand need in congested activity areas that cannot accommodate vehicle traffic.

Paratransit

Paratransit came about in the early 1970s to meet individual local transportation needs that cannot be satisfied by conventional public transportation. The elderly, the very young, the handicapped, people without access to a private automobile, and car owners who want to cut energy costs or avoid urban congestion are the main users of paratransit in areas where public transit does not exist or does not meet their transportation needs.

Paratransit vehicles can carry from 6 to 20 passengers and are not restricted to fixed routes or time schedules. A wide range of movement types are classified as paratransit including dial-a-bus, dial-a-ride, shared-ride taxis, car pools, van pools, subscription buses, school buses, and regular taxi service. When paratransit operates in cities with conventional transit, it should try to complement rather than compete with the conventional transit by providing special travel services that otherwise would not be available.

Paratransit can be divided into two main groups, namely, demand-responsive paratransit and prearranged ridesharing. Demand-responsive includes services like dial-a-ride and shared-ride taxis, and is most responsive to groups like suburban shoppers, the elderly, and rural residents. School children and commuters are the main groups requiring prearranged ridesharing which includes car pools, van pools, and subscription buses as forms of transportation. Operation of paratransit services are handled by a variety of public and private profit and non-profit groups. Local governments and regional transit authorities are the main public sector operators while private sector operators include [18]:

- non-profit social service agencies

TABLE 26. Limited Mobility User Characteristics for Special Dial-A-Ride Transportation.

Variable		Cleveland Ohio (Mixed)	Syracuse New York (DAB)	Dade County Miami Florida (Mixed)	Baton Rouge Louisiana (DAB)
Sex:	Male	8.0%	14.9%	N/A	48.0%
	Female	92.0	83.9	N/A	52.0
Age:	Under 18	0.0	N/A	2.9 (Under 21)	43.4 (Under 21)
	18–44	5.5 (20–59)	13.1 (Under 60)	25.0 (21–49)	26.4 (21–39)
	45–64	6.0 (60–64)	6.8 (60–64)	28.8 (50–64)	23.9 (40–64)
	65 & over	88.5	80.1	43.3	6.3
Cars in Household:	0	N/A	59.9	N/A	12.7
	1	N/A	N/A	N/A	42.4
	2	N/A	N/A	N/A	37.4
	3+	N/A	N/A	N/A	7.5
Purpose of Trip:					
	Work	N/A	2.3	12.7	N/A
	School/Soc.Serv.	N/A	N/A	11.4	N/A
	Medical	49.5	46.8	24.0	87.0
	Shopping	47.5	0.9	19.0	N/A
	Recreation	32.0	7.3	17.7	N/A
	Other	14.0[a]	42.7[a]	15.2	13.0
Alternate Mode:					
	Bus	33.0	11.1	9.0	20.4
	Walk	42.0	0.6	N/A	2.5
	Taxi	8.0	21.9	28.0	2.5
	Drive	7.0	1.2	N/A	5.7
	Be Driver	38.5	15.8	49.0	45.2
	No Trip	10.0	22.3	N/A	2.5
	Other	8.0	16.1	N/A	18.5

[a]Figure includes multi-purpose trips.
Source: J. W. Billheimer, et al., *Paratransit Handbook, A Guide to Paratransit Implementation*, 2 Volumes, Systan, Inc., Prepared for U.S. Department of Transportation, Transportation Systems Center, Cambridge, Massachusetts, January 1979.

- profit-making, nonsubsidized organizations
- profit-making transportation providers that have local government contracts and subsidies
- employers and employee organizations

Table 25 gives a summary of guidelines for providing demand responsive services to the public. Regional transit authorities and other local government entities operate demand responsive paratransit services in small cities. Under private sector operations, the most common operation is demand-responsive offered by social service agencies to various types of clients with most operations being door-to-door personalized service. Dial-a-ride is the most common form of paratransit with women and senior citizens being the more frequent users. Table 26 shows the characteristics of four different dial-a-ride systems that serve the general public. Shared-ride taxi services usually take the form of municipality subsidized service from a private, profit-making taxi company.

Transportation brokerage projects, which have been instituted in a number of communities recently, attempt to match the targeted demand for services with the most effective provider. Projects are usually classified into four brokerage types: commuter, elderly and handicapped, decentralized, and integrative. The impact of these projects on target markets has ranged from slight improvements in service delivery to regulatory changes that could lead to future transportation improvements. The greatest degree of brokerage effectiveness occurs when a broker is able to fulfill a true need and is accepted by the local transportation environment.

Prearranged ridesharing service is usually offered by employers and employee organizations. Typically the employer or employee organization will lease the equipment to the workers, provide insurance coverage, and offer certain incentives to employees who use car pools, van pools, or subscription bus services. Of the 77 million people who commuted to work on a regular basis in 1977, 15 million

TABLE 27. Spectrum of Urban Transportation Modes and Service Characteristics.

	Service Characteristics			
Mode	Origins-Destinations Capability	Vehicle Acquisition	Waiting Required?	Parking Required?
Private				
Automobile	unlimited	driver initiated	no	yes
Motorcycle	unlimited	driver initiated	no	yes
Truck (e.g., pickup truck)	unlimited	driver initiated	no	yes
Other	unlimited	driver initiated	no	yes
Paratransit				
Daily and short-term rental car	unlimited	hire and drive	minor	yes
Taxicab	unlimited	hail or phone	yes	no
Dial-a-ride	unlimited[a]	hail or phone	yes	no
Jitney	partially fixed	hail or phone	yes	no
Car pool	partially fixed	pre-arranged	yes	yes
Subscription bus	partially fixed	pre-arranged	yes	no
Conventional mass transit				
Commerical bus	fixed	scheduled	yes	no
Trolley coach	fixed	scheduled	yes	no
Rail				
Subway and elevated	fixed	scheduled	yes	no
Streetcars	fixed	scheduled	yes	no
Commuter rail	fixed	scheduled	yes	no
Other fixed guideway	fixed	demand & scheduled	yes	no
Other transit (e.g., air, water)	fixed	demand & scheduled	yes	no
Dual or mixed mode	partially fixed	demand & scheduled	yes	no

[a]Some operations have fixed origins or destinations.
Source: E. Weiner, "The Characteristics, Uses and Potentials of Taxicab Transportation," Article in *Urban Transportation Prospectives and Prospects*, Eno Foundation for Transportation, Inc., Westport, CT, 1982.

used car pools compared to 4 million who rode public transportation. As of 1980, the National Association of Van Pool Operators estimated that there were 8,500 employer operated van pools [18].

The National Ridesharing Demonstration Program (NRDP), established in 1979 by the U.S. Department of Transportation, just completed an evaluation of projects in 17 cities where comprehensive innovative approaches to ridesharing had been implemented. The evaluation also included an in-depth analysis of five case study sites. Principal findings of the study are as follows [19]:

- *Project design and implementation*—careful design and adequate funding appear to be essential for satisfactory program implementation.
- *Primary ridesharing market*—the primary market for commuter ridesharing appears to be multi-worker households with at least one car, located at relatively long distances from their workplaces.
- *Car pool formation and composition*—most commuter car pools seem to consist of informal arrangements between household members or fellow workers.
- *Employer involvement with ridesharing*—the proportion of employees ridesharing and car pool size appear to increase with firm size.
- *Ridesharing program outreach*—although area ridesharing programs have contacted a high proportion of firms, the share of employees actually receiving program materials appears to be much lower.
- *Ridesharing program impact*—the impact of ridesharing programs on commuter travel behavior cannot be conclusively determiined from NRDP data. The direct impact appears small, but indirect effects may be much larger.
- *Other ridesharing strategies*—neighborhood-based ridesharing promotion does not appear to be an effective alternative to employer marketing programs.

Taxicabs are private sector operations that offer service somewhere between the private automobile and mass transportation. Table 27 compares the exclusive use taxi characteristics with other forms of urban transportation. Taxicabs are mentioned here only because they are one of the older enduring forms of transportation and in 1970 they operated three times as many vehicles over twice as many revenue miles and collected more passenger revenue than the entire transit industry. The industry operated 7,200 fleet operations in 1974 that were franchised in about 3,300 communities as well as several thousand individual operations [20].

ORGANIZATION AND FINANCES

Urban transit systems are fundamentally a public sector responsibility. Even where systems are privately operated, they are publicly regulated to some extent.

As ridership declines and operating deficits grow larger, organizations responsible for public transportation are continuously striving to improve transit performance, productivity and cost effectiveness. Recent declines in levels of federal operating subsidies have put even more emphasis on the urban transportation industry to analyze their management goals, operating functions, and financial responsibilities.

Management and Operating Functions

In 1980, over 94% of the transit riders in the United States rode on publicly owned transit systems as compared with 50% who rode on publicly owned systems in 1967 [10]. The move away from private transit systems started long before World War II as systems became unprofitable. Regulatory agencies often prohibited the usual remedies of raising fares and pruning unproductive transit services. Rather than let unprofitable private systems go out of business, many municipalities took over those operations, usually with federal funding assistance.

As operating deficits have steadily increased, there has been increasing acceptance of a concept which separates managing and operating functions. Although local governments usually accept responsibility for providing public transit services and retain control of such management functions as fare setting and capital funding, there is more and more involvement by private and voluntary sectors in the operation of transit. Private firms and local or community based service organizations are more attuned to the needs of the transit user and may be more qualified and/or efficient at performing transit services.

With a variation in population density, travel patterns, and other service requirements in large regional areas, decentralization of large systems is often desirable. Large scale operations tend to have high overheads, overstaffed organizations, formalized labor-management relations, huge employee pension commitments, and large fleets of expensive equipment. Therefore, they are generally less efficient and more costly to operate than small scale systems that respond to the changing travel needs of a multitude of population groups. Small scale service districts are more flexible in adapting to changes in service requirements. Over the last several years there have been a variety of innovative arrangements started by private providers of transit services, including commuter bus subscription services, employee-organized van pooling arrangements, door-to-door services for senior citizens and handicapped persons and suburban feeder services operated under contract by private taxi companies.

Institutional Roles and Responsibilities

The Federal Government's role in public transportation is one mainly of establishing policy and providing financial support. Although numerous funding programs for transportation are administered by various federal agencies, the Urban Mass Transportation Administration (UMTA), established in 1964, represents the basic source of capital funding for virtually all transit systems in America. The majority of federal financial assistance is used for capital purchases of buses and bus facilities or for modernization of rail facilities. Federal transit assistance programs were sharply curtailed in 1981 with substantial reduction and eventual termination proposed for operating assistance activity.

The Surface Transportation Assistance Act of 1982 eliminated the operating assistance program and replaced it with a block grant program that could be used for both capital and operating programs. Federal operating subsidies are less than those allocated in 1982 and could eventually be eliminated. The authorization legislation that was passed in 1982 provides funds for three years (1984–1986) of approximately $3.5 billion each year in capital assistance with an additional $875 million available annually for operating subsidies [7]. The 1982 UMTA policy toward capital investment directs funds toward transit systems that are cost-effective from a ridership, travel time, and operating cost perspective and which can show reliable sources of operating funds when the capital improvements are completed. Capital assistance also requires that grant applicants must provide some percentage of a project's total cost as a local share. Under the 1982 authorization, that share is around 25 percent in the discretionary grant program funded with Highway Trust Fund revenues.

State by state involvement in public transit is not consistent with states having varying degrees of responsibility in such functions as policy direction, planning, funding, regulating, and in some cases, actually owning and operating transit systems. The Department of Transportation's (DOT) goal in many states is a balanced, multimodal transportation system in which DOT coordinates state involvement, eliminates duplicate efforts by separate modal agencies, and communicates with federal agencies. States often have a public utility commission or other regulatory body that oversees policy and budget considerations for public transit operations within a given state.

States will become more involved in transit as federal funds decrease and the state is put in the position of having to provide a stable, predictable source of transit deficit-financing. In many areas, state enabling legislation is required if localities are to undertake certain transporation-related activities, such as capital bonding, tax-increment financing, and various public incentives for private actions.

Local governments have traditionally been the planners and providers of public transportation. In 1974, a Federal DOT regulation required that each urbanized region must have a metropolitan planning organization (MPO) that is responsible for planning and coordinating transportation activities in a region. In some areas counties are taking on a stronger role in the allocation of resources and delivery of public transportation services where transit service operations encompass an area too large for local governments, or outside urban jurisdiction. There are numerous pros and cons for both regional and local roles in the planning, coordination, and delivery of transit services in a given area. As local governments become more responsible for operating deficits there is more argument by urban and suburban jurisdictions to decentralize transit systems based on growing evidence that centrally-run region-wide systems are not necessarily the most efficient.

The private sector is increasingly providing more and more urban transportation services including:

- urban transit systems (the few that exist are usually providing service under contract to public agencies)
- taxicab companies (sometimes under contract to publicly-owned transit systems)
- private and private non-profit, client oriented transportation programs (local, community-based organizations that often consist of a coalition of business, labor, neighborhood groups, and government agencies; these organizations combine resources and capabilities of public, private and voluntary sectors, and address local needs in ways that neither sector would have the capability to address alone

Use of private sector and voluntary associations is based on the realization that competition and the profit motive may lead to reduced public sector costs and improved service quality. Popularity of these arrangements is based on the belief that private enterprise is likely to deliver service more effectively and at lower cost; partly because available public transportation service is inadequate or nonexistent, and partly because private operators are felt to be more consumer-oriented and better motivated to respond to demands of local markets.

SCHEDULING AND CONTROL

For any given system, two basic aspects of transit operations are the movement and control of vehicles. Schedules are made up by reviewing fundamentals. The physical nature of the route and the operating characteristics of the vehicles, or trains, govern running time. The overall round trip time for any vehicle includes the running time for each direction, the dwell times at passenger stops, and layovers or turn-around times required at the ends of the route.

The frequency of service during peak periods is usually based on the number of passengers to be handled past a maximum load point. During off-peak periods, headway is often a policy determination based on desirable trip frequencies and costs. A common rule for the effective use of on-board personnel is to have them all work at least one peak period. Service levels and capacities can be maintained by using the maximum number of cars per train during peaks and shorter trains or basic units during day and evening off-peak periods.

Control systems are important in maintaining schedules and headways safely and effectively. The purpose of control and communications systems is to ensure that the transit system operates safely, reliably, and effectively within its inherent operational limits. Each powered vehicle should operate smoothly and all trains should operate in a coordinated manner on all parts of the system.

LABOR

Labor costs represent about 70 percent of total operating expenses for public transportation systems in the United States. About 88 percent of U.S. transit systems have union agreements that specify wage rates, benefits, conditions, and work rules. Although union representatives would prefer a system of straight eight hour shifts, passenger demand calls for peak employee needs which cannot be totally satisfied with scheduling techniques, such as split shifts.

Labor output in place-miles per worker is a measure of what transit workers produce and is based on annual car miles and car sizes. Running empty transit vehicles will generate place-miles, but no revenue. Therefore, passenger-miles, which is place-miles utilized must also be considered. Figure 19 gives the labor output of various transit systems and demonstrates that the output per employee of rapid transit exceeds that of bus by a wide margin.

Most rapid transit systems, shown in Figure 19, produce between 1 and 2 million place-miles of service per worker a year while the bus range, with mostly local service, is between 0.5 and 1 million [2].

Factors that can improve output per worker on rail systems are:

1. Higher speeds
2. Larger vehicles opeated in trains whenever possible
3. Self-service fare collection (enables one-man train operation)
4. Close coordination with feeder buses (to enlarge the tributary area of a high-capacity line and use each mode in a setting to which it is best suited

Labor requirements per place-mile is the reciprocal of labor output in place-miles per employee and can be converted directly into labor costs. Figure 20 shows the relation between the labor required to produce a million place-miles of service annually related to opeating speeds. Labor requirements in cost per place-hour are equal to the cost per place-mile multiplied by operating speed in miles per hour and falls in a very narrow band for buses, rapid transit, and light rail. The effect of speed on labor costs is pervasive and not limited to the ability to get an extra run out of a peak hour vehicle by increasing operating speed. The effect of service volume on labor cost can also be observed using labor requirements per place-hour. Employee requirements in administration do not change perceptibly with volume of service per mile of line. Vehicle maintenance requirements also fail to show any change with rising volume of service.

FARE SYSTEMS

Fare pricing levels and the fare structure are established by management on the basis of a multitude of complex factors, with government regulatory approval usually required before they can be implemented. Since most transit operations are subsidized, fares can be established in accordance with the following factors:

1. Requirement to produce a specific sum or a given proportion of operating costs, as determined by availability of subsidies
2. Goal to maximize transit use and reduce automobile travel
3. Social equity considerations applicable to the total ridership or various subgroups
4. Competition from other modes or systems, if any
5. Willingness of passengers to pay premium fares for special services
6. Operational constraints, such as the disadvantages of collecting multiple-coin fares on board vehicles

The equitable fare system is one graduated by distance traveled. It is mostly used in rapid rail systems where use of automated ticket systems in stations allows for accurate monitoring of distance traveled by passengers. The graduated system is not practical on one operator buses or similar transit modes that use manual fare collection methods since checking passengers at the beginning and end of a trip is time-consuming, cumbersome, and, therefore, often inaccurate.

The simplest and least expensive system to administer is the uniform fare which charges all passengers the same rate. From the passenger standpoint, it tends to favor those on longer trips while it penalizes the shorter passenger trip.

Zone fares fall between the uniform and graduated systems and relate trip length with fare paid. As a zone fare becomes more fine grained, the closer it approaches the equity of graduated systems while at the same time it becomes more cumbersome and expensive to administer with manual fare collection. They are most often used on long route sections that extend beyond the larger control areas of a uniform fare system.

Transit fare charges are often implemented in connection with other transit service improvements or reductions, and are frequently accompanied by marketing activities. A wide range of individual results are reported in response to fare charges; but consistent patterns are evident [21]. The effect of bus fare increase and decrease average roughly 0.4 percent ridership loss or gain per 1 percent fare change. Ridership responds least to fare changes on rapid transit, in large cities, and where transit service is exceptionally good or the cost of alternate modes is high. Patronage losses observed for rapid transit are 0.14 percent per 1 percent fare increase for the average case. Ridership is most sensitive to fares on small city and suburban bus operations and wherever transit service is light. The results of bus frequency changes in connection with fare changes is not real conclusive, but overall

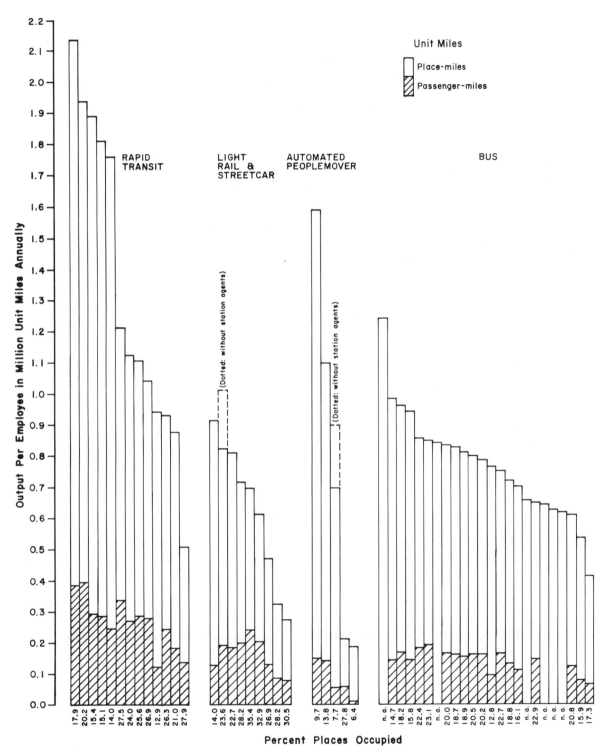

FIGURE 19. Labor output of transit operations. Notes: n.a. = not available. Based on actual data from operational transit systems that are representative of the respective modes. Source: B. S. Pushkarev and J. M. Zupan, *Urban Rail in America: An Exploration of Criteria for Fixed-Guideway Transit*; Report UMTA-NY-06-0061-80-1, Urban Mass Transportation Administration, Washington, DC, Nov., 1980.

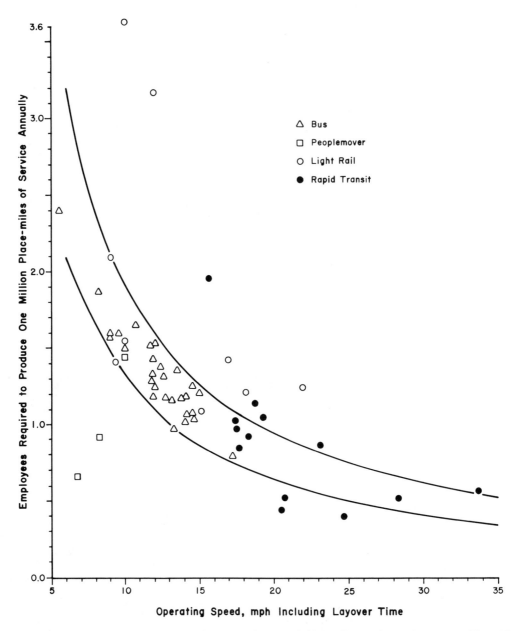

FIGURE 20. Labor requirement related to operating speed. Notes: Range shown by curves: 13 to 19 employees per million place-hours (place-miles × mph). Based on actual data from operation transit systems that are representative of the respective modes. Source: B. S. Pushkarev and J. M. Zupan, *Urban Rail in America: An Exploration of Criteria for Fixed-Guideway Transit*; Report UMTA-NY-60-0061-80-1, Urban Mass Transportation Administration, Washington, DC, Nov., 1980.

service improvements, including both coverage and frequency improvements, have been estimated to have greater impact than fare reductions. There is also a limited, but strong indication that the typical commuter railroad patron is much more influenced by service frequency than by fares.

Fare collection procedure should balance adverse transit system performance considerations against patron convenience and operation cost of collection system. There are a wide variety of fare collection systems in use which can be classified as either manual or automatic type systems. The trend today is toward systems such as prepayment and automatic payment of fares which are less labor intense than manual systems.

Financial Aspects and Trends

Prior to 1961, most transit systems were privately owned with system revenues exceeding operating costs. In the early 1950s ridership and revenues started declining rapidly as urban population became less transit oriented with the growth of the automobile. As the trend continued, private investment and ownership became more unprofitable and was replaced with public ownership. Federal assistance began in 1961 with $75 million authorized by the Federal Housing Act for demonstration projects and capital improvements for transit. Federal assistance increased substantially thereafter and by 1974, they also began providing operating assistance to transit systems. As government budgets get tighter at all levels, it is becoming increasingly difficult to adequately finance capital and operating costs of transit systems at a time when energy concerns and other factors have increased ridership in many cities. Concerns in finding new funding sources, especially with the future of federal funding uncertain, are justified with operating costs of U.S. transit systems rising at a rate somewhat higher than the consumer price index. Figures show that total transit operating deficits were about $300 million in 1970 compared to $2.2 billion in 1977, a dramatic increase.

TABLE 28. Trend of Transit Revenues.[a]

Calendar Year	Operating Revenue			Non-Operating And Auxiliary Revenue	Operating Assistance			Total Revenue
	Passenger	Other	Total		State & Local	Federal	Total	
Dollars (millions)								
1940	$ 701.5	$ 35.5	$ 737.0	b	b	b	b	b
1945	1,313.7	66.7	1,380.4	b	b	b	b	b
1950	1,386.8	65.3	1,452.1	b	b	b	b	b
1955	1,358.9	67.5	1,426.4	b	b	b	b	b
1960	1,334.9	72.3	1,407.2	b	b	b	b	b
1965	1,340.1	103.7	1,443.8	b	b	b	b	b
1970	1,639.1	68.3	1,707.4	b	b	b	b	b
1975	1,860.5	141.9	2,002.4	$ 40.6	$1,106.0	$ 301.8	$1,407.8	$3,450.8
1976	2,025.6	135.5	2,161.1	75.0	1,224.5	422.9	1,647.3	3,883.4
1977	2,157.1	122.9	2,280.0	73.6	1,319.5	584.5	1,904.1	4,257.7
1978	2,271.0	110.1	2,381.1	68.8	1,542.1	689.5	2,231.7	4,681.5
1979	2,436.3	87.9	2,524.2	123.6	2,054.6	855.8	2,910.4	5,558.2
1980	2,556.8	105.9	2,662.7	142.4	2,611.2	1,093.9	3,705.1	6,510.2
1981	2,701.4	82.3	2,783.7	261.5	3,225.7	1,095.1	4,320.8	7,336.0
1982	3,077.0	75.0	3,152.0	305.0	3,582.0	1,005.4	4,587.4	8,044.3
1983[c]	3,171.6	59.7	3,231.2	272.8	4,194.6	827.0	5,021.6	8,525.7
Percent of Total Revenues (percent)								
1975	53.9	4.1	58.0	1.2	32.1	8.7	40.8	100.0
1976	52.2	3.5	55.7	1.9	31.5	10.9	42.4	100.0
1977	50.7	2.9	53.6	1.7	31.0	13.7	44.7	100.0
1978	48.5	2.4	50.9	1.4	33.0	14.7	47.7	100.0
1979	43.8	1.6	45.4	2.2	37.0	15.4	52.4	100.0
1980	39.0	1.7	40.7	2.0	40.0	17.3	57.3	100.0
1981	36.7	1.1	37.8	3.5	43.8	14.9	58.7	100.0
1982	38.3	0.9	39.2	3.8	44.5	12.5	57.0	100.0
1983[c]	37.2	0.7	37.9	3.2	49.2	9.7	58.9	100.0

[a]Table excludes automated guideway transit, commuter railroad, and urban ferry boat.
[b]Data not available.
[c]Preliminary
Source: American Transit Association, *Transit Fact Book*, 1985.

TABLE 29. Trend of Transit Expenses by Function Class.[a]

Calendar Year	Operating Expense					Depreciation And Amortization	Other Reconciling Items	Total Expense
	Transportation	Maintenance		General Administration	Total			
		Vehicle	Non-Vehicle					
Dollars (millions)								
1940	b	b	b	b	b	b	b	$ 660.7
1945	b	b	b	b	b	b	b	1,231.7
1950	b	b	b	b	b	b	b	1,385.7
1955	b	b	b	b	b	b	b	1,370.1
1960	b	b	b	b	b	b	b	1,376.5
1965	b	b	b	b	b	b	b	1,454.4
1970	b	b	b	b	b	b	b	1,995.6
1975	$1,876.5	$814.4[d]		$ 846.4	$3,537.3	$121.0	$ 94.2	3,752.5
1976	2,033.4	894.1[d]		929.9	3,857.4	136.3	88.9	4,082.6
1977	2,219.8	972.7[d]		928.5	4,121.0	161.4	84.2	4,366.6
1978	2,508.7	$ 776.6	$292.1	961.7	4,539.1	149.6	100.2	4,788.9
1979	2,735.0	1,070.2	$398.8	1,027.7	5,231.7	253.4	126.3	5,611.4
1980	3,248.2	1,274.3	$499.7	1,224.3	6,246.5	277.6	186.5	6,710.6
1981	3,596.5	1,397.8	$547.9	1,482.1	7,024.3	386.3	211.1	7,621.7
1982	3,882.3	1,555.8	$611.8	1,503.0	7,552.9	507.1	254.3	8,314.3
1983[c]	3,930.8	1,696.6	$694.9	1,633.7	7,956.0	472.5	307.2	8,735.7
Percent of Operating Expense (percent)								
1975	53.1	23.0[d]		23.9	100.0			
1976	52.7	23.2[d]		24.1	100.0			
1977	53.9	23.6[d]		22.5	100.0			
1978	55.3	17.1	6.4	21.2	100.0			
1979	52.3	20.5	7.6	19.6	100.0			
1980	52.0	20.4	8.0	19.6	100.0			
1981	51.2	19.9	7.8	21.1	100.0			
1982	51.4	20.6	8.1	19.9	100.0			
1983[c]	49.4	21.3	8.8	20.5	100.0			

[a]Note: Table excludes automated guideway transit, commuter railroad, and urban ferry boat.
[b]Data not available.
[c]Preliminary
[d]Vehicle Maintenance and Non-Vehicle Maintenance combined.
Source: American Transit Association, *Transit Fact Book*, 1985.

REVENUES AND EXPENSES

Table 28 shows the dollar trend in transit revenues and the trend in revenues as a percent of total revenue. Likewise, the dollar trend for expenses is shown in Table 29 which also gives the trend expenses by function class as a percent of operating expense. Although total passenger revenue income has increased in recent years, it now pays less of a percentage of total revenues required to offset increasing operating expenses than it did in previous years. Government assistance from federal, state, and local sources now supplies the majority of transit operating revenues. As federal and state sources of operating assistance have dropped in the last decade, local assistance has increased with the total percentage of assistance not showing that much of an increase on a yearly basis. In Figure 21, the trend of passenger revenues and operating subsidies as a percentage of total expense is based on data in Tables 28 and 29. Between 1970 and 1980, transit fares fell from providing 84 percent of operating expenses to only 42 percent. During this same period, the average fare per unlinked transit passenger trip went from 22.4 cents to 31.0 cents [10]. Revenues and expenses of transit are negatively affected by inflation. Transit expenses can increase more rapidly than inflation while transit revenues increase at a slower rate (Figure 22). Between 1970 and 1980 the price of consumables also rose at a faster rate than the U.S. Department of Labor Consumer Price Index. During this period, diesel fuel cost for buses rose 630% or almost 5 times the 131% inflation rate of consumer prices. In addition, the cost of producing a revenue mile of transit service rose 200 percent from $1.05 in 1970 to $3.14 in 1980. As Figure 22 shows, transit fare increases lag behind the consumer price index. To encourage ridership, transit fare increases do not keep pace with inflation, mostly due to policy stipulations in the industry. During times of high inflation, the transit industry is severely affected by delays in passing cost increases through to the consumer.

CAPITAL COST

The need for and the cost of capital improvements, including modernization of older systems, gets more expensive each year. Over $1.5 billion in federal funds has been allocated in fiscal year 1985. Table 30 shows the history of capital fund grants. Since 1974, federal funds may be used to cover 80% of capital expenses with matching funds from state and local sources having to cover the balance. Table 31 gives a breakdown of the larger federal capital grants received by major urban areas for rail and bus improvements. Between 1964 and 1980, the federal government has participated, in part, in the funding of the following equipment and facilities [10]:

42,692	motor buses
678	trolley coaches
3,218	heavy rail cars
497	light rail cars
1,720	commuter railroad cars
96	commuter railroad locomotives
16	ferry boats
2	inclined plane cars
2	automated guideway transit
12	miles of commuter railroad lines
23	miles of busways (planned mileage not included)
240	miles of heavy and light rail lines (planned mileage not included)

plus bus garages, office buildings, passenger shelters, communications systems, and many other items

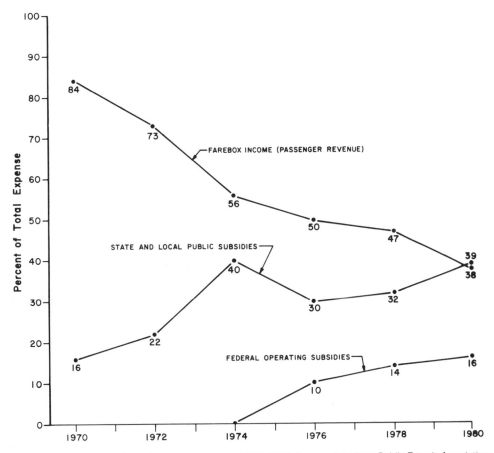

FIGURE 21. Trend of transit revenue sources, 1970-1980. Source: American Public Transit Association.

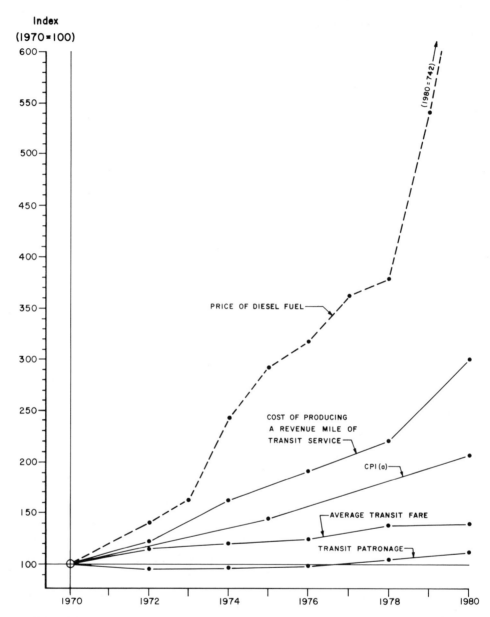

FIGURE 22. Vital revenue related trends, 1970–1980. Note: a. CPI = United States Department of Labor Consumer Price Index. Source: American Public Transit Association.

Construction costs in 1977 dollars for light and heavy rail construction are presented in Table 32. The aggregate average cost, by construction type, is based on actual construction contract documents from various cities around the world. With inflation having averaged 6.7 percent annually over the 1970's, construction costs have obviously escalated considerably since 1977, but the data in Table 32 is representative for comparison purposes.

The saving in light rail track and station cost is partly offset by closer station spacing. While less elevated and tunnel routing saves money, the savings are somewhat nullified by the lower performance and benefits derived from grade level trackage. Other capital cost elements, not shown in Table 32, are track right-of-way, land for facilities, parking lots, shops, storage yards, crossing protection, vehicles, and other miscellaneous costs.

TABLE 30. United States Government Capital Grant Approvals for Mass Transportation by Use ($\times 10^6$).[a]

Federal Fiscal Year	Bus[b]	Rapid Transit[c]	Commuter Rail	Other[d]	Total
1965–1976[e]	$1,960.1	$3,370.1	$ 937.3	$186.3	$ 6,453.8
1977	483.6	1,001.1	232.0	7.0	1,723.7
1978	598.5	1,162.9	271.7	3.8	2,036.9
1979	554.6	1,318.7	232.6	5.7	2,101.6
1980	935.8	1,474.3	340.4	36.6	2,787.1
1981	994.3	1,546.1	373.5	31.8	2,945.7
1982	854.4	1,307.1	323.0	59.6	2,544.1
1983	1,138.4	1,455.5	465.4	102.3	3,161.6
Cumulative Total	$7,509.6	$12,635.9	$3,175.9	$433.2	$23,754.6

[a]Net amounts exclude cancelled and reduced projects. Includes funding from Section 3, Section 5, Section 9A, and Section 16 (b) (2) of the Urban Mass Transportation Act of 1964, as amended. Urban Systems and Interstate Transfers Sections of the Federal Aid Highway Act of 1973, as amended, and funding from Section 14 of the National Capital Transportation Act of 1969, as amended.
[b]Motor bus and trolley coach.
[c]Heavy rail and light rail.
[d]Urban ferry boat, cable car, inclined plane, and automated guideway transit.
[e]Twelve-year total.
Source: U.S. Department of Transportation, Urban Mass Transportation Administration.

TABLE 31. Distribution of UMTA Capital Grants.[a]

Total		New Rail Construction and Modernization Only		Bus and Bus Facilities Only	
City[b]	Dollars ($\times 10^9$)	City[b]	Dollars ($\times 10^9$)	City[b]	Dollars ($\times 10^6$)
New York	$ 4.2	New York	$ 3.6	Los Angeles	$532
Washington	2.3	Washington	2.1	New York	521
Boston	2.0	Boston	1.8	Chicago	332
Chicago	1.6	Chicago	1.3	Pittsburgh	215
Philadelphia	1.2	Philadelphia	1.0	SF-Oakland	215
Atlanta	1.0	Atlanta	0.9	Washington	197
SF-Oakland	0.9	Baltimore	0.6	Seattle	194
Baltimore	0.7	SF-Oakland	0.6	Philadelphia	189
Los Angeles	0.7	Pittsburgh	0.2	Detroit	161
Pittsburgh	0.5	Cleveland	0.2	Boston	145
Total ($\times 10^9$)	$15.0	Total ($\times 10^9$)	$12.3	Total ($\times 10^9$)	$2.7
Percent of Total UMTA Program (%)	75	Percent of Total UMTA Program (%)	60	Percent of Total UMTA Program (%)	14

[a]As of 9/30/83, Capital Grants totaled $20,675,595,886. Includes funding from Section 3, Section 5, Interstate Transfer, Federal Aid Urban Systems and Urban Initiatives.
[b]Cities ranked by descending dollar volume of grants received in each category.
Source: U.S. Department of Transportation, Urban Mass Transportation Administration.

TABLE 32. Summary of Capital Costs for Construction of Double Track Systems.

	Millions of Constant 1977 Dollars[a]							
	At Grade		Cut and Cover		Elevated		Underground	
Cost Elements	Light Rail	Heavy Rail	Light Rail	Heavy Rail	Light Rail	Heavy Rail	Light Rail	Heavy Rail
Route Construction	$2.00	$ 3.20	$ 8.84	$ 8.80	$20.29	$20.30	$40.23	$40.20
Track and Electrical Work	2.90	4.20	3.00	4.20	3.10	4.30	3.21	4.30
Stations (#/ml)	0.06[2]	2.90[1]	4.78[2]	3.90[1]	4.35[2]	3.20[1]	6.90[2]	7.70[1.5]
Total	$4.96	$10.30	$16.62	$16.90	$27.74	$27.80	$50.34	$52.50

[a]All costs are averages of high and low observations: they exlude shops, yards, and land acquisition.
Source: Thomas K. Dyer, Inc.; converted from 1974 constant $$ by ENR cost index 1.28.

Land acquisition costs are site-specific and time-specific and represent extremes from lower cost suburban land to expensive central city sites. To show order of magnitude, parking lots might require 1 to 3 acres for small lots and 5 to 10 or 15 acres for large lots. Storage yards may range from 3 to 10 acres and shop areas from 4 to 12 acres.

TABLE 33. Trend of Energy Consumption by Transit Passenger Vehicles.[a]

Year	Electric Power Consumed (kilowatt hours in millions	Fossil Fuels Consumed (gallons in thousands)	
		Gasoline[b]	Diesel
1940	6,334	[c]	[c]
1945	7,033	510,000	11,800
1950	5,251	430,000	98,600
1955	3,530	276,000	172,600
1960	2,908	191,900	208,100
1965	2,584	124,200	248,400
1970	2,561	68,200	270,600
1975	2,646	7,576	365,060
1976	2,576	6,163	389,187
1977	2,303	9,273	402,842
1978	2,223	9,331	422,017
1979	2,473	8,973	423,212
1980	2,446	11,400	431,400
1981	2,655	13,950	445,950
1982	2,722	11,670	455,590
1983[d]	2,930	9,460	450,260

[a]Table excludes automated guideway transit, commuter railroad, and urban ferry boat.
[b]Includes propane.
[c]Data not available.
[d]Preliminary
Source: American Transit Association, *Transit Fact Book*, 1985.

Rail vehicles cost can be a major consideration when buying current generation rolling stock. Rolling stock costs may add between 8 percent and 13 percent, or higher to the construction costs of a new rapid transit system. Standard production model rail vehicles averaged between $300,000 and $600,000 in 1975 [17]. There are numerous variations possible utilizing standardized modules and components. The cost of purchasing motor buses went from $20,500 in 1972 to $122,200 in 1980, a 202% increase [10].

MISCELLANEOUS ASPECTS

Energy and Environmental Concerns

From the energy viewpoint, the key feature of the various modes of urban transit is the type of propulsion. The differences between electric and liquid fuel propulsion (typical of most free-wheeled vehicles) concerns both the quality of the environment and the quantity of the primary energy resources consumed.

ENERGY CONSUMPTION

Transportation is the largest single consumer of energy in the United States, using 25% of the total national energy budget for fuel. The automobile uses about 40% of the total transportation energy budget for urban passenger movements, or around 10% of the total national energy budget [22]. Table 33 shows the yearly trend in energy consumption for passenger vehicles. The rapid rise in diesel fuel consumption is due, in part, to the large increase in motor bus usage over the years.

Figure 23 shows the relationship of energy use by urban transportation modes. It can be seen that a fully loaded modern heavy rail car is the most efficient form of urban transportation. A rail car with 250 passengers is about 53 times more fuel efficient than an average automobile. Transit buses are up to 15 times more fuel efficient than the automobile.

Figure 24 compares vehicle energy requirements of various transportation systems. Data for operational transit systems is used to establish lines of best fit for rail rapid transit and buses, respectively. Energy requirements in BTU's per passenger place mile are derived for both electric and diesel oil vehicles by equating both to the common denominator of gross energy input needed for vehicle operation.

The average gross energy requirements for rapid rail systems for which comparable data are available is about 670 BTU per passenger place mile. In general, old heavy rail systems built before 1950 tend to be more energy efficient than newer systems. This is due in part to fewer power consuming amenities on older vehicles and higher ridership. The energy requirement is nearly independent of speed since the energy needed for frequent acceleration on slower systems pretty much balances out that needed to achieve high speeds on faster systems. Energy requirements for buses also average around 670 BTU per passenger place mile at 12 mph, which is the average speed at which most urban buses operate. This figure rises to 800 BTU in slow speed downtown areas and decreases to 415 BTU on higher speed freeways where speed is more constant. Figures on light rail are scarce, but they seem to require about 785 BTU per passenger place mile [2].

Other studies conducted on vehicle energy consumption show that mileage figures for motor buses is around 3.88 to 5.5 miles per gallon (mpg) compared to fuel consumption for a standard size automobile of about 9.0 to 15.1 mpg. On a 53 passenger bus, energy efficiency can be high.

ENVIRONMENTAL IMPACT

The National Environmental Policy Act of 1969 established a national environmental protection policy that along with legislation in many states has heightened environmental awareness and concern in decisions regarding transportation and land use alternatives and in decisions taken to identify and mitigate impacts. In general, environmental input issues can have consequences adverse to environmental goals including air, water, and noise pollution, undesirable land use patterns, population displacement, damage to life systems, and threats to health. There are many diverse environmental factors that can affect the detailed location and design of transportation facilities in urban areas. Only in scattered instances, however, do they ever result in project abandonment, significant relocation, or redesign.

Transit contributes positive effects to the urban environment as a result of moving people more efficiently than the automobile. In addition to energy savings, less land is used for streets and parking, air pollution is lowered, and property values are raised.

Air Pollution

Table 34 compares rush hour emissions of motor buses and heavy rail with those of automobiles. In an effort by transit systems to reduce emissions, many motor bus systems use No. 1-D diesel fuel, a more refined grade of fuel that produces fewer emissions than regular grades of diesel fuel.

Air pollution associated with electric propulsion depends in large part on the means of electric power generation. Electric rail transit has the advantage of power generation plants located away from central city areas, thus removing pollutants associated with power generation away from the passenger and away from already polluted urban areas.

Noise and Vibration

One of the basic goals of bus and rail transit noise and vibration control efforts is to provide patrons with an

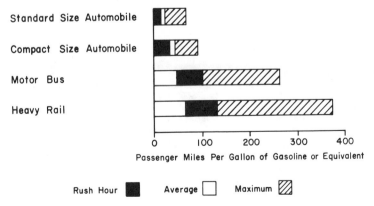

FIGURE 23. Energy use by urban transportation modes. Source: American Transit Association, *Transit Fact Book*, 1981.

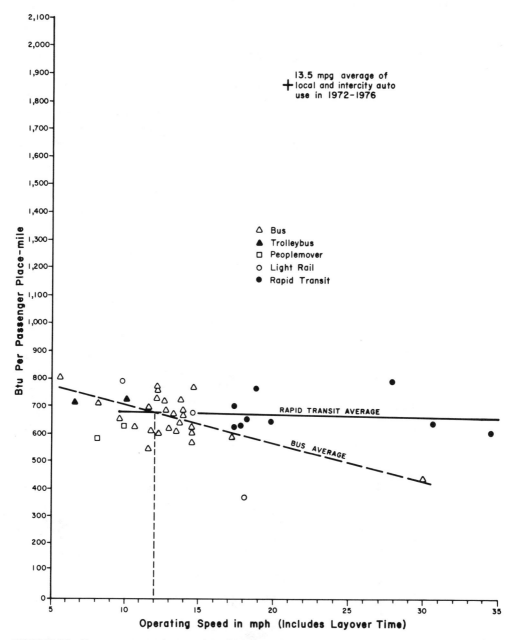

FIGURE 24. Gross energy requirements for vehicle operations. Note: Based on actual data from operational transit systems that are representative of the respective modes. Assumption: 1 KWH DC = 1,775 BTU gross input. 1 KWH AC = 11,467 BTU gross input. 1 gal deisel = 164,560 BTU gross input. 1 gal gasoline = 151,190 BTU gross input. Source: B. S. Pushkarev and J. M. Zupan, *Urban Rail in America: An Exploration of Criteria for Fixed-Guideway Transit*; Report UMTA-NY-60-0061-80-1, Urban Mass Transportation Administration, Washington, DC, Nov., 1980.

TABLE 34. Rush Hour Pollutant Emissions by Vehicle Type.

	Pollutant Emitted (grams per 100 passenger miles)		
Vehicle Type	Hydrocarbons (HC)	Nitrogen Oxides (NOx)	Carbon Monoxides (CO)
Automobile	70	220	650
Motor Bus	20	30	50
Heavy Rail	1	15	0

Source: American Transit Association, *Transit Fact Book*, 1981.

accoustically comfortable environment by maintaining noise and vibration levels in vehicles and stations within acceptable limits. In rail operations, adverse effects on the community should be reduced by minimizing transmission of noise and vibration to adjacent properties. Community acceptance requires control of airborne noise and vibration from surface and aerial operations and from yard operations. The design may also need to control ground-borne noise and vibration from subway operations.

Offending sources of noise and vibration are more of a problem for rail transit and can be divided into four general categories [17]:

1. *Vehicles*—running noise, including wheels rolling and sliding on rails; under car equipment noise from trucks, traction motors, brake shoes and other equipment; car interior noise from air conditioning, ventilation and other auxiliary equipment; and rattling noise from doors, seats, and windows
2. *Station equipment*—including ventilating fans, transformers, ducts, and maintenance facilities
3. *Trackwork*—includes track joints and irregularities, curves, and restraining rails
4. *Line structures*—bridges or aerial girders and decks as well as ground-borne radiation sources such as subways

Many of the above noise and vibration sources that pertain to vehicles and station equipment are of concern in motor bus operations. In addition to vehicle running noise, buses are also subject to vibration that is characteristic of reciprocating internal combustion engines. Vehicle interior noises from equipment as well as rattling noises can also be present in buses.

Interactions Between Transportation and High-Rise, High Density Living

To a degree, city densities have dictated transportation needs. Conversely, transportation often has inffluenced the patterns of urban growth and development. Understanding interactions between transportation and urban growth and density and the many alternatives that can occur is a key to sound overall planning, and especially to transportation planning.

High land values can dictate high-density living. In some cities, such as Hong Kong and Singapore, land values probably caused by shortages, make high-rise ("elevator") office buildings and residential units economical—and even essential.

Residences, shops, and work places are clustered along the radial transit lines, and near rapid transit and commuter rail stations. In American cities rapid transit and commuter rail lines concentrated commercial development in the city center while dispersing the residential growth along the transit arteries.

The automobile dramatically and undeniably altered urban development patterns by increasing accessibility and enabling people to live, shop, and work in the interstices between existing transit lines. This "filling in" of outward expansion between transit corridors accentuated the decentralization process already started by street railway, rapid transit, and commuter rail lines. Time and convenience became the controlling factors in choice of travel mode.

In most instances density patterns of today will largely prevail for the next twenty years, or longer; so, for transport planning purposes, drastic changes need to be anticipated. Exceptions can occur in countries with unique political philosophies and in areas where living standards are being rapidly upgraded.

Fixed transit corridors such as those provided by rail transit tend to have much stronger influences on land use and density than corridors served by buses. Because of its great flexibility, bus transportation is adaptable to almost any type of residential development. It is obvious that most private developers do not usually make investments based on the presence of bus service, if, indeed, transportation is a factor in their decisions. The presence of a fixed guideway, such as rail, imparts a permanence to the transportation system. The rapid transit systems in cities with high residential densities have a significant trip generation base in place and on which to plan. This creates a situation somewhat different from that in most cities in North America. Again, it is difficult to discern whether transportation is a key factor in effecting development, or whether transportation merely follows or serves, development. Transportation is not a principal determinant in residential densities; however, densities directly affect travel demands and patterns.

The effects of transport investment on urban development vary widely from city to city. The impacts are highly differentiated—they depend on changes in accessibility brought about by the transport system, but to an even greater extent on the condition of the land—whether vacant or built-up, whether right or wrong location, as well as on the interplay of market forces.

Transit impacts are seen most dramatically when looking at how rail transit has changed the urban landscape. The modern motor bus has total flexibility. It need not precede development because it can immediately serve development, of all types, in all locations. Once a rail guideway is installed, the likelihood of its permanence, or long life, is very great. Residential densities in New York, Chicago, Philadelphia, and Boston clearly reflect the result of rail transit development during the first half of the 20th Century. In all these cities, the patterns were similar. Rail transit lines had their greatest impacts around station areas located farthest from the city center which were previously undeveloped and unserved by public transport. Settlement generally was as close to stations as land availability allowed, while impacts were much less in already built up areas.

The land use impacts of new-generation systems built since 1960 in Atlanta, Montreal, San Francisco, Toronto, and Washington have varied widely. A study of impacts of rail transit in United States cities shows that successful cases always had a variety of other factors present. These factors included land availability, its ease of assembly, the social and physical characteristics of the area, general economic conditions, community support, and public land use policies. Conversely, when these forces were absent or weak, few land use impacts, especially in high-density residential areas, were found.

The Bay Area Rapid Transit System (BART) in the San Francisco-Oakland area of California has influenced land use and urban development largely through its secondary impacts on zoning regulations, redevelopment finances and civic improvements. Effects to date have been small but not inconsequential. BART has affected locational and timing decisions for at least six major residential projects representing approximately ten percent of the 1965-1978 volume of housing built in BART corridors. Few projects have been built that could be considered high-density even though nine station areas have been specifically zoned for high or medium-density residential development. However, some communities have explicitly prohibited high-density development, in spite of BART's presence. There is strong public feeling in some outlying residential communities against permitting high-density developments, regardless of attractiveness of BART [23].

In Washington, DC, the Metro system has speeded up commercial development at some key stations. Although not strong, there has been some measurable effects on high-density residential development as well. Between 1970 and 1976 prior to Metro, 17 percent of high-density residential permits were located in station areas; between 1971 and 1980 after Metro opened, 23 percent of the high density units permitted were located in station areas. Even after Metro, most of the higher density units permitted in the region were authorized for construction in locations outside the influence of Metrorail stations [24].

Today's rail transit improvements in American cities are superimposed on an auto-oriented environment. In location, lines are fitted to major thoroughfares, freeways, topography, and open-areas—not necessarily to density patterns. Therefore, they do not provide the dramatic improvements in overall access that was common before 1930. These systems, by themselves, do not have strong impacts on residential developments, unless they are complimented by many other factors.

The relationship of rapid transit to housing is much different in cities in the Far East than in the United States. For example, when the Metro was planned for Hong Kong, much high-rise residential capacity was already in place and much more was committed. This situation became a factor in location of lines. But, there were severe restrictions, including topography, high-density industrial and commercial uses that were already in place, and the exceptionally high costs for land. In Hong Kong, some impacts of the Metro on housing are already apparent, even though the transit system is not yet complete.

One of the most interesting plans related to the Metro is on Hong Kong Island and is known as the Kornhill site. It is about five miles distant from the Star Ferry. It is now planned for completion about 1986. It will contain approximately 8,000 flats—2,000 of which will be set aside for the Government's middle income housing plan. It will be a self-contained community. Development of this project will be jointly undertaken by the MRT and a private company. A unique feature of the site is that the MRT at this location will be on reclaimed land; the fill material for the reclamation was taken from the Kornhill site. Housing was considered to be the best use of hillside platforms created in the removal of the material. Of some eight development sites proposed on the Hong Kong Island Line, only Kornhill will be developed primarily for residential use.

Looking at overall impacts of transit on residential development—past, present, and future, it is clear that rail transit caused increased residential densities in the past where it led development. Transit can increase densities in the future where it can provide a rapid service to growing still-to-be developed areas. To accomplish such development, stations should be widely spaced on lines that extend beyond the built-up area outside the "preferred" residential corridor. Success calls for—balancing services to existing markets, with those designed to catalyze and organize new residential growth. Anticipatory transit developments are limited in American cities due to limited resources, cost-effectiveness tests and political viability.

Many social, political, and economic factors influence urban residential densities, both high and low. These include: culture and tradition, topography and terrain, land values and availability, the economic return on land, car ownership levels and constraints, and government policies and attitudes. These can be added: the cost, convenience, and quality of public transport and road systems.

In countries where government policy fosters high-density residential development at specific locations, such as in new towns, or neighborhoods set apart from central cities, transit is a very effective tool, regardless of the density of development. Any new govenment decisions to foster high-density residential patterns should be accompanied by decisions to improve transit so that transit can serve as a development catalyst.

Housing policies vary widely throughout the world, especially as to relative responsibilities of public agencies and private groups. A key issue is a policy one which, obviously, must be made by Government—it has to do with the extent to which high-density development is to be encouraged, or allowed. From the private sector point of view, decisions relate more to economic factors and these are frequently impacted by the cost of land. The term "high-density" is a variable term, especially when considered on a global basis and in the context of governmental policies. What a city is to be like must be developed by planners and adopted by Government.

It is important to combine all available professional resources and experiences in the planning and developing of high-density housing. Transit and high-density living can be mutually supportive, but one does not necessarily assure the other, at least in the United States and other locations where free market real estate conditions prevail. Also needed are supportive policies for land assembly, high-density zoning incentives, and reasonable policies for automobiles. To achieve this symbiosis calls for a new look at how to manage both housing and transit system developments. High-rise, high-density housing is woven into the total urban fabric. It is an important component of the city. The underlying goals are how to best improve urban mobility, economy, and amenity through coordination of transport and urban development.

The Future of Urban Public Transportation

The automobile will continue to be the primary mode of transportation. Between now and the year 2000, transit ridership could increase substantially due to higher automobile and gasoline costs, increased urban congestion, and an aging population.

By the year 2050, it is anticipated that the United States population will increase some 34 percent over its 1983 level. The 1983 median age of 30.9 will go to 36.3 by the year 2000, while at the same time, the proportion of the population age 65 or over will go from 11.6 percent to 13.1 percent [3,25]. If the previous trends over the past two decades continue, urban areas can be expected to grow more rapidly than rural ones. The automobile population is expected to increase to 146 million by the year 2010, and areas that have geographical growth constraints, like Manhattan and San Francisco, will have major vehicular accommodation problems as population densities increase [25].

FUNDING NEEDS

Finding funding alternatives for operating costs is becoming a major problem. Escalating costs cause passenger revenue coverage to decline. Figure 25 presents data that estimates the 1989 operating deficit, assuming a continuing 8 percent annual increase in operating costs, a 2.5 percent per year increase in operating revenues as a function ridership increases and stable fares, and federal, state, and local operating assistance remaining level after 1984. Elimination of federal operating assistance would only compound the problem and would have far reaching effects on both the transit industry and the economy.

Changes in fare policy might bring in additional funds, but they would probably not be sufficient to offset future deficits. In addition, mandated fare box recovery rates become counterproductive when they arbitrarily raise fares to the point of causing ridership instability. Methods of cost control and improving productivity will have to be considered in an effort to curb increasing operating expenses. Controlling labor costs and providing realistic service levels are two areas that have good potential for meaning-results. Private sector financial participation is desirable since increasing relationships with private sector firms would be more profitable, but such participation must carefully consider effects of increased taxes, if any.

Over $8 billion in bus system capital improvements will be required over the next five years according to the American Public Transit Association (APTA). This will be split almost equally between vehicles (40 percent) and facilities (51 percent). About 82 percent of the reported bus needs will be used to maintain existing levels of service. Rail systems will require even more money with over $28 billion required between 1984 and 1988 for capital improvements in urban areas. Over half the rail needs will be used for modernization of existing systems.

RAIL SYSTEMS

New systems offer fast, frequent, and convenient service often utilizing park-and-ride stations that allow passengers to combine the best advantage of rapid transit and the automobile. New systems have higher operating voltages (up to 1,000 volts dc) on fully grade separated tracks, operate at higher running speeds and obtain higher average speeds. Labor costs are minimized by various forms of automatic train operations that allow one man train operations; and, automatic fare collection systems in stations make it possible to reduce station staffs.

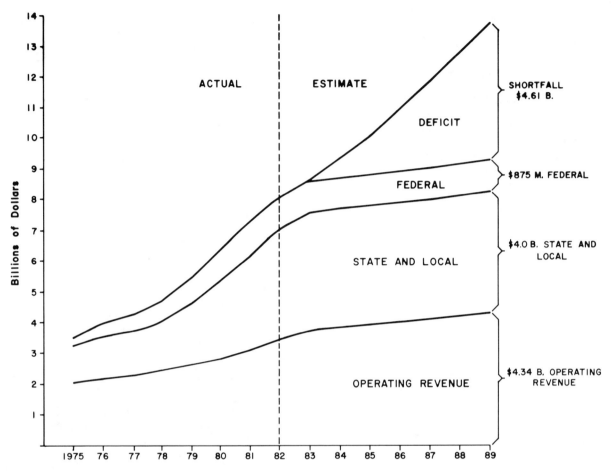

FIGURE 25. National transit cost and revenues. Source: AASHTO Report, *A Study of Future Directions of Public Transportation in the United States*, American Association of State Highway and Transportation Officials, Washington, DC, February 1985.

There has been a lot of new light rail technology in recent years. Light rail concepts have been applied in recent installations in the U.S. and Canada and are being planned or contemplated in other cities. Careful study is required to reduce capital costs by using both grade separated and mixed traffic right-of-ways without impairing reliability and safety with automobile interference. One of the better systems being built is the San Francisco Municipal Railway (Muni) which will use a subway under Market Street in the center city and grade level right-of-way in outlying residential areas. The grade-level track will use exclusive and medium strip right-of-ways, as well as in street right-of-ways with vehicular lane separation.

In 1985, one of the most ambitious, light rail systems was dedicated in Buffalo after two decades of planning, engineering, and construction. The line is 6.4 miles in length with 3.5 miles of bored tunnel (18.5 feet in diameter) and 1.8 miles is cut-and-cover tunnel design. The balance of the line operates in the central business district. The at-grade portion acts as a people mover in what will become the longest transit pedestrian mall with free rides between surface stations [26].

Fares are required on the underground portion using a concept already in use in San Diego, Edmonton and other west coast cities. There are no turnstiles, ticket takers or other collection devices which allows passengers to board trains directly subject to random proof of payment verification by inspectors. Tickets, purchased from underground and surface machines, also can be used for proof of payment as well as transfers from buses, monthly computer cards, and/or student passes.

Each of the 27 cars is 66 ft., 10 in. long, 8 ft., 7 in. wide and weighs 71,000 pounds. The cars can be operated from either end and accommodate 140 seated and standing passengers comfortably with a maximum capacity of 210 persons per car. Each car has three sliding doors on each side for boarding and alighting with boarding from floor level platforms in underground stations and retractable steps on the cars at surface stations. Overhead electrification supplies power to the car top pantographs [26].

The Niagara Frontier Transportation Authority in Buffalo is also planning a pilot project in late 1985 to test railbuses as an alternative to light rail. Although new in the U.S., the railbus is common in Europe. The experiment would test technological, operational, and maintenance aspects of the vehicle. It will operate on an unused commuter rail line. Depending on test results, the diesel power vehicles may be far less expensive to purchase and operate than light rail cars. This same vehicle was used in a recently concluded six week test in the Cleveland area. Initial results indicate that the test vehicle had a good operational and maintenance record and offers a comfortable ride at speeds of around 50 mph. The vehicle is capable of running 70 to 75 mph on longer runs without frequent stops [27].

OTHER POTENTIAL CHANGES

Service improvements that do not require expensive new facilities can help to attract riders by reducing travel time. This can be accomplished by various means, including dedicated bus lanes and express service between major traffic centers. Approaches to attracting potential riders that are not served by conventional bus and rail systems include a variety of paratransit services such as shared-ride taxis, carpools, vanpools, jitneys, subscription bus service, and others. Future paratransit will probably make extensive use of computer and microcomputer routing and dispatch services.

A new type technology with future potential in shuttle or loop type applications is a form of people mover called automated guideway transit (AGT). There are various levels of sophistication which vary from Personal Rapid Transit (PRT), using small vehicles with more complex and extensive networks and controls, to simpler kinds of shuttle and loop transit (SLT), which can carry upwards of 100 people in large vehicles. There is still a lot of disagreement about the pros and cons of automated guideway transit and whether it is reliable, safe, and economical in comparison to other more proven alternatives. There are other exotic possibilities but one more straightforward concept with promise is a dual mode bus that can operate either as a trolley bus using electric power or as a regular diesel vehicle.

Considerable progress had been made in the design of new operating support systems and more are expected in the future. Computers have made possible automation in such functions as fare collection and vehicle dispatch and control. The automation trend shows savings in labor costs and will continue in both new and old systems. The use of computers in office type functions is well established and is seeing more use in transit management for planning, modeling, and system design, and other complex areas. Innovative management will also become more commonplace in transit and will include many areas including: labor agreements; realizing cost saving potentials that make use of private sector operators working with various governments to adopt automobile disincentives; marketing programs to inform the public about available services and improve the image of transit; and periodic fare adjustments in accordance with inflation and service charges to avoid large disruptive increases. To ensure a steady increase in transit efficiency and overall productivity, in the near future, management and systems innovations need to be pursued since the rate of technological development will not be as rapid as in the past.

REFERENCES

In preparing this chapter, many relative materials were assembled and reviewed. All efforts were made to give proper credit to materials used. The following sources were referenced in the manuscript:

1. N. D. Lea Transportation Research Corp.; *Lea Transit Compendium, Vol. II,* No. 1, Huntsville, AL (1975).
2. Pushkarev, B. S. and J. M. Zupan, *Urban Rail in America: An Exploration of Criteria for Fixed-Guideway Transit,* Report UMTA-NY-06-0061-80-1, Urban Mass Transportation Administration, Washington, DC (Nov. 1980).
3. U.S. Bureau of the Census, *Historical Statistics of the United States, Colonial Times to 1970,* Series A57, Washington, DC (1975); U.S. Bureau of the Census, *Statistical Abstract of the United States,* Washington, DC (1985).
4. L. M. Schneider, *Marketing Urban Mass Transit,* Division of Research at Harvard University Graduate School of Business Administration, Boston (1965).
5. Levinson, H. S., "Modal Choice and Public Policy in Transportation," *Urban Transportation—Perspectives and Prospects,* Eno Foundation for Transportation, Inc., Westport, CT (1982).
6. Pushkarev, B. S. and J. M. Zupan, *Public Transportation and Land Use Policies,* Indiana University Press, Bloomington, Indiana (1977).
7. Cudahy, B. J., *Public Transport Administration,* Eno Foundation for Transportation, Inc., Westport, CT (1985).
8. Wilbur Smith and Associates, *The Potential for Bus Rapid Transit,* Prepared for Automobile Manufacturers Association, Inc., Detroit, MI (1970).
9. Levinson, H. S., "Parking in a Changing Time," *Urban Transportation—Perspectives and Prospects,* Eno Foundation for Transportation, Inc., Wesport, CT (1982).
10. American Public Transit Association, *Transit Fact Book,* Washington, DC (1981, 1985).
11. Institute of Transportation Engineers, *Transportation and Traffic Engineering Handbook,* W. S. Homburger, ed., Prentice-Hall, Inc., Englewood Cliffs, NJ (1982).

12. Transportation Research Board, *National Coopeative Highway Research Program Synthesis, 69,* Washington, DC (1980).
13. Levinson, H. S. and D. B. Sanders, *Reserved Bus Lanes on Urban Freeways: A Macromodel,* Transportation Research Record No. 513, Washington, DC (1974).
14. "Highway Capacity Manual," *Special Report 87,* Transportation Research Board, Washington, DC (1965).
15. Levinson, H. S., et al., *Bus Use of Highways: Planning and Design Guidelines,* National Cooperative Highway Research Program Report 155, Transportation Research Board, Washington, DC (1975).
16. "Busway Systems Expanding in the USA," *World Highways, Vol. XXXVI,* No. 6, International Road Federation, Washington, DC (July/August 1985).
17. Hintzman, K. W., et al., *Rail Transit Criteria for System Review and Preliminary Design,* Report No. UMTA/CA/PD-78/1, California Department of Transportation, Sacramento, CA (Dec. 1979).
18. *An Overview of Paratransit,* Paratransit Technology Sharing, Transportation Systems Center, U.S. Department of Transportation, Washington, DC (1981).
19. "The National Ridesharing Demonstration Program: Comparative Evaluation Report," *Paratransit, Vol. 8,* No. 1, Transportation Research Board, Washington, DC (August 1985).
20. Weiner, E., "The Characteristics, Uses and Potentials of Taxicab Transportation," *Urban Transportation—Perspectives and Prospects,* Eno Foundation for Transportation, Inc., Westport, CT (1982).
21. Barton-Aschman Associates, Inc., R. H. Pratt & Co. Division, *Traveler Response to Transportation System Changes,* 2nd Ed., Contract No. DOT-FH-11-9579, U.S. Department of Transportation, Washington, DC (July 1981).
22. McCoy, M., "Transit's Energy Efficiency," *Urban Transportation—Perspectives and Prospects,* Eno Foundation for Transportation, Inc., Westport, CT (1982).
23. Metropolitan Transportation Commission, *The Environmental Impacts of BART,* Berkeley, CA (1978).
24. Metropolitan Washington Council of Governments, *Trends Before Metrorail,* Washington, DC (July 1982).
25. ASSHTO Report, *A Study of Future Directions of Public Transportation in the United States,* American Association of State Highway and Transportation Officials, Washington, DC (February, 1985).
26. Schieber, L. "A Subway for Buffalo," *Engineering Progress of Western New York," Vol. 5,* No. 1, published by the Faculty of Engineering and Applied Sciences, State University of New York at Buffalo and the Erie-Niagra Chapter of New York State Society of Professional Engineers, Buffalo, New York (Spring 1985).
27. Simon, P., "Railbus Test in South Town May Boost Commuter Line," *The Buffalo News,* Buffalo, New York (July 18, 1985).
28. Smith, W. S., "Interaction Between Transportation and High-Rise, High-Density Living," Paper in *High-Rise, High-Density Living—SPC Convention 1983, Selected Papers,* Published by Singapore Professional Center, Singapore (August, 1984).
29. American Bus Association, *Annual Report—1984,* Washington, DC (1984).
30. Becker, R. and W. Simpson, "Getting Around New Orleans: Variety is the Spice of Life," *ITE Journal, Vol. 55,* No. 7, Institute of Transportation Engineers, Washington, DC (July, 1985).
31. Schwartz, G. G., *Where is Main Street, U.S.A.?,* Eno Foundation for Transportation, Inc., Westport, CT (1984).
32. "Transit Funding: Triumph—Then Tribulation," *Mass Transit, Vol. X,* No. 4, Washington, DC (April 1983).
33. Young, D., "Transit in Transition—Major Changes Ahead," *Mass Transit, Vol. X,* No. 3, Washington, DC (March, 1983).
34. Middleton, W. D., *The Time of the Trolley,* Kalmbach Publishing Co., Milwaukee, WI (1967).
35. Smerk, G. M., "The Streetcar: Shaper of American Cities," *Traffic Quarterly,* The Eno Foundation for Transportation, Inc., Westport, CT (October, 1967).
36. Pushkarev, B. S., J. M. Zupan, and R. S. Cumella, *Urban Rail in America: A Regional Plan Association Book,* Indiana University Press, Bloomington, Indiana (1982).
37. Spengler reported by D. E. Boyce, *The Impact of Rapid Transit on Suburban Residential Property Values and Land Development,* U.S. Department of Commerce, Washington, DC.
38. Knight, R. C., *The Impact of Rail Transit Land Use—Evidence and a Change of Perspective Transportation,* Everson Scientific Publishing Company, Amsterdam (1980).
39. Ashtakala, M., *The Impact of the Metro on an Inner City Area,* Research Project Report, McGill University, Montreal (1978).
40. *The Environmental Impacts of BART,* Metropolitan Transportation Commission, Berkeley, CA (1978).
41. *Trends Before Metrorail,* Metropolitan Transportation Commission, Berkeley, CA (1982).
42. Owen, W., *The Accessible City,* The Brookings Institution, Washington, DC (1972).
43. Levinson, H. S., "Characteristics of Urban Transportation Demand," Prepared by Wilbur Smith and Associates for Urban Mass Transportation Administration (1978).
44. Jacobs, M., R. E. Skinner, and A. C. Lemer, *Transit Project Planning Guidance—Estimation of Transit Supply Parameters,* Report No. UMTA-MA-09-9015-85-01, Urban Mass Transportation Administration, Washington, DC (1984).
45. "Highway Capacity Manual," *Special Report 209,* Chapter 12, Transportation Research Board, Washington, DC (1985).
46. Gray, G. E. and L. A. Hoel (eds.), *Public Transportation Planning, Operations, Management,* Prentice-Hall, Inc., Englewood Cliffs, NJ (1979).
47. *Bus Route and Schedule Planning Guidelines,* NCHRP, Synthesis of Highway Practice 69 (1980).
48. Rothenberg, N., *Public Transportation—An Element of the Urban Transportation System,* FHWA U.S. Department of Transportation, Washington, DC (February 1980).

49. Vuchic, V. R., *Urban Public Transportation: Systems and Technology,* Prentice-Hall, Inc., Englewood Cliffs, NJ (1981).
50. Levinson, H. S., "Analyzing Transit Travel Time Performance."
Transportation Research Record 915, Transportation Research Board, Washington, DC (1983).
51. Pushkarev, B. S. and J. M. Zupan, *Where Rail Transit Works,* Regional Plan Association, New York, NY (1978).
52. Dickey, J. W., *Metropolitan Transportation Planning* (2nd ed), McGraw-Hill Book Company, New York, NY (1983).
53. Hoey, W. F. and H. S. Levinson, "Bus Capacity Analysis," *Transportation Research Record 546,* Transportation Research Board, Washington, DC (1975).
54. Homburger, W. S., "Notes on Transit System Characteristics," *Information Circular 40,* University of California, Institute Transportation Studies, Berkeley, CA (1975).
55. Levinson, H. S. and Texas Transportation Institute, *Conceptual Planning and Design,* Lockwood Transit Center (March 1983).
56. Levinson, H. S., et al., *Bus Use of Highways: State-of-the-Art,* National Cooperative Highway Research Program Report 143, Washington, DC (1974).
57. Scheel, J. W. and J. E. Foote, *Comparison of Experimental Results with Estimated Single Lane Bus Flows Through a Series of Stations Along a Private Busway,* Research Publication GMR-888, General Motors Research Laboratories, Warren, MI (1969).
58. Hodgkins, E. A., "Effects of Buses on Freeway Capacity," *Highway Research Record No. 59,* Highway Research Board, Washington, DC (1965).
59. Kraft, W. H., *An Analysis of the Passenger Vehicle Interface of Street Transit Systems with Applications to Design Optimization,* Doctoral Dissertation, New Jersey Institute of Technology, Newark, NJ (September 1975).
60. *Capacities and Limitations of Urban Transportation Modes,* an informational report, Institute of Traffic Engineers, Washington, DC (May 1965).
61. Transportation Research Board, Circular 212—*Interim Materials in Highway Capacity,* Washington, DC (1980).
62. *Draft Alternatives Analysis Procedures and Technical Guidelines: Appendix A Estimating of Transit Supply Parameters,* Urban Mass Transportation Administration (1980).
63. Skinner, L. E., *Comparative Costs of Urban Transportation Systems,* FHWA, U.S. Department of Transportation, Washington, DC (June, 1978).
64. Levinson, H. S., *Inlet Transit Travel Times Analysis,* Prepared for Urban Mass Transportation Administration, Washington, DC (April, 1982).
65. "Transportation and Urban Development," prepared by the U.S. Conference of Mayors on Urban Transportation: New Roles for Mayors and Cities, DOT-P-30-20-33, U.S. Department of Transportation, Washington, DC (October, 1980).
66. "Transit Financing," prepared by the U.S. Conference of Mayors on Urban Transportation: New Roles for Mayors and Cities, DOT-P-30-80-34, U.S. Department of Transportation, Washington, DC (October, 1980).
67. "Auto in the City," prepared by the U.S. Conference of Mayors on Urban Transportation: New Roles for Mayors and Cities, DOT-P-30-80-35, U.S. Department of Transportation, Washington, DC (October, 1980).
68. Hutchinson, B. G., *Principles of Urban Transport Systems Planning,* McGraw-Hill Book Company, New York, NY (1974).
69. Owen, W., *Transportation for Cities,* The Bookings Institution, Washington, DC (1976).
70. Lang, A. S. and R. M. Soberman, *Urban Rail Transit: Its Economics and Technology,* Massachusetts Institute of Technology Press, Cambridge, MA (1964).
71. Kanafani, A. *Transportation Demand Analysis,* McGraw-Hill Book Company, New York, NY (1983).

CHAPTER 15

Mass Transportation: Carpool Assignment Technique Application

WILLIAM W. MANN*

INTRODUCTION

A technique to determine the usage of carpools on high occupancy vehicle (HOV) lanes has been developed by the Metropolitan Washington Council of Governments/Transportation Planning Board (MWCOG). The purpose was to be able to test the impact of HOV facilities on the entire transportation network as a part of the "Long Range Plan Alternatives Analysis" for the Washington region. The technique can be copied and applied to any other metropolitan area. It can also be accomplished in a short time. The only prerequisite is that the area's transportation planning process be able to produce zone-to-zone auto person work trips and average zone-to-zone car occupancies. Most metropolitan planning organizations already have this capability as part of their continuing travel forecasting process.

While this technique was developed for HOV facilities defined as requiring four or more occupants, it can be modified to estimate usage of HOV facilities requiring two or more occupants or three or more occupants. The application for each of these priority treatments is examined in this chapter.

It was imperative for MWCOG to have a tool to estimate HOV usage not just to test proposed HOV facilities, but to simulate the existing highway system, much of which is already devoted to preferential treatment for HOV vehicles. For instance, Shirley Highway has been operating for several years with the median lanes used exclusively as a reversible HOV (vehicles with four or more occupants) facility for cars and buses. Also, I-66, inside the Capital Beltway, opened for use in December 1982. It operates with the peak-period, peak-direction providing exclusive use for buses and cars with four or more occupants.

THEORY

The underlying theory for this approach is that there is a predictable distribution of vehicles by the number of occupants if the average car occupancy is known [5]. In other words, given the average car occupancies between zones, the distribution of vehicles by vehicle occupants can be determined based on observed vehicle occupancy distributions.

Vehicle occupancy distributions, observed in the Washington region, are shown in Figures 1 and 2. Figure 1 represents data wherer no HOV facility exists, and Figure 2 represents data where one HOV facility exists, namely, Shirley Highway, which provides preferential treatment for carpools of four or more occupants. All of the data were developed from measurements of inbound weekday traffic only from 6:30 A.M. to 9:30 A.M. and on all highway facilities crossing the Ring 1 and Beltway cordon. Ring 1 is located about 1 mile (1.61 km) from the center of the region and the Beltway is about 10 miles (16.1 km) from the center. The reason for delineating these points separately on the graphs in Figures 1 and 2 was to show that different areas of the region serving different employment densities can be explained with the same vehicle occupancy distribution model.

An example of how to read these graphs is as follows:

1. If the average car occupancy is 1.3, using Figure 1, then the percent of persons in one-occupant cars is 60%.
2. In two-occupant cars, it is 28%.
3. In three-occupant cars, it is 7%.
4. In four-or-more occupant cars, it is 5%.

Figure 2 shows the distribution of vehicle occupancy sizes

---

*Metropolitan Washington Council of Governments, National Capital Region Transportation Planning Board, Washington, DC

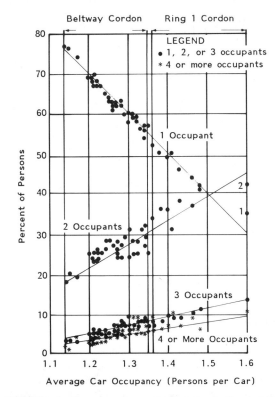

FIGURE 1. Car occupancy distributions (without high-occupancy facilities).

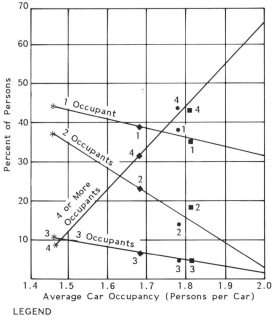

FIGURE 2. Car occupancy distributions (on high-occupancy facilities). Source: 1980 Beltway and Ring 1 cordon count data.

starting at 1.46 persons per car and increasing with increasing vehicle occupancies. The reason for not extrapolating the curves for average occupancies below 1.46 persons per car was because this is the point where the curves between HOV preferential treatment and non-HOV preferential treatment intersect.

MULTILOAD[1]

Another very important part of the process of estimating carpool sizes on various links in the network is accomplished with a unique traffic assignment computer program called MULTILOAD [4]. MULTILOAD has two significant capabilities which traditional traffic assignments do not have. They are as follows:

- MULTILOAD can load up to four trip tables, perform capacity restraint on the total volume, and report link volumes for each of the four tables (purposes or modes).
- MULTILOAD, in loading the four trip tables and performing capacity restraint with the total volume, can permit or restrict any trip table(s) from loading on specified links. For example, only HOV trips can permitted to load on HOV links; truck trips can be excluded from Parkways, etc.

Traditional traffic assignment algorithms load a single trip table to a single network without regards to specific link operational characteristics. That approach was unacceptable in the Washington, DC region because of the need to be responsive to prohibiting certain classes of vehicles from using specific roadways during portions of or all of the day. These roadways with their unique operational characteristics are:

- I-395, Shirley Highway—This facility is an eight lane radial freeway connecting the Capital Beltway in Virginia with Washington, DC and has two additional lanes in the median which operate as reversible HOV lanes.
- I-66, Custis Parkway—This facility is a four lane radial freeway connecting the Capital Beltway in Virginia with

[1]MULTILOAD is a FORTRAN computer program written for an IBM 4341 and is available for the cost of a tape copy through Metropolitan Washington Council of Governments, 1875 Eye Street, N.W., Washington, DC 20006.

Washington, DC. During morning and evening rush periods, HOV's are allowed exclusive use of the peak direction lanes only. For the minor flow direction, during peak hours, any *car*, regardless of the number of occupants, is allowed full use of the facility. Also, being classified as a parkway, trucks are prohibited from its use at all times.

- Parkways—There are numerous parkways in the Washington, DC region and trucks are prohibited from using these facilities at all times. There has always been a need in the traffic assignment process to prohibit trucks from Parkways. This need was never sufficient to solely justify building an algorithm such as MULTILOAD because of the large software development costs.

In addition to needing a better tool with which to estimate traffic on existing highway facilities, there was a more pressing need to evaluate proposed HOV facilities and proposed Parkways. MULTILOAD was designed to meet all of these needs.

MULTILOAD Inputs

There are essentially three inputs to MULTILOAD. They are:

- District[2] trip tables (any number from 1 to 4).
- One zone[2] highway network, with one out of four separate operational characteristic codes associated with each link in the network.
- Selected zones along with associated zone land activity data for the subarea where zone-level traffic assignments are needed.

The four trip tables selected most of the time, when MULTILOAD is used by practitioners in the Washington, DC region, are:

- Trucks
- LOV (low occupant vehicles) for work trips
- Non-work auto trips
- HOV (high occupant vehicles) for work trips.

And, the four operational link characteristics selected most of the time by practitioners in the Washington, DC region, when running MULTILOAD, are:

- Regular links (Type 1)
- Parkway links which prohibit trucks (Type 2)
- I-66 links: off peak hour links (Type 3)
- I-66 links: peak hour links (Type 4)
- I-395 median lanes links (Type 4)

I-66 had to be coded as two parallel roadways. One was coded as Type 3 links with a daily capacity based on use by non-work trips during all hours except rush hours. The other parallel I-66 set of links were coded as Type 4 links (HOV links) which prohibited all vehicles except carpools and buses from using the facility.

A summary of the overall process for subarea traffic assignments with MULTILOAD is illustrated in Figure 3.

MULTILOAD Procedure

The basic objective of MULTILOAD is to load four trip tables to a zone-level highway network in such a manner that trucks are prohibited from Parkways and HOVs are given exclusive access to certain links. To accomplish this, a discrete set of network links must be associated with each trip table.

The steps MULTILOAD uses to achive its objectives are summarized below:

Step   Operation

1. Read and build the entire network; speeds are read from a user supplied matrix of speeds based on ring and route type and converted to level of service (LOS) "A" speeds.
2. Read a listing of all special links (Types 2, 3, and 4), then remove these links from the network, but keep them in computer storage for later recall.
3. Build tree for zone 1 over Type 1 links. A tree specifies the minimum path from zone 1 to all other zones in the network.
4. Load trip purpose 1, origin zone 1, to the network. For example, load truck trips over regular links (Type 1 links only).
5. Update the tree built in Step 3 with Type 2 links. Tree updating isfar more efficient in terms of computer time than rebuilding the entire tree.
6. Load trip purpose 2, origin zone 1 to the network. For example, load LOV trips over both regular links (Type 1 links) and Parkway links (Type 2 links).
7. Update the tree built in Step 5 with Type 3 links.
8. Load trip purpose 3, origin zone 1 to the network. For example, load non-working trips over regular links (Type 1 links), Parkway links (Type 2 links) and the off-peak I-66 roadway links (Type 3 links).
9. Update the tree built in Step 7 with Type 4 links.
10. Load trip purpose 4, origin zone 1 to the network. For example, load HOV trips over paths using all links in the entire network.
11. Perform capacity restraint over the entire network. New speeds are calculated based on a volume-to-capacity ratio and the LOS A speed.
12. Repeat Steps 3 through 11 for zone 2 and again for each succeeding zone. Also, calculate new capacity restrained speeds for each link in the network, only after loading a specified number of zones. The capacity restraint subroutine in MULTILOAD can be applied

---

[2]Districts are defined as groups of zones. MWCOG has the region divided into 1345 zones and 200 districts.

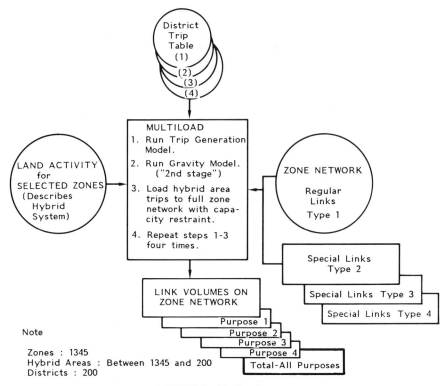

FIGURE 3. Multiload process.

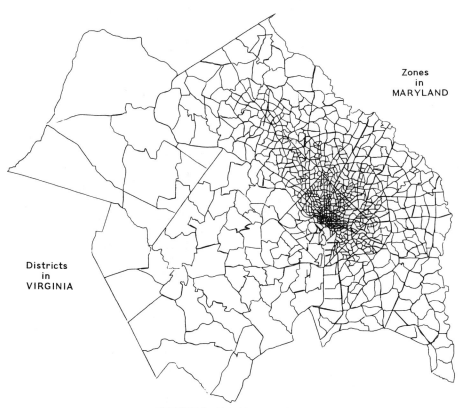

FIGURE 4. Hybrid area system.

after loading a user specified number of zones; i.e., 50. This approach of applying capacity restraint after loading 50 zones should be used only when the reverse digits loading sequence is applied. The reason is that this assures a pseudo-random loading sequence.

MULTILOAD Outputs

The output from MULTILOAD contains many of the traditional traffic assignment outputs such as link volumes, capacity restrained speeds, and volume-to-capacity ratios for each link in the network. In addition, it reports the total volume broken down into four trip categories. These are the same categories as input to MULTILOAD, i.e., truck trips, LOV work trips, non-work trips, and HOV work trips. Parkway links, of course, have no trucks assigned to them and HOV links report only HOV trips. The computer program also outputs vehicle-miles-of-travel by ring and route type for each trip category.

The "select-link" option, incorporated within the capacity restraint assignment subroutine, allows one to determine great detail about all trips passing through select-links. For example, one could determine the origin zones of all HOV trips passing through a select-link as well as each trip purpose passing through the link.

Another option available to the user is to print trees as they are built. Instead of having one tree for each origin zone, four trees are now obtained (one for each trip category) for each origin zone with each one possibly using a different path to reach the same destination zone.

MULTILOAD Subarea Applications

One very important feature of MULTILOAD is that it streamlines subarea applications. For example, suppose one wanted to create a hybrid area network with zone-level detail in the District of Columbia and Maryland suburbs and district-level detail elsewhere. This area system is illustrated in Figure 4. Hybrid area systems with great detail in a subarea and less detail elsewhere are preferred for subarea studies because they are easier to process, cheaper in terms of computer costs, and allow quicker turn-around time to test and evaluate. Sometimes zones are further subdivided into subzones where more detail is needed. (Usually, the MWCOG 4-step transportation planning process consists of using districts for trip generation and trip distribution, while zones or hybrid area systems, are used for modal split and traffic assignments.)

MULTILOAD reads up to four district level trip tables and a full zone level network for the entire region. It then expands each of these trip tables to zone level using a zone trip generation model and a modified gravity model technique, or it can expand these trip tables to a hybrid area system, specifically designed for a subarea study. If hybrid area trip tables are created by MULTILOAD, the program would then load these trip tables to the full zone network through zone centroid connectors of the selected zones.

It is this zone selection process that describes the hybrid area system to MULTILOAD. For example, if district detail is desired in Montgomery County, Maryland, the user would select one zone within each district in Montgomery County to represent the entire district. It is always best to choose a zone that is closest to the actual centroid of the district. If the user prefers a smooth transition from zone detail in the Virginia suburbs to district detail in Maryland, one might select two or three zones per district for districts in the transition area. In the subarea itself, zones or possibly subzones could be used.

Inputting land activity data for selected zones and equating these zones to their respective districts is all the user must do to create the hybrid area network and the hybrid area trip table. The program does the rest. It expands the district trip table into the hybrid area system trip table and then assigns it to the full zone network through the centroid connectors of the selected zones.

To produce the hybrid area system trip table, the user must input parameters for the trip generation and gravity models. The program first calculates hybrid area trip productions and trip attractions based on a separate model for each trip purpose. It then executes what is referred to as a "second stage gravity model" for each trip purpose (or mode) to produce the zone/district trip table for each trip purpose (or mode). The formula for the trip table expansion gravity model is shown below:

$$T_{ij} = \frac{1}{2} \times \frac{P_i}{P_I} \times \frac{A_j}{A_J} \times \frac{F_{ij} \times T_{IJ}}{\Sigma_j \frac{A_j}{A_J} \times F_{ij}}$$

$$+ \frac{P_j}{P_J} \times \frac{A_i}{A_I} \times \frac{F_{ij} \times T_{JI}}{\Sigma_j \frac{P_j}{P_J} \times F_{ij}}$$

where

T = Trips
P = Trip productions
A = Trip attractions
F = Friction factors
I,J = District designations
i = Zone designation within district I
j = Zone designation within district J

The zonal trip productions and trip attractions required in the trip table expansion model are calculated from a zonal trip generation model. The zonal trip generation model structure is illustrated below:

$$P = a_1 L_1 + a_2 L_2 + \ldots + a_n L_n$$

$$A = b_1 L_1 + b_2 L_2 + \ldots + b_n L_n$$

where

- P = Trip productions
- A = Trip attractions
- a_1 = User supplied coefficient for land-activity category 1
- L_1 = Land activity data for category 1
- n = Limited to six categories of land activity data

It is important to understand that this model only needs to determine the relative trip weights of one land activity to another for use in allocating total trips from district-level to zone-level. Therefore, a regional model is appropriate without the need for subarea correction factors.

For application in the MWCOG region, the work trip model consists of pro-rating trip productions from districts to zones by the number of households per zone and trip attractions by the number of jobs per zone. For non-work trips, trip attractions are determined using a relative weighting of 2, 5, and 1 for office employment, retail employment, and other employment, respectively. Note the heavy weighting of retail employment for the non-work model trip attraction estimation process.

APPLICATION

The procedure used by MWCOG to produce highway network traffic assignments of HOV, LOV (low-occupant vehicle), and other vehicle trips is summarized in Figure 5. A description of the steps shown in this flow chart are explained as follows.

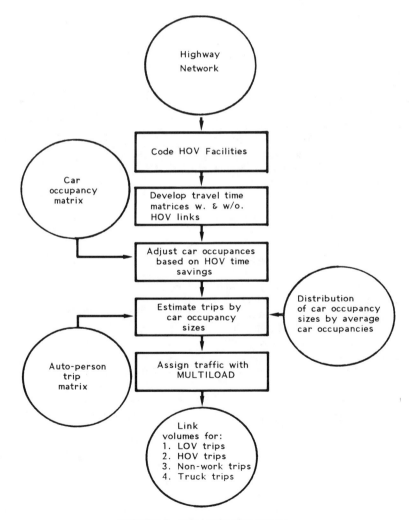

FIGURE 5. HOV estimating process.

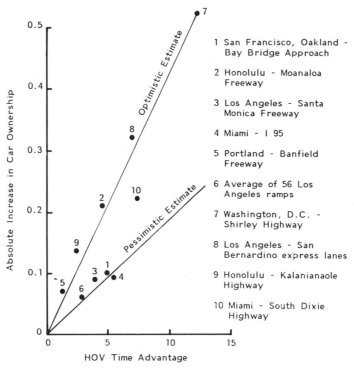

FIGURE 6. Facility impacts. Source: Correspondence with Frederick Wagner and based on HOV Training Course by JHK, et al. [2].

Step 1: Code HOV Facilities

The median lanes of Shirley Highway were coded as separate network links for loading HOV trips only. The outside lanes were coded in the usual manner. I-66 had to be coded in a unique way because it is a four-lane divided highway with the peak-period, peak-direction open to carpools only. Also trucks are prohibited at all times from using any part of I-66. In this case, parallel links were coded: one for HOVs, the other for LOVs. Both sets of links had to be coded to exclude trucks. Since separate links were coded for LOVs and HOVs, each could be given separate capacities which were needed for a more realistic capacity restraint process. This special coding of excluding trucks and allowing only HOVs to use specific links while loading multiple trip tables in a single computer execution can only be achieved with MULTILOAD.

Step 2: Develop Travel Time Matrices With and Without HOV Links

Restrained speeds on all links (HOV and non-HOV) in the network are an ouput of MULTILOAD. One could use a recent capacity restrained network to estimate LOV zone-to-zone travel times. The analyst could then code HOV links at level-of-service "A" speeds to obtain HOV zone-to-zone travel times. If one has the time and budget, one could iterate the process by taking restrained speeds on all links output from the process and using them at the beginning of the process. MWCOG, due to time constraints, did not use this recommended approach. Instead, travel time differences were estimated directly, using judgment by MWCOG staff; i.e., zones adjacent to Shirley Highway were assigned large HOV time savings to the regional core, and zones further away from Shirley Highway were assigned progressively smaller time savings.

Step 3: Adjust Car Occupancies Based on HOV Time Savings

Based on the HOV time savings from Step 2, the average zone-to-zone car occupancies output from the car occupancy model was adjusted based on the chart shown in Figure 6. For instance, if the travel-time savings from Step 2 showed an HOV time savings of 10 minutes, the average zone-to-zone car occupancy was increased by 0.38 persons per car. This was accomplished with the computer program UMATRIX [6], which is available in the Urban Transporta-

TABLE 1. Car Occupancy Distributions for Non-HOV Facilities.

% of Persons in 1-Occupant Cars = 187 − (98 × CO)
% of Persons in 2-Occupant Cars = −50 + (60 × CO)
% of Persons in 3-Occupant Cars = −20 + (21 × CO)
% of Persons in 4+ Occupant Cars = −17 + (17 × CO)

where:

CO = Average Car Occupancy and subject to:
$1.1 < CO < 1.7$

TABLE 2. Car Occupancy Distributions for HOV Facilities.

% of Persons in 1-Occupant Cars = 79 − (24 × CO)
% of Persons in 2-Occupant Cars = 131 − (64 × CO)
% of Persons in 3-Occupant Cars = 37 − (18 × CO)
% of Persons in 4+ Occupant Cars = −147 + (106 × CO)

where:

CO = Average Car Occupancy and subject to:
$1.45 < CO < 2.05$

---

tion Planning System (UTPS) battery of transportation planning software.

Step 4: Estimate Trips by Vehicle Occupancy Size

Trip tables by vehicle occupancy sizes were estimated with the curves in Figures 1 and 2, and the adjusted average car occupancies from Step 3 (Optimistic Curve). Again, the computer program UMATRIX was used. Prior to the use of UMATRIX, the data in Figures 1 and 2 had to be converted to equations. Hand-fitted lines were drawn through the data and then equations to describe the lines were developed. These equations are shown in Tables 1 and 2. If the average car occupancy was less than 1.46 persons per car, then Table 1 was used. If the average occupancy exceeded 1.46 persons per car, then Table 2 was used only if the origin and destination of the trip were in any of the HOV corridors. Zone-to-zone auto occupancies, the equations in Tables 1 and 2, and zone-to-zone auto-person trips were all used to produce person-trip tables for:

- 1-occupant cars
- 2-occupant cars
- 3-occupant cars
- 4-or-more-occupant cars

To convert these person trip tables to vehicle trip tables, each trip table was divided by its average car occupancy. For the four-or-more-occupant cars, the average-occupancy was assumed to be 4.4 persons per car except for the Shirley Highway corridor where 4.6 persons per car was used. These assumptions for the future were based on observed data.

Step 5: Assign Traffic with MULTILOAD

Four trip tables were input to MULTILOAD. The four vehicular trip tables were:

1. LOV work trips
2. HOV work trips (4 or more persons per car)
3. Non-work trips
4. Truck trips

Truck trips were treated separately because there are several Parkways in the MWCOG region where trucks are prohibited, and trucks provide an important input to the vehicle emissions calculations. Non-work trips were treated separately in the loading process because of the special treatment of I-66. Only non-work car trips were allowed to load on the regular I-66 lanes and only HOV work trips were allowed to load on the special I-66 HOV links.

RESULTS

The results of applying this technique to estimate HOV demand compare favorably with two separate cooridor studies forecasting HOV demand. One of these studies estimated HOV usage on I-66; the other estimated HOV usage on an extension of Shirley Highway about 20 miles (32.2 km) more into the Virginia suburbs. Both corridor studies were done by JHK & Associates [1,3]. Comparisons between the MWCOG regional approach and the JHK corridor approach are shown in Table 3.

APPLICATION TO OTHER METROPOLITAN AREAS

The first step in applying this technique to other metropolitan areas would be to collect local data on car occupancy distributions in the suburbs and close to the central area on radial facilities in the peak-period peak-direction. This data could then be used to develop equations similar to the ones in Table 1.

TABLE 3. HOV Volumes in Peak Period Peak Direction.

	JHK [1,3]	COG/TBP	% Difference
I-66, Close-in	5,200	4,000	23
Shirley Highway at Beltway	3,800*	3,000	21

*The peak period volume was determined by converting the reported peak-hour volume to peak period, assuming 65% of the peak-period HOV volume occurs in the peak hour.

Next, if one wanted to test an HOV facility requiring four-or-more persons per car, the MWCOG approach could be used substituting local data for Table 1.

If one wanted to test an HOV facility requiring two-or-more or three-or-more occupants per car, then techniques such as the following could be devised.

HOV Facility Requiring Two-or-More Occupants per Car

HOV travel time reductions should have very little effect on increasing three-or-more-occupant cars but should significantly reduce single-occupant cars and increase two-or-more-occupant cars. Therefore, an approach would be to estimate the car occupancy distribution "after" the HOV improvement, using the models in Tables 1 and 2. The increase in three-or-more-occupant cars could then be converted into additional two-occupant cars.

HOV Facility Requiring Three-or-More Occupants per Car

An approach for testing HOV facilities requiring usage by vehicles with three-or-more occupants would be to use the models in Table 1 if the average car occupancy is less than 1.45 persons per car. If the average car occupancy is more than 1.45 persons per car, then Table 2 should be used as follows:

- For estimates of one-occupant and two-occupant cars use the Table 2 equations without any adjustments
- For estimates of four-or-more occupant cars, use a constant 6.7 percent.

This 6.7% is the value from the model based on 1.45 persons per car, and assumes that the improved HOV time would have most of its impact on increasing three-or-more-occupant cars. The resulting equation for three-or-more-occupant cars is determined to be: The percentage of persons in three-or-more-occupant cars = $-117 + (88 \times CO)$ where: CO = Average Car Occupancy.

CONCLUSIONS

This technique is relatively easy to understand and apply. IT may also need adjustments as more data on the use of HOV facilities become available. The most significant improvement in the new technique is that this approach addresses the entire region, not just one corridor at a time. Also, with the traffic assignment algorithm MULTILOAD, the capability of determining the volumes for each carpool size for every link in the network is now possible which is another significant improvement to the state-of-the-art traffic assignment process.

REFERENCES

1. JHK & Associates, "Carpool Forecasts in the Metro K Line Corridor" for Metropolitan Washington Council of Governments (March 1978).
2. JHK & Associates, Wagner-McGee Associates, Inc., The Traffic Institute—Northwestern University, "Training Course—High Occupancy Vehicle Facility Development Operation and Enforcement," prepared for Federal Highway Administration, U.S. Department of Transportation (May 1981).
3. JHK & Associates, "Extending the Shirley Highway HOV Lanes, A Planning and Feasibility Study," Final Report for Virginia Department of Highways and Transportation (March 1982).
4. Mann, W. W., "MULTILOAD Traffic Assignment," Metropolitan Washington Council of Governments/Transportation Planning Board (August 1982).
5. Morin, D. A., "Analysis of the Distribution of Automobile Occupancies," U.S. Department of Transportation, Federal Highway Administration (November 1969).
6. U.S. Department of Transportation, "Urban Trasnportation Planning System (UTPS), Reference Manual," Urban Mass Transportation Administration and the Federal Highway Administration (April 1979).

CHAPTER 16

Transit Performance Evaluation

RICHARD P. GUENTHNER*

THE DEVELOPING NEED FOR TRANSIT PERFORMANCE MONITORING

About the time of the turn of the twentieth century, public transit was privately operated. The primary competing modes of travel in the urban setting were horseback and walking. Consequently, the street railways were a very attractive alternative. The fares were universally set at 5 cents per ride. Very substantial profits were realized from this service.

However, the emergence of the automobile coupled with the spreading of the cities into newer areas triggered a slight decline in the stability of urban transit in the mid to late 1920's. Other possible causes for this decline were overcapitalization, increasing cost of labor, and the demand for a constant fare. Most urban transit operators (many were motor coach operators by that time) survived this period. However, their profits were much less than had been realized a decade earlier. After suffering the depression years, the downward trend was interrupted during World War II. During the war years, gasoline rationing and other emergency measures forced transit riding onto the public. Record ridership was reached in 1946. However, following the war, unprecedented changes in land use were the main cause threatening the extinction of urban transit.

Post War Suburbanization Trend

During the 1950's, the increasing popularity and affordability of the automobile enabled many city residents, who were previously dependent upon public transit, to move to less densely populated suburbs or small cities. This trend is demonstrated in Table 1. The populations for 18 of the 20 largest United States cities in 1950 are shown. Note that these 18 cities all decreased in population between 1950 and 1980. The total decrease is shown as 3.6 million or 18 percent. The other two of the 20 largest cities (Los Angeles and Houston) increased in population and will be discussed later.

However, in Table 2 are shown the urbanized area populations for these same 18 cities. Note that these populations have shown an increase of 15.6 million or 40 percent during those same years that the central city populations decreased. This trend can be further demonstrated by examining the suburban population defined as the urbanized area population minus the central city population. As shown in Table 3, the resulting increase is 19.2 million or 126 percent. The conclusion which may be drawn from these statistics is that the population in the older densely populated cities has been moving out of the cities and into the suburban areas.

The population in other cities of the nation has also been increasing. In Table 4 are shown the growth of the remaining 22 of the largest 40 cities in 1980. Note a substantial increase of 7.5 million or 90.8 percent during these 30 years.

These statistics are shown to demonstrate that the primary reason for the decline in transit ridership during these post war years was not competition of the automobile. Jones [1] demonstrated that highway facilities constructed during this period did not directly compete with transit. Rather, the decline of transit was caused by a shifting in population from those areas readily served by transit (Table 1) to newer areas which were being developed to be readily served by the automobile (Tables 3 and 4). Because of the space requirements of the automobile, enough land is not usually available in densely populated cities to meet the capacity requirements by automobile. Consequently, the newer areas have been built at much lower densities.

Transit, however, cannot survive in low density land use. More vehicle-miles must be produced to attract the same

---

*Department of Civil Engineering, Marquette University, 1515 W. Wisconsin Avenue, Milwaukee, WI

345

TABLE 1. Central City Populations of Transit Oriented Cities.

City	1980	1970	1960	1950
New York	7,071,639	7,895,563	7,781,984	7,891,957
Chicago	3,005,072	3,369,357	3,550,404	2,630,962
Philadelphia	1,688,210	1,949,996	2,002,512	2,071,605
Detroit	1,203,339	1,514,063	1,670,144	1,849,568
Baltimore	786,775	905,787	939,024	949,708
San Francisco	678,974	715,674	740,316	775,357
Washington	638,333	756,558	763,956	802,178
Milwaukee	636,212	717,372	741,324	637,392
Cleveland	573,822	750,879	876,050	914,808
Boston	562,994	641,071	697,197	801,444
New Orleans	557,515	593,471	627,525	570,445
Seattle	493,846	530,831	557,087	457,591
St. Louis	453,085	622,236	750,026	856,796
Kansas City	448,159	507,330	475,539	456,622
Pittsburgh	423,938	520,089	604,332	676,806
Cincinnati	385,457	453,514	502,550	503,998
Minneapolis	370,951	434,400	482,872	521,718
Buffalo	357,870	462,768	532,759	580,132
Total	20,336,191	23,340,959	24,295,601	23,949,087

Source: U.S. Bureau of the Census.

TABLE 2. Urbanized Area Population of Transit Oriented Cities.

City	1980	1970	1960	1950
New York	15,590,274	16,206,841	14,114,006	12,296,117
Chicago	6,779,799	6,714,578	5,961,634	4,920,816
Philadelphia	4,112,933	4,021,066	3,635,228	2,922,470
Detroit	3,809,327	3,970,584	3,538,495	2,659,398
Baltimore	1,755,477	1,579,781	1,418,948	1,161,852
San Francisco	3,190,698	2,987,850	2,395,098	2,022,078
Washington	2,763,105	2,481,489	1,808,423	1,287,333
Milwaukee	1,207,008	1,252,457	1,149,997	829,495
Cleveland	1,752,424	1,959,880	1,783,436	1,383,599
Boston	2,678,762	2,652,575	2,413,236	2,233,448
New Orleans	1,078,299	961,728	845,237	659,768
Seattle	1,391,535	1,238,107	864,109	621,509
St. Louis	1,848,590	1,882,944	1,667,693	1,400,058
Kansas City	1,097,793	1,101,787	921,121	698,350
Pittsburgh	1,810,038	1,846,042	1,805,310	1,532,953
Cincinnati	1,123,412	1,110,514	993,568	813,298
Minneapolis	1,787,564	1,704,423	1,377,143	985,101
Buffalo	1,002,285	1,086,594	1,054,370	798,043
Total	54,779,323	54,759,240	47,747,502	39,225,686

Source: U.S. Bureau of the Census.

TABLE 3. Suburban Area Population of Transit Oriented Cities.

City	1980	1970	1960	1950
New York	8,518,635	8,311,278	6,332,022	4,404,160
Chicago	3,774,727	3,345,221	2,411,230	2,289,854
Philadelphia	2,424,723	2,071,070	1,632,716	850,865
Detroit	2,605,988	2,456,521	1,868,351	809,830
Baltimore	968,702	673,994	479,924	212,144
San Francisco	2,511,724	2,272,176	1,654,782	1,246,721
Washington	2,124,772	1,724,931	1,044,467	485,155
Milwaukee	570,796	535,085	408,673	192,103
Cleveland	1,178,602	1,209,001	907,386	468,791
Boston	2,115,768	2,011,504	1,716,039	1,432,004
New Orleans	520,784	368,257	217,712	89,323
Seattle	897,689	707,276	307,022	163,918
St. Louis	1,395,505	1,260,708	917,667	543,262
Kansas City	649,634	594,457	445,582	241,728
Pittsburgh	1,386,100	1,325,953	1,200,978	856,147
Cincinnati	737,955	657,000	491,018	309,300
Minneapolis	1,416,613	1,270,023	894,271	463,383
Buffalo	644,415	623,826	521,611	217,911
Total	34,479,132	31,418,281	23,451,451	15,276,599

Source: U.S. Bureau of the Census.

TABLE 4. Central City Populations of Auto Oriented Cities.

City	1980	1970	1960	1950
Los Angeles, CA	2,966,850	2,811,801	2,479,015	1,970,358
Houston, TX	1,595,438	1,233,535	938,219	596,163
Dallas, TX	904,078	844,401	679,684	434,462
San Diego, CA	875,538	697,471	573,224	334,387
Phoenix, AZ	789,704	584,303	439,170	106,818
San Antonio, TX	785,880	654,153	587,718	408,442
Honolulu, HI	762,874	630,528	294,194	248,034
Indianapolis, IN	700,807	736,856	476,258	427,173
Memphis, TN	646,358	623,988	497,524	396,000
San Jose, CA	629,442	459,913	204,196	95,280
Columbus, OH	564,871	540,025	471,316	375,901
Jacksonville, FL	540,920	504,265	201,030	204,517
Denver, CO	492,365	514,678	493,887	415,786
Nashville, TN	455,651	426,029	170,874	174,307
El Paso, TX	425,259	322,261	276,687	130,485
Atlanta, GA	425,022	495,039	487,455	331,314
Oklahoma City, OK	403,213	368,164	324,253	243,504
Fort Worth, TX	385,164	393,455	356,268	278,778
Portland, OR	366,383	379,967	372,676	373,628
Long Beach, CA	361,334	358,879	344,168	250,767
Tulsa, OK	360,919	330,350	261,685	182,740
Toledo, OH	354,635	383,062	318,003	303,616
Total	15,792,705	14,293,123	11,247,504	8,282,460

Source: U.S. Bureau of the Census.

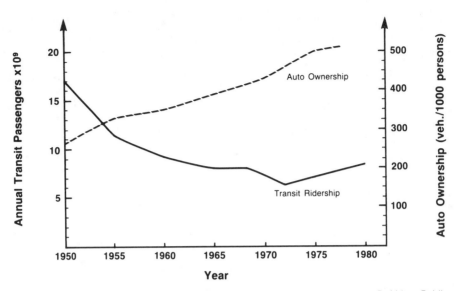

FIGURE 1. Trends in transit ridership and auto ownership. Source: Vuchic, Vukan R., Urban Public Transportation Systems and Technology, Prentice-Hall, 1981, p. 109.

ridership. This trend is shown in Figure 1. Note that while transit ridership has significantly dropped between 1950 and 1975, automobile ownership has steadily increased.

Experiencing ridership drops, especially off-peak, the transit companies suffered financial difficulties. Service cutbacks and fare increases became more plentiful. This resulting lower level-of-service discouraged more people from riding transit, thus compounding the financial problems. The downward cycle as shown in Figure 2 results. This cycle demonstrates that any cutbacks by the transit company are merely self defeating in that they only compound the initial problem rather than solving it.

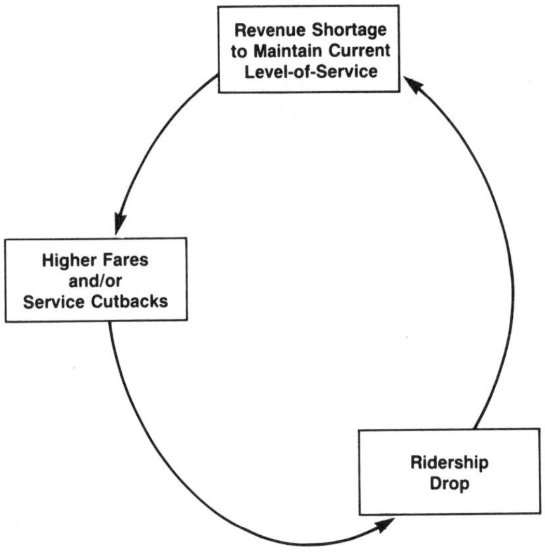

The ultimate result for many systems was bankruptcy. Many small cities were left without public transportation. However, realizing the need for public transportation in most medium to large cities, public corporations, and city and county governments have become the public transit operators. As shown in Table 5, by 1975, virtually all ridership was on publicly operated transit.

Increase in Transit Subsidy

Seeing the downfall of public transit and recognizing a need to continue the service, in 1961 the federal government provided assistance to local transit operators for capital improvements. However, these appropriations only included loans and small demonstration grants to selected properties.

The Urban Mass Transportation Act (UMT Act) was passed in 1964. This act provided significant capital improvement assistance on a federal to local matching ratio of 67 percent to 33 percent. This act was amended several times, each increasing the federal involvement in local transit service. In 1973, the federal matching rate was increased to 80 percent. In 1974, for the first time, operating assistance was available. These grants were available on a formula basis using a 50 percent to 50 percent matching ratio.

Until 1978, federal subsidy was available only to those cities classified as urbanized areas by the census bureau. Section 18 of the UMT Act was passed in 1978. This provided assistance to areas other than urbanized areas. At this

FIGURE 2. Downward transit cycle.

time many small cities and counties were able to expand their transit service.

The federal subsidy during selected years is shown in Table 6. One purpose for the federal subsidy was to encourage the states and local governments to provide subsidy to public transit. Some selected figures are also shown in Table 6 to show that the state and local support has in fact increased.

Revised Goals for Transit

While most private companies are profit oriented, the goals of public corporations or authorities are quite different. Obviously the purpose for the transit service is no longer to make a profit. Rather, because public need is the reason for existence, the goals of the system are all public service oriented. A list of possible goals for a publicly operated transit service is shown in Table 7.

The majority of the population has access to an automobile. Consequently, they have a choice between automobile and transit. They are called choice riders. However, many do not have an automobile available. These among others include the handicapped, elderly, youth, poor, and the second person in a single automobile household. These people are known as captive riders. Often transit service goals are centered around providing mobility for the captive riders.

Other goals might be stated as a minimum level-of-service to be offered. This might be expressed as a maximum headway or fare. Another goal might give a minimum percent of the population which must be served by transit. Threshold values are often stated as part of the goal. For example, "Maintain the passengers per vehicle-mile above 2.00."

Another goal of transit might relate to environmental or community needs. Included here might be reducing air pollution, conserving fuel, or revitalizing the downtown area.

Another goal of the public subsidy was to stabilize the fare levels. Because of reduced ridership and other problems facing privately owned transit during the post war years, the fares were increased substantially. Between 1945 and 1961, fares increased 165.5 percent [2], which is higher than the increase in the consumer price index. After operating subsidies were available, the goal of stabilized fares was achieved. Between 1972 and 1978 the fare only increased by 21 percent while the consumer price index rose 56 percent. Another success of the subsidy program is that while total ridership declined between 1946 and 1972, it has been since increasing every year since 1972.

However, public subsidy has not been completely successful. While the fare levels were declining in real dollars between 1972 and 1978, the cost of operating transit surpassed the inflation rate. The price of diesel fuel increased 231 percent and the cost of a new bus increased 156 percent. Labor cost per employee increased 58 percent which is slightly above the inflation rate. The result has been a sharply decreasing operating ratio (passenger revenue divided by operating cost) from 0.74 to 0.48 [2]. The conclusion is that while public subsidy has allowed a recovery in the transit industry, the industry has not stabilized. Every year, increasing subsidy is required.

Public subsidy has also allowed overcapitalization. Because capital grants have been readily available, transit operators have expanded into low density suburban areas. Local public policy has often demanded this expanded serv-

TABLE 5. Trend of Public Takeover of Public Transit.

Year	No. of Systems	Percent of Industry Vehicles	Percent of Industry Ridership
1940	20	7	—
1945	29	16	—
1950	36	28	—
1955	39	30	—
1960	58	36	—
1965	88	48	—
1970	159	66	77
1975	333	83	90
1980	576	90	94

Source: American Public Transit Association, Transit Fact Book, 1981.

TABLE 6. Transit Subsidies.

Year	Federal Capital Subsidy (millions)	Operating Subsidy Federal (millions)	Operating Subsidy State (millions)	Operating Subsidy Local millions)
1965	$ 50.7	—	—	—
1966	106.1	—	—	—
1967	120.6	—	—	—
1968	121.8	—	—	—
1969	148.3	—	—	—
1970	133.4	—	—	—
1971	284.8	—	—	—
1972	510.9	—	—	—
1973	863.7	—	—	—
1974	955.9	—	—	—
1975	1287.1	$ 301.8	$ 406.6	$ 699.4
1976	1954.8	422.9	367.1	857.4
1977	1723.7	584.5	478.4	841.1
1978	2036.9	689.5	564.3	977.8
1979	2101.6	855.8	637.7	1416.9
1980	2787.1	1093.9	820.4	1703.9
1981	1502.5	996.1	934.1	1516.5
1982	1518.8	930.9	1109.1	1596.4

Source: American Public Transit Association, Transit Fact Book, 1981. (1965–1980) and UMTA Section 15 Reports (1981, 1982).

TABLE 7. Possible Goals for a Public Transit System.

Category	Possible Goals
Transportation	Reduce travel times
	Reduce overall transportation costs
	Reduce congestion
	Reduce parking demand, esp. downtown
Public Service	Provide mobility to elderly, handicapped, and youth
	Maintain low fares
Economic	Improve accessibility to downtown
	Improve accessibility to major employers
	Revitalize downtown area
	Minimize local property taxes
Environmental	Reduce air pollution
	Reduce energy consumption
Level-of-Service	Maximize percent of population served by transit
	Maximize service frequencies

ice. In addition, the advent of grants for non-urbanized areas (Section 18) has made possible transit in very low density areas. Because transit requires high population densities to be self supporting, the resulting performance has declined as service has expanded.

With the possibility of the state and federal subsidy decreasing, many transit authorities are faced with very critical decisions. In fiscal year 1982, only 45.2 percent of the operating cost was obtained from farebox and other revenues. The remaining was subsidized with 29.1 percent local funds, 13.3 percent state funds, and 12.3 percent federal funds. This is more significant for systems in small cities which are more heavily dependent upon the subsidy than those in larger cities. For example, operators with fewer than 25 buses only recovered 23.7 percent of their expenses from farebox and other revenues. 22.5 percent of their expenses were supported by the federal government [3]. The loss of federal subsidy in these small cities could easily force the system to stop operating.

Much recent research has been conducted to develop methods to improve the performance of urban public transit [4]. Careful monitoring of performance has been demonstrated as one tool for reaching these goals.

MEASURING PERFORMANCE

One solution to help offset some of the financial problems of public transit is to maintain higher overall performance. In other words, achieve greater output for a given amount of input into the system. Two very important questions must be answered in relation to improving transit performance: 1) "How can performance be measured?", and 2) "What can be done to improve performance?"

No single method or measure to evaluate transit performance has been adopted by the transit industry. In addition, there are no set levels of what are considered "good" or "bad." The two basic reasons for this are:

1. The methods of performance evaluation are still being researched.
2. Since no two cities are exactly alike, "good" performance will have a different meaning in each city.

Defining Performance

Two basic categories of performance are generally defined: efficiency and effectiveness.

Efficiency measures demonstrate how well the operator is utilizing its resources. These measures are usually expressed as a measure of output divided by a unit of input. For example, "vehicle-miles per employee" would indicate how well the operator is using its labor. Also included in this category are overall cost efficiency and vehicle utilization [5].

Effectiveness is the degree of fulfillment of the operating objectives of the system. As the basic purpose of a transit operation is to transport passengers, most performance measures dealing with ridership are effectiveness measures. "Passengers per vehicle-mile" is a very commonly used example which indicates how well the service is being utilized. Measures of service quality and accessibility are also included here. Because system operating objectives differ depending upon local concerns, the effectiveness measures used in each system will also vary.

Numerous efficiency and effectiveness measures can be developed for different reasons. A partial list is shown in Table 8. The organizing framework, developed by Fielding [6] and shown in Figure 3, demonstrates the general form of how indicators are computed.

Selecting Performance Measures

Much effort has been made to determine a standard list of a few key performance indicators. A list of this type should enable operators to know which indicators give the most information about the system. The result would be a less costly data collection procedure since a more pointed goal would be sought. More importantly, standard indicators would provide target values for each measure by which an evaluation could be made. In other words, a comparison regarding performance could be made between cities.

However, comparison between cities is difficult because of different characteristics of each city and different com-

TABLE 8. Transit Performance Measures.

Category	Measure
Efficiency Measures	
Cost efficiency	Operator cost per vehicle-mile
	Operator cost per vehicle-hour
Labor productivity	Revenue vehicle-miles per employee
	Revenue vehicle-hours per employee
Vehicle utilization	Vehicle-miles per vehicle
	Vehicle-hours per vehicle
	Passengers per vehicle
Energy efficiency	Energy consumption per vehicle-mile
	Energy consumption per vehicle-hour
Safety	Accidents per vehicle-mile
	Accidents per vehicle-hour
Service Effectiveness Measures	
Service utilization	Passengers per vehicle-mile
	Passengers per service area population
Accessibility	Percentage of population served
Quality of service	System reliability
	Vehicle-miles per square mile of service area
Cost Effectiveness Measures	
Financial performance	Operating ratio (revenue/cost)
	Cost per passenger trip
	Cost per passenger-mile

Source: J. C. Yu, *Transportation Engineering*, Elsevier, 1982.

munity goals. For example, since transit service is best suited for large densely populated cities, the performance in cities of this type will almost always be better than that in smaller, low density cities. While city size and density are the most significant factors affecting performance, there are many others as shown in Table 9. For example, cities with grid or radial street patterns, many captive riders, realistic goals, and ample state and local financial support should expect very high performance by most standards. However, a transit manager in a city with opposite characteristics should not be expected to produce high performance.

Fielding [7] examined 48 performance indicators from 304 fixed-route bus systems. Factor analysis was employed to determine seven key dimensions of transit performance. One indicator for each dimension was determined to explain best the characteristic in question. These indicators should give maximum information about the performance of a system without repeating. They are shown in Table 10.

In addition, Fielding used cluster analysis to determine which cities could be compared with each other. Twelve peer groups were found. The common characteristics of each group are outlined in Table 11. The conclusion is that the performance can be compared between cities of the same peer group.

Data Requirements

Sinha and Guenthner [8] determined that the data requirements for careful performance monitoring are not excessive. A one page monthly report can contain enough information to compute an adequate number of performance measures for each route to define the performance of the system in most small to medium sized cities. In Figure 4 is shown an example being used by an operator in a medium sized city. From this, both monthly and route by route performance can be computed at a reasonable cost.

During the 1970s, increasing attention was given to monitoring performance. Many operators realized the importance of improved data collection. The Urban Mass Transportation Administration (UMTA) realized the need for better and more consistent data. Consequently, starting in fiscal year 1979, each operator receiving funds from UMTA was required to submit annual data in a consistent

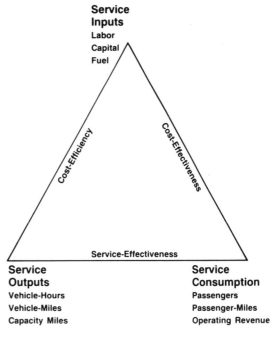

FIGURE 3. Framework for transit performance concepts. Source: G. J. Fielding, et al., "Indicators and Peer Groups for Transit Performance Analysis," University of California at Irvine, January, 1984.

TABLE 9. Urban Characteristics Which Can Affect Performance.

Category	Best Performance	Worst Performance
Density	High	Low
Urban Structure	Active Downtown Grid Pattern Radial Pattern	Dispersed Land Use Random Street Pattern Irregular Shaped City
Type of People	Captive Riders Low Income Low Auto Ownership	Choice Riders High Income High Auto Ownership
Hours	Peak Hour Service	Evenings Weekends Night Service
Community Goals	Only Basic Service Desired	Desire for High Level-of-Service Need to Serve Special Groups
Available Subsidy	Strong State and Local Support	Little State or Local Interest

TABLE 10. Key Dimensions of Transit Performance.

Factor	1st Performance Indicator	2nd Performance Indicator
1. Output per Dollar Cost	Rev. Veh.-Miles/Operating Expense	Tot. Veh.-Miles/Operating Expense
2. Utilization of Service	Total Passengers/Rev. Veh.-Hours	Total Passengers/Tot. Veh.-Miles
3. Revenue Generation/Expense	Operating Rev./Operating Expense	Rev. Veh.-Hours/Tot. Op. Assistance
4. Labor Efficiency	Tot. Veh.-Hours/Employee	Rev. Veh.-Hours/Op. Employee Hour
5. Vehicle Efficiency	Tot. Veh.-Miles/Peak Vehicle	Tot. Veh.-Hours/Peak Vehicle
6. Maintenance Efficiency	Tot. Veh.-Miles/Maint. Employee	Peak Vehicle/Maint. Employee
7. Safety	Tot. Veh.-Miles/Accident	Rev. Veh.-Miles/Accident

Source: G. J. Fielding, et al., "Performance Evaluation for Fixed Route Transit: The Key to Quick, Efficient and Inexpensive Analysis," University of California at Irvine, December, 1983.

TABLE 11. Peer Groups for Performance Comparisons.

Peer Group	City Size & Type	Area of Country	Average Bus Speed	Peak/Base Ratio
1	Smallest	All	Extremely high	Lowest
2	Small urban & suburban	All	High	Average
3	Small cities & towns	Southwest	Above average	Low
4	Small cities with suburban char.	All	High	Low
5	Below average, some private	New York & midwest	Very low	Average
6	Rural to medium	Midwest & south	Below average	Low
7	Small cities/large towns	All	Average	Above average
8	Small to medium	Midwest & eastern	Varied	High
9	Suburban, low density	Southwest	Above average	Average
10	Large urban	All except northeast	Below average	Wide range
11	Major urban areas	All	Average	Above average
12	Three largest urban areas	All	Very low	Slightly above avg.

Source: G. J. Fielding, et al., "Indicators and Peer Groups for Transit Performance Analysis," University of California at Irvine, January, 1984.

Transpo Operations Performance

19

Regular Route Weekdays	Revenue	Passengers @	Miles	Gross Cost @	Net Cost	Ratio Rev. to Gr. Cost	Hours	Passengers per Mile	Passengers per Hour	Rev. per Mile	Rev. per Hour
Madison/Bendix	$			$	$					$	$
Portage/L.W.W.											
O'Brien/S. Mich.											
N. Dame/Miami											
N. Side/West'n											
S. Side/Rum V.											
Corby/Sample											
System Sub-Total											
Saturday											
Madison/Bendix											
Portage/L.W.W.											
O'Brien/S. Mich.											
N. Dame/Miami											
N. Side/West'n											
S. Side/Rum V.											
Corby/Sample											
System Sub-Total											
Total Month											
Regular Route	$			$	$					$	$
Tripper Service											

FIGURE 4. Example of a monthly data collection. Source: South Bend Public Transportation Corporation.

fashion. This requirement was given in Section 15 of the UMT Act as amended. After implementation of the Section 15 procedure, performance could easily and readily be computed by a variety of measures. Because Section 15 specifies exactly how the data should be collected, the data are consistent and can be compared between cities.

The increased use in computer technology is allowing even small operators to record performance on a computer data base. When necessary preparation has been completed, the Section 15 reporting will be directly entered into the computer. Three levels of detail will be required with the most comprehensive for the largest systems. The result will be much more information at a reasonable cost.

Uses of Performance Monitoring

A number of uses for performance monitoring applies to different groups. A primary use is to ensure that the best use of public money is being achieved. This is the concern of the officials at all levels of government and the primary area of interest of transit boards and commissions. Much consideration is currently being given toward allocating subsidies on the basis of performance. From the standpoint of the management, the performance is generally viewed in terms of limiting the deficit so that all obligations can be met within the budget.

On the operations level, a specific use for performance measures includes continuous monitoring of system operation. This involves regular tabulation of selected performance measures with the primary purpose of examining how well the bus system is operating. Generally, the values of various indicators can be compared with themselves over time. Also, such a monitoring process can quickly detect problems in the operation. The task of determining the most productive service changes or improvements in system operation can also be performed by evaluating performance

KEY AREA	INDICATOR	OBJECTIVE
III. Service Development	D. Hours of service	1. To restructure the hours of service per route from 13.5 to 16 hours per bus per route at a cost of 120 staff hours and $1,000 within existing budget by October 1, 1979.

ACTION STEPS		DATE
a.	Analyze present schedule to determine what time service expansions are needed for	06-01-79
b.	Develop preliminary time schedules	07-01-79
c.	Present schedules for approval	07-19-79
d.	Revise and modify if necessary	08-06-79
e.	Implement printing of new schedules	08-20-79
f.	Release of information to press and public	09-17-79
g.	Service implementation	10-01-79

FIGURE 5. Example of a management by objectives report. Source: Metro Transit System, Kalamazoo, MI.

measures. At the same time, performance monitoring can also help to maintain a geographically equitable service. It can aid in determining if service is practical in locations normally found to be non-productive.

Evaluating how well the public is being served is another important use for performance monitoring. The public can be divided into three groups, captive riders, choice riders, and non-users [9]. Different performance measures should be used to demonstrate how well each group is being served. Each system should tailor its performance analysis to the relative percentage of each group involved.

Within some limits, the captive riders must ride transit regardless of the level-of-service being offered. Certain effectiveness measures should indicate how this group is being served. These would include measures indicating how well walking distances are minimized, the handicapped are being served, and in general, that their transportation needs are met.

The current and potential choice riders need to be offered service competitive with other modes, most commonly, the automobile. Measures indicating, for example, minimized transit travel times, efficient routing structures, or service reliability, demonstrate how well the system could attract choice riders.

The general public including the users as well as the non-users of the system are concerned with how well the tax dollars are being used. To this group, measures related to financial efficiencies are of primary importance [8].

By appropriately defining and measuring performance, the level of achievement of the overall goals and objectives of a transit system can be assessed. A very useful management tool for making this evaluation is Management-by-Objectives (MBO). An MBO is a device by which annual goals are specified in detail. Specific objectives are given as a means of achieving each goal. In addition, specific tasks or action steps along with the timing are given for reaching each objective. For each objective, the indicator to evaluate how well the objective has been met is set in advance. An example of one page of an MBO report is shown in Figure 5. The dates, indicator, and action steps are all shown clearly. In 1976, 41 percent of the transit managers used Management-by-Objectives. In 1982 this number had increased to 59 percent indicating the increased popularity of this method [10].

METHODS TO IMPROVE PERFORMANCE

After a performance measure approach has been implemented, decisions must be made as to where and how the performance will be improved to acceptable levels. Several techniques are available and will be discussed here.

Non-Productive Service

The first step often made by transit officials to improve overall performance is to eliminate service with poor performance. This could include nighttime, evening, or weekend service. The frequency or length of a particular route might be reduced. Sometimes an entire route might be eliminated.

Generally these decisions would be based upon criteria stating minimum threshold values for performance. If a route is found to be consistently violating the threshold values, it is "flagged" for further investigation.

Caution needs to be seriously considered if cutbacks of

this nature are made. Any cutbacks should conform to minimum level-of-service criteria. For example, a maximum headway might be set at 30 minutes. If a route is already operating at 30 minute headways, the only option available might be total elimination of the route. Furthermore, another goal might block elimination. Possible reasons might be:

1. It is the only route serving a particular corridor, or
2. A small but important group of captive riders depend upon the route.

In this situation, the transit manager must look elsewhere for cost cutbacks. Perhaps service cutbacks can more easily be implemented elsewhere in the system.

Service cutbacks are actually negative methods to improve performance. Eventually they could lead to the downward cycle shown in Figure 1. Consequently, some effort needs to be given to determine positive ways to improve performance.

One possibility is to replace the current minimum level-of-service, low productive, fixed-schedule route with some form of a demand responsive service. Dial-a-bus, subsidized taxi, subscription bus, car pool, or van pools are a few examples. In Table 12 are shown some features of selected modes. The idea behind demand responsive service is that service will be provided only when needed. A growing trend is for transit systems (known as full service transit agencies) to offer both fixed-route, fixed-schedule service and demand responsive service where appropriate.

Better Performance on Productive Service

In addition to reducing or eliminating service on low productive routes, another alternative may be found by examin-

TABLE 12. Basic Characteristics of Paratransit Modes.

Characteristic	Mode						
	Rental Car	Car pools	Van pools	Subscription bus	Taxi	Dial-a-ride	Jitney
Type of usage							
Private	X	X					
Semipublic			X	X			
Public					X	X	X
Ownership							
User		X	X				
Employer			X				
Individual	X				X		X
Public				X		X	
Routing							
Personal	X	X	X		X		
Partial				X		X	
Fixed-route							X
Method of Getting Service							
Always available	X						
Prearranged		X	X	X		X	
On street/phone					X		X
Driver							
User	X	X	X				
Partially Trained			X		X		X
Trained				X	X	X	
Vehicle Capacity							
≤ 9	X	X			X		X
10–15			X			X	X
≥ 16				X		X	
Parking at Each Trip-End							
Required	X	X	X	X			
Not Required					X	X	X

Source: V. R. Vuchic, *Urban Public Transportation*, Prentice-Hall, 1981, p. 595.

ing the operating strategies of the high performance routes. Several strategies are available which might significantly increase the performance on very high demand routes with short headways. In general, a strategy is needed which will reduce vehicle miles or vehicle hours without reducing the level-of-service. Included here are zone and skip stop service, bus priority treatment, short turns, and express bus service. In general, these strategies are most effectively used on long CBD oriented corridors.

Zone and skip stop arrangements are used to allow shorter travel times because every stop will not be made. Many variations of implementing this type service are available. The exact method chosen depends heavily upon local conditions.

For example, every other run of a zonal system would alternate between stopping at only the inner stops or the outer stops. A restricted zonal strategy would allow travel between zones by having the bus serving the outer zone stop for alighting passengers on the inbound trip and boarding passengers on the outbound trip. A disadvantage of these strategies is that while the in-vehicle travel time is reduced, either wait time or walking time is increased. Also, confusion can occur with complicated strategies.

Short turns are commonly used for routes with low demand near one end. Every other bus can turn around soon enough to reduce the total vehicle requirements.

Express bus operation could include very few or no stops for selected runs. If possible, freeways could be traversed to achieve minimum travel times.

Several bus or high occupancy vehicle (HOV) priority techniques can also be implemented. Exclusive lanes are the most commonly used. Similar in nature are separate on-ramps to freeways. Car pools are often included in these strategies. Signal preemption would allow the bus to always have a green light at signalized intersections. With an indicator on the bus, a sensor on the signal would detect when the bus is approaching. A green signal would be extended until the bus safely passes. A red signal would be truncated immediately. Bus stops in this strategy would have to be placed on the far side of each intersection.

Methods to Increase Demand

In addition to lowering the cost of supplying transit, another method to improve effectiveness is to take measures to increase the demand for transit.

Many factors outside the control of transit management can greatly increase the demand. For example, the emergency measures taken during World War II brought record ridership to the industry without any effort on the part of management. The fuel shortages of 1973 and 1979 also increased transit demand because of fear by the public that petroleum fuel would not be available. The aftereffects of the fuel shortages included a national emphasis on transit. Americans were being encouraged to save fuel by riding transit for altruistic reasons. For similar reasons, transit riding has been encouraged in cities experiencing significant air pollution problems.

Land use changes might affect the long run trend. This may apply if a major new development is constructed. Also, the demand is reduced as the population of the central city declines, as demonstrated in Tables 1 through 4.

On a shorter term basis, outside factors can also affect transit ridership. Inclement weather, for example, can cause very significant boosts especially in northern states. Employment fluctuations determine when people need to travel. Consequently, ridership peaks can be expected during periods of high employment.

Seasonal variations may also occur. School sessions dictate ridership peaks in some cities. The summertime vacation period usually causes a ridership valley (except in resort areas).

However, to some extent, transit management can encourage demand increases. Careful planning and marketing are required.

The largest element of marketing is advertising. Through the media the public must be informed of the advantages of riding transit. Also, because riding transit requires much information on the part of the rider, public information is a very important element of marketing. A good clear system map should be readily available. In addition, understandable schedules should be printed where appropriate. Most cities now incorporate telephone assisted information service. In larger cities, a computer data base can assist the telephone agent in giving information. The importance of informative bus stop signs and direction signs in rail stations (graphics) cannot be overemphasized. These have the capability of informing the unfamiliar traveler exactly where to go to reach the transit service. For example, questions of which side of the street to stand on or which direction to turn in an underground station can be answered. No other medium has this ability.

Providing good reliable service also has an effect on marketing. Reliable service will build trust in the transit system. Over time, performance will naturally improve.

Other important management tasks which will increase demand include:

- keeping the buses clean
- maintaining public relations
- providing courteous and informative vehicle operators

Fare Structure Strategies

Changes in fare structure or level may increase the fare box revenue. Also, more equitable fare arrangements may be possible. Most United States systems use a flat fare. Where inflation has made fare increases inevitable, transit authorities are looking for more flexible fare arrangements.

The most common differential fare pricing in practice today is discounts to certain sub-groups of the population. These include the elderly, the handicapped, and youth.

Half fares during off-peak hours for the elderly and the handicapped are a requirement of Section 5 of the UMT Act. This requirement is intended primarily as a public service to avoid hardship for these specific groups who might otherwise have no means of transportation. The benefit to the operator is increased ridership during the off-peak hours. During these hours, the bus capacity usually far exceeds the demand. Consequently, the marginal cost of carrying one additional passenger is zero.

The purposes for youth discounts are also public service in nature. Youths have not yet reached the age when they can legally drive. Since they are still financially dependent, the reduced fare will increase their mobility. Some systems have implemented a student fare which includes the requirement for showing a school identification card.

Multiple ride discounts are also very common practice. Their advantages are:

- convenience to the patron
- ease in collecting the fares
- prepayment to afford more cash flow to the operator
- guaranteed ridership by regular patrons

Multiple ride discounts often have a negative effect. Primarily because use of the discount is usually during the peak hour and by long distance commuters. The peak hour riders are the most expensive to service. Also, long distance riders should expect to pay more, not less.

Quality based fares involve a different fare for different type of service. For example, express or subscription service may require a premium fare.

Fare free service is normally introduced at a large cost. In this situation, other motives besides financial efficiency are needed, such as promotions or public service. Usually instituted in a downtown area, fare free may increase downtown mobility, lower air pollution in the critical downtown area, or demonstrate the convenience of transit to the public.

A somewhat less common fare structure is a time of day differential. A higher fare during peak hour is the most common form of this. In setting rolling stock and labor requirements, a transit authority will provide for the peak hour needs. Labor rules usually restrict the use of part time drivers. Consequently, during the off-peak period, the drivers as well as the buses are usually available. Off-peak service can then be offered at very little marginal cost.

Riders who have a choice are encouraged to ride during the off-peak hours. Consequently, some of the extra capacity may be used. One less rider will need service during the peak hour, which due to the capacity conditions, is the most expensive time to provide service. Also, because of the commuters, average trip lengths are longer during the peak hour. Off-peak trips are usually non-work trips. They are more elastic, indicating that a low fare will attract riders. Also, this policy allows reduced fares to disadvantaged groups who more often prefer riding during the off-peak.

Peak hour fares are also easy to administer. Those bus runs which originate during the specified time period will simply be priced at a higher fare.

The most accurate price differential technique is distance-based fares. Two methods are used: truly graduated and zonal based fares. Truly graduated pricing may be either linear or logarithmic.

Finely graduated fares are considered the most equitable of all fare structures. The main advantage is that each passenger pays very close to the actual cost of the service. This applies both to the additional vehicle-miles and the decreased load factors in providing for the longer trips. Also, those who usually make longer trips are suburbanities who can afford to pay more. Distance-based fares allow center city and downtown residents, who are more often disadvantaged, to pay a smaller fare.

The main disadvantage of distance-based fares is that they are difficult to administer and collect. In the United States they are used mainly on new rail systems where computerized fare collection is feasible. As shown in Table 13, distanced-based fares are more common in Europe. Most notably, 85 percent of the systems in the United Kingdom use distance-based fares [11].

Zonal fares are a compromise between flat and distanced-based fares. APTA [12] reported that 33.7 percent of United States cities use zone fares. However, many involve only two or three zones. Again, zones fares are more common in Europe. All cities in the Netherlands use zone fares due to a national fare policy in that country [11]. generally, the zones are arranged as concentric circles. Often cross town trips are encouraged because the trip might be completely in one zone. This, however, may be desirable

TABLE 13. Fare Structure by Country.

Country	Cities Over 200,000 Population Fare Structure			
	Graduated	Zonal	Flat	Mixed
United Kingdom	22	3	0	1
France/Belgium	5	6	11	6
Netherlands*	0	7	0	0
West Germany	3	25	8	5
Scandinavia	0	3	7	0
Switzerland/ Austria	2	2	3	1
Italy	0	1	13	3
Spain/Portugal	6	0	10	0
United States**	0	84	163	0
Canada**	0	2	9	0

*Since 1980 a national fare system has applied throughout the Netherlands.
**Not necessarily cities over 200,000 population.
Source: A. C. Dawnes and R. P. Kilvington, "The Effects of Simplified Fare Structure & Ticketing in Urban Public Transport Operations," PTRC 11th Summer Annual Meeting, July, 1983, p. 58 (Europe) and American Public Transit Assn., Transit Fare Summary, 1983 (North America).

because cross town trips more often occur during the off-peak. Four basic collection techniques are in common practice including:

- pay upon entering
- pay at zonal boundaries
- pay upon entering and receive tickets to be collected at zonal boundaries
- pay when leaving

Vehicle Maintenance

Maintenance of the vehicles is also important to performance. Poorly maintained vehicles will increase the chances of a roadcall. This will cause inconvenience to the passengers and delay to the route. Also, the cost of the roadcall will be high to the operator. In the extreme, enough vehicles may not be operational to meet the service requirements. Poor management of the maintenance department itself can affect performance because of the associated increased cost.

Improved management can, of course, improve performance. One technique which should be used is preventive maintenance of the vehicles. If done correctly, this procedure will greatly reduce the number of roadcalls because potential problems can be detected early. Purchasing buses from as few manufacturers as possible will reduce the requirements for spare parts inventory.

Another solution is to buy more reliable buses in the first place. Often more reliable buses cost more and would not be successful in a competitive bid situation. Consequently, UMTA now allows transit authorities to consider the life cycle cost in the competitive bidding process. This allows the bid to include both an initial cost and the projected maintenance cost amortized over the expected life of the vehicle. In Table 14 are shown some bids which considered life cycle costing. Note that in two of these cities the award was actually given to a different manufacturer than had life cycle costing not been considered.

TABLE 14. Awards Based Upon Life Cycle Costs.

City	Date	No. of Buses	Bidders*	Base Bid	LCC Cost	Award Price
Santa Rosa, CA	6/82	9	GFC	$152,206	$150,622	$307,482
			GMC	155,419	183,903	343,410
Reno, NV	6/82	16	GMC	142,385	39,507	186,442
			GFC	153,024	52,280	209,961
Providence, RI**	7/82	42	GFC	149,697	232,786	382,483
			GMC	145,113	241,056	386,169
Dallas, TX	7/82	89	GFC	144,050	115,174	259,224
			GMC	146,044	120,064	266,108
Buffalo, NY	8/82	40	GMC	156,800	9,874	166,674
			GFC	157,082	11,350	168,432
Madison, WI	10/82	15	GFC	144,285	184,729	329,014
			GMC	147,543	192,399	339,942
			Neoplan	148,600	207,816	356,416
Shreveport, LA	10/82	8	GFC	146,102	156,581	302,683
			GMC	152,124	157,659	309,783
Oklahoma City	11/82	15	GMC	145,969	163,876	309,845
			GFC	153,505	173,586	327,091
Houston, TX**	11/82	100	GMC	149,784	156,704	306,488
			GFC	156,957	158,549	315,506
			Neoplan	146,800	175,531	322,331
Wilmington, DE	12/82	55	Neoplan	138,600	217,402	356,002
			GMC	146,431	217,298	363,729
			GFC	152,651	217,295	369,936
Clearwater, FL	12/82	7	GMC	141,864	148,667	290,531
			GFC	149,295	152,320	301,615

*GFC = Grumman Flexible Corp.; GMC = General Motors Corp. **Award affected by life cycle cost.
Source: Metro, vol. 79, no. 1 (Jan/Feb 1983), p. 32.

Labor Relations

Forty-five percent of transit operating cost is for the vehicle operators' wages [2]. Consequently, any reduction in labor costs could significantly improve performance. However, several of the current labor practices inherently increase the cost of providing transit.

Because transit primarily serves peak hour riders, the need is to operate only during the early morning and late afternoon. The most effective use of labor is to assign split runs which typically might include working four hours, off four hours without pay, and again working four hours. However, these are undesirable working hours. Most labor contracts include a maximum of about 10 to 13 hours spread time between first reporting in the morning to release in the afternoon. Operators working beyond these hours are often paid premium wages. A maximum percent of split time may also be specified in the labor contract [13].

Also undesirable from the labor standpoint are part time drivers. Consequently, most labor contracts prohibit or strictly limit this practice. The result is less flexibility and higher cost to the operator.

Another problem is that labor resists the assigning of alternate duties to drivers. This is an especially significant problem during the off-peak hours when many drivers are idle but still being paid.

Most contracts increase driver's pay to the maximum by two years of service [13]. Cost of living increases are usually only afforded by increasing the base pay for all drivers. The major privilege given to senior drivers is first choice of assignment. Senior drivers understandably choose low demand suburban routes because the work is more relaxed. However, logic dictates that the most experienced workers should be given the most responsibility, earn the most money, and contribute the most to the employer.

A rethinking of the use of labor is needed. Low productive routes should be serviced by new drivers with corresponding entry level pay. As experienced is gained, a promotion would consist of higher pay and a more productive route. The work will then be more difficult which can be handled by the more experienced driver. Jones [1] suggested additional levels to include alternative duties such as maintenance. A plan of this type will allow drivers to advance both in pay and duties. The base pay will need only keep pace with the cost of living. The drivers would be given the incentive and ambition to progress and learn. The most experienced drivers would be utilized to their fullest potential. During the off-peak hours, further productivity can be gained from those drivers who have been promoted to a double duty position.

The driver should also be realized as a marketing tool. Many passengers turn first to the driver for information. Consequently, he/she should be polite, willing to help, and informed. Drivers meeting these qualifications should be first in line for promotions.

METHODS OF EVALUATING EFFECTS ON PERFORMANCE

The decision makers must have some idea of the possible effects before implementing a change to improve performance. Forecasting these effects requires an understanding of the components of a transit trip. Also, traveler's response to various changes in these components must be understood. Through a combination of theory and empirical evidence, these attributes have been examined. A number of analytic tools has been developed based upon these findings. In addition, many of these tools have been developed into computer packages, either for main frame or microcomputers. These packages are too numerous to list. In reference [14] is given a partial list of packages which operate on microcomputers.

Attributes of a Transit Trip

For any fare or service change, two primary outputs need to be estimated: the new operating cost, and the resulting ridership.

The operating cost must be determined from an analysis of the existing operating costs. Unit costs can be determined and applied toward the proposed changes.

Passengers may change their travel behavior if one or more of the components of the trip changes. The most significant of these components are the out-of-pocket cost, the in-vehicle travel time (IVTT), and the out-of-vehicle travel time (OVTT). The OVTT can be further subdivided into walking time and waiting time. If a transfer is required, then the time waiting for the second bus must also be considered.

Other attributes are less tangible, but can be the controlling factors in a modal decision. These may include:

- comfort
- convenience
- privacy
- security

Estimating the Attributes

The exact values of a set of the trip attributes might be different for each individual transit trip. Consequently, a methodology should be able to estimate average values for a particular type of trip; for example, all trips on a given route. If more detail is required, the estimates could be disaggregated into trip purpose (work, shop, other) or characteristics of the user (e.g. income groups).

If the average passenger trip length and the vehicle operating speed are known, then IVTT can be directly estimated. OVTT must be estimated by summing each of the components:

- access time
- waiting time
- transfer time
- distribution time

Access time is the time spent reaching the initial bus stop. Most passengers walk to the stop. Consequently, access time is often referred to as walking time. Two distances must be considered: the average distance to the route and the average distance to a stop. The minimum for both of these is zero for a person living next to a bus stop. The maximum distance to the route is one half the distance to the next route. The maximum distance to the stop is one half of the stop spacing. Consequently, these two extremes may be averaged and divided by the appropriate speed as follows:

$$ACC = [(Y + SPACE)/4]/\Omega \quad (1)$$

where

ACC = access time in hours
Y = spacing between stops in miles
$SPACE$ = distance between routes in miles
Ω = walking or driving speed in MPH

The maximum waiting time will be equal to the headway of the bus. The minimum will be zero. Similarly, these will be averaged to obtain an average waiting time of one half of the headway. However, the waiting time may be less for riders who know the schedule and plan to arrive accordingly. In general, the longer wait associated with longer headways will encourage riders to arrive in accordance with the bus schedule. By comparing previous research, Keeler [15] developed the following equation:

$$WTT = 30/X; \text{ for } X > 2 \quad (2a)$$

$$WTT = 8 + 14/X; \text{ for } X < 2 \quad (2b)$$

where

WTT = waiting time in hours
X = buses per hour

If the bus does not arrive in time, additional waiting time will be required. Furthermore, if the bus arrives before the passenger, the waiting time will increase by the amount of the headway. Regular patrons will be aware of the reliability of the bus arrival time and will leave early enough to compensate for it.

Turnquist [16] developed a model to evaluate average waiting times which also consider schedule reliability. He determined the distribution of the bus arrival time to be lognormal with a given mean and variance. The regular passengers will arrive at time TA as follows:

$$TA = \exp[-\sigma \times G(q)] + \mu \quad (3)$$

where

TA = arrival time of the passengers
μ = mean arrival time of the bus
σ = standard deviation of the arrival time
q = the probability of missing the bus
$G(q)$ = inverse of the lognormal distribution

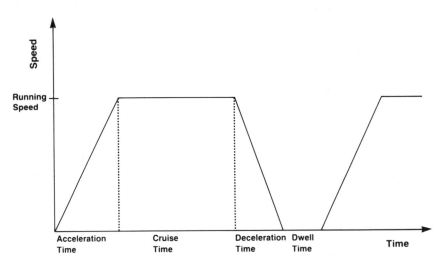

FIGURE 6. Basic relationships between bus speed and time.

The mean waiting time is then given as:

$$WTTr = a\,WN + (1 - a)\,WTT \quad (4a)$$

$$WN = q(TA - T2) + (1 - q)(T1 - TA) \quad (4b)$$

where

$WTTr$ = waiting time considering reliability
a = percent riders knowing the schedule
WN = average wait time for riders knowing the schedule
$T2$ = arrival time of the second bus
$T1$ = arrival time of the first bus

If a transfer is required, additional waiting time would be required. If no effort has been made to coordinate the two routes in question, this additional waiting time would be one half the headway of the second bus. However, routes may be coordinated such that the riders have a timed connection. In this situation, the additional wait time would be small.

The distribution time at the attraction end of the trip would normally be made by walking. The expected average distance would be divided by the appropriate speed. If shuttle bus service, taxi, or other mode is available, the appropriate speed and interface time can be estimated.

Determining Operating Speeds

The operating speed of a bus will, in general, be much lower than that of an automobile. The primary reason for this is that the bus must be continually stopping for passengers.

The running speed is defined as the speed of the bus while not stopping for passengers. Two penalties must be applied to obtain the operating speed. These are: 1) the decelerating and accelerating time at each passenger stop, and 2) the dwell time while the passengers are boarding and alighting. This movement is shown in Figure 6.

Each run of the bus will not stop at every scheduled stop along a route. Many stops will have no demand. Guenthner and Sinha [17] found that the total numbers of passengers boarding and alighting at each stop follow a negative binomial distribution. Consequently, if the demand is known, the average boardings and alightings per stop could be computed. In turn, the percentage of stops actually made is as follows:

$$SPM = Y \times (1 - p^k) \quad (5a)$$

$$p = m/s^2 \quad (5b)$$

$$k = m^2/(s^2 - m) \quad (5c)$$

where

SPM = actual stops per mile

Y = posted stops per mile
p, k = parameters of the negative binomial distribution
m = average boardings and alightings per stop
s = standard deviation (a function of m)

After the number of actual stops has been determined, a time penalty per stop must be applied. This penalty is a function of the running speed, the acceleration rate, and the traffic conditions. If a constant deceleration and acceleration is used, the total time for these maneuvers will be twice the time requirements for traveling the same distance at the running speed. Consequently, the penalty is one half the total stopping and starting time. This is equivalent to either the acceleration or deceleration. At a running speed of 30 miles per hour and an acceleration rate of 3 miles per hour per second, 10 seconds would be required.

The dwell time at each stop is a function of the number of passengers boarding and alighting as well as many other factors including:

- number and width of doors
- bus configuration
- fare collection method
- presence of elderly and handicapped
- weather
- crowding on the vehicle

No single formula has been found to adequately evaluate the dwell time. However, researchers agree that the average boarding or alighting time is about 4 to 6 seconds for the first passenger. The time per passenger declines as more boardings and alightings are at a given stop.

TABLE 15. Relative Importance of Travel Time Components.

Travel Time Component	Relative Rank*
In-Vehicle Travel Time	
Standing	6
Sitting	2
Express Service	1
Out-of-Vehicle Travel Time	
Walking to the Route	5
Walking Along the Route to the Stop	3
Waiting for the First Bus at Home	7
Waiting for the Scheduled Arrival of the First Bus	8
Waiting for a Late First Bus	11
Waiting for the Scheduled Transfer Bus	9
Waiting for a Late Transfer Bus	10
Walking to the Destination	4

*Higher numbers indicate a higher value on time.

TABLE 16. Ranges of Demand Elasticity Values.

Category	Mean	Standard Deviation
Fare Elasticities		
Estimation Method		
Quasi-experimental	−0.28	0.16
Time Series	−0.42	0.24
Type of Change		
Fare Increase	−0.34	0.11
Fare Decrease	−0.37	0.11
City Size		
Greater than 1,000,000	−0.24	0.10
500,000–1,000,000	−0.30	0.10
Less than 500,000	−0.35	0.12
Mode		
Bus	−0.35	0.14
Rapid Rail	−0.17	0.05
Route Type		
CBD Oriented	−0.40	0.04
Non-CBD Oriented	−0.62	0.09
Time period		
Peak	−0.17	0.09
Off-peak	−0.40	0.26
Trip purpose		
Work	−0.10	0.04
Shop	−0.23	0.06
Income		
Less than $5,000	−0.19	0.10
$5,000–$15,000	−0.25	0.11
More than $15,000	−0.28	0.13
Age		
Less than 17 years	−0.32	0.01
17–24 years	−0.27	0.03
25–44 years	−0.18	0.10
45–64 years	−0.15	0.03
More than 65 years	−0.14	0.02
Service Elasticities		
Total Travel Time		
Peak	−1.03	0.13
Off-peak	−0.92	0.37
In-Vehicle Travel Time		
Peak (quasi-experimental)	−0.29	0.13
Off-peak	−0.83	—
Peak (non-experimental)	−0.68	0.32
Off-peak	−0.12	—
Out-of-Vehicle Travel Time		
All	−0.59	0.59
Walk time (peak)	−0.26	—
Walk time (off-peak)	−0.14	—
Wait time	−0.54	—
Transfer time	−0.40	0.18

Source: Ecosometrics, Incorporated, "Patronage Impacts of Changes in Transit Fares and Services," Sept. 1980.

Value of Travel Time

In making modal comparisons, the tradeoff often involved is a higher out-of-pocket cost versus longer travel times. Consequently, in order to make adequate comparisons, some value needs to be assigned to a unit travel time. Many attempts have been made to evaluate the value of time. However, a variety of results has been obtained with no absolute value decided upon. However, there is an understanding of the relative values of different types of travel time.

Time values will change according to both the rider characteristics and the trip characteristics. In general, high income, choice riders will place a higher value on their time. Less busy people such as the unemployed, the retired, and the elderly who are not trying to meet a busy schedule put less emphasis on time.

Work, school, and other trips in which the arrival time at the destination is critical, would all be associated with a very high travel time value. Shopping trips can commence at any time. Consequently, the travel time would be a less important consideration in the mode choice for that trip.

In addition, within a trip, the various travel time components will have different values. In general, because progress is being made, passengers view IVTT as less costly than OVTT. Similarly, walking time is less than waiting time. In Table 15 are shown relative weights of various trip components. Again, these relative weights are only examples and may change with the riders or the trip.

Demand Elasticities

Following any change in the supply of transit service, some change in ridership will result. The demand elasticity is defined as the percent change in ridership for a unit change in a service attribute. The attribute can be either the cost, IVTT, OVTT, or any other direct or indirect cost of taking the trip. For example, if the fare changes, the change in ridership will be found from the demand elasticity with respect to fare.

The definition of a demand elasticity assumes an infinitesimally small change in service. This is known as a point elasticity and is defined as:

$$E = \frac{\partial Q}{\partial F} \times \frac{F}{Q} \qquad (6)$$

where

Q = ridership
F = value of service attribute being changed

In reality, infinitesimally small changes do not occur. Consequently, a midpoint elasticity is defined as:

$$E = \frac{Q2 - Q1}{F2 - F1} \frac{F_{ave}}{Q_{ave}} \qquad (7)$$

where

$Q1, F1$ = values before the change
$Q2, F2$ = values after the change
Q_{ave}, F_{ave} = average of before and after values

For the transit industry, a shrinkage ratio has been found to be much simpler to use:

$$SR = \frac{Q2 - Q1}{F2 - Q1} \times \frac{F1}{Q1} \qquad (8)$$

In general, ridership will decline from an increase in fare. Consequently, the elasticity value will be less than zero. An elasticity value of -1.00 will result in no change in revenue. Elasticity values less than -1.00 will result in a decrease in revenue following a fare increase. This is defined as an elastic demand. Conversely, elasticity values between 0 and -1.00 will result in increased revenue following a fare increase and are defined as having an inelastic demand. Transit has almost universally been found to have an inelastic demand.

An elasticity with respect to the attributes of a competing mode are defined as cross elasticities. These elasticities are usually greater than zero, indicating that as the price of the competing mode increases, ridership will also increase.

Estimating elasticity values requires data from before and after a change. Care must be made to avoid data that represent more than one change. However, completely isolating a change is impossible. Many factors are always influence changing ridership patterns. Time series analysis is a useful tool to separate multiple effects on ridership. The disadvantage of this method is that autocorrelation between the residuals is highly likely. Box-Jenkins time series analysis [18] may be used to reduce this problem.

The general rule for the value of the demand elasticity with respect to fare was developed by Curtin [19] as follows:

$$c = 0.80 + 0.30\,b \qquad (9)$$

where

b = percent fare increase
c = percentage ridership drop

For significant fare changes, the elasticity from Equation (9) becomes about -0.33. Consequently, the general understanding is that a fare increase of one percent will cause a ridership drop of one-third of one percent. This rule, loosely referred to as the "Curtin Rule," has been found to be quite accurate.

Further research has revealed that riders are more sensitive toward service changes than toward fare changes. Consequently, the elasticities with respect to travel time have been found to be much steeper than those with respect to fare. Also, elasticity measures for different groups, and trip types vary. For example, work trips by middle income people would be much less sensitive to a fare change than shopping trips by low income people. Some ranges of elasticity values are shown in Table 16.

REFERENCES

1. Jones, D., *Urban Transit Policy: An Economic and Political History,* Englewood Cliffs, NJ, Prentice-Hall Inc. (1985).
2. American Public Transit Association, *Transit Fact Book,* 1978–1979 edition.
3. Jacobs, M., et al., *National Urban Mass Transportation Statistics, FY 1982 Section 15 Annual Report,* Transportation Systems Center, Cambridge, MA, UMTA- MA-06-0107-84-1 (November 1983).
4. Urban Mass Transportation Research Information Service, *Transit Costs, Performance Evaluation, and Subsidy Allocation: Special Bibliography,* UMTA-DC-06-0258-83-3 (December 1983).
5. Fielding, G. J., "Issues and Challenges in Transit Performance Research," in *Innovative Strategies to Improve Urban Transportation Performance,* A. Chatterjee and C. Hendrickson, eds., American Society of Civil Engineers (1985).
6. Fielding, G. J., T. L. Babitsky, and M. E. Brenner, "Performance Evaluation for Fixed Route Transit: The Key to Quick, Efficient and Inexpensive Analysis," Institute of Transportation Studies, University of California at Irvine (December 1983).
7. Fielding, G. J., et al., "Indicators and Peer Groups for Transit Performance Analysis," University of California at Irvine, CA-11-0026-2 (January 1984).
8. Sinha, K. C. and R. P. Guenthner, "Field Application and Evaluation of Bus Transit Performance Indicators," Purdue University, CE-TRA-81-1 (March 1981).
9. Valley Transit, "Valley Transit 1979 Management Plan," Appleton, WI.
10. Miller, D. R., G. T. Lathrop, D. G. Stuart, and T. H. Poister, *Simplified Guidelines for Evaluating Transit Service in Small Urban Areas,* Transportation Research Board, NCTRP Report 8 (October 1984).
11. Dawes, A. C. and R. P. Kilvington, "The Effects of Simplified Fare Structure and Ticketing in Urban Public Transport Operations," *Public Transport Planning and Operation,* PTRC 11th Summer Annual Meeting, University of Sussex, England, pp. 55-67 (July 1983).
12. American Public Transit Association, *Transit Fare Summary* (October 1981).
13. Sinha, K. C. and D. B. Dobry, "A Comprehensive Analysis of Urban Bus Transit Efficiency and Productivity, Part 2, Labor Aspects of Urban Bus Transit Productivity," Purdue University, CE-TRA-78-3 (December 1978).
14. Urban Mass Transportation Administration, *Microcomputers in Transportation: Software and Source Book,* UMTA-URT-41-85-1 (February 1985).

15. Keeler, T. E., L. A. Merewitz, and P. Fisher, "The Full Costs of Urban Transport," Institute of Urban and Regional Development, University of California, Berkeley (December 1971).
16. M. A. Turnquist, "A Model for Investigating the Effects of Service Reliability on Bus Passenger Waiting Times," *Transportation Research Board*, TRR 663, pp. 70–73 (1978).
17. Guenthner, R. P. and K. C. Sinha, "Modeling Bus Delays due to Passenger Boardings and Alightings," *Transportation Research Board*, TRR 915, pp. 7–13 (1983).
18. Rose, G. "An Aggregate Time-Series Analysis of the Effects of Fare Changes on Transit Ridership," Master's thesis submitted to Northwestern University, Evanston, Illinois (June 1982).
19. Curtin, J. F., "Effect of Fares on Transit Riding," *Highway Research Board*, HRR 213, pp. 8–20 (1968).

# Chapter 17

Matching of Transportation Capacities and Passenger Demands in Air Transportation

Dušan Teodorović*

ABSTRACT

One of the most important air carrier's problems which has to be solved is a problem of matching transportation capacities and passengers' demand. There are many different factors which have a direct influence on this matching. Problem of matching transport capacities and passengers' demand is more complex when a whole network is considered rather than a particular route. Values of flight frequencies, as well as, distribution of departure times on certain routes represent the manner in which passengers' demand and carrier's interest is matched. Flight frequencies and departure times depend on passenger requests for transportation, as well as, on the number and type of aircraft in the fleet, number of flying and technical staff, maintenance system, etc. When matching passengers' demand and transportation capacities, after determining flight frequencies and departure times on certain routes it is necessary to make adequate aircraft routing, crew scheduling, etc.

IMPORTANCE OF FLIGHT FREQUENCY DETERMINATION PROBLEM

Over the years a certain number of methods for flight frequency determination and airline schedule design were developed. Still it is necessary to improve these methods, as well as, to create new ones. Namely, as it was mentioned earlier, flight frequency determination is very important step for organizing other activities which are necessary to be done if carriers wish to perform service on the network in the best possible way. Flight frequencies and departure times have an essential influence on the number of passengers in air traffic. Johnston et al. (1980) studied factors which have important influence on the passenger choice of a carrier in the case when competition exists between companies on some route. These authors made passenger survey on routes in Canada in order to calculate criteria weights in this decision making process. Passengers estimated the weights of the following criteria: Flight Schedules, Safety Considerations, Fare Prices, Aircraft Characteristics, Flight Related Aspects, Reservation Conditions and Auxiliary Services. The criteria Flight Schedules was composed from the following elements: On-time arrivals (departures), Frequency of flights and Non-stop flights. Final results of survey showed that for business travellers criteria Flight Schedules including Flight Frequency was the most important criteria to choose certain carrier. In the case of Vacation Travellers this criteria was on the third place of importance. Renard (1970), de Neufville and Gelerman (1973), and Ghobrial (1983) studied the influence of flight frequency on market share for routes on which competition exists. Later, these papers we will analyze in more details. It is important now to note that these authors showed the existance of great influence of flight frequencies on market share.

Flight frequency, as well as, departure times have very high influence on modal split on routes on which competition between different transportation modes exists. This influence is extremely high on short haul routes which are characterized by strong competition of different modes (bus, train, car, aircraft). Total number of passengers between two cities depends, first of all, on gross national product per capita, population of the cities, distance between cities, development of business, cultural, tourist and other connections between cities, as well as, characteristics of existing transportation system. There are many methods developed to estimate the number of passengers between two cities or between two regions. In his recent book Kanafani (1983) analyzed the most important methods for estimation of the number of passengers in air transportation. As it is usual, after estimating the total number of passengers between two cities, it is necessary to decide on modal split, i.e. to estimate number of passengers for different modes. The

*Faculty of Transport and Traffic Engineering, University of Belgrade, Belgrade, Yugoslavia

most important factors which influence passengers to use aircraft on certain route are travel costs and travel time associated with different modes, flight frequency, aircraft, train and bus departure times, schedule maintenance, comfort associated with different transportation modes etc. Number of passengers in air transportation on certain routes highly depends on flight frequency and departure times. After carefully studying passenger demand density, it is possible using appropriate frequency, to increase, up to certain limit, number of passengers who would use aircraft. On the other hand inadequate flight frequency or inadequate departure times will result in lower number of air passengers. It is important to mention that some carriers do not pay much attention to the influence which flight frequency and departure times have on number of passengers in air transportation. As Simpson (1969) mentioned "it is not usual for an airline schedule planning process to determine timetable, and return to the traffic forecasting process to regenerate new traffic data on the basis of the new timetable. It is a real feedback particularly in competitive airline markets, or short haul markets where the air system competes with automobile traffic."

DEMAND DENSITY IN AIR TRANSPORTATION

Demand density can be defined as the number of passengers per unit of time which travel from one to another city. So we can write:

$$h_{ij}(t) = \frac{\left[\begin{array}{c}\text{number of passengers who travel from city } i \text{ to} \\ \text{city } j \text{ during time interval } (t, t + dt)\end{array}\right]}{dt} \quad (1)$$

or

$$h_{ij}(t) = \frac{d\left[\begin{array}{c}\text{cumulative number of passengers who travel} \\ \text{from city } i \text{ to city } j \text{ during time interval } (0,t)\end{array}\right]}{dt} \quad (2)$$

where $h_{ij}(t)$ – demand density between city i and city j in a moment t.

It is necessary to make distinction between demand density and potential demand density. Potential demand density can be defined as the number of passengers per unit of time who would like to travel from one to another city. Very frequently there is difference (because of many factors) between real and potential demand density. For example if carrier has no competition on some route, because of low frequency value, certain number of passengers can not get vacant seats on the desired flights. In this case potential demand density is higher than real one.

Demand densities in air transportation are variable over time. Variations can be noticed by observing months in a year, weeks in a month, days in a week and finally hours in a day. It is obvious that daily and respectively weekly flight frequency on a certain route depends on demand density which means that monthly demand in the course of the year should be accompanied by corresponding variations in the flight frequency. This is particularly important for tourist services which in certain months show a considerable variation in demand. In order to solve problems involved in determining flight frequency and departure times it is extremely important to observe the daily and hourly demand density.

Variability of passengers' daily demand in the course of the week and hourly demand in the course of the day was studied by Swan (1979) and Powell (1982). Considering variability of demand "in the context of a single market with two competitors" Powell assumed that the total market size V_o is random variable which has normal distribution with mean μ and standard deviation σ. Variability of demand is caused first of all by the fact that one passenger's reasons for expressing transportation demand at a particular moment are as a rule not related to the respective reasons of other passengers. Besides that, there are a great number of different factors which have influence on the variability of demand. Because of these factors it is very logical to have normal distribution of the total market size. Number of passengers who would like to travel on a certain flight is also random variable which has normal distribution. It is very difficult to estimate the number of passengers who would like to travel by certain flight. Instead of this number of passengers it is more convenient to operate with the number of passengers in a plane. Swan (1979) and Soumis et al. (1981) showed that number of passengers in a plane on a certain flight is random variable which also has normal distribution.

Papers by Gagnon (1967), Hyman and Gordon (1967), Simpson (1969), Miller (1972), Soumis et al. (1981) and

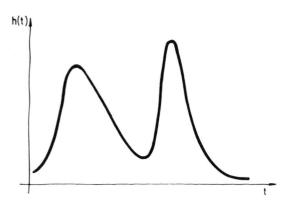

FIGURE 1. Demand density as a function of the time of a day.

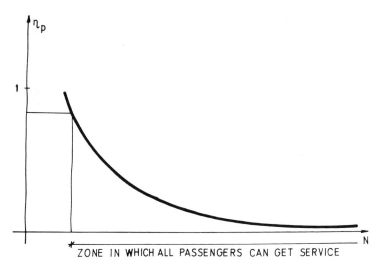

FIGURE 2. Functional relation between average load factor and the flight frequency.

Teodorović (1983) point out the characteristic form of demand density as a function of the time of a day. Figure 1 shows the characteristic functional relation between the demand density and the time of a day.

In Figure 1 $h(t)$ is used to denote demand density and T is the time interval during which the passengers' demand for transportation exists. In order to get functional relation between demand density and the time of a day it is necessary to make appropriate surveys of passengers.

LOAD FACTOR

Average load factor is defined as ratio between average number of passengers in a plane and number of seats in a plane. Swan (1979) called this load factor nominal load factor defining it by the following relation:

$$\eta_p = \frac{\mu}{n} \qquad (3)$$

where

η_p = average load factor
μ = mean number of passengers in the plane
n = number of seats in the plane

Average load factor η_p during certain period of time on some route is defined as:

$$\eta_p = \frac{\lambda}{nN} \qquad (4)$$

where

λ = average number of passengers during certain period of time
n = number of seats in the plane
N = flight frequency, i.e. number of flights during observed period of time

Through relations (3) and (4) one can get basic information about economic results achieved by the carrier. All carriers try to reach load factor not lower than so called breakeven load factor. This load factor is such one for which total revenues are equal to total costs.

Average load factor is also rough indicator for the level of service. Namely, the higher value of average load factor on some route, the greater the number of passengers refused seat on desired flight. We can conclude that average load factor depends of flight frequency and that it is an important indicator for carrier's economic results, as well as, level of service.

LOAD FACTOR AND DENIED BOARDINGS

As it was mentioned, in a case of higher average load factor values on some route a certain number of passengers cannot get a reservation for a desired flight. It is necessary to study what happens to those passengers who are denied reservation. This fraction of passengers can do some of the following:

- They do not make the trip.
- They try to get seat on the first later departure.

- They try to get seat on the first, second, third, ..., later departure.
- They try to get a seat on any flight which departs later or before desired flight.

Soumis et al. (1981) studied the way in which passengers choose certain flight. They have developed appropriate Model for Assigning Passengers to a Flight Schedule (MAPUM). They introduced into consideration attractiveness factor of a route. This factor is a function of the departure time, the number of stops and/or connections, the service quality and the aircraft type. Soumis et al. evaluated passenger's dissatisfaction as follows:

$$p(t,r) = a_1(t) + a_2(v) \tag{5}$$

where

- r = route r
- t = passenger's desired departure time
- $a_1(t)$ = penalty for the time difference between the flight time and the passenger's desired departure time
- $a_2(v)$ = penalty for the time difference between actual route time for the route r and the time for a direct flight.

Penalties $a_1(t)$ and $a_2(v)$ are calculated as follows:

$$a_1(t) = \begin{cases} k_1(u - t) & \text{,if } u > t \\ k_2(t - u) & \text{,if } t > u \\ 0 & \text{,if } t = u \end{cases} \tag{6}$$

$$a_2(v) = k_3(v - \bar{t}) \tag{7}$$

where

- u = the scheduled departure time for route r
- v = actual route time
- $\bar{t}$ = the time for a direct flight
- k_1 = the penalty factor for the passenger's waiting time
- k_2 = the penalty factor for a passenger departing earlier than desired time
- k_3 = the penalty factor for passenger who will use route r instead direct flight

The objective function in MAPUM model is the sum of the loss of revenue due to passenger overload and the cost for passenger's dissatisfaction. This model also takes into account the fact that every passenger must be assigned to one flight. We will discuss later overload situations in more details.

Attractiveness factor of a flight was introduced by Gagnon (1967) and defined in the following way. For example, let flight 1 leave at time d_1 and arrive at time a_1. Flight 2 leaves at time d_2 and arrives at time a_2 (Figure 3). Gagnon estimates that the attraction of flight 1 will end (or the attraction of flight 2 will start) at:

$$d_1^2 + \frac{a^2}{a + d} \tag{8}$$

where

$$d = d_2 - d_1 \tag{9}$$

is the time interval between two adjacent departures, and

$$a = a_2 - a_1 \tag{10}$$

is the time interval between two adjacent arrivals.

In a case when both flights have equal block-time, we have $a = d$, i.e.:

$$d_1^2 + \frac{a^2}{a + d} = \frac{d_1 + d_2}{2} \tag{11}$$

In this case passenger will choose flight which is closer to the desired departure moment. The same assumption was used by Teodorović (1983).

Every passenger independently of all other passengers makes decision about day and time for traveling. In a case that he can not get a vacant seat on a desired flight he will either not make the trip, or try to get a seat only on the next flight, or will try to get a seat on the first, second, third, ... flight etc. It is obvious that great number of factors have an influence on the total number of passengers on certain flight. From probability theory is known that in such cases it is logical to expect normal distribution of the variable which is influenced by great number of factors and when each of these factors by itself has only small influence on the variable. Soumis et al. (1981) concluded after analysis of three months data for the Air Canada network, that the distribution of the daily number of passengers using a flight

FIGURE 3. Attractiveness factor of a flight.

can be approximated by a normal distribution. Swan (1979) also showed that load factor which is random variable has normal distribution. The number of passengers on a certain flight is an excellent measure of a passenger demand for that flight. Taking into account that it is very difficult to measure passengers' demand we can conclude that known distribution of number of passengers for a certain flight is a basic information for making the analysis of demand, as well as, for calculating the probability of a person being refused space on a departure. As Swan (1979) mentioned "the best measure of the quality of scheduled service is the probability of a person being denied space on a departure." Let us calculate this probability in the following way. Let X denote random variable which represents number of passengers on a certain flight. Taking into account previous observations, probability density function $f(x)$ of this variable is:

$$f(x) = \frac{1}{\sigma\sqrt{2\pi}} \exp\left\{-\frac{(x-\mu)^2}{2\sigma^2}\right\} \quad (12)$$

where

μ = mean of the number of passengers
σ = standard deviation of the number of passengers

Let us denote by Y random variable which represents number of passengers who were not able to get a vacant seat on the desired flight. Mean of this variable is:

$$M(Y) = \int_n^\infty (x - n)f(x)dx \quad (13)$$

where n = number of seats in the plane.
After transformation we get:

$$M(Y) = \int_n^\infty (x - \mu)f(x)dx + \int_n^\infty (\mu - n)f(x)dx \quad (14)$$

i.e.:

$$M(Y) = \int_n^\infty (x - \mu)f(x)dx + (\mu - n)\left[\Phi(\infty) - \Phi\left(\frac{n-\mu}{\delta}\right)\right] \quad (15)$$

where

$$\Phi(t) = \frac{1}{\sqrt{2\pi}} \int_0^t e^{-\frac{u^2}{2}} du$$

is Laplace function, for which $\Phi(\infty) = 0.5$.

Finally we can write that the mean of number of passengers who cannot get a vacant seat on the desired flight is:

$$M(Y) = \frac{\sigma}{\sqrt{2\pi}} \exp\left\{-\frac{(n-\mu)^2}{2\sigma^2}\right\} + (n - \mu) \cdot \left[\Phi\left(\frac{n-\mu}{\sigma}\right) - 0.5\right] \quad (16)$$

In a case of aircraft DC-9-30 with 115 seats and for different values of parameters μ and σ, the calculation of mean number of passengers who can not get a vacant seat is performed. The results are shown in Table 1.

Final results shown in Table 1 were expected and they are very logical. For the same variability of demand (constant σ) with increasing the average number of passengers in the plane, the number of passengers being denied space is also increasing. For a constant average number of passengers in a plane (constant μ) with increasing the variability of demand (increasing of σ), the number of passengers being

TABLE 1. Mean Number of Passengers Who Cannot Get a Vacant Seat on the Desired Flight.

Number of seats in the plane	Mean passengers in the plane	Standard deviation of number of passengers in the plane	Number of passengers being denied space per one flight
115	70	10	<1
115	70	20	<1
115	70	30	<1
115	70	40	2.614
115	80	10	<1
115	80	20	<1
115	80	30	1.825
115	80	40	4.352
115	90	10	<1
115	90	20	<1
115	90	30	3.377
115	90	40	6.47
115	100	10	<1
115	100	20	2.625
115	100	30	5.937
115	100	40	9.56
115	110	10	1.977
115	110	20	5.723
115	110	30	9.618
115	110	40	13.58

denied space is again increasing. This conclusion is very important, because it means that during analysis of available transportation capacities use, it is necessary to carefully study load factors, but even more than that variability of demand, i.e. differences between numbers of passengers on flights and average number of passengers. Using relation (16) from which one can get an average number of passengers being denied space per one flight, it is possible to make further analysis about increasing flight frequency, changing aircraft type (changing the number of seats offered per one departure) etc. This way it is much easier to see advantages and disadvantages of different alternatives which differ one from another due to flight frequency value and/or aircraft type and to make appropriate decisions about air transport operation on some route.

FLIGHT FREQUENCY AND COMPETITION

In a case when only one carrier operates on certain route, flight frequency, as we saw, has a direct influence on the economic results of the carrier, as well as, on the level of service. Low values of flight frequency offered by monopolistic carrier will result in high average load factor values and great number of passengers being denied space. Most frequently great numbers of these passengers turn to another alternative transportation modes. In a case of two or more competitive carriers who operate on certain route it is even more important to make good estimate of flight frequency value and departure times in order to maximize market share. Let us consider one route on which operate two competitive carriers. Passenger demand density on this route is shown in Figure 4.

Two competitive carriers operate on this route. Departure times of the first carrier are x_1, x_2, x_3, x_4 and x_5. Another carrier has flights at the time points x_1, x_2 and x_4. Let us make an assumption that both carriers operate with the same aircraft type and that the air fares are also the same. Attractiveness of different flights was calculated using assumption that the passenger will choose the flight which is the nearest (in any direction) to the passenger's desired departure time. In moments x_3 and x_5 carrier 1 is without competition, so he will take all passengers who would like to travel during time intervals $(x_2 + x_3)/2$, $(x_3 + x_4)/2$ and $(x_4 + x_5)/2$, T. Only with two flights carrier 1 will take 2/5 of all passengers who would like to travel during time interval (O,T). On flights which depart in moments x_1, x_2 and x_4 carrier 1 will take $1/2 \cdot 3/5$ of all passengers who would like to travel during time interval (O,T). This means that carrier 1 who has 5/8 of all flights on this route is able to take $2/5 + 1/2 \cdot 3/5 = 7/10$ of all passengers. On the other hand carrier 2 who has 3/8 of all flights can take only $1/2 \cdot 3/5 = 3/10$ of all passengers. Even this simple example shows that carrier who operates with greater frequency usually has success to have greater percentage of passengers along a route than its share in number of total flights. The explanation for this is the fact that because of greater frequency this carrier does not have competition in some moments.

Relationship between market share and frequency share

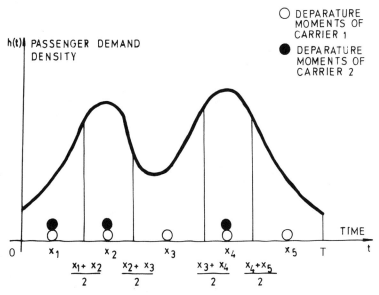

FIGURE 4. Passenger demand density on the route on which operate two competitive carriers.

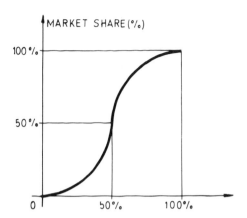

FIGURE 5. Functional dependance of market share on the frequency share.

on the routes with competition were considered by Renard (1970), Gelerman and de Neufville (1973), de Neufville and King (1979) and Powell (1982). All these authors used in their papers the following Renard's relation:

$$MS_i = \frac{FS_i^\alpha}{\sum_{i=1}^{k} FS_i^\alpha} \qquad (17)$$

where

MS_i = market share for airline i
FS_i = frequency share for airline i
k = number of airlines who operate on the route
α = empirically obtained constant which depends upon the number of competitors (Gelerman and de Neufville mentioned that this constant could be in the range $1 < \alpha < 2$)

Figure 5 shows the characteristic functional dependance of market share on the frequency share.

FLIGHT FREQUENCY AND LEVEL OF SERVICE

Level of service in air transportation consists of the sequence of different attributes such as travel costs, total travel time, flight frequency, adjustment of airline schedule to passengers' requests, maintenance of airline schedule, safety, comfort, etc. In a context of flight frequency consideration it is quite natural to consider the influence of flight frequency on level of service. In their papers Gordon and de Neufville (1973), Douglas and Miller (1974), Eriksen (1977), Swan (1979), Kanafani and Ghobrial (1982), Ghobrial (1983) and Teodorović (1983) studied the relationship between flight frequency and level of service. One of the basic measures for level of service in air transportation is schedule delay. Gordon and de Neufville defined schedule delay as "the difference in time between when a person would like to make a trip and the time he is constrained to make the trip because of the inflexibility of the airline's schedule." For example if passenger would like to make a trip at 1 P.M. and if because of airline's schedule it is possible to travel at 5 P.M., this time difference equals 4 hours. Clearly, with increase of flight frequency because of greater number of departure times which are offered to passengers, average schedule delay per passenger is decreasing. Average schedule delay can be calculated using different formulas which can be found in literature. Comparison of different schedule delay formulas will be given later.

As we discussed earlier, because of variability of demand in air transportation, number of passengers on certain flights is random variable most frequently normally distributed. Because of this some passengers can not get a vacant seat on desired flight, and as it was shown number of these passengers is increasing when load factor is increasing, i.e. when frequency is decreasing. Some fraction of these passengers will try to find place on some of next flights. So, we can conclude, that this type of passenger will have extremely great time difference between realized and desired departure. Time difference between departure moment of flight on which passenger got a vacant seat and departure moment of a flight which was first time chosen by a passenger is called stochastic delay. For a calculation of average stochastic delay, first it is necessary to calculate probability that passenger can get a vacant seat on desired flight. Time difference similar to stochastic delay can be found in a case where passenger had o.k. reservation but because of airline's mistake in overbooking he did not realize his journey. In their papers Falkson (1969), Rothstein (1971, 1975), Shlifer and Vardi (1975), Simon (1972), Vickrey (1972) and Nagarajan (1979) studied airline overbooking problem. In his research Nagarajan showed, using passengers' surveys, that because of stochastic delay due to overbooking, passengers' requests for compensation are not so high. The values of time which Nagarajan explored in a case of waiting due to overbooking probably are very similar to the values of time in a case of stochastic delay.

Flight frequency has a direct influence on the schedule and stochastic delay which are both very good indicators of level of service on some route.

FLIGHT FREQUENCY DETERMINATION ON THE SINGLE ISOLATED LINK IN A CASE OF MAXIMIZING CARRIER'S PROFIT

Flight frequency has a direct influence on the level of service, but even more on the total number of passengers carried, total carrier's revenue and total direct and indirect operating costs. Taking these facts into account de Vany and Garges (1972) have developed a model for flight frequency determination on the single isolated link. The objective function in the model was maximizing carrier's profit. The model was developed with assumption that carrier has no competition on the route. The mentioned authors expressed total profit P on a route between city i and city j as:

$$P = c \cdot \lambda_{ij}(c,N) - DOC \cdot N - L \cdot N - IOC \cdot \lambda_{ij}(c,N) - IC \quad (18)$$

where

c = fare for journey from city i to city j
N = flight frequency between cities i and j
$\lambda_{ij}(c,N)$ = average number of passengers between city i and city j due to fare c and flight frequency N
DOC = direct operating costs per one flight
L = landing fee
IOC = indirect operating costs (without landing fee) per one passenger
IC = indirect costs

The condition:

$$\frac{\partial P}{\partial N} = 0 \quad (19)$$

determines the appropriate flight frequency for which airline can maximize profit on the single isolated link. Using relations (18) and (19) we get:

$$N = \frac{c - IOC}{DOC + L} \cdot \lambda \cdot \frac{\frac{\partial \lambda}{\lambda}}{\frac{\partial N}{N}} \quad (20)$$

The expression $e = (\partial\lambda/\lambda/\partial N/N)$ is number of passengers elasticity due to flight frequency, so we can write:

$$N = \frac{(c - IOC) \cdot \lambda \cdot e}{DOC + L} \quad (21)$$

The average load factor η_p for which airline can maximize profit on the single, isolated link which operates without competition is:

$$\eta_p = \frac{\lambda}{n \cdot N} = \frac{DOC + L}{(c - IOC)e \cdot n} \quad (22)$$

For practical use of formulas (21) and (22) it is necessary to have good statistical data in order to study change of number of passengers with change of flight frequency values.

Swan (1979) also studied flight frequency for which one can maximize airline's profit on a route without competition. In order to find appropriate value of flight frequency Swan started from the following equation:

$$C = c_1 \cdot S + c_2 \cdot N + c_3 \cdot \lambda \quad (23)$$

where

C = the total expense of operating the day's schedule of services
S = the total number of offered seats per day
N = the daily frequency
λ = the daily number of passengers
c_1, c_2, c_3 = constants which have different values for different air travel markets

Swan showed that these constants for 1976 U.S. domestic air travel were:

$$\begin{aligned} c_1 &= \$3.27 + \$0.0176 \cdot d \\ c_2 &= \$379.8 + \$0.816 \cdot d \\ c_3 &= \$12.64 + \$0.008 \cdot d \end{aligned} \quad (24)$$

where d is route distance in miles.

Total airline's revenue R is:

$$R = c \cdot \lambda \quad (25)$$

and total airline's profit P is:

$$P = R - C = c \cdot \lambda - c_1 \cdot S - c_2 \cdot N - c_3 \cdot \lambda \quad (26)$$

Average load factor can be expressed as:

$$\eta_p = \frac{\lambda}{S} \quad (27)$$

so we have:

$$P = c \cdot \lambda - c_1 \cdot \frac{\lambda}{\eta_p} - c_2 \cdot N - c_3 \cdot \lambda \quad (28)$$

In his research Swan assumed the following demand function:

$$\lambda = k_1 \cdot PP^\alpha \quad (29)$$

where

λ = the total passengers' demand
PP = the perceived price of the service
k_1 = the market density constant
α = the elasticity

The total perceived price of the service could be expressed as:

$$PP = c + v[t_B + f_1(N) + f_2(\eta_p, N)] \quad (30)$$

where

v = value of passengers' time
t_B = route block-time
$f_1(N)$ = schedule delay per passenger
$f_2(\eta_p, N)$ = stochastic delay per passenger due to load factor

Using relations (28), (29) and (30) one can get:

$$P = k_1 \{c + v[t_B + f_1(N) + f_2(\eta_p, N)]\}^\alpha \quad (31)$$
$$\left[c - \frac{c_1}{\eta_p} - c_3 \right] - c_2 \cdot N$$

The condition:

$$\frac{dP}{dc} = 0 \quad (32)$$

determines the appropriate fare for which carrier can maximize profit. After we took the derivative $dP/dc = 0$ we got:

$$c = \frac{c_1}{\eta_p} + c_3 - \frac{1}{\alpha} \cdot PP \quad (33)$$

Through the relation (33) the fare for which the airline maximizes its profit is defined. This fare can be further expressed as:

$$c = \frac{\alpha}{\alpha + 1} \left\{ \frac{c_1}{\eta_p} + c_3 - \frac{1}{\alpha} \cdot v[t_B + f_1(N) + f_2(\eta_p, N)] \right\} \quad (34)$$

The condition:

$$\frac{dP}{dN} = 0 \quad (35)$$

determines the appropriate flight frequency for which carrier can maximize profit. So we get:

$$\frac{d\lambda}{dN} \left(c - \frac{c_1}{\eta_p} - c_3 \right) - c_2 = 0 \quad (36)$$

On the other hand we have:

$$\frac{d\lambda}{dN} = k_1 \cdot \alpha \cdot PP^{\alpha-1} \frac{dPP}{dN} = \frac{\alpha}{PP} \cdot \lambda \cdot \frac{dPP}{dN} \quad (37)$$

From relations (36) and (37) we get:

$$\frac{\alpha \cdot \lambda}{PP \, c_2} \left[c - \frac{c_1}{\eta_p} - c_3 \right] \cdot \frac{dPP}{dN} = 1 \quad (38)$$

From (30) we have:

$$\frac{dPP}{dN} = v \left\{ \frac{d[f_1(N)]}{dN} + \frac{d[f_2(\eta_p, N)]}{dN} \right\} \quad (39)$$

In the Equation (38) from which we will calculate appropriate flight frequency, there is expression dPP/dN. On the other hand dPP/dN depends on schedule and stochastic delay. Swan calculated schedule delay using following equation:

$$f_1(N) = \frac{5.7}{N} \quad (40)$$

Furthermore assuming that passengers' demand for transportation have been expressed during 22.8 hours per day, another assumption was made that schedule delay is equal to one quarter of the average headway. (Average headway is equal to $22.8/N$).

Swan also made an assumption that a passenger who can not get a vacant seat on the desired flight, will wait for departure one extra headway. He showed that the number of denials per customer P_u can approximately be expressed as:

$$P_u = 2.5 \left(\frac{\mu}{n} \right)^9 \quad (41)$$

where

μ = mean passenger demand
n = number of seats in a plane which operates

So the average stochastic delay is:

$$f_2(\eta_p, N) = P_u \cdot (\text{headway}) = 2.5 \left(\frac{\mu}{n} \right)^9 \cdot \frac{22.8}{N} = \frac{57 \left(\frac{\mu}{n} \right)^9}{N} \quad (42)$$

After appropriate derivations we obtain:

$$\frac{d[f_1(N)]}{dN} = -\frac{5.7}{N^2} \quad (43)$$

$$\frac{d[f_2(\eta_p, N)]}{dN} = -\frac{57 \left(\frac{\mu}{n} \right)^9}{N^2} \quad (44)$$

$$\frac{dPP}{dN} = v \cdot \left[-\frac{5.7}{N^2} - \frac{57 \left(\frac{\mu}{n} \right)^9}{N^2} \right] \quad (45)$$

Finally we have:

$$N = \sqrt{\frac{\lambda}{PP \cdot c_2} \cdot \left(c - \frac{c_1}{\eta_p} - c_3 \right) \left\{ -\alpha \cdot v \cdot \left[5.7 + 57 \left(\frac{\mu}{n} \right)^9 \right] \right\}} \quad (46)$$

Using formula (46), it is possible to calculate the flight frequency value for which airline can maximize its profit. Before using this formula for desired load factor it is first necessary to calibrate demand model (to estimate parameters k_1 and α), to estimate value of passengers' time v, parameters μ and σ, and to estimate constants c_1, c_2 and c_3.

FLIGHT FREQUENCY DETERMINATION ON THE SINGLE ISOLATED LINK FOR WHICH CARRIER MAXIMIZES NUMBER OF CARRIED PASSENGERS

Let us consider single, isolated link on which operates only one carrier without competition. Logical question about flight frequency is as follows: For which flight frequency value can carrier maximize number of carried passengers? Swan (1979) determined this flight frequency value using condition that fare equals average costs including a fair return on investment.

This means:

$$c = \frac{C}{\lambda} = c_1 \cdot \frac{S}{\lambda} + c_2 \cdot \frac{N}{\lambda} + c_3 \quad (47)$$

i.e.:

$$c = \frac{c_1}{\eta_p} + c_3 + c_2 \cdot \frac{N}{\lambda} \quad (48)$$

Taking into account relation (48) and previous formulas it is possible to express number of passengers as:

$$\lambda = k_1 \left\{ \frac{c_1}{\eta} + c_3 + c_2 \cdot \frac{N}{\lambda} \right. $$
$$\left. + v \left[t_B + \frac{5.7}{N} + \frac{57 \left(\frac{\mu}{n} \right)^9}{N} \right] \right\} \quad (49)$$

From the condition $d\lambda/dN = 0$ we finally get flight frequency for which airline can carry maximum number of passengers on the single, isolated link on which it operates without competition:

$$N = \sqrt{\lambda} \cdot \sqrt{\frac{v \left[5.7 + 57 \left(\frac{\mu}{n} \right)^9 \right]}{c_2}} \quad (50)$$

From formula (50) we can see, as Swan mentioned "that optimal frequency should rise at the square root of traffic, provided the load factor is constant."

ANALYSIS OF FLIGHT FREQUENCY INFLUENCE ON THE AVERAGE SCHEDULE DELAY PER PASSENGER ON THE SINGLE ISOLATED LINK

As we metnioned earlier there are different formulas for calculating average schedule delay per passenger. Swan (1979) used an assumption that schedule delay per passenger is equal to one quarter of the average headway (average headway was equal to $22.8/N$). Ghobrial (1983) operated with schedule delay which was equal to $4.5/N$, while some authors used an assumption that average shedule delay per passenger is equal to one half of average headway. Since schedule delay per passenger is one of the basic measures for level of service, the impression about level of service on some route, as well as making a decision about flight frequency value, will highly depend on the way in which schedule delay is defined. Detailed analysis of relationships between flight frequency and schedule delay was done by Teodorović (1983). Teodorović started from the following assumptions.

The time intervals between the moments in which the demand for transportation by particular passengers have been expressed are random variables characterized by a particular probability density function. In Figure 6, t_i ($i = 1,2,3 \ldots$) is used to denote the moments of expressed demand for transportation made by particular passengers and T_i ($i = 1,2,3 \ldots$) designates the time intervals between these demands.

Random variables $T_1, T_2, T_3, \ldots$ are not interdependent and are arranged in accordance with the same probability density function. These random variables are not interdependent because the reasons of one passenger for expressing demand for transportation in a particular moment are, as a rule, not related to the respective reasons for other passengers. Let us use $f(t)$ to mark the probability density function of the random variables T_i which present the time intervals between the demands for transportation of the particular passengers. Let us use μ and σ respectively to mark the mean and standard deviation of these random variables.

Before the flight frequency quantity on a particular route has been determined, let us consider the following facts: in the case of the fixed previously published schedule, passengers adjust their demands for the time of flight to the given schedule. The question is in what way they do that. Let us start from a simplified assumption that a passenger will choose the time of departure closest to the actually desired time of departure. The similar assumption is applied in the paper of Gagnon (1967).

The next outstanding and important question is a level of service for the determined flight frequency value and with the fixed flying schedule. It is obvious that the passengers are more satisfied with the particular value of the flight frequency and particular fixed flying schedule, if it differs, as little as possible, from their desires. Consequently, the best way to estimate the level of service with the particular value of the flight frequency and with a particular fixed schedule is to apply the total or the average time difference between the actual time of flight and the time of departure desired by all the passengers on the time interval (O,T). The quantity of the average time difference between the actual and the desired time of departure by all the passengers (hereinafter called 'time difference') depends on the flight frequency as well as on the scheduled times in the course of the day. Let us determine, for the moment, the quantity of the average time difference, only as the function of the flight frequency quantity, whereas the influence of the times of departure in the course of the day on the quantity of the average time difference will be discussed later.

In Figure 7, $x_1, x_2, \ldots, x_N$ are used to mark departure times in the time interval (O,T). Let us note the time of departure marked x_i. According to the above mentioned assumption that a passenger will choose the time closest to the desired one, the time of departure marked x_i will be chosen by all passengers who want to leave in the time interval:

$$\left(\frac{x_{i-1} + x_i}{2}, \frac{x_i + x_{i+1}}{2} \right)$$

Let us assume that the average number of passengers on each flight on the observed route is equal to $2m$, that m passengers wanted to leave in the time interval:

$$\left(\frac{x_{i-1} + x_i}{2}, x_i \right)$$

and m passengers in the time interval:

$$\left(x_i, \frac{x_i + x_{i+1}}{2} \right)$$

Moments of time in which the observed $2m$ passengers

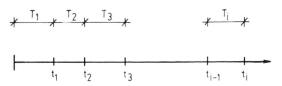

FIGURE 6. Demands for transportation expressed by individual passengers.

wanted to leave are shown in Figure 8. Let us denote the absolute value of the time difference between the actual and the desired departure time by ith passenger with W_i. Quantities W_i ($i = 1, 2, \ldots, 2m$) for the particular passengers are respectively as follows:

$$W_1 = T_2 + T_3 + T_4 + \ldots + T_m + W_m$$

$$W_2 = T_3 + T_4 + \ldots \ldots + T_m + W_m$$

$$W_3 = T_4 + T_5 + \ldots \ldots + T_m + W_m$$

W_m

W_{m+1} (51)

$W_{m+2} = T_{m+2} + W_{m+1}$

$W_{m+3} = T_{m+3} + T_{m+2} + W_{m+1}$

$W_{2m} = T_{2m} + T_{2m-1} + T_{2m-2} + \ldots + T_{m+2} + W_{m+1}$

whereas $W_m + W_{m+1} = T_{m+1}$

It is easy to show that the random variable W ($W = \sum_{i=1}^{2m} W_i$) representing the absolute value of the total time difference between the actual time and the time of departure desired by all passengers who want to leave in the time interval $(x_{i-1} + x_i)/2$, $(x_i + x_{i+1})/2$) is equal to the total m^2 of random variables T_i which are not interdependent and have the same probability density function $f(t)$ with the mean μ and standard deviation σ.

FIGURE 7. Departure times in time interval (O,T).

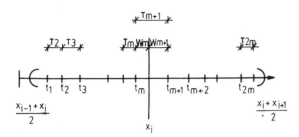

FIGURE 8. Moments of time in which observed 2m passengers want to leave.

According to the known results of the probability theory we can conclude that the random variable W has normal distribution with the mean $m^2 \cdot \mu$ and standard deviation σm. It means that we came to a conclusion that probability density function of the absolute value of the total time difference between the actual time and the desired time of departure by all the passengers who wanted to leave in the time interval $(x_{i-1} + x_i)/2$, $(x_i + x_{i+1})/2$ is equal to:

$$g(w) = \frac{1}{\sigma m \sqrt{2\pi}} \exp\left\{-\frac{[w + m^2\mu]^2}{2\sigma^2 m^2}\right\} \quad (52)$$

It has already been stated that the mean $M(W)$ of the random variable W is:

$$M(W) = m^2 \cdot \mu \quad (53)$$

The average schedule delay per passenger is equal to:

$$s_d = \frac{M(W)}{2m} = \frac{m}{2} \cdot \mu \quad (54)$$

The average number of passengers on the flight is:

$$2m = n \cdot \eta_p \quad (55)$$

where n is used to denote number of seats on a plane which operates and η_p to denote the average load factor. It is easy to show that:

$$s_d = \frac{n \cdot \eta_p}{4} \cdot \mu \quad (56)$$

Let us use N to denote the flight frequency, i.e. number of flights performed in the time interval (O,T) and λ to denote the average number of passengers expressing the demand for flight during the time interval (O,T). It is obvious that:

$$\eta_p = \frac{\lambda}{n \cdot N} \quad (57)$$

$$\mu = \frac{T}{\lambda} \quad (58)$$

On the basis of the formulae (56), (57) and (58) the following result is obtained:

$$s_d = \frac{T}{4 \cdot N} \quad (59)$$

where s_d is obtained in the same units in which T is expressed.

The formula (59) provides a very quick information about the level of the offered traffic service. For the known values of the time interval in which the demands for transportation have been expressed by the average number of passengers who want to be transported and for the particular flight frequency, it is very easy to estimate the average time difference.

In order to achieve a profitable service, the average load factor should range within the following limits:

$$\frac{K}{n \cdot c} \leq \eta_p \leq 1 \quad (60)$$

where c represents the applied fare for the transportation of one passenger, and K the total costs of one flight (direct operating, indirect operating and indirect costs).

In case the average load factor complies with (60) the flight frequency N ranges within the following limits:

$$\frac{\lambda}{n} \leq N \leq \frac{\lambda \cdot c}{K} \quad (61)$$

On the basis of the formulae (59) and (61) we have concluded that if carrier runs a profitable service, the average time difference (absolute value) does not exceed the following limits:

$$\frac{n}{4 \cdot \lambda} \cdot T \leq s_d \leq \frac{K}{4 \cdot \lambda \cdot c} \cdot T \quad (62)$$

The formulae (60), (61) and (62) provided for the relations between the flight frequency N, the average load factor η_p, used to measure the economic effects achieved by the carrier and the average time difference s_d, used to measure the level of service.

The average time difference is determined exclusively as the function of the flight frequency. It is obvious that this average time difference is the function of the flight frequency as well as of the departure times in the course of the day. If the times of departure are better adjusted to the form of the demand density, i.e. to the $h(t)$ function, the average time difference will be lower and vice versa. It is obvious that for the same value of the given flight frequency, depending on the given departure times, various values of the average time difference could appear. It is, therefore, necessary to estimate the influence of the departure times on the average time difference. Only after the quantity of this influence has been considered, it will be possible to verify the formula (59) and to come to the adequate conclusions as to how to apply it.

Let us consider first what is the lowest possible time difference between the actual and desired time of departure for a given flight frequency value. Let us then compare it with the value of the time difference given in the formula (59).

Demand density $h(t)$ is shown in Figure 9. It is obvious that the number of passengers expressing the demand for flight up to a certain moment $t \in (O,T)$ is equal to:

$$H(t) = \int_0^t h(t)dt \qquad (63)$$

The Figure 10 shows $H(t)$ function and times of departure in the course of the day $x_1, x_2, x_3, \ldots, x_N$. According to the previously accepted assumption that the passengers will choose the flight closest to the desired one, the following situation will occur: The flight at the moment x_1 will be chosen by all the passengers who wanted to leave the time interval:

$$\left(O, \frac{x_1 + x_2}{2} \right)$$

The flight at the moment x_2 will be chosen by all the passengers who wanted to leave in the time interval:

$$\left(\frac{x_1 + x_2}{2}, \frac{x_2 + x_3}{2} \right)$$

The total time difference between the actual and the desired time of departure by all the passengers in the time interval (O,T) is presented by the hatched sections on the Figure 11.

On the basis of the Figure 11 after arranging, the absolute value of the total time difference between the actual and the desired time by all passengers expressing demand for flight in the time interval (O,T) is equal to:

$$W = 2 \int_0^{x_1} H(t)\,dt + 2 \int_0^{x_2} H(t)\,dt + \ldots$$

$$+ 2 \int_0^{x_N} H(t)\,dt - 2 \int_0^{\frac{x_1+x_2}{2}} H(t)\,dt$$

$$- 2 \int_0^{\frac{x_2+x_3}{2}} H(t)\,dt - \ldots \qquad (64)$$

$$- 2 \int_0^{\frac{x_{N-1}+x_N}{2}} H(t)\,dt$$

$$- x_N H(T) - \int_0^T H(t)\,dt + TH(T)$$

where $H(T) = \lambda$

Times of departure in the course of the day $x_1, x_2, \ldots, x_N$, for which the absolute value of the time difference W is minimum will be obtained by the following terms:

$$\frac{\partial W}{\partial x_i} = O \qquad (65)$$

where $i = 1, 2, \ldots, N$

After the partial derivations have been determined, it occurs that the times of departure $x_1, x_2, \ldots, x_N$ for which the observed time difference is minimum have to comply with the following equations:

$$H(x_1) = \frac{H\left(\frac{x_1+x_2}{2}\right)}{2}$$

$$H(x_2) = \frac{H\left(\frac{x_1+x_2}{2}\right) + H\left(\frac{x_2+x_3}{2}\right)}{2}$$

$$H(x_3) = \frac{H\left(\frac{x_2+x_3}{2}\right) + H\left(\frac{x_3+x_4}{2}\right)}{2} \qquad (66)$$

$$H(x_{N-1}) = \frac{H\left(\frac{x_{N-2}+x_{N-1}}{2}\right) + H\left(\frac{x_{N-1}+x_N}{2}\right)}{2}$$

$$H(x_N) = \frac{H\left(\frac{x_{N-1}+x_N}{2}\right) + H(T)}{2}$$

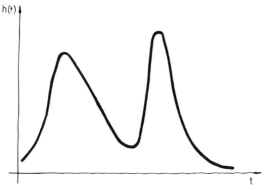

FIGURE 9. Demand density.

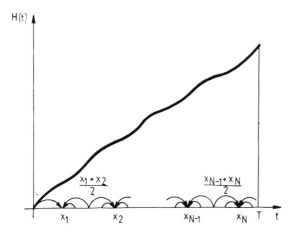

FIGURE 10. Passengers choose flight closest to the desired one.

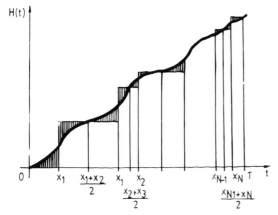

FIGURE 11. Total time difference between actual time of departure and that desired by all passengers in time interval (O,T).

On the basis of the formulae (64) and (66) it is possible for the known form of $H(t)$ function to calculate the minimum value of the total, i.e. time difference per passenger. It is also possible, on the basis of the formulae (64) to calculate the total time difference for various values $x_1, x_2, \ldots, x_N$. In this way, it is possible to quantify the influence of the schedule on the average time difference quantity and thus to verify the validity of the formula (59).

In order to verify the above results it is first necessary to study the character of the demand density, i.e. to determine the $h(t)$ and $H(t)$ function forms. In May 1980 survey was carried out among the passengers flying from Belgrade to Zagreb (Yugoslavia). The passengers were asked the following question:

If you could choose, what time of departure would you choose to fly from Belgrade to Zagreb?

After the survey's results had been classified, the interdependent relation between the demand density and particular times of the day could be shown in the following formula:

$$h(t) = -0.0802\ t^4 + 2.5827\ t^3 - 25.9952\ t^2 \\ + 79.9252\ t + 48.2199 \qquad (67)$$

i.e. that number of passengers expressing the demand for flight up to the t moment is:

$$H(t) = -0.016\ t^5 + 0.6457\ t^4 - 8.6651\ t^3 \\ + 39.9626\ t^2 + 48.2199\ t \qquad (68)$$

where 5A.M. corresponds to the $t = 0$ moment, and time 9P.M. corresponds to the $t = 16$ moment.

On the average $\lambda = 1000$ passengers fly this route. On the basis of the formula (59) it was possible to estimate the time difference per passenger, without regard to the influence of the times of departure to that difference. The formulae (64) and (66) provided for the estimation of the minimum time difference. This time difference is compared with average schedule delay per passenger which was calculated as $s_d = 1/4 \cdot \text{(headway)} = 1/4 \cdot T/N$. The results are shown in the Table 2. The results obtained by the Formula (59) were compared with the minimum time difference per passenger for the various flight frequency values.

Let us compare these results with the results which would be obtained if a few arbitrarily chosen schedules are taken into consideration. For the frequency value $N = 8$ three arbitrary chosen flying schedules have been considered and the corresponding time differences have been estimated. The results are given in the Table 3.

The formulae (59–62) provide for the relation between the flight frequency, the average load factor and level of service, measured by the total or average time difference between the actual and desired time of departure.

The average time difference has been determined without

TABLE 2. Comparison of the Average Time Difference per Passenger Without Regard to the Influence of the Times of Departure to that Difference, with the Minimum Time Difference per Passenger for the Various Flight Frequency Values.

Number of flights per day	$s_d = \frac{1}{4} \cdot \frac{T}{N}$ [min]	Minimum time difference per passenger as per Formulae (64) and (66) (min)
8	30	28.69
9	26.40	25.54
10	24	22.38
11	21.60	20.68
12	19.80	18.83
13	18	17.15
14	16.80	16.35
15	15.60	14.70

regard to the times of departure during the course of the day. The numerical example has shown that, even with this simplification, it is possible to determine satisfactorily correct values of the average time difference. The time difference resulting from the formula (59) will better correspond to the average time difference estimated on the basis of the times of departure, if the times of departure are better adapted to the demand density.

The obtained results are based on the simplified assumption that a passenger will choose the flight closest to the desired one.

The formula (59) can still offer a very useful information concerning the level of service on a particular route. When introducing new routes in air transport network and after the expected average number of passengers has been estimated, the Formulae (59–62) will offer a quick information about the level of service (s_d) and economic effects achieved by the carrier (η_p) under the particular flight frequency.

FLIGHT FREQUENCY DETERMINATION ON THE SINGLE ISOLATED LINK WITH TWO COMPETITORS

In the previous considerations we tried to determine flight frequency on the single isolated link for which it is possible to maximize airline's profit or to maximize number of car-

TABLE 3. Comparison of the Average Time Difference per Passenger for the Various Times of Departure.

Time of departure	Average time difference per passenger in minutes
$x_1 = 1$, $x_2 = 2$, $x_3 = 3$, $x_4 = 7$ $x_5 = 12$, $x_6 = 13$, $x_7 = 14$, $x_8 = 15$	36.71
$x_1 = 3$, $x_2 = 5$, $x_3 = 8$, $x_4 = 11$ $x_5 = 13$, $x_6 = 14$, $x_7 = 15$, $x_8 = 16$	45.86
$x_1 = 1$, $x_2 = 3$, $x_3 = 5$, $x_4 = 7$ $x_5 = 9$, $x_6 = 11$, $x_7 = 13$, $x_8 = 15$	30
$x_1 = 0.9215$, $x_2 = 2.2057$, $x_3 = 3.4867$, $x_4 = 4.9563$ $x_5 = 6.9615$, $x_6 = 10.6239$, $x_7 = 12.6087$, $x_8 = 14.2787$ Schedule with which the minimum time difference occurs	28.69
Value resulting from Formula (59)	30

ried passengers. One of the basic assumptions for those models, was the fact that carrier operates without competition. It is quite logical to proceed with flight frequency determination in the context of a single market with two or more competitors. Powell (1982) studied a single market with two competitors. Similarly as it was done by Renard (1970), Gelerman and de Neufville (1973), de Neufville and King (1979) and Powell also used well known market share/frequency share model as follows:

$$MS_i = \frac{N_i^\alpha}{N_1^\alpha + N_2^\alpha}, i = 1,2 \quad (69)$$

where N_i is the frequency of airline i, MS_i market share of airline i and α is a parameter.

In further discussion airline 1 will be called "the airline" and airline 2 "the competition." Total market size V_o is a random variable. Powell introduced an assumption that this random variable is normally distributed with mean μ and standard deviation σ. Earlier we showed that the number of passengers on a certain flight is normally distributed. Because of this reason the assumption that the total market size V_o has a normal distribution seems very realistic.

Let us denote by V_1 the number of passengers attracted by airline 1. Clearly it is:

$$V_1 = MS_1 V_o \quad (70)$$

On the other hand we have:

$$MS_1 = \frac{N_1^\alpha}{N_1^\alpha + N_2^\alpha} = \frac{1}{1 + N_1^{-\alpha} N_2^\alpha} \quad (71)$$

Thus we have:

$$V_1 = \frac{V_o}{1 + N_1^{-\alpha} N_2^\alpha} \quad (72)$$

in a case when number of passengers attracted by airline 1 is less or equal to the number of seats offered by airline 1.

Total number of seats offered by airline 1 is equal to $n \cdot N_1$, where n is the number of seats in the plane and N_1 flight frequency of airline 1. Finally we can write the number of passengers attracted by airline 1 can be expressed as:

$$V_1 = \begin{cases} \dfrac{V_o}{1 + N_1^{-\alpha} N_2^\alpha}, & \text{if } V_1 \leq n \cdot N_1 \\ n \cdot N_1, & \text{otherwise} \end{cases} \quad (73)$$

Total market demand V_o is a random variable with density $f(V_o)$. So, the expected profit on a route $M(P)$ is:

$$M(P) = \int_o^\infty c \cdot V_1 \cdot f(V_o) \, dV_o - N_1 \cdot K(n,d) \quad (74)$$

where

c = average yield per passenger
$f(V_o)$ = probability density function of random variable V_o
$K(n,d)$ = total cost per flight in a case of aircraft with n seats which operates on route whose length equals d

It is easily noted that:

$$P[V_1 \leq n \cdot N_1] = P\left[\frac{V_o}{1 + N_1^{-\alpha} N_2^\alpha} \leq n \cdot N_1\right]$$
$$= P[V_o \leq n \cdot N_1(1 + N_1^{-\alpha} N_2^\alpha)] \quad (75)$$

hence

$$M(P) = \int_o^{nN_1(1+N_1^{-\alpha}N_2^\alpha)} cV_1 f(V_o) dV_o$$
$$+ \int_{nN_1(1+N_1^{-\alpha}N_2^\alpha)}^\infty cV_1 f(V_o) dV_o - N_1 K(n,d) \quad (76)$$

$$M(P) = \int_o^{nN_1(1+N_1^{-\alpha}N_2^\alpha)} \frac{cV_o f(V_o)}{1 + N_1^{-\alpha} N_2^\alpha} dV_o$$
$$+ \int_{nN_1(1+N_1^{-\alpha}N_2^\alpha)}^\infty cnN_1 f(V_o) dV_o - N_1 K(n,d) \quad (77)$$

From the condition $d[M(P)]/dn = 0$ it is possible to obtain optimum aircraft size (optimum number of seats in a plane) and so to maximize carrier's profit. This number of seats must satisfy the following equation:

$$F[N_1 \cdot n^*(1 + N_1^{-\alpha} N_2^\alpha)] = 1 - \frac{\dfrac{d[K(n,d)]}{dn}}{c} \quad (78)$$

where

n^* = number of seats in the plane for which airline can maximize its profit
$F[\cdot]$ = the cumulative distribution function for V_o

Powell showed that total costs per flight $K(n,d)$ can be approximately expressed as:

$$K(n,d) = \$5.6\,n + (0.46 + 0.0093 \cdot n) \cdot d \quad (79)$$

Using this last equation finally we get that the number of seats in a plane for which carrier maximizes its profit must satisfy the following equation:

$$F[N_1\,n^*(1 + N_1^{-\alpha}N_2^\alpha)] = 1 - \frac{5.6 + 0.0093 d}{c} \quad (80)$$

In a case when random variables V_o has a normal distribution Equation (80) becomes:

$$\Phi\left[\frac{N_1 \cdot n^*(1 + N_1^{-\alpha}N_2^\alpha) - \mu}{\sigma}\right] = 1 - \frac{5.6 + 0.0093 \cdot d}{c} \quad (81)$$

where $\Phi(\cdot)$ is a Laplace function.

For different values of μ, σ, d, N_1, N_2 and α it is possible, using Equations (80) or (81) to calculate optimal aircraft size and so to maximize airline's profit.

From condition $d[M(P)]/dN_1 = 0$ we get equation:

$$F[nN_1^*(1 + N_1^{*-\alpha}N_2^\alpha)] = 1 - \frac{K(n,d) - \dfrac{\alpha \cdot N_2\,N_1^{*-(1+\alpha)}}{[1 + N_1^{-\alpha}N_2^\alpha]^2}}{c \cdot n} \quad (82)$$

where N_1 is flight frequency value for which airline 1 can maximize profit.

Through Equation (82) we established relationship between aircraft size n, flight frequency value N_1^* (for which airline 1 maximizes profit), competition's flight frequency N_2, average yield per passenger c, total costs per flight $K(n,d)$, in case of stochastic demand whose characteristics are expressed through the cumulative distribution function $F[\cdot]$. Through this model all main factors are analyzed which have influence on the flight frequency on the single isolated market with two competitors.

SIMULTANEOUS CALCULATION OF DEPARTURE MOMENTS AND FLIGHT FREQUENCY ON THE SINGLE ISOLATED LINK

Usually departure moments are calculated after the flight frequency on the single, isolated link is determined. It means that for known value of flight frequency, for some objective function (taking into account all necessary constraints) we calculate optimal departure moments. Simpson (1969) mentioned that it is possible to use one of the following objectives:

Minimize Fleet Size
Minimize Operating Costs
Maximize (Revenue−Cost) for the Carrier
Maximize (Total social benefits−costs) for operator and public

Most frequently we have some of the following model constraints: number of aircraft in the fleet, different types of aircraft in the fleet, multi-stop routes, minimum daily frequency, maximum daily frequency, location of depots in which aircraft must "sleep," daily utilization of aircraft, working times of airports, maintenance requirements, etc.

In most of scheduling and routing models in air transportation, flight frequency values on different links are input data. Only exception is Dispatching models in which flight frequency on the single, isolated link can be determined simultaneously with departure moments. The basic input data for dispatching models is functional dependence of passenger demand density on the time of a day. In order to get functional dependence of demand density on the time of a day it is necessary to count departing passengers in small time intervals in a case when airline operation is without reservations (air shuttle service), or to make appropriate passengers' surveys in a case with reservations. Simpson (1969) studied simultaneous calculation of flight frequency on the single, isolated link and departure moments for a system without reservations when functional dependence of passenger demand density is known, "so to minimize a weighted sum of costs of operating a dispatch and passenger waiting time." Simpson developed a model based on Dynamic Programming.

Let us consider route between cities A and B for which we know the functional dependence of demand density $h_{AB}(t)$ (Figure 12).

To simplify calculation it is assumed that Passenger Arrival rate $h_{AB}(t)$ is uniform. Passengers' demands for transportation have been expressed during time interval (O,T). System is without reservations and if passenger cannot get a vacant seat on the desired flight he will try to get a seat on

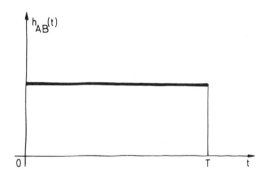

FIGURE 12. Passenger demand density between cities A and B.

some of the next departures. Let us denote by n number of seats in a plane which operates. Let us divide time intervals (O,T) into m smaller time intervals with width Δt. So we have:

$$m\Delta t = T \quad (83)$$

Let us also assume that during every time interval Δt exactly one passenger will arrive. This means:

$$h_{AB}(t) = \begin{cases} \dfrac{1}{\Delta t}, & \text{for } 0 \leq t \leq T \\ 0, & \text{otherwise} \end{cases} \quad (84)$$

Total number of passengers $H_{AB}(T)$ who would like to travel from city A to city B during time interval (O,T) equals:

$$H_{AB}(T) = \int_{o}^{T} h_{AB}(t)dt = \dfrac{T}{\Delta t} = m \quad (85)$$

Appropriate network for dispatching problem is shown in Figure 13. Time is stage variable. State varaible is number of passengers waiting in moments $0, \Delta t, 2\Delta t, 3\Delta t, \ldots, m\Delta t$.

As we can see from the Figure 13 number of waiting passengers z_i in some moment $i\Delta t$ ($i = 1,2,3, \ldots, m-1$) can be equal $0,1,2,3, \ldots$ or i passengers. Number of waiting passengers in a moment $m\Delta t = T$ must be equal to zero, because we must carry all passengers arrived until moment $m\Delta t = T$.

As it was mentioned, flight frequency and departure moments will be determined so to minimize a weighted sum of costs of operating a dispatch and passenger waiting time.

Let us consider moment $(n + 1)\Delta t$ and let the number of waiting passengers at this moment be $n + 1$. If the number of waiting passengers at the moment $(n + 2)\Delta t$ equals $(n + 2)$, it means that there was no dispatch at the moment $(n + 2)\Delta t$. In this case the length of the branch from node $(n + 1)$ in the moment $(n + 1)\Delta t$ to node $(n + 2)$ in the moment $(n + 2)\Delta t$ will be cost of waiting time of a passenger who arrived during time interval $[(n + 1) \Delta t, (n + 2)\Delta t]$. On the other hand the length of the branch from node $(n + 1)$ in the moment $(n + 1)\Delta t$ to node 2 in the moment $(n + 2)\Delta t$ will be the sum of the cost of dispatching a flight with n passengers and the cost of waiting time of a passenger who arrived during time interval $[(n + 1)\Delta t, (n + 2)\Delta t]$. Let us consider the branch (z_{i-1}, z_i) which connects number of waiting passengers z_{i-1} at the moment $(i - 1)\Delta t$ with a number of waiting passengers z_i at the moment $i\Delta t$ and let $R_i(z_{i-1}, z_i)$ denote the length of this branch. Let the length of this branch represent the weighted sum of cost of operating a dispatch and cost of passenger waiting time. Let us mark with $f_i(z_i)$ the length of the shortest path to node z_i at stage i. Since $i = 0$ is the initial stage, then $f_o(O) = O$. Thus we write the Dynamic Programming equations as:

$$f_o(O) = O \quad (86)$$

$$f_i(z_i) = \min \{f_{i-1}(z_{i-1}) + R_i(z_{i-1}, z_i)\} \quad (87)$$

$$i = 1,2 \ldots ,m$$

Using relations (86) and (87) it is possible to determine flight frequency value and appropriate departure moments, so to minimize the weighted sum of cost of operating a dispatch and cost of passenger waiting time.

Usually flight frequency value is calculated simultaneously with departure moments in a case of air shuttle service. When two cities are connected by air shuttle, many business travellers turn to air transport because it offers schedules that match better their requests. Experience shows that air shuttles have been above all introduced between cities with high passenger demands, especially business travellers. The main characteristics of air shuttle are equal intervals between two successive departures and impossibility of reserving tickets in advance. One can get a guarantee that he will depart with the first departing plane, or that within a certain probability he will depart with the first departing plane, or that he will depart with the next departing plane if he cannot get a vacant seat in the first departing plane, etc. . . . This means that passengers will get some sort of guarantee about waiting time from the carrier. These guarantees depend on the average number of passengers, demand density by time of the day, the desired average load factor, the number of engaged aircraft for air shuttle, etc. The ranking of air shuttle alternatives surely depend on several criteria. Since both carrier and passengers interests shoul be included in the development of air shuttle alternatives and their order of priority, a multiple criteria decision making method is required.

Passenger requests for transportation between cities A and B are given in time interval (O,T). With $H(t)$ we denote the average number of passengers who would like to travel in this relation up to moment t. A survey of passenger requests gives function $H(t)$. The moments of departure of first, second, third, . . . aircraft are denoted with $x, 2x, 3x, \ldots, zx$. Let us first calculate transportation capacities, which should be offered in moments $x, 2x, 3x, \ldots$ for case when at least (p) percentages of passengers could find a vacant seat in the first departing plane and when passengers had guarantee that they would get a seat in the next departing plane, if they did not get one in the first.

The passenger appearing during the time interval from $(i - 1)x$ to $ix, i = 1,2, \ldots, z$, will find a vacant seat in the plane which is to depart at the moment ix, with a probability at least equal (p). If one cannot find a vacant seat in the plane departing at the moment ix, he will find one with

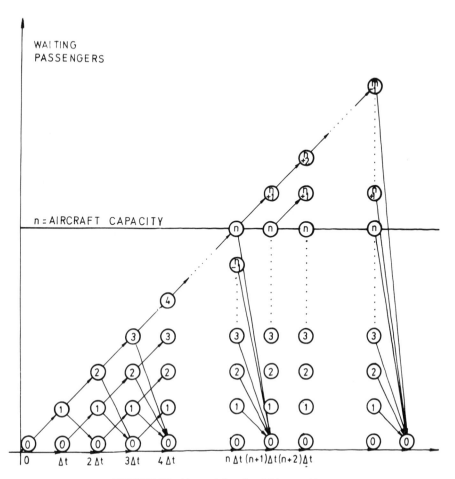

FIGURE 13. Network for dispatching problem.

probability equal to 1 in the plane departing at the moment $(i + 1)x$. This means, that the passenger has a guarantee that, with a probability at least equal to (p), he will wait for departing less than x, and with probability equal to 1 he will wait for departure less than $2x$. If t_w is the passenger's waiting time for departing, then:

$$P[t_w < x] \geq p \qquad (88)$$

$$P[t_w < 2x] = 1 \qquad (89)$$

Let us denote with $k_1, k_2, \ldots, k_z$ the minimal number of seats that should be offered in moments $x, 2x, \ldots, zx$. If one's wish is to satisfy relations (88) and (89), it is obvious that:

$$k_1 = pH(x) \qquad (90)$$

$$k_i = p\{H(ix) - H[(i-1)x]\} \\ + (1-p)\{H[(i-1)x] - H[(i-2)x]\} \qquad (91)$$

where

$i = 1, 2, \ldots, z$

and

$H(O) = O$

If the number of seats in the plane is denoted by n, then the number of departing aircraft f_i in the moments $x, 2x, \ldots, zx$ is:

$$f_i = \left[\frac{k_i}{n}\right] + 1 \quad , i = 1, 2, \ldots, z \qquad (92)$$

where $[y]$ is integer part of y.

There is a probability at least equal to p that the passenger can get a vacant seat on the first plane to depart. An interesting question regarding this probability is as follows:

Should one take p to have the same value throughout the

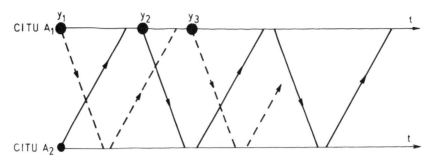

FIGURE 14. Space-time diagram.

day or is it preferable to have p different in relation to the form of $H(t)$?

Probability p has a direct influence on the average waiting time per passenger and also on the number of engaged aircraft. With different values of p during peak hours and outside of peak hours a carrier can directly change the form of $H(t)$, since regular users would become familiar with the service offered. A higher value of probability p outside of peak hours can influence the form of $H(t)$ in the same manner as a lower ticket price on planes which depart outside of peak hours. It is not easy to say which is better: to have p constant or in relation to the form of $H(t)$. The final decision concerning probability p depends on demand density by time of the day, the desired average load factor, the number of engaged aircraft for the air shuttle, etc.

The real number of offered seats $x, 2x, \ldots, zx$ is equal $n f_i$, $i = 1, 2, \ldots, z$. Knowing the number of departing aircraft f_i at $x, 2x, \ldots, zx$, it is easy to calculate the number of aircraft that should be engaged in air shuttle between two cities. Very good methods for calculating the minimum number of aircraft for a given schedule are presented by Anders Martin-Löf (1970) and A. Levin (1971). Papers by Gilett and Miller (1974) and Fisher and Jaikumar (1981) give a simple heuristic method for calculating the number of needed vehicles.

Figure 14 shows a space-time diagram where nodes are flights which have to be flown. Nodes are denoted by $y_1, y_2, y_3, \ldots$, respectively. After simple visual inspection of the space-time diagram, by connecting corresponding nodes (taking into account block time and servicing time on the ground between the two flights) it is possible to determine the route of each aircraft, as well as, the number of engaged aircraft in the air shuttle fleet. So, for example, with solid and dotted lines are determined routes of two aircraft that are engaged in air shuttle service between cites A and B.

Figure 15 shows the number of passengers $H(t)$, that

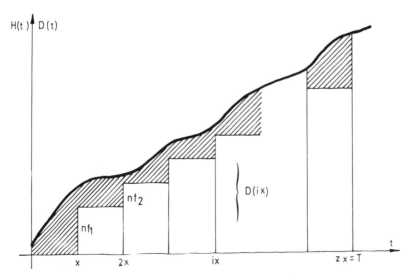

FIGURE 15. The number of passengers H(t) that wanted to travel up to the moment t and the number of already departed passengers D(t) up to the moment t.

wanted to travel up to the moment t and the number of already departed passengers $D(t)$ up to the moment t. The shaded area in Figure 15 represents the total time losses of all passengers that travel from city A to city B during time interval (O,T). These time losses are:

$$W = \int_o^T H(t)dt - x \sum_{i=1}^{z-1} D(ix) \qquad (93)$$

where

$$D(ix) = \min\left\{\sum_{r=1}^{i} sf_r; H(ix)\right\} \qquad (94)$$

Before establishing air shuttle service on a route, the carrier gets the form of $H(t)$ with a survey of passenger requests. We calculated time loss W taking into account the form of $H(t)$ from the beginning of air shuttle service establishment. After establishing the air shuttle a large number of passengers will become familiar with the service offered. Most of them will adjust their arrival at the airport to the service, trying to minimize waiting time. So, more precisely, time loss W defined by relation (93) is at a maximum unless service continuously adjusts to the changing of $H(t)$.

The average waiting time per passenger is:

$$\bar{t}_w = \frac{W}{H(T)} \qquad (95)$$

The number of passengers unable to find a vacant seat in the first plane to depart is:

$$Q = \sum_{i=1}^{z} [H(ix) - D(ix)] \qquad (96)$$

Another important characteristic of air shuttle is the probability of finding a vacant seat in the first plane to depart. This can be calculated as:

$$q = 1 - \frac{Q}{H(T)} \qquad (97)$$

where

$$q \geq p$$

Knowing the number of flights which have to be made during a day it is easy to calculate total costs of air shuttle service. If we calculate these costs, as well as, the number of engaged aircraft, average waiting time per passenger and probability of finding a vacant seat in the first plane to depart, for a few different values of time interval between two successive departures, it is possible to choose the best air shuttle alternative. Clearly, the ranking of different air shuttle alternatives and appropriate flight frequency values depends highly on criteria weights for ranking.

During the past few years air shuttle was introduced on many routes in the world. Operational research departments of air carriers have been engaged in developing theoretical air shuttle models and their practical implementations. Results of one such investigation can be found in paper by J. L. Gasco (1975). In that paper one can find experience of Spanish carrier "Iberia" about introducing air shuttle service on Madrid–Barcelona route in 1974. Average daily number of passengers during 1972 and 1973 was 1250 in a direction Madrid–Barcelona and 1280 in a direction Barcelona–Madrid. Average load factor before introducing air shuttle was 71%. First of all "Iberia" was interested in improving the level of service on this route. Gasco found that 77.6% of all passengers flying on this route are business travellers. This category of passengers usually did not cancel their tickets. Also it was found that in cases when these passengers finished their activities in a city, they were coming to the airport trying to get a vacant seat on the first departing plane. Many of these passengers were coming to the airport without reservations for certain flights having only a ticket with open date, because as a passenger with experience in air transport they had a feeling about possibility to get a vacant seat on a certain flight. "Iberia" has calculated a schedule based on the following guarantees to air shuttle passengers:

1. 100% guarantee to the passenger that he will depart if it is possible to make reservation for certain flight (for some flights it was still possible to make reservations) and if a passenger has correct reservation.
2. 90% guarantee to the passenger that he will depart with a first departing flight if for that flight it is not possible to make reservations.
3. 100% guarantee to the passenger that he will depart with the next flight if it was impossible to get a vacant seat on the desired flight.

Gasco (1975) developed simualtion model for this type of air shuttle. Simulation technique was used, first of all because of simplicity to find relationship between flight frequency, departure moments, number of engaged aircraft, total transportation costs, average waiting time per passenger and probability of finding a vacant seat in a first plane to depart in air shuttle. As it was mentioned it is very useful to apply Multiattribute Decision Making Methods for ranking of different air shuttle alternatives (different flight frequency values and different distribution of departure moments). One such attempt can be found in a paper by Teodorović (1985).

FLIGHT FREQUENCY ON ALL NETWORK BRANCHES FOR WHICH TOTAL SCHEDULE DELAY FOR ALL PASSENGERS IS MINIMIZED

In previous discussion we have considered flight frequency determination on the single, isolated link. Problem of simultaneous determination of flight frequencies on all network branches is more complex than flight frequency determiantion on the single, isolated link because it is necessary to match passenger requests between all city pairs with carrier's transport capacities. Flight frequencies on all network branches are determined simultaneously with design of air transportation network and ranking of appropriate aircraft type (aircraft size choice). Aircraft size has a direct influence on the flight frequencies on certain links, carrier's economic performance and level of service. One of the most important papers about design of air transportation networks and flight frequencies determination on all links of the network is the paper by S. Gordon and R. de Neufville (1973). As a measure for a level of service these authors used schedule delay. They started from an assumption that the total schedule delay for all passengers D can be expressed as:

$$D = a \sum_i \frac{\lambda_i}{(1 - \eta_{p_i}) \cdot N_i} \qquad (98)$$

where

a = the proportionality constant (which is assumed to be the same along every link)
N_i = flight frequency along link i
λ_i = the volume of passengers along link i
η_{p_i} = the average load factor along link i

Clearly with flight frequencies increasing along links, the total schedule delay for all passengers is decreasing. Increasing of flight frequencies can extend only to a certain limit which depends on the number of aircraft in the fleet, as well as on their types. Airline can during some time interval offer to the passengers certain number of available seat-hours. This means that number of available seat-hours S must satisfy following inequality:

$$\sum_i N_i \, n \, t_{B_i} \leq S \qquad (99)$$

where

S = number of available seat-hours which carrier can offer to the passengers during some time interval
n = number of seats in the plane
t_{B_i} = block time along link i
N_i = flight frequency along link i

Average load factor along link i is equal:

$$\eta_{p_i} = \frac{\lambda_i}{N_i \cdot n} \qquad (100)$$

So, we have:

$$D = a \cdot n \sum_i \frac{\eta_{p_i}}{1 - \eta_{p_i}} \qquad (101)$$

Using relations (99) and (100) we got:

$$\sum_i \frac{\lambda_i \, t_{B_i}}{\eta_{p_i}} \leq S \qquad (102)$$

If carrier wishes to offer to passengers the best possible level of service with existing fleet, it is necessary to minimize total schedule delay for all passengers. So, our problem can be defined as:

$$D = a \cdot n \sum_i \frac{\eta_{p_i}}{1 - \eta_{p_i}} \rightarrow \min \qquad (103)$$

subject to:

$$\sum_i \frac{\lambda_i \, t_{B_i}}{\eta_{p_i}} \leq S \qquad (104)$$

After introducing auxiliary function $F(\eta_{p_i}, \alpha)$ and La Grange multiplier α we have:

$$F(\eta_{p_i}, \alpha) = a \cdot n \sum_i \frac{\eta_{p_i}}{1 - \eta_{p_i}} + \alpha \left(\sum_i \frac{\lambda_i t_{B_i}}{\eta_{p_i}} - S \right) \rightarrow \min \qquad (105)$$

From the conditions:

$$\frac{\partial F}{\partial \eta_{p_i}} = 0 \text{ and } \frac{\partial F}{\partial \alpha} = 0 \qquad (106)$$

Gordon and de Neufville got the following relations:

$$\alpha = \frac{a \cdot n \left[\sum_i \sqrt{\lambda_i t_{B_i}} \right]^2}{\left[S - \sum_i \lambda_i t_{B_i} \right]^2} \qquad (107)$$

$$N_i = \frac{\lambda_i}{n} + \frac{S - \sum_i \lambda_i \cdot t_{B_i}}{n \cdot \sum_i \sqrt{\lambda_i \cdot t_{B_i}}} \cdot \sqrt{\frac{\lambda_i}{t_{B_i}}} \qquad (108)$$

TABLE 4. Comparison of Flight Frequency Values Which Were Calculated Using Formulae (108) and (114).

Route i	Average daily number of passengers in one direction	Block-time along the route $t_{B_i}[Hr]$	Formula (108)	Formula (114)
1	250	1.5	4.46 ≈ 4	5.22 ≈ 5
2	300	1.4	5.22 ≈ 5	5.92 ≈ 6
3	350	1.6	5.75 ≈ 6	5.98 ≈ 6
4	400	1.2	6.77 ≈ 7	7.38 ≈ 7
5	450	1.1	7.57 ≈ 8	8.17 ≈ 8
6	500	1.7	7.61 ≈ 8	6.93 ≈ 7
7	550	1.3	8.63 ≈ 9	8.31 ≈ 8
8	600	1.1	9.55 ≈ 10	9.40 ≈ 9
9	650	1.3	9.90 ≈ 10	9.04 ≈ 9
10	700	1.2	10.67 ≈ 11	9.76 ≈ 10

Using relation (108) it is possible to get flight frequencies on all links in the network so to minimize total schedule delay of all passengers with existing fleet.

We have shown earlier that an average schedule delay per passenger on link i would be approximately equal to:

$$s_{d_i} = \frac{T}{4N_i} \quad (109)$$

where

s_{d_i} = schedule delay per passenger on link i
λ_i = an average number of passengers on link i who would like to travel time interval (O,T)
N_i = flight frequency along link i
T = time interval during which passengers would like to travel

It is interesting to see which flight frequency values we will get if we use relation (109) to express schedule delay. The total schedule delay for all passengers in the whole network is:

$$W = \sum_{i=1}^{p} s_{d_i} \cdot \lambda_i \quad (110)$$

where p is a number of links in the network.

It can easily be proved that:

$$W = \frac{T}{4} \cdot \sum_{i=1}^{p} \frac{\lambda_i}{N_i} \quad (111)$$

An average time difference between between the desired and the actual departure time in the network per one passenger is as follows:

$$w = \frac{T}{4 \sum_{i=1}^{p} \lambda_i} \cdot \sum_{i=1}^{p} \frac{\lambda_i}{N_i} \quad (112)$$

There is no doubt that this time difference is one of the basic indicators of the level of service in an airline network having non-stop flights. Now, we can determine flight frequencies along links, by solving the following problems:

$$W = \frac{T}{4} \cdot \sum_{i=1}^{p} \frac{\lambda_i}{N_i} \rightarrow \min$$

$$\sum_{i=1}^{p} N_i \cdot n \cdot t_{B_i} \leq S \quad (113)$$

Solution using La Grange multiplier technique yields:

$$N_i = \frac{S \cdot \sqrt{\lambda_i}}{n \cdot \sqrt{t_{B_i}} \cdot \sum_{i=1}^{p} \sqrt{\lambda_i t_{B_i}}}, i = 1,2,\ldots,p \quad (114)$$

Using Equation (108) or (114) one can get flight frequency values along links so to minimize total schedule delay for all passengers with existing fleet. It is interesting to compare flight frequencies which were calculated using formula (108) and formula (114).

In the Table 4 are shown the values of flight frequencies on ten routes which were calculated using formulae (108) and (114). Average daily number of passengers, as well as, block times along certain routes are also shown in this table. It was assumed that the capacity of aircraft which operates is $n = 100$ seats and that the daily numbers of available seat-hours is $S = 10{,}000$ seat-hours.

The comparison of values of flight frequencies determined by the relation (108) with those determined by the formula (114) shows slight differences. The values of flight frequencies between pairs of cities with lower values of the passengers' volume determined by the relation (114) are somewhat greater than those determined by the formula (108). Formula (108) gives somewhat greater values than those determined by the relation (114) if the routes in question are used by a greater number of passengers.

Through the relations (108) and (114) we established connection between carrier's characteristics (n and S), characteristics of the network (λ_i and t_{B_i}, $i = 1,2\ldots,p$) and flight frequencies ($N_i, i = 1,2,\ldots,p$). Flight frequencies are determined so as to maximize the level of service with existing fleet. The measure for level of service was total schedule delay for all passengers. Instead of this measure for the level of service we can also use the sum of total schedule delay for all passengers and the total stochastic delay for all passengers in order to determine flight frequencies applying the same methodological approach.

FREQUENCY PLANNING FOR NETWORKS TO MINIMIZE CARRIER'S COSTS

In the previous considerations we have determined flight frequencies to maximize level of service with existing fleet Such approach is appropriate when taking into account interests of a society. On the other hand, in most countries airlines usually have not any special financial support from the governments. So, air carrier, first of all, must take care about revenue and cost. It is quite natural that most of airlines are trying to minimize total operating costs taking care about level of service up to a certain limit.

Frequencies planning for network to minimize carrier's costs is one of basic air carrier's problems which have to be solved. This problem was under the consideration of Carter and Morlok (1975). These authors developed a model for flight frequencies determination over network to minimize carrier's direct operating costs on an airline network with non-stop and one stop or two stop flights.

Let us denote by N_{ij} flight frequency in one direction (non-stop flights) between city i and city j. In a case when flights between two cities are one-stop, two-stop flights etc., it is possible that aircraft fly between these cities along different routes. Later we will discuss in more details ways in which these routes can be determined. Let us denote by r and s respectively cities between which carrier operates with one-stop and two-stop flights, and let us denote by N_{rsp} flight frequency in one direction between cities r and s along route p. Estimated daily number of passengers between cities r and s in one direction, λ_{rs} is one of basic input data for this type of analysis. Let us denote by λ_{rsp} daily number of passengers between cities r and s along route p, and by L_{rs} minimum allowed daily frequency between cities r and s. So, L_{rs} is a measure of a minimum level of services to be required between cities r and s. As we can see flight frequencies will be determined to minimize direct operating costs but simultanesouly we will take care of level of service up to a certain limit. The values of L_{rs} between all city pairs (r,s) must also be given as input. Now, we can determine flight frequencies along links, to minimize direct operating costs, by solving the following problem:

$$\sum_{i,j} c_{ij} N_{ij} \to \min \qquad (115)$$

$$CN_{ij} - \sum_{rsp \in ij} \lambda_{rsp} \geq 0 \quad ,\text{all } ij \qquad (116)$$

$$\sum_p \lambda_{rsp} = \lambda_{rs} \qquad ,\text{all } rs \qquad (117)$$

$$N_{ij} - N_{rsp} \geq 0 \quad ,\text{all } ij, \text{ for each } rsp \in ij \quad (118)$$

$$\sum_p N_{rsp} \geq L_{rs} \qquad ,\text{all } rs \qquad (119)$$

$$\text{all } N_{ij}, N_{rsp}, \lambda_{rsp} \geq 0 \qquad (120)$$

$$\text{all } N_{ij}, N_{rsp} = \text{integer} \qquad (121)$$

where

N_{ij} = flight frequency in one direction (non-stop flights) between cities i and j
c_{ij} = cost of operating a flight from i to j
λ_{rsp} = average daily number of passengers which travel from city r to city s along route p
λ_{rs} = daily number of passengers between cities r and s
C = effective seating capacity of an aircraft (the seating capacity n multiplied by the average load factor η_p)
N_{rsp} = flight frequency in one direction between cities r and s along path p

It is necessary to explain in more details relations (116–121). Relation (116) means that daily offered number of seats along every link in the network must be greater than or equal to all of the daily demand assigned to the routes using that link. The second constraint (117) requires that the sum of the passengers assigned to all of the routes connecting cities r and s, must be equal to the total number of passengers for that city pair. The third constraint (118) requires that frequency on the route is not greater than the smallest frequency along links which include that route. Constraint (119) requires that the sum of frequencies on all of the routes between each city pair must be greater than or equal to the minimum required frequency. All choice variables must be greater than or equal to 0 (Constraint 120). Constraint (121) requires that the frequency choice variables can take only integer values.

This problem is a typical integer programming problem. However, as it was mentioned by Carter and Morlok (1975) "Since real world applications of this model are likely to involve hundreds of choice variables—and in the case of large regional networks approximately 1,000 to 2,000 choice variables—it did not seem appropriate to attempt to use existing integer programming codes to solve the problem. Rather the initial approach suggested by the model was to attempt to solve it as a standard linear program, dropping the integer constraint." Carter and Morlock applied standard linear program in a case of introducing V/STOL aircraft in Appalachia. These authors used modified abstract mode model and multiple regression analysis to estimate number of passengers between city pairs. They used ATA method for calculation of direct operating costs. In a case of greater airline network there is a large number of the possible routes between two cities. Because of the efficiency of the model it is necessary to reduce the number of possible routes. Carter and Morlock reduced the number of possible routes as follows. When the distance via a route is more than double the distance of the direct route, that route is not considered. In further consideration only routes which have no more than two intermediate stops are included. Routes along which intercity distance is less than 50 miles are excluded from consideration and no other city within the circle formed by the diameter drawn as a straight line between the two cities.

Linear programming is a very useful method for solving problems of this type. The results of this particular application indicate that dropping the integer constraint does not alter much the total costs of operating the system. In this model it was assumed that all aircraft which operate have the same capacity. Since, most frequently in an airline network operate aircraft of different capacities, it is necessary to determine flight frequencies along links and routes simultaneously with assignment of different types of aircraft to particular routes. These types of models are known as Fleet Assignment Models. Simpson (1969) developed Fleet Assignment Model for solving the following problem. For a given set of non-stop routes for which market share curves relating passengers to daily frequencies are known and for given number of aircraft of particular type, assign different types of aircraft to particular routes and determine appropriate flight frequencies, to maximize income for the system. The constraints which were included in the model were: maximum permissible load factor on all routes, fleet availability, minimum daily frequency on each route, etc. Linear programming technique is also very convenient for solving this type of problem.

FREQUENCY PLANNING AND TRAFFIC ASSIGNMENT ON AIR NETWORK

Most frequently, in air transportation, it is possible to make a trip between two cities along several different routes. The manner by which the total number of passengers who would like to travel from one to another city will be distributed among different routes, depends on many factors. The most important factors are travel time, travel fares, and frequency of service along a particular route. Carrier can operate along routes by non-stop flights, direct flights with one or more stops or indirect flights. The types of the flights on particular routes also have important influence on distribution of total origin-destination demand among different routes. Taking into account the fact that the fares are very similar on different routes between certain city pairs, we can conclude that flight frequency and travel time along a route are most important factors for route choice.

Problem of frequency planning and traffic assignment on air network was studied by Kanafani (1981), Kanafani and Ghobrial (1982) and Ghobrial (1983). Let us consider the results which were obtained by Kanafani and Ghobrial.

One of the basic assumptions in models for frequency planning and traffic assignment on air network is the assumption that origin-destination demands are inelastic (fixed origin-destination matrix), as long as there is some transportation capacities to meet them. On the other hand, it was mentioned, that the distribution of this demand among the different routes is elastic and that flight frequencies are one of the most important factors which have influence on this distribution. This means that some initial flight frequencies along particular routes will have an influence on the manner by which total passenger demand will be distributed among different routes. On the other hand, the initial link and route frequencies must be adjusted in order to meet the number of passengers who would like to travel along particular routes. New, adjusted flight frequencies again have an influence on the number of passengers along different routes, etc. It is obvious that flight frequencies determination along routes and traffic assignment are parts of one iterative procedure. As it was mentioned by Ghobrial (1983) "the process is repeated iteratively until network equilibrium is reached when no change in both link frequencies and flows occurs." This approach to flight frequency determination was for the first time applied by Kanafani (1981). In the analysis of aircraft technology and network structure in short-haul air transportation, Kanafani developed iterative procedure for flight frequency determination along particular routes in the network. This procedure is based on the following multinomial logit model:

$$P(r,j) = \frac{e^{V(r,j)}}{\sum_{r \in R_j} e^{V(r,j)}} \quad (122)$$

where

$P(r,j)$ = proportion of all passengers in the jth city pair who use route r
R_j = set of feasible routes between jth city pair
$V(\cdot)$ = linear choice function

After calibration of the multinomial logit model (Kanafani and Chang (1979)), linear choice function was as follows:

$$V(r,j) = -0.0188 \cdot t_{rj} + 0.212 \cdot N_{rj} \quad (123)$$

where

t_{rj} = travel time between jth city pair along route r
N_{rj} = flight frequency between jth city pair along route r

Statistical data about traffic between 26 city pairs 500 km around the city of Atlanta were used for this calibration.

Through iterative procedure Kanafani alternatively calculated number of passengers and flight frequencies and appropriate load factors along particular routes. This iterative procedure was finished when load factors were within acceptable bounds. The lower bound on acceptable load factor on any link was a given break-even load factor. Ghobrial (1983) used similar approach for frequency planning and traffic assignment in air network. Ghobrial noted that any link in the network could be a part of more than one route between different city pairs. In Ghobrial's model the basic requirement was to ensure a conservation of flows at each node in the network. This means that the following equations must be satisfied:

$$\sum_i \lambda_{ij}^{OD} = \sum_m \lambda_{jm}^{OD} \quad , \text{if } j \neq O, D \quad, \forall O, D \quad (124)$$

$$\sum_i \lambda_{ij}^{OD} = \sum_m \lambda_{jm}^{OD} + D_{OD} \quad , \text{if } j = D \quad, \forall O, D \quad (125)$$

$$\sum_i \lambda_{ij}^{OD} = \sum_m \lambda_{jm}^{OD} - D_{OD} \quad , \text{if } j = O \quad, \forall O, D \quad (126)$$

where

λ_{ij}^{OD} = number of passengers from O to D along link (i,j)
D_{OD} = total number of passengers from O to D

Flight frequency along any route k is as follows:

$$N_k = \min_{ij \in R_k} \{N_{ij}\} \quad (127)$$

where

N_k = flight frequency along route k
N_{ij} = flight frequency along link (i,j)
R_k = set of links which are part of route k

Ghobrial (1983) started from an assumption that it is possible to get the proportion of the origin–destination passengers between certain city pairs that are flying particular route, using the following multinomial logit model:

$$\lambda_{r,j} = \frac{e^{V(r,j)}}{\sum_{r \in R_j} e^{V(r,j)}} \cdot D_j \quad, \forall j \quad (128)$$

and

$$\lambda^{(p,q)} = \sum_{r \in R^{(p,q)}} \lambda_{r,j} \quad (129)$$

where

$\lambda_{r,j}$ = number of passengers between the jth city pair who use route r
R_j = set of feasible routes between the jth city pair
$R^{(p,q)}$ = set of routes containing link (p,q)
$V(r,j)$ = linear function of the attributes of the route r between the jth city pair

$V(r,j)$ was a linear combination of the following attribute variables and dummy variables:

$$V(r,j) = a_1 T_{rj} + a_2 N_{rj} + a_3 c_{rj} + a_4 n_{rj} + a_5 G_{rj} \quad (130)$$

where

T_{rj} = travel time along route r between jth city pair
N_{rj} = flight frequency along route r between jth city pair
c_{rj} = fare along route r between jth city pair
n_{rj} = dummy variable depending upon the number of aircraft seats n
G_{rj} = dummy variable reflecting the travel pattern
$a_1, a_2, \ldots, a_5$ = parameters to be estimated

Dummy variables n_{rj} and G_{rj} are respectively defined as follows:

$$n_{rj} = \begin{cases} 0 & \text{if } n \leq 30 \\ 0.5 & \text{if } 30 < n \leq 50 \\ 1 & \text{if } n > 50 \end{cases} \quad (131)$$

$$G_{rj} = \begin{cases} 0 & \text{if it is non-stop flight} \\ 0.5 & \text{if it is direct flight with one or more stops} \\ 1 & \text{if it is indirect flight} \end{cases} \quad (132)$$

After calibration of the model Ghobrial (1983) showed that multinomial logit model is very useful for describing the distribution of total origin–destination demand among available routes. For frequency planning and traffic assignment in air network he developed the following iterative procedure:

For every city pair in the network between which carrier would like to operate, determine the set of feasible routes between them. In the following step determine initial values of flight frequencies along links, ensuring that carrier operates at break-even load factors. Taking into account initial flight frequencies, using multinomial logit route choice model, calculate the number of passengers on routes and links. For these values of number of passengers determine new values of flight frequencies. Using new values of flight frequencies and multinomial logit route choice model calculate the new values of number of passengers among routes and links etc. Finish with iterative procedure when change in neither link frequencies nor number of passengers along routes and links occurs.

In this iterative procedure for flight frequencies determination first of all care was taken about carrier's economic results. Kanafani and Ghobrial (1982) also developed iterative procedure for flight frequencies determination and traffic assignment in which both the carrier's economic interest and the level of service (society's interest) were included. These authors also used multinomial logit route choice model trying to maximize the utilization of passenger-miles per dollar of system cost. System cost in this case, was the sum of carrier's costs and passenger travel time and delay costs.

This type of iterative procedure for flight frequencies determination and traffic assignment can be repeated few times for few different aircraft of different capacities. In this manner it is much easier to make appropriate decisions about aircraft size choice and the structure of the appropriate air network.

REFERENCES

1. Carter, E. C. and E. K. Morlok, "Planning Air Transport Network in Appalachia," *Transportation Engineering Journal of ASCE*, Vol. 101, pp. 569–588 (1975).
2. de Neufville, R. and C. R. King, "Access, Fares, Frequency: Effects on Airport Traffic," *Transportation Engineering Journal of ASCE*, Vol. 105, pp. 109–125 (1979).
3. De Vany, A. S. and E. H. Garges, "A Forecast of Air Travel and Airport and Airway Use in 1980," *Transportation Science*, Vol. 6, pp. 1–18 (1972).
4. Douglas, G. W. and J. C. Miller, "Economic Regulation of Domestic Air Transport," Brookings (1974).
5. Eriksen, S., "Policy Oriented Multi-Equation Models of U.S. Domestic Air Passenger Markets," MIT Ph.D. (1977).
6. Falkson, L. M., "Airline Overbooking: Some Comments," *Journal of Transp. Econ. and Policy*, Vol. 3, pp. 352–354 (1969).
7. Fisher, M. L. and R. Jaikumar, "A Generalized Assignment Heuristic for Vehicle Routing," *Networks*, Vol. 11, pp. 109–124 (1981).
8. Gagnon, G., "A Model for Flowing Passengers over Airline Networks," *Transportation Science*, Vol. 1, pp. 232–238 (1967).
9. Gasco, J. L., "Simulation Model for MAD-BCH Air Shuttle," Paper presented at the 15th AFIFORS Symposium, Rotorua, New Zealand (1975).
10. Gelerman, W. and R. de Neufville, "Planning for Satellite Airports," *Transportation Engineering Journal of ASCE*, Vol. 99, pp. 537–551 (1973).
11. Ghobrial, A. A., "Analysis of the Air Network Structure: The Hubbing Phenomenon," Ph.D., thesis, University of California, Berkeley (1983).
12. Gilett, B. and L. Miller, "A Heuristic Algorithm for the Vehicle Dispatch Problem," *Operations Research*, Vol. 22 (1974).
13. Gordon, S. and R. de Neufville, "Design of Air Transportation Networks," *Transportation Research*, Vol. 7, pp. 207–222 (1973).
14. Hyman, W. and L. Gordon, "Commercial Airline Scheduling Technique," *Transportation Research*, Vol. 2, pp. 23–39 (1968).
15. Johnston, E. E., V. J. Jones and J. R. Ritchie, "Measuring Consumer Perceptions of Airline Competition," Presented at the World Conference on Transport Research, London (April 1980).
16. Kanafani, A., "Aircraft Technology and Network Structure in Short Haul Air Transportation," *Transportation Research*, Vol. 15A, pp. 305–314 (1981).
17. Kanafani, A. and P. Chang, "A System Analysis of Short Haul Air Transportation," Research Report, Institute of Transportation Studies, University of California, Berkeley (1979).
18. Kanafani, A. *Transportation Demand Analysis*. McGraw-Hill Book Company (1983).
19. Kanafani, A. and A. Ghobrial, "Aircraft Evaluation in Air Network Planning," *Transportation Engineering Journal of ASCE*, Vol. 108, pp. 282–300 (1982).
20. Levin, A., "Scheduling and Fleet Routing Models for Transportation Systems," *Transportation Science*, Vol. 5, pp. 232–255 (1971).
21. Martin-Lof, A., "A Branch-and-Bound Algorithm for Determining the Minimal Fleet Size of a Transportation System," *Transportation Science*, Vol. 4, pp. 159–163 (1970).
22. Miller, J. C., "A Time-of-Day Model for Aircraft Scheduling," *Transportation Science*, Vol. 6, pp. 221–246 (1972).
23. Nagarajan, K. V., "On An Auction Solution to the Problem of Airline Overbooking," *Transportation Research*, Vol. 13A, pp. 111–114 (1979).
24. Powell, W. B., "Analysis of Airline Operating Strategies Under Stochastic Demand," *Transportation Research*, Vol. 16B, pp. 31–43 (1982).
25. Renard, G., "Competition in Air Transportation: An Econo-

metric Approach," Thesis presented to the Massachusetts Institute of Technology, at Cambridge, Massachusetts (1970).
26. Rothestein, M., "Airline Overbooking: The State of the Art," *Journal of Transp. Econ. and Policy,* Vol. 5, pp. 96-99 (1971).
27. Shlifer, E. and Y. Vardi, "An Airline Overbooking Policy," *Transportation Science,* Vol. 9 (1979).
28. Simon, J. L., "Airline Overbooking: The State of the Art—A Reply," *Journal of Transp. Econ. and Policy,* Vol. 6, pp. 254-256 (1972).
29. Simpson, R. W., "A Review of Scheduling and Routing Models for Airline Scheduling," Paper presented at the AGIFORS Meeting, Broadway, England (October 1969).
30. Soumis, F., J. A. Ferland and J. M. Rousseau, "MAPUM: A Model for Assigning Passengers to a Flight Schedule," *Transportation Research,* Vol. 15A, pp. 155-162 (1981).
31. Swan, W. M., "A Systems Analysis of Scheduled Air Transportation Networks," Report FTL-R79-5, Massachusetts Institute of Technology (1979).
32. Teodorović, D., "Flight Frequency Determination," *Journal of Transportation Engineering,* Vol. 109, pp. 747-757 (1983).
33. Teodorović, D., "Multicriteria Ranking of Air Shuttle Alternatives," *Transportation Research,* Vol. 19B, pp. 63-72 (1985).
34. Vickrey, W., "Airline Overbookings: Some Further Solutions," *Journal of Transp. Econ. and Policy,* Vol. 4, pp. 257-270 (1972).

CHAPTER 18

Pedestrian Flow Characteristics

ABISHAI POLUS*

INTRODUCTION

The movement of pedestrians in the urban environment is vital for sustaining the social and economic relationships essential to city life. Walking enables individuals to have direct contact with the environment and with other people, enables the passage of people from place to place, and makes possible the access of pedestrians to areas where vehicular traffic is not possible or is not desirable for safety or ecological reasons.

To enable and encourage walking for different purposes, the physical facilities must be available to support the physiological, psychological, and social needs of pedestrians and to ensure them against overexertion, interference by other pedestrians, and accidents. Planning and implementing such facilities require an understanding of the characteristics of pedestrian traffic.

Walking, because of its infinite diversity, is sometimes the only means of transportation that satisfies many short, dispersed trip linkages required within the central business district (CBD) of a city. A study of pedestrian movement by Pfefer, Sorton, Fegan and Rosenbaum (1982) suggests that about 90 percent of all internal trips within the CBD are walking trips. In outlying areas, however, trips are divided into walking trips (such as short trips to neighbors or recreational walking); combined drive and walk trips (such as trips to and from and within shopping centers); or mainly auto trips, which involve driving for a longer distance than the usual walking distance.

The central business district is usually subject to some capacity restrictions and freedom-of-movement limitations, especially during peak-hour traffic. Often, streets and sidewalks are narrow and inadequate to handle existing volumes, since the infrastructure is a survival of an archaic street system based on land-use scales of the past. Certain sidewalks of the New York City Borough of Manhattan, for example, drain great numbers of workers as well as visitors from the many high-rise buildings, causing huge volumes of pedestrians at mid-day and the afternoon peaks and leading to the rapid deterioration in flow conditions to a low level of service. The spill-over of these pedestrians into the roadway and the congestion on crosswalks, furthermore, cause a breakdown in vehicular level of service and lead to safety problems from the increased number of conflict locations. The accident potential and its severity are increased, of course, because of the slow and relative directional freedom of pedestrian movement, compared to the rapid movement of the motor vehicle.

The following general characteristics typify pedestrian movements:

- relatively slow movement
- sensitivity to available space
- dependence on physical characteristics of pathways (slope, width, surface type)
- high directional freedom
- variability in age, sex, and physical ability or handicap
- susceptibility to emotional and psychological conditions
- sensitivity to carried loads or luggage
- great variability in purpose of trip (work, school, leisure, shopping)
- great variability in climatic and environmental conditions encountered

GENERAL CHARACTERISTICS OF PEDESTRIAN FLOW

Pedestrian flows have constituted in recent years the subject of several studies, some of which will be described

*Department of Civil Engineering and Transportation Research Institute, Technion—Israel Institute of Technology, Haifa 32000, Israel

later. Foot traffic has been found to be influenced by psychological, physiological, social and environmental factors. The most significant of these factors for defining the characteristics of pedestrian flow, the studies show, are age, gender, physical fitness, the social tie to pedestrians very close by, the purpose of the walk, and topography.

Walking-Speed

Speed is defined as the distance traversed by a pedestrian in a unit of time. It is possible to define two types of walking speeds: (1) a free walking-speed, in which the pedestrian chooses his own speed without any interruptions while walking along a level path under favorable weather conditions; (2) a forced walking-speed, which is the possible speed under the prevailing flow, geometric and weather conditions. The free walking-speed depends primarily on the pedestrian's age, physical ability and fitness, and purpose of trip. The forced walking-speeds depend mainly on the density of the pedestrian traffic (that is, the number of pedestrians per unit area), the physical characteristics of the path along which the pedestrian is moving, and weather conditions. Physical characteristics may include the width and slope of the path and the texture of the surface.

Walking-speed is sensitive to weather conditions, particularly to extreme conditions like severe cold or heat, and is also greatly influenced by the pavement condition as a result of the weather, particularly the presence of snow or rain. Typical values of average free walking-speeds for men range between 1.28 m/sec (4.2 ft./sec.) and about 1.50 m/sec. (4.9 ft./sec.). Walking-speeds for females are slightly, but significantly, lower, on the average about 7–12 percent slower. Table 1 presents some average free speeds based on typical studies of pedestrian movement.

The influence of age on walking-speed was investigated by a few researchers, among whom were Fruin (1971) in the

TABLE 1. Average Free Walking-Speed.

Walking Environment	Male m/sec (ft/sec)	Female m/sec (ft/sec)	All Pedestrians m/sec (ft/sec)
SIDEWALKS	1.28 (4.20)	1.14 (3.74)	—
Source: Polus, Schofer, Ushpiz (1983)			
CROSSING STREETS	1.36 (4.45)	1.22 (4.00)	1.28 (4.20)
Source: Traffic Engineering Handbook (1965)			
Source: ITE Technical Council Committee 5-R (1976)	1.16 (3.80)	1.00 (3.30)	1.13 (3.70)
TRANSIT TERMINALS			
*Port Authority Bus Terminal (N.Y.C.)	—	—	1.31 (4.30)
*Pennsylvania Railroad Station (N.Y.C.)	—	—	1.37 (4.50)
Source: Fruin, 1971			
CENTRAL BUSINESS DISTRICT			
*Mid-Block	1.50 (4.93)	1.41 (4.63)	1.46 (4.80)
*Intersection	1.50 (4.93)	1.38 (4.53)	1.44 (4.72)
Source: Hoel, 1968			

TABLE 2. Walking-Speed, According to Age, Group and Sex.

Age and Sex of Pedestrian	Average Walking-Speed (km/h)	Speeds on Pathway	
		(m/sec)	(ft/sec)
Men, under 55 yrs. of age	3.7	1.03	3.37
Men, over 55 yrs. of age	3.4	0.94	3.10
Women, under 50 yrs. of age	3.1	0.86	2.82
Women, over 50 yrs. of age	2.9	0.81	2.64
Women, with small children	1.6	0.44	1.46
Children, between 6 and 10 yrs. of age	2.5	0.69	2.27
Adolescents	4.0	1.11	3.65

Source: Peschel (1957).

U.S. and Peschel (1957) in Germany. The latter found that the speed of children and teenagers increases with age, whereas the walking speeds of adults and elderly people decrease with increased age. Children up to 10 years of age and women with small children are generally the slowest pedestrians; adolescents usually walk at the fastest speed. The findings of his study are summarized in Table 2.

Analysis of pedestrian speed on various slopes has shown that it is reduced considerably for slopes of 5–14 percent. For slopes up to 5%, the reduction is only about five percent relative to a level walkway. According to the Traffic Engineering Handbook (1950), pedestrian speeds on slopes tend to be relatively constant up to gradients of about 4–5%. Beyond this point, a reduction in speed of about 35% occurs as a slope increases from 4% to 14%. With further increases in the gradient of the walkway, however, a more moderate reduction in speeds takes place. Fruin (1971) analyzed the speeds of soldiers on slopes and found a similar, although smaller trend. He found an 11.5% reduction in speed when the gradient increased from 5% to 10% and only a 2.3% further reduction when the gradient increased to 20%.

Space per Pedestrian

The physical space per pedestrian is defined as the overall area needed by or available to a pedestrian under various walking or standing conditions. The space needed by a single pedestrian is defined by the shoulder breadth and body depth, with some allowance for light clothing. A general guide for physical space for standing pedestrians under various conditions, based on French guidelines, is suggested in Table 3.

Dimension values for two pedestrians with baby strollers or for a couple include an allowance for some margin between the two pedestrians. According to an ITE informational report on pedestrian characteristics (1976), the minimum space for a woman is $0.14 m^2$ ($1.51 ft^2$) and for a man $0.17 m^2$ ($1.83 ft^2$). These values, which are presented in Table 3, should be considered as absolute minimum values, not normally to be used in design. Desirable values and the effect of the functional aspects of the standing area on the space per standing pedestrian may be adopted from the study by Fruin (1971), and are presented in Table 4.

According to ITE Technical Council Committee 5-R (1976), the minimum width of a sidewalk (or any other pedestrian facility) should be approximately 1.8 meters (6 ft.) in order to allow the convenient passage of a pedestrian past a wheelchair. The limiting dimensions are as follows: 0.75 meter (2.5 ft.) for a wheelchair, 0.75 meter (2.5 ft.) for a moving pedestrian (one lane), and 0.3 meter (1 ft.) for wheelchair clearance. The old standard for a single lane was 0.56 meter (22 inches) but the new standard of 0.75 meter (2.5 ft. or 30 inches) was found to be more adequate for pedestrian needs.

In the design of ramps, one must consider the influence of slope on pedestrian volume and speed. The ITE Committee (1976) noted that slopes of 1:6 (16.7%) and 1:8 (12.5%) appeared to be the most frequently used outdoors, while a slope of 1:10 (10.0%) was recommended for ramps within buildings.

Density

In the context of pedestrian flow, density is defined as the reciprocal of space per pedestrian. Thus, density defines the number of pedestrians in a unit area, a square meter or a square foot. Adopting space values from Table 4, for example, yields a suggested density for elevators of about 5 pedestrians per square meter, or 1 pedestrian per 2 square feet.

The density value is more meaningful than space in describing pedestrian flows, as will be described in the following paragraphs. Density, therefore, will be suggested as a design measure in conjunction with the level-of-service concept that will be presented.

Speed-Density Relationships

It is generally useful to examine speed-volume-density relationships to develop an understanding of any traffic flow phenomenon because: (1) these parameters are meaningful descriptions of flow characteristics; (2) they are conceptually easy to measure and understand; and (3) past stud-

TABLE 3. Values of Physical Space for Pedestrian(s) under Various Conditions.

Pedestrian Conditions	Width m.(ft.)	Depth m.(ft.)	Space m²(ft.²)
Single pedestrian	0.66 (2.16)	0.31 (1.02)	0.21 (2.26)
One pedestrian with baby stroller	0.70 (2.30)	2.0 (6.56)	1.40 (15.06)
Two pedestrians with baby stroller	1.7 (5.58)	2.0 (6.56)	3.40 (36.59)
One pedestrian with small hand baggage (at side)	0.82 (2.69)	0.30 (0.98)	0.25 (2.69)
One pedestrian with two large suitcases (both sides)	0.93 (3.05)	0.60 (1.97)	0.56 (6.00)
One pedestrian in wheelchair	0.80 (2.62)	1.00 (3.28)	0.80 (8.61)
Two pedestrians (a couple)	1.40 (4.59)	0.43 (1.41)	0.60 (6.46)
Minimum values:			
Women	—	—	0.14 (1.50)
Mixed	—	—	0.17 (1.80)
Contact with others	—	—	0.26 (2.75)
Uncrowded	—	—	0.33 (3.50)
Dense bulk queues (at escalators and crosswalks)	—	—	0.50 (5.0)

Source: Batelle Research Center, "Synthesis of a Study on the Analysis, Evaluation and Selection of Urban Public Transport Systems," Geneva, 1974, and ITE Technical Council Committee 5-R (1976).

ies of vehicle flows have identified important requirements in the relationship between these parameters that are helpful in understanding flow mechanisms, service attributes, and design principles.

The classic linear relationship between speed and density of pedestrian flow is illustrated in Figure 1. This graph is similar in shape to speed-density relationships for vehicles. Older (1964) investigated the relationship between pedestrian speed and flow density on Oxford Street in London, and found a decreasing linear relationship. The highest speeds for the lowest densities (free-flow conditions) were 1.4m/s or less. Navin and Wheeler (1969) found a decrease in the flow rate (volume) and speed of pedestrian traffic with a rise in density and with an increase in pedestrian traffic from the opposite direction. Hankin and Wright (1958), who investigated the flow of passengers in subway stations, appear to be the first to have presented a nonlinear graph to express the speed-density relationship, which was composed of several subsections, each nearly linear in itself. Fruin (1971) explored extensions of basic vehicle-flow relationships to pedestrian volumes, average speed and density. He used the reciprocal of density, however, and, defining it in area modules of square feet per pedestrian, found that pedestrian speeds tended to have low variability, as increased crowd density restricted the ability to bypass slower-moving pedestrians and to select a desired walking-speed. Fruin used a general semi-parabolic curve to describe the increase in speed with the increase in module size (decrease in density); the rate of increase gradually decreased as the area per pedestrian (module) increased.

Polus, Schofer, and Ushpiz (1983) conducted a pedestrian-flow and level-of-service study in which data were collected by recording pedestrian sidewalk traffic with a video-tape camera. Speed-density relationships were calibrated and two types of models were obtained: a three-regime model and a linear model. The linear relationship between walking speed and average density is presented in Figure 2.

TABLE 4. Effect of Functional Aspects of the Standing Area on Space per Standing Pedestrian.

Functional Aspect	Space Area per Pedestrian	
	m²	ft²
Elevator		
mixed crowd	0.17	1.8
all female	0.14	1.5
average	0.19	2.0
for a comfortable feeling	0.28	3.0
Stairs	0.65–0.74	7.0–8.0
Escalator or crosswalk (dense bulk queue)	0.46	5.0
Ticket lines	0.74–0.84	8.0–9.0

Source: Values adopted from Fruin (1971).

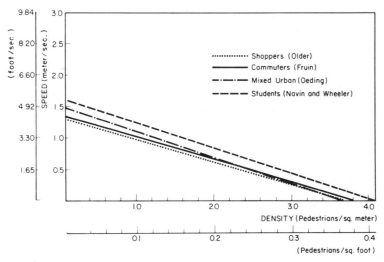

FIGURE 1. Speed-density relationships for pedestrian flow. Source: J. F. Zupan (1977).

Flow

To relate speed and space or density to pedestrian flow, the traditional vehicular traffic-flow relationship may be used:

$$V = S \cdot D$$

where:

V = pedestrian flow per minute per meter
S = speed in meters per minute
D = density in pedestrians per square meter.

If the speed were expressed in meters per second, then the product $(S \cdot D)$ should, of course, be multiplied by 60 to obtain the flow of pedestrians per minute per meter.

If a space parameter is used, the flow equation is expressed as follows:

$$V = S/SP$$

where V and S have been defined above, and SP = space area in square meters per pedestrian.

Pedestrian flow is related to speed and density in a manner similar to the relationship of vehicular traffic. Some

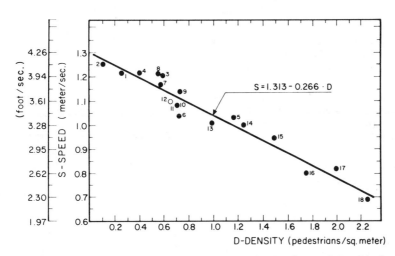

FIGURE 2. Average walking speed and average density. Source: Polus, Schofer and Ushpiz (1983).

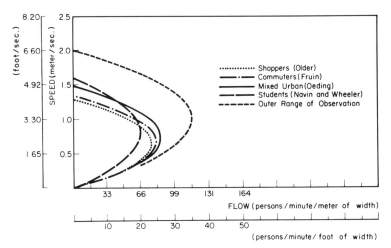

FIGURE 3. Typical speed-flow relationships. Source: Zupan (1977).

typical speed-flow relationships are presented in Figure 3. It can be noticed that the maximum observed flows ranged between 65 student pedestrians per minute per meter (20 persons per minute per foot) and some 88 pedestrians in an urban environment per minute per meter (27 persons per minute per foot). Some observations, however, yielded up to 115 pedestrians per minute per meter (35 persons per minute per foot).

LEVEL OF SERVICE

A number of studies have defined service level for pedestrians on the basis of the ranges of average area available for a single pedestrian. Fruin (1971) defined six levels of service in a somewhat similar manner to that of vehicular levels of service. He, as well as Oeding (1963), before him, described the factors that influence the quality and ease of pedestrian traffic flow at the different levels of service: the possibility of progressing at a desired normal walking-speed, the reasonable chance of conflicts between pedestrians in the main traffic flow and those in crossing directions, the chance of passing a slow pedestrian, and the existence or nonexistence of two-directional traffic.

In connection with this last factor, Fruin found that a significant difference in the behavior of pedestrians existed between one-directional and two-directional traffic when the main flow in the latter case was much bigger than that of the secondary flow. In this situation, pedestrians in the main traffic flow occupy the entire walking path space and leave the free clearance space between them to the pedestrians in the secondary traffic flow. Fruin also found that the four factors influencing service level are themselves influenced by the size of the average area available to each pedestrian in the traffic flow; furthermore, these factors differ according to the environmental attributes and characteristics of the physical facilities.

Oeding's definition for level of service for walking is essentially similar to that of Fruin's. Oeding refers to five main levels of service, which are characterized by occupancy areas of between 3.3 m^2 per pedestrian for a high level of service and 0.7 m^2 per pedestrian for a low level of service. Pushkarev and Zupan (1975) offer a similar level of service definition, but include low densities—large occupancy spaces (above 5 m^2 per pedestrian)—that are beyond the range investigated by Oeding and Fruin. Pushkarev and Zupan defined a number of levels of service for walking, beginning with open flow and unimpeded flow to dense flow and jammed flow. In the last event, progress is nil.

Based on their analysis of speed-density relationships as well as several other qualitative observations of pedestrian flow performed in their study, Polus, Schofer, and Ushpiz (1983) presented a quantitative and qualitative definition of four suggested levels of service for pedestrian flow. The parameter values are presented in Table 5 for the following levels: level of service A, which characterizes free flow; level of service B, which is typical of restricted, impeded, and unstable pedestrian flow; level of service C, which characterizes dense flow; and level of service D, which is the level obtained under very high densities and jammed flow conditions.

Referring back to Figure 2, the limits of the so defined level-of-service ranges can be plotted as shown in Figure 4.

Level of service A, with densities of fewer than 0.6 pedestrians/m^2, seems to be suitable for a design aiming at a free flow of pedestrians (e.g., on walkways in residential areas or public parks). During the research, it was observed that densities up to about 0.6 pedestrians/m^2 produce no signifi-

TABLE 5. Values of Parameters Defining Four Levels of Service for Pedestrian Flow.

Description of Flow	Level of Service	Pedestrian Density, in Pedestrians per Square Meter	Pedestrian Area Occupancy, in Square Meters per Pedestrian	Estimated Flow Volumes, in Pedestrians per Meter per Minute	Recommended Uses When Planning Walk Paths
Free flow	A	≤0.60	≥1.67	0–40	residential areas, public parks
Restricted, impeded, unstable flow	B	0.61–0.75	1.66–1.33	40–50	public buildings, commercial areas, shopping centers
Dense flow	C_1 C C_2	0.75–1.25 1.25–2.00	1.33–0.80 0.80–0.50	50–75 75–95	high-rise office buildings, sport centers, central transit stations
Jammed flow	D	>2.0	<0.5	undefined	not recommended

Source: Polus, Schofer, Ushpiz (1983).

cant deterioration in speed (to be interpreted as freedom to maneuver). Further refinement of this free-flow range, as suggested by some studies, does not seem justified for pedestrian flows because of the observed insensitivity of the flow measures (volume and speed) to variations in density in this range. Under these conditions, pedestrian-flow measures are likely to be more sensitive to esthetic and physical parameters, such as visual attractions, slope and surface smoothness.

The increasingly restricted flow represented by level of service B may be suitable for designing shopping areas and public buildings, where the monetary and land-use constraints call for some compromises in design criteria, where pedestrians are likely to have lower needs and expectations regarding flow speeds, and where merchants and owners clearly benefit from both increased density and reduced speeds.

The higher end of level C, denoted as level C_1 at densities

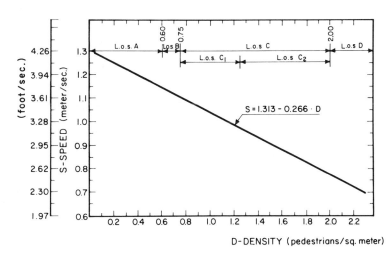

FIGURE 4. Level-of-service parameters of pedestrian flow.

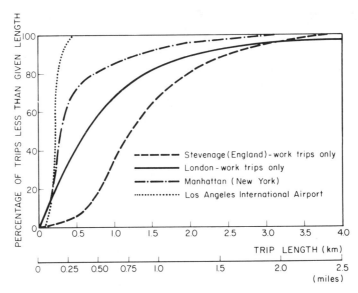

FIGURE 5. Cumulative distribution function of trip length. Source: Carter and Homburger (1978).

of 0.75–1.25 pedestrians/m², is appropriate for the design of space between high-rise office buildings or campus facilities, where peak flows occur routinely and over a relatively short period of time. The lower end, denoted as C_2, with expected densities in excess of 1.25, but fewer than 2.0 pedestrians/m² and the corresponding predicted volume approaching 85 pedestrians per meter width per minute, should probably be restricted to places where people's expectations of free flow are low because of extremely high volumes and densities (e.g., at or near sport centers, central transit stations, or similar facilities). In such high-volume cases, the costs of moving to level of service B are likely to be high and difficult to justify[1]. A fourth level of service, D, characterized by jammed flow, is observed under such conditions as the gathering of large crowds in front of a narrow gate. Fruin (1971) suggested the criterion of 2.15 pedestrians/m² as a limit above which the flow is jammed. Pushkarev and Zupan (1975) proposed a range for the jammed level of 1–5 pedestrians/m². The study by Polus, Schofer and Ushpiz (1983) suggested the value of 2.0 pedestrians/m² as the lower limit of service level D. This figure probably should be avoided as a design value although peak loads on some pedestrian facilities may fall within this range.

[1]Recent recurrent happenings in some European sport centers should rather lead to high design criteria, especially for emergency escape lanes in high occupancy stadiums.

WALKING DISTANCES

Walking trips vary considerably in length, ranging from short trips from a parked car to a near-by destination to long walks for pleasure purposes. The purpose of the trip, as well as the weather, availability and cost of alternative transportation modes determine the length of a trip that a pedestrian is willing to walk. The distance usually increases with favorable weather and as the alternative modes become more expensive, less accessible, and less reliable. Longer distances are obtained in large cities, where pedestrians may be accustomed to and accept the longer walking distances owing to the inability to park close to their destination. Transit riders, on the other hand, are normally exposed to longer distances in the smaller cities because of the lesser density of transit routes.

Acceptable walking distances are important factors in the design of transit routes and station spacing, location of parking garages and siting of commercial centers and also in the design of new neighborhoods where vehicle traffic is separated from pedestrian movement. Figure 5 presents a selected cumulative walking distance distribution.

It can be observed that acceptable trip lengths vary considerably from one city to another and that people are willing to walk longer to work than at airports, for example. Figure 6 shows the proportion of the total walking trips made to work in London as a function of the trip length. It shows a steady reduction from about 70 percent at 0.6 km (0.37 mile) to about 10 percent at 2.3 km (1.43 miles). It can

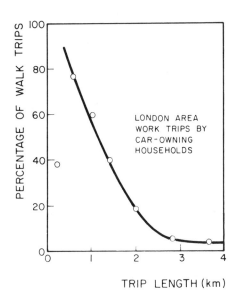

FIGURE 6. Proportion of total trips made by walking. Source: Mitchell (1973).

also be observed that 50 percent of the work trips are made by walking if the distance is about 1.15 km (0.71 mile) or less.

Carter and Homburger (1978) suggested that in auto-oriented locations, such as shopping centers and airports, walking distances should be very short. The proximity of the parking locations to the access location in these facilities is essential. They also suggested a general guide for transit planning by which service should be offered within 1/4 mile (400m.) of origins and destinations in the CBD and within 1/2 mile (800m.) elsewhere.

PEDESTRIAN CROSSINGS

Pedestrian behavior in crossing streets and roads depends primarily on the crossing facilities, such as pavement markings, road signs and particular signals. Behavior in crossing at an unsignalized location depends on a number of factors, such as the available gap in the traffic stream, type of operation (one way or two way), local geometric conditions (width, sight distance, and slope), enforcement practices, and drivers' responses to walking pedestrians.

The delay to a pedestrian waiting to cross a traffic stream is of practical interest in view of the time lost and the possibility of the crossing being associated with risk-taking and potentially higher accident rates. This issue was first treated by Adams (1936) and since then by a number of other researchers. Adams' equation for mean delay (d) was given as

$$d = \frac{1}{q}(e^{qT} - qT - 1)$$

where q is the traffic flow (in vehicles per second) and T is the critical gap (in seconds).

Goldschmidt (1977) found that vehicle-traffic flow was the only variable that significantly affected pedestrian delay. Road width, the number of lanes of traffic, and speed were rejected in his multiple regression models as insignificant variables. Most of the streets surveyed, however, were between 6 and 10 meters (19.7 and 32.8 feet) wide, and traffic speeds were concentrated between 35 and 56 km/h (22 and 35 miles per hour). A presentation of mean pedestrian delay associated with different road-crossing situations is presented in Figure 7.

At signalized intersections, pedestrian crossing opportunities are created by the signal phasing. Signalized pedestrian crossings are normally found at intersections, but sometimes at mid-block. A number of studies (e.g., McLean and Howie, 1980) have shown signalized crossings to be generally safer than zebra crossings, but the delays to pedestrians higher (note also Figure 7). In North America, pedestrian-crossing signals use words instead of symbols. Robertson (1977) reported on a U.S. Federal Highway Administration project to improve pedestrian signal displays by using symbols.

Pedestrian movements may be incorporated in the various phasing schemes. At locations with heavy pedestrian volumes and heavy turning movements, it may be desirable to provide an exclusive phase for pedestrians. A report by the U.K. Department of Transport (1981) suggested that a separate pedestrian stage or one combined with a traffic stage should be required if either (1) the flow of pedestrians across any one arm of the intersection is of the order of 300 per hour or more; or (2) the turning traffic flowing into any arm has an average headway of fewer than 5 seconds and conflicts with a pedestrian flow of at least 50 pedestrians per hour.

These flows, it is further suggested, should be taken as the average for the four busiest hours during any weekday. In special circumstances—e.g., where there is an above-average number of infirm or handicapped pedestrians—then a crossing facility may be required even though it cannot be justified in terms of the above warrants. At school crossings, special signals may also be justified. If a school crossing patrol operates at a signalized intersection for a short period of the day, then a key switch, under the control of an authorized person, may be provided to introduce an extended all-red period for the needed time.

Mid-block signalized crossings are composed of a standard three-phase lantern for vehicles which dwells in green until activated by a pedestrian and a two-aspect lantern for

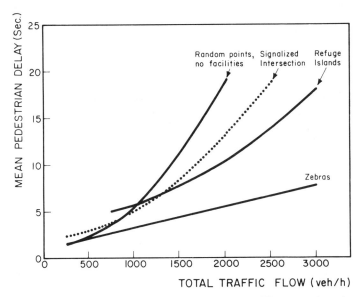

FIGURE 7. Mean pedestrian delay associated with different road-crossing situations. Source: Goldschmidt (1977).

pedestrians with words or symbols. The signals require a push-button detector, which is the common form of detection used for pedestrians. The detector is actuated by a pedestrian's pushing the button to cause a low-voltage current to flow to the controller in order to register a "demand" for pedestrian service. When the demand is received, the WALK indication (or symbol) is displayed as soon as appropriate, followed by a flashing DON'T WALK clearance display. A prescribed distance of 150 meters between a signal-controlled midblock crossing and any other signal-controlled pedestrian crossing is suggested by Ribbens (1983). He further proposes the distance of 200 meters as the distance needed between an uncontrolled mid-block crossing and any signalized intersection.

A high percentage of pedestrian violations may indicate that the signal does not adequately reflect pedestrians' needs. Good signal timing includes a minimum green time to allow pedestrians to cross the roadway safely. For example, a speed of 1.2 m/sec. (4 ft/sec) is assumed in Canada for the minimum walk phase, based on roadway width (MUTCD 1976, and UTCD for Canada, 1976). U.S. practice suggests that under normal conditions, the walk interval should be at least 7 seconds. On streets with a median of at least 6 feet in width, it may be desirable to allow only enough pedestrian clearance time on a given phase to clear the crossing from the curb to the median.

Pavement markings at pedestrian crossings are usually of the zebra or pelican type. The zebra pedestrian crossing is composed of parallel bars 0.3 to 0.6 meter (1 to 2 ft.) wide and 2.0 meters (6.5 ft) minimum length, with gaps of up to 0.2 meter (0.7 ft.) wider than the bars. Several studies have found that crossing at zebra or pelican locations is safer than crossing elsewhere, but not as safe as signal-protected crosswalks.

Zebra crosswalks can be made safer by the following means:

1. Placing guard-rails so as to force pedestrians to use the marked crosswalk.
2. Installing flashing lights and reducing vehicle speeds.
3. Adding pavement markings to prohibit passing and stopping near the crosswalk.

FIGURE 8. Typical installation of a crosswalk illumination and signing system.

The location and design of pedestrian crossings should be considered together with the road network as well as with land-use facilities that generate road-crossing by pedestrians. Special attention should be given to school crossings because of the inability of children to evaluate properly the risk of crossing the road.

Warrants for the provision of zebra pedestrian crossings are reported by Dunn (1982) for various countries. The warrants, based on delay to pedestrians and vehicles, use the hourly vehicle flow (V) and the pedestrian flow per hour (P). The following practices are reported:

New Zealand	$P \times V > 45000$
Australia	$P \times V > 90000$
England	
without a refuge island	$P \times V^2 \geq 10^8$
with a refuge island	$P \times V^2 \geq 10^8 \cdot 2$

There is an upper as well as a lower limit, however, to the use of unsignalized zebra crossings. Because of the confusion they engender over priority between drivers and pedestrians, zebra crossings are not recommended for use on arterial streets where vehicle volumes are high and speeds exceed 60 km/h (approx. 35 mph).

Symbol signs are used to designate the location of pedestrian crossings. At busier locations, the use of additional overhead signs, which can also provide some illumination on the crosswalk at night, is recommended.

The severity of the pedestrian accident problem at night has been well established. Smeed (1968), for example, found the night pedestrian accident rate to be twice as great as the daytime rate, and the night-time pedestrian casualty rate three times greater. Polus and Katz (1978) examined a special crosswalk illumination and signing system (note Figure 8) and found that a significant reduction in pedestrian night-accidents was achieved when devices shown in the figure were installed; the reduction in day-time accidents was not statistically significant. The examined system, it was further found, represented a highly worthwhile investment in terms of cost- and accident-reduction benefits. In terms of pedestrian accident countermeasures, it has become clear from numerous investigations that improved overall street lighting not only reduces motor-vehicle accidents in general, but pedestrian accidents as well.

MECHANICAL STAIRS AND MOVING RAMPS

Mechanical facilities like moving ramps or mechanical stairs help pedestrians overcome height differences or move more conveniently along an extended walk, particularly when carrying luggage as at airports or train stations. Except for some variable-speed moving ramps, these facilities are not designed to save time because they move at speeds that are equal to or slower than normal walking. In order that passengers of all ages and physical abilities may board a conveyor, its speed at the boarding point must be low (about 2½km/h or 1½mph).

Pedestrians entering or leaving moving ramps may experience some discomfort for two reasons: an abrupt change in speed; an abrupt change in inclination of plane (as they step from a level surface to an inclined moving surface or vice versa). It is important, therefore, that the initial speed of horizontal conveyors not be excessive and that mechanical stairs start out horizontally before climbing and level off toward the next floor up to allow pedestrians to step off conveniently. In the design of such facilities, it is important to provide a stationary walk or regular stairs and elevators next to the mechanical facilities, for use by handicapped persons. The stationary facilities or elevators, moreover, should be constructed to allow the easy passage of wheelchairs.

The main advantages of pedestrian conveyors are their automatic operation, non-stop journey, high capacity, absence of pollution and freedom from collision risk. Their disadvantages of high cost, modest speeds, and inflexibility (especially their inability to negotiate sharp bends) have caused transport operators to prefer other forms of transportation.

SHARED STREET FOR VEHICLES AND PEDESTRIANS

In recent years, there has been an increasing need to create new forms of residential roads. Some traffic-restraint schemes with the aim of reducing vehicle volumes and speed have been suggested for residential neighborhoods. These schemes include the introduction of one-way streets, the reduction of route continuity and even the narrowing of roads.

A residential neighborhood scheme enabling pedestrians and cars to share the street is inspired by a Dutch development termed "Woonerf." A typical such street is presented in Figure 9. The main features of this concept may be briefly summarized as follows (some of the points are suggested by Polak [1979] and Martin [1982]):

1. The residential function predominates over the traffic function; the pedestrian has rights to the whole roadway.
2. Only streets with relatively low volumes (up to 100 vehicles per hour, both ways) should be treated.
3. Adequate sight lines are required.
4. The segregation between street and sidewalk should be removed; no height difference of curb line should exist.
5. Speed-restricting devices such as humps, sharp bends, and narrow sections are recommended every 50 meters (165 ft.); cars should travel at speeds of less than 16 km/h (10 mph).
6. Only short streets of up to 250 meters (approx. 800 ft.) in length or small areas should be treated.
7. Adequate parking must be provided to handle the

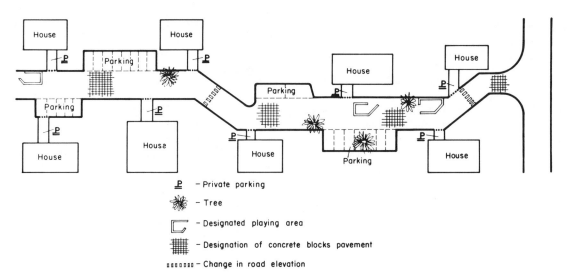

FIGURE 9. A typical scheme of a shared street for vehicles and pedestrians.

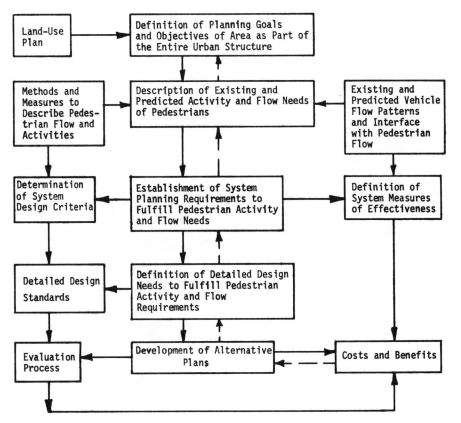

FIGURE 10. General planning process of pedestrian system.

demand of residents and visitors; whenever the expected demand is high, development of a shared street is not recommended. Most of the parking spaces should be provided off the street and within private lots. Street parking should be limited to locations where it causes no inconvenience to other street users. Parking places must be identified in the design of the pavement.
8. Special facilities, such as play areas, should be provided for children, and parking should be prohibited in these areas.
9. The streetscape of "Woonerfs" should be pleasant and appreciated by the pedestrian. Trees and plants and the use of a variety of paving materials, like various interlocking concrete blocks, are highly recommended.

The basic concept of the shared street is that pedestrians have priority over vehicles and that drivers must adjust their normal behavior pattern, particularly by reducing speed. Some willingness and cooperation on the part of the motorists are expected, therefore, and these can be found more easily in residential areas. To conclude, it should be noted that the design of shared residential streets for pedestrians and vehicles can be executed only with strict consideration of the above requirements, since the margin between a successful design and failure is rather small.

SYSTEM CONSIDERATION IN PEDESTRIAN FLOW

In recent years, there has been an increasing recognition that pedestrian movements constitute a vital part of any traffic plan. Whereas there was in the past an understanding that pedestrian flow had lower priority than vehicle flow (e.g., pedestrian delay compared with vehicle delay), traffic-engineering philosophy is today the reverse: pedestrian flow and safety receive higher priorities.

Generally, the design of pedestrian facilities must take into consideration the interrelationships of the three fundamental elements of the system: the human being, the pathway, and the environment. ITE Committee 5-R (1976) suggested that the "designer must consider physical scale, the surface and character of the pathway, physical and topographic impedences, and the need for adequate protection from adverse weather conditions and the general utility and convenience of the system. He must design adequate queuing spaces, walkways and escalators. He should also evaluate the level of service, and he may need to check the cost-effectiveness of alternative designs within given financial constraints."

The planning process of a pedestrian system at a specified urban area may contain five main steps, as described in Figure 10. The general planning and design process is suitable for implementation in any urban activity center: residential areas, shopping centers, or central business districts. Nevertheless, the local flow and activity characteristics of each area determine its specific needs, measures of effectiveness, and criteria for system design and design details.

An analysis and characterization of pedestrian flow in a central business district (CBD) may be conducted following these steps:

a) Determining land-use in and around the CBD
b) Mapping of walking paths, including major and minor flows
c) Mapping of parking spaces, bus stops, and rail transit stations
d) Identifying conflict areas between vehicles and pedestrians and physical barriers hindering flow of pedestrians
e) Mapping of exit and entrance zones of various land-uses, identifying of commercial loading and unloading zones and other service activities
f) Analyzing noise and pollution levels along walking paths
g) Analyzing climatic characteristics of the area and microclimate influences, like sun, cold or wind patterns on walking paths

The analysis methods of pedestrian-flow patterns include questionnaire-studies, observations, and counts at key points. The measures of effectiveness, level of detail, and technological tools used may vary from one study to another, based on area, size, and characteristics as well as available budget and expected results.

CONCLUSION

This chapter has described the characteristics of pedestrian flows. The following subjects were discussed: general characteristics, walking speed, space, density, speed-density relationships, the level-of-service concept and its recommended use in the design of walkpaths and walking distances for work and other trip purposes. A discussion of the pedestrian crossing followed and included an analysis and a comparison of delay at various crossing facilities, the characteristics of signalized crossings, and warrants for non-signalized crossings like zebra crossings. Next, a short presentation of moving walkways and mechanical stairs preceded a description of the new concept of the shared street (vehicles and pedestrians), which may be applicable to the design of residential streets with the aim of providing the residents with high environmental quality. System consideration in pedestrian planning, with an emphasis on the interrelationships of the three fundamental elements of the system (the pedestrian, the environment and the pathway), was the subject of the concluding discussion.

The material presented summarizes certain practical aspects of pedestrian movement. One may use the information provided for the design of pedestrian facilities, such as crosswalks, sidewalks or platforms at public transit stations; however, it is not intended to serve as a detailed design guide for all situations or locations. The information is based on various sources, mainly U.S. studies and other international research; an effort was made to distinguish between the various sources of the pertinent information. It is often left up to the design engineer to choose the appropriate values to suit the local conditions and practices.

Too often, traffic engineers apply their talents and knowledge to analyzing urban transportation problems, with an emphasis on vehicular traffic. Pedestrian flow, convenience and safety must be given higher priority than hitherto, so that future plans for the urban environment may lead to improvement in the overall quality of life. It is hoped that the foregoing discussion may shed some light on and emphasize the importance of the complex phenomena of pedestrian flow.

REFERENCES

1. Adams, W. F., "Road Traffic Considered as a Random Series," *Journal of the Institute of Civil Engineers* (1936).
2. Baerwald, J. E., *Traffic Engineering Handbook*, Institute of Traffic Engineers, Washington, D.C. (1965).
3. Carter, E. C. and W. S. Homburger, "Introduction to Transportation Engineering," Institute of Transportation Engineering, Reston Publishing Co. Inc. (1978).
4. Council on Uniform Traffic Control Devices for Canada, "Uniform Traffic Control Devices for Canada" (January 1976).
5. Department of Transport (Great Britain), Roads and Transportation Directorate, "Pedestrian Facilities at Traffic Signal Installations," Advice Note No. TA/15/81, London (1981).
6. Dun, R. C. M., "Pedestrian Crossings—A Review of Overseas Research and Practice," Department of Civil Engineering, Report No. 285, University of Auckland, New Zealand (January 1982).
7. Fruin, J. J., "Pedestrian Planning and Design," Metropolitan Association of Urban Designers and Environmental Planners, Inc., New York, N.Y. (1971).
8. Goldschmidt, J., "Pedestrian Delay and Traffic Management," Transport and Road Research Laboratory, TRRL Supplementary Report 356, Crowthorn, U.K. (1977).
9. Hankin, B. D. and R. A. Wright, "Passenger Flow in Subways," *Operational Research Quarterly*, Vol. 9, No. 2, pp. 81–88 (June 1958).
10. Hoel, L., "Pedestrian Travel Rates in Central Business Districts," *Traffic Engineering* (January 1968).
11. ITE Technical Council Committee 5-R, "Characteristics and Service Requirements of Pedestrians and Pedestrian Facilities," *Traffic Engineering*, Vol. 46, No. 5 (May 1976).
12. Maclean, A. S. and D. J. Howie, "Survey of the Performance of Pedestrian Crossing Facilities," *Australian Road Research*, Vol. 10, No. 3 (1980).
13. Martin, J. D., "Opportunities for Street Improvement as Part of a Local Area Traffic Management Package in Existing Australian Suburbia," *Australian Road Research Journal*, Vol. 11, Part 4 (1982).
14. Mitchell, C. G. B., "Pedestrian and Cycle Journeys in English Urban Areas," Transport and Road Research Laboratory, Report LR 497, Crowthorne, U.K. (1973).
15. Navin, F. P. D. and R. J. Wheeler, "Pedestrian Flow Characteristics," *Traffic Engineering*, pp. 30–36 (June 1969).
16. Oeding, D. "Verkersbelastung und Dimensionierung von Gehwegen und anderen Anlagen des Fussgaengerverkehrs," Strassenbau und Strassenverkehrstechnik No. 22, Bonn, W. Germany (1963).
17. Older, S. J., "Pedestrians," Dept. of Scientific & Industrial Research, Road Research Laboratory, Publication No. LN 275/SJ0, Crowthorne, Berkshire, U.K. (1964).
18. Peschel, R., "Studies of the Efficiency of Unprotected Pedestrian Crossings," *Strassentechnik*, 5(6), pp. 63–67 (in German) (1957).
19. Pfefer, R. C., A. Sorton, J. Fegan, and M. J. Rosenbaum, "Synthesis of Safety Research Related to Traffic Control and Roadway Elements—Chapter 16—Pedestrian Ways," Federal Highway Administration, Washington, D.C. (December 1982).
20. Polak, M., "Four Years Experience with 'Woonerf,' " Dutch Road Safety Association (April 1979).
21. Polus, A. and A. Katz, "An Analysis of Night-time Pedestrian Accidents at Specially Illuminated Crosswalks," *Accident Analysis & Prevention*, Vol. 10 (1978).
22. Polus, A., J. L. Schofer, and A. Ushpiz, "Pedestrian Flow and Level of Service," *Transportation Engineering Journal of ASCE*, Vol. 109, No. 1 (1983).
23. Pushkarev, B. and J. M. Zupan, "Urban Space for Pedestrians," A Report of the Regional Planning Association, the MIT Press, Cambridge, Mass. (1975).
24. Ribbens, H., "Revised Guidelines for the Layout, Signing, Lighting, and Siting of Midblock Pedestrian Crossings," National Institute for Transport and Road Research, CSIR, South Africa, Report RF/4/83 (April 1983).
25. Smeed, R. J., "Some Aspects of Pedestrian Safety," *Journal of Transport Economics and Policy*, Vol. 2, No. 3 (September 1968).
26. Zupan, J. M., "Pedestrian Facilities," A Draft for a Revised Highway Capacity Manual, Regional Planning Association, New York (1977).

CHAPTER 19

Simulation Model Applied to Japanese Expressway

Yasuji Makigami,* Nobuo Kumaki,** Masuo Nakajima,[†] and Kim Seil[‡]

ABSTRACT

In order to investigate countermeasures for the bottleneck sections of the often congested Japanese urban expressways, a traffic simulation model which computes traffic flow characteristics and traces congestion behavior under given traffic demands was developed. A study section was selected from the Meishin Expressway in Osaka: this section is characterized by daily congestion during the morning peak hours due to the presence of two tunnels.

An extensive traffic survey was conducted making use of aerial photography, video camera recording, and floating tests. A simulation model for traffic flow through bottlenecks was developed on the basis of the results of the traffic survey. The expressway bottleneck simulation model is a macroscopic model based upon the theory of compressible fluid. The results obtained with the simulation model have been satisfactory. The output from the model shows that the model pursues the behavior of congested areas and that the travel time through the study section computed from the model output agrees with actual measurements from floating tests. Various aspects of the model usage are also included in this paper.

INTRODUCTION

The Japanese national expressway system, constructed and operated by the Japan Highway Public Corporation since 1958, now has more than 3,000km of expressway sections that are open to traffic and more than 2,500km under construction. The traffic demands on the expressway network have increased yearly. The 1980 annual report from the Japan Highway Public Corporation showed that, on the average, more than 1,240,000 vehicles use the system each day. Naturally, this heavy use often results in traffic congestion, especially around large urban centers such as Osaka and Tokyo.

Although these overloaded sections could be improved by adding lanes or constructing parallel detours, in Japan a considerable period of time is required to acquire the right of way for these improvements. Be that as it may, it is totally unacceptable to allow such conditions to continue, therefore, various operational procedures, such as the effective use of changeable message signs and traffic control techniques, should be considered.

In order to evaluate such countermeasures, a traffic simulation model is needed to compute traffic flow characteristics such as traffic density and speed and to trace the behavior of congested areas under given traffic demands. From 1981 to 1983, a series of comprehensive traffic surveys were conducted for the purposes of developing and validating such a simulation model. Table 1 shows the date and period of time, the study section, and the objectives of the survey for each year. The traffic survey method described in this chapter and its results, are from the first year of the traffic survey when the basic structure of the simulation model was established. Although the survey methods adopted in 1982 and 1983 were essentially the same, some significant results from these surveys have also been included.

In the 1981 survey, a study section was set up in the eastbound traffic lanes on the Meishin Expressway between the Ibaraki and the Kyoto South interchanges where congestion brought on by two tunnels occurred almost every day during the morning peak period.

An extensive traffic survey was conducted throughout the study section. The three major tools of the survey were: (1) aerial photography; (2) video camera recording, and (3) test

---

*Professor of Civil Engineering, Faculty of Engineering, and Science, Ritsumeikan University, Kita-ku, Kyoto, Japan

**Chief Engineer, Asa Construction Office Japan Highway Public Corporation, Asa-ku, Hiroshima, Japan

[†]Research Engineer, Institute of Systems Science Research, Shimogyo-ku, Kyoto, Japan

[‡]Graduate Student, Transportation Engineering, Kyoto University, Sakyo-ku, Kyoto, Japan

TABLE 1. Record of Traffic Survey from 1981 to 1983.

Year	Date and Period	Traffic Direction and Section	Section Length in kilometers	Major Objectives
1981	August 26 Morning Peak Period	Eastbound Ibaraki IC. ⟨ Kyoto South IC.	24	Development of Basic structure of the Simulation Model
1982	July 27 and August 24 Morning Peak Period	Eastbound Nishinomiya IC. ⟨ Kyoto East IC.	58	Improvement of the model and Application of the model to the Eastbound traffic to investigate countermeasures
1983	August 18 Evening Peak Period	Westbound Seta IC. ⟨ Ibaraki IC.	67	Application of the Simulation Model to the Westbound Traffic to Investigate Countermeasures

car floating. A traffic simulation model was developed based on the traffic stream characteristics collected from the survey.

The basic structure of the simulation model is almost the same as that of the FREEQ Model [5,6], concerning which, one of the writers had the chance to participate in the original developmental activities under Professor A. D. May. However, with the FREEQ program there is a section length limitation due to the assumption of instantaneous demand propagation. In the simulation program developed here, traffic demands have been calculated in such a way that they reflect the time lag inherent in demand propagation. Therefore, the new simulation model, referred to as the "Expressway Bottleneck Simulation Model," can be applied to a fairly long stretch of highway.

CHARACTERISTICS OF STUDY SECTION

Figure 1 shows both the study section on the Meishin Expressway and the trunk surface highways in the vicinity. The Meishin Expressway is a divided, four-lane limited access highway. As shown in Figure 2, after its opening in July 1963, usage increased steadily each year until, in 1981 on the study section alone, the average daily traffic volume exceeded 80,000 vehicles. In addition to high traffic volume, the large proportion of trucks, an average 45%, contributes significantly to the congestion. This is especially true for the study section, which serves as a trunk highway for industrial, as well as social activities. The situation on the study section is further complicated by the presence of two tunnels: the Tennozan tunnel and the Kajiwara tunnel. The low traffic capacities of these tunnels, together with a high frequency of accidents in the tunnels, result in bottlenecks causing congestion extending upstream nearly every day. It is obvious then, that these two tunnels are the key to traffic operation and safety on the section in question.

METHOD OF TRAFFIC SURVEY AND ITS RESULTS

Method of Traffic Survey

The 1981 traffic survey was conducted between 8:30 and 11:30 A.M., on August 26th, 1981, to collect data for estimates of capacity, volume-speed relationship, travel time, etc., for both development and validation of the simulation model. The three major items of the traffic survey [2] were the following:

1. Video recording of the traffic stream at five locations.
2. Measurement of travel time and running speed using the test vehicle floating method: 12 runs.
3. Aerial photography for density measurement: 12 flights

The objectives, methods, and instruments for each item of the survey are summarized in Table 2. Traffic volume counts were also collected using ultrasonic-type detectors installed along the study section.

The survey started at 8:30 A.M. to record the initial development of congestion in the study section, and continued for 3 hours in accordance with the flight limit of the aircraft used for aerial photography. A complete set of records for the congestion development process was collected, but there was not sufficient time to record the final process of congestion dispersion. The weather conditions at the time of the survey were favorable with a completely dry road surface and no traffic tie-ups. Table 3 shows the structural charac-

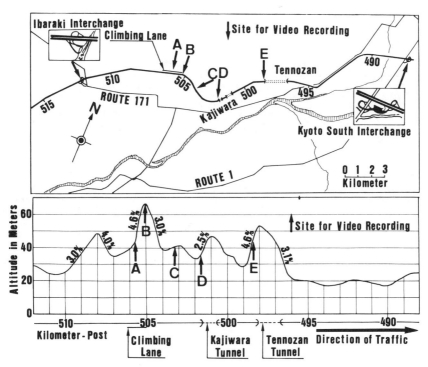

FIGURE 1. Plan and profile of study section.

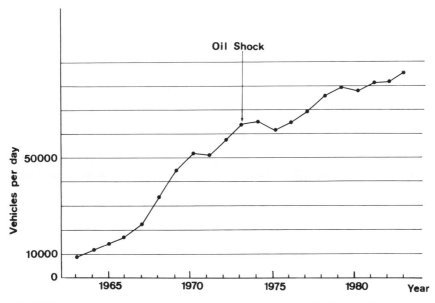

FIGURE 2. Annual average daily traffic on the study section of the Meishin Expressway.

TABLE 2. Outline of Traffic Survey.

Item (1)	Video Recording (2)	Floating Survey (3)	Aerial Photography (4)
Instruments and staff	Two engineering aids Video camera: Sony HVC 80 3 BP 60 Batteries per recording site	Three engineering aids Passenger car One tape recorder per floating crew	Pilot and photographer Fixed wing aircraft (Cessna) Hasselblad camera
Method	Record on color video tape for 50 min with 5 min pause, for 3 hrs	Record of passage time at each 200-m post with 15 min intervals	Continuous photo shooting at an altitude of 600 m
Objectives	Measurement of spot running speed Estimation of capacity of bottleneck section	Measurement of travel time Measurements of spatial speed distribution	Measurement of traffic density Pursuit of behavior of congested area
Number of measurements	Recording at five locations	12 floating surveys with five crews	12 flights

TABLE 3. Video Recording Sites Data.

Camera site (1)	Kilometer post (2)	Characteristics of road structure (3)	Objectives (4)	Camera mount position (5)
A	505.66	Midway point of climbing lane section with upgrade slope of 1.5–4.6%	Determine relationship between volume and speed on upgrade section	Over bridge
B	505.08	Merging taper end of climbing lane	Estimation of capacity at merging taper end	Over bridge
C	503.19	Flat section with a horizontal curve of radius 1,400 m	Estimation of volume speed relationship on flat section	Over bridge
D	501.56	Approach section to Kajiwara Tunnel with 2.5% upgrade	Estimation of capacity and volume speed relationship for tunnels	On the portal of Kajiwara Tunnel
E	498.20	Approach section to Tennozan Tunnel with 4.5% upgrade	Estimation of capacity and volume speed relationship for tunnels	On the portal of Tennozan Tunnel

teristics of the roadway and the study objectives for the five video recording locations.

Results of Survey and Input Data for Model

FLUCTUATION OF TRAFFIC VOLUME AND ESTIMATION OF INPUT DEMAND

Using the video recordings, five-minute manual volume counts were made of both passenger cars and trucks for each lane at the five video recording stations. Figure 3 shows the results of the manual volume counts for recording site A farthest upstream, and for site D at the portal of the Kajiwara tunnel. Replacement of batteries and tapes necessitated breaks in the recording; there were no 15-min volume measurements between a quarter and half past each measurement hour.

Comparing the counts from the two recording stations, traffic volume at site A increases gradually from the beginning and reaches a peak between 9:30 and 9:45 A.M. After that, the volume fluctuation is moderate, remaining between 700 and 750 vehicles per 15 min. In contrast, the traffic volume at site D does not fluctuate so much, but remains steady at between 700 and 770 vehicles per 15 min. This tendency also prevails on the three downstream video recording sites C, D, and E.

As shown later in the results of the aerial photography and the floating test vehicle surveys, three bottlenecks appeared during the 3-hr period of the traffic survey: (1) Video recording site B, which is the taper end of the climbing lane, (2) site D at the portal of the Kajiwara tunnel, and (3) site E at the portal of the Tennozan tunnel. Due to the capacity constraint of these bottlenecks, the traffic-volume fluctuation pattern of site A with its high peak seems to be leveled off like the fluctuation pattern for site D shown in Figure 3.

Since there was no continuous volume count, a continuous pattern of the 15 min volume fluctuations was prepared from the outputs of detectors in the study section, and rectified by the manual volume counts from the recorded video screen. Figure 4 shows the input traffic demand pattern for simulation model.

SPATIAL AND TEMPORAL DISTRIBUTION OF SPEED AND DENSITY

Figure 5 shows the spatial and temporal distribution of speed and density in the form of contour diagrams drawn from the aerial photographs and the floating test vehicle survey.

The density contour diagram is drawn on the basis of the density measurement obtained from the aerial photographs. The traffic density was measured by counting each vehicle in each lane, in uniform subsections about 500m in length.

The speed contour diagram was made from the running speed measurements that were calculated using the travel time records from each 200m post.

Both diagrams show that on the day of the traffic survey,

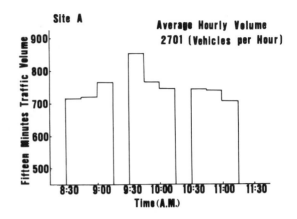
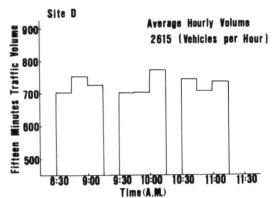

FIGURE 3. Fluctuation of fifteen minutes traffic volume with time.

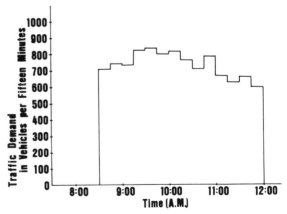

FIGURE 4. Input traffic demand pattern for simulation model.

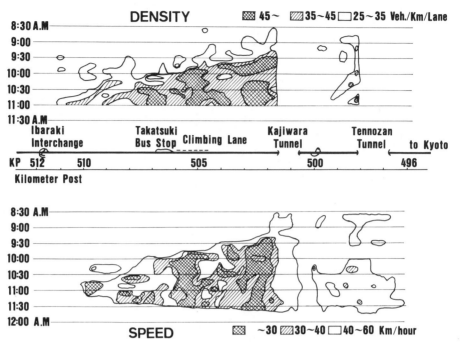

FIGURE 5. Traffic density and speed contour diagram from 1981 survey.

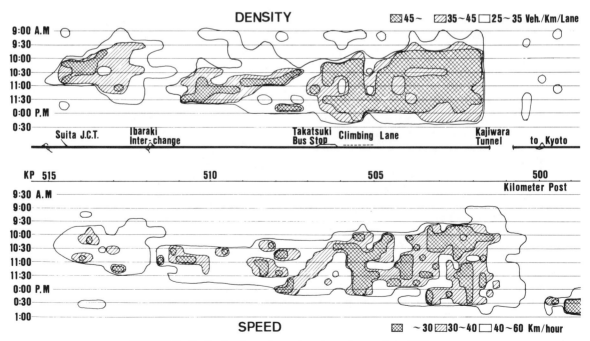

FIGURE 6. Traffic density and speed contour diagram from 1982 survey.

congestion originated from the portals of both tunnels and from the taper end of the climbing lane, with the major problem initiating at the portal of the Kajiwara tunnel and extending beyond the merging end of the climbing lane. The congestion upstream of the Tennozan tunnel was fairly light.

Figure 6 shows the speed and density contour diagrams from the 1982 survey. In that case, congestion extended beyond the Suita interchange, which is located about three kilometers west of the Ibaraki interchange. Based on these results, the Expressway Bottleneck Simulation Model was improved so as to compute traffic demands incorporating the time lag inherent in demand propagation.

Fluctuation of Travel Time

Figure 7 shows the fluctuation of travel time between the Ibaraki interchange and both the Tennozan tunnel and the Kyoto South interchange. The travel time between the Ibaraki interchange and the Kyoto South interchange varies from 22 min, between 8:30 and 9:00 A.M., to a maximum of 33 min 30 sec at 10:45 A.M., which is almost one and half times as long. In contrast, the travel time between the portal of the Tennozan tunnel and the Kyoto South interchange, which is depicted as the travel time difference between the two travel time curves, is quite stable in the range of 7.5 to 8 min.

Capacity Analysis

Three types of investigations were made to estimate the capacity of each subsection: (1) Direct estimation based on manual volume counts from the video recordings, (2) estimation from the cumulative gap distribution obtained manually from the video recordings, and (3) capacity calculation according to the Japanese manual for highway capacity (3). An outline of these analyses follows:

1. Direct estimation: The highway capacity at the bottleneck stations was estimated directly from the volume counts under saturated traffic-flow conditions when the upstream was congested.
2. Estimation from gap distributions: Figure 8 shows an example of cumulative gap distributions at recording site A. An estimate of capacity flow rate can be obtained by removing the long gaps from the traffic flow rate. This can be done by dividing 3,600 sec by the average value of gap measurements, excepting very long gaps which are in the range of the flat portion of the cumulative gap distribution curves as shown in Figure 8. This method was especially effective for recording sites A and C, where an estimated capacity value could not be obtained directly from the volume measurements.
3. Capacity calculation: The practical capacity was calculated for each subsection in the study section according to the Japanese manual for highway capacity.

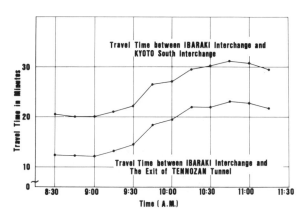

FIGURE 7. Fluctuation of travel time with time.

Table 4 shows the capacity estimations obtained for the five video stations using the three different methods just described. The values on the bottom line are the subsection capacity values used as original input data for the simulation model.

Relationship between Speed and Volume

Figure 9 shows the relationship between the 5-min average running speed and the volume capacity ratio. The 5-min average running speed is the average of the speeds of all the vehicles passing by a given video recording station in 5 minutes. The speed of each vehicle in uncongested traffic flow is measured directly from the video screen using a pair of pulse-height analyzers which measure the time taken by each vehicle to traverse two fixed points on the roadway. However, because the accuracy of the automatic speed measuring system is not adequate for a congested traffic flow, the running speed under congested traffic flow conditions was measured manually with stopwatches.

The results of the floating survey, as well as speed measurements by other studies, were used in determining the following free speeds for three structural conditions: (1) 90km/h for flat sections, (2) 75km/h for up-grade sections and tunnels, and (3) 85km/h for the merging section of the climbing lane. The average running speed during capacity flow was 55km/h [7].

COMPUTATION LOGIC OF EXPRESSWAY BOTTLENECK SIMULATION MODEL

System Description and Basic Assumptions

The basic objective of the Expressway Bottleneck Simulation Model is to simulate traffic flow and to follow the be-

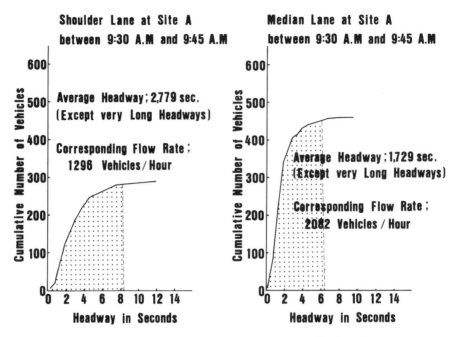

FIGURE 8. Cumulative gap distribution at Site A.

havior of traffic in a congested region with two or more bottlenecks. In order to ensure a meaningful relationship between traffic volume and the average running speed, the freeway section is divided into several homogeneous subsections with constant capacity over their length. Those subsections are numbered 1, 2, ..., n from upstream to downstream. Furthermore, a series of subsections are grouped together in such a way that the travel time through the group is approximately 15 minutes.

Traffic demands are introduced into the freeway section in the form of O–D tables. The first subsection and each on-ramp are considered as origins, and each off-ramp and the last subsection (n) as destinations. The origins and the destinations are numbered consecutively from upstream to downstream.

Considering the fact that traffic demands during a peak period usually fluctuate, the peak period should be divided into a number of smaller time intervals. It is therefore necessary to input an O–D table for each time interval.

This method of treating traffic demand, although adding complexity, yields the following desirable characteristics: (a) Actual demand patterns are more realistically simulated, (b) travel times for individual O–D movements, which are essential for evaluating the effectiveness of such improvements as ramp control, can be readily obtained, and (c) the resultant Expressway Bottleneck Simulation Model exhibits a flexibility that will facilitate considerations of network traffic movements and patterns.

Basic assumptions for the model are as follows:

1. Traffic is treated as a compressible fluid in which an individual vehicle is a part of the fluid and is not considered individually.
2. Traffic demands are introduced into the simulation sec-

TABLE 4. Capacity Estimation in Vehicles per Hour.

Estimation Method (1)	Site A (2)	Site B (3)	Site C (4)	Site D (5)	Site E (6)
Possible capacity according to manual	3,267	3,910	3,910	3,910	3,830
Estimation from volume measurement	—	3,037	—	2,915	2,910
Estimation from headway measurement	3,381	3,413	3,537	3,226	3,295
Input for model	3,380	3,040	3,600	2,915	2,910

tion every 15 min, and for a given time interval of 15 min traffic flows remain constant and do not fluctuate. Therefore, traffic demands are expressed as a step function over the entire simulation period.
3. For each subsection, the relationship between traffic volume and average running speed must be established.

Demand Calculation

As mentioned previously, an O–D table format is used to input traffic demand. It is therefore necessary to calculate total demands for on- or off-ramps and for each subsection from the O–D table.

Let $QOD(J, K/T)$ be the traffic demand (number of trips) between the Jth origin and the Kth destination at the time interval T

$$QO(J/T) = \sum_{k=1}^{ND} QOD(J, k/T)$$

$$QD(K/T) = \sum_{j=1}^{NO} QOD(j, K/T)$$
(1)

where $QO(J/T)$ is the Jth on-ramp demand and $QD(K/T)$ is the Kth off-ramp demand in time interval T. ND and NO are the number of on-ramps and off-ramps respectively.

From assumption 2, the subsection demand between the Jth origin and the Kth destination can be calculated as follows:

$$D(i/T) = \sum_{j=1}^{J} QO(j/T) - \sum_{k=1}^{K-1} QD(k/T) \quad (2)$$

where i is any subsection number between the Jth on-ramp and the Kth off-ramp.

The time lag in demand propagation is brought into the computation process by shifting the time interval at the boundaries between adjoining subsection groups as shown in Figure 10. In the simulation calculus, the subsection demand in group N in time interval T is replaced by the subsection demands calculated from Equation (2) for the time interval $[T - (N - 1)]$. Since the travel time through each subsection group is set at approximately 15 minutes, traffic demands from any given on-ramp reach downstream subsections with a reasonable time lag.

As described later, particular attention is given to the behavior of shock waves should they appear at the upstream end of any congestion. It is supposed that a shock wave can traverse the boundary of each subsection group in spite of a demand gap created by the time interval shift; thus, traffic

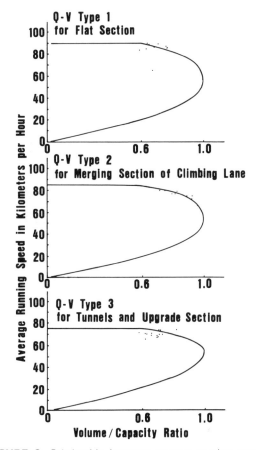

FIGURE 9. Relationship between average running speed and volume/capacity ratio.

flow characteristics can be calculated over the entire length of the simulation based on the traffic demand with the time interval shift.

Mathematical Logic [9]

The following is the mathematical logic of the model developed on the assumptions previously described. Consider subsection i at the present time interval with variables defined as follows:

1. $c(i)$ = capacity of subsection i in vehicles/hr.
2. $q(i)$ = traffic volume of subsection i in vehicles/hr.
3. $q'(i)$ = traffic volume of subsection i under congested flow conditions.
4. $v(i)$ = average running speed in subsection i in km/hr.
5. $v'(i)$ = average running speed in subsection i under congested flow conditions.

It is possible to compute the average running speed using

the relationship between speed and the volume-capacity ratio under given traffic volume, if the traffic conditions—whether congested or uncongested—are defined. Let $f(\cdot)$ and $f'(\cdot)$ be speed functions for uncongested and congested traffic flow as shown in Figure 11. Then for uncongested flow:

$$v(i) = f\left[\frac{q(i)}{c(i)}\right] \quad (3)$$

for congested flow:

$$v'(i) = f'\left[\frac{q'(i)}{c(i)}\right] \quad (4)$$

Furthermore, let's define those variables that are related to the present time interval T as follows:

1. $D(T)$ = traffic demand for time interval T in vehicles/hr.
2. Tr = average travel time in hours through the study section for time interval T.
3. TTr = total travel time in vehicle hours for the study section for time interval T.

In addition to the preceding notation: $L(i)$ = length of subsection i in kilometers; $k(i)$ = traffic density of subsection i in vehicles/km; and $k'(i)$ = traffic density of subsection i under congested traffic conditions in vehicles/km.

Suppose the traffic demand at time interval T is less than the capacity of any subsection, i.e., $c(i) \geq D(T)$ for $i = 1, 2, \ldots n$, then

$$q(i) = D(T); \; v(i) = f\left[\frac{q(i)}{c(i)}\right]; \; Tr = \sum_{i=1}^{n} \frac{L(i)}{v(i)};$$

$$TTr = \sum_{i=1}^{n} \frac{0.25 q(i) \cdot L(i)}{v(i)};$$

$$TTr = 0.25 \sum_{i=1}^{n} k(i) \cdot L(i) \because k(i) = \frac{q(i)}{v(i)} \quad (5)$$

If subsection i is fully congested throughout the time period T, then all three in Equation (3) are valid for congested traffic conditions if $q(i)$, $v(i)$, and $k(i)$ are replaced by $q'(i)$, $v'(i)$, and $k'(i)$, respectively. Problems remain when two regions of congested and uncongested traffic conditions appear at the same time. Further investigation is needed to find a way to follow the behavior of shock waves which appear along the boundaries of two such regions.

Shock Wave Behavior Follow-up

Suppose that for time period T the traffic demand in subsection j exceeds capacity for the first time since the start of simulation. In this case, for all subsections downstream of the bottleneck subsection j, $q(i) = c(j)$ for all i in which $j \leq i \leq n$, and the average running speed and the travel time can be calculated from Equation (5).

For subsections upstream of subsection j, calculations must be made for two traffic conditions: for uncongested regions

$$q(i) = D(i/T); \; v(i) = f\left[\frac{q(i)}{c(i)}\right]; \; k(i) = \frac{q(i)}{v(i)} \quad (6)$$

and for congested regions

$$q'(i) = c(j), \; v'(i) = f'\left[\frac{q'(i)}{c(i)}\right]; \; k'(i) = \frac{q(i)}{v'(i)} \quad (7)$$

The propagation speed of the shock wave at subsection i $\gamma_j(i)$, in kilometers per hour, can be calculated by

$$\gamma_j(i) = \frac{q'(i) - q(i)}{k'(i) - k(i)} \text{ (kilometers per hour)} \quad (8)$$

Assuming that the shock wave is to pass through subsection i within the time period, then time $t(i)$, i.e., the time for the shock wave to reach the upstream end of subsection i measured from the start of time period T, is given by

$$t(i) = \sum_{k=j-1}^{i} \left[\frac{L(k)}{-\gamma_j(k)}\right] \quad (9)$$

The average traffic volume for subsection i throughout time period $T[q_a(i)$ in vehicles per hour] is calculated as follows (Figure 12):

$$q_a(i) = \frac{1}{0.25}\left[0.25 + \frac{t(i+1) + t(i)}{2}\right]q'(i)$$
$$+ \frac{t(i+1) + t(i)}{2} q(i) \quad (10)$$

In the same way, the total travel time for subsection i for time period $T[TTr(i)$ in vehicle hours] is computed by multiplying the area of congested and uncongested regions in the time and space field by the corresponding traffic density as follows:

$$TTr(i) = k'(i)L(i)\left[0.25 - \frac{t(i+1) + t(i)}{2}\right] +$$
$$+ k(i)L(i)\frac{t(i+1) + t(i)}{2} \quad (11)$$

Traffic stream characteristics of subsection i for time

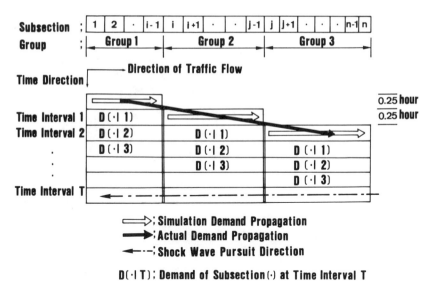

FIGURE 10. Traffic demand propagation in actual versus simulated conditions.

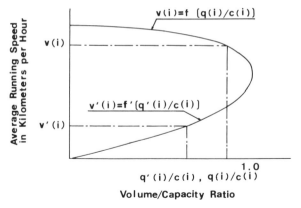

FIGURE 11. Speed function for congested and uncongested traffic flow.

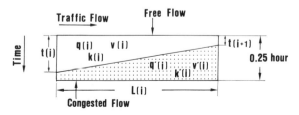

FIGURE 12. Shock wave behavior in subsection i.

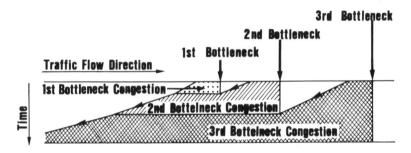

FIGURE 13. Shock wave behavior in multiple bottleneck situation.

period T are calculated on an average as follows:

average travel time: $\quad Tr(i) = \dfrac{TT(i)}{0.25\, q_a(i)}$

average running speed: $\quad V_a(i) = \dfrac{L(i)}{Tr(i)} \qquad (12)$

average traffic density: $\quad k_a(i) = \dfrac{q_a(i)}{v_a(i)}$

The same types of calculi can be applied in various cases, e.g., (1) when the shock wave stays in subsection i, (2) when the shock wave continues from one time period to the next, (3) when the shock wave propagates into other congested regions upstream, and (4) when the shock wave moves downstream toward dissolution after the peak period. However, if we pursue the behavior of two or more shock waves the logic becomes very complicated, as we must consider the shock wave velocities when they meet and pass through another congested region upstream. Therefore, in such multiple bottleneck situations, it is supposed that a shock wave from a downstream bottleneck propagates through the upstream congested region instantaneously just like demand propagation in the subsection groups. The multiple bottleneck simulation shock wave behavior is shown in Figure 13.

PROCEDURE FOR COMPUTER CALCULATION AND RESULTS OF PRESENT TRAFFIC SIMULATION

Input Data

There are three formats for the model input: (1) The number of subsections, (2) subsection parameters, and (3) traffic demand. The variable names in the simulation program and the coding formats for those input data are shown in Table 5.

Figure 14 shows the subsection parameters for simulating the traffic flow in the study section at the time of the 1981 traffic survey: the item "Key number to Q/V curves" refers to which of the three curves of speed and volume-capacity ratio is applied to calculate the average running speed in each subsection.

This simulation, hereafter referred to as the simulation of the 1981 "actual conditions," was carried out on the 16.4km section between the Ibaraki interchange and the exit portal of the Tennozan tunnel. The simulation section was divided into 28 subsections in the same way as for the aerial photography analysis.

Computer Calculation Procedure and Output Format

The flow diagram of the Expressway Bottleneck Simulation Model is shown in Figure 15. The outline of the computation process is as follows: after reading input data, the computer calculation starts by comparing input traffic demand with the capacity of each subsection, starting from upstream, in order to identify bottleneck sections. When the demand exceeds the capacity of a particular subsection that becomes a bottleneck section in the time period; then traffic demands of subsections downstream are replaced by the capacity of the bottleneck. This kind of demand replacement is also necessary during the discharging of vehicles stored in the congestion and remaining from the previous time period, even if the demand does not exceed the capacity in the present time period. After identifying the bottle-necks, if any, and making the consequewnt demand adjustments, the program calculates the traffic stream characteristics with equations that are appropriate for the traffic conditions in each subsection, starting backward from the downstream end subsection, n. The program follows each step of the described calculation from the first time period to the last.

Results from Simulation of 1981 Actual Conditions

Figure 16 shows the contour diagrams for average running speed and traffic density drawn from the computer output of

TABLE 5. Input Data.

Items (1)	Variable Name (2)	Format (3)	Card Arrangement (4)
Number of Subsection	NS	15	
Subsection data capacity (vehicles per hour)	C(I), I = 1 NS	F7.1	One for each subsection
Subsection length (meters)	L(I), I = 1 NS	F7.1	One for each subsection
Key number to Q/V curves	QVTP(I), I = 1 NS	F7.1	One for each subsection
Demand (vehicles per hour)	DU	F6.1	One for each on-ramp for each time period

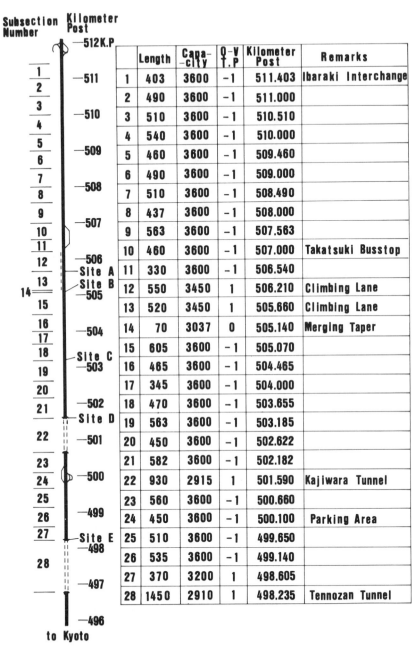

FIGURE 14. Subsection data for simulation of present conditions.

the simulation of actual conditions. Although the contour diagrams drawn from the simulation results cannot depict delicate fluctuations, when compared with the results of the traffic survey shown in Figure 5, the simulation results are satisfactory from the following viewpoints:

1. The simulation results identify the same three bottlenecks as the traffic survey. The time when the congestion starts in the simulation is also about the same as in the survey.
2. In the simulation, the congested area extends to almost

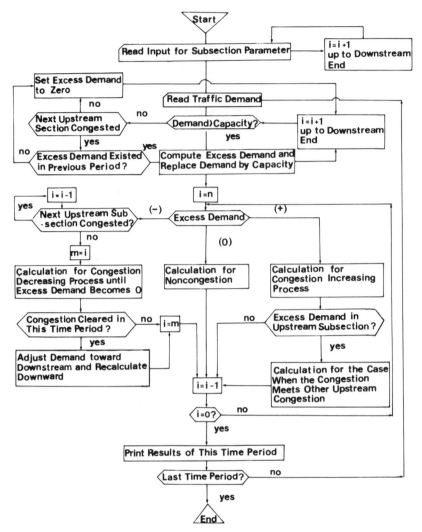

FIGURE 15. Flow diagram.

the same point as it does in the actual congested conditions, although the propagation speed of congestion in the simulation is a little slower than that of the actual congestion.

3. Comparing the actual and simulated average densities subsection by subsection for each time period, the simulation results have a tendency to higher speed and lower density in uncongested traffic and, conversely, slower speed and higher density in congested traffic. Furthermore, the simulation results exhibit a uniform value of the average running speed and the traffic density for subsections with the same capacity under given traffic demands; whereas, the actual average running speed and the actual traffic density varied slightly with the location.

However, the difference in values between the simulation results and the actual measurements is not significant.

Figure 17 shows the simulation output and the actual measurements for the travel time between the Ibaraki interchange and the exit of the Tennozan tunnel. The difference between the simulation results and the actual measurements is less than 1 min. The travel time fluctuation from the simulation results is quite similar to that of the actual measurements.

Possible Countermeasures

Here, congestion countermeasures are investigated in reference to the study results. On the study section, as pointed

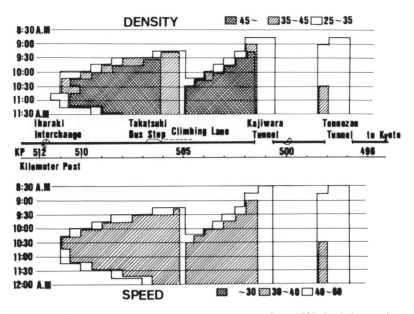

FIGURE 16. Traffic density and speed contour diagram from 1981 simulation results.

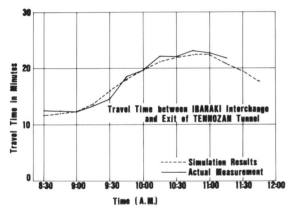

FIGURE 17. Simulation travel time fluctuation (1981 survey).

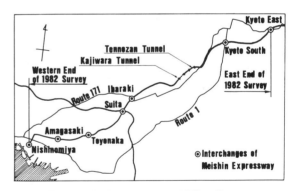

FIGURE 18. Study section in 1982 traffic survey.

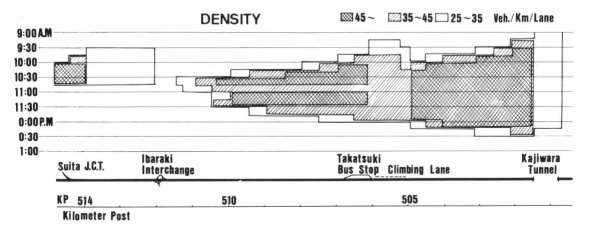

FIGURE 19a. Density contour diagram from 1982 simulation results (actual conditions).

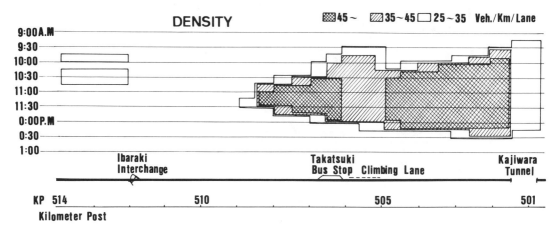

FIGURE 19b. Density contour diagram from 1982 simulation results (Case 4).

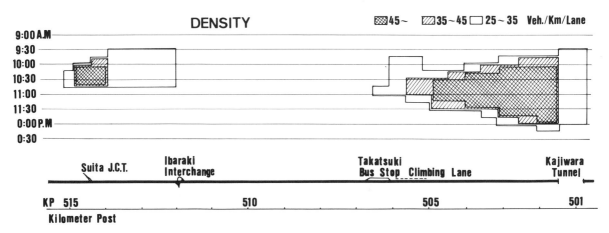

FIGURE 19c. Density contour diagram from 1982 simulation results (Case 5).

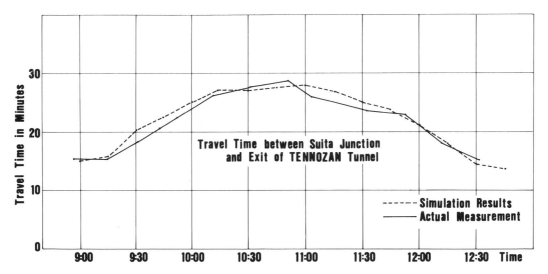

FIGURE 20. Travel time fluctuation (1982 survey).

out before, there are three bottlenecks that daily cause heavy congestion. In general, the principles for congestion countermeasures are as follows:

1. To increase the capacity of bottlenecks by adding one or more lanes.
2. To decrease the demand on the bottlenecks by controlling the traffic volume on the on-ramps.
3. To combine the previous two.

In fact, the Japan Highway Public Corporation has already determined to improve a part of the Meishin Expressway by adding a lane for both directions. The improvement is to cover a 24.5 kilometer section including the 1981 study section. However, people living along the Meishin Expressway are antagonistic to the highway because of traffic noise, vibration, air pollution, and so on. Since people are against widening the expressway, it is anticipated that some time will be required to obtain the cooperation and consent of the people and the municipal authorities; therefore, it is necessary to consider operational congestion countermeasures during the interim. In order to establish practical control plans, it is necessary to extend the simulation coverage further upstream. For this purpose, as shown in Figure 18, in summer 1982, survey coverage was extended to encompass the 58km section between the Nishinomiya interchange at the west end of the expressway and the Kyoto East interchange. The 1982 speed and density contour diagrams are already shown in Figure 6.

Density contour diagrams from the simulation of actual traffic conditions in 1982 are shown in Figure 19a. Figure 20 shows "travel time fluctuation with time" measurements obtained with the floating car method and by simulation. The travel time diagram in Figure 20 depicts travel time on an 18km section of the expressway from the Suita interchange to the exit of the Tennozan tunnel. The figures from Figure 6, 19a and 20 indicate that traffic congestion initiating at the entrance of the Kajiwara tunnel extended as far as the Suita interchange. The congestion length exceeded 13 kilometers and the congestion period lasted for three and a half hours. Meanwhile, the travel time was about fifteen minutes longer than before or after the congestion period. The excess demand on the bottleneck section of the Tennozan tunnel amounted to about 570 vehicles during the peak 2 hours and 45 minutes.

Five operational improvement plans are being considered based on the actual traffic conditions of 1982 survey. These include input demand control at several on-ramp booths and discharge control at the off-ramp nose. Strategy for each control plan is as follows:

In-put demand control:
Case 1; to eliminate excess demands by closing or reducing the number of on-ramp toll gate(s) in operation at each site, starting from the nearest upstream interchange.
Case 2; to reduce excess demands by 50% using methods described in Case 1.
Case 3; to eliminate excess demands by reducing the number of on-ramp toll gates in operation at those four upstream interchanges with identical input demand reduction rates.
Case 4; to reduce excess demands by reducing the number of on-ramp toll gate(s) in operation by one at each upstream interchange.

Discharge control:
Case 5; to reroute drivers from the Ibaraki interchange to the surface route as much as the off-ramp toll gate capacities will allow, using changeable message signs installed on the road side.

In each case, traffic would be controlled or diverted only during the 2 hours 45 minutes when the excess demand occurs on the bottleneck section. Density contour diagrams from the simulations for Case 4 and 5 are shown in Figure 19b and c: both show a reasonable reduction in the length of the congestion period. Table 6 is the summary of all the simulations, listing the average travel time between the Suita interchange and the entrance of Tennozan Tunnel, and the maximum congestion length. Naturally, no congestion occurs in Case 1 and 3, because all the excess demands are eliminated: of course, both of these plans are impractical from the view point of expressway operation, as the sizable

TABLE 6. Simulation Results for Countermeasures.

Lase	Average Travel Time*		Maximum Congestion Length in Meters
	Minutes'	Seconds"	
Actual Conditions	22'	56"	7955
Case 1	14'	42"	0
Case 2	18'	41"	6112
Case 3	14'	22"	0
Case 4	18'	36"	6403
Case 5	17'	11"	4885

*Average travel time between Suita Junction and Exit Portal of Tennozan Tunnel.

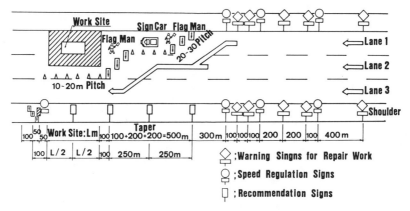

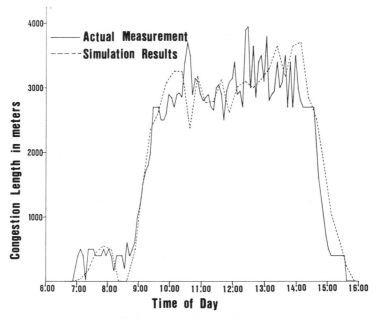

FIGURE 21. Application to maintenance works (continued).

queue which would develop in front of the on-ramp toll gate would impair effective traffic operation on the surface streets. The simulation objective of Case 1 and 3 was to evaluate the effect of complete congestion elimination by channeling traffic to the surface streets. Case 5 is also not so practical, because too much traffic diverted to the surface route is expected to damage the parallel surface roadway.

It is concluded that Cases 2 and 4 are the most promising control plans, because they reduce the average travel time by about four minutes and shorten the maximum congestion length by nearly twenty percent. From the view point of actual control operation, Case 4 is the most practical; although the reduction rate of the average travel time and the congestion length is not so great, the control is effective from the aspect of traffic safety, because it will prevent the congestion from extending upstream into the Suita Tunnel.

ANOTHER ASPECT OF MODEL USAGE

Application to Maintenance Works [11]

More than 60 percent of congestion occurs due to lane closures for maintenance, such as pavement repair and

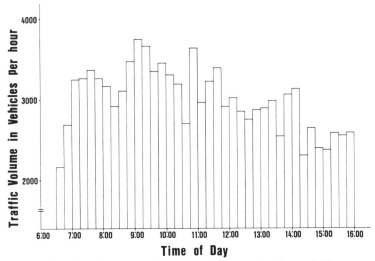

(c) 15 minute Input Traffic Demand in Hourly Rate of Flow

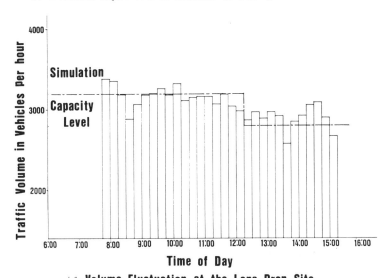

(d) Volume Fluctuation at the Lane Drop Site

FIGURE 21. Application to maintenance works.

expansion joint replacement in bridges and viaducts. Tokyo First Bureau of Traffic Operation, Japan Highway Public Corporation, is now investigating the application of the Expressway Bottleneck Simulation Model for estimation of traffic congestion caused by the lane closure operations.

If it is possible to get an estimated pattern of traffic volume fluctuation with time, for a particular scheduled maintenance day, it would be possible to predict the development of congestion: then, with such knowledge of expected congestion behavior, maintenance schedules could be adjusted as needed. For this reason, Tokyo First Bureau of Operation is looking for a way to input the Expressway Bottleneck Simulation computer program into minicomputers installed in each operation office on the Tomei Expressway.

Figure 21 shows the results of one of the maintenance work site traffic flow simulations. In this case, lanes 1 and 2 were closed for pavement repair as shown in Figure 21a. Congestion length fluctuation with time, in-put traffic demands and volume fluctuation at the lane drop site are shown in Figures 21b, c and d, respectively. The simulated congestion length is contrasted with actual measurements in Figure 21b.

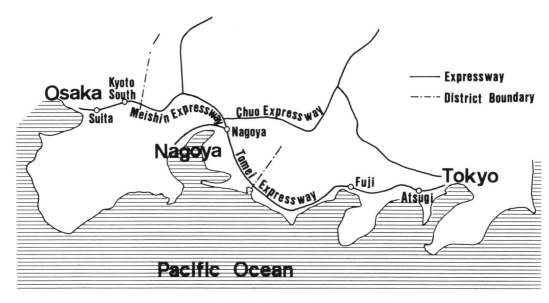

FIGURE 22. District of Nagoya Operation Bureau.

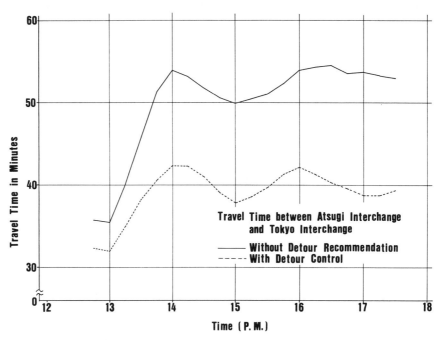

FIGURE 23. Travel time fluctuation with and without detour control.

Investigation of Diversion to Alternate Long Distance Routes [12]

As shown in Figure 22, Nagoya Operation Bureau is in charge of maintaining and operating the three major expressways that form the main corridor connecting the "big three" metropolitan areas of Tokyo, Nagoya, and Osaka. These three major expressways are 1) the Tomei expressway, 2) the Meishin expressway and 3) the Chuo expressway. The Tomei expressway is the most direct route between Nagoya and Tokyo: it carries an average daily traffic volume of more than 60,000 vehicles in the Nagoya district alone. Lane closures for maintenance and construction often cause bottlenecks and congestion: especially heavy congestion occurs in the eastbound traffic near Tokyo where extremely heavy demands exceed the capacity of several particular bottleneck sections. The Chuo expressway, branching away from the Tomei expressway east of Nagoya, runs north of Mt. Fuji and ends in the north-west suburbs of Tokyo. The traffic demand on the Chuo expressway is moderate and the flow is smooth.

Attention is being given to route guidance strategy which would advise expressway drivers leaving the Nagoya area for Tokyo, to divert to the Chuo expressway. The Expressway Bottleneck Simulation model is being used to evaluate the effect of this long diversion strategy. Figure 23 shows the fluctuation of travel time from Nagoya to Tokyo under the assumption that all the drivers bound for Tokyo divert to the Chuo expressway in the year of 1990.

REFERENCES

1. Cillers, M. P., A. D. May, and R. Cooper, "Development and Application of a Freeway Priority-Lane Model," *Transportation Research Record,* No. 722, pp. 16–26 (1979).
2. Hom, R. E., and W. E. Fillifillan, "Bay Area Freeway Operations Study, Sixth Interim Report," Institute of Transportation and Traffic Engineering, University of California, Berkeley, Calif. (March, 1969).
3. "Interpretation and Usage of Policy on Geometric Design of Highways," Japan Road Associations, pp. 47–64 (Nov., 1970).
4. Lighthill, M. J., "On Kinematic Waves; A Theory of Traffic Flow on Long Crowded Road," *Proceedings, Royal Society of London, Part A, Vol. 299,* No. 1178 (May, 1955).
5. Makigami, Y. and W. L. Woodie, "Freeway Travel Time Evaluation Technique," *Highway Research Record,* No. 223, pp. 33–45 (1970).
6. Makigami, Y., W. L. Woodie, and A. D. May, "Bay Area Freeway Operations Study, Final Report, Part I, Freeway Model," Institute of Transportation and Traffic Engineering, University of California, Berkeley, Calif. (Aug., 1970).
7. Masaki, H., et al., "On Analysis of Traffic Stream on Keio Toll Road," presented at the 22nd Symposium on Business Activities and Research in Japan Highway Public Corporation, pp. 850–854 (Nov., 1980).
8. Wang, J. J. and A. D. May, "Computer Model for Optimal Freeway On-Ramp Control," *Highway Research Record,* No. 469, pp. 16–39 (1973).
9. Makigami, Y., et al., "Simulation Model Applied to Japanese Expressway," *Journal of Transportation Engineering, ASCE, Vol. 110,* No. 1, pp. 94–111 (January, 1984).
10. Osaka Operation Bureau, Japan Highway Public Corporation, "Report of Traffic Survey on Tunnel Sections of the Meishin Expressway" (February, 1983).
11. Tokyo First Operation Bureau, Japan Highway Public Corporation, "Report on Congestion Estimation for Maintenance Works" (March, 1985).
12. Nagoya Operation Bureau, Japan Highway Public Corporation, "Report on Traffic Control and Surveillance Project for Nagoya District" (March, 1985).

CHAPTER 20

The Role of Driver Expectancy in Highway Design and Traffic Control

Gerson J. Alexander* and Harold Lunenfeld**

SCOPE

The concept of expectancy is presented within the context of the driving task and the reception and use of information by drivers. The discussion centers on driver expectancy and the way it affects and is affected by the presentation of information. It describes the ways in which expectancy and expectancy violations influence driving task peformance. It includes an historical development of the expectancy concept and its application to highway design and traffic engineering. It describes various highway and traffic situations where expectancy impacts on driver behavior. Finally, it presents a procedure for the application of the expectancy concept in the analysis of problem locations and the development of information system improvements.

INTRODUCTION

Driver expectancy is a key factor in driving task performance. It affects all aspects of driving, including pre-trip planning, hazard avoidance, lateral placement, speed control, road following, route following, and direction finding. It affects how drivers react to and handle information, how they make decisions, and how they translate their decisions into control actions and driving strategies. It affects the safety and efficiency of driving task performance, and, ultimately, the effectiveness and suitability of highway design and traffic operations [30]. Accordingly, the focus of this discussion is on expectancies, what they are, how they are structured, reinforced, or violated, and how engineers and designers can use them in the design and operation of highways.

---

*President, Positive Guidance Applications, Inc., Rockville, MD
**Engineering Psychologist, Office of Traffic Operations, Federal Highway Administration, Washington, DC

Ultimately, the success or failure of any design or operational strategy rests in its ability to be used safely and efficiently by drivers. Therefore, since expectancy is so basic to driver task performance, it and its antecedent driving task performance and information handling framework should be considered in all driver-related aspects of highway design and traffic engineering.

The material on the driving task, information reception and use, and expectancy is derived from several primary references. These include an article by the authors [5], NCHRP Report No. 123 (30), the 2nd Edition of the *Users' Guide to Positive Guidance* [32], and SAE Report SP 279(37). The reader is referred to these references for detailed discussions of specific factors.

BACKGROUND

The highway system consists of a complex array of elements: drivers; in vehicles; on roads; in traffic; in an environment. It is dynamic, with diverse subsystems (e.g., information displays, interstate roads, urban arterials, city streets, police, traffic platoons) and interactions, often of a transitory nature. As a principal controlling element, drivers are primary determining factors in the system's successful operation. Skillful driving task performance, maintenance of vehicle control, safe and efficient guidance through roads and traffic, and proper navigation using an optimum mix of routes, represent ways in which driver performance enhances system operations and safety. Highway and traffic engineers are also major determiners of the success of the highway system. Their production of designs that are compatible with the capabilities of drivers, that take human limitations in account, and that, through the highway information system, convey the operating conditions of the highway enhance optimum driving task performance.

Since the safety and operational efficiency of the highway

FIGURE 1. Levels of the driving task.

system depends, in great measure, on a driver's ability to perform in a proper, error-free manner, an appreciation of human factors is essential to highway design and traffic control. While engineers have considerable knowledge about vehicle characteristics, load factors, environmental effects on pavement, etc. [1,14], they often have only a rudimentary understanding of the motorist. They fail to account for driver error, the consequence of designs that are beyond driver capabilities, maneuvers that are unusual or unexpected, decisions that are overly complex, or information displays that are confusing or ambiguous. Driver error is one of the leading contributors to accidents and inefficient traffic operations and must be minimized for the highway system to perform its intended function, the safe and efficient movement of people and goods.

Driver Error

Driver errors occur for a variety of reasons. A leading cause that is beyond the scope of this discussion is that drivers are unable to perform due to drugs or medication, alcohol, fatigue, etc. Causes within the scope include expectancy violations; situations that place too much demand on drivers, causing overload; long periods that put too little demand on drivers, causing lack of vigilance; information displays that are deficient, ambiguous, or missing content; misplaced information; blocked or obscured information; and information that does not possess sufficient size, contrast, or target value [29,34,35,42]. These deficiencies cause drivers to miss or be unable to process traffic control information.

In cases where errors are committed due to the nature of the task, the demands of the situation, the inability of drivers to handle high information load, the inadequacy of the information being presented, or the violation of expectancies, it is the responsibility of designers and engineers to reduce the sources of error. Positive Guidance, a procedure that identifies information system deficiencies and provides suitable, expected information, when needed, where required, and in a form best suited for its intended purpose, achieves this goal [4]. The premise of Positive Guidance is that competent drivers, using properly designed roads with appropriate traffic control devices, will drive safely and efficiently. Conversely, if designs are incompatible with driver attributes, if the information displays are ambiguous or erroneous, or if expectancies are violated, drivers will commit errors, and the system will fail. Armed with the proper information, designers and traffic engineers can provide roads and information displays matched to highway users and their expectancies.

The Driving Task

In order to understand expectancies, it is necessary to understand what drivers do, and how they receive and use information to perform the driving task. The basic driving task consists of three performance levels—Control, Guidance, and Navigation [3] (see Figure 1). These levels and their associated activities and subtasks can be described according to scales of complexity and priority (Primacy). The scale of complexity increases from Control through Guidance to Navigation; priority (Primacy) decreases in the same direction.

Control: Control refers to a driver's interaction with the vehicle. The vehicle is controlled in terms of speed and path. Passenger vehicle drivers exercise control through three or four mechanisms—steering wheel, accelerator, brake, and gear shift. Information about how well or poorly the driver has controlled the vehicle comes primarily from the vehicle and its displays. Drivers receive continual feedback through vehicle response to various control manipulations.

Guidance: Guidance refers to a driver's maintenance of a safe speed and path. Control subtasks require action by the driver; guidance requires decisions involving judgment, estimation, and prediction. The driver must evaluate the immediate environment and translate changes into control actions needed to maintain a safe speed and path in the traffic stream. Information at this level comes from the highways—alignment, geometry, hazards, shoulders, etc.; from traffic—speed, relative position, gaps, headway, etc.; and from traffic control devices—regulatory and warning signs, traffic signals, and markings.

Navigation: Navigation refers to the activities involved in planning and executing a trip from origin to destination. Navigation information comes from maps, verbal directions, guide signs, and landmarks.

The three levels—Control, Guidance, and Navigation—form a hierarchy of information handling complexity. At the Control Level, performance is relatively simple and so completely overlearned by most drivers that it is performed almost by rote. At the Guidance and Navigation Levels, information handling is increasingly complex, and drivers need more processing time to make decisions and respond to information inputs.

The key to successful driving task performance is efficient

information handling. However, the total driving task does not consist of independent activities peformed independently. At any given point in time, drivers are faced with a multitude of information, transmitted from a variety of sources, and received through a number of sensory channels. They may be required to sift through this information, determine its relative importance, make proper interpretations, decide on courses of action, and take those actions in a limited time period.

When drivers are required to sift through the mass of information, both relevant and extraneous, under time pressures, they need to assign relative priority to the competing sources and therefore require a criterion upon which to base their decisions. Similarly, engineers need a basis for deciding what information to give the driver. The concept of "primacy" has been developed to deal with this problem.

Primacy

Primacy refers to the relative importance of each level of the driving task and of the information associated with a particular activity. The major criterion upon which primacy is assessed is the consequence of driver error on system performance. Since Control and Guidance Level errors often result in crashes, these levels assume a higher primacy than do errors at the Navigation Level, where the consequences of error are likely to be lost or confused drivers. There are, in addition, primacy gradients within a given performance level. At the Control Level, going into a skid is likely to be more serious than stripping a gear. At the Guidance Level, failure to avoid an immediate hazard is probably worse than driving too fast. Although failures at the Navigation Level assume a lower primacy, they can and do have an effect on system operations and safety. Erratic maneuvers and traffic conflicts are two common indicators of Navigation Level uncertainty and often cause problems for the traffic stream. In any event, primacy is an important consideration when information needs compete. Higher primacy needs should be satisfied and lower primacy information deferred.

Information Handling

Drivers use most of their sensory input channels to gather information. They hear horns, radio broadcasts, and engine noise; they feel road surface texture and raised pavement markers through the vehicle; they smell burning insulation wires; they sense changes in acceleration, pitch and yaw through the "seat of their pants;" and they see road alignment, traffic, signs, signals, and markings. Although most senses are used to gather information, drivers receive more than 90 percent of all information visually.

While driving, motorists do many things either at the same time or very nearly so. They look at traffic, follow the road, read signs and listen to the radio. To accomplish this, they gather information from many diverse information sources, both informal (e.g., the road, its alignment) and formal (e.g., signs, signals, markings); make many decisions (e.g., take a particular exit, brake for a road hazard, speed up to avoid a signal change); and perform continuous control actions (e.g., steering, speed control, gear shifting). At any point in time, drivers may have several overlapping information needs associated with each individual activity for each driving task level. To fulfill these needs, drivers must search the environment, detect information, and use it in a safe and efficient manner. Thus, information must be available when needed, where required, and in a form best suited for its intended purpose.

An important consideration in the reception and use of visually displayed information is that drivers can only attend to and process one source of visual information at a time. Drivers are "serial," rather than "parallel," processors of information. Since the driving task often requires motorists to perform more than one activity at a time or in very close temporal proximity, and since drivers do not parallel process information, they have developed a way to cope with competing visual information through an information handling "juggling act." Drivers integrate various subtasks and maintain an overall appreciation of a dynamic, ever-changing environment by sampling information in short glances and shifting attention from one source to another. They make some decisions and delay others, depending upon the primacy of the need. They rely on judgment, estimation, and prediction to fill in gaps. Such task sharing behavior enables drivers to use their limited attention span and information processing capabilities [17].

Drivers receive and handle information using a signal search, detection, recognition, and use process. In the search and detection modes, a driver scans the environment and samples available information in short glances until a potentially needed source is detected. Once detected, the source is attended to, either continuously or intermittently, until recognized. The driver then determines whether the information is needed. If the information is needed, it is read (if verbal or symbolic), or otherwise perceived, and used to make decisions and perform control actions in a feedback process.

In situations where information needs compete, unneeded and low primacy information is (or should be) shed. Relevant information that is not immediately used is stored in short-term memory for rapid access and retrieval. If this stored information is not quickly used, reinforced, or repeated, it is usually forgotten. On the other hand, if information in short-term memory is reinforced or repeated, it is stored in long-term memory for future use. This process structures expectancies. If other sources of information interpose before information in the short-term memory is used, the new information often extinguishes the information in storage. Relevant information immediately needed is attended to and is processed by comparison with *a priori* knowledge and expectancies in long-term storage; decisions are made and control actions taken (including no change in speed, path, or direction, if applicable). Once used, the driver then gathers and uses information from other sources. This process is repeated continually throughout the driving task [7].

REACTION TIME

The time it takes drivers to process information and respond is their reaction time. Reaction time, like visual acuity, color vision, and eye height, varies from individual to individual. Reaction time also varies with decision complexity, information content, and expectancy. The more complex the decision, the more information required to make the decision, the longer the reaction time, and the greater the chance for error [39].

The relationship between information content and reaction time is based on the amount of information needed to resolve uncertainty. The information needed to make a decision can be broken down into binary units called "bits," where a bit is the smallest amount of information required to decide between two alternatives. Since a bit is a binary unit, the relationship between information content and decision complexity is 2^n, where n = number of alternatives to make a decision [7]. Thus, a "zero" bit decision has only one response, a "one" bit decision has 2, a "two" bit decision 4, etc. This is an exponential relationship where a complex decision (4 bits or more) often exceeds a driver's capacity to respond, and either a very long reaction time or confusion occurs. Hence, several simple decisions are usually preferable to a single complex one.

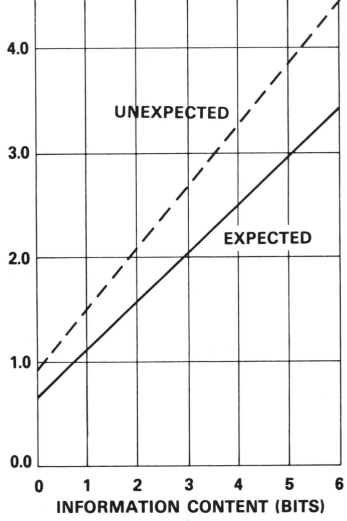

FIGURE 2. Median driver reaction time.

Whether or not a decision is expected also affects reaction time. Researchers [18] measured brake-reaction time for expected and unexpected signals. When information was expected, reaction time was, on average, 2/3 sec (ranging from less than 0.2 sec to greater than 2.0 sec). When the signal was unexpected, reaction time approached 1 second, with some drivers taking over 2.7 sec to respond. In addition, a complex, unexpected, multi-alternative decision has a considerably longer reaction time than a simple, expected one (see Figure 2). Long reaction time decreases a driver's time to sample information and leaves less time to attend to other important sources of information. This, in turn, increases the chances for missing information and committing errors.

Thus, the nature of the driving task and the process of information handling, from the detection of information through its use in making decisions, can contribute to driver error. Information may be missing or ambiguous. It may not be visible, legible, or conspicuous enough and may be missed. Drivers may not have sufficient time to handle it, or its content may require overly long reaction time or lead to confusion. In addition, there may be too few or too many sources to handle and a driver may become inattentive or overloaded. In each of these conditions, reinforced driver expectancies regarding the presence, form, or location of information play a key role in rapid, error-free information handling and task performance. Conversely, violated expectancies worsen the situation and result in slower, less appropriate responses.

THE EXPECTANCY CONCEPT

The nature of the driving task and the driver's information handling characteristics point to the importance of expectancies. For example, a driver's reaction time to an unexpected situation or source of information is longer than when the situation or information is expected. Conversely, drivers are less likely to become confused or commit errors when their expectancies are reinforced. Since the key to safe and efficient driving task performance is rapid, error free information handling, what a driver expects and does not expect, has a major impact on task performance, particularly under time pressures and/or high information loading. Because the expectancy concept is such an important consideration in driver task peformance and information processing, it is one that engineers, designers, and operations pesonnel should understand and use. Expectancies affect all levels of the driving task and should be accounted for in highway design, traffic operations, and traffic control device applications.

The expectancy concept was first identified by psychologists [19,41] over 50 years ago. It was not until the 1960's, however, that the concept found its way into highway applications [27,38]. Since then, a number of highway researchers and practitioners have recognized the importance of expectancy and used it in diagnostic and design activities [2,8–11,13,15,25,28,36,40]. The landmark work in expectancy was accomplished by NCHRP Project 3-12 [30] and refined by the Federal Highway Administration (FHWA) [5]. Since then, the expectancy concept was further developed and applied in a number of additional FHWA efforts [20–22]. Its use and application in highway design and traffic control culminated in the development of Positive Guidance [4,32,33] and its demonstration in FHWA Demonstration Project No. 48 [6,23,31].

Definitions

Expectancy relates to a driver's readiness to respond to situations, events and information in predictable and successful ways. It influences the speed and accuracy of driver information processing and is one of the most important considerations in the design and operation of highways and the presentation of information. Configurations, geometrics, traffic operations, and traffic control devices that are in accordance with and/or that reinforce expectancies aid drivers and help them respond quickly, efficiently, and without error. On the other hand, configurations, geometrics, traffic operations, and traffic control devices that are counter to and/or violate expectancies lead to longer reaction time, confusion, inappropriate response, and driver error.

Expectancies operate at all levels of the driving task. At the Control Level, expectancies relate to vehicle handling characteristics, the placement of controls and displays, and vehicle response to control manipulations. At the Guidance level, expectancies relate to highway design and traffic operations. They affect how a driver negotiates the road, responds to traffic, selects a safe path, perceives hazards, and avoids them. At the Navigation Level, expectancies relate to drivers' trip plans, their use of route markers and guide signs, their selection of exits at interchanges and streets at intersections, how they locate destinations, and use services. They affect route choice, in trip–route diversion, and, ultimately, whether or not motorists arrive at their destinations with a minimum of inefficiency and confusion.

There are two kinds of driver expectancy. The first are long-term, *a priori* expectancies that drivers bring to the task based on past experience, upbringing, culture, and learning. The second are short-term, *ad hoc* expectancies that drivers formulate from site-specific practices and situations encountered in-transit. Both types affect driving task performance and should be accounted for in highway design and traffic control.

A PRIORI EXPECTANCIES

Because things are designed to operate in standard, consistent ways, and are applied nationwide, certain expectancies are structured over a lifetime. Whether it's the typewriter keyboard, the direction of movement of a clock's hands, or the placement of HOT and COLD shower knobs, the intent of consistent, uniform design is to foster rapid, error-free operation. Red is used to signify danger and green to signify safety in a similar manner.

Thus, people expect things to operate predictably. For example, when entering a room all light switches are expected to be toggles, and are expected to operate UP for ON and DOWN for OFF. It is also expected that there will be no difference between wall switches in Minot, North Dakota, and Selma, Alabama. As a result, if these expectancies are fulfilled, user performance is rapid and error free. However, if a light switch is not placed on the wall adjacent to the door, but is on a wall behind the door and, if instead of a toggle, is a push button or a toggle installed upside-down, then expectancies would be violated. The results of these "surprises," or expectancy violations, range from taking longer to figure out how to turn on the light to frustration, anger, inappropriate action, and the increased possibility of accident involvement.

In designing highways and traffic control devices, it is necessary to understand the nature of *a priori* expectancies. For example, because most freeway exits are on the right, drivers expect *ALL* exits to be on the right. Unexpected left exits often have serious consequences. However, not all *a priori* expectancies are held by the entire driving population. There are regional and local differences. Thus, if most interchanges in a given area contain left exits, then drivers in that area would expect to exit on the left, rather than the right. This expectancy aids performance in the area a driver is familiar with, where interchanges are as expected. However, outside the area, the same drivers' response would be inappropriate. In a similar manner, if most signalized intersections in a Central Business District use an all-red phase for pedestrian movements, pedestrians will come to expect this treatment to be used elsewhere—sometimes with disastrous results.

AD HOC EXPECTANCIES

In designing and operating highways, it is as important to recognize and understand the nature of short term, *ad hoc* expectancies structured in response to in-transit, site-specific situations. Drivers form initial expectancies from their trip plan and experience. At the Guidance Level, these relate to what roads and traffic will be like. At the Navigation Level, initial expectancies relate to information (e.g., freeway guide signs, route markings, destinations signed for) service availability, land use, etc. As drivers traverse an unfamiliar area, the geometry of the routes, the traffic control devices, and the traffic patterns structure *ad hoc,* site-specific expectancies. For example, when driving on a rural road, if several relatively sharp curves are preceded by curve warning signs, an *ad hoc* expectancy is structured that similar curves will be similarly signed. If a sharp downstream curve is not preceded by a curve warning sign, thereby violating the *ad hoc* expectancy, drivers may not respond properly. Unfamiliar drivers may misinterpret the sharpness of the curve, take it too fast, and run off the road. In a similar manner, if the upstream road geometry provides a 70 MPH design speed with clear sight lines and adequate stopping sight distances, then strangers will expect these design standards to continue, and if services and rest areas are readily available, motorists will expect services and rest areas to continue to be available, etc.

Thus, not only does a driver bring a set of previously held *a priori* expectancies into the driving task, but he or she is constantly formulating new *ad hoc* expectancies based on what is encountered in transit. The engineer and designer must understand both type of expectancies and account for each.

EXPECTANCIES AT WORK/VIOLATIONS

Since expectancies affect all aspects of the reception and use of information by drivers, this section presents and discusses examples of expectancies and expectancy violations to illustrate how they are structured, how they are violated, and how they can be restructured to aid driver task performance. The material is based on a series of lectures and training courses on driver expectancy presented by the authors to engineers and technicians at the federal, state, and local levels throughout the United States (Human Factors Symposium, 1973-1974; Positive Guidance in Traffic Control, 1977-1979; Seminar on the New Positive Guidance Procedure, 1985-1987).

General Expectancies

The first group of expectancies are general in nature, and are taken from everyday experience. They are designed to illustrate how expectancies are formed, what they are like, and what occurs when they are violated. To gain a full appreciation of these general expectancy examples, it is recommended that the reader follow along and solve the problem or answer the questions prior to reading the explanation.

EXPECTANCY RELATED TO SERIES AND SETS

In Figure 3, three sets of numbers are presented. The problem is to supply the missing number signified by the question mark (?) and to identify the process by which it is derived.

- In the first example (3-6-9-?), the correct answer is "12." The process is to add three to each number.
- In the second example (3-9-27-?), the correct answer is "81." The process is to multiply each number by three.

In this *ad hoc* expectancy example, three sets of numbers are used. Two expectancies are structured, both seemingly

3 - 6 - 9 - ?

3 - 9 - 27 - ?

3 - 14 - 159 - ?

FIGURE 3. Expectancies related to series and sets.

Mac Duff

Mac Donald

Mac Hinery

FIGURE 4. Word expectancies.

designed to enhance rapid problem solution. While they succeed in the first two sets, they inhibit a solution in the third. First, the number 3 seems to play a role, since there are three cases of three numbers, each beginning with the number 3. An expectancy is structured that the number 3 is somehow involved in the solution to all sets. This is not correct for the third set. Secondly, each of the first two sets is written in the form of a mathematical progression "3-6-9-?; 3-9-27-?." Because the third set is also written in the same form, the expectancy is structured that all are mathematical progressions. Again, this is not true for the third set.

In order to solve this problem, one needs to take the numbers 3, 14, and 159 out of the progression context and write them as "314159?". Does that make it easier? If that does not help, try putting a decimal point after the 3.

Out of over 5,000 people who have been exposed to this problem, only 3 were able to identify the third case as pi before any clues were given. Hence, once an expectancy is established, it is very hard, if not impossible for individuals to change their information handling behavior, even when given what should be familiar information. The implications for design and traffic control are clear. The engineer should determine what expectancies exist and/or are being established by the road, its environment, and its complement of traffic control devices prior to presenting the driver with additional information. Further, the engineer should recognize that information being presented, while correct and accurate, may lead to an expectancy violation if the form of the information presentation is unusual or unique.

WORD EXPECTANCIES

In Figure 4, three sets of words are presented. The problem is to identify the three "MACs."

- The 1st, "MAC DUFF," is a character from Shakespeare's MacBeth.
- The 2nd, "MAC DONALD," is the farmer of note.
- The 3rd, "MAC HINERY," is actually a noun, the word "machinery."

In this example, an expectancy was structured that all three were proper names that began with "MAC." This was violated by the third case, a noun, "machinery," rather than a name. In most instances, people have solved this correctly, particularly in light of the "Series and Set" example, which structured an expectancy that there will be a "catch" in the third case. However, here again it can be seen how expectancies are structured and how a violation can affect the way words are perceived. Consider the use of cardinal directions on guide signs. When drivers read the message "East St. Louis," do they read the message as "EAST to St. Louis" or the city of "East St. Louis, Illinois?" Obviously, any time a potential destination has a cardinal direction associated with it, such as "West Springfield" (Massachusetts), "South Saint Paul" (Minnesota), "East Dubuque" (Illinois), and "North Little Rock" (Arkansas), there is a potential for misinterpretation. The fact that cardinal directions on guide signs are all upper case letters, while destinations are upper and lower case, is probably not understood by most drivers. How drivers plan their trips and what destinations they expect to set on guide signs also affect how word messages in general and cardinal directions in specific are perceived.

SPATIAL EXPECTANCIES

Figure 5 presents a matrix of nine dots. The problem is to connect all dots with four straight, connected lines. Problems of spatial relationships, such as this example, are usually solved within some kind of visual framework. The framework provided here, the eight dot perimeter, is almost always seen as the limiting boundary. As in the case of the "Series and Sets" problem, the expectancy thus created inhibits correct problem solution. Remember, there is a visual framework, but it is not the eight dot perimeter—it is the whole two-dimensional surface of the page. The solution can be found at the end of this chapter (see Figure 32).

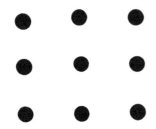

FIGURE 5. Spatial expectancy.

MECHANICAL EXPECTANCIES

Figure 6 shows a car radio panel to illustrate a mechanical expectancy.

In the case of the radio panel, the problem is determining which knob controls volume (left or right?) and which controls tuning. In the vast majority of American automobiles, the left knob is the volume control and the right knob is the tuning control. However, in Japanese and some other "foreign" vehicles, the left knob is for tuning and the right knob controls volume. Anyone who does not drive a foreign car and is not used to this configuration can attest to the difficulty in trying to change stations and or control volume. Usually, the result is very high or low volume and an unwanted station. It is very hard to break old habits.

The effects of mechanical expectancies and their violation are most applicable to vehicle design and control placement and operation. When people drive unfamiliar vehicles, finding an emergency brake, for example, that is expected to be on a console on the right, when it is on a floor pedal on the left, or locating a horn on a lever rather than on the steering wheel, could be critical in an emergency. Standardization of vehicle controls and displays would help eliminate this violation. Similarly, standardization would also help eliminate many design and traffic control device expectancy violations.

On-The-Road Expectancies

The next group of expectancies and violations relate to roadway design. Some are more common than others, but all surprise motorists.

LEFT EXITS

Because most exits are on the right, unfamiliar drivers usually expect to exit from the right hand lane of a freeway. In the absence of advance warning, unfamiliar motorists desiring to exit at a downstream interchange will move to the right lane. If the exit ramp is on the left, as shown in Figure 7, drivers in the right lane desiring to exit may either miss their exit or perform a hazardous late lane change to get to the left lane. Such a maneuver often results in traffic conflicts or collisions.

The conventional guide sign treatment applied to left exits has not proven as effective as the diagrammatic treatment contained in the *Manual on Uniform Traffic Control Devices* (MUTCD) [24] and shown in Figure 8. Studies [26] have shown that diagrammatics are effective in providing advance notice of an unexpected highway feature. It is important to understand, however, that diagrammatics are applicable in a limited number of cases, primarily when an off-route movement is to the left of a through-route movement.

INTERCHANGE LANE DROPS

At most freeway exits, motorists must move into a deceleration lane to leave the facility. It is therefore an expectancy violation when a lane which had been a through-lane becomes a deceleration lane and then exits the freeway (see Figure 9). Instead of having to change lanes to leave the freeway, drivers in the dropped lane have to change lanes to stay on the freeway.

The standard MUTCD device applied at interchange lane drops is the "EXIT ONLY" panel (see Figure 10). Other messages such as "MUST EXIT" and "ONLY" have also been used, but "EXIT ONLY" has been shown to be the most effective [21]. The "EXIT ONLY" panel, by virtue of its black-on-yellow color scheme, has the requisite target value, when placed on the white-on-green freeway guide sign, to gain a driver's attention. Its "EXIT ONLY" message, by virtue of its clarity and uniform application at interchange lane drops, serves to structure the appropriate expectancy.

LEFT LANE DROP

In Figure 11, a multiple expectancy violation is shown—the combination of a left exit and an exit lane drop. As might

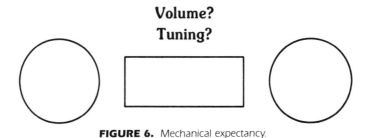

FIGURE 6. Mechanical expectancy.

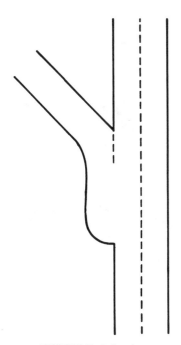

FIGURE 7. Left exit.

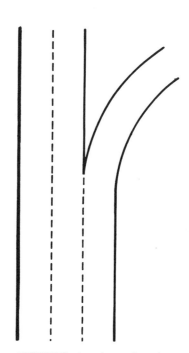

FIGURE 9. Interchange lane drop.

be expected, this is a far more serious problem than either of its individual components. More drivers are affected, interactions in the traffic stream are more turbulent, and the potential for confusion and accidents is substantially greater. Wherever a left-exit lane drop is located, it has been a recognized source of operational problems. In at least one of these locations, a number of redundant information sources were used in an attempt to overcome substantial operational problems. Included were a median mounted regulatory sign, "LEFT LANE MUST TURN LEFT"; a black-on-yellow "ONLY" panel on the overhead guide sign; the word "ONLY" with an arrow painted in the dropped lane; and a different color and texture on the dropped lane. However, diagrammatic signs with a black-on-yellow EXIT ONLY panel in accordance with the MUTCD are recommended for this location (see Figure 12).

TANGENTIAL OFF-RAMPS

Figure 13 shows a schematic of a freeway tangential off-ramp. Unless a driver is alert, he or she could be unintentionally "pulled off" a freeway by following the heretofore straight roadway alignment on to the exit ramp. A tangential off-ramp is thus both an unexpected feature and one that creates perceptual problems. To date there is no traffic control device that can adequately warn drivers about tangential exit ramps. Solutions to this expectancy problem appear to rest in changing the geometric design itself.

Tangential off-ramps are best treated by so configuring the

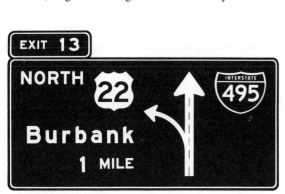

FIGURE 8. Diagrammatic treatment for left exit.

FIGURE 10. Interchange lane drop "EXIT ONLY" treatment.

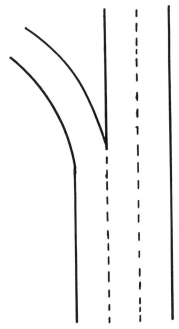

FIGURE 11. Left lane drop.

ramp terminal that the peception of a continuous tangential movement is visually disrupted. If the ramp terminal could be located as little as 100 feet up or downstream, the desired effect might be achieved.

PARALLEL ROADSIDE FEATURES

Rural road situations similar to the freeway tangential off-ramp are parallel roadside features and tangential roads intersecting at the point of curve of a turning road. A line of utility poles, trees, or railroad tracks running parallel and adjacent to a long section of tangent rural two-lane road structures an expectancy for this condition to continue, with the road remaining tangent and following the off-road feature. Such situations are fairly common in States where rural two-lane roads are built on section lines. When the road curves away from the parallel feature (see Figure 14) or a tangent road begins at the point of curve (see Figure 15), an unexpected and hazardous condition is created. The expectancies structured by the parallel roadside feature or tangential road are analogous to the freeway tangential off-ramp. In these situations, drivers may inadvertently take the tangent road, or in the case of utility poles, trees, or tacks, run off the road or cross into the path of oncoming vehicles. The best way to overcome these kinds of expectancy violations, short of removing the offending feature, is to inhibit the structuring of the expectancy by obstructing the tangential line of sight; by using landscaping that follows the curve; or by strategically placing warning signs, delineators, or markings that emphasize the true alignment.

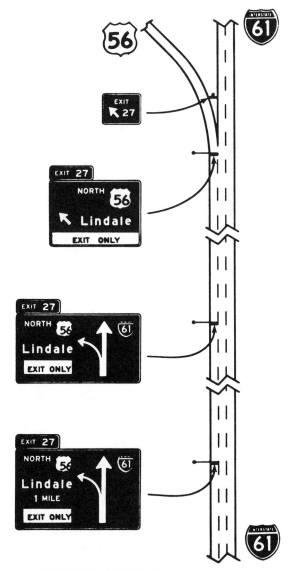

FIGURE 12. MUTCD left lane drop treatment.

FREEWAY SPLIT

A common freeway design feature with the potential for violating expectancies is the split. Several aspects of its design can lead to violations. The first is the split's geometrics, particualarly when an optional lane is used (see Figure 16). The second is when the off-route movement is to the left of the through-route movement.

The optional lane design creates expectancy problems for drivers in the optional lane. They do not expect to have to make a lane or directional choice by staying in lane. This leads to a classical dilemma—the choice between two equal

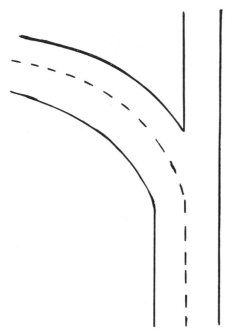

FIGURE 13. Tangential off-ramp.

FIGURE 15. Tangential road at point of curve.

FIGURE 14. Parallel roadside features.

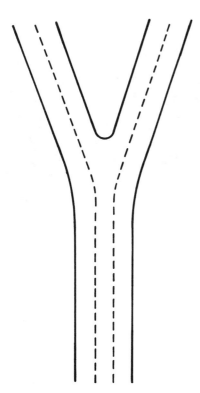

FIGURE 16. Optional lane split.

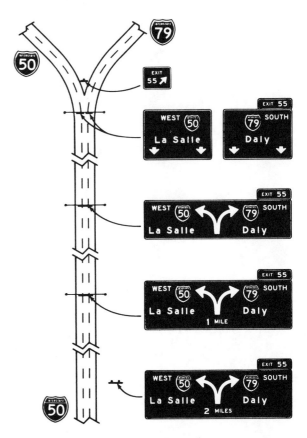

FIGURE 17. Optional lane split treatment.

alternatives. If drivers have directional uncertainty—because of an imprecise trip plan or ambiguous guide signs—many will be unable to resolve the dilemma and will perform a dangerous maneuver such as stopping in the gore or weaving across several lanes of traffic.

An effective way to avoid this expectancy problem is to eliminate the option by adding a full lane one-quarter to one-half mile upstream of the gore area, and eliminate the optional feature of the middle lane by striping. Failing that, a diagrammatic treatment can be used. As in the case of the left exit, diagrammatic signs in accordance with the MUTCD are recommended for splits and two lane exits (see Figures 17 and 18).

CONSTRUCTION JOINTS THAT DO NOT FOLLOW LANE MARKINGS

Figure 19 shows a location where a lane is added to a facility on a curved bridge approach, with the striping reflecting the lane configuration. However, as Figure 20 shows, the construction joints do not follow the lane markings. This is the case where a lane was added by construction on the left side, and the lane was added by markings on the right. Drivers expect lane markings and construction joints to lie parallel and adjacent to each other. When they don't, drivers often have problems, particularly at night in the rain, when painted lines are nearly invisible. In this case, drivers expecting lane markings and pavement joints to coincide, may follow the joints, often with catastrophic results. Solutions to this problem involve making the joints and markings coincide or making the correct lane markings visible under adverse weather. In the latter case, raised pavement markers have proven effective. In the former case, if it is not possible to use raised pavement markers, some jurisdictions have used asphalt overlays and "artificial" joints cut into the road to make them appear to follow the markings.

DIPS

Figure 21 shows a dip on a rural two-lane road. Several expectancy violations can occur at this kind of location. One concerns whether or not the road is continuous, two separate roads, or intersecting roads. Another expectancy, brought about by the broken line striping, is that the dip is too shallow to hide a car. However, with lower seated eye heights and smaller vehicles, it may be unsafe to pass. To resolve the intersecting road violations, engineers should provide route information. No solution exists short of providing better sight distance if there are two separate roads. Finally, engineers should assure that the striping reflects that latest 3.5-foot seated eye height of the "Green Book" [1] so that no vehicle can be hidden in the dip.

NARROW BRIDGES

Any reduction in the width of the road represents an expectancy violation and a hazard to the driver. Such situations as lane drops, construction zones, and narrow bridges are common sources of pavement width reduction. While all are expectancy violations, the narrow bridge situation is one that is particularly difficult because of the many configurations that a narrow bridge can take. Narrow bridges come in a variety of shapes and sizes, from those that are short box culverts (Figure 22) to long bridges with trusses. "Narrowness" of narrow bridges ranges from a loss of shoulder, a situation that often occurs on freeways, to a narrowing of lane width, to a one-lane bridge. Narrow bridges range from a loss of shoulder, a situation that often occurs on freeways, to a narrowing of lane width, to a one-lane bridge. Narrow bridges also occur on curves or dips, making them very difficult to perceive (Figure 23). Thus, not only are narrow bridges unexpected, but they may also be hard to recognize, detect, and negotiate in the presence of oncoming traffic.

Narrow bridges, depending upon their configuration, require a variety of treatments, both to warn motorists of the unexpected feature or features, thereby restructuring their expectancies, and to make the bridge and its approach more visible. In one-lane situations, a Positive Guidance [32] treatment may be necessary to accomplish the aforementioned goals, and to assign right-of-way on the single lane span (see the following section for a discussion of Positive Guidance). Positive guidance treatments for a variety of narrow bridge configurations are set forth in Appendix A of the "Yellow Book" [12]. Figure 24 shows a recommended treatment for a one-lane bridge on a curve with restricted sight distance.

Traffic Control Devices

Traffic control devices structure expectancies about downstream features. They also structure expectancies concerning information treatments at similar locations. The key to effective expectancy structuring is uniformity and standardization, using standard MUTCD devices consistently applied. If devices are inconsistently applied, drivers experience problems. For example, if upstream curve warning signs tend to underestimate maximum safe speed, then drivers expect similar underestimations for similar curves downstream. However, when a downstream curve is more realistically signed, then expectancies are violated, and drivers may be unprepared or unable to respond properly.

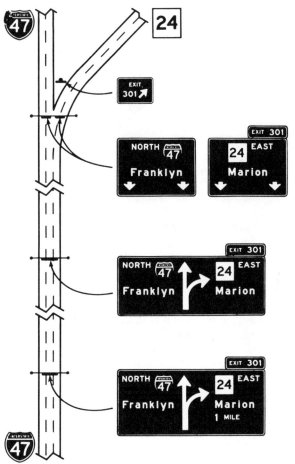

FIGURE 18. Two-lane exit treatment.

Thus, not only do traffic control devices serve to structure expectancies, they also serve to violate expectancies when misapplied, inconsistently applied, unique to a given location, and/or ambiguous.

TRAFFIC SIGNALS

In Figure 25, the signal indication is changing from green-to-yellow in the cross street signal head, resulting in main street drivers expecting their signal to change from red-to-green. However, in the event of a lagging green or a protected turn phase, the signal indication in the driver's direction will not immediately turn green. Thus, drivers expecting a green signal may inadvertently enter the intersection on a red indication. One way to resolve this problem is to mask the cross street signal indication so drivers cannot see it, thereby inhibiting the expectancy from being structured.

Another example of an unexpected signal indication is the

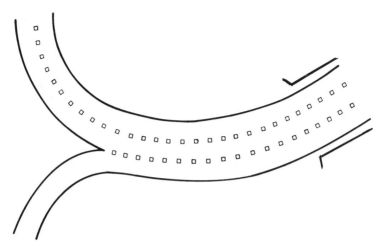

FIGURE 19. Lane added by striping.

midblock signal shown in Figure 26. Drivers do not, in most instances, expect a signal anywhere but at an intersection. When a midblock signal is used, they may not be prepared, without advance warning, and may not react in time, or may rear-end another vehicle stopped at the crosswalk. The key in this situation is to provide conspicuous advance warning.

SIGNS

Signs which provide information at the Guidance Level (regulatory and warning signs) and at the Navigation Level (guide signs) also have the potential to structure or violate expectancies. With regard to guide signs, drivers, by virtue of their trip plan, formulate *a priori* expectancies relative to what route, direction, and destination information will be displayed. These are often violated in transit. There are also sign-related expectancies that are violated at the Guidance Level, such as the sign shown in Figure 27.

The figure shows a standard curve warning sign with a speed advisory plate. Automobile drivers generally expect to be able to exceed the advisory speed by a substantial margin. This is based on their experience with such signs, which are often understated when it comes to a "safe" speed, at least for passenger vehicles. Truck drivers, on the other hand, may not be as likely to exceed the advisory speed, as their vehicles do not track curves as well as cars. On wet pavement, when curve tracking capabilities deteriorate for most vehicles, auto drivers still tend to overdrive curves. There are locations where the advisory speed must be adhered to, even in dry weather. Here, since drivers tend to exceed the advisory speed limit, a "We Really Mean It"

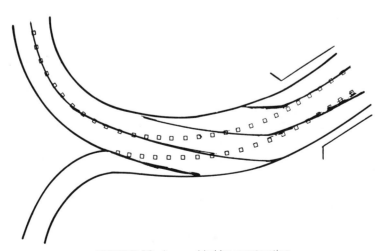

FIGURE 20. Lane added by construction.

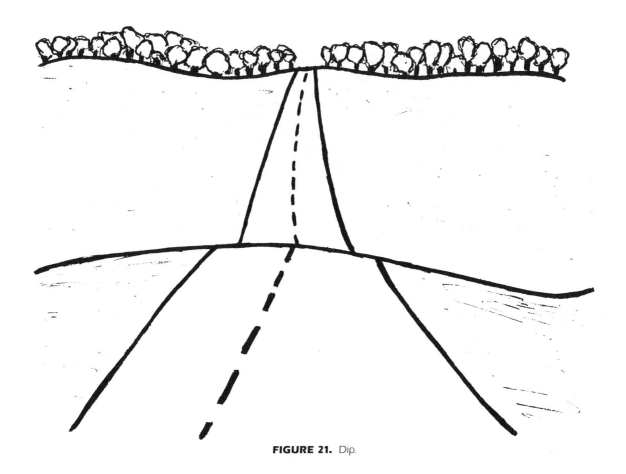

FIGURE 21. Dip.

FIGURE 22. One lane bridge.

FIGURE 23. Narrow bridge on curve.

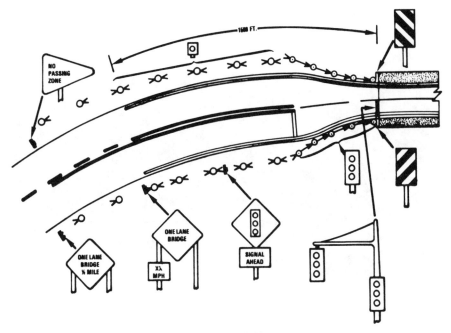

FIGURE 24. One-lane bridge treatment.

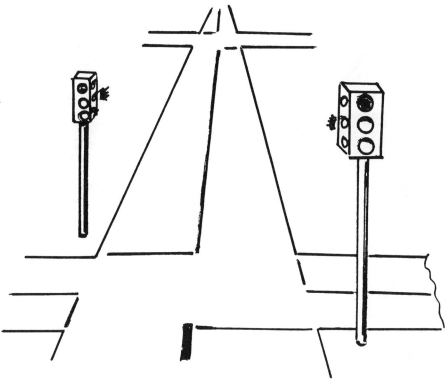

FIGURE 25. Cross street indication.

FIGURE 26. Midblock signal.

445

FIGURE 27. Curve warning sign.

technique is often used. Jurisdictions often employ various display treatments including oversized signs, Chevrons, and flashing beacons. An example is shown in Figure 28 (see Reference 6).

It is difficult, at the Navigation Level, to anticipate exactly what drivers expect. For example, some drivers may expect their specific destination to be signed for at their exit. While this is a reasonable expectancy if the destination is a major traffic generator, there are so many potential destinations that this expectancy will be violated most of the time. One way to overcome this problem is through education and training relative to trip planning and destination finding, thereby assuring that unreasonable expectancies are not formed. Another way to aid is through supplemental signing of potential destinations. Various innovative aids to navigation are in the developmental stage to help in trip planning and destination finding, including Highway Advisory Radio and in-vehicle systems.

Finally, the two common problems of freeway names versus numbers and facility-route continuity are shown in Figure 29. Local drivers generally refer to freeways by name, while unfamiliar drivers often expect route numbers to be used. When route numbers are signed, local drivers, who often do not know them, may experience expectancy violations. The converse is true about strangers who have planned their trip using route numbers. Many jurisdictions attempt to solve this problem by displaying both the route name and number on signs.

The facility-route continuity problem is more complex, with no clear-cut way to solve it. Most drivers expect the facility they are on to carry the through-route, and to exit onto a ramp that leads to the off-route. That is, people do not expect to have to "exit" to stay on their route. When this occurs, there is invariably driver uncertainty and confusion. Although many "solutions" have been tried, ranging from diagrammatics to different colored pavements, few seem to work.

Key Considerations

The development of appropriate designs, the display of needed information, the operation of traffic in accordance with driver expectancies, and the restructuring of expectancies that are violated through the use of standard traffic control devices are primary ways to aid performance and enhance safety and efficiency. Unusual, ambiguous, or nonstandard designs or information displays should be avoided, and traffic control devices should be consistently applied throughout the system. Attention should be given to ensure design consistency from one segment of roadway to another. When drivers get the information they expect from the highway and its information system, driver response tends to be rapid and error free. When drivers do not get what they expect, or get what they do not expect, then slow response, confusion, and errors occur.

Key considerations about expectancies include the following:

- Expectancies are associated with all levels of the driving task and all phases of the driving situation.
- Drivers experience problems and commit errors when their expectancies are violated.
- Drivers should not be surprised.
- Drivers tend to anticipate upcoming situations and events that are common to the route they are driving.
- The more predictable the design, information display, or traffic operation, the less likely will be the chance for driver error.
- Drivers, in the absence of information to the contrary, assume that they will only have to react to "standard," i.e., expected, situations.
- The roadway, the information system, and the environment upstream of a location structure expectancies of downstream conditions. Drivers experience difficulty in transition locations, and places with inconsistencies or unexpected features in design and/or traffic operations.
- The objective in helping drivers overcome the effects of expectancy violation is to structure the appropriate expectancy through advance warning. When it is not possible to give drivers what they *do* expect, it is essential to tell them what they *should* expect.

FIGURE 28. Curve warning treatment.

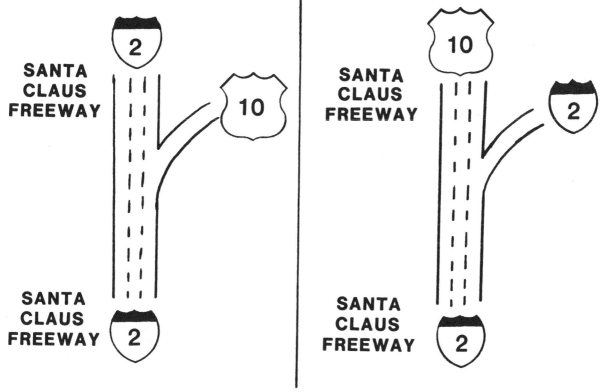

FIGURE 29. Route continuity and name vs. number.

IDENTIFYING EXPECTANCY VIOLATIONS USING POSITIVE GUIDANCE

Positive Guidance

The Expectancy Violation Analysis and Review is derived from the *2nd Edition* of the *Users' Guide to Positive Guidance* [32]. The Users' Guide presents a conceptual development of Positive Guidance and contains a step-by-step description of the "Engineering and Human Factors Procedure," the heart of the process.

Positive Guidance is an approach to enhance the safety and operational efficiency of problem locations. It joins the highway engineering and human factors technologies to produce an information system matched to the characteristics of a location and the attributes of drivers. It is designed to provide high-payoff, short-range, low-cost solutions to safety and/or operational problems. It is based on the premise that a driver can be given sufficient information to avoid accidents and/or drive efficiently at problem locations.

Since few locations are identical, each is individually analyzed to develop improvements tailored to the particular site. Positive Guidance in general and the Expectancy Violation Analysis and Review in particular are tools to analyze a site, identify its problems, develop information system improvements, and determine their effectiveness. Using Positive Guidance procedures, a site is reviewed and analyzed in a number of ways including a drive-through from a "driver's eye" point of view, analysis of films, slides, video, and/or photologs, and the collection of performance data. Information gained from these sources is used to perform the various steps in the Engineering and Human Factors Procedure, develop improvements, and evaluate their effectiveness. Results of various projects using Positive Guidance are contained in a number of reports [6,23,31]. Since the thrust of this discussion is on expectancies, the Expectancy Violation Analysis Step is presented in detail.

Expectancy Violation Analysis and Review

The Expectancy Violation Analysis and Review is designed to identify expectancy violations, pinpoint their sources, and develop information displays to restructure violated expectancies or structure appropriate ones. The Analysis and Review is initiated by first reviewing the area upstream and downstream of a problem location (if already identified) or assessing a road segment as part of general surveillance. This General Review provides an understanding of the land-use, geometric design, traffic, operational procedures, and traffic control devices which serve to structure driver expectancies. Once this understanding is obtained and/or unidentified problems are found through routine surveillance, a Detailed Analysis is then performed to zero in on specific expectancy violations and locate them on the road.

GENERAL REVIEW

A General Review can be performed using a variety of approaches and sources. One approach is to drive through a location and record observations in the field using audio tape or pencil and paper. Another way is to film, videotape, or photograph the site and analyze the data in the office. Existing photologs can also be used. Finally, a combination of approaches and sources can be employed. In any event, the reviewer should always drive through the site and obtain a "driver's eye" view in real time. The most important activity to perform during the General Review is to obtain a "feel" for the site. Hence, this part of the analysis should be fairly informal. If possible, help should be obtained from someone who is unfamiliar with the road or area, because familiarity often causes a reviewer to miss an expectancy violation that is quite apparent to a stranger. In performing the review, it is useful to generate a list of conditions upstream of the location as well as informally locating features for further assessment during the Detailed Analysis. Table 1 lists factors to consider in the General Review. Changes from upstream to downstream factors should also be noted.

DETAILED ANALYSIS

Going from the General Review to the Detailed Analysis links expectancy violations to specific features. This is accomplished by identifying specific expectancy violation(s), their source(s), their effect(s) on driver behavior, and driver information needs brought about by the violation(s). As in the General Review, data are obtained from an in-field drive-through and photographic sources. Since this analysis is keyed to specific site features, a schematic or plan view of the site (locating all traffic control devices and

TABLE 1. Factors to Consider in the General Review.

- Land Use
- Road Type
- Road Surface
- Cross Section
- Terrain
- Geometry
- Sight Distance
- Weather
- Lighting
- Traffic
- Signals
- Markings
- Warning & Regulatory Signs
- Guide Signs & Route Markers
- Missing Information

other relevant features) helps to locate source of the violation and where it operates. A checklist, such as the one shown in Figure 30, also can be used. Its inputs are essentially derived from the General Review, with specific details entered on the form.

Using the General Review as a point of departure, the Detailed Analysis begins by selecting a convenient starting point upstream of a problem location. This starting point should be "typical" of upstream conditions in terms of traffic, geometrics, land use, etc., since it structures a driver's *ad hoc* expectancies. Having located a starting point, the analysis proceeds toward and through the problem location. A Detailed Analysis is performed for all applicable directions. In performing the Detailed Analysis, a target population is assumed, usually unfamiliar drivers (the "stranger with a map"). There are times when locals or commuters are the target group. When this occurs, strangers should also be considered.

Identify Navigation Expectancies

Assuming unfamiliar drivers as the target group, the first activity is to identify potential Navigation Expectancies. In doing so, the assumption is made that strangers have con-

```
            DETAILED EXPECTANCY CHECKLIST - 1.

Reviewer: _____   Date: _____
Location: _____


1.  Upstream Land Use: _____  Have Changes Occurred? _____
    Where: _____   What: _____

2.  Upstream Road Type: _____  Have Changes Occurred? _____
    Where: _____   What: _____

3.  Upstream Road Surface: _____  Have Changes Occurred? _____
    Where: _____   What: _____

4.  Upstream Cross-Section: _____  Have Changes Occurred? _____
    Where: _____   What: _____

5.  Terrain:  Do Terrain Features or Manmade Elements Provide False
    Cues? _____
    Where: _____   What: _____

6.  Geometry: Does Geometry or Geometric Inconsistencies Surprise
    Drivers? _____
    Where: _____   What: _____

7.  Sight Distance: Does Poor Sight Distance Cause Drivers to Miss
    Unexpected Features? _____
    Where: _____   What: _____
```

FIGURE 30. Detailed expectancy checklist (continued).

```
┌─────────────────────────────────────────────────────────────────┐
│              DETAILED EXPECTANCY CHECKLIST  —  2.               │
│                                                                 │
│  8.  Weather: Are Temporary Weather Features Involved? _____  │
│                                                                 │
│      What: _____                              │
│                                                                 │
│                                                                 │
│  9.  Lighting: Does Lighting (Including Natural Light) Contribute to │
│      Expectancy Violations? _____                             │
│                                                                 │
│      Where: _____   What: _____             │
│                                                                 │
│ 10.  Traffic: Do Any Unusual Traffic Patterns or Mixes Exist (Including │
│      Pedestrians)? _____                              │
│                                                                 │
│      Where: _____   What: _____             │
│                                                                 │
│ 11.  Signals: Are Any Signals, Signal Configurations, and/or Signal │
│      Patterns Confusing or Unusual? _____                     │
│                                                                 │
│      Where: _____   What: _____             │
│                                                                 │
│ 12.  Markings: Are Any Markings (Delineation) Confusing or Unexpected? │
│      _____                                            │
│                                                                 │
│      Where: _____   What: _____             │
│                                                                 │
│ 13.  Warning & Regulatory Signs:  Are Any Warning and/or Regulatory │
│      Signs Surprising, Confusing, Obsolete and/or Nonstandard? _____ │
│                                                                 │
│      Where: _____   What: _____             │
│                                                                 │
│ 14.  Navigation: Are any Guide Signs, Directional Signs, and/or Route │
│      Markers Surprising, Confusing, Obsolete and/or Nonstandard? ____ │
│                                                                 │
│      Where: _____   What: _____             │
│                                                                 │
│ 15.  Missing Information: Is any Needed Information Missing? ____ │
│                                                                 │
│      Where: _____   What: _____             │
│                                                                 │
│ 16.  Others: Is There Anything else About the Site or Location  │
│      Surprising or Confusing? _____                           │
│                                                                 │
│      Where: _____   What: _____             │
│                                                                 │
└─────────────────────────────────────────────────────────────────┘
```

FIGURE 30. Detailed expectancy checklist.

sulted a map and prepared a trip plan. It is likely, particularly at choice points, that strangers will be both looking for and expecting information relating to route following and direction finding. The reviewer should obtain maps of the area and identify major destinations, routes and traffic generators. The following should be kept in mind about Navigation Expectancies: Drivers expect all nodes (intersections, interchanges, choice points, etc.), major routes, and cardinal directions to be identified; drivers expect their destination, if major, to be signed for; drivers expect well-known traffic generators, such as colleges, stadia, shopping centers, etc., to be indicated; and drivers expect service availability information. These and other route-specific expectancies are most common in urban areas, but often are a major determinant of driver task performance in rural locations. Using maps and the drive-through, the reviewer should determine if any Navigation Expectancy violations occur. These are noted on Figure 30.

Identify Guidance Expectancy Violations

The next activity is to identify and locate Guidance expectancy violations. Using the data generated by the drive-through and photographic means, the reviewer should search for expectancy violations, note them on a plan or site diagram, and enter them on the checklist (Figure 30). Expectancy violations can be numbered for convenience. Identifying an expectancy violation may require, in addition to the criteria presented herein, considerable engineering judgement. Another aid is an accident review to see if there are location patterns. It may also be useful to talk to maintenance pesonnel, police, and operational personnel. The accident review and discussions can serve to identify situational problems (e.g., at night, in rain). The basic aim is to determine whether any aspect of the site and its traffic control and operations is *surprising* to the target group(s). As an aid, particularly when a reviewer is familiar with the site, the following questions should be considered.

1. Is the location one that exhibits features or attributes that drivers may find unusual or special?

    Discussion: Table 2 presents a list of "special features" to serve as a guide. Each one is a potential expectancy violation. Since there are usually regional differences, the reviewer may have additional special features to add to this list.

2. Is the feature a "first of a kind?"

    Discussion: Even though a feature may not be unusual per se, it may be the first one encountered on a road. For example, if all freeway exits are cloverleafs, then the first diamond would be unexpected. If the feature is both a "special feature" and a "first of a kind," then it probably will be a major problem for an unfamiliar driver.

3. Are there changes in site characteristics?

    Discussion: Drivers may be surprised by changes in geometrics, design, or operational characteristics. Changes such as different cross sections, different land use, differences in terrain, differences in road surface, closer interchange spacing, and new vehicle/pedestrian mixes may violate expectancies. It is the transition that surprises.

4. Are there changes in practices?

    Discussion: Operating practice changes, though often subtle, can violate expectancies. Differences in speed zoning, no passing zoning, or signal timing can vary from jurisdiction-to-jurisdiction. Sign placement and location can be different, curves that are signed in one place may not be signed in another. Once drivers get used to a specific practice, they expect it to continue. Locals are more affected by operations that vary from the usual such as maintenance, changes in railroad operations, and new traffic control devices.

5. Are there sight distance restrictions?

TABLE 2. Special Features (NCHRP 123 (30)).

Unusual Intersections—Circles; squares; leading/lagging green; 4 way stops.

Unusual Interchange Design—**Exits:** Bifurcation; double exit; exit on horiz curve; exit on vert curve; exit on combined horiz/vert curve; lane drop at exit; left exit; missing or short exit decl lane; tangent off-ramp; two (or more) lane exit; exit to collector-distributor road; unusual ramp and/or ramp terminus features. **Entrances:** Double entrance; entrance on horiz curve; entrance on vert curve; entrance on combined horiz/vert curve; lane addition; left entrance; missing or short accel lane; two (or more) lane entrance; unusual ramp geometrics; metered ramps; extremely high volume entrances. **Exits/Entrances:** Multilevel exit/entrance; common accel/decel lane; inadequate weaving sect. **Misc:** At-grade crossing (on freeways and expressways); restricted interchanges (by type of traffic or time of day); uncontrolled access; very long/very short interchange spacing.

Extremes in Roadway Geometry—Steep hills; extreme horiz curves; combined curves; dips; bumps; improper superelevation.

Unusual Maneuvers—Weaves; stops on exit ramps; stops on entrance ramps; discontinuous route; through route on off-ramps; off route on through lanes; U turns; left turns from "Jughandle."

Changes—**Cross Section:** Lane drops; lane additions; shoulders; medians; lane width. **Roadway Environment:** Urban-rural; rural-urban; trees, foliage, etc.; surface; elevated-depressed; freeway, arterial, two lane. **Legal:** Speed limit; lane restrictions; pedestrian zones; bicycle zones; diamond lanes; turn restrictions.

Off-Line Restrictions: Abutments; piers; underpasses; culverts; cuts; curbs; guardrail ends; luminares; sign supports; parked vehicles.

Sight-Line Restrictions: Horizontal, vertical; combined.

Environmental Problems: Freezing roadways; fog; high background lighting; sun.

Traffic: Heavy traffic; congestion; large proportion of trucks, RV's, etc.; bicycle traffic; pedestrians.

Miscellaneous: Construction zones; maintenance zones; fallen rock zones; animal crossings; narrow bridges; tolls; railroad crossings; poor road surfaces; school zones; bus stops; billboards, tree lines and/or telephone poles that deviate from road align; pavement joints that deviate from road align; route changes; discontinuities in street grids; one-way roads; parking restrictions; hidden driveways; farm vehicles; underpasses; special use/reversible lanes.

```
             SITE IMPROVEMENT FORM (Use a Separate Sheet for Each Violation)


Engineer: _____        Date: _____
Location: _____


Expectancy Violation: _____
Expectancy Violation Source: _____
Expectancy Violation Location: _____
Affected Driver Performance:  Speed _____  Path _____  Direction _____
Driver Information Need(s) _____
Potential Improvement(s): Standard Device(s): _____


Non-Standard Device(s) :_____
If Nonstandard Devices Are Used, Do They Result in Expectancy Violations?


Comments: _____


                        Sketch of Location and Change(s)
```

FIGURE 31. Site improvement form.

Discussion: Drivers have difficulty preparing for unexpected features that they cannot see. They must have sufficient time to see and respond. Thus, any unseen feature, be it a standard intersection beyond a crest vertical or a lane drop beyond a horizontal curve, is unexpected.

6. Is it signed for?

   Discussion: One main reason for signing is to provide advance warning of an unexpected situation or event. If an expectancy has been violated, the reviewer must ask, "Was there advance warning—was it signed for?" Even when there is advance warning, the reviewer should recognize that the sign itself could result in an expectancy violation. This is particularly true of Navigation information.

7. Is the signing adequate?

   Discussion: The reviewer should assess each traffic control device to determine the expectancies the signs markings, or signal(s) structure; the effectiveness of the device(s) in providing advance warning; and the possibility of the device violating an expectancy.

Determine Affected Driver Performance

Having identified expectancy violations and their sources and located them, the next activity is to gauge their affects on driver task performance, specifically on speed, path and/or direction. This will provide an understanding of the consequences of the expectancy violation and on the identification of an associated information need.

Identify Information Needs

The reviewer should identify driver information needs associated with each expectancy violation. These represent information required to structure and/or restructure a driver expectancy at the site. Hence, these information needs are ultimately translated into traffic control device improvements.

Assess Safety and Operational Consequences

If a large number of expectancy violations are identified, the reviewer should assess each in terms of its consequences on the safety and/or operations of the location. A determination may have to be made on which violations can be restructured if a number are found. A priority estimation, based on primacy, will aid in this determination.

Develop Site Improvements

Data generated by the Detailed Analysis are ultimately translated into information system improvements using the Form(s) shown in Figure 31. Improvements may include removing sources of expectancy violations as well as enhancing traffic control devices at the problem location. Standard traffic control devices, applied in accordance with the MUTCD, should be used to the greatest extent practicable. It must be born in mind that any nonstandard device or unusual application has a potential of leading to subsequent expectancy violations. Accordingly, the location should be reassessed after changes are made (particularly nonstandard ones), to assure no new expectancy violations have been structured. This assessment should be conducted upstream of the problem location as well as at the hazard zone.

SUMMARY

Expectancy relates to a driver's readiness to respond to situations, events and information in predictable and successful ways. As such, it is a key factor in driving task performance since it affects the speed and accuracy of a driver's

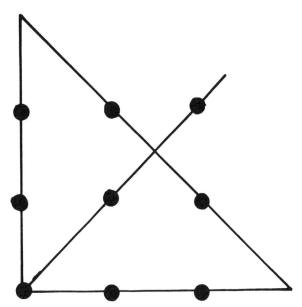

FIGURE 32. Solution to spatial expectancy problem.

response. Prevalent expectancies that are reinforced aid drivers, while expectancies that are violated increase reaction time and driver error. Expectancy and expectancy violations operate at all levels of the driving task, from vehicle Control through Guidance in the traffic stream to Navigation of the road network; they include *a priori* (brought by the driver as a result of culture and experience) and *ad hoc* (caused by exposure to a set of site-specific practices) aspects; and they encompass all phase of geometric design and traffic control. Included in the design category are such features as left exists, interchange lane drops, tangential off-ramps, parallel roadside features, optional lanes, construction joints that do not follow markings, dips, and narrow bridges. Traffic control device examples include lagging green and midblock signals, curve warning signs that lack credibility, freeway names versus numbers, and route discontinuity. Because of these considerations, engineers and designers need to maintain an appreciation of the expectancy concept, prevalent *a priori* expectancies, *ad hoc* expectancies structured in transit by a location's design and traffic control device complement, and the adverse effects of expectancy violations. They should ensure that their designs and traffic control devices do not violate expectancies by surprising the driver. Finally, they should rectify expectancy violations through information system improvements, including removal of the expectancy violation, based on Positive Guidance.

To emphasize the point, when drivers get the information they expect to from the highway and its traffic control devices, performance tends to be error free. When they don't get what they expect, or get what they don't expect, errors and system failures are the usual result.

REFERENCES

1. *A Policy on Geometric Design of Highways and Streets* ("Green Book"), American Association of State Highway and Transportation Officials, Washington, D.C. (1984).
2. AASHTO Subcommittee on Design, Region 2, *Designing for the Stranger,* Troy, N.Y. (June 18–20, 1980).
3. Alexander, G. J. and H. Lunenfeld, "A Users' Guide to Positive Guidance in Traffic Control," in R. Easterby and H. Zwaga (eds.), *Information Design,* John Wiley, Chichester (UK), pp. 351–383 (1984).
4. Alexander, G. J. and H. Lunenfeld, *Positive Guidance in Traffic Control,* Federal Highway Administration, Washington, D.C. (April 1975).
5. Alexander, G. J. and H. Lunenfeld, "Satisfying Motorists' Need for Information," *Traffic Engineering,* Vol. 42, No. 1, pp. 46–70 (October 1972).
6. Barsness, J. and M. Nesbitt, *Application of Positive Guidance at a Reverse Curve/Narrow Bridge Site in Washington State,* Report No. FHWA-DP-48-2, Federal Highway Administration, Washington, DC (April 1982).
7. Cumming, R. W., "The Analysis of Skills in Driving," *Australian Road Research,* Vol. 9, pp. 4–14 (1964).
8. *Driver Expectancy Checklist,* American Association of State Highway Officials, Washington, D.C. (1972).
9. Ellis, N. C., *Driver Expectancy: Definition for Design,* Report No. 606-5, Texas Transportation Institute, College Station, Texas (June 1972).
10. Gordon, D. A., *Psychological Contributions to Traffic Flow Theory,* Report No. FHWA-RD-74-53, Federal Highway Administration, Washington, D.C. (June 1973).
11. Graham, J. L., D. J. Migletz, and J. C. Glennon, *Guidelines for the Application of Arrow Boards in Work Zones,* Report No. FHWA-RD-79-59, Federal Highway Administration, Washington, D.C. (1978).
12. *Highway Design and Operational Practices Related to Highway Safety* ("Yellow Book"), American Association of State Highway and Transportation Officials, Washington, D.C. (2nd Edition) (1974).
13. Hill, B. L., "Vision, Visibility, and Perception in Driving," *Perception,* Vol. 9, pp. 183–216 (1980).
14. Homburgh, W. S. (ed.), *Institute of Transportation Engineers, Transportation and Traffic Engineering Handbook (2nd Edition),* Prentice-Hall, Englewood Cliffs, N.J. (1982).
15. Hostetter, R. S., K. W. Crowley, G. W. Dauber, L. E. Pollack, and S. Levine, *Determination of Driver Needs in Work Zones,* Report No. FHWA-RD-82-117, Federal Highway Administration, Washington, D.C. (September 1982).
16. Hulbert, S. F., "Human Factors in Transportation," In W. S. Homburgh (ed.) *Transportation and Traffic Engineering Handbook,* Prentice-Hall, Englewood Cliffs, N.J. (1982).
17. Hulbert, S. F. and A. Burg, "Application of Human Factors Research in Design of Warning Devices for Highway Rail Grade Crossings," *NCHRP Report No. 50,* "Factors Influencing Safety at Highway-Rail Grade Crossings," Highway Research Board, Washington, D.C. (1968).
18. Johannson, C. and K. Rumar, "Driver's Brake Reaction Time," *Human Factors,* Vol. 13, No. 1, pp. 22–27 (1971).
19. Kimble, G. A., *Hilgard and Marquis' Conditioning and Learning,* Appleton-Century-Crofts, New York (1961).
20. Lunenfeld, H., *Improving the Highway System by Upgrading and Optimizing Traffic Control Devices,* Report No. FHWA-TO-77-1, Federal Highway Administration, Washington, D.C. (April 1977).
21. Lunenfeld, H. and G. J. Alexander, *Signing Treatments for Interchange Lane Drops,* Report No. FHWA-TO-76-1, Federal Highway Administration, Washington, D.C. (June 1976).
22. Lunenfeld, H. and G. J. Alexander, "Human Factors in Highway Design and Operations," *Journal of Transportation Engineering,* Vol. 110, No. 2, pp. 149–158 (March 1984).
23. Lunenfeld, H. and R. D. Powers, *Improving Highway Information at Hazardous Locations,* Report DOT-I-85-16, U.S. Department of Transportation, Washington, D.C. (March 1985).
24. *Manual on Uniform Traffic Control Devices,* U.S. Department of Transportation, Federal Highway Administration, Washington, D.C. (1978).
25. Markowitz, J. and C. W. Dietrich, *Investigation of New Traffic Signs, Markings, and Signals,* Report No. BBN-1762, Bolt, Beranek, and Newman, Inc., Cambridge, Mass. (1972).
26. Mast, T. M. and G. S. Kolstrud, *Diagrammatic Guide Signs for Use on Controlled Access Highways,* Report No. FHWA-RD-73-21, Federal Highway Administration, Washington, D.C. (December 1972).
27. McGill, W., "Population Expectancies and Traffic System Design," *Australian Road Research,* Vol. 2, No. 7, pp. 19–42 (1966).
28. Messer, C. J., J. M. Mounce, and R. Q. Brackett, *Highway Geometric Design Consistence Related to Driver Expectancy,* Report No. FHWA-RD-81-035, Federal Highway Administration, Washington, D. C. (April 1981).
29. Michaels, R. M. "Human Factors in Highway Safety," *Traffic Quarterly,* Vol. 15, No. 4, pp. 586–599 (Oct. 1961).
30. *NCHRP Report No. 123,* "Development of Information Requirements and Transmission Techniques for Highway Users," Highway Research Board (1971).
31. Opland, W., *Application of Positive Guidance at a Freeway Split in Michigan,* Report No. FHWA-DP-48-1, Federal Highway Administration, Washington, D.C. (April 1982).
32. Post, T. J., G. J. Alexander, and H. Lunenfeld, *A Users' Guide to Positive Guidance (2nd Edition),* Report FHWA-TO-81-1, Federal Highway Administration, Washington, D.C. (December 1981).

33. Post, T. J., H. D. Robertson, G. J. Alexander, and H. Lunenfeld, *A Users' Guide to Positive Guidance (1st Edition),* Federal Highway Administration, Washington, D.C. (June 1977).
34. Reason, J., *Man in Motion: The Psychology of Travel,* Weidenfeld and Nicholson, London (1974).
35. Rockwell, T. H., "Driver-Sensory Load," in *Proceedings of the National Conference on Highway Operations in the 1980's,* T. C. Helvey (ed.), University of Tennessee at Nashville (September 24–27, 1973).
36. Sanders, J. H., G. S. Kolsrud, and W. G. Berger, *Human Factors Countermeasures to Improve Highway-Railway Intersection Safety,* Report No. HS-800-888, National Highway Traffic Safety Administration, Washington, D.C. (1973).
37. Schmidt, I. and P. L. Connolly, *Visual Considerations of Man, the Vehicle, and the Highway,* Publication SP279, Society of Automotive Engineers, New York (March 1966).
38. Shore, R. E., "Shared Patterns of Nonverbal Expectations in Automobile Driving," *Journal of Social Psychology,* Vol. 62 (first half), pp. 155–163 (1964).
39. Taylor, J. I., H. W. McGee, E. L. Seguin, and R. S. Hostetter, *NCHRP Report 130,* "Roadway Delineation Systems," Highway Research Board, Washington, D.C. (1972).
40. Taylor, J. I. and H. T. Thompson, *Identification of Hazardous Locations on Highways,* Report No. FHWA-RD-77-83, Federal Highway Administration, Washington, D.C. (1977).
41. Tolman, E. C., *Purposive Behavior in Animals and Men,* Appleton-Century, New York (1932).
42. Versace, J., "Factor Analysis of Roadway and Accident Data," *HRB Bulletin 240,* pp. 24–32 (1960).

CHAPTER 21

Technique for Identifying Problem Downgrades

RONALD W. ECK*

INTRODUCTION

Statement of the Problem

Highway construction in mountainous terrain is extremely costly due to the large volumes of earth and rock which must be moved. The total cost of a construction project in rugged topography may be several times that of a similar project in a level environment. In the past, engineers attempted to minimize construction costs in mountainous areas by designing highways with sharper horizontal curvature and steeper grades than normally desirable. These minimum geometric design standards often have a detrimental effect on traffic operations and safety.

Highways in mountainous terrain pose a number of special problems for motor vehicle operators. These problems may be especially critical for large commercial vehicles. Problems include maneuverability, poor performance traveling upgrade, and braking problems when descending grades.

The effects of steep positive grades on traffic flow have been well-documented and translated into criteria for the design and location of truck climbing lanes. Since a positive grade for one direction is a negative grade for the other direction, the downhill situation may also be significant in mountainous areas. On long, steep downgrades there is the possibility of vehicle brake failure and/or inattentive drivers failing to downshift properly. In such situations, trucks frequently accelerate uncontrollably down the steep grades endangering not only the lives of the truck drivers but also the occupants of other vehicles on the highway. In addition, residences and business enterprises at the foot of long steep downgrades may be damaged or destroyed by runaway vehicles. Due to the high speeds involved, a large percentage of runaway vehicle accidents result in fatalities.

There are a variety of countermeasures available for reducing runaway truck accidents and other operational problems associated with long steep downgrades. The most desirable approach would be to flatten the grades and straighten out winding alignments. However, due to financial constraints, this is not usually possible. As a result, various other types of runaway vehicle countermeasures have been developed by highway agencies. These include reduced speed limits, improved signing, alternative routings for trucks of specified sizes, brake check areas at summits, and truck escape facilities to "catch" out-of-control trucks. While warrants and guidelines have existed for a number of years, relative to erecting signs on problem grades, they have not been applied uniformly. Similarly, other than identifying high accident locations, there have not been, up until recently, general guidelines that engineers could follow for installing the more costly facilities such as brake check areas and escape facilities. Total reliance on accident records may not always be appropriate for identifying downgrades prone to runaway vehicle accidents since truck-involved accidents may be difficult to track and runaway truck accidents are relatively rare events. As competition for scarce highway funds becomes more intense, engineers must have a rational procedure by which they can justify the installation of the different downgrade accident countermeasures.

This chapter reviews and compares several approaches for identifying problem downgrades. Emphasis is placed on a simple yet reliable technique that can be readily used by technicians and engineers.

Review of Previous Approaches

Probably the first and still most common type of downgrade accident countermeasure is the use of warning signs to

---
*West Virginia University, Morgantown, WV

alert motorists to the existence of the grade. Nationwide surveys made in the late 1970s provide insight on the use of these signs. An American Association of State Highway and Transportation Officials (AASHTO) study (Wagner, 1976) found wide variations in application of *Manual on Uniform Traffic Control Devices* (MUTCD) recommendations on signing long, steep downgrades. A number of recommendations relative to downgrades were presented in Wagner's (1976) report. It was recommended that the grade length criteria which had appeared in the 1961 MUTCD should be reintroduced. The 1961 MUTCD criteria are presented later in this chapter. Further, advisory speed plates were recommended for use on hills that have curves which restrict speeds below the legal limit. The use of flashers was encouraged for downgrades where accident rates continue above normal.

In an effort to determine state practices regarding runaway truck countermeasures such as described above, the Pennsylvania Department of Transportation conducted a survey in December, 1977. It was found that reduced gear zones and speed limits are the most common types of truck restrictions used on long, steep downgrades. In general, states had not developed warrants for establishing truck restrictions on downgrades. Any restrictions were usually established based on engineering judgment rather than specific warrants.

Questionnaire results indicated some cases where state laws provided authority to enact truck restrictions on downgrades. For example, the Pennsylvania code listed elements to be considered in studies dealing with reduced speed limits on downgrades. PennDOT criteria indicated that a reduced speed limit for trucks could be established where one or more of the following conditions exist: a) the percent and length of grade exceed the values set forth in Table 1; b) an accident has occurred on the downgrade which can be attributed to the speed of a truck of specified weight; and c) receipt of a verified report of a truck losing control on the downgrade. A second criterion indicated that the reduced speed limit used should be consistent with the speed at which a specified truck can climb the hill.

Wagner's study (1976) also addressed the topic of truck turnouts. The American Trucking Associations (ATA) had indicated in the past that there was no justification for requiring trucks to stop at the tops of grades less than 5 or 6 percent and less than one mile long. While an AASHTO committee in 1961 likewise agreed there was no need for general provisions of truck turnouts, several states have used them successfully. One of Wagner's recommendations was that truck brake check areas be considered for long, steep down-grades where brake failure is a cause of accidents. He suggested that, where truck turnouts do not reduce the accident rate, consideration be given to truck escape ramps where terrain permits. Johnson, DiMarco, and Allen (1982) noted that escape ramps are generally constructed as a last-resort effort to reduce severity of damage, injury, and loss of life where signing and turnouts alone do not reduce accidents.

Highway departments in the moutainous western states were apparently the first public agencies to use truck escape ramps. In general, these ramps consisted of a roadway at a very steep positive grade composed of layers of loose gravel or uncompacted sand. Based on actual experience with existing truck escape ramps, revisions, primarily in the types of materials used, were made to the original designs. Until the mid-1970s, little formal research and development had taken place relative to the design and construction of truck escape ramps. Since that time, however, there has been increasing interest in all facets of steep highway downgrades and truck escape facilities.

Four basic types of truck escape facilities can be delineated (Eck, 1980). These are: (1) ascending grade, (2) horizontal grade, (3) descending grade, and (4) sandpile. The basic truck escape ramp designs are shown in Figure 1. Note that no single ramp design criterion has been adopted on a nationwide basis. This is primarily due to the fact that each ramp location possesses unique problems depending upon topography, land use, and length and steepness of grade.

Williams (1979) compiled a comprehensive state-of-the-art survey of truck escape facilities with emphasis on ramp design and construction. Williams found, as has been indicated previously, that there are no formal criteria used in planning escape ramps. He noted that roadways with long descending grades which can cause truck brakes to become overheated showed the greatest need for escape ramps. Traffic accident records can help identify specific areas that have a high incidence of runaway trucks. He stated that the

TABLE 1. Minimum Length and Steepness of Grade for Establishing Hazardous-Grade Speed Limit in Pennsylvania (Source: Eck, 1980).

Average Grade (Percent)	Length of Grade (Feet)	
	Condition A[a]	Condition B[a]
3	20,000	—
4	8,000	16,000
5	5,000	10,000
6	3,000	6,000
7	2,000	4,000
8	1,800	3,600
10	1,500	3,000
12	1,250	2,500
15	1,000	2,000

[a]Condition A applies if trucks are required to stop at or before the bottom of the hill, if traffic is required to reduce speed at or before the bottom of the hill due to horizontal alignment, if critical sight distance limitations exist, or if there is a reduced speed limit of 35 mph or below or an urban area at the foot of the hill. Condition B pertains at all other locations.

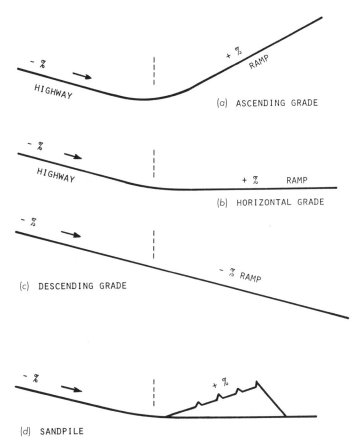

FIGURE 1. Profile views for the four basic types of truck escape facilities.

best location is prior to a critical curve or the bottom of the grade. Eck (1980) found that, in general, ramps near the crest of downgrades were not used by vehicles since truck drivers would rather stay with the out-of-control vehicle and attempt to bring it under control.

Johnson, DiMarco, and Allen (1982) indicated that in addition to factors such as grade, curvature, truck volumes, and accident experience, countermeasure development appears to be a function of public and/or political pressure. They noted that conditions at the bottom of the grade seem to dictate the nature of measures taken to prevent accidents having catastrophic results. For example, in eastern states, many downgrades terminate in small towns. These communities are frequently the sites of accidents caused by a truck losing control on a downgrade which forms a T-intersection with a main downtown street.

This brief review of countermeasures for the downgrade accident problem has given particular emphasis to locational and operational aspects. Available information indicates that while truck escape ramps are cost-effective in many instances, they are not a panacea for the problem of downhill accidents. Other solutions to the problem may be more appropriate in certain situations. Thus, the engineer involved in selecting countermeasures should examine a number of variables in addition to highway parameters. For example, the engineer should obtain data on the types (delivery, dump, flatbed, etc.) of trucks using a given downgrade. The condition of the vehicles should be observed since this may provide clues to vehicle maintenance policies. An effort should be made to determine the types of drivers operating over a given grade (common carrier regular route, common carrier irregular route, independent, etc.). This may provide information on driver familiarity with the route, likelihood of fatigue, as well as vehicle maintenance condition information. Analysis of such information will be useful as a guide in selecting the appropriate countermeasure, e.g., signing, truck turnout, or escape ramps. Where escape ramps appear to be the best solution, the information presented later in this chapter should be consulted for more detail.

While design and construction of escape ramps have advanced in recent years, there are still no criteria for determining escape ramp need. The recently revised *Policy on Geometric Design of Highways and Streets* (AASHTO, 1984) presents considerations to be used in design and construction of escape ramps but points out that specific guidelines for the design of escape ramps are lacking at this time. Ramps are installed on what might be called a "seat-of-the-pants" basis. Similarly, ramp location is usually based on finding a convenient site that will minimize earthwork and construction costs. The AASHTO design policy (1984) notes that selection of escape ramp locations on existing highways is usually based on accident experience. As previously noted, runaway truck accidents are relatively rare events and a concentration of accidents on a particular downgrade may not always be immediately apparent.

Since only limited published information was available on the subject of downgrade countermeasures, the author felt that correspondence with state highway agencies would be a source of data and insight concerning the topic of downgrade countermeasures. A questionnaire (Eck, 1979) was sent to 24 highway agencies known to have long, steep downgrade sections on their highway system. Agencies were asked to identify variables considered in determining whether an escape facility should be installed. Figure 2 shows the factors cited by the 19 responding agencies along with the frequency of citation. It should be noted that each factor listed by the states was given equal weight in making the plot of Figure 2. Several states noted certain factors as being more important than others but lack of additional data made it difficult to assign a numerical weight to the importance of each factor. The prominence of three factors which have been repeatedly mentioned in this discussion is apparent, namely, runaway truck accident rate and length and steepness of grade.

Several states added comments that might be useful to engineers involved with escape ramps. One state highway agency noted that percent grade and length of grade were factors that should be considered in designing the length and grade of the escape ramp rather than in determining escape ramp need. It was felt that the horizontal alignment of the roadway should be considered in designing the width and alignment of the escape ramp. Another engineer noted that there were three criteria used to determine if an escape facility should be installed: (1) Is there a runaway truck problem? (2) Can the problem be corrected by signing or delineation? (3) If the problem cannot be corrected by signing, where should the facility be built to best fit conditions?

State highway agency engineers were asked to specify factors which enter into the decision of where to locate escape facilities. As shown in Figure 3, a number of factors were cited by the states. The most frequently mentioned factor was topography due to its obvious effect on construction costs. Horizontal alignment and accident location were the second most frequently mentioned factors. Several respondents noted that the ramp should be located up-grade from sharp horizontal curvature since runaway trucks would not be able to traverse the curves at high speeds. Accident location is probably closely related to horizontal alignment since runaway vehicle accidents are usually run-off-the-road type caused by failure to negotiate horizontal alignment. Right-of-way availability, truck driver input, speeds of out-of-control vehicles, and length of grade were each cited once.

Questionnaire respondents were asked to indicate the number of escape facilities in each of several locations, such as in a cut, on a fill, in the median, etc. Results for this question are shown in Table 2. An unknown number of ramps were double counted. For example, a ramp located in a cut section on a tangent would receive two citations.

The greatest number of ramps are located on curves and/or in cuts, with ramps on tangents also receiving wide usage. Although ramps on tangents may be more desirable from the viewpoint of the driver, the wide usage of ramps on curves is not surprising. A large percentage of highway mileage in mountainous terrain consists of horizontal curvature. The fact that a roadway is in a cut makes it easier to achieve the desired uphill gradient for the escape ramp.

Fifteen escape facilities were located on fills; a large number of these were sandpile ramps. Sandpile ramps were specifically developed for use in locations where inadequate space existed for a typical escape ramp. A common location for sandpile ramps is on small fills where the downgrade lane(s) is on the outside of the mountain.

Two states had escape facilities at the end of freeway off-ramps. California and Utah had ramps in the median. Only one state had a ramp on the left-hand side of a two-lane highway. Reluctance of state highway agencies to use left-hand ramps is probably due to serious questions about liability (should a runaway truck heading for an escape ramp strike an on-coming vehicle), signing problems, and truck driver reaction to this type of ramp.

With regard to the question on ramp locations, states were asked whether any of the six locations cited were unsatisfactory in terms of operational or accident experience. Only a few states mentioned problems that were strictly locational in nature. Problems included poor visibility and vehicles at night mistaking the escape ramp for the through roadway (this particular ramp left the roadway at the right-hand side where the mainline curved to the left).

The review of published literature and questionnaire results tended to confirm that most downgrade countermeasures are installed on the basis of judgment rather than formal engineering analysis. As problems are encountered with specific countermeasures and their designs or locations, improvements and modifications are being made. In this manner, the state of the art is being advanced. While it is true that each escape facility location and physical condition should be considered individually, there is a need to develop methodologies by which an optimum use and location of downgrade countermeasures can be determined. Thus, research was undertaken to develop, analyze, and compare techniques for identifying problem downgrades.

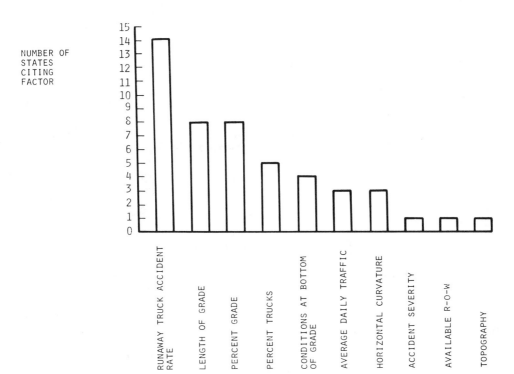

FIGURE 2. Factors considered by state highway agencies in determining need for truck escape facilities.

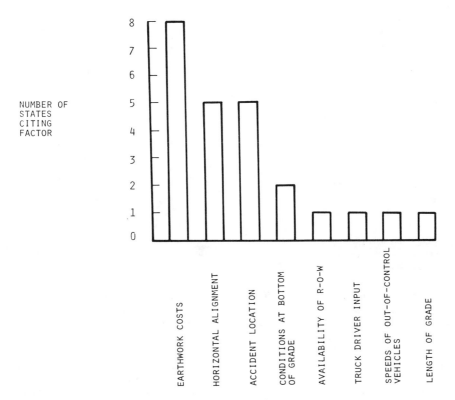

FIGURE 3. Factors considered by state highway agencies in determining location of truck escape facilities.

TABLE 2. Number of Escape Facilities in Various Locations, as Determined by Eck (1980).

Escape Facility Location	Number of Facilities
On Tangent	25
On Curve	29
In Cut	29
On Fill	15
In Median	1
Left Side of Highway	1
End of Off-Ramp	2

DESCRIPTION OF THE METHOD

Data Collection

Data from fifteen downgrades in West Virginia were used to develop and test accident prediction techniques. Accident data for each site consisted of milepost location of accident, calendar date of accident, time, light condition, pavement condition, accident type, and severity code. Runaway truck accidents were not identified explicitly in the accident listing. Thus, a critical element in the data collection task was the identification of runaway vehicle accidents. A rational approach to identifying runaway vehicle accidents was developed. Three years of accident data were obtained for each downgrade. Of the approximately 600 total accidents, 105 were identified as runaway truck accidents. Geometric and environmental data collected for each site included percent grade, length of grade, pavement width, horizontal curvature, speed limit, signing, land use characteristics, and vertical alignment prior to the downgrade.

Methodology

Data collected in the study were analyzed in a number of ways to try to determine which geometric and site variables were strongly related to runaway truck accidents. Accident prediction models were developed using a regression approach as well as Bayesian statistics. Due to the small sample size used, there was poor agreement between predicted and observed runaway truck accident rates using these two approaches.

Due to the lack of success in formulating statistically based models, an heuristic technique was developed (Eck, 1983) based on the idea that the interaction between the absolute magnitude of the grade (in percent) and the cumulative degree ($D = 5,729.58/R$) of horizontal curvature (measured from the crest of the downgrade) would influence runaway vehicle accidents. For example, on a given downgrade, a series of curves, each of decreasing radius, would be expected to place greater demands on vehicle brake systems than a series of curves, each of increasing radius. While not a predictive "model" in the formal sense, the technique recognizes that the arrangement of horizontal curves has an effect on runaway vehicle accidents.

Plots, such as that shown in Figure 4, were made of cumulative degree of curve versus cumulative distance from the crest of grade for each downgrade. A concave upward curve means that a vehicle driver experiences continually sharper curvature while descending the grade. A convex upward curve means that curves are becoming less sharp as one proceeds downgrade. It was hypothesized that downgrades having generally convex shapes would be safer, in terms of runaway truck accident rate, than those having concave shapes.

The plot described above can also be used to indicate consistency of horizontal curvature. A relatively smooth plot indicates that most horizontal curves are of similar order of magnitude. An isolated sharp curve appears as a discontinuity on the plot. A route with wide variation in curvature demonstrates an irregular appearance. One might expect runaway vehicle accidents to occur at, or just beyond, these points of sharp curvature.

Validation

To examine the above hypotheses, the cumulative number of runaway vehicle accidents by location on the downgrade was plotted on the drawing described above; this was done for the 15 sites studied. Three plots (Figures 4, 5, and 6) are presented here to illustrate some of the more important features as they relate to the location characteristics of runaway truck accidents.

Figure 4 depicts cumulative degree of curve versus distance from top of grade for US Route 50 at Cheat Mountain, West Virginia. Note that the graph is quite steep, indicating a large number of horizontal curves per mile. Much of the plot is concave upward, indicative of steadily increasing curvature. Accidents are roughly equally distributed throughout the length of the grade.

Figure 5 represents the degree of curve versus location plot for US Route 50 at the Allegheny Front. Although this curve is not as steep as the Cheat Mountain case, there are more discontinuities or irregularities. This is indicative of isolated sharp curves occurring in the alignment. Most of the curve is concave upward. The plot of accidents versus distance from top of grade indicates that most accidents occur either at sharp curves or a short distance downstream from a concave upward section of the curve.

Figure 6 shows the plot for WV Route 42 at the Allegheny Front. In terms of physical characteristics, Route 42 is very similar to US 50 at the Allegheny Front. However, Route 42 has a substantially lower accident rate. The lower rate may be partially explained by the plot, which is flatter and smoother than that for US 50. While there are concave upward sections along the downgrade, these are separated by

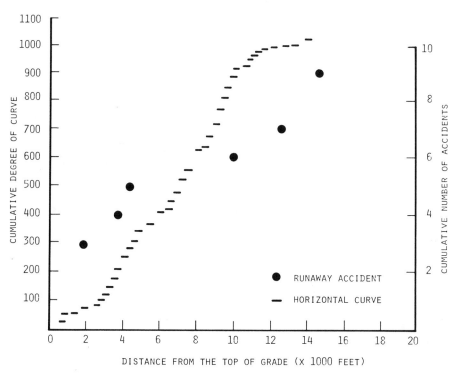

FIGURE 4. Cumulative degree of horizontal curvature plot for US 50 Cheat Mountain.

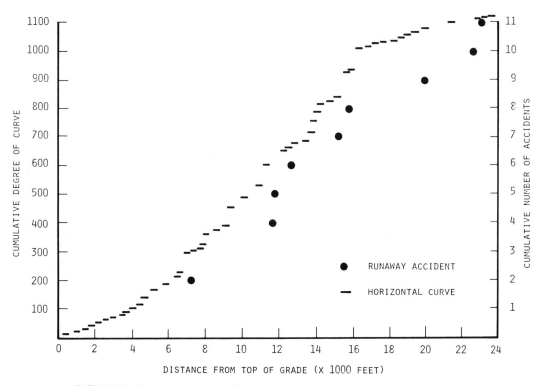

FIGURE 5. Cumulative degree of horizontal curvature plot for US 50 Allegheny Front.

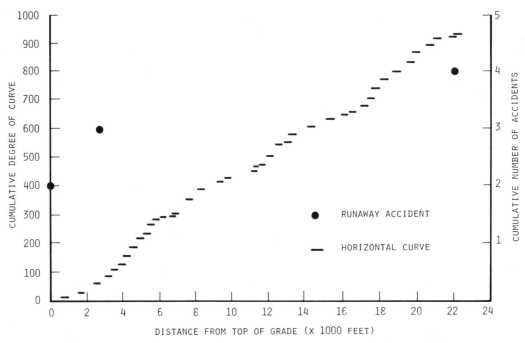

FIGURE 6. Cumulative degree of horizontal curvature plot for WVA 42 Allegheny Front.

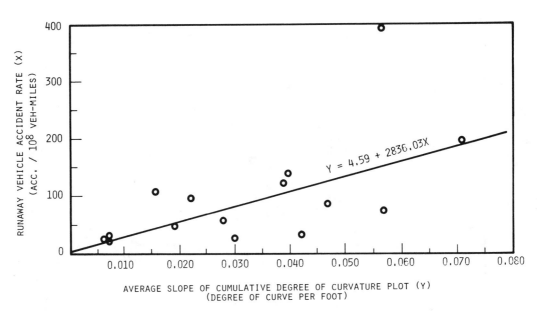

FIGURE 7. Relationship between runaway truck accident rate and average slope of cumulative degree of curvature plot.

convex upward sections. Accidents at this location bear no apparent relationship to roadway geometry.

Although a formal evaluation of the cumulative degree of curve approach could not be made, attempts were made to assess this technique as a tool for identifying problem downgrades. Observed accident rates on the 15 downgrades were correlated with the average slope of the cumulative degree of curvature plot. The correlation coefficient between the observed runaway truck accident rate and slope of the plot was 0.60. The relationship between runaway truck accident rate and average slope of the cumulative degree of curvature plot is shown in Figure 7. The plot indicates a trend toward increasing accident rate as the average slope increases.

An average slope over the entire downgrade might not be as significant as the variation in horizontal curvature at specific subsections of the downgrades. To verify this idea, an additional analysis was undertaken. Collision diagrams were plotted for all 15 downgrades. Any location having more than one runaway truck accident during the three year study period was arbitrarily defined as being "accident prone." These sites were examined in detail to determine any common characteristics. Variables examined included total number of curves traversed prior to the accident, total degree of curve negotiated, distance from adjacent upstream curve, and degree of adjacent upstream curve. Overall, there were 12 accident prone sites, accounting for 30 of the 105 runaway truck accidents.

Ten of the 12 accidents were on horizontal curves. The plots indicated that 10 of the accident prone locations occurred just downgrade from sections of increasing horizontal curvature. A possible reason for this is that the increasingly sharp horizontal curvature requires vehicles to make frequent brake applications. Thus, brakes are in degraded condition when the sharp horizontal curve is encountered. The sharp curve requires severe braking; the brake system cannot meet the demands imposed and failure results. Unable to reduce vehicle speed to negotiate the curve, the driver loses control of the vehicle and either runs off the road or strikes another vehicle.

Although a quantitative evaluation was not performed, examination of the plots tended to verify the aforementioned reasoning. For example, the plots for two downgrades were sharply concave upward near the bottom of the grade. It was here that most accidents occurred. Another downgrade was over 6 miles long. Because this is a relatively long downgrade for West Virginia, it was expected to have a high accident rate; however, the downgrade had a low runaway vehicle accident rate. This can be partially explained by the plot which, except for a short section which was concave upward, had a generally smooth convex upward slope.

AVAILABLE COUNTERMEASURES

A wide variety of countermeasures to prevent or reduce the severity of downgrade truck accidents has been developed and utilized. These can generally be broken down into three major categories which are described here in order of increasing costs: signing, brake check areas, and truck escape ramps. Each of the countermeasures will be discussed in more detail below.

Signing

One of the simplest yet most important ways to assist truckers on steep descending grades is the effective use of warning and informational signing. Both research and accident analyses (Carrier and Pachuta, 1981; Eck, 1980; Eck and Lechok, 1980; and Johnson, DiMarco, and Allen, 1982) have indicated that driver inexperience and unfamiliarity with the route are directly related to major downgrade truck accidents. The MUTCD (Federal Highway Administration, 1978) includes a warning sign to provide some of this information. The Hill sign (W7-1), shown in Figure 8, is intended for use in advance of a downgrade where the length, percent grade, horizontal curvature, or other physical features require special precautions on the part of drivers. When percent grade is shown within the Hill sign, this message shall be placed below the inclined ramp/truck symbol. The MUTCD states that the word message HILL may be used as an alternate legend.

The Hill and Grade signs should be used in advance of downgrades for the following conditions:

- 5% grade and more than 3,000 feet long
- 6% grade and more than 2,000 feet long
- 7% grade and more than 1,000 feet long
- 8% grade and more than 750 feet long
- 9% grade and more than 500 feet long

(a) HILL SIGN

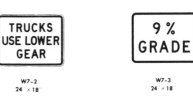

(b) SUPPLEMENTAL PLAQUES

FIGURE 8. Downgrade warning sign with supplemental plates as shown in MUTCD (Source: Federal Highway Administration, 1978).

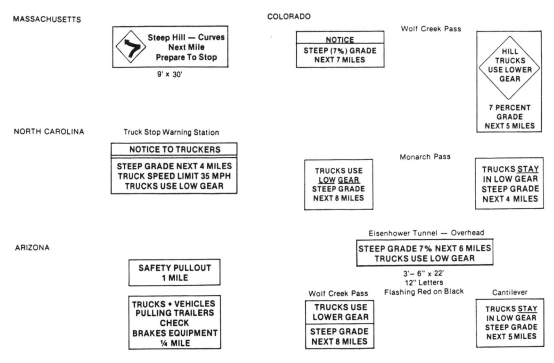

FIGURE 9. Examples of special downgrade warning signs (Source: Johnson, DiMarco, and Allen, 1982).

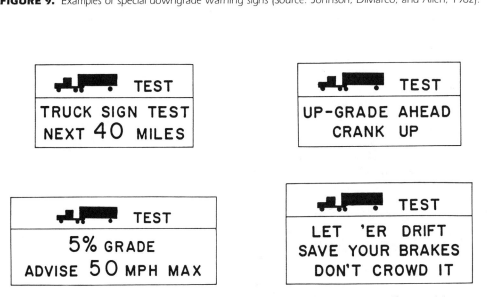

FIGURE 10. Examples of downgrade signing used in Donner Pass experiment (Source: Johnson, DiMarco, and Allen, 1982).

These signs should also be installed for steeper grades or where accident experience and field observations indicate a need.

The supplemental plaques shown in Figure 8 (W7-2 and W7-3) or other appropriate legends and larger signs should be used for emphasis or where special hill characteristics exist. On longer grades, the use of the mileage plaque at periodic intervals of approximately one mile spacing should be considered.

Wagner (1976) presented a survey of practice of selected states in implementing the MUTCD signing requirements on long, steep downgrades, based on driver needs and a human factors evaluation of typical driver information requirements. State experiences with signing strategies for downgrade trucks were reviewed. Primary techniques included providing drivers with a more thorough description of the (a) length and percent of downgrade, (b) need for the use of low gears, (c) nature and severity of the downgrade ahead, and (d) selection of alternative routes. Several of these techniques are shown in Figure 9. Flashing lights are also used to increase the effectiveness of these signs.

Johnson, DiMarco, and Allen (1982) described an innovative approach to downgrade signing that was tested on Interstate 80 between Donner Pass and Colfax, California. The 40-mile section of roadway contained several of the characteristics typical of problem downgrades across the country. State officials, in cooperation with the California Trucking Association, erected 27 signs along the route in order to "talk down" drivers in the same general manner that air traffic controllers bring aircraft into an airport. In addition to isolated circumstances along the grade, an important concern was that unfamiliar drivers over-utilized their brakes near the top of the hill, where grades are not severe. By the time the steeper grades were reached in the last 10 miles, some trucks did not have any brakes left. The "total concept" signing plan was developed so that drivers would feel an experienced expert was in the cab telling them how to drive each section of the road. The signs, examples of which are shown in Figure 10, are expressed in distinctive trucker language.

Brake Check Areas

Some downgrades have potentially hazardous conditions such as a stop condition, railroad grade crossing, sharp curvature, or a community that may not be readily apparent to the unfamiliar driver. A truck turnout at the crest of the hill and a special trucker information diagrammatic sign may be necessary for these situations. Such turnouts, which may be referred to as safety pullouts, brake check areas, and brake test areas provide drivers with the opportunity to check the vehicle's brakes and other equipment prior to descending the grade. Perhaps more importantly, the truck begins the descent in a low gear which makes it easier for the driver to control downgrade speed.

Johnson, DiMarco, and Allen (1982) noted that the method by which states approach the testing or checking of brakes varies. In some cases, informal areas such as scenic overlooks and rest stops are used by truckers to inspect their vehicles. Elsewhere, such as in Parley's Canyon, Utah, an extended shoulder is designated as a point where trucks can pull out of the traffic flow in safety, while the vehicle's brakes are checked by the driver. Many such voluntary brake check areas have been constructed at or near the summits of downgrades throughout the country.

Mandatory brake check areas can take several forms according to Johnson, DiMarco, and Allen (1982). In California, brake inspections by State Highway Patrol officers are conducted on a random basis at truck weighing stations. In other states, for example, New York, Pennsylvania, and West Virginia, mandatory turnouts are located at the summits of several downgrades where all trucks are required to stop. The testing of brakes is usually accomplished by the fact that a truck is able to stop at a line controlled by a stop sign. The thinking is that if a truck can stop, the brakes have been tested. A number of states, including New York, Pennsylvania, and Idaho, have erected informational signing, which shows road alignment on the downgrade, percent grade, and length of downgrade section ahead, at these turnouts. An example of this type of signing is shown in Figure 11.

Truck Escape Ramps

The most positive countermeasure for the reduction of truck accidents on long, steep downgrades is the truck escape ramp. While the countermeasures just discussed at-

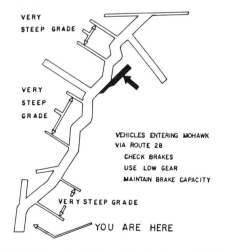

FIGURE 11. Example of information signing erected at brake check area, New York Route 28 at Vickerman Hill (Source: Johnson, DiMarco, and Allen, 1982).

tempt to warn and assist truckers in negotiating the entire downgrade, escape ramps are a last resort to reduce the severity of accidents which do occur.

The types of escape ramps were outlined earlier in this chapter. All escape ramps should include the considerations noted by Versteeg and Krohn (1977):

1. *Length* of ramp is dependent on ramp geometry and aggregate utilized.
2. *Width* of ramp should be sufficient to accommodate more than one vehicle.
3. *Aggregate* utilized for the ramp should be free-draining and clean (e.g., pea gravel).
4. *Anchors* (deadmen) are required to secure tow trucks when removing vehicles from the bed.
5. *Surfaced road* is needed adjacent to the ramp for use by tow trucks and maintenance vehicles.
6. *Advance signing* is important so that drivers are aware of the ramp. Access to the ramp itself must be signed so that it will not be missed by drivers of runaway vehicles.
7. *Entrance* to ramp should be designed so that vehicles traveling at high speeds can enter safely. This includes provision so that all wheels on any axle (or both steering wheels) enter the arrester bed at the same time.

In 1979, the Institute of Transportation Engineers (ITE) established a committee (Technical Committee 5B-1, 1982) to develop a set of recommended practices for the design of truck escape ramps. The Committee's report is now being considered for adoption as a Recommended Practice of the Institute.

Johnson, DiMarco, and Allen (1982) concluded that the feasibility, type of design, and location of truck escape ramps are based primarily on engineering judgment. They emphasize that escape ramps have not been constructed randomly; however, the current practice and experience of agencies is due to the fact that each situation presents a unique set of design requirements, depending in part on the following factors:

- nature of terrain along the route
- degree of slope and roadway alignment
- availability of sites adjacent to the highway
- environmental impact
- logical site distance below the summit
- maximum potential speed of runaway trucks

The four basic types of escape ramps were shown in Figure 1. Johnson, DiMarco, and Allen (1982) noted that each of these designs has been effective in reducing the severity of truck accidents in specific situations. Although they differ in technique, each having advantages and disadvantages, each provides a specific solution to a specific set of problems. Much of the discussion which follows comes from Ballard (1983); other excellent overviews of escape ramp technology have been prepared by Bullinger (1980) and Erickson (1980).

An ascending grade ramp consists of a ramp with a positive grade that has a bed of arresting material (usually sand or gravel). The arresting material and gravity contribute to the deceleration of a vehicle that enters the ramp. Ascending grade ramps exist with lengths from 330 to 1,560 feet long. Baldwin (1974) has given a detailed description of the design, construction, and surveillance of an ascending grade ramp in Utah.

A special case of the ascending grade ramp is the gravity ramp. A gravity ramp consists of a hard-surfaced lane that is on an ascending grade that may or may not have a small aggregate bed near the top. Vehicles that enter gravity ramps are decelerated primarily by the force that results from gravity acting opposite the direction of movement. The purpose of the bed is not to contribute significantly to the deceleration of the vehicle but to keep the vehicle in place once it has stopped. If an aggregate bed is not present, there is the possibility that a truck may roll backward and jackknife. Gravity ramps are typically longer than other ramp types, on the order of 1,200 to 1,600 feet in length.

Horizontal grade escape ramps are arrester beds that are approximately level. In these ramps, the deceleration of vehicles is the result of rolling resistance provided by the aggregate.

Descending grade ramps are facilities on which the vehicle is decelerated by the arresting material. The resistance provided by this material must also counteract the effect of the descending grade. These ramps will generally be longer than the other types discussed due to the adverse effect of negative grade. Descending grade ramps are typically used as roadside arrester beds. A roadside arrester bed is parallel and adjacent to the mainline and has provisions whereby a vehicle may enter from the side as well as the upstream end of the arrester bed. Allison, Hahn, and Bryden (1979) and Young (1979) described the results of field testing conducted to evaluate ramp effectiveness.

Sandpiles are masses of arresting material placed on the roadside such that the top surface is approximately level or at a slightly ascending grade. The surface of the sandpile may or may not be covered with transverse ridges. When a vehicle enters a sandpile escape ramp, the arresting material increases rolling resistance against the tires and, if the vehicle sinks in the sand far enough, against the undercarriage. Sandpiles are the shortest of all truck escape ramps; they are usually less than 400 feet long. Brittle (1977), Crowe (1977), and Rhudy (1978) described specific applications of sandpile ramps.

In designing truck escape ramps, regardless of the type, the length should be determined by analytical techniques. A design formula reported by FHWA (1979) is given in Equation (1):

$$L = \frac{V_i^2 - V_f^2}{30 \, (R + G)} \qquad (1)$$

where

L = length of ramp (ft)
V_i = velocity at beginning of ramp (mph)
V_f = velocity at end of ramp (mph)
R = rolling resistance (divided by 100) expressed as an equivalent percent grade
G = percent grade divided by 100

The suggested values for rolling resistance are 0.15 and 0.25 for sand and pea gravel, respectively.

The material used in arrester beds is independent of the grade of the ramp; that is, ascending grade, descending grade, and horizontal grade ramps all use approximately the same aggregate types. The most common aggregates are pea gravel, although angular aggregates have been used where rounded gravel is not available at reasonable cost. Pea gravel is the most desirable because of the high percentage of voids, which provides better drainage than angular aggregate.

Signing on the approaches to and along truck escape ramps should follow the provisions of the *Manual on Uniform Traffic Control Devices* (Federal Highway Administration, 1978). The Manual indicates that the signing, shown in Figure 12, shall be black on yellow with the message "Runaway Truck Ramp." A supplemental panel may be used with the words "Sand," "Gravel," or "Paved" to describe the ramp surface. These advance warning signs should be located in advance of the gore approximately one mile, one-half mile, and then one at the gore. A regulatory sign near the entrance should be used containing the message "Runaway Vehicles Only" to discourage other motorists from entering the ramp. "No Parking" signs may be placed as required near the ramp entrance.

To indicate the alignment of the ramp, delineators (light-retroreflecting devices) may be used. When used, delineators should be red in color and should normally be placed on both sides of truck escape ramps. The delineators should be spaced at 50-foot intervals for a distance sufficient to identify the ramp entrance. Delineator spacing beyond the ramp entrance should be adequate for guidance in accordance with the length and design of the escape ramp.

All escape ramps having arrester beds require maintenance after each use. When a vehicle enters a ramp, its wheels create ruts in the aggregate. These ruts must be eliminated and the shape of the bed restored before the next vehicle enters the bed. Note that the arresting material must be replaced after it has accumulated too many fine particles. Frequency of aggregate replacement is currently not well defined and is an area where additional research is needed. Sandpiles and arrester beds may require a de-icing agent if the facility is in an area prone to freezing.

Ballard's (1983) review of escape ramps throughout the country found that a variety of bed depths, as well as depth tapers, are currently in use. Although beds only 6 inches

FIGURE 12. Signing for truck escape ramps, as shown in MUTCD (Source: Federal Highway Administration, 1978).

deep have been reported, most arrester beds are 18 to 24 inches deep. Research that defines the optimum depth of aggregate for various types of ramps and aggregates is lacking.

Several states have used variations on standard runaway vehicle escape facilities. Both Virginia and California have placed escape facilities at the bottom of interchange ramps. California also has a somewhat unique arrester bed design in the metropolitan Los Angeles area ("Roadside Bunkers . . . ," 1972). A typical escape ramp could not be used since the highway in question became a city street with homes on both sides of the route. To solve this problem, three gravel arrester beds were constructed in the center (median) of the roadway. An important advantage of the median arrester bed was that no additional right-of-way was required.

CONCLUDING REMARKS

This chapter has discussed a number of issues relative to highway safety in mountainous terrain. The lack, heretofore, of easy to apply general guidelines for identifying problem downgrades and locating countermeasures was noted. The large number of human, vehicular, and roadway factors which play a role in downgrade accident occurrence and their complex interactions have precluded development of a simple yet reliable technique for predicting downgrade accidents. Thus, most downgrade countermeasures are installed on the basis of judgment rather than formal engineering analysis.

Data indicate that lack of driver familiarity with a route is a contributing factor in runaway vehicle accidents. This demonstrates the importance of truck drivers receiving adequate information about length, steepness, and horizontal alignment of unfamiliar grades in order that they can select the proper gear to avoid overloading the brake system.

Plots of cumulative degree of curve versus cumulative distance from crest of grade appear to be a useful tool in identifying problem downgrades. Concave upward curves are good predictors of high runaway truck accident rates. Irregular curves with frequent and/or sharp discontinuities can be used to identify runaway truck accident locations since trucks usually encounter problems at points of sharp horizontal curvature. Such points stand out dramatically on the plots. Conversely, concave downward curves, or smooth curves with few discontinuities, generally have a smaller number of runaway truck accidents. Although they do not provide quantitative accident prediction, the plots alert engineers to downgrades (and locations along a downgrade) where runaway truck accidents would be expected to occur frequently. Additional, more detailed study could be performed to evaluate countermeasures for the problem downgrades identified by the technique.

Perhaps the main advantage of the technique is its simplicity. When highway plans and drawings are available, the plot can be constructed in the office by plotting the summation of degree of curve versus distance from the summit of the grade. Where plans are not available, a stringlining technique, such as that described by Hickerson (1964), can be used in the field to estimate degree of curve and a plot can be prepared from these data.

Although a variety of countermeasures exist for the downgrade accident problem, perhaps the most effective are truck escape ramps. Several types of ramps are used, the specific type for a given application depending primarily on site conditions.

A multi-stage decision process is recommended for those downgrades where ramp feasibility has been established. The first step would be to determine accident prone locations. This could be accomplished in several ways: (1) utilize existing accident records system, (2) utilize cumulative degree of curvature plot, and (3) combination of (1) and (2). Engineering judgment must then be used to predict the economic benefit of installing a ramp at an accident prone location in terms of estimating the reduction in runaway vehicle accidents. Computing the net present worth of the facility will indicate if an escape facility is still feasible. Where a ramp is shown not to be feasible, the engineer selects another location and repeats the procedure. In those cases where no ramp location is feasible, other runaway vehicle countermeasures should be investigated.

Important characteristics of escape ramps include physical dimensions, gradient, arresting material, drainage, signing, delineation, and maintenance practices. The individual truck escape ramps in the United States illustrate a wide variety of approaches to the characteristics just listed. As Ballard (1983) noted, this may indicate that optimum designs have not yet been identified and that there is still opportunity for further advancement in the whole area of highway downgrade countermeasures.

REFERENCES

1. Allison, J. R., K. C. Hahn and J. E. Bryden, "Performance of A Gravel-Bed Truck-Arrester System," *Transportation Research Record 736*, pp. 43–47 (1979).
2. American Association of State Highway and Transportation Officials, *A Policy on Geometric Design of Highways and Streets*, American Association of State Highway and Transportation Officials, Washington, D.C., pp. 293–303 (1984).
3. Baldwin, G. S., "Truck Escape Lanes on Mountain Roads," *Civil Engineering*, 44 (7), pp. 64–65 (July 1974).
4. Ballard, A. J., "Current State of Truck Escape Ramp Technology," *Transportation Research Record 923*, pp. 35–42 (1983).
5. Brittle, W. J., Jr., "Truck Escape Ramps and Sandpiles in Virginia," paper presented at the Southeastern Association of State Highway and Transportation Officials Meeting, Asheville, North Carolina, 8 pages (1977).
6. Bullinger, M. J., *Truck Escape Ramps: Operating Experience and Design Considerations*, Department of Civil Engineering, Iowa State University, Ames, Iowa (October 1980).
7. Carrier, R. E. and J. A. Pachuta, "Runaway Trucks in Pennsylvania," *SAE Technical Paper 811262*, Society of Automotive Engineers, Warrendale, Pennsylvania, 10 pages (1981).
8. Crowe, N. C., Jr., *Photographic Surveillance—A Study of Runaway Truck Escape Ramps in North Carolina*, Traffic Engineering Branch, North Carolina Department of Transportation, Raleigh, North Carolina, 9 pages and appendix (September 1977).
9. Eck, R. W., *Development of Warrants for the Use and Location of Truck Escape Ramps*, Department of Civil Engineering, West Virginia University, Morgantown, West Virginia, 532 pages (February 1980).
10. Eck, R. W., "State Practice and Experience in the Use and Location of Truck Escape Facilities," *Highway Research Record 736*, pp. 37–42 (1979).
11. Eck, R. W., "Technique for Identifying Problem Downgrades," Technical Note, *Journal of Transportation Engineering*, 109 (4), pp. 604–610 (July 1983).
12. Eck, R. W. and S. A. Lechok, "Truck Drivers' Perceptions of Mountain Driving Problems," *Transportation Research Record 753*, pp. 14–21 (1980).
13. Erickson, R. C., Jr., "A History of Runaway Truck Escape Ramps in Colorado," paper presented at the Joint Meeting of the Intermountain and Colorado-Wyoming Sections of ITE, Jackson, Wyoming, 28 pages (May 16–17, 1980).
14. Federal Highway Administration, "Interim Guidelines for Design of Emergency Escape Ramps," FHWA Technical Advisory T 5040.10, Washington, D.C. (July 5, 1979).
15. Federal Highway Administration, *Manual on Uniform Traffic Control Devices for Streets and Highways*, U.S. Department of Transportation, Washington, D.C. (1978).
16. Hickerson, T. F., "String-Lining Railroad Curves," in *Route Location and Design*, 5th Edition, McGraw-Hill Book Company, New York, pp. 346–349 (1964).
17. Johnson, W. A., R. J. DiMarco and R. W. Allen, *The Development and Evaluation of a Prototype Grade Severity Rating System*, FHWA/RD-81/185, Federal Highway Administration, Washington, D.C., 184 pages (March 1982).
18. Rhudy, H. C., "Truck Sandpiles," paper presented to Subcom-

mittee on Traffic Engineering, American Association of State Highway and Transportation Officials, Scottsdale, Arizona (June 11–14, 1978).
19. "Roadside Bunkers to Trap Runaway Vehicles," *Public Works*, *103* (7), p. 55 (July 1972).
20. Technical Committee 5B-1, "Proposed Recommended Practice—Truck Escape Ramps," *ITE Journal*, *52* (2), pp. 16–17 (February 1982).
21. Versteeg, J. H. and M. Krohn, "Truck Escape Ramps," report presented at the Western Association of State Highway and Transportation Officials Meeting, Colorado Springs, Colorado, 20 pages (June 1977).
22. Wagner, D. C., "Signing for Long, Steep Downgrades," paper presented to the AASHTO Operating Subcommittee on Traffic Engineering, Kansas City, Missouri, 21 pages (June 29, 1976).
23. Williams, E. C., Jr., *Emergency Escape Ramps for Runaway Heavy Vehicles*, FHWA-TS-79-201, Federal Highway Administration, Washington, D.C., 68 pages (March 1979).
24. Young, J., "Field Testing A Truck Escape Ramp," *Highway Focus*, *11* (3), pp. 43–55 (September 1979).

CHAPTER 22

Transportation Planning and Network Geometry: Optimal Layout for Branching Distribution Networks

PRAMOD R. BHAVE* AND CHAN F. LAM**

TRANSPORTATION PLANNING

A modern society depends on several types of networks for communication of messages and signals, and transportation and distribution of gas, water, materials, energy, etc. Some of these networks are highway networks, urban transportation networks, telephone networks, gas pipeline networks and water distribution networks. The networks include connected lines and points. The lines may be roads, power lines, telephone wires, railroad tracks, airline routes and the gas or water mains. The points may be highway intersections, power stations, telephone exchanges, railroad yards, airline terminals and the gas depots or the water reservoirs; in general the points where a flow originates, is relayed or terminates. Each one of these networks has its own characteristics and forms a field of study by itself. We will, however, briefly discuss the planning of public transportation in general and of water distribution in particular.

Public transportation, including highways and urban roadways, is a very complex system. Short term and long term planning are needed, the former one being less complex. The short term planning, in general, is concerned with obtaining maximum capacity or optimal operation from existing facilities. Because of the huge financial expenditures and large construction programs, long term planning should rely on a systems approach, which is composed of systems analysis and systems engineering [1]. Systems analysis is an evaluation of all the components and subcomponents of the whole system and the forces or inputs to the system to achieve the desired outputs or objectives. Systems engineering is the organizing and scheduling of strategies for the solution of the problems. For detailed planning procedures, one can refer to Paquette, et al. [1]; Stopher and Meyberg [2]; and Ashford and Clark [3].

The planning process of public transportation systems can be used in the planning of water distribution systems as well. In addition to the study of the population growth pattern, the demand of the quantity of water, and the topology of the area such as hills or other physical barriers, one needs to consider the water pressures or the hydraulic gradient levels at the demand locations. The size and length of the pipes in a water distribution system dictates its cost and also the quantity (amount) and the quality (water pressure) of the supplied water. Various design and optimization procedures, which help in planning the layout of transportation systems in general and water distribution systems in particular, are considered in this chapter.

NETWORK GEOMETRY

A large number of transportation systems, including water distribution systems, are studied, planned and designed through networks represented by linear graphs, or simply graphs. Graph theory [4–6] is a mathematical discipline for the analysis of such systems.

Terminology

For ease of presentation some terminology from graph theory is introduced herein and explained with the help of Figure 1. Wherever necessary, the terminology is slightly modified to suit water distribution networks.

A *linear graph* or simply a *graph* is a graphical representation of a network and consists of links (also called elements, arcs or edges) and nodes (also called vertices).

---

*Department of Civil Engineering, Visvesvaraya Regional College of Engineering, Nagpur 440 011, India
**Department of Biometry, Medical University of South Carolina, Charleston, SC

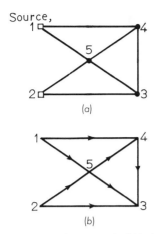

FIGURE 1. (a) Distribution network; (b) its linear graph.

Figure 1(a) shows a water distribution network, represented by a graph in Figure 1(b). A *link* is a line segment in the system graph, whereas a node is an end point of a link. In Figure 1(b), 1, 2, etc., are the nodes and 14 (to be read as one-four), 15, 54, etc., are the links. In a *directed graph* the links have arrows indicating direction of flow, whereas the direction is not indicated in an *undirected graph*. A *source node* or simply a *source* is a node with all connected links having outward flow (nodes 1 and 2 in Figure 1). An *intermediate node* has some connected links with inward flow and some with outward flow (nodes 4 and 5); a *sink node* or simply a *sink* has all connected links with inward flow (node 3). The *degree of a node* is the number of links incident at the node. (In Figure 1, the degree of nodes 1 and 2 is 2, of nodes 3 and 4 is 3, and of node 5 is 4.) A *loop*, also called a *circuit*, is a subgraph that is connected such that every node of this subgraph has a degree 2. (In Figure 1, links 15, 54, and 41 constitute a loop.) By following the links of a loop one can get back to whichever node one begins with. A *tree*, also known as a *spanning tree*, is a connected subgraph that contains all the nodes of a graph but has no loop (Links 15, 25, 53, and 54 give a tree).

A graph is simply a graphical representation, and, therefore, it is not possible to tell from the graph whether it represents an electrical, a highway or a water distribution network. Similarly, the links and the nodes are simply the lines and points, respectively. For ease of presentation and interpretation we shall show the graphs in which the source nodes will be shown by squares and the intermediate nodes and sinks by circles.

Number of Trees

The number of trees available for a distribution network can be obtained with the help of graph theory [4,7]. Let A be a matrix such that its diagonal elements (a_{ij} with $i = j$) represent the number of links meeting at node i (the degree of the node i), whereas the nondiagonal elements (a_{ij} with $i \neq j$) represent the negative number of links connecting node i with node j. (In a distribution network, two nodes, if directly connected, are usually connected with one link only. Therefore, a_{ij} ($i \neq j$) $= -1$ if the nodes are connected, or $= 0$ if they are not connected.)

It is observed that A is a particular type of square matrix whose all row sums and column sums are zero. Therefore, according to the matrix-tree theorem from the graph theory [4], the cofactors of all the elements a_{ij} of matrix A are equal, and their common value gives the number of trees for the network. (The cofactor of an element a_{ij}, i.e., CF_{ij} in matrix A is $(-1)^{i+j}$ times the determinant obtained by deleting the ith row and the jth column from A.)

For the 5-node, 6-link distribution network of Figure 2(a), the matrix A is given by

$$A = \begin{array}{c} \text{Node} \\ 1 \\ 2 \\ 3 \\ 4 \\ 5 \end{array} \begin{array}{c} \begin{array}{ccccc} 1 & 2 & 3 & 4 & 5 \end{array} \\ \begin{bmatrix} 3 & -1 & -1 & 0 & -1 \\ -1 & 2 & 0 & -1 & 0 \\ -1 & 0 & 2 & -1 & 0 \\ 0 & -1 & -1 & 3 & -1 \\ -1 & 0 & 0 & -1 & 2 \end{bmatrix} \end{array} \quad (1)$$

The number of trees T given by the cofactor of any element, say a_{23} is given by

$$T = CF_{23} = (-1)^{2+3} \begin{vmatrix} 3 & -1 & 0 & -1 \\ -1 & 0 & -1 & 0 \\ 0 & -1 & 3 & -1 \\ -1 & 0 & -1 & 2 \end{vmatrix} = 12 \quad (2)$$

List of Trees

The listing of all the trees of a small network such as in Figure 2(a) can easily be done by inspection. As the network in Figure 2(a) has 5 nodes, the number of links in a tree must be $5 - 1$, i.e., 4 out of the total 6 links. Therefore, showing a loop-forming and therefore non-permissible 4-link combination within parentheses, for the network of Figure 2(a), the 12 trees obtained by inspection are (1234), 1235, 1236, 1245, 1246, 1256, 1345, 1346, 1356, (1456), 2345, 2346, (2356), 2456 and 3456. All these 12 trees are shown in Figure 2(b).

Several algorithms are available for systematically listing all the trees of a network. An algorithm developed by Maxwell and Kline [8] is herein described. This algorithm incorporates the principles of alphanumeric multiplication and Wang algebra [9]. Alphanumeric multiplication of 2 and 4 is 24 (read as two-four) rather than 8. In Wang algebra, the sum or product of two or more identical constants (or sym-

TRANSPORTATION PLANNING AND NETWORK GEOMETRY: OPTIMAL LAYOUT FOR BRANCHING DISTRIBUTION NETWORKS   475

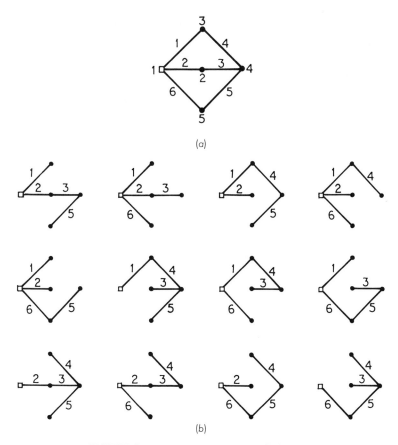

FIGURE 2. (a) Distribution network; (b) its trees.

bols) is zero, i.e.,

$$\sum_{i=1}^{N} C_i = 0 \quad \text{for } n > 1$$

$$\prod_{i=1}^{N} C_i = 0 \quad \text{for } n > 1$$

(3)

For example, carrying out alphanumeric multiplication and using Wang algebra, $(C_1, C_2)(C_1, C_2, C_3) = C_1C_1 + C_1C_2 + C_1C_3 + C_2C_1 + C_2C_2 + C_2C_3 = C_1C_3 + C_2C_3$.

The terms C_1C_1 and C_2C_2 are discarded because of the product principle and the terms C_1C_2 and C_2C_1 are discarded because of the addition principle of the Wang algebra. The algorithm, based on these principles, for the generation of all the trees is as follows:

1. Inscribe circles about all but one nodes of the graph. (The results are independent of the one node being ignored.)
2. For each circled node, list the branches cut by the circles.
3. Multiply alphanumerically the listing obtained in step 2, applying the principles of Wang algebra during all the steps of this multiplication.
4. The remaining terms give the listing of the trees.

For example, the network of Figure 2(a) is shown in Figure 3 in which node 2 is ignored and circles are drawn around nodes 1, 3, 4 and 5 (step 1). The cut branches (step 2) are 1, 2 and 6 for node 1; 1 and 4 for node 3; 3, 4 and 5 for node 4; and 5 and 6 for node 5. Multiplying alphanumerically and applying the principles of Wang algebra (discarded terms are shown within parentheses) yields (step 3): (1, 2, 6)(1, 4) = (11), 12, 16, 14, 24, 46; (12, 16, 14, 24, 46)(3, 4, 5) = 123, 136, 134, 234, 346; 124, 146, (144), (244), (446); 125, 156, 145, 245, 456; (123, 136, 134, 234, 346, 124, 146, 125, 156, 145, 245, 456)(5, 6) = 1235, 1356,

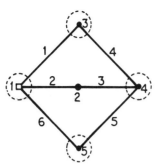

FIGURE 3. Distribution network with circles around all nodes but one.

1345, 2345, 3456, 1245, (1456), (1255), (1556), (1455), (2455), (4556); 1236, (1366), 1346, 2346, (3466), 1246, (1466), 1256, (1566), (1456), 2456, (4566).

Thus, the listing of the trees is 1235, 1356, 1345, 2345, 3456, 1245, 1236, 1346, 2346, 1246, 1256, 2456, which are the 12 trees shown in Figure 2(b).

It is, however, difficult to computerize this technique because the digital computer generally does not handle alphanumeric operations. Utilizing prime numbers, a completely algebraic method of enlisting all the trees of a graph, has been proposed by Henry et al. [10]. Herein the real numbers are replaced by the corresponding prime numbers. Thus, the real numbers 1, 2, 3, 4, 5, 6, etc., are replaced by the corresponding prime numbers 2, 3, 5, 7, 11, 13, etc., and the link numbers are denoted by these prime numbers. Algebraic operations are then carried out instead of the alphanumeric operations, and the Wang algebra principles are used.

For the network of Figure 3, using the corresponding prime numbers, the cut branches (1, 2, 6) (1, 4)(3, 4, 5)(5, 6) are changed to (2, 3, 13), (2, 7), (5, 7, 11)(11, 13). Therefore, (2, 3, 13)(2, 7)(5, 7, 11)(11, 13) = (4, 6, 26, 14, 21, 91)(5, 7, 11) (11, 13) = (20, 30, 130, 70, 105, 455; 28, 42, 182, 98, 147, 637; 44, 66, 286, 154, 231, 1001)(11, 13) = 220, 330, 1430, 770, 1155, 5005, 308, 462, 2002, 1078, 1617, 7007, 484, 726, 3146, 1694, 2541, 11011; 260, 390, 1690, 910, 1365, 5915, 364, 546, 2366, 1274, 1911, 8281, 572, 858, 3718, 2002, 3003, 13013.

Terms having the same values are deleted due to the summation principle of Wang algebra. Thus, the two 2002 terms are deleted. The remaining terms are then factorized into prime numbers and the terms containing duplicate prime numbers are deleted due to the product principle of Wang algebra. For example, $220 = 2 \times 2 \times 5 \times 11$ and is therefore deleted because of the duplication of prime number 2. However, $330 = 2 \times 3 \times 5 \times 11$, which constitutes a tree 1235.

Lam [11] has given a computer program for listing all the trees of a network.

WATER DISTRIBUTION NETWORKS

Water distribution networks consist of sources, demand nodes and links, which are the commercially available pipes. The networks are branching or tree-like for small communities and rural areas, but are looped for large communities and urban areas.

Head Loss and Cost Functions

During the planning and designing of water distribution systems, it is common to assume that the link diameter is a continuous variable [12–16] and that the unit capital cost of a link varies nonlinearly with its diameter [13,15,16]. Thus,

$$C = K_1 L D^m \quad (4)$$

in which C = capital cost of a link of length L and diameter D, K_1 = a link cost constant, which depends upon the units of C, L, and D, the pipe class and the geographic location; and m = an exponent which generally lies between 1 and 2.

The general head-loss relationship for a link can be expressed as

$$H_L = \frac{K_2 L Q^p}{D^r} \quad (5)$$

in which H_L = head loss through a link of length L and diameter D while carrying a discharge Q; K_2 = a coefficient that depends upon the pipe material, diameter and pipe conditions, type of flow, and the units of other terms; p = an exponent lying between 1.7 and 2; and r = an exponent lying between 4.7 and 5.

Combining Equations (4) and (5) and eliminating D,

$$C = K_1 K_2^{m/r} L^{1+m/r} Q^{pm/r} H_L^{-m/r} \quad (6)$$

or

$$C = K L^{1+m/r} Q^{pm/r} H_L^{-m/r} \quad (7)$$

in which $K = K_1 K_2^{m/r}$

Two-Link Loop

Consider a simple two-link looped network shown in Figure 4. Let A be the source and B the demand node, connected by two links 1 and 2. Let the demand at B be Q and the discharges in the links 1 and 2 be Q_1 and Q_2, respectively, so that $Q_1 + Q_2 = Q$. Now, as the exponent of Q in Equation (7), i.e., pm/r, is almost always less than 1, C is concave with respect to Q. Therefore, the cost of the looped network of Figure 4 is minimum when the entire demand Q at node B is transported either through link 1 or through link 2; i.e., $Q_1 = Q$ and $Q_2 = 0$ or $Q_1 = 0$ and $Q_2 = Q$ [7].

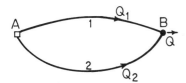

FIGURE 4. Two-link loop.

Thus, from Equation 7, the link having the minimum $KL^{1+m/r}$ value will yield minimum-cost network.

If K values are the same for both the links, as is usually assumed in network planning, the entire flow should be transported through the shorter link for the network to be optimal. Such a link, which would carry the entire (or most of the) discharge, is termed *primary link* herein. The other link, which is required to maintain the geometry of the looped network, is termed *nonprimary* link. It is generally of the minimum or some specified diameter (D-specified condition), or it carries the minimum or some specified discharge (Q-specified condition).

Multiple-Link Loop

Consider a multiple-link loop connecting N nodes as shown in Figure 5(a). The number of links is N. It can be seen that $N - 1$ links are necessary to connect all the nodes, and, therefore, in this network $N - 1$ links will be the primary links and only one link will be nonprimary link. The combination consisting of the primary links only is a spanning tree termed *distribution tree* herein for water distribution networks. For the network of Figure 5(a), there will be N distribution trees, two of which are shown in Figure 5(b) and Figure 5(c), respectively.

Multiple-Looped Network

Consider a multiple-looped network as shown in Figure 6. Let the network have N nodes (labeled $j, j = 1 \ldots N$), C circuits or loops (labeled $c, c = 1 \ldots C$) and E elements or links (labeled $e, e = 1 \ldots E$). According to the Euler's theorem on connected graphs [4], the number of links E can be expressed as

$$E = N + C - 1 \qquad (8)$$

(For the network of Figure 6, $E = 17$; $N = 12$; and $C = 6$).

Out of these E links, $N - 1$ links will be the primary links and the remaining C links will be the nonprimary ones. Each combination of primary links will constitute a distribution tree. Each network corresponding to a distribution tree and the accompanying nonprimary links can be optimally designed, but the solution will be a local optimum solution. Usually only one of all these local optimum solutions is the global optimum solution. As of 1985 no method is available that can identify the distribution tree leading to the global optimum solution. As shown earlier, it is possible to find out the number of all the distribution trees and also enlist them. Therefore, even though it is theoretically possible to obtain the global optimum solution by comparing all the local optimum solutions, each corresponding to a different distribution tree, the total number of distribution trees would be too large for practical networks to allow the use of this approach. Thus, obtaining the global optimum solution has remained elusive in practice. However, different approaches have been suggested to select, a priori, a branching layout which would lead to a fairly good but still local optimum solution.

SELECTION OF BRANCHING LAYOUT BY SIMPLE METHODS

Using Equation (7), the total capital cost C_T of a distribution network having E links can be expressed as

$$C_T = \sum_{e=1}^{E} (K L^{1+m/r} Q^{pm/r} H_L^{-m/r})_e \qquad (9)$$

Thus, the capital cost of a distribution network is deter-

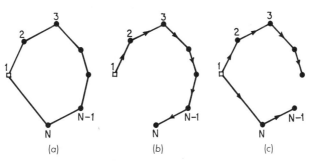

FIGURE 5. Multiple-link loop.

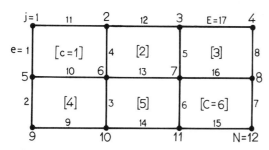

FIGURE 6. Multiple-looped network and its labeling.

mined by the combined effect of the length, discharge and available head loss in each link.

Minimal Spanning Tree

In Equation (9), considering the link length L to be the predominant parameter, it is reasonable to assume that the spanning tree, giving the minimum total length of the links (i.e., the minimal spanning tree) would give a layout having quite less, if not the minimum, cost of transportation. The algorithm for obtaining the minimal spanning tree, as sug-

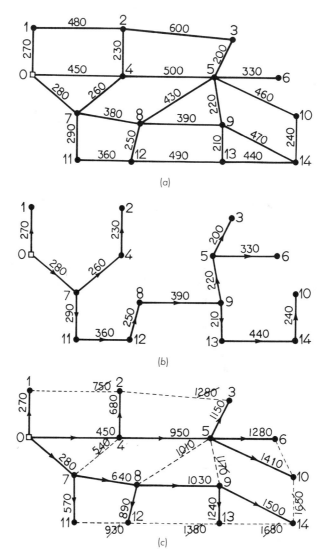

FIGURE 7. (a) Distribution network; (b) distribution tree according to minimum spanning tree concept; and (c) distribution tree according to path concept.

gested by Prim [17,18], is briefly as follows:

Starting at any node of the network, select from amongst the links connected to this node, the link with the least length. Call the combination of the link and the two end nodes as partial spanning tree. Next, from amongst all the links connected to the nodes of the partial spanning tree, select a link of minimum length connecting one of the remaining unconnected nodes. Continue this procedure until all the nodes are connected to the partial spanning tree. The resulting tree connecting all the nodes is the minimal spanning tree.

As an illustration, consider the single-source network of Figure 7(a), in which the source is labeled as 0 and the demand nodes are labeled 1 ... N. The number of links in the distribution tree will also be N. Let the link-lengths be as shown along the links in the figure. Following the above procedure, the minimal spanning tree is obtained for this network as shown in Figure 7(b).

Spanning Tree by Path Concept

In a looped network, water from the source can reach a demand node through several link combinations. It is therefore reasonable to assume that a distribution tree in which all demand nodes are connected to the source so that water has to travel minimum distance in reaching a node would also give a tree having quite less, if not the minimum, cost of transportation. This concept is termed *path concept* by Bhave [7,16,19]. The path concept and the algorithm to obtain the spanning tree based on path concept are now described.

SINGLE-SOURCE NETWORKS

In Figure 7(a), a link-combination starting from the source 0 and terminating at a demand node j ($j = 1 ... N$) is termed *route* to node j, i.e., R_j. Some of the routes from source 0 to demand node 9 are 0-7-8-9, 0-1-2-3-5-9, 0-7-11-12-13-14-10-5-9. Of these various routes, the shortest route is termed *path* P_j. Thus,

$$P_j = \text{shortest } R_j \quad (10)$$

$$L_{P_j} = \text{minimum } L_{R_j} \quad (11)$$

in which L_{P_j} = length of path P_j; and L_{R_j} = length of a route R_j.

The algorithm for determining the path-lengths and hence the path from source 0 to the various demand nodes, total number N, is given by

$$L_{P_j} = \text{minimum } (L_{P_i} + L_{ij}), j = 1 ... N \quad (12)$$

in which L_{P_i} = length of path P_i estimated earlier; and L_{ij} = length of link ij.

The path concept is applied to the network of Figure 7(a) as shown in Figure 7(c). The route lengths are shown near the respective nodes. The longer route lengths are cancelled and the minimum route length to each node, i.e., the path length, is retained. The distribution tree obtained by the path concept is shown by thick lines.

When two or more paths are available to a demand node (this happens in rectangular grid type distribution networks), the path in which the flow gets concentrated is selected because of the concave behavior of the flow in the link-cost function, Equation (7).

MULTIPLE-SOURCE NETWORKS

For multiple-source networks, assume for simplicity that the available hydraulic gradient level (HGL) values at the source nodes are the same. Further assume that the supplies at all the source nodes are unrestricted. (General case in which these assumptions do not hold good is considered later.) Herein, all the demand nodes are free to receive their requirements from any of the source nodes. Using the path concept, it is assumed that a demand node receives its requirement from the "nearest" source node.

To obtain the branching configuration for the entire network, obtain the distribution tree for each source node, superimpose them and then allot a node to the nearest source. In the network of Figure 8(a), nodes 1, 2, and 3 are

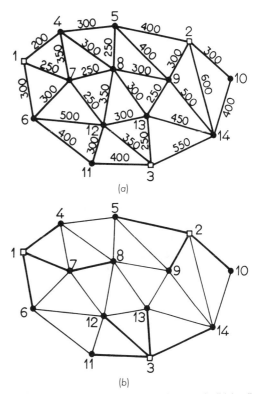

FIGURE 8. (a) Multiple-source looped network; (b) its distribution trees.

the source nodes and nodes 4–14 are the demand nodes. The branching configuration (shown by thick lines) consists of three distribution trees, one from each source as shown in Figure 8(b).

The branching layouts obtained by the minimum spanning tree concept and the path concept give fairly good layouts. However, because of the assumptions on which these concepts are based, the obtained layouts do not necessarily lead to the global optimum solutions.

OPTIMAL LAYOUT FROM LINEAR TRANSPORTATION MODEL

Selection of the optimal layout for a general multiple-source network is developed through the classical transportation problem principles in which the transportation of goods from the manufacturing plants to the warehouses is considered [20,21]. The methodology as developed by Bhave [22,23] is based on linear programming (LP) formulation.

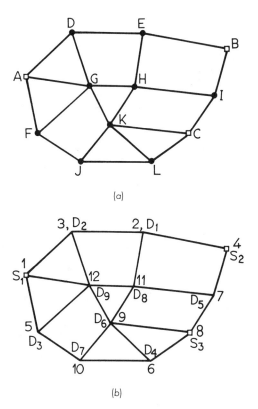

FIGURE 9. (a) Multiple-source looped network; (b) node labeling by CLP and SLP procedures.

Node-Labeling Procedure

Let a multiple source network contain M source nodes and N demand nodes (demand can be zero). These source and demand nodes are labeled consecutively from 1 to $M + N$ in the decreasing order of their hydraulic gradient level (HGL) values so that

$$H_j \geq H_{j+1}; j = 1 \ldots M + N - 1 \quad (13)$$

in which H_j = provided HGL value at the source nodes or the minimum required HGL value at the demand nodes. When some source and demand nodes have the same such HGL values, the demand nodes are labeled first and the source nodes thereafter. This labeling procedure is termed "combined labeling procedure" (CLP).

The source and demand nodes are also labeled separately as S_m, $m = 1 \ldots M$ and D_n, $n = 1 \ldots N$, respectively, in the increasing order of their CLP labels. This labeling procedure is termed "separate labeling procedure" (SLP).

Both these labeling procedures are illustrated through a 3 source-node (labeled A, B, and C), 9 demand-node (labeled D-L) network of Figure 9(a), as shown in Figure 9(b) and Table 1. Note that 1) the source and demand nodes are intermingled in the CLP; 2) always the node 1 in the CLP is S_1 in the SLP, and node $M + N$ in the CLP is node D_N in the SLP; and 3) the source node C and the demand node I have the same HGL value (170.00 m); and, therefore, demand node I is labeled earlier than source node C (7 and 8, respectively, in the CLP).

Design Feasibility Constraints

For planning and designing of water distribution networks, certain constraints termed "design feasibility constraints" must be satisfied. In the classical transportation problems involving transportation of goods from the manufacturing plants to the warehouses, only the network demand constraint must be satisfied for the design feasibility. In water distribution systems, however, the HGL values at all the nodes need be considered; and, therefore, the group demand constraints and the transportation constraints must also be considered in checking the network design feasibility.

NETWORK DEMAND CONSTRAINT

Consider the demand for the entire network:

$$\sum_{m=1}^{M} Q_{S_m} \geq \sum_{m=1}^{M} q_{S_m} + \sum_{n=1}^{N} q_{D_n} \quad (14)$$

in which Q_{S_m} (>0) = available supply at source node S_m; q_{S_m} (≥0) = demand at S_m; and q_{D_n} (≥0) = demand at D_n. This constraint, Equation (14), is "network demand con-

TABLE 1. Node Labeling by Combined and Separate Node Labeling Procedures, for Figure 9.

Node	Nature	HGL, (m)	HGL[a] arranged according to Eq. (13) (m)	Node	Nature	CLP	Node label according to SLP source node	Node label according to SLP demand node	Group source node	Group demand node
A	Source	200.00	200.00	A	Source	1	S_1	—	1	—
B	Source	180.00	190.00	E	Demand	2	—	D_1	—	1
C	Source	170.00	185.00	D	Demand	3	—	D_2	—	1
D	Demand	185.00	180.00	B	Source	4	S_2	—	2	—
E	Demand	190.00	175.00	F	Demand	5	—	D_3	—	2
F	Demand	175.00	172.00	L	Demand	6	—	D_4	—	2
G	Demand	155.00	170.00	I	Demand	7	—	D_5	—	2
H	Demand	158.00	170.00	C	Source	8	S_3	—	3	—
I	Demand	170.00	165.00	K	Demand	9	—	D_6	—	3
J	Demand	160.00	160.00	J	Demand	10	—	D_7	—	3
K	Demand	165.00	158.00	H	Demand	11	—	D_8	—	3
L	Demand	172.00	155.00	G	Demand	12	—	D_9	—	3

[a]Provided for the source nodes, and minimum required for the demand nodes.

straint" and ensures the total supply for the entire network to be at least equal to the total demand. This constraint is sufficient when

$$H_{S_m} > H_{D_n}^{min}; m = 1 \ldots M, n = 1 \ldots N \quad (15)$$

in which H_{S_m} = provided HGL at S_m; and $H_{D_n}^{min}$ = minimum required HGL at D_n.

GROUP DEMAND CONSTRAINTS

When Equation (15) is not satisfied (source and demand nodes are intermingled as in Table 1), not all the demand nodes are free to receive their requirements from all the source nodes. For example, in the network of Figure 9 and Table 1, demand nodes 9–12 can receive their requirements from source nodes 1, 4 or 8, whereas demand nodes 2 and 3 must receive their requirements from source node 1; and demand nodes 5–7 must receive their requirements from source nodes 1 or 4. Thus, a demand node can receive its requirement only from a source node, label of which in the CLP is less than that of the demand node under consideration. Therefore, let the source and demand nodes be divided into G groups, each group consisting of source or demand nodes having consecutive labels in the CLP, as shown in the last two columns of Table 1. Let S_{a_g} and D_{b_g} be the SLP labels of the last source node and the last demand node in their respective source node group and demand node group g, $g = 1 \ldots G$, respectively. In the network of Figure 9 and Table 1, for the source nodes in groups 1, 2, and 3, i.e., for $g = 1, 2,$ and 3; $a_1 = 1, a_2 = 2,$ and $a_3 = 3$, so that $S_{a_1} = S_1, S_{a_2} = S_2,$ and $S_{a_3} = S_3$. For the demand nodes, however, for $g = 1, 2,$ and 3, $b_1 = 2, b_2 = 5,$ and $b_3 = 9$

so that $D_{b_1} = D_2, D_{b_2} = D_5$ and $D_{b_3} = D_9$. Thus, from the group demand considerations

$$\sum_{m=1}^{a_g} Q_{S_m} \geq \sum_{m=1}^{a_g} q_{S_m} + \sum_{n=1}^{b_g} q_{D_n}; g = 1 \ldots G \quad (16)$$

that gives a set of G "group demand constraints." The last group demand constraint naturally reduces to the network demand constraint of Equation (14).

TRANSPORTATION CONSTRAINTS

When transportation of water from a source node group to a demand node group is considered, the HGL values at the intermediate nodes become effective. Assume that a particular demand node D_n is assigned to a particular source node S_m to satisfy the group demand constraints. Let the route for transportation of water from S_m to D_n pass through an intermediate node D_i. Then, from transportation point of view,

$$H_{S_m} > H_{D_i}^{opt} > H_{D_n}^{opt} \quad (17)$$

in which $H_{D_i}^{opt}, H_{D_n}^{opt}$ = optimal HGL values (finally obtained in the optimal design) at the demand nodes D_i and D_n, respectively.

Now, to meet the HGL constraints at D_i and D_n, respectively,

$$H_{D_i}^{opt} \geq H_{D_i}^{min} \quad (18)$$

and

$$H_{D_n}^{opt} \geq H_{D_n}^{min} \qquad (19)$$

in which $H_{D_i}^{min}$ = minimum required HGL at D_i.

In a particular case, it may be that $H_{D_n}^{min} > H_{D_i}^{min}$. Therefore, combining Equations (9), (10), and (11),

$$H_{D_i}^{opt} > H_{D_n}^{opt} \geq H_{D_n}^{min} > H_{D_i}^{min} \qquad (20)$$

However, to satisfy the group demand constraints, considering the D_i^{min} value, D_i might have been assigned to a particular source node group. But as D_i^{opt} is now a higher value, it may be necessary to assign D_i to a higher source-node group.

Further, it might also happen that some downstream demand nodes, routes to which pass through D_i while satisfying the group demand constraints, may now be cut off from their respective source nodes, if no alternate routes are available. Therefore, they must now be supplied from higher source-node groups. Thus, from the transportation point of view,

$$\sum_{m=1}^{a_g} Q_{S_m} \geq \sum_{m=1}^{a_g} q_{S_m} + \sum_{n=1}^{b_g} q_{D_n}$$

$$+ \sum_{\substack{\text{routes} \\ \text{selected}}} q_{D_i} + \sum q_{D_c}; \, g = 1 \ldots G \qquad (21)$$

in which q_{Di} = demand at D_i transferred from the lower source-node group; and q_{Dc} = demand at a downstream node D_c, completely cut off from the lower source-node group to which it was assigned to satisfy the group demand constraints. Because of the transportation constraints, the discharge requirements increase at the higher source-node groups, but they decrease by the same amount at the lower source-node groups, Herein, also, the last transportation constraint reduces to the network demand constraint.

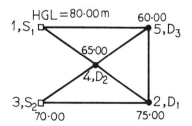

FIGURE 10. Network illustrating design feasibility constraints.

Illustration Explaining Design Feasibility Constraints

To understand the various design feasibility constraints, consider the two source, three-demand node network of Figure 10. For the HGL values as given in the figure, the nodes are labeled by both the node-labeling procedures. For the different nodal supplies and demands as shown in Table 2 ($q_{S_1} = q_{S_2} = 0$ for simplicity), the various design feasibility constraints are checked in Table 2.

The cases 1 and 2 are simple, and the constraints can be easily verified. For case 3, the transportation constraint cannot be satisfied, as both the demand nodes D_2 and D_3 have to be supplied from source node S_2. This cannot be done, as the HGL value at either of the demand nodes D_2 or D_3 has to be more than 75.00 m so that water can be transported through it from S_1 to D_1. In case 4, the design is feasible, as S_1 can supply to D_1 and D_3, and S_2 can supply to D_2. In case 5, also, the design is feasible, as S_2 can supply to D_2 and S_1 can supply to D_1 and D_3 and partly to D_2.

Selection of Design Paths and Distribution Trees

PROBLEM FORMULATION

When a network satisfies all the design feasibility constraints, the network design is feasible. From the several feasible routes the shortest route, i.e., the path P_{mn}, should be selected, as described earlier, from each feasible source node S_m to each demand node D_n. From all these paths, one from each feasible source node, the least costly one, termed "design path" is determined as follows: Consider a demand node D_n of a multiple source system. Let S_m, $m\epsilon M$ be the source nodes which can supply to D_n after satisfying all the transportation constraints. All the paths P_{mn} having lengths L_{mn} from the feasible source nodes S_m to the demand node D_n are considered in determining the design path to D_n. Using Equation (6), the transportation cost from S_m to a particular node D_n can be written as

$$C_{mn} = (K_1 K_2^{m/r} L^{1+m/r} Q^{pm/r} H_L^{-m/r})_{mn}; \, m \in M$$

or

$$C_{mn} = (K_1 K_2^{m/r})_{mn} L_{mn}^{1+m/r} Q_{mn}^{pm/r} (H_{S_m} - H_{D_n}^{min})^{-m/r}; \, m\epsilon M \qquad (22)$$

in which Q_{mn} (≥ 0) = allocated discharge from the source node S_m to the demand node D_n.

Now assume that the transportation cost varies linearly with Q_{mn} (an assumption made for the selection of the design path only). Then Equation (22) can be written as

$$C_{mn} = (K_1 K_2^{m/r})_{mn} L_{mn}^{1+m/r} (H_{S_m} - H_{D_n}^{min})^{-m/r} Q_{mn} \qquad (23)$$

Now for a path P_{mn}, the values of L_{mn}, H_{S_m}, and $H_{D_n}^{min}$ are

TABLE 2. Checking Design Feasibility Constraints for Network of Figure 10.

	Source supply (L/min)		Node demand, (L/min)			Constraints satisfied?			Network design feasible?
Case	Q_{S_1}	Q_{S_2}	q_{D_1}	q_{D_2}	q_{D_3}	network demand constraint	group demand constraint	transportation constraint	
1	5,000	2,000	6,000	1,000	1,000	No	—	—	No
2	5,000	3,000	6,000	1,000	1,000	Yes	No	—	No
3	5,000	3,000	5,000	2,000	1,000	Yes	Yes	No	No
4	5,000	3,000	4,000	3,000	1,000	Yes	Yes	Yes	Yes
5	5,000	3,000	3,000	4,000	1,000	Yes	Yes	Yes	Yes

For simplicity, $q_{S_1} = q_{S_2} = 0$

known. Further, assuming that K_2 is the same for all the paths, Equation (23) can be written as

$$C_{mn} = c_{mn} Q_{mn} \quad (24)$$

in which c_{mn} = transportation cost, per unit quantity of water, from source node S_m to demand node D_n and is given by

$$c_{mn} = (K_1 K_2^{m/r})_{mn} L_{mn}^{1+m/r} (H_{S_m} - H_{D_n}^{min})^{-m/r} \quad (25)$$

The design paths to all the demand nodes of the entire multisource network can now be obtained by solving the following linear transportation problem:

$$\text{Minimize:} \quad \sum_{m=1}^{M} \sum_{n=1}^{N} c_{mn} Q_{mn} \quad (26)$$

$$\text{subject to:} \quad \sum_{n=1}^{N} Q_{mn} \leq Q_{S_m} - q_{S_m} \quad (27)$$

for all source nodes m;

$$\sum_{m=1}^{M} Q_{mn} = q_{D_n} \quad (28)$$

for all demand nodes n; and

$$Q_{mn} \geq 0 \quad (29)$$

Equation (26) is the objective function in the linear transportation problem. It contains only the feasible paths, the decision variables Q_{mn} for nonfeasible paths being discarded. Equation (27) gives a set of M linear constraints, one for each source node, and ensures that the total allocation from a source node S_m to the various demand nodes is less than or equal to the net available supply at S_m. Similarly, Equation (28) gives a set of N linear constraints, one for each demand node, and ensures that the total supply to a demand node D_n from the feasible source nodes is equal to the discharge requirement at D_n. Equation (29) gives the usual non-negativity constraints for all the decision variables Q_{mn}.

As this transportation problem is used to obtain the design paths only, the actual value of the objective function is not necessary. Further, as all the c_{mn} terms contain the same values of $K_1 K_2^{m/r}$, the exact values of K_1 and K_2 need not be known. Therefore for simplicity, c_{mn} values in Equation (26) can be taken as

$$c_{mn} = L_{mn}^{1+m/r} (H_{S_m} - H_{D_n}^{min})^{-m/r} \quad (30)$$

SOLUTION

Consider a 2-source node distribution network of Figure 11 [13,23]. The HGL values provided at the two source nodes A and B are 54.90 m and 51.90 m, respectively, while the minimum required HGL values at all the other nodes are 42.70 m. The supplies at A and B are 530 L/s and 150 L/s, respectively, and the discharge requirement at all the nodes

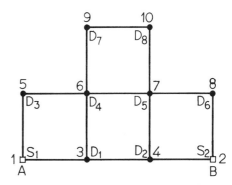

FIGURE 11. Illustrative network for selection of design paths and distribution trees.

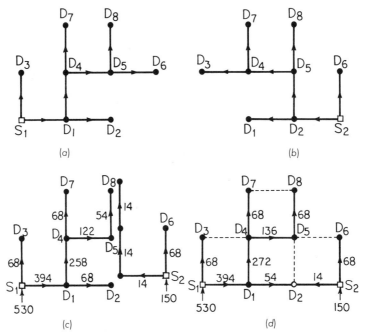

FIGURE 12. Selection of distribution trees for illustrative network: (a) distribution tree for source node S_1; (b) distribution tree for source node S_2; (c) superimposition of distribution trees; and (d) selected distribution trees.

(source as well as demand) is 68 L/s. Each link is 850 m long.

Considering the given HGL values, all the nodes are labeled according to the combined and separate node labeling procedures (Figure 11). As Equation (15) is satisfied, only the network demand constraint, Equation (14), must be satisfied. This constraint is satisfied with an equality sign giving the total supply equal to the total demand for the network.

At each source node, $Q_{s_m} > q_{s_m}$ and, therefore, each source satisfies its own demand. The feasible paths to various demand nodes from source nodes S_1 and S_2 are obtained as shown in Figures 12(a) and 12(b), respectively. In Figure 12(a), the paths S_1D_1, $S_1D_1D_2$ and S_1D_3 are obvious as these are the shortest routes from source S_1 to demand nodes D_1, D_2 and D_3, respectively. Two routes $S_1D_1D_4$ and $S_1D_3D_4$ of equal length are available from S_1 to D_4. However, as the discharge in link S_1D_1 is already greater than that in link S_1D_3, to concentrate the discharge further for optimality, route $S_1D_1D_4$ is selected as path instead of the route $S_1D_3D_4$. Similarly, the other paths in Figure 12(a) and all the paths in Figure 12(b) are selected from the two source nodes S_1 and S_2, respectively.

Considering the appropriate values when the transportation problem given by Equations (26-29) is solved, the node demand allocations are obtained. This leads to the link discharges as shown in Figure 12(c). (For clarity, the paths from source node 2 are slightly shifted.)

The network of Figure 12(c) contains a loop (loop $D_1D_2D_5D_4$) which must be eliminated. In this loop, link D_2D_5 carries the minimum discharge (14 L/s). Therefore, link D_2D_5 is converted to a nonprimary link by superimposing a clockwise discharge of 14 L/s on this loop. This modifies the link discharges as shown in Figure 12(d), which also shows the selected distribution trees and the primary links (thick lines), nonprimary links (thin lines) and the tree-connecting node D_2 which receives its requirement from both the sources. The nonprimary links can be selected according to either the D-specified condition or the Q-specified condition. Due to the flow in the nonprimary links, the actual discharge allocations from the two source nodes and the link flows are slightly modified as explained by Bhave [23].

General Comments

NODE LABELING PROCEDURES

Two node labeling procedures are introduced herein to explain the theory and develop expressions for the various design feasibility constraints. These procedures consider the available HGL values at the source nodes and the minimum required HGL values at the demand nodes. However, barring the sources and sink nodes, the actual HGL values would be different for the other nodes. The node labels given earlier are, however, retained. The node labeling procedures suggested herein help in checking the various design feasibility constraints. Therefore, in cases where only the network demand constraint is applicable (as in the illustrative example), the nodes need not be labeled using procedures suggested herein, but can be labeled in any convenient manner.

BRANCHING CONFIGURATIONS

For the illustrative example Cembrowicz and Harrington [13] have estimated the total number of branching configurations as 208, and the total number of feasible flow combinations as 2,944. (The total number of feasible flow combinations is obtained by considering both the flow directions in links.) The procedure suggested herein has given a unique branching configuration as shown in Figure 12(d). Although this may lead to a fairly good solution, because of the several assumptions (especially the linear variation of the transportation cost with Q_{mn}), the solution would still be a local optimum and not necessarily the global optimum one.

OPTIMAL LAYOUT FROM NONLINEAR PROGRAMMING MODEL

Rowel and Barnes [24] suggested the formulation and solution of a nonlinear programming model to obtain the branching configuration for single and multiple-source water distribution networks. The link-cost function, Equation (6), is rewritten as

$$C = \frac{K_1 \, K_2^{m/r} \, L \, Q^{pm/r}}{(H_L/L)^{m/r}} \quad (31)$$

It is assumed that the hydraulic gradient, i.e., the head loss per unit length of link, given by H_L/L has the same value on all links. Therefore, taking H_L/L also as constant in addition to K_1 and K_2, Equation (31) can be expressed as

$$C = K' \, L \, Q^{pm/r} \quad (32)$$

in which

$$K' = K_1 K_2^{m/r} \, (H_L/L)^{-m/r}$$

The nonlinear optimization model uses Equation (32) as the objective function in addition to the usual constraints. As $pm/r < 1$, the nonlinear minimum cost flow model involves minimizing a concave function over a convex set. The nonlinear objective function may be approximated, to any degree of desired accuracy, with piecewise linear functions. However, since the resulting linear program involves minimizing a strictly concave function, the simplex method must be modified slightly to ensure that only adjacent points of the piecewise linear approximation function are chosen.

The solution of the nonlinear model automatically assigns each demand node to a particular source for multiple source systems. Thus, the network is automatically partitioned into a set of disconnected trees, each rooted at a source. Nonprimary links are then added to obtain a desired degree of reliability of water distribution.

OPTIMAL LAYOUT BY DYNAMIC PROGRAMMING MODEL

An approach based on dynamic programming is developed by Bhave [25] to obtain an optimal layout to connect several distribution reservoirs to the source of water for a large water supply system.

Assumptions

1. The layout is of the branching type without any loops.
2. The link diameter is a continuous variable and the link cost can be expressed by Equation (7).
3. The required HGL values at the distribution reservoirs are known and the provided HGL values at these reservoirs are equal to the required values. For reservoirs forming the end nodes of the branching system, this assumption leads to the optimal layout. However, for intermediate reservoirs, in some cases, providing HGL values greater than the required ones might prove to be optimal.
4. Any reservoir can be supplied directly from the source or via any other reservoir having higher HGL value.

Theory

The source is labeled as 0 and the distribution reservoirs are labeled in the decreasing order of their HGL values, according to Equation (13). Because of this labeling procedure, for a reservoir with label j, total j possibilities of getting water from 0, 1, . . . , $j-1$ nodes may exist. However, the least costly possibility should be selected.

Now let subscripts gj and ij represent two links gj and ij ($g < i < j$) connecting reservoirs g and i to reservoir j. Therefore, from Equation (7), for demand Q_j at node j,

$$C_{gj} = K_{gj} \, L_{gj} \, (L_{gj}/H_{L_{gj}})^{m/r} \, Q_j^{pm/r} \quad (33)$$

and

$$C_{ij} = K_{ij} \, L_{ij} \, (L_{ij}/H_{L_{ij}})^{m/r} \, Q_j^{pm/r} \quad (34)$$

Assuming K values the same for both the links, as $g < i$, $(HGL)_g \geq (HGL)_i$ and therefore $H_{L_{gj}} \geq H_{L_{ij}}$. Thus when $L_{ij} \geq L_{gj}$, $C_{ij} \geq C_{gj}$. Therefore, all the possibilities of con-

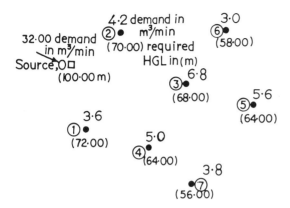

FIGURE 13. Source and distribution reservoirs.

necting reservoir j to reservoir i for which $L_{ij} \geq L_{gj}$ can be discarded at a glance. Further, even when $L_{ij} < L_{gj}$, if $L_{ij}(L_{ij}/H_{L_{ij}})^{m/r} \geq L_{gj}(L_{gj}/H_{L_{gj}})^{m/r}$, $C_{ij} \geq C_{gj}$. These possibilities also can be discarded.

For the remaining possibilities, $C_{ij} < C_{gj}$, i.e., it is cheaper to carry water to j from reservoir i rather than from reservoir g when links gj and ij only are considered. However, because of the discharge requirement of Q_j at j, the requirements at g or i must increase from Q_g and Q_i to $Q_g + Q_j$ and $Q_i + Q_j$, respectively. This naturally would increase the cost of the link carrying the increased discharge. Such increase must also be considered before making the final choice of the layout.

Procedure and Illustrative Example

A source and 7 distribution reservoirs are located as shown in Figure 13. The HGL values and the water demands are as shown in the figure. The various link lengths are given in Table 3. In Equation (7), $K = 16.29$, $m/r = 0.283$ and $pm/r = 0.524$.

Considering the link lengths columnwise (Table 3), the lengths greater than or equal to the previous lengths are discarded because of the reasons stated earlier. The discarded lengths are shown within parentheses. The remaining possibilities are considered in Table 4. For reservoirs 1 and 2, as seen from Table 3, there are no other alternatives except routes 01 and 02; and, therefore, the cost of these links is calculated as shown in column 5, Table 4, and the table is completed up to column 7. For reservoir 3, however, all the routes 03, 013 and 023 must be considered. Therefore, considering these routes in Table 4, $L^{1.283} H_L^{-0.283}$ values for these routes are evaluated (column 4) for links 03, 13 and 23. As these values (column 4) for links 13 and 23 are more than that for 03, they are discarded (shown within parentheses) and the table is completed up to column 7 for link 03 only, which is the selected alternative. Proceeding in the same manner for reservoir 4, alternatives with links 24 and 34 are discarded (Table 3). However, from column 4 in Table 4 it is clear that both the alternatives 04 and 014 must be considered. The costs of links 04 and 14 are calculated (Rs. 483,900 and Rs. 408,300, respectively) and shown in column 5. For alternative 014, the link 01 has now to carry a discharge of 3.6 + 5.0, i.e., 8.6 m³/min. Its revised cost is determined (Rs. 358,000) and shown in column 5. The excess cost for link 01 is Rs. 131,400 (358,500–227,100) as shown in column 6. Thus, the cost of carrying water to reservoir 4 via route 014 becomes Rs. 539,700 (Rs. 408,00 for link 14 and Rs. 131,400, excess cost for link 01). Therefore link 04 is selected. (Discarded route 014 is shown within parentheses in column 7.)

Table 4 is completed for all the reservoirs up to column 7. All the selected routes, i.e., paths, are shown underlined in column 2. The finally selected links and their final costs are shown in columns 8 and 9, respectively. For example, the finally estimated cost of link 03 is Rs. 1,002,500 as obtained in column 5 for path 036, which is the last path containing link 03. The cumulative cost of the finally proposed layout is Rs. 3,504,900 (columns 7 and 9). The finally selected layout is shown in Figure 14.

TABLE 3. Link Lengths in Meters for Different Connections for Figure 13.

Upstream end	Downstream end						
	1	2	3	4	5	6	7
0	2,100	1,800	3,800	3,500	5,500	5,500	6,000
1	—	(3,200)	3,500	2,200	5,200	(6,000)	4,500
2	—	—	2,500	(3,700)	4,500	3,800	(6,000)
3	—	—	—	(2,500)	2,500	2,500	3,700
4	—	—	—	—	(3,500)	(4,800)	2,700
5	—	—	—	—	—	(2,500)	(3,500)
6	—	—	—	—	—	—	(5,500)

TABLE 4. Cost Estimation and Layout Selection for Figure 13.

Reservoir 1	Route 2	Link 3	$L^{1.283} H_L^{-0.283}$ 4	C (Rs.) 5	ΔC (Rs.) 6	Cumulative cost (Rs.) 7	Selected link 8	Cost of selected link (Rs.) 9
1	01	01	—	227,100	—	227,100	01	227,100
2	02	02	—	198,100	—	198,100	02	198,100
3	03 013 023	03 13 23	14,686 (23,804) (18,809)	653,200	—	653,200	03	1,002,500
4	04 014	04 14 01	12,782 10,783 —	483,900 408,300 358,500	— 131,400	483,900 (539,700)	04	650,800
5	05 015 025 035	05 15 25 35 03	22,827 (32,513) (29,299) 15,458 —	917,100 621,100 894,900	— 241,700	(917,100) 862,800	35	621,100
6	06 026 036	06 26 02 36 03	21,852 19,384 — 11,927 —	633,000 561,500 262,800 345,500 1,002,500	— 64,700 107,600	(633,000) (626,200) 453,100	36	345,500
7	07 017 037 047	07 17 01 37 03 47 04	24,114 22,197 — 18,732 — 14,023 —	790,700 727,800 331,300 614,200 1,125,300 459,800 650,800	— 104,200 122,800 166,900	(790,700) (832,000) (737,000) 626,700	47	459,800
				Total	3,504,900	—		3,504,900

OPTIMAL LAYOUT BY BRANCH EXCHANGE METHOD

Rathford, et al. [26] have proposed a *branch exchange* method for bringing gas from offshore fields to the onshore separation and compressor plant by a branching layout. Their method incorporates 1) The selection of optimal diameters in a given pipeline network to minimize the sum of investment and operation costs; and 2) the selection of the optimal layout.

Assumptions

1. The average daily production rates for the gas fields are specified.
2. Because of the availability of the drilling platforms,

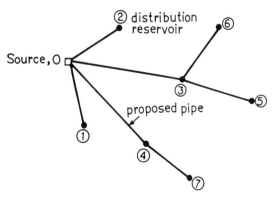

FIGURE 14. Optimal layout.

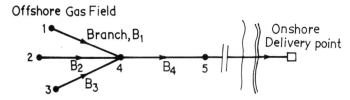

FIGURE 15. Part of network for illustrating partial assignment method.

TABLE 5. Pipe Sizes and Corresponding Pressure-Squared Differences for Figure 15.

Pipe size[a]	Pressure-squared difference for branch		
	B_1	B_2	B_3
1	830	1030	650
2	760	960	590
3	630	810	550
4	450	600	420
5	370	520	380
6	280	480	330
7	210	460	220

[a]Pipe size successively increases.

TABLE 6. Partial Assignments for Three-Branch Part-Network of Figure 15.

Partial assignment number	Pressure-squared difference for branch			Selected pipe size for branch		
	B_1	B_2	B_3	B_1	B_2	B_3
1	830	1,030[a]	650	1	<u>1</u>[a]	1
2	830	<u>960</u>	650	1	2	1
3	<u>830</u>	810	650	<u>1</u>	3	1
4	760	<u>810</u>	650	2	3	1
5	<u>760</u>	600	650	<u>2</u>	4	1
6	630	600	<u>650</u>	3	4	<u>1</u>
7	<u>630</u>	600	590	<u>3</u>	4	2
8	450	<u>600</u>	590	4	<u>4</u>	2
9	450	520	<u>590</u>	4	5	<u>2</u>
10	450	520	<u>550</u>	4	5	<u>3</u>
11	450	<u>520</u>	420	4	<u>5</u>	4
12	450	<u>480</u>	420	4	<u>6</u>	4
13	450	<u>460</u>	420	4	<u>7</u>	4

[a]Critical branches and pipe sizes are underlined.

only the gas-field locations are used as junction points for connecting links.

3. The flow of gas through a pipeline is a function of the internal diameter, the length, the pressure at the end points, the elevation profile of the pipeline, the flowing temperature, and the physical characteristics of the gas. The steady-state gas flow is described by the standard "Panhandle" equation expressed as

$$Q = K[(P_1^2 - P_2^2)/L]^{0.5394} D^{2.6182} \qquad (35)$$

in which Q = flowing volume (m³/d); P_1 = input absolute pressure (kPa); P_2 = output absolute pressure (kPa); D = internal pipe diameter (mm); L = length of pipe (km); and K = constant. (The technique, however, can be used for any other form of the flow equation.)

4. The pipelines are available in a finite selection of standard diameters. For a particular diameter, the cost depends on the depth of the water at which the pipe is to be installed. For example, one value for a range of water depth 0–27 m, and a higher value for a range of 27–60 m. Total capital pipe costs are converted to the annual costs for comparison.

5. The power required by the compressor is a function of the amount of gas that flows through the compressor and the input and output pressures.

6. The maximum permissible pressure in a pipeline and the minimum required pressure for delivery of the gas on the onshore are specified.

Pipe-Size Optimization

The pipe sizes are optimized by examining combinations for one set of branches at a time. An assignment of diameters to such a set is termed *partial assignment*. This method considers only those partial assignments which can be in the optimal assignment for the entire network and discards the rest of the partial assignments.

The path from the delivery point to the node of greatest pressure is termed *critical path*. The sum of the pressure-squared differences along the critical path determines the pressure at the delivery point and hence the cost of compression needed to meet the minimum required pressure at the delivery point.

To understand the selection of the partial assignments that

can be in the optimal solution, consider a part of a gas pipeline network as shown in Figure 15. The downstream ends of the three branches B_1, B_2, and B_3 meet at gas field 4. Let the available pipe-size set consist of seven sizes 1–7, increasing in size from 1 to 7. For each of these pipe sizes, let the required pressure-squared difference to maintain the gas flow (Equation (35)) in each branch be as given in Table 5.

Initially, the minimum pipe size, i.e., size 1, is assigned to all the branches. As the pressure-squared difference is maximum for branch B_2, this branch lies on the critical path. Therefore, when B_2 has the minimum pipe size, branches B_1 and B_3 also must have minimum pipe sizes (as already selected), because adopting larger pipe sizes for B_1 and B_3 would increase the pipe costs without reducing the compression cost. Therefore, all the combinations with size 1 for B_2 and sizes 2–7 for B_1 and B_3 can be discarded. Thus, pipe sizes 1 for all the three branches belongs to the set of partial assignments, which could be optimal. This is shown as partial assignment 1 in Table 6.

Next, pipe size 2 is selected for B_2. Herein also, as the pressure-squared difference for B_2 is still more than the values for B_1 and B_3, B_2 lies on the critical path. Therefore, the partial assignment consists of pipe size 2 for B_2 and pipe size 1 for B_1 and B_3, (partial assignment 2).

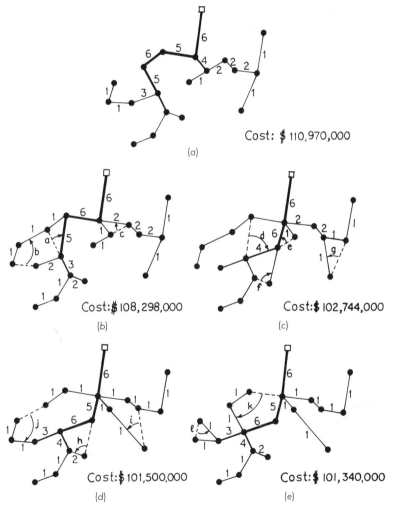

FIGURE 16. Layouts for a gas network: (a) initial layout with optimized pipe diameters; (b) intermediate layout with branches a, b and c replaced; (c) intermediate layout with branches d, e, f and g replaced; (d) intermediate layout with branches, h, i and j replaced; (e) optimal layout obtained after replacing branches k and l. The numbers beside branches indicate relative pipe diameters for the network.

Next, when pipe size 3 is selected for B_2, keeping pipe size 1 for B_1 and B_3, branch B_1 determines the critical path, and the partial assignment is pipe size 1 for B_1 and B_3 and pipe size 3 for B_2 (partial assignment 3).

This procedure is continued. At each stage, the pipe having the largest pressure-squared difference and therefore deciding the critical path is located, and its size is increased. The process terminates when the largest pressure-squared difference occurs across a pipe with the largest available diameter. Further consideration of partial assignments is unnecessary as this would increase pipe cost without possible saving in pipe cost. Thus, out of 7^3, i.e., 343 possible partial assignments (seven pipe sizes for each of the three branches), only 13 partial assignments as listed in Table 6 need be considered for optimal assignment. The pressure-squared difference and the pipe size for critical paths are shown underlined in Table 6.

The selected partial assignment can be viewed as an equivalent branch connecting a combination of gas fields 1, 2, and 3 to gas field 4. Thus, the branches B_1, B_2, and B_3 are merged in parallel and therefore this procedure is termed *parallel-merge* procedure.

A procedure, similar to the above procedure and termed *serial-merge* procedure is then applied to serially merge the equivalent branch and branch B_4, to obtain a new equivalent branch for branches B_1–B_4. The parallel-merge and serial-merge techniques are continued to obtain a single equivalent branch for the entire network. The cost of diameter assignment corresponding to each entry in the final list is then evaluated by summing the pipe cost and the cost of compression associated with the pressure-squared difference. The diameter assignment with the smallest cost is the optimal-diameter assignment. The size of the intermediate and final lists have been observed to have not more than 1000 possible diameter assignments for trees of 20 nodes [26].

Layout Optimization

Usually, the minimum spanning tree is selected initially and the pipe sizes are optimized by the procedure described earlier. This layout is then optimized by the *branch-exchange* method, which repetitively removes a branch from the initial tree and replaces it by a new branch.

Any node in the initial tree is selected and is connected directly to one of its three nearest neighbors (not already directly connected) by a new branch to form a loop. Then each branch of this loop is omitted and the optimal diameters are calculated for the entire new branching layout. This process is repeated for the other two nearest neighbors. The new branch yielding the maximum reduction in total costs is retained and replaces the old branch. The entire process is repeated for all nodes in the layout. During this process it might happen that a branch replaced in an earlier iteration might enter into the layout in a subsequent iteration. The process of replacing old branches with new branches and optimizing the pipe diameters is continued until no further saving in cost occurs. This procedure gives a good optimal layout but the final layout cannot be claimed to be the globally optimum one.

In a particular study of a gas pipeline network for the Gulf of Mexico [26], the initial network as shown in Figure 16(a) was adopted. This network with optimized pipe diameters had a 20-year cost of $110,970,000. This layout was successively improved and some intermediate layouts with their optimal costs as shown in Figures 16(b), 16(c) and 16(d) were obtained [26,27]. Finally selected layout with a 20-year cost of $101,340,000 is shown in Figure 16(e). This layout, though quite economical, cannot be guaranteed to be the global optimal one.

OPTIMAL LAYOUT HAVING STEINER POINTS

The networks considered initially had fixed branches and the optimal branching layout was obtained using only these branches. For example, for the looped network in Figure 7(a), optimal branching layouts were obtained, using only the branches of Figure 7(a), by the minimal spanning tree approach in Figure 7(b), and the path concept in Figure 7(c). It was not allowed to consider an ungiven branch, e.g., a branch joining nodes 4 and 8. Later, the branches were not specified and the optimal branching layout was obtained to connect the given nodes. However, the junction points of the branches needed to coincide with the given nodes. If the choice of the junction points is not restricted to the given nodes, additional junction points can be selected to obtain more economical layout as shown by Bhave and Lam [28].

Three-Node Network

LAYOUT CONSIDERATION

Consider a three-node network having one source node labeled 1 and two demand nodes labeled 2 and 3 (Figure 17) for which an optimal branching layout is required. Using the approaches considered so far, the optimal layout would be the one from the possible 3 layouts: (1) 12, 13; (2) 12, 23; and (3) 13, 32. However, the general case of the branching layout would be to carry the combined discharge requirements of nodes 2 and 3 from source 1 to a common junction point J, through a link $1J$, and then carry the individual demands for nodes 2 and 3 through links $J2$ and $J3$ respectively. The junction point, J, can lie anywhere in the space (including points 1, 2, and 3) and the objective would be to obtain the optimal branching layout such that the total capital cost of the branching layout consisting of links $1J$, $J2$, $J3$ is the minimum.

Even though J can lie anywhere in the space, it is obvious that for the layout cost to be the minimum, J should lie on the plane of nodes 1, 2, and 3. This is because instead of the layout $1J$, $J2$, and $J3$; a layout $1J'$, $J'2$ and $J'3$ (J' is the projection of J on plane 123) would be more economical. Further, when the point J lies on the plane 123, it can lie in one of the seven regions shown in Figure 17. Region 1 consists of the three sides, 12, 23 and 31 of the triangle 123 and the area within it; and the other six regions are outside the triangle 123 as shown in the figure. It can be easily shown that when J lies in any one of the outside six regions, i.e., regions 2–7, there is always a point on 12, 23 or 31 in region 1, which is a better location for J. For example, when J lies in region 2, instead of the layout $1J$, $J2$, and $J3$, the layout 12 and 23 (J merging with 2) is obviously more economical. Similarly, if J lies in region 3, instead of the layout $1J$, $J2$ and $J3$, the layout $1J'$, $J'2$ and $J'3$ (J' being the intersection point of $1J$ and 23) is more economical. Therefore J must lie in region 1 for optimality. Further, for simplicity, it is assumed that the plane of region 1 is horizontal.

FORMULATION OF OPTIMIZATION PROBLEM

In Figure 18, let 1 be the source node with X_1, Y_1 as the cartesian coordinates. Similarly, let 2 (X_2, Y_2) and 3 (X_3, Y_3) be the demand nodes, and J (X_j, Y_j) be the junction node. Let the available HGL value at the source node be H_1 and the minimum required HGL values at the two demand nodes 2 and 3 be H_2 and H_3, respectively. As the junction point J is not a demand node, there is no HGL constraint for J. Let L_{1j}, L_{j2} and L_{j3} be the lengths of the links $1J$, $J2$, and $J3$, respectively. Let the demands at nodes 2 and 3 be Q_2 and Q_3, respectively so that the discharges in links $1J$, $J2$, $J3$ become Q_{1j} ($=Q_1 + Q_2$), Q_{j2} ($=Q_2$) and Q_{j3} ($=Q_3$), respectively.

For optimal branching layout, the objective function of the optimization problem can be written as:

Minimize

$$C_T = C_{1j} + C_{j2} + C_{j3} \qquad (36)$$

in which C_T = total cost of the branching layout; and C_{1j}, C_{j2}, C_{j3} = cost of links $1J$, $J2$, $J3$, respectively.

Using Equation (7) and proper suffixes, Equation (36) can be written as

$$C_T = K L_{1j}^{1+m/r} Q_{1j}^{pm/r} (H_1 - H_j)^{-m/r}$$
$$+ K L_{j2}^{1+m/r} Q_{j2}^{pm/r} (H_j - H_2)^{-m/r} \qquad (37)$$
$$+ K L_{j3}^{1+m/r} Q_{j3}^{pm/r} (H_j - H_3)^{-m/r}$$

Expressing Equation (37) in a concise form

$$C_T = \sum_i K L_{ji}^{1+m/r} |Q_{ji}|^{pm/r} |H_i - H_j|^{-m/r} \qquad (38)$$

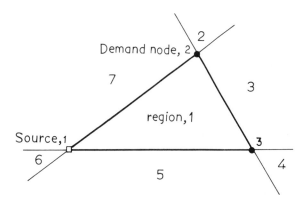

FIGURE 17. Three-node network and various regions.

or

$$C_T = \sum_i K[(X_j - X_i)^2 + (Y_j - Y_i)^2]^{(1+m/r)/2} |Q_{ji}|^{pm/r} |H_i - H_j|^{-m/r} \qquad (39)$$

OPTIMALITY CRITERIA

The objective function of Equation (39) is unconstrained nonlinear function having X_j, Y_j, and H_j as variables. It is strictly convex and therefore has a unique minimum value. Therefore, differentiating Equation (39) partially with respect to X_j, Y_j, and H_j and simplifying

$$\frac{\partial C_T}{\partial X_j} = \sum_i \frac{C_{ji}}{L_{ji}^2} (X_j - X_i) = R_{X_j} = 0 \qquad (40)$$

$$\frac{\partial C_T}{\partial Y_j} = \sum_i \frac{C_{ji}}{L_{ji}^2} (Y_j - Y_i) = R_{Y_j} = 0 \qquad (41)$$

and

$$\frac{\partial C_T}{\partial H_j} = \sum_i \frac{C_{ji}}{H_j - H_i} = R_{H_j} = 0 \qquad (42)$$

Equations (40–42) are three independent equations representing the X, Y, and H optimality criteria, respectively. Their solution would yield the optimal values of X_j, Y_j, and H_j and thereby the optimal layout.

CORRECTION TERMS

The suggested procedure is iterative in which the values of X_j, Y_j, and H_j are initially assumed. As the optimality cri-

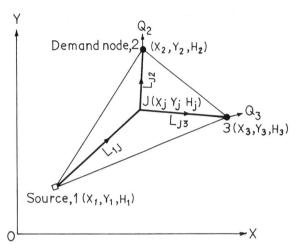

FIGURE 18. Definition sketch.

teria would rarely, if ever, be satisfied for these assumed values, the residues R_{x_j}, R_{y_j}, and R_{H_j} would not be zero. Therefore, let ΔX_j, ΔY_j, and ΔH_j be the *additive* corrections to the assumed values of X_j, Y_j, and H_j, respectively. Then these corrections are given as

$$\left[\sum_i \frac{C_{ji}}{L_{ji}^2} + \left(\frac{m}{r} - 1\right) \sum_i \frac{C_{ji}}{L_{ji}^4} (X_j - X_i)^2 \right] \Delta X_j$$

$$+ \left(\frac{m}{r} - 1\right) \left[\sum_i \frac{C_{ji}}{L_{ji}^4} (X_j - X_i)(Y_j - Y_i) \right]$$

$$\times \Delta Y_j = - R_{x_j} \qquad (43)$$

$$\left(\frac{m}{r} - 1\right) \left[\sum_i \frac{C_{ji}}{L_{ji}^4} (X_j - X_i)(Y_j - Y_i) \right] \Delta X_j$$

$$+ \left[\sum_i \frac{C_{ji}}{L_{ji}^2} + \left(\frac{m}{r} - 1\right) \sum_i \frac{C_{ji}}{L_{ji}^4} (Y_j - Y_i)^2 \right]$$

$$\times \Delta Y_j = - R_{y_j} \qquad (44)$$

and

$$\Delta H_j = \frac{R_{H_j}}{\sum_i \frac{C_{ji}}{(H_j - H_i)^2}} \qquad (45)$$

Simultaneous solution of linear Equations (43) and (44) gives the values of ΔX_j and ΔY_j; and Equation (45) gives the value of ΔH_j. For rapid convergence in the iterative procedure, it is suggested that for each iteration, ΔX_j and ΔY_j should be calculated, the values of X_j and Y_j corrected; and then ΔH_j calculated and H_j corrected (using the corrected X_j and Y_j values) to complete one iteration.

SELECTING INITIAL X_j, Y_j, AND H_j VALUES

Even though the final optimal layout can be obtained from any set of reasonable X_j, Y_j, and H_j values, it would be preferable to select their initial values, which would reduce the number of iterations.

X_j and Y_j Values

In the total cost function, Equation (37), the index of L is numerically much larger than the indices of the other terms. Thus, L is more influential than other terms. Therefore, it would be logical to start the iterative procedure with the location of the junction point J given by the minimum spanning tree layout; i.e., $L_{1j} + L_{j2} + L_{j3}$ is minimum. This is achieved when the links $1J$, $J2$, and $J3$ subtend an angle of 120° between them. The location of J can be easily obtained graphically as follows:

As shown in Figure 19, draw the perpendicular bisectors of any two sides, say 12 and 13; and take points A and B on them, outside the triangle 123, such that $\angle A12 = \angle B13 = 30°$. Taking A and B as centers and radii equal to $A1$ and $B1$, respectively, draw arcs of circles to intersect at a point within the triangle 123. This point is the initial position of J.

When the angle subtended at one of the nodes is equal to or more than 120°, J coincides with this node. If this node is the source node 1, J would always coincide with 1 in the optimal solution, giving 12, 13 as the optimal layout. However, if such a node is the demand node 2 or 3, J may coincide with this node. Therefore, it is preferable to locate J slightly away from this node and allow the iterative procedure to decide the optimal location of J.

H_j Value

It is obvious that H_j should lie between H_1 and the larger of the two values H_2 and H_3. Therefore, the initial value of H_j can be suitably assumed or selected by using the critical path concept [16,19], also described by Bhave and Bhole in *Civil Engineering 5*.

ILLUSTRATIVE EXAMPLE [28]

It is desired to connect the demand nodes 2 and 3 to source node 1 (Figure 19). The details are [28]: $X_1 = 0$ m; $Y_1 = 0$ m; $X_2 = 383$ m; $Y_2 = 321.4$ m; $X_3 = 600$ m; $Y_3 = 0$ m; $L_{12} = 500$ m; $L_{13} = 600$ m; $H_1 = 100.00$ m; $H_2 = 90.00$ m; $H_3 = 80.00$ m; $Q_2 = 3.00$ m³/min; and $Q_3 = 2.00$ m³/min. In the cost function, Equation (4), $K_1 = 0.0809$, $m = 1.394$, and C, L, and D are in rupees, meters and millimeters, respectively.

Hazen-Williams head-loss formula is used with coefficient

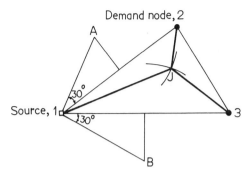

FIGURE 19. Graphical location of initial position of junction point J.

$= 100$. For H_L and L in meters, Q in cubic meters per minute, and D in millimeters in Equation (5), $K_2 = 4.3734 \times 10^8$, $p = 1.85$, and $r = 4.87$. Therefore, Equation (7) becomes

$$C = 24.061 \ L^{1.2862} \ Q^{0.5295} \ H_L^{-0.2862} \quad (46)$$

From the graphical method, initial position of J is obtained as $X_j = 370$ m; and $Y_j = 165$ m. H_j is taken as 92.8 m. The optimal location of J ($X_j = 262$ m; $Y_j = 135$ m; $H_j = 94.35$ m) is obtained in four iterations as shown in Table 7. The optimal layout is shown in Figure 20, and its cost is Rs. 112,937. Note that because of the influence of Q and H_L terms (Equation (46)), the total link length in the optimal layout is more (880.9 m) than that in the minimum length layout (845.1 m).

For comparison, the minimum costs of networks 12, 13 (junction point at 1); 12, 23 (junction point at 2); and 13, 32 (junction point at 3) are also estimated. These values are Rs. 121,100, Rs. 124,790 and Rs. 187,820, respectively. It is seen that the network 12, 13 is least costly but its cost is approximately 7% more than that of the optimal network shown in Figure 20.

Multiple Node Networks

The procedure of starting with a minimum-length layout and then optimizing it can also be applied to multiple node networks.

SELECTION OF INITIAL LAYOUT

As was done for the three-node network, for multiple node networks also the initial layout is selected such that the total length of the layout is minimal. This layout also provides the general nature of the optimal layout giving the number of the junction points (this depends upon whether a junction point is separate or merges with a given node) and the nodes to which each of them is connected.

TABLE 7. Iteration Details for 3-Node System of Figure 20.

Iteration	X (m)	Y (m)	H (m)	Total L (m)	Total C (Rs)
Initial	370	165	92.80	845.1	117,656
Iteration 1	301	176	93.36	862.6	113,705
Iteration 2	282	151	93.94	870.0	113,129
Iteration 3	269	140	94.22	876.9	112,984
Iteration 4	262	135	94.35	880.9	112,937

The minimum length layout for a multiple node system is obtained by treating it as a *Steiner problem*. The Steiner problem [29–32] is to find a layout connecting all the given nodes so that the total length of the layout is minimal. Such a layout is termed *minimum Steiner tree* and the junction points are termed *Steiner points*.

The minimum Steiner tree for a multiple-node network can be obtained by applying computerized algorithms [33–36]. A simple practical method based on soap film techniques is also available. It has been shown that the geometry of soap films is governed by the area minimizing principle [37–39]. The soap film procedure is as follows:

Take a transparent glass, Perspex or Plexiglas sheet and, adopting some convenient scale, mark the positions of all the given nodes on the sheet. Fix nails or Plexiglas rods of equal length and perpendicular to the sheet at all the marked positions of the nodes. Attach another sheet firmly on top of the nails or rods to form a sandwich with empty space and nails or rods in between. Immerse this sandwich in a soapy solution and take it out. A soap film connecting all the nails or rods will be formed in the sandwich space. The film is traced on the sheet giving a layout. Since the area of the soap film is minimal, its cross section and therefore the traced layout is a minimum length layout. This is the minimum Steiner tree; giving the location of the Steiner points.

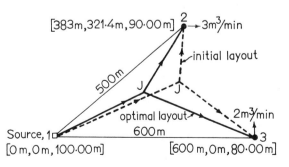

FIGURE 20. Initial and optimal layouts for 3-node network of illustrative example.

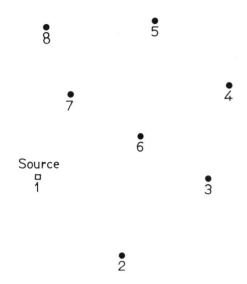

FIGURE 21. 8-Node system.

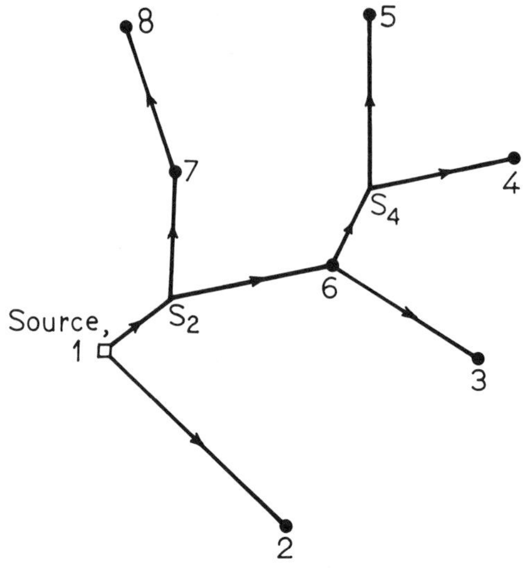

FIGURE 23. Optimal layout for 8-node system.

OPTIMIZATION OF SELECTED LAYOUT

The minimum Steiner tree gives the initial layout and thus the X_j and Y_j values for all the Steiner points. These values are already known for the given nodes. The starting H_j values for all the Steiner points and junction nodes can be suitably assumed or obtained by the critical path concept as stated earlier.

At each Steiner point the X, Y, and H criteria, Equations (40–42) must be satisfied, while at the junction node, only the H criterion, Equation (42) must be satisfied. Equations (43–45) give the iterative corrections for the assumed X, Y, and H values for the Steiner points. Equation (45) gives the correction to the assumed HGL values for the junction nodes, the corrections being limited to conform with the minimum required HGL value at the junction nodes. For some junction nodes, termed *critical* nodes, optimal HGL values might be equal to the minimum required HGL values.

ILLUSTRATIVE EXAMPLE [28]

The 8-node system [28] shown in Figure 21 has node 1 as source node and nodes 2–8 as demand nodes. Their locations (X and Y coordinates), the available HGL value for the source node and the minimum required HGL values for the demand nodes, and the nodal demands are given in Table 8. The minimum Steiner tree is shown in Figure 22. It has S_1, S_2, S_3, and S_4 as Steiner points and nodes 6 and 7 as junction nodes. The locations of the Steiner points are also given in Table 8.

The optimal layout obtained from an iterative procedure, outlined earlier, is shown in Figure 23, and the details are given in Table 8. Steiner points S_1 and S_3 have merged with nodes 1 and 6, respectively, and the junction node 6 becomes a critical node as seen from Table 8. The total link length in the minimum Steiner tree is 3,761 m, which increases to 4,063 m in the optimal layout. The cost of the initial network is Rs. 884,390 which reduces to Rs. 770,100 in the optimal layout. Thus, even though the total link length increases approximately by 8%, the cost decreases approximately by 13%.

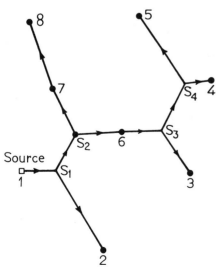

FIGURE 22. Minimum Steiner tree for 8-node system.

TABLE 8. Data and Solution of 8-Node System of Figure 21.

Node	Nature of node	Location Coordinates			Q (m³/min)	Optimal Solution			Remarks
		X_i or X_j (m)	Y_i or Y_j (m)	H_i or H_j (m)		X_i or X_j (m)	Y_i or y_j (m)	H_i or H_j (m)	
1	Source	0	0	100.00	−25	0	0	100.00	
2	Demand	550	−500	90.00	4	550	−500	90.00	
3	Demand	1,120	0	80.00	3	1,120	0	80.00	
4	Demand	1,250	590	70.00	2	1,250	590	70.00	
5	Demand	770	1,000	75.00	4	770	1,000	75.00	
6	Demand	670	260	90.00	5	670	260	90.00	Critical junction node
7	Demand	200	530	85.00	4	200	530	87.72	Noncritical junction node
8	Demand	40	950	80.00	3	40	950	80.00	
S_1	Steiner point	230	10	97.00	—	—	—	—	merges with node 1
S_2	Steiner point	360	240	93.00	—	196	159	96.72	
S_3	Steiner point	930	270	85.00	—	—	—	—	merges with node 6
S_4	Steiner point	1,080	570	80.00	—	785	492	85.23	

Application to Large Networks

The methodology can be applied to large networks having even multiple sources. The minimum Steiner tree would have several Steiner points. Naturally, therefore, the number of iterations required may be large. However, each Steiner point will be connected to two or three Steiner points so that these Steiner points form a group. Thus, all the Steiner points will be divided into several groups, and all these groups will be connected to one another through the given nodes with fixed positions. For example, in Figure 22, Steiner points S_1 and S_2 form one group and are connected to nodes 1, 2, 6, and 7 of fixed positions. Similarly, S_3 and S_4 form another group having connection with nodes 6, 3, 4, and 5.

Now, when the X, Y, and H values of Steiner points in a group change, the values for Steiner points in another group are also affected. However, such an effect is transferred only through the variation of the HGL value of the connecting nodes. Further, if the connecting nodes happen to be the critical nodes, the transfer is completely absent. In Figure 22, the group of Steiner points S_1 and S_2 is connected to the group of Steiner points S_3 and S_4 through only node 6, which also happens to be critical. Therefore, changes in the positions and HGL values of S_1 and S_2 have no effect on S_3 and S_4, and vice versa. Thus, even for large networks, the convergence is fast and the number of iterations comparatively small.

GENERAL COMMENTS

Several procedures for obtaining optimal branching layouts for connecting the given nodes have been described. These procedures deal with three situations: 1) The links are specified; 2) the links are not specified; and 3) additional junction points and corresponding links are acceptable.

Links Specified

The number of links required to connect N nodes through a branching layout is $N − 1$. When the number of given links is more than $N - 1$, loops are formed. However, the choice for selecting the branches of the layout is restricted to this given set of links.

In the analysis and design of water distribution systems, it is assumed for convenience that the demands are concentrated at the nodes. However, in practice as the actual water supply to the consumers takes place along the links, all the links must be provided. The links lying on the distribution trees are the primary links and the remaining links are secondary links, having the minimum or specified size, or carrying the minimum or specified discharge. The secondary links also increase the reliability of water distribution.

The methods described herein for obtaining the branching layouts are direct; i.e., they explicitly give the optimal distribution trees. The "minimum spanning tree" and "path concept" approaches are simple to understand and easy to apply. The minimum-spanning-tree approach has been applied to water distribution systems by Ridgik [40] and Ridgik and Lauria [41]. However, this approach has several limitations [42]. Similarly, even though the path concept is found to give good optimal layouts [7], as stated by Rowell and Barnes [24], better layouts may exist.

The approaches based on "linear programming" and "nonlinear programming" concepts are more complex. However, they provide better distribution trees.

Once the optimal branching layout, i.e., the distribution tree, is identified, the optimal pipe sizes can be obtained. However, as the objective herein has been to describe the methods of obtaining optimal branching layouts, optimization of pipe sizes has not been considered.

Several approaches which do not explicitly find the optimal branching layouts but carry out simultaneously the conversion of the looped networks to branching ones and the optimization of the pipe sizes are available [12–15, 43–46]. Naturally, these approaches also provide optimal branching layouts. However, as stated earlier, none of the presently available techniques can guarantee the distribution tree, either obtained directly or indirectly, to be the global optimum one.

Links Unspecified

When the links are not specified, the connection of a node with every other node must be considered. According to graph theory, for N nodes N^{N-2} possible branching configurations exist. If direct connection between two nodes is infeasible or undesirable, excessively large lengths between them is assumed to avoid this connection in the optimal layout.

Two methods, one based on dynamic programming approach and another based on branch-exchange concept, are described. Both these methods simultaneously obtain the optimal branching layout and the optimal pipe sizes. The dynamic programming approach as described herein assumes the pipe diameter to be a continuous variable. It successively connects the nodes and at each stage provides the optimal pipe sizes and the optimal branching layout. In the branch-exchange method all the nodes are initially connected (usually by the minimum spanning tree), the pipe sizes are optimized and the layout is improved by exchanging the branches. Complete branching layout is available at every stage of the procedure.

Additional Junction Points Acceptable

When additional junction points, i.e., the Steiner points, are acceptable, the branching layout becomes less costly. However, in practice, there may be some limitations for the adoption of such a layout. For example, in the gas pipeline networks, construction of special offshore platforms at Steiner points will be necessary. The economics of constructing such platforms must be considered. Because of the additional platforms, the branching layout with Steiner points may be costlier and the branching layout using the given nodes only may be optimal. Further, the branching layout using Steiner points connects the various points by straight lines. In practice such straight-line connections may not be possible because of hills and depressions, intervening structures, property right of way, river, railway and highway crossings, etc. However such an optimal layout would provide better insight and would help the planner in adopting better physical designs.

REFERENCES

1. Paquette, R. J., N. Ashford, and P. H. Wright. *Transportation Engineering—Planning and Design.* 2nd ed., John Wiley and Sons (1982).
2. Stopher, P. R. and A. H. Meyburg. *Urban Transportation Modelling and Planning.* Lexington Books, Lexington, Massachusetts, U.S.A. (1975).
3. Ashford, N. J. and J. M. Clark, "An Overview of Transport Technology Assessment," *Transportation Planning and Technology, 3* (1975).
4. Harary, F., "Graph Theory," Addison-Wesley Publishing Co. (1969).
5. Chan, S-P., "Introductory Topological Analysis of Electrical Networks," Holt, Rinehart and Winston Inc. (1969)
6. Seshu, S. and M. B. Reed, "Linear Graphs and Electrical Networks," Addison-Wesley Publishing Co. (1969).
7. Bhave, P. R., "Selection of Optimal Distribution Tree for Optimization of Single-Source Looped Networks," *Ind. Jour. Env. Hlth, 21:* 220 (1979).
8. Maxwell, L. J. and J. M. Kline, "Topographical Network Analysis by Algebraic Methods," *Proc. Inst. of Elect. and Electronics Engrs., 113:* 1344 (1966).
9. Wang, K. T., "On a New Method for the Analysis of Electric Networks," *Academica Sinica, Inst. Engrs. Memoirs, 2* (1934).
10. Henry, D., P. Arouhine, and A. Stoudinger, "On Using Primes

to Find Trees," Midwest Circuits Symposium, Denver, Colorado, U.S.A. (1971).
11. Lam, C. F., "Techniques for the Analysis and Modelling of Enzyme Kinetic Mechanisms," Research Studies Press (John Wiley & Sons), New York (1981).
12. Jacoby, S. L. S., "Design of Optimal Hydraulic Networks," *Jour. Hyd. Div.*, Am. Soc. of Civ. Engrs., *94:* 641 (1968).
13. Cembrowicz, R. G., and J. J. Harington, "Capital Cost Minimization of Hydraulic Networks," *Jour. Hyd. Div.,* Am. Soc. of Civ. Engrs., *99:* 431 (1973).
14. Watanatada, T., "Least Cost Design of Water Distribution Systems," *Jour. Hyd. Div.*, Am. Soc. of Civ. Engrs., *99:* 1497 (1973).
15. Deb, A. K., "Optimization of Water Distribution Network Systems," *Jour. Env. Engrg. Div.*, Am. Soc. of Civ. Engrs., *102:* 837 (1976).
16. Bhave, P. R., "Noncomputer Optimization of Single-Source Networks," *Jour. Env. Engrg. Div.*, Am. Soc. of Civ. Engrs., *104:* 799 (1978).
17. Prim, R. C., "Shortest Connection Networks and Some Generalizations," *The Bell System Technical Journal, 36:* 1389 (1957).
18. Epp, R. and A. G. Fowler, "Efficient Code for Steady-State Flows in Networks," *Jour. Hyd. Div.*, Am. Soc. of Civ. Engrs., *96:*43 (1970).
19. Bhave, P. R., "Selecting Pipe Sizes in Network Optimization by LP," *Jour. Hyd. Div.*, Am. Soc. of Civ. Engrs., *105:* 1019 (1979).
20. Smythe, W. R., Jr. and L. A. Johnson, "Networks and Flows," *Introduction to Linear Programming with Applications*, Prentice-Hall Inc., Englewood Cliffs, NJ, U.S.A. (1966).
21. Wagner, H. M., *Principles of Operations Research,* Prentice-Hall of India Pvt. Ltd., New Delhi, India (1973).
22. Bhave, P. R., "Optimization of Gravity-Fed Water Distribution Systems: Theory," *Jour. of Env. Engrg.*, Am. Soc. of Civ. Engrs., *109:* 189 (1983).
23. Bhave, P. R., "Optimization of Gravity-Fed Water Distribution Systems: Application," *Jour. of Env. Engrg.*, Am. Soc. of Civ. Engrs., *109:* 383 (1983).
24. Rowell, W. F. and J. W. Barnes, "Obtaining Layout of Water Distribution Systems," *Jour. Hyd. Div.*, Am. Soc. of Civil Engrs., *108:* 137 (1982).
25. Bhave, P. R., "Optimal Layout for Connecting Distribution Reservoirs to the Source," *Jour. Indian Wat. Wks. Assoc.*, *9:* 37 (1977).
26. Rothfarb, B., et al., "Optimal Design of Offshore Natural-Gas Pipeline Systems," *Oper. Res.*, *18:* 992 (1970).
27. Frank, H. and I. T. Frisch, "Network Analysis," *Scientific American, 223:* 94 (1970).
28. Bhave, P. R. and C. F. Lam, "Optimal Layout for Branching Distribution Networks," *Jour. of Trans. Engrg.*, Am. Soc. of Civil Engrs., *109:* 534 (1983).
29. Garey, M. R., R. L. Graham, and D. D. Johnson, "The Complexity of Computing Steiner Minimal Trees," *SIAM Jour. of App. Maths, 32:* 835 (1977).
30. Gilbert, E. N. and H. O. Pollak, "Steiner Minimal Trees," *SIAM Jour. of App. Maths, 16:* 1 (1968).
31. Graham, R. L., "Some Results on Steiner Minimal Trees," *Bell Lab. Tech. Memorandum,* Murray Hill, N.J. (1967).
32. Shamos, M. I. and D. Hoey, "Closet-Point Problems," *16th Ann. Symp. on Foundations of Computer Science,* Inst. of Electrical and Electronic Engrs., 151 (1975).
33. Boyce, W. M. and J. B. Serry, "Steiner 72, An Improved Version of Cockayne and Schiller's Program STEINER for the Minimal Network Problem," *Bell Lab. Techn. Memorandum,* Murray Hill, N.J. (1973).
34. Boyce, W. M., "An Improved Program for the Full Steiner Tree Problem," *Bell Lab. Tech. Memorandum,* Murray Hill, N.J. (1975).
35. Chung, F. R. K. and R. L. Graham, "Algorithm Aspects of Combinations—Steiner Trees for Ladders," *Annals of Discrete Maths,* 173 (1978).
36. Cockayne, E. J. and D. G. Schiller, "Computation of Steiner Minimal Trees," *Combinatiories,* D. J. A. Welsh and D. R. Woodall, eds., Inst. of Maths and Applications, 53 (1972).
37. Almgrem, F. J., Jr., "Existence and Regularity Almost Everywhere of Solution of Elliptic Variation Problem with Constraints," *Mem. of Am. Math. Soc., 4* (1976).
38. Almgrem, F. J., Jr. and J. E. Taylor, "The Geometry of Soap Films and Soap Bubbles," *Scientific American,* 82 (July 1976).
39. Taylor, J. E., "The Structure of Singularity in Soap-Bubble-Like and Soap-Film-Like Minimal Surfaces," *Annals of Maths, 103:* 489 (1976).
40. Ridgik, T. A., "Design of Water Distribution Networks Using Linear and Heuristic Programming," Report submitted to Univ. of North Carolina in Partial Fulfillment for MS, Chapel Hill, N.C. (1981).
41. Ridgik, T. A. and D. T. Lauria, discussion of "Optimization of Gravity-Fed Water Distribution Systems: Theory," by P. R. Bhave, *Jour. of Env. Engrg.*, Am. Soc. of Civ. Engrs. *110:* 504 (1984).
42. Bhave, P. R., closure to "Optimization of Gravity-Fed Water Distribution Systems: Theory," *Jour. of Env. Engrg.*, Am. Soc. of Civ. Engrs., *110:* 509 (1984).
43. Lam, C. F., "Discrete Gradient Optimization of Water Systems," *Jour. Hyd. Div.*, Am. Soc. of Civ. Engrs., *99:* 863 (1973).
44. Rasmusen, H. J., "Simplified Optimization of Water Supply Systems," *Jour. Env. Engrg. Div.*, Am. Soc. of Civ. Engrs., *102:* 313 (1976).
45. Alperovits, G. and U. Shamir, "Design of Optimal Water Distribution Systems," *Wat. Res. Research, 13:* 885 (1977).
46. Quindry, G., E. D. Brill, and J. C. Liebman, "Optimization of Looped Water Distribution Systems," *Jour. Env. Engrg. Div.*, Am. Soc. of Civ. Engrs., *107:* 665 (1981).

CHAPTER 23

Factors Affecting Driver Route Decisions

Manouchehr Vaziri*

This chapter describes various factors affecting driver route decisions. These factors are associated with the characteristics of driver, vehicle, and route. In highway planning and transportation planning, information about these factors is essential to predict traffic flows in different components of transportation systems. The predicted flows can be used to analyze current conditions, and to evaluate alternative transportation proposals.

INTRODUCTION

The trip-making process is essentially a complex choice process which involves a number of interrelated decisions about which not much is well known. Route choice is a part of this complex individualized decision process which reflects how a trip-maker who is often faced with a choice among a number of alternative routes executes his (her) route decision.

An understanding of driver route choice behavior has always been of interest to transportation engineers and planners. Previous research has provided important insights to route choice and the results have been instrumental in developing procedures for urban transportation planning and highway planning. Indeed, information on route choice behavior has proven to be essential to traffic assignment procedures and transportation systems management schemes. In examining route choice one should determine:

1. What alternative route choices drivers perceive
2. What factors and consequences of these alternatives drivers consider important
3. Based on the perceived alternatives and their consequences, how driver route decisions are reached

*Department of Civil Engineering, University of Kentucky, Lexington, KY

The state of the art of route choice behavior has evolved from findings of studies of the above three considerations. With relevant knowledge of these three dimensions of the problem, one can predict route choice decisions.

Previous studies have shown large variations in these three dimensions; major determinants of these variations are found to be factors that are related to drives and vehicles.

In this chapter the effects of the factors related to drivers and vehicles will be discussed first. Then the findings from research on each of these three facets of route choice will be reviewed. At the end, current applications will be explained.

FACTORS RELATED TO DRIVER AND VEHICLE

Different drives have different route choice behaviors partially due to factors that are related to drives and vehicles. For a driver, route choice decision processes and the perceived alternative routes and their attributes are found to be related to the following:

1. Driver trip related choices and decisions
2. Driver information about alternative routes
3. Driver socio-economic characteristics
4. Vehicle characteristics.

Driver Trip Related Choices and Decisions

There is a set of interrelated decisions that are made about a trip. A person decides whether to make a trip, where to make it, at what time, by what mode, and by what route. Trip purpose has been found to be an important attribute that reflects the limits of these interdependences [25,46,52, 85]. Indeed, trips are of different types that have specific temporal and spatial characteristics, and the first activity of

499

any trip study is often to identify and classify them. In many transportation planning and route choice studies, trips are classified by their purpose and then are separately analyzed [46,52,85,115]. Classifications used for home-based trips are: (1) work trips, (2) shopping trips, (3) business trips, (4) school trips, (5) social trips, and (6) recreational trips. Other trip types identified are: non-home-based trips, commercial vehicle trips, and taxi trips. For a home-based work trip, perceived alternative routes are interrelated to time and mode choices, assuming the person has to make the trip with fixed residence and work place. For a shopping trip, perceived alternative routes depend on time and mode choices plus perceived alternative shopping destinations. Previous route choice research has only focused on home-based trips because they constitute the larger portion of trips, especially work trips in which travel patterns are rather homogeneous on all working days.

With the hypothesis that the options one perceives about different choices of a trip are interrelated, the next face of the trip study problem is to determine whether these dependent choices are made simultaneously or sequentially. Simultaneous choices imply no hierarchy among trip choices and that attributes of all trip options affect the choice of any one of them. Evidence of both assumptions has been reported in the literature; however, most of the urban trip studies have hypothesized that a hierarchy exists and that trip making decisions are made sequentially [24,42,46,48,49,83,84,115]. Research to date, however, has not yet clearly determined the structure of the hierarchy and interdependencies of trip choices. Perceived alternative route choices have often been investigated as being dependent on other trip related decisions. That is, the route choices are determined when their origin and destination, mode, time, frequency, and purpose are already known and are assumed as given [24,42,46,48,52,83,84,115].

Driver Information about Alternative Routes

With the hypothesis of route choice being the last stage of a sequence of trip choices and within the context of available information about existing possible routes, the domain of perceived alternative routes can be determined. Past studies have often postulated that a driver has a complete and perfect knowledge of alternative routes and their attributes. It has been shown, however, that driver information is limited due to poor dissemination of information and lack of experience with the network [2,4,87,97,113,114]. For long trips, a driver often has high preference for highways due to lack of information about local streets and the easy availability of maps and signing upon highways [87,97]. Drivers are often unable to meet route choice objectives due to poor knowledge of route characteristics [11,22]. Sometimes a driver selects a particular route because of lack of awareness of alternative routes [113,114]. Improved information dissemination systems, suggested by transportation systems management, ameliorate the driver's lack of knowledge of alternative routes. Development of new and automated systems such as ALI [71], Driver Guidance and Information System, and IMIS [116], Integrated Motorist Information System, are promising steps toward improving a driver's knowledge of route characteristics. The extent of a driver's experience with the existing network also rectifies the driver's perception of alternative routes. Driver information will change over time due learning from travel experience in the network. However, once a given route becomes habitual for a frequently made trip, the driver most often will not consider alternatives. Indeed, trip frequency is an important attribute that reflects the effect of driver experience with a particular type of trip. Smith [81] showed that a car driver, 89% of the time, uses a single route as his habitual route of work. Other studies also have shown that for trips made almost every day, such as work trips, the selected route has the tendency to become habitual [36,86,105,106].

For infrequent trips such as the ones that might occur less often than once a month, even with a very limited knowledge of street and highway network, a driver might think of different alternative routes [3,87]. Wachs' [107] study showed that length of residence, a proxy for the knowledge of network, is a determinant of avoiding congested routes. Other studies have shown that for many drivers the route choice reasons for always using a particular route are perceiving a route as the only route and not knowing about alternative routes [105–107].

Driver Socio-Economic Characteristics

Route choice decision processes, perception of alternative routes and their attributes vary among drivers due to personal characteristics. In the previous route choice research, a wide range of driver attributes has been found to be the route choice determinants. These major socio-economic characteristics are listed in Table 1.

Benshoof [3] showed that route choice decision processes are related to the age and sex of the driver. His study revealed that older men tend to select a route before getting into the car, whereas younger men select their route later, most probably during the course of the journey. Thomas [89,90] concluded that high income drivers consider travel time savings as the most important factor in route choice as compared to medium and low income drivers who are less time conscious. Wachs [107] showed that most driver socio-economic characteristics listed in Table 1 have relationships with route choice reasons; however, it was also found that such relationships are difficult to interpret. He concluded that older drivers have a preference of less congestion and more pleasant scenery. He found that higher income and better educated drivers have a lower preference for safety. He also concluded that residents of Central Business Dis-

TABLE 1. Driver Attributes That Influence Route Choice.

Age
Life cycle
Sex
Income level
Education level
Household structure
Number of persons in the household
Number of employees in the household
Household car ownership
Race
Profession
Length of time at current occupation
Residence type
Residence ownership status
Length of time at previous occupation
Residence type
Residence ownership status
Length of time lived at current address
Length of time lived at previous address
Place of birth
Miles driven last year
Years having driving license
Number of drivers in the household

tricts are less concerned with congestion. Female drivers were believed to have a high preference for shortest route and commercial development along the route.

Vehicle Characteristics

The literature contains little documentation of studies of route choice related to vehicle cahracteristics; however, there is evidence showing that vehicle characteristics can affect driver route decisions [89,90]. The driver of a commercial vehicle may not have access to all routes open to a passenger car driver. Indeed, commercial vehicle trips are of a different type and mode that should be analyzed separately from passenger car trips. Automobile characteristics such as age, type, and performance influence driver route decision. Thomas [89,90] concluded that model year of a car, a variable reflecting the age of the car, is among the statistically important variables of a route choice model. He showed that the older the car, the lower the value of travel time in the making of route choices. Once a person acquires a new automobile, he is often very conscious of it and avoids roads with bad surface conditions. A driver naturally selects routes on which his vehicle can properly perform. If the driver has an old and poorly maintained car, perhaps there is a lower preference for high speed routes. As a minimum, different cars have different meaning, purpose, and value to

persons of different socio-economic classes. This is very clearly spelled out in research by Roberts [76] in assessing the used car domain especially with an urban black perspective in classifying social data related to trip purpose in older, heirloom quality cars. Expressive travel is often via high visibility routes.

ALTERNATIVE ROUTE CHOICES

For a resourceful urban commuter, the accessible streets and highways provide numerous alternative routes between his (her) origin and destination. A driver is in a good position to explore the many feasible routes available in search of the "best route." A driver, however, perceives a limited number of these as viable alternatives for consideration due to the effect of the four groups of determinants explained above. A study at the Technical University of Helsinki [36] showed that for a passenter car driver, the number of comparable alternative routes is between three and five. Smith [81] concluded that work trips are often via habitual route. Many drivers select a particular route because that is the only one of which they are aware [113,114]. For long trips, drivers are inclined to limit their routes to highways due to lack of information about local streets and easy availability of maps and signing upon highways [87,97]. Vaziri [105,106] showed that for many drivers the reasons for selection of the preferred route include: driver has always used a particular route; driver perceived it as the only route; driver did not know about other alternative routes.

ROUTE CHARACTERISTICS

Route characteristics that influence a driver to select a certain route are numerous. These characteristics basically reflect the general concept of level of service of a route that a driver experiences during his trip. Table 2 lists major route attributes that previous research has shown to be among the reasons or factors of driver route choice. These factors can be classified into three groups:

1. Roadway attributes which reflect transportation supply variables such as number of highway lanes, lane width, design speed, etc.
2. Traffic attributes which reflect transportation demand variables such as traffic volume, percentage of trucks, etc.
3. Service attributes which incorporate both the transportation demand and supply variables to reflect the performance and operational condition of the route such as travel time, operating speed, etc.

Different drives consider different route characteristics to be important in their route choice, reflecting the influence of

TABLE 2. Route Characteristics That Influence Driver Route Choices.

Roadway attributes:
  Number of lanes
  Lane width and lateral obstructions
  Travel distance
  Design speed
  Speed limit
  Traffic control devices such as signs, signal markings
  Intersections
  Number of route segments
  Other roadway design elements such as curves, slopes
  Parking along the route
  Type of development and land-use along the route
  Parking at origin and destination
  Aesthetics and scenery
  Walking distance
  Weather and temperature
  Law enforcement
  Direct charges such as tolls
  Pavement condition and rideability

Traffic attributes:
  Traffic volume
  Traffic mix and composition
  Traffic volume of opposite direction
  Traffic mix of opposite direction

Service attributes:
  Travel time
  Reliability and variation in travel time
  Operating cost
  Operating speed
  Safety and probability of accident
  Congestion and bottlenecks
  Ease of driving
  Pedestrian crossing
  Ease of pick-up and drop-off passengers en route
  Pleasant and relaxing
  Security and privacy
  Status

the four types of determinants explained before. Driver route choice reasons vary not only in which route attributes drives consider important, but also in relative values they place on each important attribute [36,52,105,106]. Some of the attributes such as comfort, safety, status, and security cannot readily be quantified, and their practical applications for route choice and traffic assignment studies are very difficult. Even such a quantifiable attribute such as travel time is for itself quite complex and consists of different components that have different influences in driver route choice behavior [42,52,89,90].

In almost all of the past studies, travel time and/or distance minimization were found tobe the most important factors for work trip route choice decisions [2,4,13,32,34,36, 51,64,70,72,73,75,81,89,90,95,97,105-107]. Number of stop signs and traffic signals have also been found to be among the important factors for work trip route choice behavior [97,107]. For social and recreational trips scenery and aesthetic appeals were found to be among the important factors of route choice [74,87,93,98,107]. Even for work trips the important factors in trips to work can be different from trips from work [105,106]. Safety and road information signs are found to be among the important factors for long trips, especially for female drivers [64,107,112]. Vaziri [105,106] indicated that the average number of reasons influencing a driver in his route is five. Statistical method of factor analysis has been used to determine the major dimensions and statistical structure of route choice reasons. Past studies have shown that there are between five and seven groups of statistically similar reasons for route choice. They include reasons related to shortest route, overall roadway quality, congestion, safety, pleasant driving, development along the route, delays, and reliability [105-107].

DRIVER DECISION PROCESS

The driver will make a route decision based on perceived alternative routes and their attributes. Different hypotheses have been made about the driver decision process. Benshoof [3] revealed that some drivers, especially older men, tend to select their route before getting into the car and they make their decisions before starting a trip. He also showed that for some other drivers, especially younger men, route selection has more characteristics of a Markov decision process and it takes place during the course of the trip. This decision process hypothesizes that as a driver reaches different points (most probably at intersections) of his route, he makes decisions that are independent of his prior decisions. Hamerslag [33] showed that for some drivers, route selection has both characteristics of preplanning and some immediate decisions. Some drivers start from their origin on local streets at the lowest level in the road network, ascending in hierarchy to expressway, and then descending into local streets in the vicinity of their destination [8,39].

Whether a driver preplans or makes intermediate decisions, his decision process is most probably a rational choice behavior. The theory of rational choice behavior, which originated from econometric and psychometric sciences, has seen significant progress in recent years and, generally speaking, asserts that a driver can rank alternative routes in order of their perceived merits and demerits, and will select the option which provides most satisfaction. For a driver, d, an alternative route, r, of M perceived routes has a level of satisfaction, U_{rd}, comprising of perceived characteristics of the route. The level of satisfaction can be expressed as:

$$U_{rd} = f_{rd}(PR_{rd}) \qquad \text{for } r \in M \qquad (1)$$

where U_{rd} is expressed as a function, f_{rd}, of PR_{rd}, a vector representing perceived characteristics of route r by driver d. The value of U_{rd} can be interpreted as a measure reflecting the degree of satisfaction, generally called "utiles." As it was disscussed in previous sections of this chapter, there are many factors related to drivers and vehicles which influence the set of perceived routes, M. Furthermore, these factors affect the consequences of alternatives considered to be important. In other words, for relation (1) they influence the functional form and parameters of function f_{rd}, and the set of perceived important characteristics of route r, PR_{rd}, to be included in the relation. The decision process was also found to be dependent to these factors. Relation (1) is a model of driver's economic consumer behavior which represents his perceived route choice utility function. The economic theory of travel behavior postulates that a driver d who is faced with a set of perceived alternative routes M, each described by the U_{rd}, will choose the alternative with highest utility value. The psychologist's approach postulates that the driver d will make a probabilistic choice which is proportional to the value of U_{rd}. Whether one adopts an economist's approach or psychologist's approach, the practical application of models such as relation (1) is very difficult if not impossible [46,52,85]. This is mainly because, for a transportation analyst, it is very difficult to comprehend and measure the heterogeneous driver route choice perceptions.

Our knowledge of driver route choice and the important factors affecting route decisions is far from perfect and much more research is needed to provide further insight to this problem. Probably the most important achievements are the Wardrop principles. Wardrop [108] suggested two criteria to simulate driver route choice behavior. These critera which are called Wardrop principles are the basis of most existing route choice and traffic assignment analysis. Wardrop's original article elaborates his principles:

> The first criterion is quite a likely one in practice, since it might be assumed that traffic will tend to settle down into equilibrium situations in which no driver can reduce his journey time by choosing a new route. On the other hand, the second criterion is the most efficient in the sense that it minimizes vehicle-hours spent on the journey.

The first criterion postulates that driver route choice is based on individual travel time and/or cost minimization, whereas the second criterion postulates that driver route choice should be based on total travellers' time and/or cost minimization. The second criterion results in system optimal assignment which can only be achieved through traffic control measures.

APPLICATIONS

In transportation planning, route choice information is essential to predict flows in different components of any transportation system. In the conventional urban transportation planning process, the final stage of travel demand analysis often deals with the assignment of interzonal trips to different routes [7,24,42,49,84,101,103,115]. Once the estimations about the interzonal trips (origin-destination matrix, O-D matrix) are made, route choice information will be used to determine the flow in the network links. During the last three decades, techniques for performing traffic assignment have evolved from the use of ad hoc manual methods to the use of complex iterative computer programs. Route choice information and predicted link volumes can be used:

1. To conduct diagnostic studies of existing transportation facilities
2. To develop alternate transportation system proposals
3. To determine the efficiency and effectiveness of alternate transportation systems proposals
4. To determine design hourly volumes
5. To determine short-range transportation development priority programs
6. To formulate control and management strategies within transportation system mangement
7. To provide inputs and feedbacks to other stages of the transportation planning process

During the massive construction of urban freeways in the early 1950s, the earliest attemts to predict freeway flows were made. During this period, diversion curves which predict the percentage of freeway users were developed and used extensively. The diversion curve method is basically a two-route traffic assgnment method; it has many shortcomings and disadvantages. One of the major shortcomings of this method is that it is unable to reproduce and predict the traffic volumes on the entire network of major links. In order to overcome some of the shortcomings of diversion curves, network assignment procedures were developed. A number of network assignment procedures have been developed; their major differences are due to their assumptions about:

1. Driver route choice decision process
2. Important route characteristics and their dependency on traffic flows

In all the network assignment procedures, drivers are assumed to perceive all possible routes between their origins and destinations. Most network assignment procedures have assumed that drivers will select the "best route." As discussed in the previous section, the notion of best route reflects the route which maximizes driver satisfaction and results in largest value of his utility function. Most of the procedures have assumed that the shortest route, in the

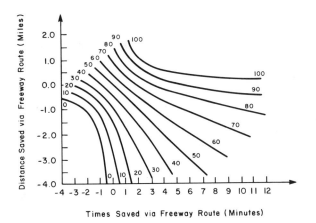

FIGURE 1. California diversion curve.

sense of travel time and/or distance, will always be chosen by drivers. In other words, it is assumed that drivers are following the Wardrop first principle. These procedures seldom have made any references to the rather modern utility theory of route choice behavior, but one can conclude that they, intentionally or unintentionally, have assumed that the drivers all have similar route choice perceptions and preferences. Few procedures have assumed that driver route selection is probabilistic and not necessary that a driver will always choose the shortest route [2,4,15,22,23,46,52,58,62,63,78,79]. These studies have incorporated the recently developed stochastic route choice models into their assignment methods and have relaxed the assumption that all drivers have similar route choice perceptions and preferences. The optimal assignment method, the second Wardrop principle, was adopted when, by means of traffic control measures, the drivers can be forced or motivated to choose the routes which will result in optimum total travel cost or total travel time [68,78,83]. Table 2 lists the important route characteristics that drivers were found to perceive in their route choice. Except for a very few practical studies, travel time, distance, and/or costs were the only route consequences that drivers were assumed to consider important. The exception to this is the very few recent works on stochastic route modelling which have included other factors such as road quality, commercial development, and scenery in traffic assignment models [2,4]. With respect to flow dependency of route characteristics, the developed procedures can be classified into two groups: those that assume travel time and cost are flow independent and those that assume travel time and cost increase as flow increases. Except the diversion curves which consider only two alternative routes, all other methods postulate that drivers consider all possible routes between origin and destination.

Diversion Curve

Diversion curves allocate trips between two possible routes. The two possible routes between origin-destination pairs are often assumed to be the fastest all-surface route and the fastest route partially or totally consisting of freeway. The principle behind diversion curves is to determine the shortest and next-shortest routes through the network for origin-destination pairs, and then assign the traffic between these two routes [5,66,100]. Figure 1 presents the California Diversion Curve which provides the percentage of freeway usage as a function of savings in travel time and distance [66]. The family of hyperbolic curves shown in Figure 1 are graphical representations of the following mathematical model:

$$P = 50 + \frac{50 (\Delta d + .5\Delta t)}{[(\Delta d - .5\Delta t)^2 + 4.5]^{.5}} \quad (2)$$

where P is the percentage of trips using freeway, Δd is the distance saved in miles by freeway and Δt is the time saved in minutes by freeway. When there is no time and distance savings by freeway, Δd and Δt are equal to zero and the model predicts that 50% of the trips will use the freeway. Figure 2 presents the Federal Highway Administration's diversion curve which was widely used in the 1960s. As this figure shows, the percentage of traffic using the freeway is assumed to depend on the ratio of travel time via freeway to travel time via quickest alternate arterial route [100].

Deterministic Network Assignment

In conventional transportation planning the study area is often divided into zones each presented by a centroid. The

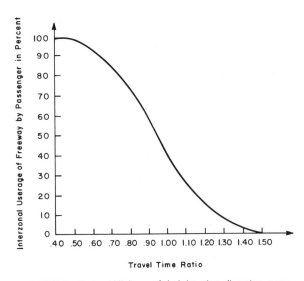

FIGURE 2. Federal Highway Administration diversion curve.

centroid is a single node, the center of activities of zone, which is assumed to be the origin and destination of all interzonal trips. Major intersections and interchanges are also presented as nodes. Major roads connecting these nodes are then presented as links. Collection of minor roads and local streets may also be presented as single links. In this way, the actual transportation network, which is often very complex to analyze, will be approximated by a manageable set of nodes, N, and links, L. The network assignment problem then is to simulate the traffic behavior when trips are using the links to reach from their origin nodes to their destination nodes. The Q_{ij} trips from a node i to a node j are distributed along M routes, all possible routes, connecting i to j, and the rth route carries Q_{rij} in such a way that:

$$\sum_{r \in M} Q_{rij} = Q_{ij} \quad \text{for all } i \text{ and } j, i, j \in N \quad (3)$$

If a link (l,m), as identified by its nodes l and m, is a link of rth route connecting i to j, then it should also carry Q_{rij}. The link (l,m) flow resulting from Q_{ij} will be the sum of flows of all M routes that have link (l,m) as one of their links. The total link (l,m) flow, V_{lm}, then will be the link (l,m) flow resulting from each Q_{ij} summed over all i and j of the network.

$$\sum_{i \in N} \sum_{j \in N} \sum_{r \in M} Q_{rij} = V_{lm} \quad (4)$$

for all (l,m), $(l,m) \in L$ and $(l,m) \in r$

Most of the network assignment procedures assume that driver route choice behavior is deterministic and is based on shortest route. For most of the procedures, shortest route is assumed to be the route with minimum travel time. Minimum travel cost and travel distance have also been used as the criteria of the shortest route. As indicated above, network assignment procedures are basically of two types: procedures that have postulated that travel time and cost are independent of flow and procedures that have postulated that they are flow dependent. When travel time or travel cost are independent of flow, the solution of traffic assignment will be the same whether it is based on the first or second Wardrop principle. The route travel time or cost, which is the summation of all its link travel time or travel cost, will be constant. Different shortest path algorithms have been developed that can efficiently find the shortest path (route) between node pairs through the network [26,42,63,68,83]. Once the shortest path between a node pairs ij is known, then the path is assumed to carry all the Q_{ij}, whereas other routes between node i and node j will assume to carry zero amount of Q_{ij} flow; this is called the all-or-nothing assignment. The total link flow of any link (l,m), V_{lm}, will be the sum of all route flows of all node pairs that are using the link

(l,m), and can be computed from relation (4). Although it is not an efficient procedure for assignment, linear programming can be used to find the link flows [6,68]. The mathematical model can be expressed as:

$$\text{Min} \sum_{(l,m) \in L} c_{lm} V_{lm}$$

$$\text{s.t.} \sum_{(i,x) \in L} V_{ix} - \sum_{(x,i) \in L} V_{xi} = \sum_{x \in N} Q_{ix} \quad (5)$$

for all i, $i \in N$

$$V_{lm} \geq 0 \quad \text{for all } (l,m), (l,m) \in L$$

Where c_{lm} is the travel time or travel cost of link (l,m). For mathematical model (5) relations are all linear and the only unknowns are V_{lm}'s. This linear programming model reflects the second Wardrop principle when the total travel time or total travel cost should be minimized, whereas shortest path algorithms reflect the first Wardrop principle when each driver selects the shortest route from origin to destination. The major disadvantage of these procedures is that they will often overload few links, and allow the rest to carry zero flows. This is mainly because the travel time and cost of the link are assumed to be constant no matter how much traffic flow they carry, when it is obvious that once the link flow reaches capacity the travel time or travel cost on the link will drastically increase. To overcome this problem, procedures that consider travel time and travel cost as flow dependent have been developed [10,16,18,21,27,37,38,42–44,46,48, 49,67–69,82,104]. The general mathematical form of the network assignment problem for the case that drivers select their routes on the second Wardrop principle can be stated as follows:

$$\text{Min} \sum_{(l,m) \in L} c_{lm}(V_{lm}) V_{lm}$$

$$\text{s.t.} \sum_{(i,x) \in L} V_{ix} - \sum_{(x,i) \in L} V_{xi} = \sum_{x \in N} Q_{ix} \quad (6)$$

for all i, $i \in N$

$$V_{lm} \geq 0 \quad \text{for all } (l,m), (l,m) \in L$$

The solution to the above nonlinear programming model will provide optimal traffic assignment for link flows, V_{lm}'s, when link travel time or travel cost, c_{lm}, is assumed to be a function of link flow, V_{lm}. Different functional forms have been used which are directly related to assumptions about speed flow relation. As the link flow increases the speed decreases and consequently the travel time increases. The literature of traffic engineering contains many graphs of speed versus flow, presenting mathematical models of speed as a function of flow. The Federal Highway Administration suggests the following relation to compute travel time as a function of link flow.

$$c_{lm} = c_{lm0}\left[1 + .15\left(\frac{V_{lm}}{C}\right)^4\right] \quad (7)$$

Where c_{lm0} is the free flow travel time, and C is link capacity flow. Relation (7) adjusts the travel time, c_{lm}, to reflect assigned flow V_{lm}. Other functional forms can be found in the Schneider model [77] and the Smok model [82]. The general mathematical model pertaining to drivers selecting their route based on the first Wardrop principle can be stated as follows:

$$\text{Min} \sum_{(l,m) \in L} \int_0^{V_{lm}} c_{lm}(x)dx$$

$$\text{s.t.} \sum_{(i,x) \in L} V_{ix} - \sum_{(x,i) \in L} V_{xi} = \sum_{x \in N} Q_{ix} \quad (8)$$

for all i, $i \in N$

$$V_{lm} \geq 0 \quad \text{for all } (l,m), (l,m) \in L$$

Efficient algorithms for solving of relationships of (6) or (8) have been developed [18,19,50]. These algorithms will iteratively evaluate link flows based on estimates of link travel time or costs which are modified and reevaluated at each iteration. When either or both travel time or cost are assumed to be flow dependent, the assignment based on the first Wardrop principle will be different from assignment based on the second Wardrop principle.

Stochastic Network Assignment

Recently, due to advancements in theory of travel choice behavior, a new generation of network assignment procedures has been developed. For most of these procedures, the relation (1) has been taken into account. The stochastic network assignment is similar to deterministic assignment except the drivers are assumed to have different route choice perceptions and preferences [2,4,46,52,54,56,78,79]. Different assumptions about the form of the utility function and important route characteristics have been made. To account for the effects of unobserved driver, vehicle, and route characteristics, the utility of each alternative route is assumed to be a random variable. The utility function is often assumed to consist of two parts, a deterministic component of measured utility and a random error term reflecting the influence of factors unobserved by the analyst. The relation (1) is then modified to a random utility model which can be expressed as:

$$U_{rd} = u_r(R,D) + e_r \quad \text{for all } r \in M \quad (9)$$

where U_{rd} is the perceived utility of alternative route r followed by driver d. The measured utility of alternative route r is $u_r(R,D)$ which is a function of a vector of observed route characteristics R and a vector of observed driver and vehicle attributes D. The random error term of utility of alternative route r is e_r which cannot be observed by the analyst. The set of alternative routes M will include all possible routes. Once the distribution of error term is specified, with proper assumption of functional form of $u_r(R,D)$, and the observed values for R and D, the distribution of random variable U_{rd} can be determined. This results in assigning a probability to any driver route decision. For a utility maximizer driver the probability of selecting a route r among M possible routes between his origin and destination is given as:

$$P_{rd} = Pr[U_{rd} > U_{kd} \text{ for all } k, k \neq r \text{ and } k \in M] \quad (10)$$

where P_{rd} is the probability of driver d selecting route r, Pr stands for probability, U_{kd} is the utility of route k for driver d. With proper aggregation procedure, the drivers' route choice behavior of each origin-destination pair can be estimated and route flows can be predicted [2,4,46,52,54,57,78,79]. Then, the total link flow resulting from each route flow can be computed, as it was shown in relation (4). Replacing U_{rd} and U_{kd} by their components of relation (9) results in:

$$P_{rd} = Pr[u_r(R,D) + e_r > u_k(R,D) + e_k \\ \text{for all } k, k \neq r \text{ and } k \in M] \quad (11)$$

The deterministic component of the utility function, $u_r(R,D)$ is often postulated to have a linear or a product form of variables R and D, and the error term, e, is often postulated to have a normal or Gumbel distribution. If for alternative routes, the error terms of the choice utility function are all independent and have identical Gumbel distribution, the model is referred to as the multinomial logit [2,4,46,52,78,85] and can be expressed as:

$$P_{rd} = \frac{e^{u_r(R,D)}}{\sum_{k \in M} e^{u_k(R,D)}} \quad (12)$$

where e is the natural logarithms base, 2.71828. If for alternative routes, the error terms of the choice utility function have a joint density function of multivariate normal, the model is referred to as multinomial probit [46,52,79,85].

CONCLUSION

The empirical evidences clearly indicate that there are many factors affecting driver route decisions. These factors are related to the characteristics of route, driver, and vehicle. For transportation system evaluation, the analyst must have adequate information about these factors in order to predict traffic flows. Furthermore, the information can be used to formulate control and management strategies within Transportation System Management (TSM) programs. To accomplish these tasks, often a network consisting of highways and major streets is defined, and its traffic flow on a link-by-link basis is predicted. By comparing capacity to predicted flow then, performance of each transportation system component will be determined.

The few existing stochastic network assignment procedures that can incorporate a large number of factors affecting driver route decisions are currently at research stage. Deterministic network assignment procedures will continue to dominate the practice until stochastic network procedures prove reliable and easy to apply. For these procedures, travel time, distance and/or costs are the only factors assumed affecting driver route decisions. The flow independent procedures are easier to apply, but the capacity-restrained techniques provide more reliable predictions and will be more employed in practice.

REFERENCES

1. Andrew, C., "An Interview Survey of Motorway Driver Information," *Transport and Road Research Laboratory*, No. 742, Berkshire, England (1977).
2. Ben-Akiva, M., M. J. Bergman, A. J. Daly, and R. Ramaswamy, "Modelling Inter Urban Route Choice Behavior," *Ninth International Symposium on Transportation and Traffic Theory*, VNU Science Press, pp. 299-330 (1984).
3. Benshoof, J. A., "Characteristics of Drivers' Route Selection Behavior," *Traffic Engineering and Control*, 606-609 (1970).
4. Bergman, M. J., M. Ben-Akiva, L. A. Silman, and S. B. Pitschke, "An Analysis of Interurban Route Choice in the Netherlands," *Proceedings of the PTRC Summer Annual Meeting*, University of Warwick, United Kingdom (1982).
5. Bevis, H. W. and A. M. Voorhees, "Estimating A Road-User Cost Function from Diversion Curve Data," *Highway Research Record*, 100, 47-54 (1965).
6. Blunden, W. R., "Some Applications of Linear Programming to Transportation and Traffic Problems," *Institute of Traffic and Transportation Engineering*, University of California, Berkeley, California (1956).
7. Blunden, W. R., *The Land-Use/Transportation System*, Pergamon Press (1971).
8. Bovy, P. H. L., "Het Kortste—Tijd Routekeuzecriterium: Een Empirische Toetsing," Colloquium Vervoersplanologisch Speurwerk, The Netherlands (1979).
9. Branston, D., "Link Capacity Functions: A Review," *Transportation Research*, 10, 223-236 (1976).
10. Brown, C. and Y. Gur, "Adaptable-Zone Transportation Assignment Package," *Transportation Research Record*, 637, 57-62 (1977).
11. Burrell, J. E., "Multiple Route Assignment and Its Application to Capacity Restraint," in W. Leutzbach and P. Baron, eds., *Fourth International Symposium on the Theory of Traffic Flow*, Karlsruhe, Germany (1968).
12. Campbell, E. W. and R. S. McCargar, "Objective and Subjective Correlates of Expressway Use," *Highway Research Board*, Bulletin No. 119, 17-38 (1956).
13. Carpenter, S. M. *Driver's Route Choice Project—Pilot Study*, Research Report, Transport Studies Unit, Oxford University (1979).
14. Carr, S. and D. Schissler, "The City as a Trip: Perceptual Selection and Memory in the View from the Road," *Environment and Behavior*, 1, 7-35 (1969).
15. Colony, D. C., *An Application of Game Theory to Route Selection*, University of Toledo (1969).
16. Dafermos, S. C., "An Extended Traffic Assignment Model with Application to Two-Way Traffic," *Transportation Research*, 5, 336-389 (1971).
17. Dafermos, S. C. and F. T. Sparrow, "The Traffic Assignment Problem for a General Network," *Journal of Research*, National Bureau of Standards, No. 73B, 91-118 (1969).
18. Daganzo, C. F., "On the Traffic Assignment Problem with Flow Dependent Costs," *Transportation Research*, 11, (6), 433-438 (1977).
19. Daganzo, C. F. and Y. Sheffi, "On Stochastic Models of Traffic Assignment," *Transportation Research*, 11, (3), 253-274 (1977).
20. Daly, A. J., "Estimating Choice Models Containing Attraction Variables," *Transportation Research*, 16B (1), 5-15 (1982).
21. Davidson, K. B., "A Flow-Travel Time Relationship for Use in Transportation Planning," *Proceedings, Australian Road Research Board, Vol. 3*, Melbourne, 183-194 (1966).
22. Dial, R. B., *Probabilistic Assignment: A Multipath Traffic Assignment Model which Obviates Path Enumeration*, Ph.D. dissertation, University of Washington, Seattle (1970).
23. Dial, R. B., "A Probabilistic Multipath Traffic Assignment Model which Obviates Path Enumeration," *Transportation Research*, 5, 83-111 (1971).
24. Dickey, J. W., *Metropolitan Transportation Planning, Second Edition*, McGraw-Hill (1983).
25. Domencich, T. A. and D. McFadden, *Urban Travel Demand, Behavioral Analysis*, North-Holland/American Elsevier (1975).
26. Dreyfus, S. E., "An Appraisal of Some Shortest and Path Algorithms" The Rand Corporation RM-5433-PR, Santa Monica, Calif. (1967).

27. Eash, R. W., B. N. Janson, and D. E. Boyce, "Equilibrium Trip Assignment: Advantages and Implications for Practice," *Transportation Research Record, 728,* 1-8 (1979).
28. Freeman Fox and Associates, "Speed/Flow Relationships on Suburban Main Roads," *Road Research Laboratory,* Department of the Environment, London (1972).
29. Fox, P. D., "The Value of Time for Passenger Cars, A Behavioral Study of Driver Route Choice, Stanford Research Institute, Stanford, Calif. (1965).
30. Griep, D. J., "An Analyses of the Driving Task: Analytical Points of View of the System," *OSCE Symposium,* Rome, Italy (1972).
31. Golob, T. F. and R. Dobson, "Assessment of Preferences and Perceptions Toward Attributes of Transportation Alternatives," *Transportation Research Board,* Special Report No. 144, 58-78 (1974).
32. Haefner, L. E. and L. V. Dickinson, "Preliminary Analysis of Disaggregate Modeling in Route Choice," *Transportation Research Record, 527,* 66-72 (1974).
33. Hamerslag, R., "Onderzoek naar Routekeuze met Behulp van een Gedisaggregeerd Logitmodel," *Verkeerskunde, 8,* 377-382 (1979).
34. Haney, D. B., "The Value of Time for Passenger Cars, Theoretical Analysis and Description of Preliminary Experiments," Stanford Research Institute, Stanford, Calif. (1967).
35. Hansson, A., "Studies in Driver Behavior, with Applications in Traffic Design and Planning: Two Examples," Tekniska Hoegskolan, Bulletin 9, Lund, Sweden (1975).
36. Helsinki Technical University, *Factors Governing Choice of Route,* Helsinki, Finland (1969).
37. Huber, M. J., H. B. Boutwell, and D. K. Withford, "Comparative Analysis of Traffic Assignment Techniques with Actual Highway Use," *National Cooperative Highway Research Program Report, 58* (1968).
38. Hymphrey, T. F., "A Report on the Accuracy of Traffic Assignments When Using Capacity Restraint," *Highway Research Board, 191,* 53-75 (1967).
39. Hidano, N., "Driver's Route Choice Model: An Assessment of Residential Traffic Management," WCTR, Hamburg, Germany (1983).
40. Holden, S. M. and C. C. Wright, "Route Choice and Traffic Control in Central Urban Areas," *Proceedings of the International Symposium of Traffic Control Systems,* Institute of Transportation Studies, *Vol. 2D,* University of California, Berkeley, Calif., 223-250 (1979).
41. Hornbeck, P. C., *Visual Values for the Highway User: An Engineer's Notebook,* U.S. Department of Transportation, Federal Highway Administration (1973).
42. Hutchinson, B. C., *Principles of Urban Transport Systems Planning,* McGraw Hill (1974).
43. Irwin, N. A., A. N. Dodd, and H. G. Von Cube, "Capacity Restraint in Assignment Programs," *Highway Research Board,* Bulletin No. 297, Washington, DC (1969).
44. Irwin, N. A. and H. G. Von Cube, "Capacity Restraint in Multi-Travel Mode Assignment Programs," *Highway Research Board,* Bulletin No. 345, Washington, DC (1962).
45. Johnson, M. A., "Attribute Importance in Multiattribute Transportation Decisions," *Transportation Reserach Record,* No. 673, pp. 15-21 (1978).
46. Kanafani, A., *Transportation Demand Analysis,* McGraw Hill (1983).
47. Kansky, K. J., "Travel Patterns of Urban Residents," *Transportation Science," 1* (4), 261-285 (1967).
48. Kruecheberg, D. A. and A. L. Silvers, *Urban Planning Analysis: Methods and Models,* John Wiley (1974).
49. Lane, R., T. J. Powell, and P. P. Smith, *Analytical Transport Planning,* John Wiley (1972).
50. Le Blanc, L. J., E. K. Morlok, and W. P. Pierskalla, "An Efficient Approach to Solving the Road Network Equilibrium Assignment Problem," *Transportation Research, 9* (5), 309-318 (1975).
51. Lessieu, E. J. and M. M. Zupan, "River Crossing Travel Choice: The Hudson River Experience," *Highway Research Record, 322,* 54-67 (1970).
52. Manheim, M. L., *Fundamentals of Transportation Systems Analysis Volume I: Basic Concepts,* MIT Press (1979).
53. Manski, C. F., *The Analysis of Qualitative Choice,* Ph.D. dissertation, Department of Economics, MIT (1973).
54. Manski, C. F., "The Structure of Utility Models," *Theory and Decision, 8,* 229-254 (1977).
55. Mast, T. M. and J. A. Ballas, "Diversionary Signing Content and Driver Behavior," *Transportation Research Record, 600,* 14-20 (1976).
56. May, A. D. and H. L. Michael, "Allocation of Traffic to Bypasses," *Highway Reserach Board,* Bulletin No. 61, Washington, DC (1955).
57. McFadden, D., "Modelling the Choice of Residential Location," *Transportation Research Record, 673,* 72-77 (1978)
58. McLaughlin, W. A., *Multi-Path System Traffic Assignment Algorithm,* Department of Highways, Research Report No. RB 108, Ontario (1966).
59. Michaels, A. J., "Attribute Importance in Multiattribute Transportation Decisions," *Transportation Research Record, 673,* 15-21 (1978).
60. Michaels, R. D., "Attitudes of Drivers Toward Alternative Highways and Their Relation to Route Choice," *Highway Research Record, 122,* 50-74 (1966).
61. Michaels, R. M., "The Effect of Expressway Design on Driver Tension Responses," *Public Roads, 32* (5) (1962).
62. Minty, G. J., "A Comment on the Shortest Route Problem," *Operations Research, 5* (5), 724-728 (1967).
63. Moore, E. F., "The Shortest Path Through a Maze," *Proceedings of the International Symposium on the Theory of Switching,* Harvard University, Cambridge, Massachusetts (1968).
64. Morisugi, H., N. Miyatake, and A. Katoh, "Measurement of Road User Benefits By Means of a Multi-Attribute Utility Function," *Papers of the Regional Science Association, 46,* 31-43 (1981).

65. Mosher, W. W., "A Capacity Restraint Algorithm for Assigning Flow to a Transportation Network," *Highway Research Record, 6,* Washington, DC (1963).
66. Moskowitz, K., "California Method of Assigning Diverted Traffic to Proposed Freeways," *Highway Research Board,* Bulletin No. 130, Washington, DC (1956).
67. Newell, G. F., "The Effect of Queues on the Traffic Assignment to Freeway," *Proceedings of the Seventh International Symposium on Transportation and Traffic Flow Theory,* Kyoto, Japan, 311–340 (1977).
68. Newell, G. F., *Traffic Flow on Transportation Networks,* MIT Press (1980).
69. Nguyen, S., "An Algorithm for the Traffic Assignment Problem," *Transportation Science, 8,* 203–216 (1974).
70. Orman, J. C., "Study of Driver Route Choice," Graduate Report, No. 1967-6, Institute of Transportation Studies, University of California, Berkeley, Calif. (1967).
71. Ottenroth, K., "Das Autofaher-Leit-und Informationsystem. ALI. Neue Möglichkeiten der Verkehrsbeeinflüssung auf Autobahnen und Schnellstrassen," *ATZ Automobiltechnische Zeitschrift, 81* (1), 3–7 (1979).
72. Outram, V. E., "Route Choice," *Proceedings of the PTRC Summer Annual Meeting,* University of Warwick, United Kingdom (1976).
73. Outram, V. E. and E. Thompson, "Drivers' Perceived Cost in Route Choice," *Proceedings of the PTRC Summer Annual Meeting,* University of Warwick, United Kingdom (1978).
74. Parody, T. E., "Technique for Determining Travel Choices for a Model of Nonwork Travel," *Transportation Research Record, 673,* 47–53 (1978).
75. Ratcliffe, E. P., "A Comparison of Driver's Route Choice Criteria and Those Used in Current Assignment Processes," *Traffic Engineering and Control,* 526–530 (1972).
76. Roberts, J. M., M. D. Williams, and G. D. Poole, *Used Car Domain: An Urban Black Perspective in Classifying Social Data,* Jossey Pass Inc. (1985).
77. Schneider, M., "A Direct Approach to Traffic Assignment," *Highway Research Record, 6,* 71–75 (1963).
78. Sheffi, Y., *Urban Transportation Networks, Equilibrium Analysis with Mathematical Programming Methods,* Prentice-Hall Inc. (1985).
79. Sheffi, Y. and W. Powell, "A Comparison of Stochastic and Deterministic Traffic Assignment over Congested Networks," *Transportation Research, 15B* (1), 53–64 (1980).
80. Shortreed, J. H. and J. Wilson, *A Minimum Path Algorithm,* Department of Civil Engineering, University of Waterloo, Waterloo, Ontario (1968).
81. Smith, R. W., *Attitude Toward Route Choice in Urban Transportation Systems,* Damas and Smith Limited, Toronto, Canada (1969).
82. Smock, R. B., "An Iterative Assignment Approach to Capacity Restraint on Arterial Networks," *Highway Research Board,* Bulletin No. 347, Washington, DC (1962).
83. Steenbrink, P. A., *Optimization of Transportation Network,* John Wiley & Sons, Inc. (1974).
84. Stopher, P. R. and A. H. Meyburg, *Urban Transportation Modeling and Planning,* Heath and Company (1975).
85. Stopher, P. R. and A. H. Meyburg, *Behavioral Travel-Demand Models,* Heath and Company (1976).
86. Surti, V. H. and E. F. Gervais, "Peak Period Comfort and Service Evaluation of an Urban Freeway and an Alternative Surface Street," *Highway Research Record, 157,* 144–178 (1967).
87. Tagliacozzo, F. and F. Pirzio, "Assignment Models and Urban Path Selection Criteria: Results of a Study of the Behavior of Road Users," *Transportation Research, 7,* 313–329 (1973).
88. Taylor, W. C., "Optimization of Traffic Flow Splits," *Highway Research Records,* No. 230, Washington, DC (1958).
89. Thomas, T. C., "The Value of Time for Drivers of Passenger Cars, An Experimental Study of Commuters Values," Stanford Research Institute, Stanford, Calif. (1967).
90. Thomas, T. C., "Value of Time for Commuting Motorists," *Highway Research Record, 245,* 17–35 (1968).
91. Tomlin, J. A., "A Linear Programming Model for the Assignment of Traffic," *Proceedings of Australian Road Research Board,* Melbourne, 263–269 (1966).
92. Toronto Metropolitan and Region Transportation Study, *Calibration of a Regional Traffic Prediction Model for the A.M. Peak Period,* Toronto, Ontario (1967).
93. Trankle, V., "Precaution versus Enjoyment: A Significant Dimension of Motivation of Traffic Behavior," *Psychologic and Praxis, 25,* 105–121 (1981).
94. Transporation Planning Associates, *Inter Urban Route Choice Study—Driver Interview and Journey Time Surveys in Gloucester,* Department of the Environment, Economics, Highway and Freight Division, London, England (1975).
95. Trueblood, D. L., "Effect of Travel Time and Distance on Freeway Usage," *Highway Research Record,* Bulletin No. 61, 18–35 (1952).
96. Tversky, A., "Elimination by Aspects: A Theory of Choice," *Psychological Review, 79,* 281–299 (1972).
97. Ueberschaer, M. H., "Choice of Routes on Urban Network for the Journey to Work," *Highway Research Record, 369,* 228–238 (1971).
98. Ulrich, R. S., "Environment as a Factor in Route Choice," *Michigan Geographical Publication,* No. 12.
99. U.S. Department of Commerce, Bureau of Public Roads, *Traffic Assignment Manual,* Washington, DC (1964).
100. U.S. Department of Transportation, Federal Highway Administration, *Traffic Assignment,* Washington, DC (1973).
101. U.S. Department of Transportation, Federal Highway Administration, *PLANPAC/BACKPAC, General Information Manual,* GPO Stock No. 050-011-00125-0, Washington, DC (1977).
102. U.S. Department of Transportation, Federal Highway

Administration, *Traveler Response to Transportation System Changes, 2nd ed.*, Washington, DC (1981).

103. U.S. Department of Transportation, Urban Mass Transit Administration, *UTPS Reference Manual*, Washington, DC (1977).

104. Van Vliet, A., "Road Assignment," *Transportation Research, 10*, 137–157 (1976).

105. Vaziri, M., "Driver Perceptions of Route Factors for Automobile Work Trips," M.S. Thesis, University of California, Davis, Calif. (1976).

106. Vaziri, M. and T. N. Lam, "Perceived Factors Affecting Driver Route Decision," *Journal of Transportation Engineering, ASCE, 109* (2), 297–311 (1983).

107. Wachs, M., "Relationship Between Driver's Attitudes Toward Alternative Routes and Driver and Route Characteristics," *Highway Research Record, 197*, 70–87 (1967).

108. Wardrop, J. G., "Some Theoretical Aspects of Road Traffic Research," *Proceedings of the Institute of Civil Engineers, Vol. 1*, part 2, London, 325–378 (1952).

109. Whiting, P. D. and J. A. Hillier, "A Method for Finding the Shortest Route Through a Road Network," Research Note RN/3337, Road Research Laboratory, Ministry of Transport, London (1958).

110. Wilson, A. G., *Urban and Regional Models in Geography and Planning*, John Wiley (1974).

111. Wohl, M. and B. V. Martin, *Traffic System Analysis for Engineers and Planners*, McGraw Hill (1967).

112. Wootton, H. J., M. P. Ness, and R. S. Burton, "Improved Direction Signs and the Benefits for Road Users," *Traffic Engineering and Control* (1981).

113. Wright, C. C., "Some Characterists of Driver's Route Choice in Westminster," *Proceedings of the PTRC Summer Annual Meeting*, University of Warwick, United Kingdom (1976).

114. Wright, C. C. and H. C. Orrom, "The Westminster Route Choice Survey," *Traffic Engineering and Control, 17* (8/9), 348–354 (1976).

115. Yu, J. C., *Transportation Engineering, Introduction to Planning, Design and Operation*, Elsevier (1982).

116. Zove, P. and C. Berger, *Integrated Motorist Information System (IMIS) Feasibility and Design Study*, Sperry Rand Corporation, New York (1978).

SECTION FOUR
Energy

CHAPTER 24 Cooling Water Supply for Energy Production 513
CHAPTER 25 Storage Treatment of Salt-Gradient Solar Pond 535
CHAPTER 26 Economics of Industrial Fluidized Bed Boilers 557

CHAPTER 24

Cooling Water Supply for Energy Production

BENJAMIN F. HOBBS*

INTRODUCTION

The 1970s were the decade of the energy crisis; many observers believe that in the 1980s, water problems will come to the fore [e.g., 1]. Droughts in many regions of the U.S. have received front-page coverage, and political quarrels over who should be able to use our increasingly scarce water resources are commonplace. Electric power and synthetic fuel plants, while helping to solve energy problems, unfortunately exacerbate water conflicts because of the huge amounts of water they require. For example, a 600 megawatt (MW) coal fired power plant with evaporative cooling towers and sulfur scrubbers can consume up to 12 million gallons per day (mgd) [2]. Although electric utilities evaporated merely 1% of the water consumed by all users in the U.S. in 1975, that fraction may increase to as much as 8% by the year 2000 [3]. This increase will cause controversy out of proportion to the actual amount of water used because individual plants are large water users, have many environmental and social impacts, and are often the latest arrivals in basins whose waters are already fully allocated.

Energy production requires water for a number of purposes. Among these are fossil fuel mining and land reclamation, direct generation in hydroelectric plants, process water in synfuel production, and transportation of coal via slurry pipelines. But the largest consumer (via evaporation) of water is thermal power production. Nuclear and fossil steam plants need boiler makeup water, condenser cooling water, potable water, and plant service water; in addition, coal plants often require water for ash handling and flue gas desulfurization [2]. The most important use is cooling. Thermal plants are only 30% to 40% efficient, meaning that approximately two-thirds of the heat value of the fuel must be rejected into the environment. This is commonly accomplished by passing the steam through the turbines and then through condensers, where heat is rejected to a secondary water loop which then transfers the heat to the atmosphere by evaporation, radiation, or conduction or to a water body by conduction (Figure 1). Most new plants use evaporative cooling towers, in which cooling water is passed through an air stream, cooled, and recycled to the condensers (Figure 1b). Water consumed in cooling can account for more than two-thirds of the total consumption of a power plant.

Growth in energy demand is only partially to blame for increased conflict between energy and other water users. Until recently, the cooling technology of choice for the power industry has been once-through cooling, in which cool water is withdrawn from a water body, passed through the condensers and heated, and then immediately returned to the water body (Figure 1a). If ample water is available, this cooling method is the least expensive because of its low capital and energy requirements and high thermal efficiency. More of the heat is dissipated via convection and radiation than with other cooling methods; as a result, once-through causes significantly less evaporation. But concerns about the resulting thermal pollution in natural water bodies contributed to the passage of the Federal Water Pollution Control Act Amendments [5]. That law mandates, with a few exceptions, that all new power plants adopt the "best available control technology" (BACT), subsequently defined by the U.S. Environmental Protection Agency as closed cycle evaporative cooling tower systems. Exceptions are permitted, if no significant impact to aquatic ecology would result [6]. BACT, while solving water *quality* problems, worsens water *quantity* difficulties by causing more evaporation.

A second trend that has aggravated water supply conflicts is the efforts by states, notably California, to discourage

---

*Department of Systems Engineering, Case Western Reserve University, Cleveland, OH

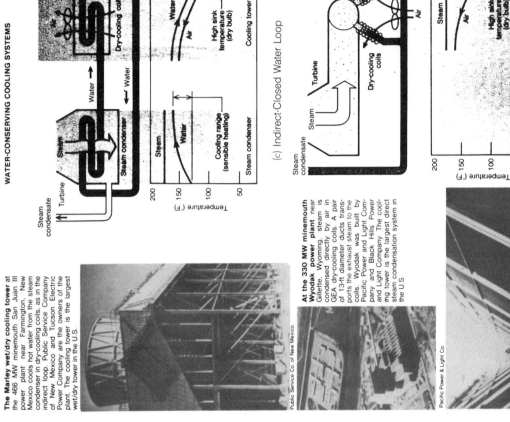

FIGURE 1. Schematics of various types of conventional cooling systems with corresponding plots of fluid temperature variations in the heat exchangers (source: Reference [4]. Reprinted with the permission of the Electric Power Research Institute and Mechanical Engineering). (a) Once-through, (b) evaporative, (c) indirect-closed water loop and (d) direct steam condensation.

facility siting along their coasts. This forces utilities to look inland for sites and, therefore, water supplies. A final trend has been a slowdown in improvements in power plant heat rates. This is due to stricter pollution regulations and the adoption of atomic power. As a result, new plants must reject more heat and, ultimately, evaporate more water.

Energy production's large appetite for water presents problems in both the arid western U.S. and the humid East—but for different reasons. It is an unfortunate coincidence that in the West, rainfall and streamflows are relatively small in precisely those regions which contain the bulk of the nation's oil shale, tar sands, and thick low-sulfur coal seams. Often, what streamflows there exist are over appropriated, meaning that water diversions to energy production will necessarily be at the expense of other (probably agricultural) users. California, Montana, and a few other states, concerned about the social changes that would result from such reallocations, have effectively banned transfers of water rights from irrigated agriculture to industry—even though the economic value of water to farmers is two orders of magnitude smaller than its worth to energy producers.

In the relatively wet eastern U.S., there is plenty of extra water available for energy development, *on average*. But specific subregions are nevertheless chronically water short. In theory, there exist sufficient flows and reservoir sites to accommodate all possible needs. For example, the U.S. Army Corps of Engineers [7] has estimated that 3246 cfs is the "maximum practical" reservoir development in the Delaware River basin, a level which far exceeds conceivable demands. But the obstacles to obtaining sufficient water for energy development in that basin are not hydrologic in nature, but legal and institutional. Despite repeated droughts that have stressed the many public water supply systems dependent on that basin (including New York City's), environmental and social concerns have killed the proposed Tocks Island and Trexler dams. Further, the Delaware River Basin Commission, which controls water allocation in that basin, has stated that:

> To continue to evaporate waters of the Delaware River Basin in large quantities in the cooling of electrical generating stations appears to be inconsistent with the doctrine of equitable apportionment. Therefore, it has been assumed for purposes of this staff report that no new quantities of water will be available for the electric utility industry beyond that which will be required for power installations to be operational by the year 1982 [8, p. 1-66].

As a result, the basin's electric utilities have been forced to plan their own small high-flow skimming reservoir and to accept a lower level of water supply reliability for the Limerick nuclear plant. In general, in eastern river basins which are already heavily used for municipal and industrial water supply, the environmental and social impacts of further water supply development will make it difficult for energy industries to obtain the water they need. Unhappily, these problems are likely to be worse in precisely those basins that would be most attractive for power development because of their proximity to energy consumers.

This chapter is addressed to civil engineers who must help secure water supplies and design water conservation measures for energy firms and those in government agencies who must regulate the industry's use of water and write comprehensive water or energy plans. The purpose is to present an overview of alternatives for solving water supply problems for new energy facilities. One possibility is lowering energy's water requirements. Because the greatest opportunities for water conservation in energy production lie in condenser cooling, technologies for decreasing cooling water evaporation are summarized. Next, supply alternatives are discussed. These include not only the traditional option of surface water supply development, but also groundwater, wastewater, brackish supplies, and other innovative sources. Because water laws and government institutions play such an important role in allocating water, they will be discussed along with the engineering and economic aspects of water acquisition. All costs are expressed in 1980 dollars. Finally, methods for consistent evaluation of conservation and supply measures are discussed.

CONDENSER COOLING TECHNOLOGIES

Table 1 summarizes some data on conventional cooling technologies, including the extent of their use in 1980, capacity penalties, and amount of evaporation. Conventional technology relies entirely on water as a heat transfer medium between the condensers and the heat sink. Cooling systems using other media, especially anhydrous ammonia, are now in the demonstration phase, but have yet to be adopted commercially. This section gives a brief overview of alternative cooling methods and their economics; the reader is referred elsewhere for detailed information on cooling system design [see especially 12,13; also 14,15,16,17].

As Table 1 shows, once-through technology still cools most steam-electric capacity in the U.S. But because of thermal pollution regulations, less than 20% of capacity installed or under construction since 1978 adopts this approach [4]. These new plants are generally found on the Great Lakes and along the coasts where the impacts of heated water can sometimes be shown to have no significant impact. The economic advantages of once-through relative to the other technologies should be obvious from the second row of Table 1, given that new steam capacity generally costs $2000/KW or more.

An additional advantage of once-through cooling is that water treatment costs are much less than for systems which require recirculation of water. Recirculation treatment systems generally include ion exchange, clarification, and addition of biocides (see [13]). The cost of such treatment is generally between $0.20 and $0.50 per 1,000 gallons of

TABLE 1. Conventional Power Plant Cooling Technologies.

	Once-through	Evaporative			Dry Cooling		Mixed wet/dry
		Mechanical draft	Natural draft	Cooling pond	Mechanical draft	Natural draft	
Percentage of total 1980 U.S. steam generation capacity[a]	60.2	----------------19.0----------------		9.9	----------------0.07[b]----------------		0[c]
Percentage capacity penalty[d]	0	2.7	3.0	3.6	16.7	14.4	[e]
Water evaporated, (gal/10^6 BTU rejected)[f]	130	180	no data	190[g], 240[h]	0	0	[e]

[a]Total steam-electric power plant capacity in 1980 = 461,099 MW. Percentages exclude the 10.8% of steam capacity which used a combination of cooling methods. Source: Sonnichsen et al. [9].
[b]Consisting of the 330 MW dry-cooled, coal-fired Wyodak unit in Wyoming.
[c]The only wet-dry plant in the U.S. is San Juan 3, a coal-fired unit in New Mexico whose cooling system consumes less than 35% of the amount of water of an equivalent evaporative system. This unit was brought online after 1980.
[d]Penalty relative to once-through cooling, including loss of capacity due to thermal inefficiencies (elevated pressures at turbine outlets), and pumping and fan power requirements. This data is for a 1200 MW nuclear power plant sited along the mid-Atlantic coast. The penalties for coal fired power plants, which have higher heat rates, would be smaller. Source: Meier et al. [10, derived from data in 11].
[e]Between the values for wet and dry cooling, the exact value depending on the design.
[f]Evaporation due to condenser cooling, assuming ambient wet-bulb temperature of 50°F. Source: Probstein and Gold [12, Fig. 4-5].
[g]Pond area = 1 acre/MW.
[h]Pond area = 2 acre/MW.

makeup water [18]. But because of the volume of water required in once-through cooling, water transport costs can be prohibitively high if the plant is not sited next to the water source. Beyond a certain distance, economics may favor recirculation. A final consideration in once-through systems is that if seawater is used, a capital cost penalty of 20%, relative to the cost of a freshwater once-through system, is incurred (due in large part to the use of corrosion-resistant titanium in the condenser tubing) [16,18].

Of the two major types of evaporative recirculation systems shown in Table 1, cooling towers are most often used. Cooling ponds suffer from the disadvantages of requiring more land (1 to 5 acres/MW, depending on the rate of evaporation at the site) and evaporating more water. Nevertheless, if water is very cheap, the low capital and maintenance costs of cooling ponds may still make them attractive. They may also be advantageous if the utility in any case would need to build a small storage reservoir near the plant. Probstein and Gold [12] report that cooling ponds are attractive only if the cost of cooling water saved is less than $0.10 to $0.25 per 1000 gallons. Cooling ponds tend to be located in the southern U.S. [9], because higher evaporation rates there minimize the amount of land required.

Evaporative cooling towers are of two kinds: mechanical towers, in which fans force air upwards through the stream of water, and natural-draft towers, in which a 200 ft or taller hyperbolic shell induces the upward air flow. The choice between them depends on the relative cost of capital (more of which is required for natural-draft towers) and energy (which is used more intensively by mechanical towers), and on whether climatic conditions are favorable for natural-draft towers. The actual costs of either type of system depend strongly on climatic factors. Generally, the capital cost of a system is determined by the "design" environmental temperature and humidity, which represent the most severe conditions under which the plant is expected to operate, the desired cooling water temperatures at the tower inlet and outlet, and the rate at which water is recirculated (see [12,13,15,16]). The lowest temperature to which evaporative towers can force cooling water is the wet-bulb temperature. Thus, on hot, humid days, cooling towers exact an efficiency penalty compared to cooling ponds and once-through cooling, whose water temperatures are often lower than the wet-bulb temperature at such times. Unfortunately, it is during those periods that power demands are at their highest, due to air conditioning loads.

Dry cooling solves the problem of water consumption by rejecting heat via conduction and radiation only (Figures 1c and 1d). This technology has been applied in Europe for some time, and is now used at a mine-mouth power plant in the U.S. Dry cooling, while easing water problems, is unfortunately very costly—three to five times as much as evaporative cooling [4]. Figure 2 shows that for an indirect conventional system, dry cooling costs $0.004 per kilowatt-hour (kWh), while the expense of 100% evaporative cooling is only slightly more than $0.001/kwh. Indirect systems contain a secondary water loop which picks up heat at the condensers and rejects it at the towers (Figure 1c); in direct dry cooling systems, the condensing steam from the turbines is passed directly through the heat exchangers (Figure 1d).

One-third to one-half of the cost of dry cooling is due to the severe capacity penalties that result from its inability to lower the temperature of cooling water below the dry-bulb temperature, which in dry climates can be as much as 40°F higher than the wet-bulb temperature. Most of the remainder of the cost is for the huge heat-exchangers made necessary by the relative inefficiency of radiation and convection compared to evaporation in removing heat. As in evaporative systems, these exchangers have to be sized to accommodate the extreme "design" temperature conditions, even if during most of the year far smaller exchangers would suffice.

To justify choosing conventional dry cooling over evaporative towers, water would have to cost upwards of $10/1000 gallons [19]. This far exceeds typical water acquisition costs for utilities, and is even an order of magnitude higher than municipal water rates. Therefore, only if water is simply unavailable (perhaps for political reasons) would dry cooling be attractive. In the case of the mine-mouth Wyodak dry-cooled unit in Wyoming, it was decided that it would be cheaper to build dry towers than to either ship the coal from the mine to a site closer to a water source or to transport water a long distance.

The high costs of dry cooling and the need to conserve water have motivated development of combined wet/dry cooling towers. One such system has two sets of cooling tower sections, one dry and one wet. This is the system used at the third unit at the San Juan power station in New Mexico. In such a system, dry sections are sized so that they can handle the heat load for much of the year, but not so large as to be able to reject all of the heat under the hottest conditions. During hotter periods, the dry sections are used to the maximum extent possible, and the remaining heat load is rejected through the wet sections. Only as many wet sections as are needed for maintaining the desired water temperature are turned on. The system's capital cost is much less than for 100% dry systems, while consuming only a fraction of the water that a 100% evaporative system would use. The economic mix of wet and dry sections and their sizes depends on water costs and site-specific climatic factors.

Other wet/dry systems include water deluge of dry heat exchangers during warm periods and presaturation of air before it contacts heat exchangers [4]. A fourth type of wet/dry system, called the binary tower, allows water to be diverted either inside or outside of plastic heat exchangers, depending on environmental conditions [20]. One propo-

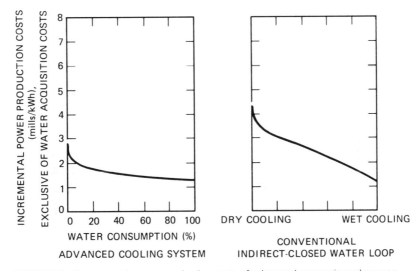

FIGURE 2. Incremental power production costs of advanced ammonia and conventional indirect cooling systems. (Source: Reference [19], based on Reference [4]. Copyright© 1980, Electric Power Research Institute. EPRI P-3647, final report, August 1984, A Method for Evaluating Regional Water Supply and Conservation Alternatives for Power generation. Reprinted with permission.)

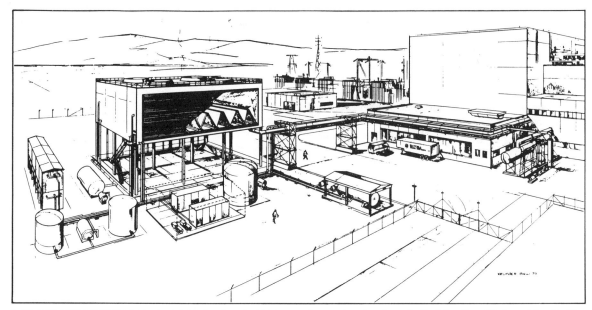

A 10 MW demonstration of an advanced water-conserving cooling system has been built at the Kern Station of Pacific Gas and Electric Company near Bakersfield, California.

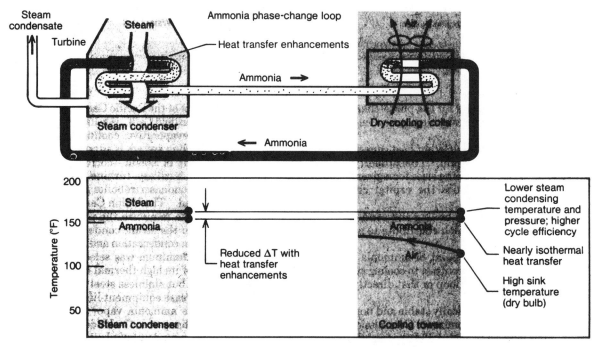

FIGURE 3. An advanced ammonia-based water-conserving dry-cooling system (source: Reference [4]. Reprinted with the permission of the Electric Power Research Institute and Mechanical Engineering).

nent of this system claims that it would cost significantly less than the first of the wet/dry systems described above [20].

One way to characterize wet/dry systems is to describe them in terms of the amount of water they consume in an average year compared to a 100% evaporative system. Figure 2 shows the costs of conventional wet/dry systems with an indirect dry cooling loop as a function of percentage wet cooled (see also [21]). Costs decrease rapidly from the 0% to the 10% level, and decline in a fairly linear fashion from that point on. Water would have to cost approximately $3/1000 gallons to justify anything except 100% wet cooling. At that price, a 10% wet/dry conventional system would be attractive. Only an expense of about $20/1000 gallons would make it worthwhile to lower consumption down to 5% of that of a completely evaporative system, while a price of roughly $25/1000 gallons would be needed before a 100% dry system would be economic compared to wet/dry cooling [19, based upon Figure 2]. These estimates are of course dependent upon a number of assumptions; the reader may wish to refer to [13,16] for details as to how cooling system costs can be estimated.

Other improvements of cooling technology are in the development and demonstration phase [22]. One which holds particular promise is an advanced system which uses anhydrous ammonia rather than water in its secondary loop (Figure 3). Its cost may be as small as one-half to two-thirds that of a conventional indirect system (Figure 2), and the incremental cost of saving the first 60% of the water used may be only one-sixth that for the conventional system [19, derived from Figure 2]. The advanced ammonia system's advantages include: high latent heat capacity per unit volume compared to water, which permits smaller equipment and pipes; nearly isothermal behavior, which means that a greater proportion of the heat is transfered via phase changes rather than temperature changes, permitting lower temperatures and back-turbine pressures (Figure 3); a low freezing point; and chemical stability [4]. A 10 MW version of this system is currently being demonstrated at the Kern fossil fuel-fired plant in California.

Another ammonia based cooling system is being installed by Electricité de France as part of a bottoming cycle of a 20 MW power plant. The heated ammonia will be expanded through a turbine, generating up to 5 MW of power, before being condensed in a dry cooling tower [23].

Other innovations now being investigated could potentially lower the costs of wet/dry and dry cooling even further. These include nonmetallic heat exchangers [21,24] and capacitive cooling systems [25]. The capacitive system permits a smaller dry cooling system by eliminating the need to design the system for hottest hours of the day. Instead, during the day, a "peak shaving" heat exchanger rejects a portion of the heat to a water storage tank. During the cooler evening hours, the water from the tank can be cooled, for example, by direct air cooling or an ammonia dry cooling system. For ammonia based dry cooling systems, use of capacitive cooling may yield cost savings of 25% to 40% [25].

But for utilities who must install new capacity within the next 15 years, the range of cooling options is probably restricted to conventional evaporative, wet/dry, and dry cooling and, if the Kern plant demonstration continues to work successfully, an anhydrous ammonia wet/dry or dry system. Power companies will have to weigh the costs of dry or wet/dry cooling, described above, against the expense and political uncertainty of water supply.

WATER SUPPLY

The main problem of wet/dry and dry cooling is its expense; the difficulties of water supply acquisition, in contrast, are predominantly political and institutional in nature. In much of the U.S., there are more demands for water than supply. These demands include those of irrigated agriculture; municipal water supply systems; self-supplied industry; and instream uses, including fish and wildlife, navigation, recreation, and water quality preservation. To obtain a reliable water supply, energy industries will often have to take water away from these other uses.

The resulting conflicts with agricultural, municipal, and other users of water can be resolved by four different institutions [18]: the marketplace and the legislative, judicial, and administrative branches of government. It is possible that all four will be involved in a given instance. For example, a utility may buy a water right from an irrigator and get the necessary approval from the state engineer, only to find itself sued by a downstream user who believes that his property rights to the return flows are being compromised. The state legislature may become a party to the controversy by passing a law restricting the circumstances under which water can be transferred to industrial use. Thus, hydrologic availability alone is not enough; permission from one or perhaps more of these institutions is required. For excellent overviews of institutional constraints which energy industries must face when acquiring water, refer to Weatherford et al. [26] or Trelease [27]. Other analyses of alternative water sources for energy include [28,29,30].

Fresh Surface Water

Fresh surface water is used by most power plants, although new facilities are increasingly turning to the other sources discussed later in this chapter. (Seawater is not considered here.) Energy companies wishing to secure a surface water supply must first obtain the right to use the water. Then, if the source is not sufficiently reliable, storage will have to be provided. Finally, the water must be treated; this expense is a significant one when waters of low quality are used.

There are several ways to obtain surface water rights. These include: exercise of riparian or appropriation rights, acquisition of a state water use permit, purchase of water from federal water projects, and, finally, leasing, purchase, or condemnation of rights from other users, considered later in this chapter. The riparian rights system is followed by many eastern states, and descends from English common law. The western states, except for Hawaii, rejected the riparian system early on, and follow the appropriation system. A number of eastern states have also jettisoned the riparian system in favor of a permit system administered by the state. Weatherford et al. [26] present a thorough discussion of water rights acquisition, including a number of illuminating case studies.

Riparian water rights are attached to real property which borders natural water courses. The right entitles owners of land adjacent to streams to use flows for any beneficial purpose, as long as the use is reasonable with respect to the needs of other riparian owners and does not unreasonably interfere with other legitimate water uses. "Reasonable use" is defined by the courts. During periods of drought, users may have to cut back in proportion to their use, or the state may impose a priority system. Such a priority system is unlikely to favor industrial uses.

Under the appropriation system, water does not have to be used on riparian lands, and is allocated to beneficial uses by the "first in time, first in right" rule. When streamflows are low, senior rights holders get priority and junior rights holders may get no water at all. There are three common exceptions to this rule, the major one being that certain preferred users, such as households, receive their full appropriation regardless of the date of first use [26]. Priority users can also condemn water rights applied to lower priority purposes.

Water permit systems have superceded the riparian system in many eastern states and are used by all western states, except Colorado, to administer the appropriation rights system. Under a permit system, potential water users apply to the appropriate state or local agency or, in a few cases, to an interstate river basin commission for approval of the proposed use. The agency will consider a number of criteria in deciding whether to grant a permit. These include: 1) physical availability of water; 2) beneficial use; 3) non-injury to other water users; and 4) the "public interest" [26]. Public hearings are often held, after which a decision is rendered. Applicants may appeal decisions to the courts. Often, permits will be heavily qualified with restrictive conditions. An example of such a condition is the one imposed upon large users by the Delaware River Basin Commission. New users must provide storage, either by purchase or by building their own reservoir, which will augment the river's flow during droughts by an amount equal to the amount of consumption. Permit conditions may also be changed even after the granting of the permit by successful law suits filed by other water users or environmentalists [18].

The final means of obtaining surface water rights, other than purchase of an existing right, is the acquisition of water from large federal water resource projects [28,31]. The U.S. Army Corps of Engineers and Bureau of Reclamation are authorized to allocate water to industry from the reservoirs they construct, although the marketing is often done in cooperation with the state through a permit system. Federal reservoirs will probably continue to be an important source of water for energy projects, particularly in the western states [31]. An attractive feature of this source is that the water is likely to be inexpensive, perhaps as low as a few pennies per 1000 gallons [18].

The acquisition of a right by any of the above means does not guarantee that the right is a secure one. There still remain important legal uncertainties that, once resolved, could drastically alter the allocation of water. These uncertainties include [26,27,28]: 1) unsettled Indian water rights; 2) the question as to whether water rights are implictly reserved for national forests and other federal lands; and 3) the extent to which increased protections will be established for instream uses of water.

Once a right is obtained, it may be necessary to provide storage to ensure that the supply is reliable. For example, any unappropriated water rights in the West will likely be junior rights, over which senior rights holders will have priority in dry years. Storage may be provided as part of a large multipurpose reservoir built by a state or federal agency. However, energy industries have increasingly found it necessary to provide their own storage, primarily because public concern about the environmental and societal impacts of large reservoirs now makes it very difficult for public agencies to build new dams.

Because the amount of storage needed to firm up a power plant's water supply is not large (e.g., 20,000 to 25,000 acre-ft for a 2200 MW nuclear plant on the Delaware River; 1 acre-ft = 326,000 gallons), small pump-in reservoirs are often adequate [32]. Such reservoirs are usually sited on a small tributary close to a major river. Water is pumped from the river into the reservoir during times of high flow and released during dry periods. Advantages of small pump-in reservoirs include: fewer people and facilities must be relocated; minimal spillways are needed because of the watershed's small size; river handling during construction is not a serious problem; and downstream impacts are minor [32,33].

The size of reservoirs is usually determined by the traditional mass or Rippl curve technique [34] (for an example, see [35]). This method determines the minimum amount of storage necessary so that the reservoir would not fail to provide the required yield if the historical streamflow record repeated itself. The resulting yield is sometimes misleadingly termed the "firm yield." Since the historical record is just one sample of the universe of possible streamflows and will not be repeated, such a reservoir will, in general, not be perfectly reliable in the future. More sophisticated reservoir design methods which explicitly consider the random nature of flows and the probability of failure are available [34], but are not as yet widely used by the energy industry. Such methods would permit comparison of the contribution that water supply uncertainty makes to power system reliability problems with other sources of such problems (such as power plant forced outages) so that the most cost-effective means of enhancing system reliability can be adopted.

The cost of building a reservoir with a given "firm yield" depends on many site specific factors, among which are the variability of flows, the topography and geology of the site, and the amount of existing development that would have to be removed. Nevertheless, some rough generalizations can be made. Typical costs for small utility-built reservoirs are \$400,000/ft^3/sec (cfs) of firm yield in the northeastern U.S. [10,36], \$650,000/cfs in Wyoming and Montana [37], \$1,100,000/cfs in Kentucky [33], \$900,000/cfs in the Pacific Northwest [37], and over \$1,000,000/cfs in the Texas-Gulf region [19]. Some of these cost differences can be accounted for by the different sizes of the reservoirs; for example, those of the northeast are larger, and therefore cheaper per unit, than those in Texas [19]. A statistical analysis of the cost of water from reservoirs as a function of storage size was estimated from cost and firm water yield data for 14 dams in Texas [19]. The costs include only those expenses allocated to water supply. The yields varied from 5 to 550 cfs. The resulting equation is:

$$\text{Cost (\$/cfs)} = 9{,}100{,}000 \, [\text{Yield (cfs)}]^{-0.5937} \quad (1)$$

The R^2 of the log-log form used to estimate the equation was 0.998. For typical utility reservoir yields of between 40 and 200 cfs, this equation yields costs between \$400,000 and \$1,000,000/cfs.

Water treatment requirements for cooling water, boiler makeup water, and other energy facility uses are discussed in [2,12,13]. Less pure water sources, such as saline lake water [38], must be subjected to additional treatment if used as makeup water for closed cycle cooling systems. Israelson et al. [39] (see also [40]) estimated treatment costs for brackish and saline water. They assumed that reverse osmosis is to be used and that water circulates through the cooling system with a total dissolved solids (TDS) concentration of 24,000 mg/l. The resulting costs vary from \$0.30 to \$3/1000 gallons, approximately, for treating water with total dissolved solids concentrations of 1000 mg/l to 10,000 mg/l, respectively.

Groundwater

To obtain groundwater, one must first secure the right to it, and then pump and treat it. If there do not exist excess groundwater rights, then rights must be purchased or condemned from other users. If there are excess rights, then how an energy facility obtains the right to pump water depends on the particular groundwater law system that applies. There are four principal systems of groundwater rights in the U.S. [26]. At one extreme is the doctrine of absolute ownership, in which the landowner can withdraw any amount without any liability for damages caused by the pumpage. Texas, Illinois, and Connecticut adhere to this system. Priority appropriation, adopted by most western states, represents the other extreme. There, the state grants use permits with priority accorded to early users. Some states have discriminated between critical and noncritical groundwater withdrawal areas, and regulate pumpage only in the former. Critical areas are usually those in which withdrawals have exceeded the sustainable yield of the aquifer, resulting in falling water tables or land subsidence. In granting a permit, the state may consider not only the effects of withdrawals upon other groundwater users, but also their impacts on flows in streams and from springs.

Between the absolute ownership and appropriation doctrines lie two somewhat similar systems, the reasonable use and correlative rights doctrines. In the former, which is the more widespread of the two, land owners have coequal rights of nonwasteful use on overlying lands, but water may not be used off of those lands. Owners may be liable for unreasonable lowering of water tables.

The cost of groundwater withdrawal depends on the number of wells needed to obtain the required water (a function of the expected yield of each well), the depth of the aquifer, and the geologic medium. The following function describing the cost of well drilling and construction was obtained by regression analysis for a sample of wells from across the U.S. between 85 and 1000 ft deep [41]:

$$\text{Cost (\$/well)} = 25{,}000 + 0.111[\text{DEPTH (ft)}]^2 + \\ + 0.0582 \, \text{GEOL}[\text{DEPTH (ft)}]^2, \quad R^2 = 0.78 \quad (2)$$

where DEPTH is the depth of the well and GEOL is a dummy variable which equals 0 if the well is drilled only through sand and gravel and 1 if the well is drilled through a consolidated geologic medium. For very deep wells (over 1000 ft), a cost figure of \$25 to \$100 per foot of depth might be used instead [42].

The cost of well construction, multiplied by the number of wells required, must be added to any land acquisition expenses and the cost of pumping energy. Land costs can be large if wells are of low yield and must be widely spaced; in the case of at least one energy project, that expense was prohibitive [43]. Energy costs are calculated using the engineering formula:

$$\text{Cost (\$/1000 gal)} = \frac{0.00315[\text{DEPTH (ft)}][C_p \, (\$/\text{kwh})]}{E}$$

(3)

where C_p is the cost of electric power and E is pump efficiency (which recent surveys of electric high yield pumps in Texas found to be approximately 0.5 [19]).

In regions of the country where surface water is scarce, there is often ample groundwater [e.g., 44], although much of it may be deep and saline [42]. For example, the upper Colorado River basin holds an immense amount of groundwater in the upper 100 ft of saturated rock—perhaps one-half trillion gallons [45]. Most of that water is saline, which means that there are likely to be few conflicts with other users. Nevertheless, energy firms have been reluctant to pursue this option. This is due to several factors [38], including exploration costs, possible land subsidence from large-scale pumping, treatment expenses, the cost of corrosion resistant materials, and possible legal risks associated with groundwater pumpage by energy projects. But the absence of competing uses and the fact that use of saline groundwater is cheaper than developing surface supplies in many parts of the West [42] means that this source will probably be used more often in the future by the energy industry.

Water Rights Transfer

Several authors [e.g., 18] have maintained that the "water for energy" literature overemphasizes governmental forums for resolving water conflicts and the problems they present while failing to note the many opportunities that the market presents for securing water supplies. Weatherford et al. [46] argue that although formal market mechanisms are rudimentary or lacking in most areas of the West, markets that promote economically efficient reallocations of water supply will nevertheless evolve. The absence of unappropriated surface and groundwater supplies in many basins will force energy firms to enter the market and purchase water rights owned by other parties.

Indeed, they are already doing so. Abbey and Loose [47] list 12 recent transfers of water rights to energy firms in the West, totaling 86 billion gallons per year. Most transfers were from irrigators, although a few consisted of sewage effluent owned by cities. At least two included transfers of groundwater rights. Transfers are also occurring in the East. For example, Pennsylvania utilities have been negotiating with the U.S. Army Corps of Engineers to reallocate storage in a flood control reservoir to water supply.

Perhaps the most celebrated example of a successful transfer is that made by the Intermountain Power Project in Utah [48]. The project purchased 15 billion gallons per year of rights from irrigation companies in the Sevier River basin at a price of $5.40/1000 gallons/year (or, at a discount rate of 8%/year, about $0.45/1000 gallons). Groundwater constitutes 23% of the water transferred. The price paid is well above the marginal worth of that water for agriculture, which is perhaps $0.10/1000 gallons [e.g., 49]. Other examples include energy firms in Texas who purchased surface water rights from municipalities and irrigators, and a utility which bought the groundwater rights of a large amount of ranchland in the High Plains region [19].

This is not to say, however, that water markets are working well and that energy firms can depend on them. As Trelease [27] points out, western water rights were established primarily to create a stable agriculture, not to facilitate transfers of water from agriculture to other uses. In most states, markets do not exist—willing buyers cannot easily purchase rights from willing sellers. This is because, in essence, many states consider water too "important" to be treated as a mere commodity. Instead, several states have erected hierarchies of priorities, with domestic and agricultural uses on top and industry on the bottom. Transfers of rights from "high" to "low" uses is often difficult or even prohibited (as it is, for example, in Montana and North Dakota [26]), even though the economic value of water may be much greater for the so-called "low" uses. The result is waste—for instance, water being used to irrigate cattle fodder, while power plants are forced to build multimillion dollar water conservation systems.

In states such as Texas and Utah, where rights transfers from agriculture to industry are encouraged, the cost of water rights may in some circumstances merely represent the value of water in irrigation, or in others the higher value of water to energy firms. Which prices result depend on the relative bargaining power of farmers and companies. In one analysis of water supply for Texas utilities, it was assumed that farmers would charge industry twice what the water was worth to them for raising crops [41]. The worth of water for irrigation was estimated using the agricultural sector's demand curve for water, which shows how much farmers are willing to pay for different quantities of water. The demand curve used in the Texas study is shown in Figure 4. It was derived from estimates of the response of irrigators to energy price increases in the year 2000 [50]. Assuming that the least valued water is sold first, the cost to utilities of rights transfers was related to the integral of the demand curve between the amount of water that would be consumed by agriculture with the transfer and the amount consumed without the transfer. This integral is an estimate of the farmers' willingness to pay for water, which is assumed to be the same as the economic value of that water in irrigated agriculture.

There are many variations of rights transfers that can be pursued even in states where outright transfer of title is difficult. One approach is for energy firms to fund water conservation efforts by irrigators and then use a portion of the saved water for energy facilities. Utilities, for example, have financed modifications to irrigation reservoirs [51], lining of canals, and more efficient water application methods [52]. A different approach that might minimize political opposition is to negotiate with farmers to purchase their water only in abnormally dry years. This type of agreement has been successfully implemented [28].

Yet another approach is to purchase water entitlements from a water management organization, who would still

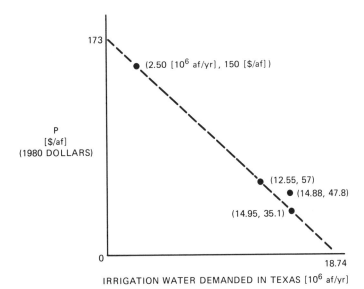

FIGURE 4. Demand Curve for Irrigation Water, Texas, year 2000. (Source: Reference [19]. Copyright© 1984, Electric Power Research Institute. EPRI P-3647, final report, August 1984, A Method for Evaluating Regional Water Supply and Conservation Alternatives for Power Generation. Reprinted with permission.)

hold the actual appropriative right [28]. The energy firm could then avoid many of the problems that arise in transferring rights, such as possible requirements for state approval and public hearings. Such organizations could include conservancy districts, mutual water or ditch companies, and irrigation districts. Together, they deliver most of the water in the West to the final user. Transfer of water from such organizations occurs with relatively little difficulty, often via purchases of shares in the organization [28]. Shafer et al. [53] describe a successful transfer of this type in which a complicated borrowing agreement between an irrigation company, a small city, and a utility resulted in the provision of a reliable supply for a power plant and operational benefits for the other parties.

There seem to be two keys to such successful agriculture-to-energy transfers [51]. One is to avoid injuring third parties. Thus, for example, only the net amount of water consumed by agriculture should be transferred, not the entire quantity that was diverted. This will help to prevent suits by downstream users who depend on return flows. A second key is be flexible and imaginative in assembling water rights and dealing with those who own them [48].

Water Reuse

This source consists of effluents from other uses, including wastewater within a power or synfuels plant that can be reused for other purposes, irrigation runoff, and sewage plant effluent. With proper treatment, this source can be used for cooling tower makeup water, dust control, sulfur scrubbing, and other purposes.

In many water-scarce basins, irrigation return flows are usually depended upon and, indeed, owned by other users downstream. But in some places, runoff has high levels of salts and is not a desirable commodity. In that case, energy firms can help to solve a pollution problem and can, at the same time, avoid conflicts with other users by diverting this effluent. For this reason, California law explicitly encourages the use of return flows, and two proposed nuclear plants in that state, now cancelled, were to have used that source [38]. Yet no plants in the U.S. presently use drainage water [30].

In contrast, there is over two decades of experience in the U.S. and abroad with the use of treated sewage as a water source [30]. A recent example is the purchase of effluent from the City of Phoenix, Arizona for use in the Palo Verde Nuclear Station for less than $0.10/1000 gallons [18]. The proposed, but now dropped Warner-Allen project in Nevada would have used one-half of Las Vegas's effluent at a cost of $0.26/1000 gallons [44].

Both irrigation drainage and sewage effluent have to undergo additional treatment before being used in cooling systems. Tests have shown that wastewaters with TDS concentrations of 1600 to 7000 mg/l can be treated for satisfactory cooling system operation [38]. But the cost of such treatment can be the major expense involved in using this source. The incremental cooling cost is on the order of $0.15 to $0.25/1000 gallons [18,19]. These sources have the additional problems of potential environmental impacts from cooling tower aerosol emissions and disposal of blowdown [30,38]. Blowdown can be disposed of in evaporation ponds, but that can cost as much as $8 to $20/1000 gallons [18].

The final type of reuse is the utilization of otherwise discarded water from within the power plant. This is motivated not only by the scarcity of water supplies, but also by water pollution control laws which discourage release of liquid effluents. Maximizing their reuse of water, at least 30 power plants in the U.S. are "zero-discharge" plants, in which all water is reused or evaporated [54]. In-plant alternatives for reuse include [30]: 1) increased cycles of concentration in cooling tower systems, which decreases blowdown; 2) cascading uses of water in a series of different processes requiring decreasing water quality; and 3) storage and treatment of water for reuse in the plant at a different time. In-plant reuse increases the cost of treatment and may worsen the environmental problems of disposal, but, as noted, this particular water "source" has nonetheless become popular.

Water Transport

Energy facility siting is a multiobjective problem in which not only water availability must be considered, but also the

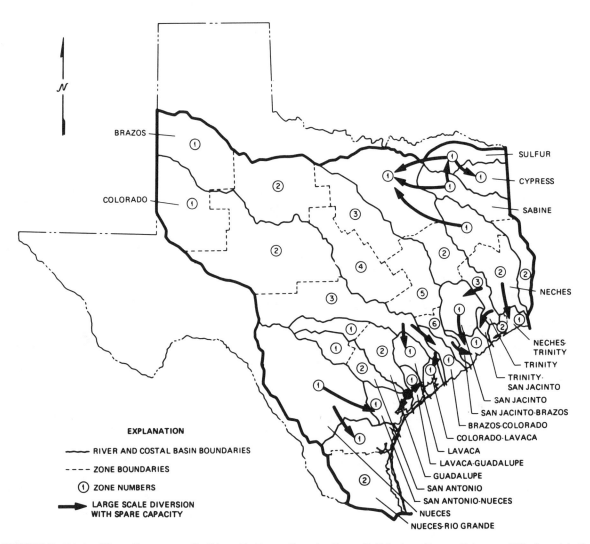

FIGURE 5. Existing Water Conveyance Facilities with Excess Capacity, Texas-Gulf Region. (Source: Reference [19]. Copyright© 1984, Electric Power Research Institute. EPRI P-3647, final report, August 1984, A Method for Evaluating Regional Water Supply and Conservation Alternatives for Powe Generation, Reprinted with permission.)

cost of power and fuel transport, various environmental and social impacts, site geology, and numerous other factors [55]. Therefore, an energy facility will frequently be located far from its water source. For this reason, and because several alternative water sources at different distances may be available, it is important to analyze the economics—and politics—of water transport.

An energy firm can transport water long distances using either existing diversion facilities owned by a public agency, if available, or a pipeline or canal constructed especially for the facility. In regions such as Texas or California, where there are large aqueducts which transport water to irrigation projects and cities, there often exists spare capacity that can be used by energy firms [19]. Figure 5 shows the general location of existing and proposed interbasin large-scale water conveyance facilities in the Texas-Gulf region which were assumed to have extra capacity in one water for power study [19]. Use of publicly owned aqueducts is desirable, as it is usually inexpensive compared to construction of a water transmission system dedicated to the energy facility.

The cost of constructing and operating a new water conveyance facility depends on distance traversed and the height differential. Gold and Goldstein [56] developed an engineering cost equation for pipes and pumps based on an optimization of pipe diameter and water velocity. The equation is:

$$\text{Cost (\$/yr)} = 105.12 C_{pipe} K_c Q^{0.5} L$$
$$\times \left\{ \frac{f[N K_p + 0.0268 C_{pump}(+ 1.84E/Q)]}{(E C_{pipe} K_c)} \right\}^{1/6}$$
$$+ \left(\frac{1150}{E} \right) H Q [K_p N + 0.0268 C_{pump} (1 + 1.84E/Q)] \quad (4)$$

in which

C_i = Capital recovery factor (1/yr) for item i
E = Pump efficiency (suggested value [56] = 0.8)
f = Manning's coefficient of roughness (suggested value = 0.016)
K_c = Cost of pipe, in \$/inch(diameter)-miles (suggested value = \$37,000/inch-mile)
K_p = Cost of electricity, in \$/kWh
H = Height (ft) water is raised. If negative, set to zero
L = Length (mi) of pipe
N = Capacity factor of system (average pumpage divided by peak pumpage)
Q = Peak quantity pumped (10^6 gallons/day)

As an example, a construction and operating cost of \$0.04/1000 gallons/mile results if Q = 20,000,000 gallons/day, N = 0.6, H = 0, K_p = \$0.192/kWh, and CRF_{pump} = CRF_{pipe} = 0.2/year. This cost is similar to estimates (in 1980 dollars) in or obtainable from other sources [18,28,57,58].

But economic and hydrologic practicability of water transport does not imply political feasibility. The public and local governments often oppose proposals to transfer water out of their local area, especially when the water would be diverted to another basin. For example, according to laws and state constitution of Texas [59], a river basin can be legally protected from diversions to other basins until its needs are assured for the next half-century, as determined by the state's water plan. Energy firms would be well advised to avoid interbasin transfers, if at all possible, because of the political difficulties that are likely to arise.

INTEGRATED WATER SUPPLY ASSESSMENTS

The increasingly complex institutional environment of water planning and the larger range of source and conservation options available mandates *integrated* water for energy assessments. The costs and political feasibilities of surface, ground, and waste water sources must be compared not just with each other, but also with those of different condenser cooling designs. This is true whether the objective of the study is to identify the best alternative for a particular facility or to determine whether the energy needs of a river basin, state, or nation are in conflict with its water objectives. To do otherwise is to risk overlooking cost-effective options and making poor predictions of how energy-water conflicts will be resolved.

Water Studies for Energy Facility Planning

An integrated water supply study for a particular facility can be conducted using either a cost-minimizing/engineering economy approach or a multiple objective framework. A cost-minimizing study would choose the mix of supply sources and water conservation measures that yields the lowest possible overall expense of water provision and use. But even for a profit-making corporation, consideration of cost alone is generally not enough. Environmental and social impacts are also important, if only because more severe impacts increase the chance of law suits by citizen groups and delay in project completion. Important, too, are the legal and institutional uncertainties concerning whether water supplies will really be available in the amounts and at the costs assumed. In general, there are tradeoffs among these objectives that should be considered. For example, the alternative that minimizes expected cost for a particular plant might consist of the purchase of water rights from irrigated agriculture. But that alternative might also be risky, in that the state engineer might not approve the transfer or third parties may sue to prevent the sale. A less risky, but much more costly option in that instance might be to install dry or wet/dry cooling towers instead. Many methods are available for considering these tradeoffs and making multiobjective decisions in a systematic and defensible fashion [e.g., 55,60,61,62].

In undertaking water supply planning, utilities need estimates of:

- how much water will be available
- from what sources
- at what reliability
- where
- at what cost

They also need to know the magnitude of the uncertainties of these estimates, and the impact of different possible resolutions of those uncertainties. These information needs are discussed in turn below.

How much water will be available? As pointed out, not only hydrologic but also institutional factors determine how much water will be available to utilities. Indeed, legislation and lawsuits often prevent utilities from either obtaining rights to excess water or purchasing rights from people who are willing to sell them. To assume, for example, that the only limits to water supply are those posed by hydrology and physical availability of reservoir sites inevitably results in overly optimistic estimates of the amount of surface water that utilities can acquire.

Water-for-energy studies should start by estimating the physical supply available. Then, existing and potential claims on the supply (including instream uses) should be tallied. The legal mechanisms (if any) by which electric utilities can either obtain rights to the unclaimed water or purchase rights already owned by other users must then be examined to determine how much water could, in reality, be acquired for power plant use. Finally, institutional obstacles to exercising rights thus obtained must then be considered. These include, for example, the political and legal difficulties associated with acquiring reservoir storage, transporting water from source to plant, and purchasing land for wells. The result of such an analysis will probably be an estimate of water availability that is much less than that which would result from consideration of hydrology only.

From what sources? Each of the possible sources described earlier in this chapter—unappropriated surface and groundwaters, rights transfers, waste waters, brackish and saline supplies, and interbasin transfers—should be considered. Utilities, particularly in the west, frequently turn to "alternative" sources, and to disregard any of them risks unnecessary expense and political opposition. Further, a power project will sometimes have to draw upon several different sources of water, each having a different priority. Combining different sources so that the plant has an economic and reliable water supply can be a challenging task.

At what reliability? "Firm" water supplies, in the sense of having zero probability of failure, do not exist. Because not all sources of water supply are equally reliable and because utilities generally are very conservative regarding what probabilities of failure they are willing to accept, it is important to consider source reliability in water-for-energy studies. The picture is further complicated by the ability of utilities to combine two or more unreliable sources of water into a single reliable supply. An example is conjunctive ground and surface water management.

Hydrologic reliability is not the only type of reliability that should be considered. Institutional reliability can be just as important. Although a utility may be able to obtain the necessary permissions to obtain water rights, later court suits may overturn the permits, perhaps resulting in expensive construction delays. (The Wheatland power station in Wyoming is a case in point [18].) Thus, utilities may prefer to build a costly wet/dry cooling system rather than, say, purchase water rights from irrigated agriculture and run the risk of lawsuits and delay.

Methods are becoming available for explicitly considering the stochastic nature of streamflows when deciding upon the optimal mix of water sources and conservation measures for a particular facility. Traditionally, the randomness of streamflows have been acknowledged only in calculating the "firm yield" of reservoirs, without directly analyzing the impact of that randomness upon optimal cooling systems design. One more advanced method has been developed by Shaw et al. [63], whose approach is applied to the situation in which a utility purchases water from different users, depending on the level of streamflow. An example of this would be where a utility has obtained the option to use a farmer's senior water rights during exceptionally dry years. The method considers the resulting randomness in the cost of water, the expense of on-site storage, and the possibilities of evaporative and wet/dry towers and cooling ponds. Palmer and Lund [64] present another such method. This approach includes the options of storage and purchases during droughts of either power from other utilities or water rights.

Where? Because long distance transport of water is costly and often politically difficult, it is important to consider the location of water supplies. It makes no sense to say, for example, that the Northeast is water rich and thus utilities will have no trouble gaining water supplies when that region contains both the bounty of Great Lakes and the water shortages of the Delaware River basin. Thus, there can still be a demand for advanced cooling methods in regions with relatively ample supplies.

At what cost? Just because water is available from both a physical and institutional standpoint does not mean that it should be used. Its expense must be balanced against that of water conservation measures, especially wet-dry and dry cooling. Further, if there are several alternative sources of water, cost will be a major consideration in determining which, if any, should be acquired. Estimates of water costs can be made using the methods described earlier in this chapter.

Sources which are unreliable from an institutional point of view (because of the possibility of law suits and delays) should be assessed a cost penalty. This penalty should be at least equal to the costs of delay and, perhaps, acquisition of other water supplies times the probability of a successful legal action. Doing this would at least begin to capture the conservative behavior of utilities who, all things else being equal, prefer water sources that are unlikely to cause political or legal controversy.

What would be the consequences of alternative possible resolutions of the uncertainties? When conducting water for energy studies and interpreting their results, analysts and users should keep the political and institutional uncertainties of water supply well in mind. These uncertainties are so large in many cases that an investigation's conclusions will depend on what assumptions are made about their resolution. Therefore, studies should emphasize sensitivity analysis. Economic uncertainties are also important and should likewise by analyzed. Consider, for example, how unpredictable energy prices and demands have been over the last 15 years. Sensitivity analyses should be conducted of energy and water demand growth rates and costs of alternative water sources.

Water Studies for Policy Analysis and River Basin Planning

Analyses of water supply and conservation options performed for industrial organizations, such as the Electric Power Research Institute, or government agencies have a variety of aims. In the mid-1970s, many studies were carried out to examine whether water supply constraints would prevent achievement of the energy goals of Project Independence. For example, the U.S. Department of Interior examined the upper Colorado and Missouri basins to see if sufficient water was available to accommodate the huge increases in fossil fuel production forecast for those regions [65,66]. Likewise, the U.S. Nuclear Regulatory Commission sponsored analyses of possible water supplies for nuclear energy centers [11].

The conclusions of many of these studies [e.g., 67,68] were unduly pessimistic, in part because energy supply needs were overestimated and also because energy firms have proven to be creative in solving water problems. For example, Wright et al. [69] quote one government report on the Colorado River:

> When energy requirements for water use are added to non-energy requirements for the year 2000, the total exceeds minimum availability estimates by as much as one million acre-feet per year [70].

But now, there is apparently enough water for new energy facilities in the region [69]. In general, energy firms have exploited sources of water, such as groundwater, sewage plant effluent, and transfers from irrigated agriculture, that many water for energy studies have failed to account for adequately [18]. One reason for this is the difficulty faced in national-level studies in obtaining data on a uniform basis on water sources other than streamflows [67,68].

Many later studies made more sophisticated analyses of water/energy policy issues in particular basins. Examples include investigations of water rights transfers in the Yellowstone River basin [71], impacts of water use by the coal industry in the Tongue River basin [72], instream water requirements [73,74], and salinity in the upper Colorado basin [75]. Other studies have examined the impacts of different regulations concerning thermal pollution control and utility water use upon future power plant siting patterns and costs [10,36]. The purpose of a few recent analyses has been to assist research and development planning. These include studies that attempted to project the demand for dry cooling [e.g., 68] and to estimate the benefits of improvements in wet/dry cooling technology [41].

A variety of methodologies has been used in these studies. Some have simply compared projected demands for water with an arbitrary standard based on streamflow levels. For example, several studies have been based on the premise that if total water use is more than 10% of the lowest 7 day average streamflow expected once every decade, then water-energy conflicts are likely [e.g., 11,67,76]. Another, more sophisticated study derived supply curves for surface water for every major basin in the U.S. based on the cost of reservoir construction, and then used mathematical programming methods to obtain an equilibrium for the entire economy [77]. However, that model considered neither nonsurface water supplies nor institutional constraints on reservoir construction.

Concerning traditional water-for-energy studies which determined whether water shortages would occur by simply summing up future "requirements" for water and then comparing them to available surface supply, Abbey [45, p. 2] writes:

> It is a fundamental economic tenet that, as a resource becomes more scarce, substitutes are found and conservation is practiced. A long tradition of water resources research has argued and demonstrated that this principle applies to water as it does to any good. Over the long run, notions of shortages and constraints are inconsistent with acceptance of the substitution principle.

Indeed, as previously noted, utilities are following the substitution principle. They are turning to sources other than unallocated surface water and, in rare cases, are investing in wet/dry and dry cooling, which substitutes capital and energy for water. Cost is the major factor in deciding what mixes of sources and water conservation will be adopted for particular plants. Thus, water-for-energy assessments should consider the relative cost of different sources, not merely their availabilities. Further, those costs should be compared to the expense of wet/dry and dry cooling.

More recent studies have considered the possibility of substitution. Several have used linear programming to calculate equilibria for the water and energy sectors of the economy and have considered, in addition to the cost of reservoirs, the expense of groundwater, water rights transfers, wastewater, and wet/dry and dry cooling towers [41,47]. These studies explicitly included institutional restrictions on certain transactions. For example, water rights transfers would be assumed impossible where state law prohibits them.

Another important issue that analysts should consider is the degree to which they disaggregate a study area. Too little disaggregation, and subareas in which water problems are critical might be hidden by more optimistic region-wide averages. Subareas should be small enough so that within each subarea water supply conditions are fairly uniform and/or water transport is feasible and not too costly. Unfortunately, some compromise must often be made because of mathematical tractability and data availability considerations. For these reasons, some studies have considered only large basins, treating, for example, the upper Colorado basin as one homogeneous region. The results of such studies, all else being equal, are likely to be overly optimistic concerning the likelihood of water shortages.

When conducting water for energy studies and interpreting their results, analysts and users should keep the political and institutional uncertainties of water supply well in mind. These uncertainties are so large in many cases that an investigation's conclusions will depend on what assumptions are made about their resolution. Therefore, studies should emphasize sensitivity analysis [69,78]:

> (I)t is suggested that rather than focus on prediction, the objective of water availability assessment should be to acknowledge this uncertainty and play out the consequences of some of the ways that unpredictable political, judicial, and administrative decisions may affect water availability [69].

But at the same time, analysts should not underestimate the imagination of energy firms and the ability of water markets, imperfect as they are, to adapt and to accommodate changing water needs [18,42,46].

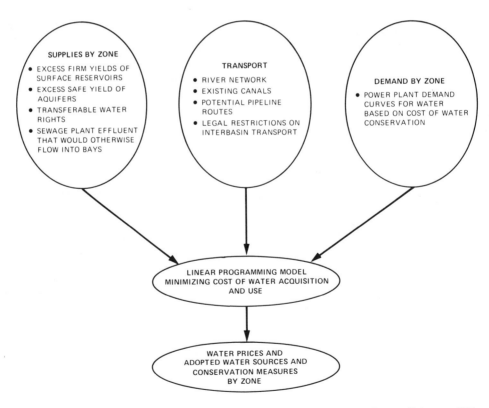

FIGURE 6. Framework for Water for Power Analysis, Texas-Gulf Region. (Source: Reference [41]. Reprinted with the permission of the Electric Power Research Institute and the American Society of Civil Engineers.)

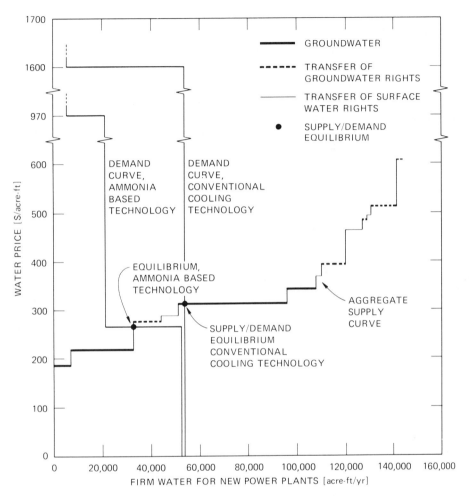

FIGURE 7. Supply and Demand Functions for Cooling Water, Brazos River Basin, zones 1-5, year 2000. (Source: Reference [19]. Copyright© 1984, Electric Power Research Institute. EPRI P-3647, final report, August 1984, A Method for Evaluating Regional Water Supply and Conservation Alternatives for Power Generation, Reprinted with permission.)

An Example

Figure 6 presents a schematic representation of a water for power investigation for the Texas-Gulf region [41]. The purpose of the study was to project the sources and amounts of water used by the power industry in the years 2000 and 2030 in order to assess the market potential for conventional and ammonia-based wet/dry and dry cooling. The costs for a wide range of water sources in different zones of different basins were described in terms of supply curves, while the expense of wet/dry cooling was portrayed using demand curves. Figure 5 shows the zones used in that study. For example, the unappropriated groundwater in a particular zone might be given as a stepped supply curve, with each step representing the cost and available amount of water from a particular aquifer in that zone. The demand curve for water in each zone would show the marginal worth of water to the power industry as a function of the amount provided, assuming a fixed amount of generation capacity. (It is also possible to optimize the locations of energy plants in this framework; see [10,36,37,79].) The worth of water was based upon the avoided expense of conservation—i.e., wet/dry or dry cooling. Supplies could be transferred between different zones, where legally permissable, using rivers, existing large canals, or new pipelines. A linear program was used to calculate the supply-demand equilibrium in each zone, which yielded the least cost combination of supply, conservation, and water transport. A supply-demand equilibrium occurs at the values of price and quantity at which the supply and demand curves (include net imports and exports) cross. This represents the least cost solution because the marginal cost of water will have been equated with its marginal worth.

Figure 7 shows the nature of the equilibria calculated in that application. Using a standard result of microeconomics [80] and, for the purpose of presentation, ignoring the cost of water transport, the aggregate supply curve for the Brazos River basin, Zones 1–5 in the year 2000 can be defined as the horizontal summation of the supply curves for the individual sources of water in those zones (the upward sloping step function in Figure 7). Different segments of the supply curve represent different sources of water with different costs. The curve shows that groundwater constitutes most of the supply below 110,000 acre-ft/year.

Figure 7 also displays two demand curves, derived from the generic curves of Figure 8 (which are based on Figure 2). One of the two represents the marginal value of water if new power plants use conventional wet/dry cooling technology, and the other shows the value of water if the advanced ammonia technology is available. The generic curves are converted to those of Figure 7 by, first, creating a curve for each plant by multiplying the X axis of Figure 8 by the capacity of the plant and dividing the Y axis by the number of acre-ft of cooling water consumed per MW per year under 100% evaporative cooling. Then, the demand curves of the individual plants in each zone are summed horizontally to obtain the zone's demand function [80].

The demand curves can be interpreted as follows. For example, if the price of water is very low, cost-minimizing utilities would adopt 100% evaporative cooling and consume 55,000 acre-ft of water per year. Indeed, this is the solution in the conventional technology scenario. Higher prices can induce wet/dry cooling. For example, a price of $1600/acre-ft or more would mean that 10% wet/dry conventional cooling would be less expensive than 100% wet cooling, while a price of $270/acre-ft would render 40% wet/dry ammonia based cooling economic. Comparison of the two equilibria in Figure 7 shows that the availability of ammonia wet/dry cooling towers would save utilities some money in the year 2000, but not a large amount. The amount saved equals the area between the two equilibria under the supply curve (the marginal cost of water) but above the ammonia technology demand curve (the marginal worth of water). The savings are small because little wet/dry cooling is adopted in either case. For the entire Texas-Gulf region, the savings would be approximately $1,000,000/year, and roughly 28,400 acre-ft/yr of water would be conserved.

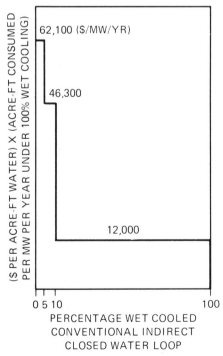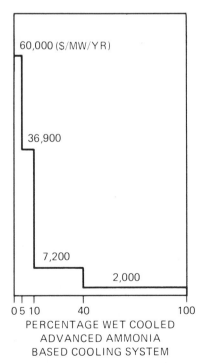

FIGURE 8. Generic Demand Functions for Cooling Water, Conventional and Advanced Ammonia-Based Cooling Technologies. (Source: Reference [41]. Reprinted with the permission of the American Society of Civil Engineers.)

CONCLUSION

Alarmist projections that water shortages would stand in the way of necessary energy development now appear to be overly pessimistic. Technological improvements in dry and wet/dry cooling offer the prospect of less costly energy conservation, and energy industries have proven adept at developing and using nontraditional sources of water, such as groundwater, wastewater, and rights transfers. There nevertheless remain large political and institutional uncertainties that can make water acquisition an expensive, time-consuming, and frustrating process. Integrated analysis of the options available, combined with explicit recognition of legal and institutional uncertainties, can help identify cost-effective and politically feasible alternatives.

ACKNOWLEDGEMENT

Helpful comments by J. A. Bartz and financial support by the Electric Power Research Institute for the case study are gratefully acknowledged.

REFERENCES

1. Powledge, F., *Water: The Nature, Uses, and Future of Our Most Precious and Abused Resource*, Farrar, Straus, and Giroux, New York (1982).
2. Shorney, F. L., "Water Conservation and Reuse at Coal-Fired Power Plants," *J. Water Res. Plan. Manage.*, 109(4): 345–359 (1983).
3. U.S. Water Resources Council, *The Nation's Water Resources, 1975-2000*, U.S. Government Printing Office, Washington, D.C., 4 Vols. (1978).
4. Bartz, J. A. and J. S. Maulbetsch, "Are Dry-Cooled Power Plants a Feasible Alternative?" *Mechanical Engineering*, 103(10): 34–41 (1981).
5. U.S. Code Annotated, Sections 1251-1378, Federal Water Pollution Control Act of 1972.
6. U.S. Code Annotated, Section 1326(a).
7. U.S. Army Corps of Engineers, *North Atlantic Regional Water Resources Study*, North Atlantic Division, New York (1972).
8. Delaware River Basin Commission, *Water Management of the Delaware River Basin*, Trenton, N.J. (1975).
9. Sonnichsen, J. C., et al., *Steam-Electric Power Plant Cooling Handbook*, Report HEDL-TME 81-53, UC-11,12, Hanford Engineering Development Laboratory, Richland, WA (1982).
10. Meier, P., B. Hobbs, G. Ketcham, M. McCoy, and R. Stern, *The Brookhaven Regional Energy Facility Siting Model (REFS): Model Development and Application*, Report BNL 51006, Brookhaven National Laboratory, Upton, NY (1979).
11. U.S. Nuclear Regulatory Commission, *Nuclear Energy Center Site Survey*, 5 Vols., Report NUREG-0001 (1976).
12. Probstein, R. F. and Gold, H., *Water in Synthetic Fuel Production: The Technology and Alternatives*, MIT Press, Cambridge, MA (1978).
13. Cheremisinoff, N. P., and P. N. Cheremisinoff, *Cooling Towers—Selection, Design, and Practice*, Ann Arbor Science, Ann Arbor, MI (1981).
14. Giaguinta, A. R., T. E. Croley II, and T. D. Hsu, "Hybrid Cooling System Thermodynamics and Economics," *J. Energy Div. (ASCE)*, 106(EY1): 89 (1980).
15. Hu, M. and G. Englesson, *Wet/Dry Cooling Systems for Fossil-Fueled Power Plants: Water Conservation and Plume Abatement*, Report EPA-600/7-77-137, U.S. Environmental Protection Agency, Washington, D.C. (1978).
16. Hu, M., G. Pavlenco, and G. Englesson, *Water Consumption and Costs for Various Steam Electric Power Plant Cooling Systems*, Report EPA-600/7-78-157, U.S. Environmental Protection Agency, Washington, D.C. (1978).
17. Lum, L. and L. W. Mays, "Optimizing Cooling Policy for Power Plant Systems," *J. Energy Div. (ASCE)*, 105(EY1): 137 (1979).
18. Abbey, D. and F. Lucero, *Water Related Planning and Design at Energy Firms*, Report DOE/EV/10180-1, U.S. Department of Energy, Washington, D.C. (1980).
19. Hobbs, B. F., D. J. Skolits, and B. B. Turner, *A Method for Evaluating Regional Water Supply and Conservation Alternatives for Power Generation*, Report EPRI P-3657, Electric Power Research Institute, Palo Alto, CA (1984).
20. Sanderson, W. G., "Wet/Dry Cooling with the Binary Cooling Tower and the Water Economizer Cooling Tower," in *Proceedings of an EPRI Workshop on Water-Conserving Cooling Systems* (J.A. Bartz, ed.), Electric Power Research Institute, Palo Alto, CA (1982).
21. Bartz, J. A., *Cooling Tower Research and Development at EPRI*, Electric Power Research Institute, Palo Alto, CA, Paper presented at the Third Workshop of the International Association for Hydraulic Research (October 12–15, 1982).
22. Parkinson, G., "Water-Saving Recipes Set High Conservation Standards," *Chemical Engineering*, pp. 18–20 (December 26, 1983).
23. Electric Power Research Institute, *Binary Ammonia Cycle Pilot Plant at the Gennevilliers Power Station: Summary of Design and Operating Principles*, Report EPRI CS-3254-SR, Palo Alto, CA (1983).
24. Dynatech R&D Company, *Nonmetallic Heat Exchangers: A Survey of Current and Potential Designs for Dry-Cooling Systems*, Report EPRI CS-3454, Electric Power Research Institute, Palo Alto, CA (1984).
25. Melin, K., M. Husain, K. Primavera, "Enhancement of Dry Cooling CBI Capacitive Cooling System," in *Proceedings of an EPRI Workshop on Water-Conserving Cooling Systems* (J. A. Bartz, ed.), Electric Power Research Institute, Palo Alto, CA (1982).
26. Weatherford G., et al., *Acquiring Water for Energy: Institutional Aspects*, Water Resources Publications, Littleton, CO (1982).
27. Trelease, F. J., "Legal Problems in the Allocation and Transfer

of Water to Electric Utilities," in *Proceedings: Workshop on Water Supply for Electric Energy*, Report EPRI WS-79-237, Electric Power Research Institute, Palo Alto, CA (1980).
28. Hendrickson, P. L., *An Overview of Issues Affecting the Demand for Dry and Wet/Dry Cooling for Thermal Power Plants*, Report BNWL 2268 Rev, Battelle Pacific Northwest Laboratory, Richland, WA (1978).
29. Jury, W. A., G. Sinai, and L. H. Stolzy, "Future Sources of Cooling Water for Power Plants in Arid Regions," *Water Res. Bull.*, 15(5) (1979).
30. Tucker, R., E. Altouney, R. Ehrhardt, and T. Fabian, "Alternate Sources of Water for Power Plant Cooling," in *Water and Energy: Technical and Policy Issues* (F. Kilpatrick and D. Matchett, eds.), American Society of Civil Engineers, New York, pp. 202–207 (1982).
31. Fred C. Hart Associates and Water Purification Associates, *Reservoir Water Availability for Western Energy Resource Development*, Report Prepared for G. Krug, U.S. Department of Energy, Washington, D.C. (1981).
32. O'Brien, E., C. N. Freeman, and J. H. Dixon, "Water and Power in the Northeast," *J. Power Div. (ASCE)*, 102(PO2): pp. 195–208 (1976).
33. Rajagopal, H. Y., C. T. Crawford, and R. E. Hughes, "Water Supply and Conservation in an Energy Facility: Design and Operation of an Off-Stream Reservoir," Paper presented at American Society of Civil Engineers Conference on Water and Energy: Technical and Policy Issues, Pittsburgh, PA (May 24–26, 1982).
34. Loucks, D. P., J. R. Stedinger, and D. A. Haith, *Water Resource Systems Planning and Analysis*, Prentice-Hall, Englewood Cliffs, N.J. (1981).
35. United Engineers and Constructors, *Statewide Site Selection Survey*, Prepared for the New York Power Pool, Schenectady, N.Y. (1977).
36. Hobbs, B. F. and Meier, P. M., "An Analysis of Water Resources Constraints on Power Plant Siting in the Mid-Atlantic States," *Water Res. Bull.* 15(6): pp. 1666–1676 (1979).
37. Eagles, T. W., J. L. Cohon, and C. ReVelle, *Modeling Plant Location Patterns: Applications*, Report EPRI EA-1375, Electric Power Research Institute, Palo Alto, CA (1982).
38. MacDonald, T. K., "Power Plant Cooling in California: Issues and Alternatives," in *Conservation and Utilization of Water and Energy Resources*, American Society of Civil Engineers, New York, pp. 271–278 (1979).
39. Israelson, C., et al., *Use of Saline Water in Energy Development*, Report UWRL/P-80/04, Utah Water Research Laboratory, Utah State University, Logan, UT (1980).
40. Milliken, J. G. and L. C. Lohman, *Feasibility of Financial Incentives to Reuse Low Quality Waters in the Colorado River Basin*, Report OWRT/RU-81/7, Denver Research Institute, University of Denver, Denver, CO (1981).
41. Hobbs, B. F., "Water Supply for Power in the Texas-Gulf Region," *J. Water Res. Plan. Manage.*, 110(4): pp. 373–391 (1984).
42. Roach, F., "Groundwater Alternatives and Solutions," in *Groundwater and Energy*, Report CONF-800137, U.S. Department of Energy, Washington, D.C. (1980).
43. Hughto, R. J., B. M. Harley, D. A. Woodruff, and H. F. Mulligan, "Water Resources for a Coal Gasification Facility," in *Water and Energy: Technical and Policy Issues* (F. Kilpatrick and D. Matchett, eds.), American Society of Civil Engineers, New York, NY, pp. 164–169 (1982).
44. Gertsch, W. D., *Assessment of Water Resources in Utah and Nevada for a Proposed Electric Power Generating Station*, Report LA-6670-MS, Los Alamos Scientific Laboratory, Los Alamos, NM (1977).
45. Abbey, D., *Water Supply/Demand Alternatives for Electric Generation in the Colorado River Basin*, Report LA-7662-MS, Los Alamos Scientific Laboratory, Los Alamos, NM (1979).
46. Weatherford, G., L. Brown, H. Ingram, and D. Mann, eds., *Water and Agriculture in the Western U.S.: Conservation, Reallocation, and Markets*, Westview Press, Boulder, CO (1982).
47. Abbey, D. and V. Loose, *Water Supply and Demand in a Large Scale Optimization Model of Energy Supply*, Report DOE/EV/10180-2, U.S. Department of Energy, Washington, D.C. (1979).
48. Clark, R., "An Energy Industry View," in *Groundwater and Energy*, Report CONF-800137, U.S. Department of Energy, Washington, D.C. (1980).
49. Gisser, M., et al., "Water Tradeoff Between Electric Energy and Agriculture in the Four Corners Area," *Water Res. Research*, 15(3): pp. 529–538 (1979).
50. Texas Department of Water Resources, *Texas Water Planning Projections for Texas—1980–2030*, Review Draft, Austin, TX (1982).
51. Humm, W. and E. Selig, *Water Availability for Energy Industries in Water-Scarce Areas: Case Studies and Analyses*, prepared for G. Krug, U.S. Department of Energy, Washington, D.C. (1978).
52. El-Ashry, M., "Testimony before the U.S. Senate Committee on Energy and Natural Resources," *Water Availability for Energy Development in the West*, Hearings before the Subcommittee on Energy Production and Supply, 95th Congress, Second Session, U.S. Government Printing Office, Washington, D.C. (1978).
53. Shafer, J. M., J. W. Labadie, and E. B. Jones, "Analysis of Firm Water Supply under Complex Institutional Constraints," *Water Res. Bull.*, 17(3): 373–379 (1981).
54. Friedlander, G. D., "Zero-Discharge Stations Gain Favor," *Electrical World*, pp. 103–109 (March 1982).
55. Hobbs, B. F., "Multiobjective Power Plant Siting Methods," *J. Energy Div. (ASCE)*, 106(EY2): pp. 187–200 (1980).
56. Gold, H. and Goldstein, D., *Water Related Environmental Effects in Fuel Conversion*, Report EPA 600/7-7-1976, U.S. Environmental Protection Agency, Washington, D.C. (1976).
57. Gottlieb, P., J. H. Robinson, and D. R. Smith, *Preliminary Assessment of Nuclear Energy Centers and Energy Systems Complexes in the Western United States*, Report ORNL/Sub-7272/1, Dames and Moore, Los Angeles, CA (1978).

58. Rogers and Golden, Inc., *Eastern Shore Power Plant Siting Study. Maryland Major Facilities Study*, Vol. 2, Maryland Power Plant Siting Program, Annapolis, MD (1977).
59. Constitution, State of Texas, Article III, Section 49-D.
60. Chankong, V., and Y. Y. Haimes, *Multiobjective Decision Making: Theory and Methods*, North-Holland, Amsterdam (1983).
61. Cohon, J., *Multiobjective Programming and Planning*, Academic Press, New York (1978).
62. Goicoechea, A., D. Hansen, and L. Duckstein, *Introduction to Multiobjective Analysis with Engineering and Business Applications*, J. Wiley, New York (1982).
63. Shaw, J. J., E. E. Adams, D. R. F. Harleman, and D. H. Marks, *A Methodology for Assessing Alternative Water Acquisition and Use Strategies for Energy Facilities in the American West*. Energy Laboratory Report No. MIT-EL 81-051, Massachusetts Institute of Technology, Cambridge, MA (1981).
64. Palmer, R. N., and J. R. Lund, "Drought and Power Production II: Risk Analysis Planning," *J. Water Res. Plan. Manage.*, forthcoming.
65. U.S. Department of the Interior, *Report on Water for Energy in the Upper Colorado River Basin*, Water for Energy Management Team, Denver, CO (1974).
66. U.S. Department of the Interior, *Report on Water for Energy in the Northern Great Plains Area with Emphasis on the Yellowstone River Basin*, Water for Energy Management Team, Denver, CO (1975).
67. Dobson, J. E., and A. D. Shepherd, *Water Availability for Energy in 1985 and 1990*, Report ORNL/TM-6777, Oak Ridge National Laboratory, Oak Ridge, TN (1979).
68. Sonnichsen, J. C., *An Assessment of the Need for Dry Cooling, 1981 Update*, Report HEDL-TME 81-47, UC-11,12, Hanford Engineering Development Laboratory, Richland, WA (1982).
69. Wright, K. R., L. M. Eisel, and R. J. Johnson, "Water for Energy in the Upper Colorado Basin," in *Water and Energy, Technical and Policy Issues* (F. Kilpatrick and D. Matchett, eds.), American Society of Civil Engineers, New York, pp. 328–333 (1982).
70. U.S. Environmental Protection Agency, *Energy from the West: Policy Analysis Report*, Washington, D.C. (1979).
71. Boris, C. M. and J. V. Krutilla, *Water Rights and Energy Development in the Yellowstone River Basin: An Integrated Analysis*, The Johns Hopkins University Press, Baltimore, MD (1980).
72. Hickcox, D. H., "The Impact of Energy Development in the Tongue River Basin, Southeastern Montana," *Water Res. Bull.*, *18*(6): pp. 941–948 (1982).
73. Spofford, W. O., Jr., A. L. Parker, A. V. Kneese, eds., *Energy Development in the Southwest: Problems of Water, Fish, and Wildlife in the Upper Colorado River Basin*, Resources for the Future, Washington, D.C. (1980).
74. Shupe, S. J., "Instream Flow Requirements in the Powder River Coal Basin," *Water Resources Bulletin*, *14*(2): pp. 349–358 (1978).
75. Flug, M., W. R. Walker, and G. V. Skogerboe, "Energy-Water-Salinity: Upper Colorado River Basin," *J. Water Res. Plan. Manage. Div. (ASCE)*, *105*(WR2): pp. 305–315 (1979).
76. Fuessle, R. W., R. M. Lyon, E. D. Brill, G. E. Stout, and K. F. Wojnarowski, "Power Development and Water Allocation in Ohio River Basin," *J. Water Res. Plan. Manage. Div. (ASCE)*, *104*(WR1): 193–209 (1978).
77. Buras, N., *Modeling Water Supply for the Energy Sector*, Report EPRI EA-2559, Electric Power Research Institute, Palo Alto, CA (1982).
78. Harte, J., "Water Constraints on Energy Development: A Framework for Analysis," *Water Res. Bull.*, *19*(1): pp. 51–57 (1983).
79. Lall, U., and L. W. Mays, "Model for Planning Water-Energy Systems," *Water Res. Research*, *17*(4): 853–865 (1981).
80. Mansfield, E., *Microeconomics, Theory and Applications*, Second Edition, W. W. Norton, New York (1975).

CHAPTER 25

Storage Treatment of Salt-Gradient Solar Pond

KOICHI KINOSE *

INTRODUCTION

One type of solar pond is known as "shallow solar pond"—a horizontal plastic bag filled with water. The theory and technology is that of conventional flate collectors. This paper will concentrate on the salt-gradient, non-convecting pond since this is the area where most practical progress is reported.

A salt-gradient solar pond dealt with herein, as shown in Figure 1, is a solar energy system which has both solar collection and heat storage in the one system. In the upper layer, a gradient is established by increasing salt concentration with depth, from very low value at the surface to near saturation at a depth of, usually, 1 – 2 m. The density gradient, which is caused by the salt concentration gradient, prevents thermal convection. Therefore, the upper layer is called the non-convective zone and plays an important role in insulating heat in the lower layer because of the low thermal conductivity of the zone. In the lower layer, which forms a zone for storage and extraction of heat, the salt concentration gradient is zero; thus, thermal convection can occur. This layer is called the convective storage zone. Incident solar radiation is absorbed in the saline water in the pond and is trapped as sensible heat by a selective absorption and, consequently, the temperature of the pond is raised.

The idea of creating artificial solar ponds was hinted by natural salt-gradient lakes—which exhibit a temperature gradient, i.e. show an increase in temperature and salt concentration with depth—have existed in nature. In limnology, these lakes are called meromictic lakes and the salt concentration gradient, the halocine. If the halocine is maintained sufficiently by surrounding geographical or geological conditions, then incident solar radiation can cause a considerable temperature rise above ambient in the lake. As a result, the temperature gradient parallels the halocline. The first record to a natural solar lake was that of Kalecsinsky [1] who reported the Medve Lake in Transylvania (42° 44′N, 28° 45′E). The temperature of this lake was about 70°C at a depth 132 cm at the end of summer. After this several meromictic lakes were found in various countries [2,3]. A natural solar lake which existed in the most severe environment is the Vanda Lake (77° 35′S, 161° 39′E) in Victoria Land (the Antarctic Continent) [4]. This lake showed temperatures increasing up to 25°C at a depth of more than 50 m in the summer. It was confirmed by measuring electric conductivity of the lake water that the temperature distribution of the lake parallels the halocine [5]. It has been considered that this heat storage was done for very a long duration, i.e. more than 1,000 years.

The investigation of an artificial solar pond was initiated in Israel in 1954 [6]. The Israel efforts were continued through to 1966, but stopped for economic cost reasons when fuel-oil was cheap. After the energy crisis of 1973, the researches of the theory and technology of the solar pond were revived in a number of countries.

THEORY AND TECHNOLOGY OF THE SOLAR POND

Thermal Convection and Heat Trap

Because the solar radiation is absorbed in the upper layer of the lake or pond, the depth of the solar pond must be shallow, say 1 – 3 m deep. The solar radiation will penetrate the water and be absorbed until reaching the bottom by a selective absorption. The solar radiation which reaches the

---

*Department of Hydraulic Engineering, National Research Institute of Agricultural Engineering, Yatabe, Tsukuba, Ibaraki, Japan

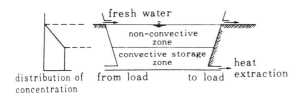

FIGURE 1. Schematic representation of solar pond.

bottom will be almost absorbed, if the bottom is black or dark, and the water temperature will increase. Due to buoyancy caused by the heated water, the hot water of the lower layer will rise to the surface where the heat is rapidly dissipated into the atmosphere. This phenomenon is a thermal convection. However, if the water at the lower layer of the pond were heavier than the upper layer, the thermal convection could not occur. The weight difference of the water per unit volume at each water level, i.e. the density gradient, is produced by the salt concentration gradient. The heat stored in the lower layer cannot radiate to the surface of the pond due to the selective absorption, because the heat is long wavelength energy—which is the reason that the solar pond can function as a heat trap.

Solar Radiation Absorption

In this way, the solar radiation is absorbed not only in the convective storage zone where we want it, but also in its way through the non-convective zone where it is less effective for getting heat. The absorption of the radiation as it passes through water can be described by an exponential series, because different wavelengths differ widely in their absorption coefficients. As shown in Table 1 [7], the short wavelength region of the sun's spectrum penetrates meters and ten of meters, while the near infrared region (long wavelength) is absorbed within a few centimeters—this phenomenon is a selective absorption as previously stated. By Rabl's expression [7], the radiation reaching the depth of z in the

TABLE 1. Energy Distribution in Percent in Sunlight Spectrum after Passing through Water Layers of Different Thickness [7].

Wavelength	Thickness of Water				
	0	1 cm	10 cm	1 m	10 m
0.2–0.6	23.7	23.7	23.6	22.9	17.2
0.6–0.9	36.0	35.3	30.5	12.9	0.9
0.9–1.2	17.9	12.3	0.8	—	—
Over 1.2	22.4	1.7	—	—	—
Total	100.0	73.0	54.9	35.8	18.1

pond can be represented as follows:

$$H(z) = \tau H_s \sum_{n=1}^{4} \eta_n e^{-\mu_n z} \quad (1)$$

in which

$$\eta_1 = 0.237 \quad \mu_1 = 0.32 \times 10^{-3} \text{ cm}^{-1} \quad (2a)$$

$$\eta_2 = 0.193 \quad \mu_2 = 4.5 \times 10^{-3} \text{ cm}^{-1} \quad (2b)$$

$$\eta_3 = 0.163 \quad \mu_3 = 0.03 \text{ cm}^{-1} \quad (2c)$$

$$\eta_4 = 0.179 \quad \mu_4 = 0.35 \text{ cm}^{-1} \quad (2d)$$

τ = the coefficient of transmission; and H_s = the full radiation incident upon the water surface. If the sunlight ray incident on the horizontal pond surface at an angle of i to normal is refracted to an angle of γ according to Snell's law,

$$\frac{1}{n} = \frac{\sin \gamma}{\sin i} \quad (3)$$

in which n = the refractive index of the pond water. For practical calculation, the absorption coefficients of $\bar{\mu}$ given by Equation (4) must be used instead of μ:

$$\bar{\mu}_n = \frac{\mu_n}{\cos \gamma} \quad (4)$$

Salt-Gradient Creation

The salt concentration gradient in solar ponds must be created by dissolving low cost salts of which the solution has a high degree of clearness. Usually two salts are considered most likely to be used in solar ponds, i.e. NaCl or $MgCl_2$.

Several theories for stability criteria against thermal convection have been proposed. According to Nielsen and Rabel [8], the following equation gives the criteria,

$$\alpha_T \frac{\partial T}{\partial z} < \alpha_s \frac{\partial C}{\partial z} \cdot \left(\frac{D}{k_w}\right)^{1/3} \quad (5)$$

in which $\alpha_s = (1/\varrho)(\partial \varrho/\partial C)$ and $\alpha_T = (-1/\varrho)(\partial \varrho/\partial T)$ = the saline and the thermal expansion coefficients, respectively. The value of $\partial T/\partial z$ depends upon the insolation and other factors. According to Tabor's experience [6], the maximum value of the local temperature gradient in midsummer was considered to reach above 400°C/m and the following Table 2 was given.

The pond is filled in layered sections one after the other, each layer having a slightly different salt concentration. The pond is poured from the bottom upwards—the densest bot-

TABLE 2. Necessary Quantity of Salt for Maintaining Stability of Solar Pond [6].

	Pond Surface	Pond Bottom 20°C	Pond Bottom 90°C	
$MgCl_2$				
C	20	300	300	(kg/m³)
$\partial\varrho/\partial C$	0.75	0.65	0.68	
$-\partial\varrho/\partial T$	0.3	0.25	0.45	(kg/m³°C)
$\partial C/\partial z$	$>0.46\partial T/\partial z$	$>0.44\partial T/\partial z$	$>0.75\partial T/\partial z$	
	i.e. > 230	(same as when hot)	>300	(kg/m³)
	for $\partial T/\partial z$ 500		for $\partial T/\partial z$ 400	
NaCl				
C	20	260	260	(kg/m³)
$\partial\varrho/\partial C$	0.8	0.62	0.52	
$-\partial\varrho/\partial T$	0.3	0.5	0.51	(kg/m³°C)
$\partial C/\partial z$	$>0.44\partial T/\partial z$	$>0.92\partial T/\partial z$	$>1.18\partial T/\partial z$	
	i.e. > 220	(same as when hot)	>230	
	for $\partial T/\partial z$ 500			

tom layer is filled first and successively lighter layers are floated upon the denser layer below. As a result of this, the pond will have an approximately stepwise distribution of density, as shown in Figure 2. The initial stepwise distribution becomes smooth with time due to the diffusion of the salt. The diffusing process can be calculated by the present writer [9] as follows,

$$\frac{C}{C_0} = \sum_{n=1}^{\infty} B_n \cos\left[\frac{(2n-1)\pi}{2}\frac{(L-z)}{L}\right]$$

$$\exp\left[-D\left(\frac{2n-1}{2L}\right)^2 \cdot t\right]$$

$$B_n = \frac{4}{(2n-1)\pi}\left\{\sum_{m=M'}^{M} 2C'_m \cos\right.$$

$$\left[\frac{(2m-1)(2n-1)\pi}{4M}\right]\sin\left[\frac{(2n-1)\pi}{4M}\right] \quad (6)$$

$$\left. + \sin\left[\frac{2n-1}{2M}\pi(M'-1)\right]\right\}$$

$$C'_m = \frac{1}{M}\left[(M'-1) + \frac{1}{M-M'}\sum_{m=M'}^{M}(M-m)\right]$$

$$M = \frac{L}{\Delta z}, M' = \frac{l_c}{\Delta z}$$

in which l_c = the thickness of the convective storage zone; and Δz = the thickness of each solution layer established when the pond was filled saline water as shown in Figure 2. This equation was derived on the following conditions; $C = 0$ at the pond surface ($z = 0$), $\partial C/\partial z = 0$ at the bottom ($z = L$), and the kinetic energy of water injection into the pond during the filling process was neglected. The kinetic energy effect reduces the time for getting the smooth distribution, because it has a mixing effect. But if the kinetic energy is very large—say the flow velocity is too large, a large scale mixing can occur. The limit of the injection velocity is about 0.12 m/sec [10]. Figure 3 shows one of our experimental results for the timewise development of concentration profiles—which were gotten from an outdoor test of the heat storage by use of a small size test pond [9].

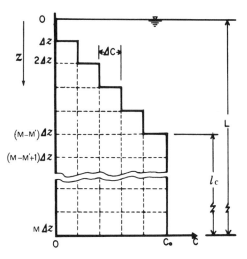

FIGURE 2. Initial concentration distribution in solar pond.

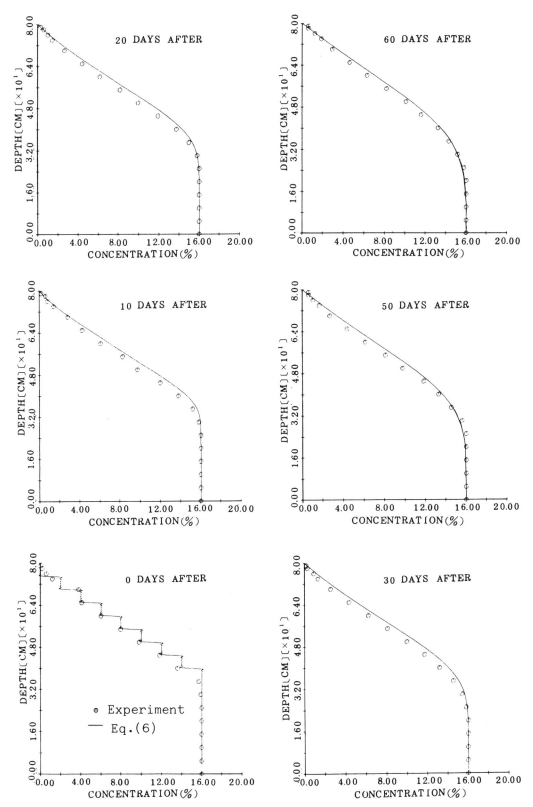

FIGURE 3. Timewise development of concentration profiles in pond no. 1 [9].

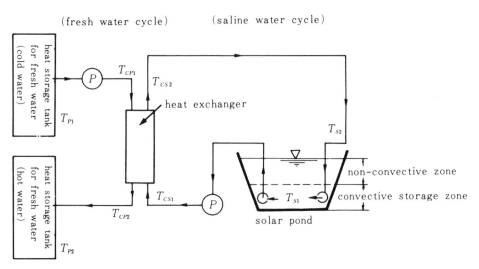

FIGURE 4. System of heat extraction.

Usually maintaining the salt concentration is effected by controlling the salt concentrations at the top and the bottom of the pond. At the top, the control is done by holding the water surface level; because the water level variation takes place due to evaporation and rainfall, the evaporated water was replaced by fresh water and the rain was drained, based on the use of a water level controller. By the water level control, the fresh water floats at all times on the water surface. At the bottom, the salt concentration is controlled when the storage heat extracts from the storage convective zone.

Heat Extraction

One method of extracting the stored heat from the storage convective zone is by doing heat exchange in the pond, and another, out of the pond. The former is done by setting up a heat-exchanger near the bottom, for example in the form of a series of parallel pipes. This method has been described in the references [11] and [12]. In this method, the construction cost for heat extraction is relatively high because a large heat-exchanger is needed to extract efficiently the heat from the pond. The latter is given by recycling the hot saline water through a heat exchanger on the outside of the pond. This heat extraction system has been introduced in Israel [13–16]. The outline of the writer's heat extraction system [17] is shown in Figure 4. In this system, the hot saline water in the convective storage zone is pumped from one end of the pond, passed through a heat-exchanger, and returned to the other end of the pond. As a result of the heat extraction by the heat-exchanger, the water which returns to the pond is cooled. Then a procedure is followed in order to maintain concentration of the convective storage zone at a required value, if necessary. By means of water circulation, a horizontal flow takes place in the convective storage zone. It is necessary for the velocity of the horizontal flow to be small enough to prevent the disturbance of the concentration gradient in the non-convective zone. The extent of the disturbance can be estimated by analyzing a problem of a stratified flow in hydraulics.

Before the heat extraction is done, the stratified fluid has a density distribution as shown by the dashed line in Figure 5. However, for our analysis [17], the thickness of the convective storage zone, d, is defined as shown by the straight line in Figure 5. Because the density gradient in the non-convective storage zone is zero, the flow velocity which takes place in the heat extraction is very small, the flow can be analyzed as a problem of potential flow. Considering two-dimensional flow, the flow in the convective storage zone is generated by a line sink and a line source. From this

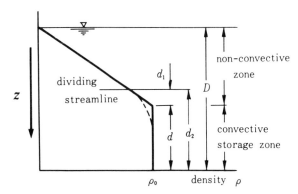

FIGURE 5. Simulation model of density distribution.

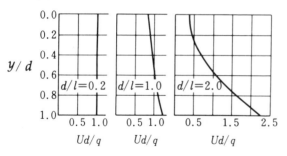

FIGURE 6. Velocity distribution at center of convective storage zone (y = distance from border line between non-convective zone and convective storage zone).

analysis, the velocity distribution at the center of the pond were obtained when the aspect ratio of the storage convective zone, is generated by a line sink and a line source. From this analysis, the velocity distribution at the center of the pond were obtained when the aspect ratio of the storage convective zone, d/l (d = the thickness of the convective storage zone; and l = the half width of the pond), were 0.2, 1.0, and 2.0, respectively as shown in Figure 6. The dependence of the center velocity on the upper border plane, between non-convective and convective storage zones, on the aspect ratio, d/l, was estimated as shown in Figure 7.

From Figure 7, the following facts can be read: The velocity distribution of the flow in the convective storage zone is uniform when the non-dimensional value, Ud/q is equal to 1, and the vertical distribution is uniform in the case when aspect ratio, d/l, is under 1/3. Since the velocity is not zero on the border plane, the water in the non-convective zone, which has a linear density gradient, should be entrained near the border plane by the flow in the convective storage zone, in which q = the discharge per unit width; and U = the velocity of the horizontal direction. The ratio of entrainment can be estimated use of the analysis for the stratified fluid with a stable linear density distribution. The ratio of entrainment, Δq, by the flow in convective storage zone is given by the following equation (17):

$$\Delta q = \frac{q}{0.248\alpha_o l^{2/3}} \sum_{n=1}^{\infty} \frac{\sin(N)}{\sinh\left(\frac{Nd}{l}\right)}$$

$$= Q\left[1 - \frac{1}{1 + \frac{1}{(0.284\alpha_o l^{2/3})}\sum_{n=1}^{\infty}\frac{\sin N}{\sinh\left(\frac{Nd}{l}\right)}}\right] \quad (7)$$

$$N = \frac{(2n-1)\pi}{2}, \quad n = 1, 2, 3 \ldots$$

$$\alpha_0^6 = \frac{\epsilon q}{D\nu}, \quad \epsilon = -\frac{1}{\varrho_0}\frac{\partial \varrho}{\partial y}$$

in which Q = the total discharge of the heat extraction per unit width and equals the sum of q and Δq. The extent of the disturbance in the non-convective zone is given the following procedure. The disturbance region is bounded in two regions by a dividing streamline. The upper region is essentially stagnant and very thick, and in the lower region the discharge is concentrated. From an experimental equation for the stratified flow with the linear density distribution on the whole depth, the level of the dividing streamline, d_1, from the border plane of the non-convective and the convective storage zones can be obtained as follows [17],

$$\left(\frac{d_1}{D-d}\right)^2 = \frac{3.95\ \Delta q}{(D-d)^2\sqrt{g\epsilon}} \quad (8)$$

The level of the dividing streamline, d_2, from the pond bottom (Figure 5) is

$$d_2 = d_1 + d \quad (9)$$

From the preceding analyses it becomes clear that the flow in the convective storage zone has nearly a uniform vertical velocity distribution by the line sink and the line source. The region of the disturbance induced by the flow in the non-convective zone is separated by the dividing streamline. In order to extract almost all of the heat from the convective storage zone, it is necessary that the transverse distribution of the velocity should be as uniform as possible. This condition can be made clear by the following consideration.

The inlet pipe and outlet one with a thin slit for the line sink and the line source are used (Figure 8). The flow in the pipe has a pressure distribution owing to the intake from the center of the pipe as shown in Figure 8. The difference between the pressure at the end and that at the center of the

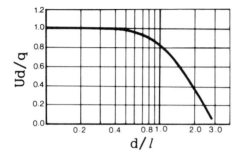

FIGURE 7. Dependence of center velocity on upper border plane of convective storage zone aspect ratio, d/l.

pipe is given by the energy balance in pipe flow:

$$\Delta P_1 = \frac{V_1^2}{2g} = \frac{8Q_i^2}{g(\pi D_s^2)^2} \quad (10)$$

in which ΔP_1 = the differential pressure; Q_i = the half discharge of the water intake; V_1 = the velocity at the center of the pipe; and D_s = the inner diameter of the pipe. This pressure difference in the pipe induced a non-distribution of the flow velocity is non-uniform.

The entrance head through the slit into the pipe is

$$\Delta P_2 = \frac{(1+f)V_2^2}{2g} = \frac{(1+f)Q_i^2}{(2gb^2L_D^2)} \quad (11)$$

in which ΔP_2 = the entrance head; V_2 = the velocity through the slit; f = the coefficient of inlet loss; L_D = the half length of the pipe; and b = the thickness of the slit. In order to decrease the influence of the pressure difference, ΔP_1, on the velocity non-uniformity, the entrance head, ΔP_2 and ΔP_1, must satisfy the following relation:

$$\frac{\Delta P_1}{\Delta P_2} = \frac{16}{\pi^2} \frac{1}{1+f} \left(\frac{b}{D_s}\right)^2 \left(\frac{L}{D_s}\right)^2 < < 1 \quad (12)$$

Therefore, the condition for the uniform transverse distribution of the velocity in the convective storage zone can be obtained by determining the geometrical configuration of the inlet pipe. Though the flow direction in the outlet pipe is opposite to that in the inlet pipe, both flows have the same pressure relation. So, this relation can also be applied to the outlet pipe. Figure 9 shows the effect of the geometrical configuration of the inlet pipe on the pressure difference.

In the next figures the results of our heat extraction test [17] will be shown. The outlines of our test pond are shown in Figure 10. To obtain a uniform transverse velocity in the pond, the inlet and the outlet pipes for the heat extraction were designed under the condition, $\Delta P_1/\Delta P_2 = 0.01$, which satisfies Equation (12), as shown in Figure 11. The pipes were installed in the test pond (Figure 10). The pond was waterproofed with two vinyl chloride sheets of 0.1 cm in thickness. Figures 12–13 show the examples of the observation data from the pond. Figure 12 shows the observed distributions of the storage temperature and the salt concentration just before the extraction was started and after four hours. The value of the level of the dividing streamline, d_2, could be obtained as shown in Figure 12. The timewise development of the level is shown in Figure 13; the straight line is the computed value, and the dotted marks are the measured value. The discharge for the heat extraction can be given by considering the following conditions: the storage temperature, the necessary quantity of heat for heat usage, and the efficiency of the heat-exchanger.

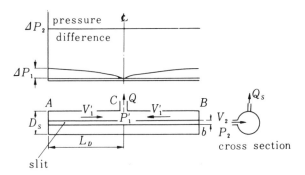

FIGURE 8. Pressure distribution in inlet and outlet pipes.

Equations for the Temperature of the Pond

In order to understand quantitatively the storage temperature of the solar pond, it is necessary to estimate the heat flux from the pond to the surrounding ground (or from the surrounding ground to the pond). The storage temperature in the solar pond, which changes with time, is dependent on weather conditions during the daytime, night, and seasons. Due to the storage temperature variation, heat transfer to the surrounding soil occurs. Then, the heat transfer must be

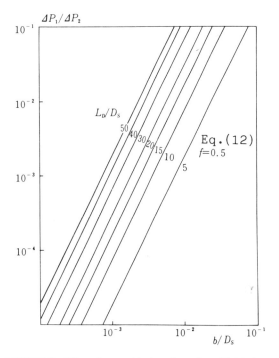

FIGURE 9. Effect of geometrical configuration of inlet pipe on pressure difference.

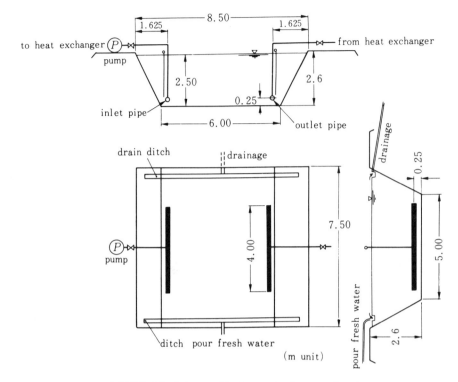

FIGURE 10. Solar pond for heat extraction experiment.

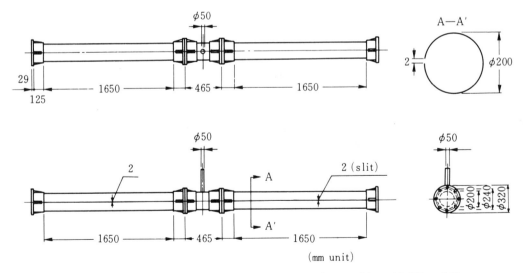

FIGURE 11. Structure of inlet and outlet pipes under design conditions, $\Delta P_1/\Delta P_2 = 0.01$.

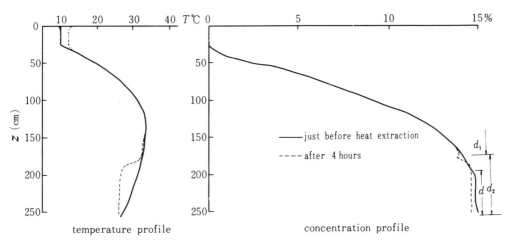

FIGURE 12. Deformation of distribution of storage temperature and concentration owing to heat extraction.

treated as unsteady boundary problems of heat conduction to a semi-infinite body.

Supposing the pond is bounded by a homogeneous soil, the next solutions [9] describing heat conduction into the surrounding homogenous ground are obtained form the thermal conduction equation. Taking the time interval of Δt small enough in comparison with the temperature change, the storage temperature variation of the pond is:

$$T(z)_1 = \text{const.} \quad 0 < t \leq \Delta t \tag{13a}$$

$$T(z)_2 = \text{const.} \quad \Delta t < t \leq 2\Delta t \tag{13b}$$

$$\vdots \qquad \vdots$$

$$T(z)_N \equiv \text{const.} \quad (N - 1)\Delta t < t < N\Delta t \tag{13c}$$

in which $T(z)_N$ = the storage temperature at the depth of z in the pond. According to this approximation, the surrounding ground temperature, $T_s(x)$, at the point from the side wall and the bottom of the pond can be obtained:

$$T_s(x) = \sum_{N=1}^{N} [T(z)_N - T(z)_{N-1}]$$

$$\times \left[1 - \text{erf}\left(\frac{x}{2\sqrt{k_s(t - (N - 1)\Delta t)}} \right) \right] + T_{s\infty} \tag{14}$$

in which erf($\bullet$) = the error function; x = the coordinate taking the side wall or the bottom of the pond; and $T(z)_0$ = the initial temperature of the pond water. Then, at the depth of z in the pond, the heat flux from the pond to the

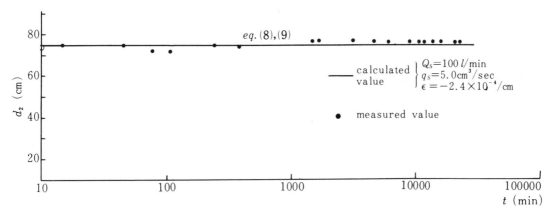

FIGURE 13. Timewise development of dividing streamline.

surrounding ground (or reverse transfer) is represented by

$$q_s(z) = -K_s \left. \frac{\partial T_s}{\partial x} \right|_{x=0}$$
$$= K_s \sum_{N=1}^{N} \frac{T(z)_N - T(z)_{N-1}}{\sqrt{\pi k_s [t - (N-1)\Delta t]}} \quad (15)$$

in which K_s = the thermal conductivity of soil; k_s = the thermal diffusivity of the soil; and $T_{s\infty}$ = the temperature of the soil at $x = \infty$.

Calculating one-dimensional heat balance for a thin layer with the thickness of Δz, the sectional area of $A(z)$, and the lateral area of $A_B(z)$ at the depth of z in the pond, the storage temperature results in

$$\varrho_w \frac{\partial T}{\partial t} = K_w \left[\frac{1}{A(z)} \frac{\partial A(z)}{\partial z} \frac{\partial T}{\partial z} + \frac{\partial^2 T}{\partial z^2} \right]$$
$$- \left[\frac{\partial H(z)}{\partial z} + \frac{H(z)}{A(z)} \frac{\partial A(z)}{\partial z} \right] - q_s(z) \frac{A_B(z)}{A(z)\Delta z} \quad (16)$$

in which T = the storage temperature at the depth of z; ϱ_w = the heat capacity per unit volume; K_w = the thermal conductivity of the pond water; and $H(z)$ = the solar radiation reaching the depth of z. In this equation, the terms with $\partial A/\partial z$ show the amount of heat induced by the change of cross-sectional area in z direction; the terms with $\partial H(z)/\partial z$ and $H(Z) \cdot \partial A(z)/\partial z$ are the heat quantity absorbed into the pond water and the side wall, respectively, where $H(z)$ is given in Equation (1).

Two boundary conditions, i.e., the surface and the bottom conditions, are required to solve Equation (16). Heat losses in the pond surface are composed of the heats of radiation, convection, and evaporation. The writer has calculated this heat balance using empirical methods and compared the calculated results with observation data obtained from a solar pond test [18]. As a result, it became clear that a certain degree of approximation can be obtained even when the calculation is carried on under the following condition:

$$T(0) \approx T_a \quad (17)$$

in which T_a = the ambient air temperature. Owing to the occurrence of the heat convection in the convective storage zone, the zone has uniform temperature distribution. Then, the following bottom boundary condition is obtained from the heat balance for the convective storage zone:

$$\varrho_w l_c \frac{\partial T}{\partial t} = H(l_i) - q_s(l_i + l_c) \frac{A(l_i + l_c)}{A(l_i)} - \frac{A_B(l_c)}{A(l_i)} q_s(l_c)$$
$$- K_w \left. \frac{\partial T}{\partial z} \right|_{z=l_i} - \frac{Q_L}{A(l_i)} \quad (18)$$

in which Q_L = the heat load (quantity of heat extraction from the pond); and l_i, l_c = the thickness of the non-convective and convective storage zones, respectively.

Influence of Pond Scale and Heat Insulation on Storage Temperature

Here, we will discuss the influence of the pond scale and the effect of the heat insulation which covers the side and the bottom of the pond on the storage temperature, using the yearly average meteorologial data. The analysis using the yearly average data gives the yearly average temperature of the pond, because the average value is the stationary component in yearly climatic variations.

Under the conditions, i.e., non-change of the cross sectional area with the depth ($\partial A/\partial z = 0$); the temperature does not change with time (a steady-state condition; $\partial T/\partial t = 0$), the temperature distribution in the non-convective zone is given by:

$$\bar{T}(z) = \bar{T}_a + \frac{z}{\ell_i} \left[\bar{T}_c - \bar{T}_a - \sum_{n=1}^{4} \frac{\bar{H}_s}{\mu_n K_w} (1 - e^{-\mu_n \ell_i}) \right]$$
$$+ \sum_{n=1}^{4} \bar{H}_s/\mu_n (1 - e^{-\mu_n z}) \quad (19)$$

in which $\bar{T}$, $\bar{H}(z)$ = the yearly average values of temperature in the pond and solar radiation, respectively; and $\bar{T}_c$, $\bar{T}_a$ = the yearly average values of the temperature of the convective storage zone and ambient air temperature.

The more simple expression [19] for the heat flux from the pond to the surrounding ground is used. Namely, supposing that the shape of the pond is hemisphere and the heat transfers hemispherically, the steady-state heat flow from the pond to the surrounding ground is given from Equation (20).

$$\bar{Q}_s = \bar{A}_B K_s / r_0 (\bar{T}_c - \bar{T}_{s\infty}) = \bar{A}_B U'(\bar{T}_c - \bar{T}_{s\infty})$$
$$U' = K_s/r_0 \quad (20)$$

Where r_0 is

$$2\pi r_0^2 = \bar{A}_B \quad (21)$$

in which $\bar{Q}_s$ = the total quantity of heat flow from the pond to ground; $\bar{A}_B$ = the sum of the lateral area and the bottom one of the pond; and U' = U-factor of thermal conductance. Using the U-factor, the effect of the heat insulation on the heat transfer from the pond to the ground can be expressed by

$$1/U = d/K_D + r_0/K_s \quad (22)$$

TABLE 3. Quantities for Computation.

Thermal conductivity	
K_w	0.542 kcal/mhr°C (40°C)
K_s	0.828 kcal/mhr°C
K_D	0.042 kcal/mhr°C
Meteorological data	
$\bar{H}_s$	122.92 kcal/mhr°C (Tokyo)
$\bar{T}_a$	15.0°C (Tokyo)
$T_{s\infty}$	15.0°C

in which d and K_D are the thickness and the thermal conductivity of the heat insulator, respectively. From the heat balance of the convective storage zone, the temperature of the steady-state of $\bar{T}_c$ in the zone is obtained by

$$\bar{T}_c = \frac{K_w/\ell_i \left[\bar{T}_a + \tau \bar{H}_s/K_w \sum_{n=1}^{4} \eta_n/\mu_n (1 - e^{-\mu_n \ell_i}) \right]}{K_w \ell_i \bar{A}_B/s}$$

$$+ \frac{\bar{A}_B/A_s \, U \, \bar{T}_{s\infty}}{K_w \ell_i + U \, \bar{A}_B/A_s} \qquad (23)$$

in which A_s = the surface area of the pond.

Figures 14–16 show the influences of the variations on the thickness of the non-convective zone, the heat insulator thickness on the storage temperature estimated from Equation (23). The quantities for the estimations are in Table 3. The meteorological data in Tokyo (Japan) are the average values for 30 years. The influence of variation on the thickness of the non-convective zone is shown in Figure 14. In the figure, (a) and (b) show the results calculated by using two values of the thermal conductivities, i.e. 0.542 kcal/mhr°C (fresh water) and 1.1 kcal/mhr°C (double value of fresh water), respectively. Figure 15 shows the effect of variation in the insulator thickness, d, when $K_D = 0.042$ kcal/mhr°C on the storage temperature under the conditions were $l_i = 1.2$ m; the total depth, L, of the pond = 3.0 m; and $A_s = 160$ m². Figure 16 shows the influence of the surface area, A_s, on the storage temperature when $d = 0.0$ m; $l_i = 1.2$ m; and $L = 2.0$ m.

Seasonal Variation of Storage Temperature

The seasonal variations of the storage temperature are estimated by use of the monthly average meteorological data. Rabel [4] has strictly analyzed the response of the storage temperature to a single sinusoidal function expressing the yearly variation. When the variational characteristic of the weather condition is more complex than that in Japan, the variant weather condition must be expressed by finite Fourier Series functions. In the expression using monthly average meteorological data, solar radiation and ambient air

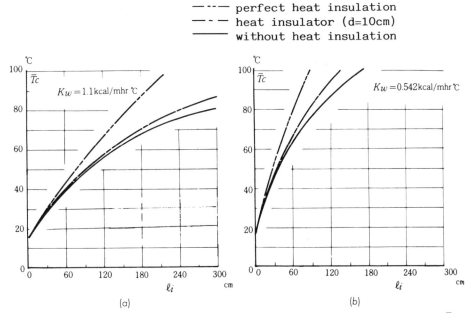

FIGURE 14. Relation between thickness of non-convective zone, l_i and storage temperature $\bar{T}_c$.

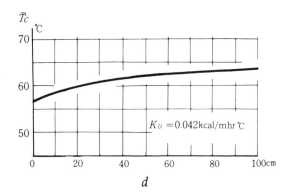

FIGURE 15. Influence of heat insulator thickness, d, on storage temperature.

temperature can be denoted by the following

$$T_a(d_a) = \bar{T}_a + \sum_{k=1}^{5} (\tilde{T}_{a0})_k \cos[\omega_k d_a - (\phi_a)_k]$$

$$+ (\tilde{T}_{a0})_6/2 \cdot \cos(\omega_6 d_a) \quad (24)$$

$$H_s(d_a) = \bar{H}_s + \sum_{k=1}^{5} (\tilde{H}_{s0})_k \cos[\omega_k d_a - (\phi_a)_k] +$$

$$+ (\tilde{H}_{s0})_6/2 \cdot \cos(\omega_6 d_a)$$

in which $(\tilde{T}_{a0})_k$, $(\tilde{H}_{s0})_k$ = the kth mode amplitudes of am-

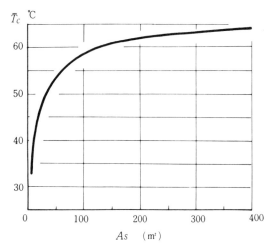

FIGURE 16. Influence of pond surface area, A_s, on storage temperature.

bient air temperature and solar radiation, respectively; $(\phi_a)_k$, $(\phi_s)_k$ = the kth mode phase lags of them; $\bar{T}_a$, $\bar{H}_s$ = the yearly average values of ambient air temperature and solar radiation, respectively; d_a = the date; and ω_k = the kth mode frequency ($\omega_k = 2\pi k/365$). According to thermotics, the response of the storage temperature to the above weather variation can be considered to have the same frequency as each sinusoidal term of Equation (24). Therefore, the yearly variation of the storage temperature can be given by the same Fourier series function:

$$T_c(d_a) = \bar{T}_c + \sum_{k=1}^{5} (\tilde{T}_{c0})_k \cos\{\omega_k d_a - [(\phi_s)_k + (\phi_t)_k]\}$$

$$+ (\tilde{T}_{c0})_6/2 \cdot \cos\{\omega_6 d_a - (\phi_t)_6\} \quad (25)$$

in which $(\tilde{T}_{c0})k$, $(\phi_t)k$ = the amplitude and the phase lag of kth mode of the storage temperature, respectively; $\bar{T}_c$ = the average storage temperature which can be obtained from Equation (23). According to Rabl's solution [7] for the response to the single sinusoidal function of the weather variation, the amplitude and the phase lag of each term on Equation (25) are given by the following equations.

$$(\phi)_k = (\phi_a)_k + \arctan \frac{R\beta + \alpha}{R\alpha - \beta}$$

$$(\tilde{T}_{c0})_k = \frac{\alpha \cos[(\phi)_k - (\phi_a)_k] + \beta \sin[(\phi)_k - (\phi_a)_k]}{(K_s/\sigma_s) + (K_w/\sigma_\omega)G_+} \quad (26)$$

Where

$$\sigma_s = \sqrt{2(K_s/\varrho_s)/\omega_k}, \quad \sigma_\omega = \sqrt{2(K_w/\varrho_s)\omega_k}$$

$$R = (K_s/\sigma_s) + (K_w/\sigma_\omega)G_+/[(K_s/\sigma_s) + (K_w/\sigma_\omega)G_- + \varrho_w \ell_c \omega_k]$$

$$\alpha = C \cos(\phi_a)_k + S \sin(\phi_a)_k + 2(\tilde{T}_{a0})K_w/\sigma_\omega F_+$$

$$\beta = -C \sin(\phi_a)_k + S \sin(\phi_a)_k + 2(\tilde{T}_{a0})K_w/\sigma_\omega F_-$$

with

$$C = \tau(\tilde{H}_{s0})_k \sum_{n=1}^{4} \eta_n/[1 + 1/4(\bar{\mu}_n\sigma_\omega)^4]\{e^{-\mu_n \ell_i}$$

$$+ (\bar{\mu}_n\sigma_\omega)[(\bar{\mu}_n\sigma_\omega)^2/2 \, F_+ + e^{-\mu_n \ell_i}/2(G_- - (\bar{\mu}_n\sigma_\omega)^2/2 \, G_+)]\}$$

$$S = \tau(\tilde{H}_{s0})_k \sum_{n=1}^{4} \eta_n/[1 + 1/4(\bar{\mu}_n\sigma_\omega)^4]\{(\mu_n\sigma_\omega)^2/2 \, e^{-\mu_n \ell_i} +$$

$$+ (\mu_n\sigma_\omega)[(\mu_n\sigma_\omega)^2/2 \, F_- - F_+ e^{-\mu_n \ell_i}/2((\mu_n\sigma_\omega)^2/2 \, G_- + G_+)]\}$$

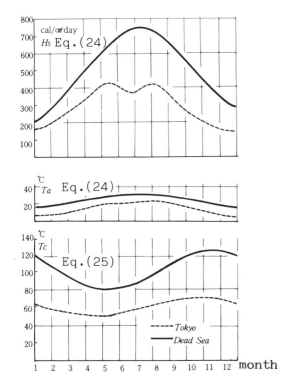

FIGURE 17. Comparison of storage temperature at Tokyo and Dead Sea (Israel).

and

$$F_\pm = [\sin(d')ch(d') \pm sh(d')\cos(d')]/[ch(2d') - \cos(2d')]$$

$$G_\pm = [sh(2d') \pm \sin(2d')]/[ch(2d') - \cos(2d')]$$

$$d' = \ell_i/\sigma_\omega$$

The storage temperature variations on the ponds built in Tokyo (Japan) and the Dead Sea (Israel) are shown in Figure 17. In this figure, the values of the solar radiation of H_s and the mean air temperature of T_a in the Dead Sea are quoted from the data by Weinberger [20], and those values of them in Tokyo are the average data of 30 years. Both ponds have the same dimension; $A_s = 160$ m, $l_i = 1.5$ m and $l_c = 1.5$ m.

Numerical Simulation for Temperature Variation by Heat Storage and Heat Extraction

When the data of ambient air temperature and solar radiation are given by meteorological observation near the solar pond with a time interval of one hour, the variation of the storage temperature due to the weather condition can be analyzed numerically by Hull [21], Akabarzadeh and Ahmadi [22], Hawlader and Brinkworth [23], and the writer [9]. In the writer's analysis [6], in order to get the storage temperature from Equation (16) under the surface boundary condition, Equation (17), and the bottom one, Equation (18), a numerical integration with an explicit finite-difference scheme was used. Equation (6) is used to get the concentration distribution when no thermal convection occurs. In a solar pond, the local temperature gradient in the pond does not satisfy the stability criteria. Consequently, thermal convection occurs locally in the pond. This phenomenon is found not only in the convective storage zone, where there is no density gradient, but in the vicinity of the border, between the non-convective zone and the convective storage zone, or the pond surface. When the concentration and temperature gradients calculated by Equations (6), (16) and do not meet the stability condition (Equation (5)), the distributions of the temperature and the concentration must be corrected by the following procedure. The thickness of the zone in which convection occurs should be found by substituting the calculated values of both gradients into Equation (5). The values of temperature and concentration in the zone are the average values of these values, which are obtained by using Equations (6) and (16).

Thereafter, the value of the concentration must be calculated from the solution of a diffusion equation Equation (27) for the concentration by use of the numerical integration instead of Equation (6).

$$\frac{\partial C}{\partial t} = D \frac{\partial^2 C}{\partial z^2} \quad (27)$$

The diffusion coefficient of the salt in water, D, may be proportional to the absolute temperature, according to Chepurniy and Savage [24]. Therefore, D changes with the depth of the pond as the temperature varies with depth. But, to simplify the calculation, it is assumed that D is constant, and the value is calculated from the average temperature in the temperature profile of the pond, as follows:

$$D = D_{20}\left(\frac{273 + T_{\text{mean}}}{273 + 20}\right) \quad (28)$$

in which D_{20} = the diffusion coefficient at 20°C; T_{mean} = the average temperature in the pond. The value of D_{20} is about 1.5×10^{-5} cm/s.

Figure 18 shows the calculated and the observed temperature profiles every 10 days in pond No. 1. Figure 19 shows the temperature changes with time. The calculated values of the temperature are compared with the measured ones at the depths of 2 cm, 20 cm, and 70 cm, in pond No. 1. In addition, heat storage in pond No. 1 was started on January 12,

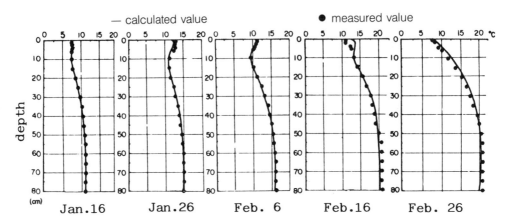

FIGURE 18. Typical comparisons between observed and computed values of storage temperature in pond no. 1.

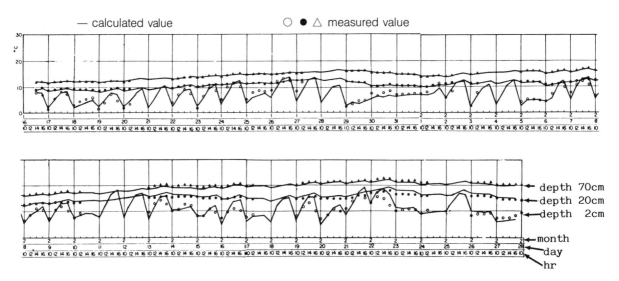

FIGURE 19. Comparison between observed and computed values of storage temperature in pond no. 1 for comparatively long period.

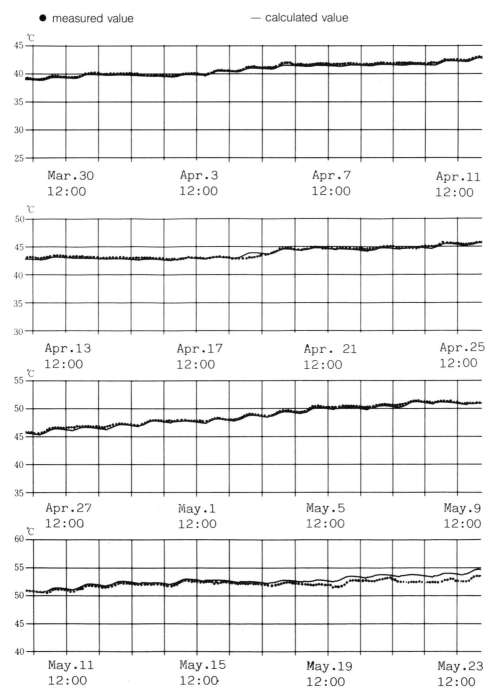

FIGURE 20. Comparison between observed and computed values of storage temperature at depth, 225 cm, in pond no. 2.

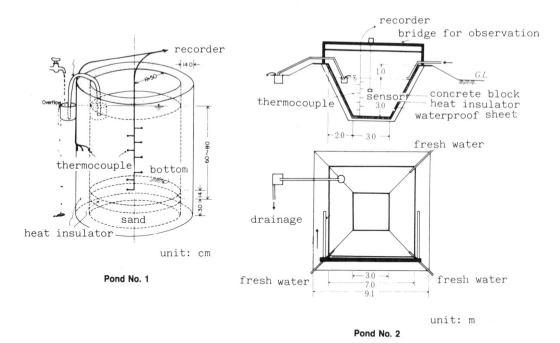

FIGURE 21. Test ponds for heat storage experiment.

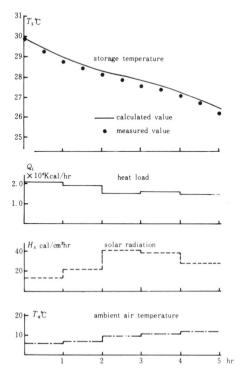

FIGURE 22. Temperature variation of convective storage zone by heat extraction.

1979; and the observed values of the temperature at the time when the pond was filled were used as the initial temperature profile of the pond for the calculation. Figure 20 shows the calculated values and measured ones at the depth, 225 cm, in pond No. 2. The heat storage in pond No. 2 was started on January 16, 1980. The temperature profile observed on the first of March was used as the initial temperature profile of pond No. 2 for calculation. The outlines of test ponds, No. 1 and No. 2, were shown in Figure 21. The calculation and the experiment of the above mention were done under the condition without heat extraction from the ponds.

The variation of the storage temperature owing to the heat extraction can be estimated by the above simulation method [17]. Figure 22 shows the temperature variation of the convective storage zone in the pond on the heat extraction experiment. The curved line and the dotted marks in Figure 22 explain the calculated values and the measured ones, respectively. The meteorological data and heat load used for the the calculation are shown in the same figure. For these calculations, the data observed just before the heat extraction were used as initial distributions of the storage temperature and the concentration. Figure 23 shows the calculated and the measured changes of the storage temperature distribution which occurred by the heat extraction.

Next, an example of the storage temperature variation will be shown, assuming a heat extraction for a large solar pond, using the above simulation method. Based on these calcula-

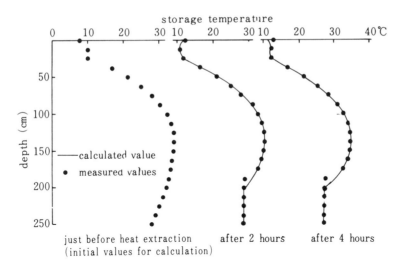

FIGURE 23. Change of distribution temperature after heat extraction.

tions, the values obtained from the monthly average meteorologial data in Tokyo were used as the weather conditions. It is assumed that the heat storage in the pond, with $A_s = 1,000$ m², $l_i = 1.5$ m, and $l_c = 1.5$ m, was started on the first of April, and the heat was extracted from the first of November in the following year to the end of February in the year after that. The ambient air temperature on April 1 was used as the initial temperature in the pond. Figure 24 shows the calculated results of the storage temperature variation, the meteorological data used, and the heat load on the assumed heat extraction. It is understood from Figure 24 that the storage temperature variation will approach the stationary variation after about one year. If the amount of heat usage in winter is limited within that the calculation condition, the storage temperature cooled by the heat extraction is considered to be recovered until the heat extraction in the following year.

PROBLEMS OF SOLAR POND TECHNIQUE

Mixed Layer at the Surface

In a real pond, there is a mixed layer at the surface caused by wind action and heat transfer. The main factor of the formation of the mixed layer is the surface wave generated by the wind. Hull [25] has analyzed the instability near the pond surface induced by the wind. The mixed layer has a uniform profile of the concentration and the temperature. The layer can have a considerable deleterious effect upon the temperature and the efficiency of the pond, because the thickness of the non-convective zone is reduced by the mixed layer. Tabor and Matz [26] have found that a wave amplitude of 2 cm can cause mixing to a depth of 20 cm and Tabor [27] reported mixing to a depth of 50 cm after strong winds. Namely, the formation of the mixed layer has a bad influence upon the thermal insulation which is the most important role of the non-convective zone. Weinberger [20] has shown that a 20 cm mixed layer at the surface of 120 cm deep pond would cause an output loss of 30%. It is therefore desirable to keep the mixed layer thickness as small as possible, preferably in the 10–30 cm range [6] if possible.

A number of proposals have been made as to how to reduce the effect of the wind economically and effectively without simultaneously attenuating the incident solar radiation. For example, floating plastic pipe or grids were experimentally used in the Israeli solar pond in order to absorb the wind wave.

Evaporation at the surface of the pond, which can be as high as 10 mm on a clear day (Tabor [27]), causes a local concentration of salt and thus disturbs the stability of the pond. This is one of the factors of the mixed layer formation, also. So at the surface the concentration of the pond water must be controlled by use of the fresh water.

Fouling of Pond Water

The water fouling caused by dirt entering and growth of undesirable algae will decrease the transparency of the pond water and the value of the solar radiation penetration. The dirt particles which float on the surface may be blown to the side by the breeze or removed when the water level is controlled by fresh water. The other part of the dirt will generally settle to the bottom—the speed of the settling is much slower than the usual lake or pond owing to the salt concentration. According to Tabor [6], at the bottom it has negligible effect on the pond performance.

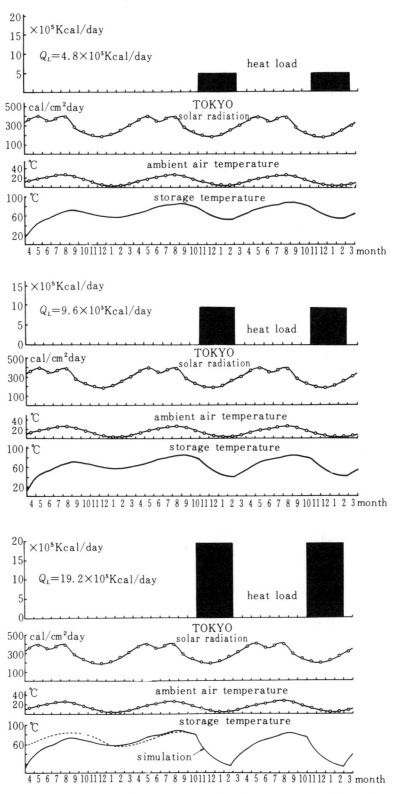

FIGURE 24. Examples of calculated values of storage temperature variation owing to assumed heat extraction.

The main factor of the fouling water is the growth of algae. Algae development is highly site-sensitive and the appropriate chemical treatment has to be applied in each area where algae may appear. By Davey [28] hypochlorite was found to be effective in order to disinfect the algae. In the writer's solar pond experiment, cupric sulfate ($CuSO_4 \cdot 5H_2O$) was used.

Efficiency of the Pond

The efficiency of a pond is limited by some physical properties. Firstly, there are reflection losses at the surface of the pond. Then, by the selective absorption as above mentioned, the insolation is attenuated about 50 percent in the first few cm of water, i.e. the heat absorbed near the surface is lost by heat radiation, evaporation and convection in the mixed layer near the surface. As a result, the temperature at the surface of the pond is nearly equal to the ambient air temperature. The other heat loss is caused by the heat transfer to the surrounding ground of the pond. Additionally, the thermal efficiency depends on the situation of the pond, i.e. the water cleanliness, the bottom reflection, and the steady or unsteady state.

By Weinberger [20], a steady state efficiency of 20 to 30% is predicted for a pond of infinite lateral dimensionals and a gradient of $1m$. Recently, the effects of bottom reflection on solar pond thermal efficiency assuming specular or diffusely reflecting bottoms have been calculated by Hawlader [29], Kooi [30] and Hull [31]. Figure 25 shows the results of the writer's experiment by use of Pond No. 2. Where each dotted mark shows the average value of every 10 days; T_{mean} = the temperature of the convective storage zone averaged every 10 days' data; H_s = the average incident solar radiation; and T_a = the average temperature of ambient air temperature. By Tabor [6], the practical values for ponds with about a $1m$ convective storage zone thickness are 15−25%; this agrees well with Figure 25.

Leakage of the Pond Water

The leakage of pond water brings both a loss of storaged heat and of salt. If the pond site is in an area where there is underground fresh water near the pond bottom, this water could become polluted. A natural areas such as salt lakes may be selected for artifical solar ponds, then the areas may also be impervious. However, for almost all the solar ponds constructed to date, a plastic or elastomor lining has been employed to ensure no leakage. The lining has to satisfy a number of conditions; the thermal resistance (until 100°C), the basic resistance, ease of sealing and a minimum mechanical strength (at the lowest estimate, to support a workman during the construction process). Alternatively, some thermal insulation may be added under the lining.

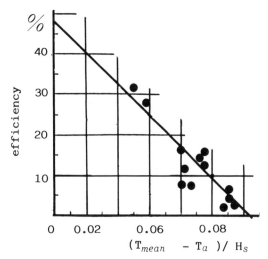

FIGURE 25. Thermal efficiency of pond no. 2.

Application for the Solar Ponds

The more obvious applications for solar ponds are for heating and cooling of buildings or green houses [7,10,32], power production [6,16] and desalination [33]. There are also application industries requiring a large amount of low-temperature heat.

NOTATION

The following symbols are used in this paper:

$A(z)$ = cross-sectional area at depth, z, of pond
$A_B(z)$ = lateral area of thin layer with thickness, z, at depth, z, in pond
$\bar{A}_B$ = sum of lateral area and bottom one of pond
A_s = surface area of solar pond
b = thickness of slit
C = salt concentration of water
D = diffusion coefficient of salt in water
D_s = inner diameter of inlet pipe
d = thickness of convective storage zone
d_1 = level of dividing streamline from border plane between non-convective and convective storage zones
d_2 = level of dividing streamline from pond bottom
f = coefficient of inlet loss
$H(z)$ = solar radiation reaching depth, z, of pond
H_s = solar radiation incident at pond surface
$\bar{H}_s$ = yearly average value of solar radiation
$(\tilde{H}_{s0})_k$ = kth mode amplitude of solar radiation
i = angle of incidence
K_D = thermal conductivity of heat insulator

K_s = thermal conductivity of soil
K_w = thermal conductivity of pond water
k_s = thermal diffusivity of soil
L = depth of pond
l = half width of pond
l_i = thickness of non-convective zone ($= L - d_2$)
l_c = thickness of convective storage zone ($= d_2$)
n = refraction index
ΔP_1 = pressure difference in inlet pipe
ΔP_2 = entrance head through slit of inlet pipe
Q = total discharge per unit width for heat extraction
Q_i = half discharge for heat extraction
Q_L = heat load
Q_s = total quantity of heat flow from pond to ground
q = discharge per unit width of convective storage zone
$q_s(z)$ = heat flux from pond to surrounding ground at depth, z, of pond
Δq = ratio of entrainment
$T(z)$ = storage temperature at depth, z, of pond
$T(z)_0$ = initial temperature of pond
T_a = ambient air temperature
$\bar{T}_a$ = yearly average value of ambient air temperature
$(\tilde{T}_{a0})_k$ = Kth mode amplitude of ambient air temperature
$T_c(d_a)$ = temperature of convective storage zone at d_a day
$\bar{T}_c$ = yearly average temperature of convective storage zone
$(\tilde{T}_{c0})_k$ = kth mode of amplitude of temperature on convective storage zone
$T_s(x)$ = temperature of surrounding ground at point, x, apart from pond wall
t = time
x = distance from side wall or bottom of pond
y = distance from border line between non-convective zone and convective storage zone
z = distance from pond surface
α_s = saline expansion coefficient of water
α_T = thermal expansion coefficient of water
γ = angle of refraction
η_n = fraction of solar radiation having absorption coefficient
μ_n = absorption coefficient for nth portion of solar spectrum
ϱ_w = heat capacity per unit volume of pond water
τ = coefficient of transmission.
$(\phi_a)_k$ = kth mode phase lag of ambient air temperature
$(\phi_s)_k$ = kth mode phase lag of solar radiation
$(\phi_t)_k$ = kth mode phase lag of temperature on convective storage zone
ω_k = kth mode frequency ($\omega \times_k = 2\pi k/365$)

REFERENCES

1. Kalecsinsky, A., "Ungarische Warme und Heisse Kochsalzeen," *Ann. D. Physik, Vol. 7,* No. 4, pp. 408–416 (1902).
2. Anderson, C. G., "Limnology of a Shallow Saline Meromitic Lake," *Liminology and Oceanog, Vol. 3,* pp. 259–269 (1958).
3. Cohen, Y., W. Krumbein, and M. Shilo, "Solar Lake (Sinai)," *Limnology and Oceanog, Vol. 22,* pp. 609–634 (1977).
4. Wilson, A. T. and H. W. Wellmann, "Lake Vanda, An Antarctic Lake," *Nature, Vol. 196,* pp. 1171–1173 (1962).
5. Nakao, K., Y. Nishizaki, and K. Nakayama, "Sedimentary Structure near the Saline Lake in Three Ice-Free Valleys, Victoria Land, Antarctica," *Report of the Japanese Summer Parties in Dry Valleys, Victoria Land,* XI, pp. 89–104 (1972).
6. Tabor, T. "Solar Pond," *Solar Energy, Vol. 27,* No. 3, pp. 181–194 (1981).
7. Rabel, A. and C. E. Nielsen, *Solar Ponds for Space Heating, Vol. 17,* No. 1, pp. 1–12 (1975).
8. Nielsen, C. E. and A. Rabl, "Salt Requirement and Stability of Solar Ponds," *Proceedings of the Joint Conference of the American Section,* International Solar Energy Society and the Solar Energy Society of Canada, pp. 183–187 (Aug., 1976).
9. Kinose, K., "Storage Temperature of Salt-Gradient Solar Pond," *Journal of Energy Engineering, ASCE, Vol. 110,* No. 2, Paper No. 18924, pp. 113–123 (1984).
10. Tabor, H. and H. Z. Weinberger, Chapter on: "Non-Convecting Solar Ponds," *Solar Energy Handbook,* Edited by Kreider, McGraw-Hill (1980).
11. Hipsher, M. S. and R. F. Bodeehm, "Heat Transfer Considerations of a Non-convecting Solar Pond Exchanger," *ASME, 76-WA/Sol* 4 (1976).
12. Harris, M. J. and Wittenberg, L. J., "Heat Extraction from a Large Salt-Gradient Solar Pond," 2nd Ann. Solar Heating & Cooling Conference, Colorado Springs (Nov., 1979).
13. Elata, C. and O. Levin, "Selective flow in a Pond with Density Gradient," *Hydraulic Laboratory Report,* Technion, Haifa, Israel (1962).
14. Daniels, D. G. and M. F. Merriom, "Fluid Dynamics of Selective Withdrawal in Solar Ponds," International Solar Energy Society Congress, Session 35, Los Angeles, Calif. (1975).
15. Elata, C. and O. Levin, 11th Congress of the Int. Assoc. for Hydraulic Research, Leningrad (1965).
16. Tabor, H., "Non-convecting Solar Ponds," *Phil. Trans. R. Soc. London,* A295, pp. 423–433 (1980).
17. Kinose, K., "Heat Extraction from Salt-Gradient Solar Pond," *Journal of Energy Engineering, ASCE, Vol. 109,* No. 3, Paper No. 18223, pp. 152–164 (1983).
18. Kinose, K. and K. Sakurai, "Study on Collection and Storage of Solar Heat by Solar Pond with Vertical Gradient of Salt Concentration (I)," *Bulletin of the National Research Institute of Agricultural Engineering* (in Japan), No. 19, pp. 65–84 (1980) (in Japanese).
19. Kinose, K. and K. Sakurai, "Study on Collection and Storage of Solar Heat by Solar Pond with Vertical Gradient of Salt Concentration (II)," *Bulletin of the National Research Institute of Agricultural Engineering* (in Japan), No. 21, pp. 203–225 (1981) (in Japanese).
20. Weinberger, H., "The Physics of the Solar Pond," *Solar Energy, Vol. 8,* No. 2, pp. 45–80 (1964).

21. Hull, J. R., "Computer Simulation of Solar Pond Thermal Behavior," *Solar Energy,* pp. 33–40 (1980).
22. Akbrazadeh, A. and G. Ahmadi, "Computer Simulation of the Performance of a Solar Pond in the Southern Part of Iran," *Solar Energy, Vol. 24,* pp. 143–151 (1980).
23. Hawlade, M. N. and Brinkworth, B. J., "An Analysis of the Non-Convecting Solar Pond," *Solar Energy, Vol. 27,* No. 3, pp. 195–204 (1981).
24. Chepurniy, N. and S. B. Savage, "Effect of Diffusion on Concentration Profiles in Solar Pond," *Solar Energy, Vol. 17,* pp. 203–205 (1975).
25. Hull, J. R., "Wind Induced Instability in Salt Gradient Solar Ponds," *Proceeding of American Section of International Solar Energy Society,* Phoenix, pp. 371–375 (1980).
26. Tabor, H. and R. Matz, "Solar Pond Project," *Solar Energy, Vol. 9,* p. 177 (1965).
27. Tabor, H., "Solar Ponds," *Science Journal,* p. 66 (June, 1966).
28. Davey, T. R. A., "The Aspendale Solar Pond," Report R15 CSIRO (Australia) (1968).
29. Hawlader, M. N. A., "The Influence of The Extinction Coefficient on the Effectiveness of Solar Ponds," *Solar Energy, Vol. 25,* pp. 461–464 (1980).
30. Kooi, C. F., "Salt Gradient Solar Pond with Reflective Bottom: Application to the 'Saturated Pond'," *Solar Energy, Vol. 26, pp. 113–120 (1980).
31. Hull, J., "Calculation of Solar Pond Thermal Efficiency with a Diffusely Reflecting Bottom," *Solar Energy, Vol. 29,* No. 5, pp. 385–389 (1982).
32. Bryant, H. C. and Colbeck, I., "A Solar Pond for London," *Solar Energy, Vol. 19,* No. 3, pp. 321–322 (1977).
33. Tabor, H., "Solar Ponds as Heat Source for Low-Temperature Multi-Effect Distillation Plant," *Desalination, Vol. 17,* pp. 289–302 (1975).

CHAPTER 26

Economics of Industrial Fluidized Bed Boilers

JOHN D. KEENAN* AND BRYAN E. KEENAN**

The concept of fluidized bed combustion (FBC) has been discussed with great interest over the past decade as a means of combusting coal and a variety of non-conventional fuels [1]. The promise shown by FBC technology includes advantages associated with greater thermal energy transfer and reduced emission of air pollutants. Compared to conventional fixed bed units, an FBC boiler provides a higher volumetric energy release. In addition, a fluidized bed system facilitates the direct removal of sulfur oxides during the combustion process, thus negating the need for expensive flue gas desulfurization systems. An FBC boiler is well-adapted for the direct injection of a chemical to react with and remove the oxides of sulfur as they are formed. The best such sorbents include limestone and dolomite, which are readily available and inexpensive.

The FBC process is, therefore, attractive to both the industrial decision-maker seeking to use cheaper fossil fuels and to the governmental policy-maker wishing to encourage the use of solid fuels over liquid and gaseous fuels, but not at the expense of degraded air quality. In spite of the recent plentiful supply of natural gas and petroleum, coal has historically been cheaper in terms of dollars per million Btu. During the environmental epoch since the mid-1960's, coal is less desirable on air quality grounds. Thus, the FBC process gives the industrial planner a mechanism for overcoming this environmental impediment. Coal utilization is also supported on the basis of policy considerations because of the great abundance of United States reserves compared to other fossil fuels.

The purpose of this chapter is to update previous work [2-3] and to review the factors affecting FBC process economics, to develop a cash flow model to predict the economic consequences of an investment in FBC technology, and to use the model to explore the impact of governmental policy decisions on FBC investments. The principal parameters influencing process economics are capital costs, reduced operating costs associated with the substitution of coal for either natural gas or fuel oil, and tax-related factors, such as depreciation schedules and tax credits. These are each quantified and incorporated into a cash flow model which is then applied to the question of what is needed to encourage an expansion in the use of FBC technology.

FLUIDIZED BED TECHNOLOGY

Fluidization is established when the upflowing air passing through the perforated distributor plate of a boiler achieves sufficient velocity to suspend the particles of coal. The three major regimes of fluidization are fixed bed, fluidized bed, and pneumatic transport. The particular regime may be defined in terms of the Froude and Reynolds Numbers. In general, it may be said that pneumatic transport occurs at air velocities high enough to transport particles out of the reactor, or, in this case, the boiler. This flow regime is applicable for transportation systems, but not for combustion applications because of the high particulate emissions. When the air velocity drops very low, the bed slumps, the particles are no longer suspended, and the regime is said to be fixed bed. At intermediate air velocities, the fluidized bed flow regime prevails. One of the major advantages of the fluidized bed is the intimate mixing between the two principal reactants, coal and oxygen. This mixing results in a much higher volumetric energy release for the fluidized bed (500,000 Btu/hr-sq ft (5.68 GJ/hr-sq m)) than for the conventional fixed bed coal-fired boiler (19,000 Btu/hr-sq ft (0.22 GJ/hr-sq m)). Because of the advantages of the fluid-

---

*Department of Civil Engineering, University of Pennsylvania, Philadelphia, PA
**Currently Mr. Keenan is associated with the law firm of Schlusser & Reivers, Wilmington, DE

ized regime for combustion purposes, it is the subject of the remainder of this discussion.

It is sometimes necessary to distinguish sub-categories within the fluidized regime. Incipient fluidization is associated with air velocities only slightly higher than the lowest necessary to suspend the coal particles. This sub-category is also known as minimum fluidization. As the air velocity increases, bubbles of gas form in the bed, rise, and burst through the bed surface. This is referred to as aggregative fluidization. With further increases in the upflow velocity, the occurrence of gas bubbles becomes more frequent, and the bubbles tend to coalesce to the point where their diameter approaches that of the vessel itself. This is termed slugging. The aggregative fluidization regime is preferred, because it provides the degree of particle mixing and uniformity of bed temperature required for optimum combustion and heat transfer.

Even when the pneumatic transport flow regime is avoided, there is still a substantial loss of particulates from the combustion chamber under typical operating conditions. As a consequence, the flue gases from fluidized bed boilers contain a high concentration of particulate matter which is removed by conventional air cleaning techniques, usually cyclones and baghouses. The collected fly ash is relatively high in unburned carbon, and, therefore, it is recycled either to a carbon-reinjection cell or to the combustion chamber in order to enhance overall efficiencies. A more detailed description of the process of fluidization is beyond the scope of this chapter, but is readily available in the literature [4–8].

FBC and Air Pollutant Emissions

In addition to the advantages from the perspective of thermal energy transfer, the fluidized bed technique is attractive in terms of air pollution control. This results from the unique ability to remove oxides of sulfur during the combustion process, without the need for expensive add-on flue gas desulfurization (FGD) systems.

In the FBC process, a sorbent, typically dolomite or limestone, is added to the combustion chamber along with the coal. These are thoroughly mixed during the combustion process, and several chemical reactions take place on the sorbent surface, leading to the removal of the sulfur oxides from the off-gases [9–12]. The removal chemistry is termed sulfation, and it is a function of temperature, sulfur dioxide concentration, particle size, and even the limestone source [13].

The sulfation process is a surface phenomenon, with the result that a dense layer of calcium sulfate forms on the surface of the limestone particle, leaving a non-reacted core behind. As a consequence, increasing the sorbent utilization is an area of interest, and the most promising approach seems to be recycling fines back to the combustion chamber [14].

An alternative for sulfur removal is FGD, particularly, wet scrubbing. FGD is attractive because the requirement for sorbent is about one-third that for the FBC process. On the other hand, limestone and dolomite are plentiful and inexpensive. Additionally, the massive capital investment required for FGD equipment is avoided. Therefore, for the scale of operations under consideration here, the FBC process allows the use of relatively inexpensive coal without the onerous burden associated with FGD systems.

The chemistry of sulfation in the FBC boiler is dependent upon a number of factors. Among the important determinants of sulfur removal are the type and quality of sorbent, fluidization regime, operating temperature and pressure, and mean particle diameter. The dosage of limestone required for varying degrees of sulfur has been observed empirically, and is expressed as a ratio of calcium to sulfur. This in turn may be expressed as a limestone dosage using the following expression

$$D = \frac{(m)(n)}{32} \qquad (1)$$

where D is the limestone dosage; m, the molar ratio of Ca/S; and n, the sulfur content of the coal.

In addition to reducing sulfur dioxide emissions, the FBC system also performs well in terms of oxides of nitrogen (NO_x) emissions. NO_x emissions are typically low because of the relatively cool FBC boiler operating temperature required for optimal sulfur removal (approximately 1500°F). At low combustion temperatures such as this, significant amounts of NO_x are neither released from the coal-bound nitrogen nor formed from the excess air.

However, particulates are a problem with FBC systems. This is not only a result of the fly ash normally associated with coal combustion but also a consequence of elutriation of particles from the fluidized bed itself. The control strategy used to capture these particles usually occurs in two stages. First, the escaping particles are collected from the main bed flue gas and then injected into carbon burnup cells. The function of the cells is to enhance the burnup of carbon, that is, to improve thermal efficiency. The carbon burnup cells are followed by a final particulate removal scheme, typically employing a mechanical collector/baghouse combination.

The main bed mechanical collectors are designed to capture as much unburned, elutriated carbon as possible. The usual operating efficiency of the main bed collectors is 90 to 92 percent. The gas phase leaving the main bed collectors moves to the baghouse, operating at about 99 percent efficiency. Emissions of particulates are typically in the range of 0.016 to 0.027 lb particulate/million Btu heat input (6.89 to 11.6 g/GJ). These are within the EPA guideline of 0.03 lb particulate/million Btu [14]. Particulate control is generally achieved with baghouses, rather than electrostatic precipitators, because of concerns about precipitator efficiency with low sulfur, high particulate flue gas streams [15].

DEVELOPMENT OF COST MODEL

A cash flow model was developed to provide an opportunity for investigating the role of various factors in the investment decision-making process. The basic premise of the modelling effort is that it deals with industrial scale boilers for which the decision to invest in an FBC system would allow the substitution of coal for fuel oil and/or natural gas without the expense of elaborate FGD technology. The model incorporates investment costs, depreciation, pre-tax savings, and income tax. The investment costs include the capital cost of the boiler, working capital, and salvage value. Pre-tax savings refer to the net reduction in operating costs due to the fuel oil and natural gas savings minus the coal and other operating expenses. All costs are expressed as March 1985 dollars.

The development of the cash flow model is described in detail in the paragraphs below. Calculations were made for each of six boiler capacities in the range of 50,000 through 500,000 lb of steam/hr. The computer program was designed to calculate net cash flow, internal rate of return, and the payback period. It was programmed in the Pascal language to be interactive, and was run on an Apple //e microcomputer. A copy of the source code is available upon request.

Installation Costs

A detailed estimate of the costs of installation for a 50,000 lb/hr boiler was made earlier [2] and updated using the appropriate construction cost index [16]. The installation costs for this size unit are summarized in Table 1. These costs were scaled up using the following economy of scale relationship:

$$I = (I_R)(C/C_R)^{0.875} \qquad (2)$$

where I is the capital investment for the boiler of capacity C, and I_R is the capital cost of the reference boiler of capacity C_R.

The cost of the building to house the boiler is depreciated differently than the cost of the equipment. Therefore, it was necessary to segregate the building cost from the total investment determined with Equation (1). Building area as a function of boiler size is estimated as

$$B = (30)(C) + 3000 \qquad (3)$$

where B is the building area for the boiler and

$$I_B = (B)(C_B) \qquad (4)$$

where I_B is the building investment cost, and C_B is the building unit cost.

TABLE 1. Installation Estimate—50,000 lb/hr FBC Boiler.

Item	Amount
Construction of new boiler house	$225,000
Stacks and breeching	287,000
Baghouse	219,000
Boiler	1,130,000
Boiler feedwater equipment	130,000
Boiler blowdown equipment	45,000
Ash handling system	360,000
Limestone handling system	253,000
Coal handling system	462,000
Utilities	237,000
Emissions monitoring	124,000
Subtotal	3,472,000
Engineering Fee (Five percent)	173,600
Subtotal	3,645,600
Contingency (Twenty percent)	729,120
Total	4,374,720

FBC Savings and Operating Costs

The purpose of this section is to enumerate the expenses and savings associated with the conversion to a coal-fired FBC unit. The annual energy use of the boiler is:

$$E = (C)(D_v)(k_1) \qquad (5)$$

where E is the annual energy consumption; D_v, diversity; and k_1, a conversion factor. The value of k_1 is 1.031×10^{10}, based on 8760 hr/yr and on steam at 1000 Btu/lb produced at a thermal efficiency of 85 percent. Diversity is the ratio of actual energy utilization to the maximum possible, and it is an indicator of the extent of equipment utilization.

The displacement of fuel oil and natural gas results in savings which may be represented as

$$A_G = (E)(f_G)(C_G) \qquad (6)$$

and

$$A_O = (E)(f_O)(C_O) \qquad (7)$$

where A_G and A_O are the annual savings in replaced gas and oil; f_G and f_O, the fraction of energy requirement satisfied by gas and oil; and, C_G and C_O, the unit cost of gas and oil.

The annual fuel cost incurred for purchasing coal, A_C, is

$$A_C = (E)(C_C) \qquad (8)$$

where C_C is the unit price of coal.

TABLE 2. Additional Labor Requirements.

Boiler Capacity, 1000 lb/hr	Additional Operators
50* 99	3
100*399	4
400*500	5

The cost for limestone is

$$A_L = \frac{(E)(D)(C_L)}{H_C} \quad (9)$$

where A_L is the annual limestone cost; C_L, the limestone unit cost; and H_C, the heat value of coal.

The coal-fired FBC unit replacing the gas- or oil-fired boiler will require somewhat more energy, because of the power requirements of the coal, limestone, and ash handling systems and because of the larger induced draft fans. As an approximation, annual electrical costs are expressed as a function of the total energy requirement, E:

$$A_E = (E_d)(E)(k_2)(C_E) \quad (10)$$

where A_E is the annual electrical cost; E_d, electrical demand; k_2, conversion factor (2.93×10^{-4}); and C_E, electricity unit cost.

Ash disposal costs for an FBC system are significant because of the addition of limestone and because of the ash native to the coal. These costs are expressed as:

$$A_A = \frac{(E)(A + D)(C_D)}{(H_C)} \quad (11)$$

for A_A, the annual ash disposal cost; A, ash content of coal; and C_D, unit ash disposal cost.

Additional manpower will be required for a coal-fired facility due to the increased workload. Table 2 presents a summary of the additional labor requirement as a function of boiler size. The annual cost of the extra labor is given as

$$A_{LC} = (N_1)(C_{LC}) \quad (12)$$

where N_1 is the number of additional operators; and C_{LC}, the unit labor cost.

The additional costs incurred for increased property tax, A_T; insurance, A_I; and maintenance, A_M, are related to the boiler investment as shown below:

$$A_T = (R_T)(I) \quad (13)$$

$$A_I = (R_I)(I) \quad (14)$$

$$A_M = (R_M)(I) \quad (15)$$

Cash Flow Model

Cash flow analysis is performed over some future span of time. It is, therefore, necessary to consider appropriate rates of inflation for the various parameters. The rates used are summarized in Table 3. These inflation rates are applied to the various unit costs (C_G, C_O, C_C, C_L, C_E, C_D, C_{LC}, and C_R) and rates (R_I, R_M, and R_T). For example,

$$(C_G)_i = (C_G)(1 + r_G)^i \quad (16)$$

where $(C_G)_i$ is the cost of gas in the ith year after construction begins; and r_G, the inflation rate for natural gas.

Several assumptions were made in the analysis. Capital expenses were assumed to be evenly spread over the two year construction period, at the end of which the boiler is placed in service. The building (Class 6 property) is depreciated differently than the boiler and its ancillary equipment (Class 4 property) ([19] Section 168). For the latter, the depreciable life for tax purposes is taken to be five years, with depreciation calculated, using the half-year convention, as 150 percent of the declining balance with conversion to straight line when advantageous. The Class 4 property qualifies for the 10 percent investment tax credit, and, as a result, only 95 percent of this part of the investment is eligible for depreciation ([19] Sections 38, 46, and 48). The assumption is made that the corporation's taxable income is sufficient to absorb all of the investment tax credit in the year the property is placed in service. For the building, the depreciable life is 18 years, and the depreciation method is 175 percent of declining balance, with conversion to straight line when advantageous. In this case, the mid-month convention rule is used. The corporate federal plus state tax rate is 50 percent ([19], Section 11).

Working capital, WC, is the amount of money needed to stock the new facility with a sixty day supply of coal and a thirty day supply of limestone:

$$WC = \frac{(60)(C_C) + (30)(D)(C_L)/(H_C)}{(E)/(365)} \quad (17)$$

This expense is incurred at the beginning of the first year of operation, and is assumed to be recovered, at inflated prices, at the end of the period of analysis.

As noted earlier, the output from the interactive computer program consists of a cash flow table, the payback period, and the internal rate of return. The cash flow table (for example, see Table 6) has seven columns of data. Column 1, investment, includes the capital outlay for each year, including working capital and salvage value. The second column lists the amount of depreciation taken in each year, calculated as described above. Column 3 is pre-tax savings. This amount is calculated as the fuel oil and natural gas savings, minus the coal and all other additional operating expenses. Column 4, income tax, is computed as the

TABLE 3. Inflation Rates Used in Cash Flow Analysis.

Parameter	Annual Percent Change	References
Fuel Oil	8.4	17
Natural Gas	6.2	17
Coal	2.6	18
Limestone	8.4	2
Electricity	11.1	17
Ash Disposal	2.2	2
Labor	4.7	16
Property Tax	10.2	2
Insurance	8.4	2
Maintenance	9.3	2

difference between pre-tax savings and depreciation multiplied by the corporate tax rate. The post-tax savings, column 5, represent the difference between pre-tax savings and income tax. The net cash flow, column 6, is the post-tax savings minus the investment. And column 7 presents the cumulative cash flow, determined as the sum of the net cash flows through that year.

The payback period is evaluated as the time required for the cumulative savings to equal the investment, using the first investment year as year 1. The payback period is considered a crude and unsophisticated investment decision-making tool. Consequently, the internal rate of return (IRR) is used. The IRR is the rate which makes the present value of the cash outflows equal to the present value of the cash inflows. This value, denoted here as R, is calculated by an iterative procedure such that

$$\sum_{i=1}^{K} ((1 + R)^{-1})(F_i) = 0 \quad (18)$$

where K is the period of analysis; and F_i, the net cash flow in year i. The internal rate of return is often deemed the most valid predictor of a project's feasibility, because it recognizes both the timing and the magnitude of cash flows.

APPLICATION OF CASH FLOW MODEL

The cash flow model described in the previous section was used to study the impact of various parameters on the decision to invest in a coal-fired FBC unit. These computer runs occurred in several phases. First, the so-called base case was evaluated, in which all the variables took on their nominal or reasonably expected values. Second, a sensitivity analysis was undertaken to isolate those variables with a profound effect on the results (as measured by payback period and IRR), and those with little or no effect. The third phase of the effort was to determine the impact a number of policy instruments would have on FBC investment decisions under the assumptions of the cash flow model.

Response to Base Case

The rate of return and payback period were determined for a series of six boiler sizes under the assumption of a set of base case conditions. The boiler sizes investigated were 50,000, 100,000, 200,000, 300,000, 400,000, and 500,000 lb/hour. The base case scenario is presented in Table 4. All costs reflect March 1985 prices. The inflation rates of Table 3 were applied to the base case analysis. The various cost elements calculated for the 300,000 lb/hr base case are summarized in Table 5. The boiler investment, I, was calculated using Equation (2); the building area, B, using Equation (3); and, the building investment, I_B, using Equation (4).

The various energy data are also shown in Table 5 for the 300,000 lb/hr FBC base case. The annual energy usage was determined with Equation (5) assuming 60 percent diversity in all cases. It was assumed that natural gas and residual fuel oil supplied 65 and 35 percent, respectively, of the original fuel requirement. That is, these were the fuels replaced by coal upon completion of the FBC unit. The natural gas and fuel oil consumption and the coal expenses and fuel savings were calculated using Equations (6) through (8). Additional determinations are also presented in Table 5.

The cash flow table for the 300,000 lb/hr base case boiler is shown in Table 6. For this investment, the IRR is 19.0 percent, and the payback period is 6.9 years. The other base case results are summarized in Table 7.

TABLE 4. Base Case Data for Cash Flow Analysis.

Analysis Input	Base Case Value
Building cost, C_B, $/sq ft	55
Energy Diversity, D_V, percent	60
Fraction gas, f_G	0.65
Fraction fuel oil, f_O	0.35
Cost of natural gas, C_G, $/10^6$ Btu	5.47
Cost of fuel oil, C_O, $/10^6$ Btu	4.67
Cost of coal, C_C, $/10^6$ Btu	2.12
Cost of electricity, C_E, $/Kwh	0.057
Limestone dosage, D, lb/lb coal	0.33
Cost of limestone, C_L, $/ton	40
Heat value of coal, H_C, 10^6 Btu/ton	25
Ash content of coal, A, fraction	0.10
Ash disposal costs, C_D, $/ton	30
Labor cost, C_{LC}, $/man-year	29,400
Property tax, R_T, fraction of investment	0.01
Insurance, R_I, fraction of investment	0.01
Maintenance, R_M, fraction of investment	0.02
Electrical demand, E_d, Fraction of energy consumption	0.01

TABLE 5. Base Case Results. Results for 300,000 lb/hr Fluidized Bed Boiler. Unless Otherwise Indicated, All Values Are in the Units of $1000. Operating Savings and Costs Refer to the First Year of Operation.

Parameter	Value
Boiler Capacity, C, 1000 lb/hr	300
Investment, I	20,981
Building Area, B, sq ft	12,000
Building Investment, I_B	660
Annual Energy Consumption, E, 10^9 Btu	1,855
Natural Gas Use, 10^9 Btu	1,206
Fuel Oil Use, 10^9 Btu	649
Natural Gas Savings, A_G	7,900
Fuel Oil Savings, A_O	3,862
Coal Expense, A_C	4,248
Electrical Expense, A_E	425
Limestone Expense, A_L	1,249
Labor Costs, A_{LC}	135
Ash Disposal Costs, A_A	1,023
Property Tax, A_T	281
Insurance, A_I	267
Maintenance, A_M	548
Net Savings	3,587

Sensitivity Analysis

An investigation of the sensitivity of the model to changes in key parameters was made. The variables studied included the investment cost, thermal efficiency, and various unit costs and associated inflation rates. The cash flow model was used to determine the individual effect of changes in these parameters on the payback period and the internal rate of return. In general, the data are shown here in terms of the IRR only.

The results of the sensitivity analysis are presented in several ways. First, the effect of a change over a range in the parameters was determined. This is shown in Tables 8 and 9 for investment cost and natural gas price, respectively. Table 8 clearly indicates the economies of scale. Because the pattern is the same in all cases, the results of the additional sensitivity analyses are shown for only the 300,000 lb/hr boiler size.

The impact of a change of plus or minus 25 percent in each variable was determined. These results are shown in Table 10. Also determined was the change in an individual parameter needed to change the solution by 10 percent, that is, to change the IRR from 19.0 percent to 17.1 and 20.9 percent. These results, which are shown in Tables 11 (impact of unit prices) and 12 (effect of inflation rates), clearly show

TABLE 6. Cash Flow Results. Data Shown are for 300,000 lb/hr Boiler. Base Case Assumptions Used Throughout. Except for the Years, All Values are in Units of $1000.

Year	Investment	Depreciation	Pre-Tax Savings	Income Tax	Post-Tax Savings	Net Cash Flow	Cumul. Cash Flow
1	10,490.7	0.0	0.0	0.0	0.0	*10,490.7	*10,490.7
2	10,490.7	0.0	0.0	*2,032.1	2,032.1	*8,458.5	*18,949.2
3	800.9	2,957.3	3,586.6	314.7	3,272.0	2,471.1	*16,478.1
4	0.0	4,981.0	4,007.4	*486.8	4,494.2	4,494.2	*11,983.9
5	0.0	3,656.2	4,458.6	401.2	4,057.4	4,057.4	*7,926.5
6	0.0	3,200.6	4,941.9	870.7	4,071.3	4,071.3	*3,855.2
7	0.0	3,196.0	5,459.5	1,131.7	4,327.7	4,327.7	472.5
8	0.0	1,615.2	6,013.2	2,199.0	3,814.2	3,814.2	4,286.7
9	0.0	34.9	6,605.3	3,285.3	3,320.1	3,320.1	7,606.8
10	0.0	31.5	7,238.0	3,603.2	3,634.7	3,634.7	11,241.5
11	0.0	29.2	7,913.6	3,942.2	3,971.4	3,971.4	15,213.0
12	0.0	29.1	8,634.7	4,302.8	4,331.9	4,331.9	19,544.8
13	0.0	29.1	9,403.7	4,687.3	4,716.4	4,716.4	24,261.3
14	0.0	29.1	10,233.4	5,097.1	5,126.3	5,126.3	29,387.5
15	0.0	29.1	11,096.6	5,533.7	5,562.9	5,562.9	34,950.4
16	0.0	29.1	12,026.1	5,998.5	6,027.6	6,027.6	40,978.0
17	0.0	29.1	13,014.9	6,492.9	6,552.0	6,552.0	47,500.0
18	0.0	29.1	14,066.2	7,018.6	7,047.7	7,047.7	54,547.7
19	0.0	29.1	15,183.2	7,577.0	7,606.2	7,606.2	62,153.8
20	0.0	29.1	16,369.1	8,170.0	8,199.1	8,199.1	70,352.9
21	*3,363.8	1.2	17,627.3	8,813.0	8,814.3	12,178.1	82,531.0

the critical role played by the investment cost, thermal efficiency, fossil fuel cost, and limestone cost. The most important determinants of investment characteristics are the investment cost itself, thermal efficiency, and the cost of displaced natural gas. By the same token, the price of coal and residual fuel oil have a strong effect on process economics. Except for limestone, most other variables exert a moderate effect on the IRR; variations in the labor cost have a negligible change in the payback and the return on investment. It may be that a crew as small as two would be sufficient to operate an FBC facility. The results of the sensitivity analysis indicate that a reduction such as this in the workforce would not noticeably alter the long-term investment characteristics.

The effect of thermal efficiency at first seems contrary to intuition. As shown in Tables 10 and 11, the IRR and the thermal efficiency of the boiler are inversely related. That is, as the energy conversion efficiency gets poorer, the investment becomes more favorable. In this study, the displaced boiler and the FBC boiler were assumed to have the same efficiency. The relationship between thermal efficiency and IRR is really a measure of the replacement of expensive gaseous and/or liquid fuel by cheaper solid fuel. Hence, the lower the thermal efficiency, the greater the net fuel savings.

EFFECT OF POLICY INCENTIVES ON CASH FLOW MODEL

The significance of the results shown in Table 7 is that, under the various assumptions implicit in the cash flow model, the choice of an FBC boiler as a replacement for an existing oil- or gas-fired unit is not an overly attractive investment. For the base case, such an investment will provide a rate of return of 14–20 percent, depending on the scale. This rate is generally considered to be insufficient to promote the widespread use of the FBC technology in industrial applications. On the other hand, a policy objective may be to encourage the use of the FBC technology, either to promote the use of sulfur-bearing coal, or to discourage the use of liquid and gaseous fuels, or both. In order to achieve this type of an objective, it may be necessary to implement some form of an incentive. The impact of such incentives is discussed in the following paragraphs.

It is interesting to note that the predicted IRR has improved over the past few years [2,3]. A comparison of this study with the previous one indicates that this change is primarily due to the relative price changes within the fossil fuels which have occurred during the first half of the 1980s. This, in turn, is largely an effect of the deregulation of natural gas prices, which itself is a result of governmental policy.

TABLE 7. Base Case Results Summary.

Boiler Capacity, lb/hr	Internal Rate of Return, percent	Payback Period, years
50,000	14.4	8.3
100,000	16.4	7.5
200,000	18.0	7.1
300,000	19.0	6.9
400,000	19.6	6.8
500,000	20.0	6.7

TABLE 8. Sensitivity Analysis. Effect of Investment on Internal Rate of Return.

	Boiler Capacity, 1000 lb/hr					
Investment	50	100	200	300	400	500
	Internal Rate of Return, %					
Base case	14.4	16.4	18.0	19.0	19.6	20.0
+ 5%	13.6	15.7	17.3	18.2	18.8	19.3
+10%	13.0	15.0	16.6	17.5	18.0	18.6
+15%	12.4	14.4	15.9	16.8	17.4	17.9
+20%	11.8	13.7	15.3	16.2	16.8	17.2
+25%	11.2	13.2	14.7	15.6	16.1	16.6
* 5%	15.1	17.2	18.8	19.8	20.3	20.9
*10%	15.8	18.0	19.6	20.6	21.2	21.8
*15%	16.7	18.8	20.5	21.5	22.1	22.7
*20%	17.6	19.8	21.5	22.5	23.2	23.7
*25%	18.5	20.8	22.5	23.6	24.3	24.8

TABLE 9. Sensitivity Analysis. Effect of Natural Gas Price on Internal Rate of Return.

	Boiler Capacity, 1000 lb/hr					
Gas Price	50	100	200	300	400	500
	Internal Rate of Return, %					
Base case	14.4	16.4	18.0	19.0	19.6	20.0
+ 5%	15.4	17.5	19.1	20.0	20.6	21.2
+10%	16.4	18.4	20.1	21.1	21.7	22.2
+15%	17.3	19.4	21.1	22.1	22.7	23.3
+20%	18.2	20.3	22.0	23.1	23.8	24.3
+25%	19.1	21.3	23.0	24.0	24.7	25.3
* 5%	13.3	15.4	16.9	17.8	18.4	18.9
*10%	12.2	14.2	15.7	16.7	17.3	17.7
*15%	11.0	13.1	14.6	15.5	16.0	16.5
*20%	9.8	11.9	13.4	14.3	14.8	15.3
*25%	8.5	10.6	12.1	13.0	13.5	13.9

TABLE 10. Sensitivity Analysis. The Effect of a 25 Percent Change in Each Variable on the Internal Rate of Return (IRR).

Case and Change		IRR,%	Case and Change		IRR,%
Base Case		19.0	Base Case		19.0
Gas price	*25%	13.0	Gas price	+25%	24.0
Thermal efficiency	+25%	15.5	Thermal efficiency	*25%	23.8
Investment	+25%	15.6	Investment	*25%	23.6
Oil price	*25%	15.7	Oil price	+25%	21.8
Gas inflation	*25%	16.0	Gas inflation	+25%	21.7
Coal price	+25%	16.4	Coal price	*25%	21.5
Oil inflation	*25%	16.8	Oil inflation	+25%	21.2
Limestone price	+25%	17.9	Limestone price	*25%	19.9
Limestone inflation	+25%	18.1	Limestone inflation	*25%	19.6
Ash disposal cost	+25%	18.4	Ash disposal cost	*25%	19.5
Coal inflation	+25%	18.4	Coal inflation	*25%	19.4
Maintenance cost	+25%	18.5	Maintenance cost	*25%	19.4
Electricity cost	+25%	18.5	Electricity cost	*25%	19.4
Electrical demand	+25%	18.5	Electrical demand	*25%	19.4
Electricity inflation	+25%	18.5	Electricity inflation	*25%	19.3
Maintenance inflation	+25%	18.5	Maintenance inflation	*25%	19.3
Property tax inflation	+25%	18.7	Property tax inflation	*25%	19.2
Property tax rate	+25%	18.7	Property tax rate	*25%	19.2
Insurance cost	+25%	18.7	Insurance cost	*25%	19.2
Insurance inflation	+25%	18.8	Insurance inflation	*25%	19.1
Ash disposal inflation	+25%	18.8	Ash disposal inflation	*25%	19.1
Labor cost	+25%	18.9	Labor cost	*25%	19.0
Labor inflation	+25%	18.9	Labor inflation	*25%	19.0

TABLE 11. Sensitivity Analysis. Change in Unit Costs Required to Affect 10 Percent Change in Solution Internal Rate of Return.

	Percent change needed	
Parameter	+10%	*10%
Natural gas price	+9	*8
Thermal efficiency	*11	+12
Investment cost	*12	+13
Fuel oil price	+16	*15
Coal price	*19	+18
Limestone price	*49	+44
Ash disposal cost	*84	+78
Maintenance cost	<*100	+96
Electrical demand	<*100	+110
Electricity cost	<*100	+110
Property tax rate	<*100	+177
Insurance cost	<*100	+209
Labor cost	<*100	+515

TABLE 12. Sensitivity Analysis. Change in Inflation Rates Required to Affect 10 Percent Change in Solution Internal Rate of Return.

	Percent change needed	
Parameter	+10%	*10%
Natural gas	+17	*17
Fuel Oil	+22	*21
Limestone	<*100	+48
Electricity	<*100	+68
Maintenance	<*100	+75
Coal	<*100	+78
Property tax	<*100	+99
Insurance	<*100	+137
Ash disposal	<*100	+304
Labor	<*100	+395

Impact of Taxation Policies

The effect of several types of policy instruments were investigated. The incentives are reduced corporate tax rates, investment tax credits, energy tax credits, subsidies, and accelerated depreciation schedules. They were selected because they represent the principal methods available to the government to influence the IRR of such a project. Other types of policy incentives, such as providing bond revenues, which would provide additional or lower cost capital resources, were not considered, because the existence of a supply of capital was assumed.

The cash flow model was run by varying the incentives over a range of values. Energy tax credits and investment tax credits of 10, 20, 30, and 40 percent of eligible costs were used. Subsidies were taken at 10, 20, 30, 40, and 50 percent of total investment, with the subsidized amount considered ineligible for tax credits and depreciation. The net effect of the subsidy was to reduce the investment by the amount of the subsidy. Accelerated depreciation was handled by changing the depreciable life of the Class 4 property to 2, 3, and 4 years. Corporate federal plus state tax rates were varied over the range of 30 to 50 percent.

The results are presented in Table 13. The impact of an accelerated depreciation schedule on the investment decision is marginal at best, with a change in depreciable life from five to two years increasing the return on investment from 19.0 to 20.2 percent. Rather substantial decreases in the corporate tax rate alter the investment characteristics only slightly. Dramatic changes in the rate of return are produced only by substantial changes in policy regarding subsidies and tax credits. It is clear from Table 13 that hefty subsidies or tax credits are required to yield a rate of return exceeding 25 percent.

Effect of Proposed Tax Changes

An issue of considerable current importance concerns possible revisions to the tax code. In a preliminary attempt to determine the effect of the changes proposed by the President [20], the cash flow model was run under the following conditions: 37 percent corporate federal plus state tax rate, 0 investment tax credit with 100 percent of the investment depreciable. The depreciable lives of the equipment and building are 7 and 28 years, respectively. The depreciation is calculated as a fixed percent (which is less than 100 percent) of the declining balance with conversion to straight line when advantageous. The rates are 22 and 4 percent of the declining balance for the Class 4 and 6 properties, respectively. That is, an accelerated cost recovery system is not part of the proposed plan. The mid-month convention is used throughout, and the half-year convention is used in the year that the switch to straight line is made. Given these assumptions, the return on investment increased to 20.9 percent, but the payback period increased to 7.2 years. The impact of the proposed changes is thus to delay the time when the cumulative cash flow turns positive, but to improve the investment picture because of the impact of the positive net cash flows after that point. In other words, there appears to be a positive impact apparently because, over the long term, the effect of the reduced tax rate outweighs the influence of the lost investment tax credit and the changes in the depreciation procedure.

SUMMARY AND CONCLUSIONS

The industrial application of fluidized bed technology for coal-fired boilers offers many advantages both in terms of resource utilization and air quality. Essentially, the use of coal burning FBC boilers will allow the substitution of relatively inexpensive and abundant coal for oil and natural gas, without the heavy capital expense of flue gas desulfurization. The FBC technology employs the use of a sorbent, such as limestone or dolomite, which is added with the coal to the

TABLE 13. Impact of Policy Instruments. Effect of Changes in Policy on the Internal Rate of Return (IRR) and the Payback Period in Years.

Policy Instrument		IRR	Payback
Investment Tax Credit	0%	17.5	7.4
	10%	19.0	6.9
	20%	20.7	6.4
	30%	22.7	5.9
	40%	25.1	5.4
	50%	28.0	5.0
Energy Tax Credit	0%	19.0	6.9
	10%	20.7	6.4
	20%	22.7	5.9
	30%	25.1	5.4
	40%	28.0	5.0
	50%	31.5	4.4
Subsidy	0%	19.0	6.9
	10%	20.6	6.8
	20%	22.5	6.0
	30%	24.8	5.9
	40%	27.6	5.5
	50%	31.2	5.1
Corporate tax rate	50%	19.0	6.9
	45%	19.7	6.8
	40%	20.4	6.8
	35%	21.1	6.8
	30%	21.7	6.7
Equipment Life	2yr	20.2	6.5
	3yr	19.7	6.5
	4yr	19.3	6.7
	5yr	19.0	6.9

boiler. The sorbent reacts with the sulfur from the coal, thus reducing emissions of sulfur oxides.

In order to assess the economic consequences of an investment in an industrial coal-fired atmospheric fluidized bed combustion boiler, a cash flow model was developed. The model was developed as an interactive computer program which was used to determine the payback period and the internal rate of return for a coal conversion project under a given set of conditions. Base case determinations were made for a series of FBC boilers in the size range of 50,000 through 500,000 lb steam/hr. For these boilers, the internal rate of return ranges from 14.4 to 20.0 percent, and the payback period from 6.7 to 8.3 years.

A sensitivity analysis was undertaken to investigate the effect of input parameters on the solution. The most important parameters include the investment itself and the relative value of the natural gas and fuel oil as compared to that of the coal which displaces them. Changes in most of the other parameters did not cause a significant change in the internal rate of return.

The potential for encouraging the use of this technology by means of policy instruments was analyzed. The impact of tax rates, tax credits, subsidies, and accelerated depreciation schedules on the rate of return was evaluated using the cash flow model. It was apparent that rather substantial tax credits and subsidies are needed to increase the return on investment to at least 25 percent.

NOMENCLATURE

A = Ash content of coal, percent
A_A = Annual cost of ash disposal, \$/yr
A_C = Annual cost of coal, \$/yr
A_E = Annual cost of electricity, \$/yr
A_G = Annual savings in replaced natural gas, \$/yr
A_I = Annual cost of insurance, \$/yr
A_L = Annual cost of limestone, \$/yr
A_{LC} = Annual cost of additional labor, \$/yr
A_M = Annual cost of maintenance, \$/yr
A_O = Annual savings in replaced fuel oil, \$/yr
A_T = Annual cost of property tax, \$/yr
B = Building area for boiler, sq ft
C = Boiler capacity, 1000 lb steam/hr
C_B = Unit cost of building, \$/sq ft
C_C = Unit cost of coal, \$/Btu
C_D = Unit cost of ash disposal, \$/ton ash
C_E = Unit cost of electricity, \$/Kwh
C_G = Unit cost of natural gas, \$/Btu
C_L = Unit cost of limestone, \$/ton
C_{LC} = Unit cost of labor, \$/person-yr
C_O = Unit cost of fuel oil, \$/Btu
C_R = Capacity of reference boiler, 1000 lb/hr
D = Limestone dosage, lb/lb coal
D_V = Diversity, percent

E = Energy consumption of boiler, Btu/yr
E_d = Electrical demand, Btu/yr
F = Net cash flow, \$
H_C = Heat value of coal, Btu/ton
I = Boiler capital investment, \$
I_B = Building investment, \$
I_R = Capital investment for reference boiler, \$
K = Period of analysis, yr
N_1 = Number of additional operators
R = Internal rate of return, percent
R_I = Insurance expense as fraction of I
R_M = Maintenance expense as fraction of I
R_T = Property tax as fraction of I
W_C = Working capital, \$
f_G = Fraction of E satisfied by natural gas
f_O = Fraction of E satisfied by fuel oil
k_1 = Conversion factor
k_2 = Conversion factor, 2.93×10^{-4}
m = Ca/S molar ratio
n = Sulfur content of coal, percent
r = Inflation rate, percent per year

REFERENCES

1. Shagian, J., in *Energy Progress*, 1, 51 (1984).
2. Keenan, J. D. and M. Maguire, in *Journal of Energy Engineering, ASCE*, 109, 113 (1983).
3. Maguire, M. and J. D. Keenan, in *Journal of Environmental Systems*, 13, 59 (1983).
4. Walsh, P. M., A. K. Gupta, and J. M. Beer, in *Proceedings of the DOE/WVU Conference on Fluidized Bed Combustion System Design and Operation*, Morgantown, West Virginia, pp. 455-509 (1980).
5. Wen, C. Y., J. Shang, and D. F. King, in *Proceedings of the DOE/WVU Conference on Fluidized Bed Combustion System Design and Operation*, Morgantown, West Virginia, pp. 164-217 (1980).
6. Rao, C. S. R., *Environmental Science and Technology*, 16, 215 (1977).
7. Fitzgerald, T. J., in *Proceedings of the DOE/WVU Conference on Fluidized Bed Combustion System Design and Operation*, Morgantown, West Virginia, pp. 8-57 (1980).
8. Daman, E. L., *Journal of the Technical Association of the Pulp and Paper Industry*, 62, 47 (1979).
9. Park, D., O. Levenspiel, and T. J. Fitzgerald, *Chemical Engineering Science*, 35, 295 (1980).
10. Keairns, D., in *Proceedings of the DOE/WVU Conference on Fluidized Bed Combustion System Design and Operation*, Morgantown, West Virginia, pp. 58-135 (1980).
11. Sarofim, A. F. and J. M. Beer, in *Proceedings of the DOE/WVU Conference on Fluidized Bed Combustion System Design and Operation*, Morgantown, West Virginia, pp. 136-163 (1980).

12. Heitner, K. L., *Journal of the Air Pollution Control Association*, 27, 1173 (1977).
13. Johnson, I., in *Proceedings of the DOE/WVU Conference on Fluidized Bed Combustion System Design and Operation*, Morgantown, West Virginia, pp. 332–362 (1980).
14. Curtis, R. E. and P. S. Dzierlanga, in *Journal of the Energy Division, ASCE*, 106, 131 (1980).
15. Pruce, L. M., in *Power*, 124, 5 (1980).
16. Anonymous, in *Engineering News Record*, p. 77 (April 4, 1985).
17. Lawrence, G. H. and M. L. Barcella, in *Energy Progress*, 4, 117 (1984).
18. U. S. Department of Energy, *Annual Energy Outlook 1983*, Energy Information Administration (1984).
19. Internal Revenue Code of 1954, as amended, Title 26, U. S. Code.
20. The Office of the President, *The President's Tax Proposals to the Congress for Fairness, Growth and Simplicity: General Explanation*, pp. 136–141, U. S. GPO (1985).

SECTION FIVE
Economics/Government/Data Acquisition

CHAPTER 27	Managing Public Involvement	571
CHAPTER 28	Municipal Service Distribution: Equity Concerns	589
CHAPTER 29	Engineering Economic Evaluation	601
CHAPTER 30	Microprocessor-Based Data Acquisition Systems	651
CHAPTER 31	BASIC Programming for Civil Engineers on Micro-Computers	669

CHAPTER 27

Managing Public Involvement

RICHARD F. ASTRACK,* NANCY A. BAUMANN,** AND GERALD L. REYNOLDS†

INTRODUCTION

Was the Colosseum of ancient Rome an early scene of public involvement as the citizens of Rome decided the fate of the gladiators?

With their thumbs turned downward, the spectators expressed their desire for the death by sword of the conquered gladiator [10]. Likewise, today's various natural resource projects may be terminated by a thumbs down action of the public.

Public involvement now plays a more critical role than ever as all levels of government attempt to provide public services with dwindling resources. Difficult decisions and choices must be made in allocating limited resources as federal, state, county, and city governments cut back and challenge deficit budgets. The public now has the opportunity and the obligation to help decide how to allocate available resources. Faced with limited resources, it is truer than ever that public support, or opposition and complacency, can make or break a proposed action. Even a highly efficient project can be lost without public support.

Dramatic changes have occurred in the role that the public may assume in decisions relating to water resource management. Only a decade ago, public involvement meant review at a public hearing or, at most, consultation with a few influential groups, often just prior to decision implementation [32]. Currently, most governmental units are aware of the need and often the legal requirement to broaden the opportunities for involvement by the general lay public. In many areas, citizens are demanding a more active role. For example, the U.S. Army Corps of Engineers has been one agency that has placed an emphasis on greater citizen participation as an integral part in discharging its responsibilities for water resource planning [3,20]. The Corps has been given specific and general directions in broadening this involvement, e.g., EM 1120-2-101 and ER 360 and 1165 series.

Public involvement emerged during the 1970s by means of the Principles and Standards [34] and Principles and Guidelines [35] of the U.S. Water Resources Council (WRC) to the extent that federal law now requires public involvement as a part of the planning for water resources development. The importance of general public participation in planning is found in the Principles and Guidelines, which state:

> Interested and affected agencies, groups, and individuals should be provided opportunities to participate throughout the planning process. The responsible federal planning agency should contact and solicit participation of: other federal agencies; appropriate regional, state, and local agencies; national, regional, and local groups; other appropriate groups such as affected Indian tribes; and individuals. A coordinated public participation program should be established with willing agencies and groups. (WRC, p. 3)

Additional justification for public participation in decision making can be found in publications by Pierce and Doerksen [23], Rosenbaum [26], Sewell and Coppock [31], Priscoli [25], and others. During the past decade, there has been considerable research, numerous publications, and much discussion on the role of the public in water resource planning.

In this evolution, new methods and approaches must be sought whereby inputs from the public can be creative and constructive rather than negative and obstructive.

EXPERIENCES IN PUBLIC INVOLVEMENT

Numerous experiments in the past can offer some insights into how the public, the planning officials, and the policy

---

*U.S. Army Corps of Engineers, St. Louis District, St. Louis, MO

**Planning and Managing Consultants, Ltd., 808 W. Main, P.O. Box 927, Carbondale, IL

†Department of Geography, University of Central Arkansas, Conway, AR

makers respond to particular techniques of public involvement. Three more enlightening studies are reviewed.

Delaware Estuary Comprehensive Study (DECS)

This study was undertaken in 1961 to 1966 at the request of various state and interstate pollution control agencies to determine the range of acceptable levels of water quality and review the control alternatives.

The main structure of participation was three committees of over 200 participants from 100 organizations. The Policy Advisory Committee sought the representation of public agencies in devising a water pollution control plan. The Technical Advisory Committee offered specific expertise on technical aspects of the pollution problem. The Water Use Advisory Committee was divided into four subcommittees, representing the main interest groups in the area: industry; local government and planning agencies; recreation, conservation, and wildlife; and the general public. The functions of this last committee were to indicate the needs and desires of people regarding water quality, to act as public relations group for the DECS and to assist in the nontechnical phase of the study [5].

Sewell [29] reviews important lessons from the DECS experience. The first is that it provided a continued mechanism for resolution of differing perceptions of the problem of environmental management.

Another lesson is that it offered an opportunity for interaction among different water users, all of whom had to learn to understand each other's goals and problems and who collectively recognized the need for a solution. Each group shifted its views from its initial position in this study.

The Susquehanna River Basin Study

This study was undertaken in 1968 to design an approach for improving communications between the public and government agencies involved in water resource planning in the basin, covering parts of New York, Pennsylvania, and Maryland.

The public involvement process model consisted of five phases, each with a focal concern: determination of goals; data collection; interim discussions of needs and possible plan alternatives; presentation and discussion of tentative plan alternatives; and formal plan presentation. Various techniques were used to obtain inputs from the public: questionnaires, content analysis of press items, workshops, public meetings, public hearings, and referenda. An important segment of the study included evaluation of the effectiveness of each technique [4].

A number of important findings should be noted. The researchers emphasized that the success of a communication-citizen involvement program hinges on the development of mutual confidence and respect [30]. Although not easily attainable, it requires sufficient time to establish this necessary support: each stage must not be rushed.

A second lesson is that different techniques for obtaining input are more effective at some stages than at others [8]. For example, interviewing key persons, followed by an open meeting, appears to be more effective in identifying goals at public hearings, which are viewed mainly as opportunities for adversary speeches. Workshops/seminars appear to be valuable approaches in discussing plan outlines and reviewing tentative alternatives.

The Susquehanna study also pointed to the difficulty of sustaining the interest and commitment of everyone throughout the entire planning period. Competing time and work requirements of the actors make continued involvement impossible. If individuals have some exposure to the process, it is often sufficient to breed necessary confidence and credibility [30]. One St. Louis area study concluded that public involvement programs for water resources should not be predicted on the assumption of a very active or interested public [14]. Other research has stressed that the traditional core of pressure groups stems from those already enjoying social and political privilege—the middle-income, well-educated, and politically articulate people [32].

The Brandywine

The Institute for Environmental Studies at the University of Pennsylvania developed the plan and program for the Brandywine River. The study received and recommended actions for water resources, land use, and future urban development. A minimum of damage and disruption to the water, scenic, and other natural resources of the area was the aim [18].

Throughout the three-year development of the plan, the Institute was keenly aware of the need to involve the relevant public; but, in the end, the plan was defeated. Apparently, not all the relevant public had in fact been involved, and the techniques employed were either inadequate or ineffective because of their timing.

An additional study had been conducted at the outset of the project so the researchers could assess the public's goals and perceptions. Other studies have indicated stable answers about preferences among goals, and alternative solutions will not be obtained if a well-informed public being interviewed does not know about the problem being discussed [8].

Another important lesson was drawn from the use of public hearings. The principal method of communication with the general public, after the initial survey, was a series of public hearings. These were not well attended initially, and by the time they were, the plan had been formed and those in attendance were very much opposed to what they perceived to be the plan.

Other water resource studies have provided valuable insight and guidance for determining how to develop a public involvement program and what methods to use. Success results only if a program involves the public *and* the study goals and objectives are met [27].

GOALS AND OBJECTIVES

Early in any planning process, establishing goals and objectives is important to guide the public involvement. Then, it is much easier to select techniques to work to accomplish one's goals and achieve superior results with less resources.

Study Purpose and Public Involvement

If a goal is to inform the public (information giving) one does not select techniques such as small group workshops, which emphasize receiving public input. If an objective is to both provide information to the public and receive input at one point during a study, techniques can be tailored to accomplish the objective if one has identified the objective. A public involvement activity such as a workshop could include a short general presentation (information giving), followed by small group sessions (receiving input).

In the development of a public participation program, there are two primary goals to be considered: involve the public in the decision-making process and meet the goals of the study. Applying Rosener's (1978) Participation Evaluation Matrix, public participation programs are incomplete if they succesfully involve the public but fail to accomplish the study goals; if they fail to involve the public but accomplish goals; or if they fail to do both. Public involvement should never be considered an end in itself.

The Planning Framework

It is critical that the public involvement program be designed around the framework of the particpular planning process for the sponsoring agency. For federal resources projects, particularly those assigned to the Corps of Engineers, this planning process now consists of six major steps, as identified in the Principles and Guidelines (1983). Currently, these steps include:

1. Specification of water and related land resources problems and opportunities
2. Inventory, forecast, and analysis of resource conditions relevant to the identified problems and opportunities
3. Formulation of alternative plans
4. Evaluation of the effects of plans
5. Comparison of alternative plans
6. Selection of a recommended plan

The earlier planning framework for the Corps of Engineers consisted of three stages—Stage 1: Reconnaissance Study; Stage 2: Development of Intermediate Plans; Stage 3: Development of Detailed Plans. Within each of these stages were the four functional planning tasks of problem identification, formulation of alternatives, impact assessment, and evaluation. The similarities between the two processes are clearly evident.

Both the current and previous planning processes are dynamic, meaning that steps or tasks may be iterated one or more times. As a public involvement program is developed, each of these steps must be considered in establishing the general objectives of a public involvement program.

Realizing the variability of study goals and planning frameworks, it becomes apparent that the objectives of a public involvement program will vary considerably. However, based on previous experiences in several Corps of Engineers' studies, several general public involvement objectives were consistently established. These included:

1. To identify the affected, concerned, and interested public within the study area
2. To provide the public throughout a study area with complete and factual information on water resources needs and planning efforts
3. To provide and maintain two-way communication between the Corps of Engineers and the public during the planning process
4. To provide the public with structured opportunity to a) identify and evaluate problems and needs in the study area, b) influence the formulation of alternative plans, and c) influence the formulation and selection of the final plan

PUBLIC INVOLVEMENT TECHNIQUES

General

It is critical when developing any public involvement program to carefully assess each technique being used, focusing on its purpose and expectations. A method which has proved successful in one public involvement program may not be successful in another. Also, a method successful in one stage may be less successful in another. For example, the public was highly cooperative in identifying the water resource problems in the study [1]. However, in evaluating the options to deal with those problems, there was considerably less interest. Flexibility is the key to developing any type of public involvement program.

Active public involvement is a necessary element in successfully completing each of the tasks. Different methods for obtaining input are more effective at some stages than at others [8], and the best method is dependent on the situation and the objectives sought [15]. The selection of methods to involve the public actively in this study was based on such factors as 1) the public involvement objectives; 2) the study history; 3) the study area characteristics; 4) preliminary and potential problems; 5) costs; 6) scheduling.

Public Involvement Techniques

As evidenced by Table 1, many public involvement techniques have been generated and applied during this era of public involvement. These techniques have focused on the identification of not only problems but also people, on communication between not only the public and lead agency but also other agencies and other publics, on interpretation and evaluation of problem solutions, and on resolutions to the resource problem.

There will probably never be a public involvement program that will include and implement all of the techniques in Table 1. The Issue Specific Reputational Survey [21,28] and Delphi [9] are effective in identifying and developing clientele prior to initial public meetings. Public information newsletters, community displays, and media exposure are effective in informing and educating the public. Workshops and public meetings are effective not only in obtaining public input but also in information sharing and interaction between the lead agency, other affected agencies, and the public. Committees, questionnaires, and surveys are effective in determining the concerns and values of the public.

As previously stressed, techniques should match the goals and objectives of the study. If the objective is to both provide information to and receive information from the public, a workshop format will probably prove more successful than a public hearing format. The less formal structure of the workshop encourages interaction more than the formality of the public hearing. Likewise, if the objective is only to inform the public, the public hearing will probably be more effective than the workshop.

Timing is a key consideration in the selection and implementation of a public involvement technique, especially as the techniques are incorporated into the planning framework. Similar techniques may be repeated again in the various stages or steps. Avoid assuming that successful techniques in one stage will again be successful in later stages. For example, limited distribution of public information newsletters and moderate media exposure can result in high attendance at initial public meetings focusing on problem identification. As the study progresses to the alternative evaluation stage, public interest may wane. Newsletters may be supplemented by other more creative techniques such as community displays and advertisements to achieve higher levels of participation.

The demographic and geographic features of the study area are key elements in technique selection. Information dissemination and information-gathering techniques will vary considerably for a rural and urban environment, a lower- or middle-to-upper socioeconomic area, and a well educated or low-educated community. Extensive notification efforts may be necessary for areas isolated by topographical features. Variation in these features applies especially to media exposure and range.

The study topic cannot be overemphasized in the selection of public involvement techniques and extent of the public involvement program. A study involving site measures (e.g., levees, floodwalls) for flooding problems will require considerably different technqiues than one of greater magnitude that may require a reservoir to solve the flooding problem. If there is a strong agreement within a study area on problems and alternatives, there will likely be no need for consensus-building techniques such as Delphi and role playing. However, if the study area is polarized these techniques would be stressed.

An additional consideration in the selection and implementation of public involvement techniques is the experience and skill of the agency staff on the various techniques. If agency skills for given techniques are not apparent, either avoid those techniques or seek the assis-

TABLE 1. List of Techniques

Booth/Display
Charrette
Citizens' Committees, etc.
Citizen and Technical Review Panel
Computer Based Techniques
Contest
Delphi
Fact Sheet (Public Information Newsletter)
Field Office
Fishbowl Planning Process
Hot Line (800)
Identify Influentials (Issue Specific Reputational Survey)
Interviews
Mailing List
Newspaper, Radio, and TV Advertisement
Ombudsman/Community Advocate
Participatory Television (QUBE)/Radio
Presentation
Press Release
Public Hearing
Public Meeting
Public Service Announcement
Reports, Brochures, etc.
Role Playing
Simulation Game
Site Visit—Self Guided Tour
Small Group
Speakers Bureau
Survey (Questionnaire)
Task Force
Technical Assistance
Training Program for Citizens
Workshops

tance of public involvement specialists who will advise, even train, agency personnel to interact with the public.

The following narrative provides valuable information on technique definitions, purposes, and advantages and disadvantages for the techniques listed in Table 1.

BOOTH/DISPLAY

Prepare a booth, display, or poster to attract public interest and attention at a county fair, travel show, shopping center, or similar location. Increased awareness, interest, and attention should result. The booth or display could be used at multiple locations if the study area is large. The display may be updated and reused as appropriate and as the study progresses.

—Advantages: Identify new publics not previously identified. Educate broader public.
—Disadvantages: Time invested for preparation and fair time. Create future expectations—if not fulfilled this would lead to resentment.

CHARRETTE

A Charrette is a highly intense meeting or series of meetings, to obtain public input and consensus, used particularly at a major decision point of a study. Useful in a crisis situation to obtain agreement. Typically held over an entire weekend, an entire week, or over a series of weekly meetings. All major publics must be present so decisions reached will be accepted on a consensus basis. The expectation is that an acceptable plan will be developed.

—Advantages: Achieve consensus among conflicting groups/interests. Result in commitment by all to support agreed-upon plan. Result in deeper understanding of positions held. Develop feeling of teamwork/cooperation.
—Disadvantages: *All* major publics must be willing to actively participate. *All* participants must be willing to attempt mutual problem solving and the agency must be willing to accept the results. It is time-consuming and difficult to get participation of decision makers for the length of time required.

CITIZEN ADVISORY COMMITTEE

The committee is established for the life of the study. It meets periodically and acts as a sounding board. The committee provides visibility for the study. It is usually a large group which, due to its size, is not effective in problem solving. To be effective, any citizen/advisory committee should follow these guidelines: a) clearly define limits of authority at the beginning of its existence; b) be representative of full range of values in the study area; c) life of committee should be limited; d) committee members should maintain regular communication with the constituencies they represent; e) membership selection must be fair and balanced.

—Advantages: Effective sounding board. Provide visibility and credibility. Assist in public involvement program design and execution. Creat emotional/vested interest in study results.
—Disadvantages: Not an effective working group—too large. Waste too much time on organizational details rather than focusing on study. Membership is time-consuming. Membership can be frustrating because of differences of expertise of members. The committee may assume an "adversary" relationship with the agency. May become "rubber stamp" without substantial technical assistance.

CITIZENS AND TECHNICAL REVIEW PANEL

The panel is established for the duration of the study to openly discuss the study, identifying problems, needs, concerns, and conflicts from the area public's perspective, and review the study with the study personnel. The panel membership, composed of area citizens, representatives of major interests affected by the study, and technical personnel with expertise in study components, also provides a channel of communication to and from other area citizens and groups.

—Advantages: Provide for continuous two-way communication between agency and study area citizens. Assist in gaining public interest in a study. Provide insight into planning process not revealed at larger public gatherings.
—Disadvantages: Purpose of panel sometimes mistaken by area citizens. May be seen as decision-making body. Difficulties will be encountered in scheduling a meeting.

COMPUTER BASED TECHNIQUES

Typical computer-based techniques include:

Tele-Conferencing. Public in various locations meet by computer rather than physically.

Polling. Public can respond with immediate results/feedback which can be useful to identify areas of consensus/disagreement.

Interactive Graphics. Visually display alternatives in response to changing priorities/preferences.

Simulation Games. See this technique presented separately.

—Advantages: More convenient—widely separated people can be brought together. More access to technical data as well as being able to immediately see implications.
—Disadvantages: Turns some people "off," as this conjures up imagery of machines over man. Fascination with computer may detract from the real purpose. Expensive and still at developmental stage.

CONTEST/EVENT

Sponsor a contest or an event which will attract attention and interest. Usually the theme is related to study topics.

For example, offer a canoe trip down river/stream of a water resources study area or have an essay contest on a related topic. If newsworthy, this can generate considerable interest and publicity. It is important for follow-up activities to capitalize on interest created, or resentment occurs.

— Advantages: Generate substantial interest. Identify persons interested in issues addressed.
— Disadvantages: Result may not be directly applicable. Expectation created which may turn to resentment if not fulfilled.

DELPHI

The Delphi process was designed to obtain a consensus on forecasts by a group of experts while minimizing group dynamics among members. The Delphi participants do not meet. A series of open-ended questionnaires are submitted individually to each participant. After each round, the answers are analyzed statistically and the statistical summary of the previous round is presented with each new round as participants respond again. Participants whose responses vary significantly from the average are also asked to state their reasoning. A new round presents the statistical analysis and the reasoning of those outside the average range. When participants no longer move toward agreement, the last results are statistically analyzed and summarized as the forecast.

— Advantages: Provide a method to obtain immediate feedback on issues. Effective method to achieve consensus on forecasts among experts. Minimize disadvantages of group dynamics—dominance by a single personality, for example.
— Disadvantages: Time consuming—with several mailed rounds. Public may prefer person-to-person interaction. Requires skillful panel coordinator. Requires commitment of panel members.

FACT SHEET (ALSO KNOWN AS NEWSLETTER, BULLETIN)

Provides information to the public. It is typically two to six pages in length. A fact sheet is useful in maintaining regular contact with public. Requires quality preparation—content, form and continuity. Distribution may be by mailing list, handout, newspaper insert, left at public places or other appropriate means.

— Advantages: Inform public of important data, events. Reach a large number of people at relatively low cost. Provide a visible representation of ongoing study.
— Disadvantages: May be used just because "it's time" rather than waiting for worthwhile information to share—which can turn public off. Mailing list, if used, may not be representative. Highly technical information may be overly simplified and misunderstood.

FIELD OFFICE

If a main agency office is away from the study area, or if there is a large seasonal influx of users, for instance, a ski resort or reservoir, a field office or local office on site in the study area may be considered. This allows more informal communication and interactions. A highly visible location—shopping center or storefront should be selected to maximize traffic and contact with the public.

— Advantages: Provides a visible means of informal interaction and encourages public participation. Communicates the value that agency places on community/study area. Staff obtain deeper/fuller understanding (sensitivity) of the study area.
— Disadvantages: Costly in staff and operation. Study must be of considerable local interest to justify cost. Several offices may be required for several communities if a large study area exists, resulting in some loss of central control. Local officials may feel an undercutting of their position.

FISHBOWL PLANNING PROCESS

Everyone can view the entire process and see how a decision is developed. Repetitive rounds of public meetings, brochures, workshops and citizen committee meetings are carefully documented in cumulative brochures which describe the entire process.

— Advantages: Process is visible—allows public to clearly see impact of their input. Encourages open communication. No special status for one person or group over others. If successful, it can be assumed that a consensus has been formed.
— Disadvantages: As the agency prepares the brochures, these can be abused if not prepared in an unbiased manner. Public response is "pro or con" when using a brochure—general comments not heard. Brochure must be in lay language to be understandable by the public. Later brochures are large, expensive.

HOTLINE (800 NUMBER)

Single number for the public to call with questions or comments about the study. This is especially useful if the agency is located away from the study area and a toll-free 800 number is used which allows access to the agency.

— Advantages: Any citizen can participate. Allows public to locate staff best able to answer questions. Can be used to provide information on future meetings, important future events.
— Disadvantages: The agency person must be careful in the response, or this may produce negative reaction of commercialization. Agency/staff must respond quickly.

IDENTIFY INFLUENTIALS

With limited resources, orient the public involvement program toward key/influential persons. In any study, one does not want to overlook key persons and they should be identified. One method is an Issue Specific Reputational

Survey (ISRS). A structured group of persons is surveyed as to whom they believe is influential in the study area. Key influentials will be most frequently mentioned.

- Advantages: Ensures that key persons are a part of the public involvement program. Identifies reputational influentials and not just positional influentials.
- Disadvantages: There may be tendency to ignore the general public when the key influentials are included.

INTERVIEW

Interview persons representing the full range of publics to assess the public attitude. Particularly useful at beginning or when long-time lapse has occurred in the study. A statistically chosen general public interview should provide typical results.

- Advantages: Provides a quick picture of political context of study. Can provide information on how the various interests want to particpate in the study. Useful in building personal relationships. Establishing communication means that persons are more likely to participate.
- Disadvantages: Key persons may not be representative of public sentiment. A negative impression can be created if the interview is pooly handled. Sample selection requires expertise.

MAILING LIST

Used to distribute information about the study. Used to distribute public input devices—surveys, questionnaires, also. Must continually work to make the mailing list representative and not restricted to institutions, agencies, public officials. Another technique should be used to help build the mailing list so that it is more representative. Newspaper insert, handouts, or a similar method to distribute fact sheets with mail back if persons want to be on the mailing list are ways to build a useable list.

- Advantages: Provides a simple method to distribute study information and some public input devices.
- Disadvantages: Mailing lists emphasize institutions and are not representative of the public. Require considerable time and effort to keep current.

NEWSPAPER, RADIO, TV ADVERTISEMENT (PAID)

Press releases are typically used announcement technique; however, they may not reach a significant percent of the public, as the agency has little say in when or how it is used. Paid announcement is more effective because the desired public can be targeted, and the agency has control over the presentation and its timing.

- Advantages: Paid avertisement reaches large population. It communicates a genuine desire to reach the public.
- Disadvantages: Use of funds for this purpose is not always acceptable to some publics. Can be expensive.

OMBUDSMAN/COMMUNITY ADVOCATE

One person acts as an ombudsman representing community and public interests by responding to public concerns and questions. Public will have one main point of contact. Creating such a position would require a major project/study with a considerable duration to be effective, unless an agency/proponent already occupies this position similar to newspaper ombudsman positions.

- Advantages: One contact point for public concerns, questions, and input at all times. Increase credibility and public confidence. Provide two-way communication.
- Disadvantages: Person must be impartial—agency employee must have free hand and not be subject to agency pressures, or public mistrust will result. Expensive because a new, additional position is required.

PARTICIPATORY TELEVISION/RADIO

A call-in show is an example of participatory television/radio. One option is to present information and then ask a mail response using perhaps a preprinted form in the newspaper. Cable TV with QUBE (Warner Communications) can allow interactive viewer response from home and may be used with a computer to provide immediate results and feedback. Educational TV may be available but is typically nonrepresentative.

- Advantages: Reaches largest audience. Convenient for public. Good educator of the public, public is comfortable with TV and radio.
- Disadvantages: Audience not representative. Appearance of vote may stifle the agency's selection of a different result. Requires time to prepare the program.

PRESENTATION

This is a formal speech/talk, typically using audiovisual aids, to present information to the public in a clear, understandable manner. For instance, it may well be a series of presentations oriented toward different groups or interests, general public, technical groups, business and recreation interests.

- Advantages: Present important information to public in a relatively inexpensive manner. Prepared presentation ensures uniformity.
- Disadvantages: Careful preparation is required, as image and credibility are at stake. Can't just "throw something together."

PRESS RELEASE/PUBLIC SERVICE ANNOUNCEMENT

To be used by media, the release/announcement must present a newsworthy, understandable story. A good point to present is how the given issue touches people's lives.

- Advantages: Will reach a large audience *if* printed or read on radio. No cost to agency.

—Disadvantages: May not be printed or if so, buried inside the newspaper. Subject to editorial distortion of the issue.

PUBLIC HEARING

A formal public meeting usually required by regulation/policy used to present proposed plans or discuss other related issues. Generally, it is not effective except for meeting a legal regulatory requirement.

—Advantages: Accepted by public due to its long history as a traditional technique. Provides an opportunity for public to challenge agency.
—Disadvantages: At end of study, too late for much change. Tend to intimidate most public. Encourage polarization on issues. Public feels it is a formality. Does not allow discussion, negotiation, or dialogue among opposing viewpoints. May limit other public involvement techniques because of sole requirement.

PUBLIC MEETING

Public meetings are held while work is in process so public input can affect the results. Small groups subdivision within the meeting allow more opportunity to obtain everyone's input.

—Advantages: Obtain input while work is underway. Less intimidating because it is informal. Can inform public and provide information. Can reach a large number of persons.
—Disadvantages: Large meeting may not allow much interaction. Encourages polarization. Results in negative reaction to agency if not well led/prepared.

REPORT, BROCHURE

Issue brochures and reports to present progress, the current status, and general information of the study. Brochures are longer in length typically than fact sheets and are prepared to present the results of significant events or progress of the study.

Some guidelines for preparation include: Keep the material simple, avoid jargon, do not get too technical, identifying the reader's stake in the study and use effective graphics.

—Advantages: Direct means for large amount of information to the public. Provides a permanent record of the study. Can reach a large number of the public.
—Disadvantages: Requires unique skills to prepare a successful publication, and the agency image is at stake. Chances of reaching all of intended audience are slim, since not all will read. Provides only one way of communication.

ROLE PLAYING

Individuals play the role of decision makers or citizens to increase the sensitivity of the public and planners to the varied aspects of resource decision-making.

—Advantages: Involves the public in the experience of decision-making and its varied complexities. Provides additional insight and understanding for the decision makers and the public.
—Disadvantages: Requires considerable skill to implement.

SIMULATION GAME

A simulation game, which could be on a computer, provides feedback on the most likely results as policies or decisions are considered. It provides a "risk free" opportunity to evaluate choices. Simulation games can be useful to educate the public and gain their enthusiasm in the study.

—Advantages: Provide information on consequences of various decisions/policies. Provide understanding of study dynamics—trade-offs of one aspect versus another. Enjoyable experience that helps develop relationships which may last through the study.
—Disadvantages: Must pick most suitable game for study, as many are complex and confusing. Educational but do not provide an effective opportunity for direct comment on the study. Participants maybe so engrossed in the game they forget the issues.

SITE VISITS (SELF-GUIDED/GUIDED TOURS)

Trips to the sites of problem or potential impacted areas provide the public and planners with increased understanding of study and its significance. May consist of guided tours, self guided tours (with brochure-map), or both.

—Advantages: Provide additional information regarding the study not possible through other techniques.
—Disadvantages: Time consuming and somewhat expensive if study area is large.

SMALL GROUP

Limited public group size allows for maximum public input, as structured small group techniques can be applied to achieve input. Brainstorming and the nominal group process are two small-group processes, for example. Best group size is 7 to 10 persons, with each group requiring a facilitator and a recorder. Effectiveness decreases as size increases, but satisfactory results can be obtained from groups of up to 20 persons. A large public meeting could be divided into small groups to achieve the desired group size.

—Advantages: High opportunity for public input. Structural process increases input.
—Disadvantages: May require special facilities, several small rooms, and several agency persons to lead the groups. One small group may not be representative, so may have to have several different groups from one large meeting or different small groups at varying time and place.

SPEAKERS BUREAU

A group of persons available to make presentations upon request to interested groups. To ensure uniformity, prepared speeches, typically with audiovisual aids, should be used by each speaker.

- Advantages: Provide consistent information presentation to public groups. Does not overburden one person and ensures more opportunity to respond positively to requests by having a group of persons available.
- Disadvantages: A detailed presentation must be developed. Training may be required of persons who do not typically make presentations.

SURVEY (QUESTIONNAIRE)

The survey must be a representative sample of the study area to be a valid assessment of public values and positions. Surveys may be conducted by phone, mail, interview, small group, or even newspaper, magazine and television with a return ballot. Preparers must be experienced in survey design, which usually requires professional assistance. The following three areas should be addressed before initiating a survey: determine what information is needed; determine how the information is to be used; and determine if the information you desire has already been collected.

- Advantages: Result is received from a more objective general public than those already participating. Assist in evaluating whether active participants are representative.
- Disadvantages: Expensive and requires experts. Usually surveys do not provide for interaction. If the subject of the survey is not of broad public interest, substantial number of respondents may be uninformed. Cannot substitute for political negotiation. They must be carefully designed and conducted.

TASK FORCE

A task force comprised of citizens is organized to study concerns on a specific problem. Size is limited to a maximum of 15 members to be effective. At the beginning, clearly define limits of authority of the task force and make it representative of the full range of values of the study area. Task force members must maintain contact with constituents to ensure they are representative and do echo public sentiment.

- Advantages: Effectively works on one specific problem/objective because of its small size. More successful and thus creates sense of satisfaction. Result is more credible than if just achieved by the agency alone. Provide in-depth information on particular issues.
- Disadvantages: Difficult to be representative because of its small size. Limited mandate. Does require a considerable investment of time and staff. Task force is usually only advisory.

TECHNICAL ASSISTANCE

Provide direct technical assistance to public so the public is able to fully develop alternatives to same level of detail as the agency. This assistance could be provided by: all agency staff as part of their job; specific staff persons assigned to different groups; and/or specially hired independent consultants.

- Advantages: Public will feel less intimidated by staff professionals if they have technical assistance. Public alternatives created are of comparable detail to agency alternatives. Independent consultants are more acceptable to the public.
- Disadvantages: Staff could be placed in position of divided loyalties. More assitance is usually provided to most active groups, thus biasing the results. Staff assistance can be misused to manipulate the public.

TRAINING PROGRAM FOR PUBLIC

This is to improve the public's understanding of how the study is conducted so that they comprehend the overall process and, therefore, are better able to participate effectively. Training may be on the overall process or limited to specific areas such as environmental impact assessment as appropriate.

- Advantages: Increase effectiveness/impact of public upon study. Public feels less intimidated. Valuable contribution to public involvement program.
- Disadvantages: Public may resent need for training. Raise the issue of who gets training/who does not. Special skills are required to ensure effectiveness of the training.

WORKSHOP

Typically a smaller scale, informal working session which allows participants to identify problems and assess solutions. While informal, the workshop process is structured to maximize public input.

- Advantages: Achieve significant public input through two-way communication. Can be very educational to public.
- Disadvantages: Does require care in preparation. Does require skilled personnel.

PUBLIC INVOLVEMENT PLANS

During the time of ancient Rome, gladiator events were announced several days before the event by placing bills (notices) on the walls of houses and public buildings [10]. The bills included the names of the competitors, the date of the event, the sponsor, and the kinds of combat. Over 2,200 years later, many resource agencies use these same approaches to public participation, perceiving that the key to public involvement is simply notifying the public of various opportunities to provide input. It is inaccurately assumed that once the public knows of a given event, if the issue is of importance to them, they will become involved in the decision process.

This misconception continues to plague public agencies that attempt to include the public in the planning process. A public involvement program includes more than simply soliciting public attendance. A comprehensive program also includes the components of collecting and analyzing information, soliciting public response, and disseminating information.

Public involvement is not spontaneous. It does not simply evolve as the study progresses. Public interest may be very high during the initial stages of a study, with only announcements of involvement opportunities. However, to maintain this continued level of involvement throughout the study, a comprehensive public involvement program should be designed and implemented.

The magnitude of a study on project is a decisive factor in developing a public involvement program and selecting activities. An extensive, multi-issue, controversial study requires a correspondingly detailed, somewhat complex public involvement program, whereby a small scale, single issue study in which there is perceived public consensus may require a very simple public involvement program.

The funds available for public participation also dictate the extensiveness of the public involvement program. The U.S. Environmental Protection Agency [33] suggestst that 10% of a project budget might be allotted to public involvement for effective water quality planning programs. Controversial, multi-issue projects could require as much as 20% to 30% of a total budget, while a single-issue, no-controversy project could require much less than 10%. It becomes apparent that no standard formula can be provided for public involvement program expenditures.

Additional considerations in the development of a public involvement program for a given study include time constraints, the authority of the lead organization and its staff capabilities, and the level of citizen interest and involvement (e.g., coalition formation).

Corps Public Involvement Program

To provide additional insight into technique selection and application, selected techniques of several Corps of Engineers, St. Louis District, public involvement programs are described. These public involvement programs were designed and implemented through the St. Louis District from 1978 to 1982.

Two of the public involvement programs, the Cahokia Canal and harding Ditch, were components of the East St. Louis and Vicinity, Illinois Interior Flood Control Project. The goal of this project was to alleviate water-related problems, primarily substantial and repetitive flood damages, in the project area. The third public involvement program was a component of the Pine Ford Study, a reformulated multi-purpose lake project, originally proposed to serve the purposes of flood control, water supply, fish and wildlife conservation, area redevelopment, water quality control, recreation, and flow augmentation for navigation.

Each of these public involvement programs was designed around the previously mentioned three-stage planning framework of the Corps of Engineers, which consisted of water resource problem identification and evaluation, development of intermediate plans, and development of detailed plans. Activities within these stages can be differentiated into three major categories. These include collecting and analyzing information, soliciting public response, and disseminating information.

The Public Involvement Model

The model presented in Figure 1 represents the public involvement plan for the Cahokia Canal study and is intended primarily to illustrate the general approach to the public involvement component of the study rather than provide a detailed description. It does not include every specific public involvement activity. For example, only three general public newsletters are shown. However, for each major contact with the public, such as a public meeting, a fact sheet is issued to announce the event and provide background information. Immediately following the meeting, another fact sheet is prepared and distributed to present the results. Also, additional fact sheets were prepared and distributed as important work efforts were completed on the study.

The public involvement plan illustrated employs several public involvement techniques—Delphi panel, Issue Specific Reputational Survey (ISRS), public information fact sheets, press releases, public meetings, questionnaires, and public workshops. The implementation of these methods will be examined by stages of the planning process.

Identifying Problems and Needs

Key public involvement activities in the identification of water-related problems and needs focus on data collection, identifying the target population, disseminating study information, and obtaining public response. Through these efforts a thorough understanding of the resource problems, concerns, and needs can be established.

Phase One: Information Gathering

During the initial phase of the public involvement programs, data collection was a major activity. The information collected and analyzed focused on the study, the study area, the people within the study area, and the people's response to the study. Socioeconomic, demographic, and geographic data were obtained and reviewed in order to establish a solid base in understanding the study area and the people residing in the study area. Additional data gathered dealt with the identification of existing and potential conflicts related to the plan of study. Sources for the varied information included technical reports, review of newspaper article files and discussions with informed individuals, and visits to city halls and county courts. Planning commissions within and adjacent to the study are also provided worthwhile information on the study area as well as influentials.

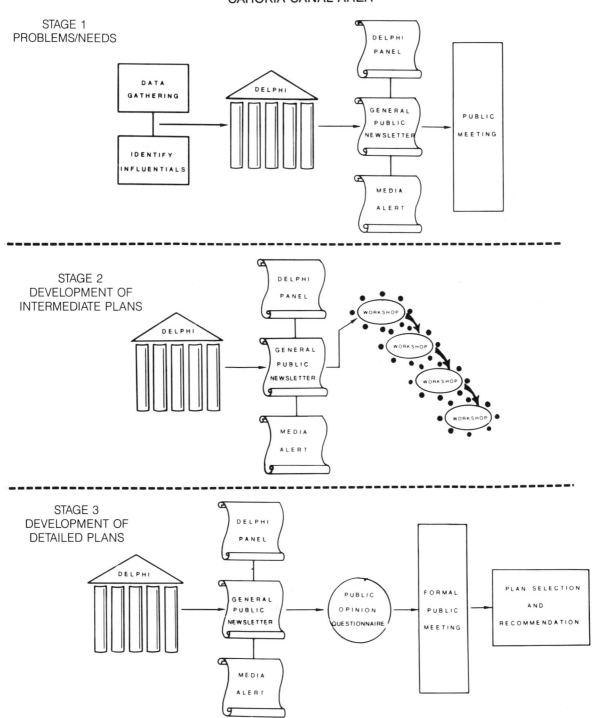

FIGURE 1.

Targeting Population. Probably the most important activity in the initial phase of the public involvement program is Targeting the Population [38]. It is through this activity that the people who will be either affected by, concerned with, or interested in the study and its outcome are identified. Influentials within the study area are also identified. The targeted population may be organized, unorganized, belong to a special interest group, or be public servants.

The Issue Specific Reputational Survey (ISRS) [6,7,38] was implemented during the initial phase of the Cahokia Canal, Harding Ditch, and Pine Ford studies to identify those people within the study area who would be influential in making decisions pertaining to the study. The ISRS was essential in forming Delphi panels in the Cahokia Canal and Harding Ditch studies and a Citizens and Technical Review Panel in the Pine Ford Study.

The potential [2] and reputational [28] approaches were highly successful in identifying influentials. The positional approaches identifies those persons who occupy important official positions (e.g., major county board chairman, major industry president, bank president, and area development director) in study area communities. The reputational approach identifies people in the study area reputed to be the most informed regarding a paticular issue, such as city engineers, public works and building superintendents, and private citizens. Panelists were selected based on their identified reputation as indicated during the ISRS [13]. Those area citizens identified during the ISRS can be especially influential in making decisions regarding a study, therefore having a significant effect on the study results.

Existing mailing lists for the various studies were updated and expanded to include a greater representation of the general public. Mailing lists of federal agencies frequently include a greater percentage of politicians/public administrators, agency representatives, and special interest groups than the nonaffiliated general public. Indeed, all study area residents cannot be entered on the list, but broad representation cannot be overstressed. Commercial interests within the study area should also be included in the mailing lists.

An additional method for identifying affected, concerned and interested parties within a study area was the third-party approach. Individuals aware of the studies identified other people through letters and newsletter reponse forms. Individuals also directly identified themselves, requesting that they be added to the mailing lists. Media coverage was also effective in this self-identification process.

Information Dissemination. Providing the public with information concerning the study and public involvement opportunities during the initial segment is essential for an effective public involvement program. There must be clear identification of how the study can affect the area people and an explicit statement on how participation can influence the outcome of the study [36]. During the Cahokia Canal, Harding Ditch, and Pine Ford studies, the techniques employed for the purpose of providing information during the problem identification stage included public information fact sheets, press releases, and media contacts.

The public information newsletter (fact sheet) was the primary method for communicating study information and involvement opportunities to the public. Research findings indicate that the most successful tools for generating attendance at public meetings at direct mailings such as newsletters and newspaper coverage [11].

Fact sheets were distributed to those people on the revised and expanded mailing list and were also made available at various locations throughout the study areas, including such locations as city halls, county courthouses, post offices, banks (where approval granted), and libraries.

Fact sheets were distributed approximately three weeks prior to the initial public meetings. Material included in the fact sheets focused on thevarious components of the study (e.g., history, purpose, scheduling, and the Corps' planning process) and detailed the upcoming public meeting and its significance. Whenever possible, technical jargon was avoided.

During the initial phase of one study, the fact sheet served as a feedback mechanism for the identification of additional people to be included on the mailing lists. It also served as a platform for general comments on water resource problems in the study area.

Other agency newsletters, including planning commissions, soil conservation service, and economic development commissions, were identified and contacted. When possible, these newsletters provided space to present study information and public involvement opportunities.

Prior to the initial Pine Ford public meetings, a multicolor poster was developed that briefly described the study, its purpose, and announced the public meetings and their purpose. These were distributed and posted at various public facilities throughout the study area. Additional media contacts also proved effective during the problem identification phase of these studies. Not only were the media used to announce public involvement opportunities; they also provided additional study information, somewhat of a media blitz. Study-related news items, especially relating to water resource problems in the study area newspapers, increased the public awareness on study topics. To assure media coverage of study activities, contact was made and maintained with assigned reporters. All media contacts were coordinated through the Corps' Public Affairs office.

Although not implemented in these three studies, paid advertisements in newspapers and flyers in Sunday supplements were considered to improve notification. This approach was especially appealing for the rural Pine Ford Study area, but was not implemented because of budget constraints.

An additional method for disseminating information during the Pine Ford Study was a booklet that provided additional background information on the study, the somewhat delayed, interrupted study process, and key supporting studies. This booklet was considered necessary to bring the public up to date and provide answers to frequently asked questions regarding this much delayed study.

Public Response. A key objective of any public involvement program is providing the opportunity for citizens to contribute their varied opinions and ideas during the planning process. During the initial phase of the Corps' studies, several methods of public response were provided. This included Delphi Panels, Citizens and Technical Review Panels, Public Response Forms in Fact Sheets, and Public Meetings. Additionally, public letters and phone calls are always encouraged as means of public response.

In the Cahokia Canal and Harding Ditch Study, initial public input was through a Delphi panel consisting of 35 members. These included public officials, engineering and management experts, and knowledgeable interested citizens in the study area [1]. Numerous studies have discussed the use of the Delphi for eliciting and refining local judgements in the development of water resource plans [2,16,17,19,24]. Following panel selection, there were iterative administrations of questionnaires for identifying, assessing, and evaluating water resource problems in the study area.

The major objective of the Delphi was to obtain informed local opinion about area water resource problems and, during later study phases, optional courses of action. The information provided by the panel was especially beneficial in guiding the planning staff in preparing for public meetings.

The Citizens and Technical Review Panel, implemented during the Pine Ford Study, was composed of 22 study area citizens representing the private and public sectors. Although similar in formation to the Delphi, the purpose of this panel was more varied, including not only expressing views concerning problems and issues and increasing the Corps' sensitivity to public preferences but also strengthening the public information dissemination network.

One of the more effective methods for soliciting public response was the public meeting. The public meetings should not be confused with the public hearings that are highly formalized. The public meetings held during the problem identification stage of the Cahokia Canal, Harding Ditch, and Pine Ford studies, were less formal and structured to facilitate public response.

Following brief opening sessions in which pertinent study information was provided, participants were divided into small groups (20 or less was preferred). Each group was assigned a group leader or "facilitator" whose role was to assist each group and assure that each group member had the opportunity to express concerns. Prior to public meetings, facilitators (Corps personnel) were trained in their roles and responsibilities.

The format of the small groups allowed each member to express a concern which was then listed and discussed. This process continued in a round robin manner until all ideas and concerns were listed and discussed and were then reviewed and ranked. The information obtained at these initial meetings was especially critical in defining planning alternatives and establishing priorities during Stage 2.

Well over 100 participants attended each of the Cahokia Canal and Harding Ditch Meetings. The more controversial Pine Ford Study had alsmot 250 people attending two meetings held. The participants at these meetings included primarily residents and business owners, with lesser numbers of public officials/representatives, agency and interest group representatives, and media personnel. At the Cahokia Canal and Harding Ditch public meetings, nine specific problem issues were identified ranging from priority topics of flooding and water quality to lessor important water supply and recreation [1].

The problems, needs, and concerns identified at the Pine Ford meetings were categorized into 14 topics. Water quality and erosion control were viewed as high priority items; flood control, water supply, and economic development were assessed as moderate priority items; while navigation, cultural resources, and hydropower were lowest priority items.

Although there were minor discrepancies between public input at meetings and panel input, the consistently identified priority problems provided the Corps with ample public feedback to develop alternative solutions to the water resource problems.

Phase Two: Alternative Planning

This second phase of the planning process, following the identification of concerned, interested, and affected publics and their perceived water-related problems, emphasized information dissemination and public response regarding alternative solutions to problems. Participant disagreement was much more evident during the Pine Ford public meetings than during the Cahokia Canal and Harding Ditch meetings. Because of this strong disagreement, essentially polarization, it was of utmost importance that clear, factual information be disseminated throughout the study area.

Although information gathering and targeting population continues to be practiced throughout a study, their importance diminishes as studies advance from initial to intermediate and final stages.

Information Dissemination. As with identifying problems, the Public Information Fact sheet again was the principal method for disseminating study information. Fact sheets for all studies recapped the activities of Stage 1 and presented detailed information on specific options to reduce or solve the identified problems. Question and answer segments were also included in response to key questions asked during Stage 1.

During the Cahokia Canal and Harding Ditch Studies, the identified problems were categorized to four primary planning objectives including flood control, sedimentation control, environmental quality, and water-related recreation.

The fact sheets provided details on alternatives (referred to as options) for meeting these objectives.

Because the Pine Ford Study was more complex than either of the two other studies, the alternatives were much more numerous. The alternative measures were categorized into the four topics of environmental/water quality, water supply, flooding, and recreation. Nineteen measures for solving or reducing these water related problems were identified and briefly presented in the fact sheet prior to the series of public workshops to assess the alternatives. Because of reduced budgets and an accelerated schedule, fewer fact sheet distributions were implemented during the Pine Ford Study than during the Cahokia Canal and Harding Ditch Studies.

Additional methods of workshop notification included media coverage, press releases, and telephone contacts to key influentials.

A somewhat unique method for information dissemination was implemented during Stage 2 of the Pine Ford Study. To enable the public to view the various key sites within the study area that related to water resource problems, alternatives, and impacted areas, a Self-guided Tour Pamphlet was developed and distributed. Although most of the identified sites related directly to the study, historical, scenic, and recreational sites were also included. Examples of sites were flood-prone areas, floodproofing examples, the proposed damsite, and river access points.

An additional method for disseminating information during the Pine Ford Study was the Speakers Bureau. Speakers were provided by the Corps' Public Affairs office to make informal presentations to public and private gatherings within the study area. These gatherings included various civic organizations and provided the opportunity for direct, informed interaction between the public and the Corps.

Public Response. The Delphi, Citizens and Technical Review Panel, and public workshops were techniques for assessing public views on alternatives to water-related problems. During this stage of each of the studies, the public was encouraged to comment on and evaluate the specific options for achieving each of the planning objectives.

The Delphi panelists participated in two rounds of option evaluation. The purpose of the second round was to more clearly define the options of agreement and disagreement. On-site control was the preferred alternative to sedimentation problems, while floodplain evacuation, flood-proofing, and floodwall construction were clearly identified as desirable measures for flood control. Measures of considerable disagreement identified included the importance of cultural resources, nonstructural options for flood contol, and recreational measures.

The Citizens and Technical Review Panel continued to function during the alternative evaluation stage of the Pine Ford Study. Although a meeting was scheduled prior to the public workshops, for the purpose of discussing and assessing the alternatives, it was canceled because of heavy snowfall. Mail responses were therefore solicited, and with only a moderate response, the panel's effectiveness prior to the workshops was considerably less than prior to the inital public meetings.

The public workshops held during the second stage of the Cahokia Canal and Harding Ditch studies indicated several key factors in terms of public involvement. Possibly due to the extended time frame of almost one year since the initial public meetings, the attendance at the public workshops was less than half that of the meetings. The effort to solicit participation had been equal to that of previous meetings, but the interest seemed to wane.

Those who did participate in the workshops, which consisted of Flood/Sedimentation Control and Environmental Quality/Recreation work groups, indicated agreement with the Delphi panelists. Although there were areas of disagreement, these pertained primarily to measures to solve less important problems such as recreation.

Although the public workshops to evaluate alternatives of the Pine Ford Study were held within six months of the initial meetings, attendance was also well below expectations. An attendance drop of over 50% was attributed primarily to heavy snowfall and subfreezing weather in the study area. The weather factor must continually be considered, the snow and cold of winter, the rains and winds of spring, the heat of summer. Although there is no best time to hold meetings and workshops, experience in these studies indicate that mid-fall and mid-spring are most favorable. When possible, avoid scheduling in winter and summer.

At the Pine Ford workshops, small group working sessions were also conducted, with group size ranging from 15 to 25. The participants were provided with information on measures such as descriptions, locations, benefits, advantages, and disadvantages. Similar to the other studies, measures were grouped by categories of Environmental/Water Quality, Water Supply, Flooding, and Outdoor Recreation. Participants were encouraged to identify additional advantages and disadvantages. Using 9-point rating scales ranging from not preferred to highly preferred, participants evaluated each measure by category. A similar procedure had been followed during the Cahokia Canal and Harding Ditch workshops.

As expected, due to the nature of the study, considerable disagreement was apparent in the preferred alternative measures. Many of the participants preferred no Corps action, while an almost equal number preferred structural measures, especially reservoirs. Just as the participants had perceived the problems differently, they likewise perceived the solutions differently.

Although polarization and conflict in a given study aid in maintaining interest, it does require a more intensive effort in implementing a public involvement program. During the alternative evaluation, it is especially important that information pertaining to measures be as accurate and clear as possible to reduce confusion and misinformation. This stage also requires the participants to more seriously study the measures so that they can better understand and more accurately assess them. Frequent complaints expressed by the public during the assessment of alternatives is that the information is too technical and requires too much thought. This increased degree of technicality cannot always be avoided and must be dealt with on a study-by-study basis.

Phase Three: Evaluation of Plans

The final stage of the studies consisted of the evaluation of the recommended plans. These recommended plans or plan were based primarily on public, technical, and economic assessment of the alternative measures of Phase Two.

Information Dissemination. During this final segment of the public involvement programs, the fact sheets continued to be the primary mechanism for information dissemination. For the benefit of more recent participants, the fact sheets developed during this stage included a complete recap of study activities as well as ongoing study activities and future public involvement opportunities. Question and answer segments continued to be included.

Fact sheets during this stage emphasized "how we got here" more than "where we are going." The importance of continued public participation was stressed.

Other methods of information dispersal included press releases and media coverage consisting of related news items on TV and radio and in newspapers. During this stage, contact continued to be maintained with the media by the Public Affairs office to assure not only coverage but also accuracy in reporting.

Public Response. The primary method for assessing public response to the recommended plans is the public meeting. The purpose of this meeting for each of the studies was to present the conclusions and recommendations of the study team and to provide the participants the opportunity to evaluate the recommendations.

The public meeting for assessing the study recommendations for Pine Ford was held within six months of the public workshops to evaluate measures. Even with this expedited scheduling, attendance was less than 25% of the initial public meeting. With the majority of public involvement programs, even those controversial studies, the public can quickly lose interest.

The Cahokia Canal and Harding Ditch studies are ongoing and the Stage 3 meeting on recommendations has not been held. It therefore is not discussed here.

At the Pine Ford meeting the large group format was maintained, and participants were invited to make verbal comments as well as written. Several recommendations were presented to the Pine Ford meeting participants. These consisted primarily of deauthorizing the proposed reservoir, placing other proposals in inactive status, and additional studies on flood control and water supply. Using a five-point scale, participants indicated general agreement with the recommendations, with strong support for deauthorization.

The Citizens and Technical Review panel continued to function during the final stage of the Pine Ford Study with similar assessments of the recommendations.

Although not implemented at this time, a general public opinion questionnaire mailed to randomly selected people within the study area is planned. This will serve as a check on representation at public meetings. As with all public meetings, it cannot and should not be assumed that those attending are a representative sample of the entire study area. For this reason additional methods of obtaining public response (i.e., surveys, panels) should be considered during evaluation segments of public involvement programs.

Technique/Program Evaluation

An additional technique of public involvement that is often overlooked in the documentation of public involvement activities and processes is the evaluation of the process and/or technique. In the Corps of Engineers public involvement programs discussed, this evaluative response was included. Participants at each public gathering were provided the opportunity to evaluate the activity using structured response forms. Additional non-structured response was also encouraged.

Study personnel were also provided the opportunity to evaluate the activity, identifying strengths and weaknesses. These evaluative sessions were generally held immediately following each public gathering.

Both the participants' evaluations and study personnel evaluations provided timely information on the effectiveness of the activity and additional information on the overall program effectiveness. From this information, modification could be made in ongoing programs and future programs.

CONCLUDING REMARKS AND SUMMARY

As evidenced by current resource decision-making, agencies are committed to public participation. The public are being provided increasing opportunities to express their views. This level of commitment, however, is as varied as the resource topics. Studies are still conducted with public involvement being only a public hearing, with little additional effort beyond the press release to notify the public. Although the study goals may be achieved, effective public involvement is not. Likewise, an agency can focus its efforts primarily on public involvement, and not achieve the goals of the study.

Although there is no standard formula for developing public involvement programs, public involvement experiences have provided the following guidelines for developing an effective public involvement program:

1. Emphasis should be placed on identifying the affected, concerned, and interested public. These individuals should be sought out with attempts made to involve them [22, 26].

2. Conflict reduction should be considered a major element in any resources planning effort [37]. Consensus is the key to successful planning, and planning efficiency is directly related to the degree of conflicts.
3. Directly related to conflict reduction, the public should be involved in the initial stage of the planning process, or at the earliest possible moment [12]. Absence of this early involvement negates full participation and additional conflicts could result as the study progresses.
4. Efforts should be made to inform the affected, concerned, and interested publics of study activities [26]. If the public does not know of the involvement opportunities, they will not participate. This can result in false representation of public preferences.
5. Mutual confidence and respect must be developed and nutured between the planning agency and the public [30]. A lack of public confidence in the lead agency, especially relating to skepticism over input effectiveness can result in low public response to involvement efforts.
6. Provisions should be made for interaction among differing resource users [29]. The development and implementation of an effective public involvement plan depend on the resource user's ability to learn and understand the others' goals and problems and collectively recognize the need for a solution.
7. The public involvement techniques selected and applied should match the goals and objectives of the public involvement program [12,25]. This guideline, as previously mentioned, cannot be overemphasized.

Following these guidelines does not guarantee an effective and successful public involvement program; however, their exclusion in the program development will generally result in an unsuccessful program. Effective public involvement programs in resources planning have generally incorporated a mixture of different techniques at specific stages of a study. In their extensive investigation of public involvement activities, Ertel and Koch [12] indicated that a varied, multifaceted public involvement program is essential for program success.

As previously mentioned, the selection and implementation of public involvement methods in a public involvement program should be based primarily on 1) the public involvement objective, 2) the planning framework, 3) the study history, 4) the study area characteristics, 5) preliminary and potential problems, 6) costs, and 7) scheduling. Each public involvement program should be tailored to individual studies/projects.

REFERENCES

1. Astrack, R., N. Baumann, and G. Reynolds, "Managing A Public Involvement Program," *Journal of Water Resources Planning and Management*, Vol. 110, No. 2 (April 1984).
2. Baumann, N., O. Ervin, and G. Reynolds, "The Policy of Delphi and Public Involvement Programs," *Water Resources Research*, Vol. 18, No. 4, pp. 721–728 (August 1982).
3. Bishop, A. Bruce, "Public Participation in Water Resources Planning, IWR Report 70-7, U.S. Corps of Engineers, Institute for Water Resources, Fort Belvoir, VA (1970).
4. Borton, T. E., et al., "The Susquehanna Communication Participation Study: Selected Approaches to Public Involvement in Water Resource Planning," IWR Report 70-6, U.S. Army Corps of Engineers, Institute for Water Resources, Fort Belvoir, VA (1970).
5. Chevalier, M. and I. G. Cartwright, "Public Involvement Planning: The Delaware River Case," in W. R. D. Sewell and Ian Burton, eds., *Perceptions and Attitudes in Resource Management*. Ottawa: Information Canada (1971).
6. Clark, T., "Present and Future Research in Community Decision-Making," *Community Structure and Decision-Making: Comparative Analysis,* Terry Clark, ed., Chandler Publishing Co., Scranton, PA, pp. 463–478 (1968).
7. Clark, T., *Community Power and Policy Outputs: A Review of Urban Research,* Chandler Publishing Co., Scranton, PA (1973).
8. David, E. L., "The Role of the Public in Decision-Making," in *Priorities in Water Management,* Francis M. Leversedge, ed., Victoria, British Columbia: University of Victoria Press (1974).
9. Delbecq, A. L., A. H. Van de Ven, and D. H. Gustafson, *Group Techniques for Program Planning,* Scott, Foresman and Company (1975).
10. *Encyclopedia Britannica* (1968).
11. Ertel, M. O., "A Survey Research Evaluation of Citizen Participation Strategies," *Water Resources Research, Vol. 15,* No. 4 (August 1979).
12. Ertel, O. and G. Koch, *Public Participation in Water Resources Planning: A Case Study and Literature Review,* Water Resources Research Center, University of Massachusetts, Amherst, Mass., Publication No. 89 (July 1977).
13. Ervin, L., "A Delphi Study of Regional Industrial Land-Use," *The Review of Regional Studies, Vol. 7,* No. 1, pp. 42–58 (1977).
14. Fleishman-Hillard Opinion Research, "Attitudes of Residents and Leaders of the St. Louis Metro. Area Toward the Management of Water Resources," St. Louis, Missouri (1973).
15. Glass, J., "Citizen Participation in Planning: The Relationship Between Objectives and Techniques," *Journal of the American Planning Association,* pp. 180–189 (Apr., 1979).
16. Harman, J. and S. J. Press, "Collecting and Analyzing Expert Group Judgement Data," Rand Corp., Santa Monica, CA., P-5467 (July 1975).
17. Helmer, O., "The Delphi Method of Systematizing Judgements About the Future," Report MR-61, University of California, Los Angeles, CA (1966).
18. Institute for Environmental Studies, *The Plan and Program for the Brandywine,* University of Pennsylvania Press (1968).
19. Linstone, H. A. and M. Turoff, eds., *The Delphi Method: Techniques and Applications,* Addison-Wesley Publishing Co., Inc., Reading, MA (1975).

20. Mazmanian, D. A. and J. Nienaker, "Prospects for Public Participation in Federal Agencies: The Case for the Army Corps of Engineers," in *Water Politics and Public Involvement*, John C. Pierce and Harvey R. Doerksen, ed., Ann Arbor, MI: Ann Arbor Science Publishers, Inc. (1976).
21. Nix, L., "Concepts of Community and Community Leadership," *Sociology and Social Research*, Vol. 53, pp. 500–510 (1969).
22. O'Riordan, J., "The Public Involvement Program in the Okanagan Basin Study," *Natural Resources Journal*, Vol. 16, pp. 177–196 (Jan. 1976).
23. Pierce, C. and H. R. Doerksen, eds., *Water Politics and Public Involvement*, Ann Arbor Science Publishing Inc., Ann Arbor, MI (1976).
24. Pill, J., "The Delphi Method: Substance, Context, A Critique and an Annotated Bibliography," *Socio-Economic Planning Science*, Vol. 5, No. 1 (1971).
25. Priscoli, J. D., "Public Involvement and Social Impact Analysis: A Union Seeking Marriage," paper presented at AWRA annual meeting, Tucson, AR (Nov. 1977).
26. Rosenbaum, M., "Citizen Involvement in Land Use Governance-Issues and Methods," The Urban Land Institute, Washington, D.C. (1976).
27. Rosener, J. B., "Citizen Participation: Can We Measure Its Effectiveness?" *Public Administration Review*, pp. 457–463 (Sept./Oct. 1978).
28. Sanders, I. T., *Allocation of Power, The Community: An Introduction to a Social System*, 2nd ed., Ronald Press, New York, N.Y. (1966).
29. Sewell, W. R. D., "Broadening the Approach to Evaluation in Resources Management Decision-Making," *Journal of Environmental Management*, Vol. 1, pp. 33–60 (1973).
30. Sewell, W. R. D., "Perceptions, Attitudes and Public Participation in Countryside Management in Scotland," *Journal of Environmental Management*, Vol. 2, pp. 235–257 (1974).
31. Sewell, W. R. D. and J. T. Coppock, eds., *Public Participation in Planning*, John Wiley and Sons, Inc., New York, N.Y. (1977).
32. Sewell, W. R. D. and T. O'Riordan, "The Cultural of Participation in Environmental Decision-Making," *Natural Resources Journal*, Vol. 16 (1976).
33. U.S. Environmental Protection Agency, *Public Participation Handbook for Water Quality Management* (Jan. 1976).
34. U.S. Water Resources Council, "Principles and Standards for Planning Water and Related Land Resources" (1973).
35. U.S. Water Resources Council, "Economic and Environmental Principles and Guidelines for Water and Related Land Resources Implementation Studies" (March 10, 1983).
36. Wagner, T. P. and L. Ortolano, "Field Evaluation of Some Public Involvement Techniques," *Water Resources Bulletin*, Vol. 13, No. 6, pp. 1131–1140 (1977).
37. Wengert, N., "Public Participation in Water Planning: A Critique of Theory, Doctrine, and Practice," *Water Resources Bulletin*, Vol. 7, No. 1, pp. 26–32 (1971).
38. Willeke, G. E., *Identification of Publics in Water Resources Planning*, Department of City Planning and Environmental Resources Center, Georgia Institute of Technology, Atlanta, GA (Sept. 1974).

Municipal Service Distribution: Equity Concerns

M. C. IRCHA*

INTRODUCTION

During the past three decades of almost continual urbanization, North American cities have become dispersed, low-density suburban agglomerations. Urban sprawl was aided both by widespread automobile ownership and by government housing policies which favored suburban development. Urban planners, municipal engineers and others involved in civic administration encouraged urban sprawl through the improvement and extension of municipal utility services to underdeveloped suburban areas. The urban transportation network was similarly upgraded and expanded to meet the demands of a growing, automobile dominated suburban population.

Many municipalities focussed their attention on the problems of new development and growth, often to the detriment of existing areas. As new and better services were being provided in the suburbs, scant attention was paid to the deterioration of existing facilities and services in the inner city areas. During the past several decades, movement to the suburbs has been dominated by middle- and upper-income families. This meant the abandonment of a growing number of urban poor within the older central parts of cities. This "flight of the middle classes" further exacerbated the problems of spatial segregation of urban groups. Individuals are not randomly scattered throughout an urban area, but rather tend to be clustered into identifiable socio-spatial groupings (neighborhoods) on the basis of socioeconomic status, ethnicity, and stage in the life-cycle. Each neighborhood, therefore, has differing municipal service needs and the provision of facilities and services should respect these variations.

Providing various municipal services to urban residents is generally considered to be the *raison d'etre* of local government. Indeed, modern urban systems have become highly dependent upon the services being provided. Public safety is ensured through police and fire protection, public health, and building inspections. Convenient and safe urban transportation is provided by improved roadways, automated traffic signals, and public transit. Potable water is delivered, and liquid and solid wastes are sanitarily removed and disposed. In a variety of ways, the quality of urban life is defined by the nature of the municipal services being provided.

The essence of municipal service distribution lies in the political science dictum "who gets what, when and how" [26]. Concerns for equity underlie this dictum. During the past decade, there has been an increased interest in determining equitable service distribution. In the U.S., various judicial proceedings have forced municipal consideration of equity in service delivery. In 1968, the Kerner Commission (tasked with determining the cause of urban riots in the mid-1960s) reported that one of the most serious complaints of urban ghetto residents was the apparent inadequacy of municipal services [41]. During the 1970s, lawsuits flourished as civil rights lawyers challenged municipalities to justify apparently inequitable service distribution patterns. In a classic case, *Hawkins v. Shaw* established that it was illegal to discriminate on the basis of race in providing basic municipal services [16]. This case and others like it have had a significant impact on the service distribution policies of U.S. municipalities.

In a parallel process, the whole field of public administration at all levels of government has shifted to a more socially equitable stance. During the 1970s, the public was continually exposed to problems in public administration. From Vietnam to Watergate, it became increasingly apparent that the dichotomy between the policy making process at the political level and the implementation of those policies by the administration no longer existed. Example after example of administrative involvement in the political policy making

---

*Department of Civil Engineering, University of New Brunswick, Fredericton, New Brunswick, E3B 5A3 Canada

process demonstrated the pervasive power of the bureaucracy. Growing public distrust of governmental systems led to the emergence of "New Public Administration" [12]. A process which entailed the addition of a fourth "E," equity to the traditional public administration values of efficiency, economy, and effectiveness.

DISTRIBUTING MUNICIPAL SERVICES

Equity is defined as the quality of being just, impartial, and fair. However, equity is an elusive concept. A number of questions may be raised in trying to determine an equitable situation:

- What is to be equalized? Inputs? Activities? Outputs? Outcomes?
- What is the basis for equalizing? Needs? Preferences? Effort? Willingness to pay?
- What is the appropriate unit? Individuals? Neighborhoods? Jurisdictions?
- What is the level of equalization? A minimum standard? Uniformity? [32]

Equality tends to be a complex and an uncertain guide in devising public policy. Equality (providing equal levels of service to all) does not always result in equity. Unequal needs demand unequal service provision—the needs of the handicapped often differ from others in our society. Equal resource inputs in differing circumstances can lead to varied outputs—equal amounts of police patrols in crime-ridden, inner-city neighborhoods and in suburban areas ignore differing crime rates and yield varied outcomes.

On what basis can service delivery be equalized? Willingness to pay (reflected in water rates, transit fares, recreation facility use) tends to be inequitable when considering ability to pay. Antunes and Plumlee found that, in Houston, municipal engineers provided street maintenance through local improvement taxes, the underlying assumption being that those who had poor streets did not wish to have them improved as they were unwilling to pay for it [2]. Preference or demand for service provision tends to be inequitable owing to the varying abilities of individuals and groups to voice their concerns.

Defining equity on the basis of need implies that unequals should be treated unequally. Those neighborhoods or groups of citizens having a greater need for a specific service should have their needs satisfied. In this manner municipal services will tend to be differentially distributed focussing on defined target groups. Providing services on the basis of need is the fairest approach of all. However, care must be taken in defining need—how is need established? Who is involved in determining it? The essential thrust of this chapter is that the distribution of municipal services on the basis of need is the most equitable approach.

Causes of Inequities

In the small but growing body of literature on the equity of municipal service provision, three fundamental causes of current inequities are typically cited. Early in the 1970s, the focus of attention rested on the "underclass" hypothesis—the quality and quantity of service provision being directly related to the socioeconomic status of the area receiving it. In other words, poorer neighborhoods received inferior service delivery due to their relative political powerlessness and their inability to effectively voice their demands. Using this perspective, the author demonstrated that several municipal services in Kingston were biased in favor of higher socioeconomic status areas [22]. Similarly, Levy and others found that class differences affected service delivery in Oakland: "by following the course of least resistance, local agencies reinforced a class bias—the more you have, the more you get" [28].

In his testing of the "underclass" hypothesis, Lineberry found that despite the intuitive simplicity of this approach, other complications affected service delivery patterns [29]. One of these complexities was an "ecological" concern: the age of the neighborhood and its population density determined differing needs. For example, older, inner-city neighborhoods usually lack the abundance of open space found in newer suburban areas, underground utilities are older and more prone to collapse, and roads tend to be congested with commuting suburbanites and often are in poor condition. These conditions in older areas imply greater needs, hence a greater share of municipal resources should be devoted to resolving these needs. A situation which occurs infrequently as civic attention typically tends to be focussed on new development and expanding residential areas.

An additional complexity in service delivery is the role of the appointed bureaucrats and their use of simplifying decision–rules. Mladenka argues that: "the bureaucracy may well be the prime determinant of who gets what at the distribution stage" [38]. Typically, bureaucracies employ well established rules to provide services. These bureaucratic decision–rules have been defined as a "rough admixture of professional norms, rules and regulations . . . loose perceptions of both needs and demands and a search for economizing devices" [29]. Such rules are used in libraries when "good" books are provided rather than information on job training, in road departments when most of the funds are used to maintain high-speed expressways which carry the majority of traffic rather than upgrading local streets. Decision–rules or "standard operating procedures" have evolved as a means of simplifying the complexities of the urban system. In seeking simplicity these rules tend to focus on economy and efficiency (allocative mechanisms) rather than on the equity (distributive) implications of service provision.

Operational stability is reinforced through the establishment and enforcement of decision–rules designed to limit the discretionary powers of lower-level bureaucrats. Some authors have suggested that these rules enable bureaucrats to resist interference in service provision by both municipal councillors and citizen interest groups [1,43].

Another cause of inequity in service delivery which relates to the role of the bureaucrat has been suggested by Kemper [24]. As a sociologist, he argues that the apparent decline of services within inner-city areas is due, in part, to changes in the status differences between the consumers and the providers of the service. His basic thesis is that white civic workers providing essential services (street maintenance, garbage collection, police and fire protection) in minority areas tend to do so "reluctantly, indifferently, dilatorily, and often enough, if one can judge from observing how ancient the trash is, rarely." Kemper's assertions have recently been reinforced by a major survey of six urban services across 30 central cities and 20 school districts:

> Regardless of whether you consider critical services such as education and police services, or less critical shopping and street services, it is evident that minorities are systematically treated differently and adversely [31].

Evaluating Distributional Equity

All three major causes of service distribution inequities can be examined in a municipal setting. The socioeconomic status (relative wealth) of neighborhoods or other defined areas can be estimated from census data. Ecological parameters can be established from both census material and a review of the municipality's historic development pattern. Bureaucratic decision–rules can be identified through in-depth interviews with the various bureaucratic and political actors involved providing municipal services.

The socioeconomic status (SES) of census enumeration areas (the smallest areal unit in which discrete data can be provided due to confidentiality requirements) can be obtained through the use of data provided in the national decadal census publications.

Socioeconomic status is generally considered to be related to income level, educational attainment, prestige, and occupation. In the Canadian context, Blishen and McRoberts developed a socioeconomic status (SES) scale using these four parameters based on reported male occupations in the 1971 census [6]. Some 480 occupations were ranked from 23.02 for farmers to 75.28 for administrators in the teaching field. Blishen and Carroll further refined this data by providing an additional SES scale for reported female occupations [5]. Census data provided for each enumeration area classifies occupations into 14 major groups. Dividing the 480 occupations reported by Blishen and McRoberts into the 14 major occupational groupings provided by the census publication permits the derivation of

TABLE 1. Occupational SES (Male) In Census Enumeration Areas [18].

Occupational Group	Mean SES
Managerial, administrative	63.61
Teaching and related	61.23
Medicine and health	56.94
Technological, social, artistic and related	57.83
Clerical and related	48.64
Sales	47.68
Transport equipment operating	41.89
Construction trades	36.64
Machining, fabrication, assembly	36.46
Processing	35.80
Service	34.25
Other primary occupations, fishing . . .	31.13
Farming, horticulture and animal husbandry	27.87
Unclassified occupations	41.60

a mean value of SES for each grouping. These values are shown in Table 1.

To obtain the SES for a neighborhood or other defined area, one begins by determining the census enumeration areas which fit within the established neighborhood boundaries. Some overlaps will inevitably occur as the census enumeration area boundaries tend to be randomly established and do not necessarily conform to traditional neighborhoods. Hence, a judgement must be made as to whether or not to include enumeration areas bisected by the neighborhood boundary. Estimating whether the majority of the population of the enumeration area resides within the neighborhood generally suffices to add or omit an enumeration area in the calculations.

Within the enumeration areas considered to be in the neighborhood or defined area, the number of individuals reported within each of the 14 major occupational groupings can be used to drive a mean area SES. A further refinement entails determining similar occupational values for females and weighting the mean area SES with these values. The mean SES of the enumeration areas within the neighborhood can thus be proportioned by their population relative to that of the total neighborhood to obtain a mean SES for the whole of the defined area.

Having obtained mean SES values for either the enumeration areas or the neighborhood, it is then possible to prepare an SES map for a given municipality. The example shown in Figure 1 is based on the mean SES of census tracts — the collation of enumeration areas into larger, defined areal units. This type of map provides a visual presentation of the SES area within a city.

The ecological parameters of neighborhood age and population density can also be derived. In the first case, a

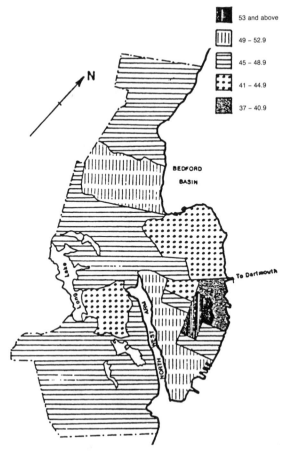

FIGURE 1. Distribution of SES in Halifax [18].

review of city development patterns from municipal records and, in particular, older civic maps indicates the age of neighborhoods. The number of years of service of the neighborhood's infrastructure identifies various needs (unless, of course, some major remedial work has been undertaken, thus reducing the age of the given infrastructure).

Population density is often directly related to the age of an area—older neighborhoods tend to contain fewer open spaces and more crowded housing stock. Census data at the enumeration area level provides a measure of population density. Again, higher concentrations of people may define differing needs than low density, suburban circumstances.

The third cause of distributional inequities is bureaucratic decision-rules. These standard operating procedures may distort the distribution process. In some cases these rules may result in greater equity (e.g., uniform water distribution), in other cases they may not. Structured interviews with the bureaucratic and political officials involved in service delivery may yield evidence of the use of decision-rules and their effect on equity.

An example of using both SES and ecological data in evaluating the characteristics of specific neighborhoods was reported earlier by the author [18]. In this case, specific portions of municipalities were designated for special remedial funding under Canada's Neighborhood Improvement Program (similar in many respects to the U.S. Community Development Block Grant Program). Typically, the neighborhoods selected were older, inner-city areas whose housing stock had notably deteriorated. As shown in Table 2, the five municipalities examined in this study were generally older, contained a significantly higher population density than the overall municipality, and housed a population whose socioeconomic status was below that of the whole municipality.

TABLE 2. NIP Area Characteristics [18].

Municipality (1)	1976 population (2)	1976 population density in persons per acre (3)	1976 mean SES (4)	NIP areas (5)	1976 population density in persons per acre (6)	1976 mean age in years (7)	1976 mean SES (8)
Halifax, N.S.	95,289	6.15	46.30	NIP II*	28.7	100+	47.12
				Nip III	30.6	100+	41.41
Saint John, N.B.	85,956	1.08	44.60	South End	24.8	100+	43.95
				North End	29.6	100+	43.69
Moncton, N.B.	55,934	1.59	45.89	------------	84.8	100+	44.47
				North Side	10.0	64	44.50
Fredericton, N.B.	45,248	1.42	46.40	South Side	17.4	100+	47.60
				West Highlands	6.3	60	42.40
Amherst, N.S.	10,263	2.92	43.80	Southeast	12.4	100+	41.39

*Data for NIP I not available.

MUNICIPAL SERVICES: EXAMINING INEQUITIES

Equity studies have been carried out on a wide range of municipal services including: police and fire protection, building inspection, road construction and maintenance, transit service, sanitation, recreation, zoning, education, and library services. Many of these case studies have shown contradictory results, a finding which Lineberry termed "unpatterned inequality" [29]. That is, specific services may be equitably distributed in one jurisdiction but not in the next. Rich's analysis of equity research tends to partially refute the problems of unpatterned inequality [42]. In his analysis, urban services were divided into two generic categories: economic and social. Economic public services, aimed at supporting the circulation of capital (property assessment and taxation) and the production processes of the urban system (maintaining the physical intrastructure, providing public health, fire, and police protection), are generally considered to be essential and were found, in most cases, to be equitably distributed. On the other hand, social services aimed at class reproduction and social control (education, library services, housing, and welfare) tended to be targeted at specific groups rather than being generally distributed. Considerable case study evidence has shown that social service provision tends to be inequitably distributed, particularly in education and library services.

Given the apparent weaknesses of generalizing service distribution across a number of municipal units rather than focussing on a specific jurisdictional situation, the following examples of inequities in municipal service distribution are provided to illustrate the difficulties that different groups in our urban areas are having in obtaining their fair share of municipal resources.

Transit Services

During the post-World War II era, North Americans have become highly dependent upon the personal mobility provided by automobiles. From 1951 to 1981, while Canada's urban population more than doubled in size, private automobile ownership grew five-fold. This rapid rise in personal mobility diminished the need for elaborate and extensive urban public transportation systems. Public policy shifts in favor of the automobile resulted in the development of an extensive urban highway network. As pointed out by Tait, automobile ownership today can be considered a basic need in our society:

> . . . without a car, people cannot get what they require to pursue a career, develop their personalities or lead fulfilling lives. If mobility is defined as "access to opportunities," the partially auto-induced, spatial segregation of jobs, health care, shopping and recreation in cities means that those without cars may be denied benefits available to car owners [47].

On the surface, with 80 percent of Canadian households owning one or more automobiles, it appears that there is little need for public transit for the mass of our society. In reality, of course, public transit services are required for a substantial proportion of society who are "transit captives"—the poor, elderly, young, and handicapped (essentially the non-drivers). In 1982, sixty percent of Canadian households earning less than $8000 did not possess a car; similarly, twenty-one percent of elderly households did not possess a car [46].

The widespread dispersal of urban activities accompanying suburbanization resulted in altered transportation requirements. Meyer has argued that there are four distinct urban transportation needs: traditional public transit, long distance commuting, cross-commuting, and inside-out or reverse commuting [35]. Meyer went further to show that transportation engineers tend to deal with long-distance commuting (commuter rail systems and urban expressways) and cross-commuting (ring road development) to the detriment of the other needs (both of which reflected the transportation requirements of poorer, inner-city residents).

In studies of urban transit, the author found a number of inequities in transit services being provided [21]. In an examination of two Maritime cities he found that although the transportation needs of the elderly tended to be accommodated (through appropriate routing and reduced fares), youthful patrons (5–19 years) were not so well served. Low income families were also poorly serviced both through inadequate route distribution and the application of a flat fare structure. In another study, evidence was found of the regressive nature of flat fare and simple zone fare systems [13]. In each case, long distance, suburban riders were cross-subsidized by lower-income, inner-city, off-peak users. Similarly, despite higher vehicle occupancy during peak periods, suburban fares recovered only about 57 percent of the associated higher costs of providing the service compared to the off-peak average recovery rate of about 62 percent. Although this difference is not large, it does indicate cross-subsidization within transit systems. As pointed out by Cervero in his examination of U.S. urban transit systems:

> Short distance, off-peak users pay disproportionately higher fares to offset losses incurred in serving long-haul, peak hour trips. On the whole, those highly dependent on transit and least able to pay lose the most from cross-subsidization [10].

Recreation Services

Recreation services have been shown in a number of studies to have a direct relationship with the socioeconomic status of the recipient neighborhood. The need for specific services tends to vary by social class. Foresta has shown that "city people or poor people or poorly educated people are completely indifferent to nature-based uses of open space while urban, affluent and well educated people are equally indifferent to its social and intensive recreation uses" [11].

Needs for different types of recreational services tend to vary by social class and these needs are changing over time as individuals within our western society are experiencing an increasing amount of leisure time. It has been estimated that Canadians will have between 20 and 50 percent more discretionary (free) hours by the year 2001. Although leisure provides the time for human development and life enrichment in our technological, work-oriented society, the major concern of growing numbers of people will be how to utilize this time.

The recreationist of the future will:

> be virtually overwhelmed by demand for his services. During the day he will be busy with early retirees, the unemployed and school children participating in physical activity and outdoor education programs. Evening programs, both educational and recreational . . . will be increasingly popular as "baby boom" adults, unencumbered by family responsibilities and locked into "slow advancement careers," seek a positive self-image outside the work spheres [3].

Today's urban recreation problems of inadequate parks and programs which do not serve the needs of many of their client groups such as children, the elderly, the poor, and the handicapped are bound to become more severe during the coming decade. Inner-city recreation facilities and programs provide a significant service to many of these urban residents, as often these are the only facilities to which they have access. "Because they are less mobile and have less amenities among their own resources, low-income families . . . have a greater need for community supported recreation services" [45]. Gold reinforces this perspective by pointing out that a significant portion of the urban population (more than 30 percent) either cannot drive or do not have access to a car and thus depend on their local parks for recreational activities. As most Americans live in urban areas, the accessibility problem leads to "severe inequities" which are becoming worse [14].

Several studies of the distribution of recreational services have shown that in some cities (Houston, San Antonio, Detroit) there is a general equality in their allocation in different neighborhoods [14,29,37]. Although the physical service may be equal, there are some concerns about the qualitative differences among neighborhoods. For example, Gold found that the age of recreational centres in Detroit tended to vary inversely with the socioeconomic status of the neighborhood. Poorer residents tend to dwell in older parts of the city and use aged facilities built for an earlier era when higher-class families lived in the neighborhood [14].

Equal distribution of recreational services to all areas of the urban centre may in itself be inequitable if the need for these services varies. As discussed by Merget and Wolff, "frequently inner-city neighborhoods, because of their age, density or population characteristics require different levels of inputs to produce some minimum level of service" [33]. Mladenka in his study of park distribution in Houston showed an equal distribution of facilities and services; however, he cautions: "if the concept of need is employed as a test for equity, the results of this study indicate that park acreage and facilities are inequitably distributed" [38].

Some of the inequities in recreational service distribution in Canadian municipalities were ameliorated in the 1970s through the implementation of the federal government's Neighborhood Improvement Program (NIP). NIP funding was biased in favor of social-recreational programs, and designated neighborhoods usually contained lower-income families living in generally deteriorated housing. This often meant that older, inner-city areas were selected for NIP, enabling historic inequities to be reduced. In an examination of the impact of NIP in a selected sample of four Canadian

TABLE 3. Recreational Facilities—NIP Areas [17].

Municipality (1)	1976 Population (2)	Municipal Socio-economic status (3)	Nip Area Socio-economic status (4)	Nip Area Population (5)	Total Nip Funding (000) (6)	% NIP Funds for Social/Rec (7)	Developed Park Space (acre/1000)	
							Before NIP (8)	Post NIP (9)
Amherst, N.S.	10,263	43.8	42.4	2400	$624	43.1%	0.69	10.40
Fredericton, N.B.	45,248	46.4	44.5(N) 47.6(S)	3345 3735	$371 $702	64.7% 75.8%	10.21 3.41	10.21 3.41
Kingston, Ont.	56,032	46.2	41.7	6500	$1,290	57.7%	0.49	1.21
Vanier, Ont.*	19,812	45.3	44.4	7146	$2,044	69.9%	0.79	1.90

*Vanier is an inner city within the Ottawa-Hull metropolitan area whose urban population was 602,487 in 1976.

municipalities, the author found that designated NIP areas were generally older and of a lower socioeconomic status than the rest of the municipality [17]. In all cases, as shown in Table 3, the majority of available NIP funding was used to develop or expand social and recreational amenities in the designated neighborhood. These amenities included the construction of recreation centres and gymnasia, tennis courts, playgrounds, and in one case, a community swimming pool. Further, NIP funds were used to acquire additional parkland within the NIP area. In four of the municipalities examined, available park space was increased by almost 75 percent.

The Neighborhood Improvement Program, initiated by the federal government, forced many local governments to examine the needs of their previously neglected areas. Through the injection of considerable funds into these areas, many of the physical ecological problems caused by neglect were alleviated. In addition, many of the designated NIP areas lost their earlier political powerlessness through the program's required resident participation process. In a number of cases, neighborhood residents learned how to organize and make themselves effectively heard within city hall. New municipal politicians, arising from the ranks of NIP residents, ensure these areas will not be forgotten in the future. In a similar vein, the municipal bureaucracy, having been exposed to the problems in these poorer areas, will adjust their decision-rules to better meet their demands. As shown by Mladenka in Chicago the civic bureaucracy responds to articulated demands and adjusts their routinized distributional mechanisms accordingly [36].

Urban Roadway Maintenance

> The politics of streets involve who gets streets, whose streets are repaired and where streets are built. It means having sidewalks to walk on and curbs, gutters and storm drains to keep one's house from flooding [28].

In their extensive review of Oakland's Street and Engineering Department, Levy and others determined that the Department's primary goal was to provide a fast, safe, and efficient traffic circulation system. Achieving this goal meant a concentration on the construction and maintenance of major arterials and urban expressways designed to carry motorists long distances quickly. In turn, this implied:

> maximizing traffic flow in Oakland to provide well-to-do commuters (including those who live outside Oakland) with routes from their homes to their offices. The poorer citizens were left with transportation that did not improve over time [28].

Given this primary goal in Oakland (in reality, a reflection of the professional norms guiding traffic engineers), relatively few budget funds were left for the maintenance of local neighborhood streets.

Little objective or scientific testing was carried out in Oakland to determine road repair needs on local streets. Establishing priorities for the expenditure of the few funds available was left to the subjective judgement of district supervisors. A study of four New Brunswick municipalities of varying size has shown similar results [44].

In Oakland, the criteria used by these supervisors in judging street maintenance needs included the degree of deterioration, traffic volumes, levels of citizen complaints, and the need to avoid utility repairs for at least five years after resurfacing. In applying these decision-rules, the supervisors tended to favour upper income areas because they knew how to register the complaints effectively, and since their utility lines were generally newer, they were in less need of repair. Similar funding biases in favor of arterials and high-traffic streets were found in Houston [2].

While employed as a municipal engineer in Nova Scotian municipality several years ago, the author noted significant inequities in the street resurfacing program—most of the work tended to be done in the wealthier areas rather than in other, more deteriorated, neighborhoods. To objectively define street resurfacing needs, a modified riding comfort test was carried out on all municipal streets. This approach yielded a street resurfacing program ranked in order of objectively measured needs. The elected officials were thus able to direct alloted funds towards the streets in most need of repair. A more sophisticated technique of determining roadway repair needs has evolved in Ontario. "The Road Needs Study" approach involves an annual objective appraisal of a number of roadway deficiencies including structural failures, riding quality, roadway capacity, and safe driving speeds [34]. Assessing this information for each section of urban street permits priorities to be set annually on needed repairs, as well as aiding longer term planning and programming.

During the past several years, additional developments in the roadway maintenance field have aided in surmounting some of the equity problems identified in earlier studies. *A Training Manual for Setting Street Maintenance Priorities* [4] provides a useful methodology for municipalities seeking to objectively prioritize repairs and reconstruction. Similarly, Haas's text on *Pavement Mangement Systems* [15] provides useful information on the need for maintenance and provides methods for evaluating this need in a scientific and rational manner.

Land Use Planning

Current urban literture stresses the problems of suburbanization. The "flight to the suburbs" of wealthy, white citizens has been well documented. Abandoned central cities have become increasingly populated by the black and the poor. In Cleveland, Krumholtz found that during the past three decades the proportion of the city's black population rose from 16 to 44 percent; in the 1960s, the city lost 25 percent of its wealthier families to the suburbs [25]. Jobs were being lost in Cleveland (110,000 from 1958 to 1977) but were being gained by the city's suburbs (210,000 over the same period).

Middle class abandonment of central cities with the resultant diminution of the urban tax base does not necessarily imply reduced costs to the municipality. Most older cities are unable to cut back due to the costs of maintaining an established infrastructure and high cost labour unions [23]. Indeed, Muller demonstrated that costs per capita for municipal services are higher in declining cities than in growing municipalities [39].

Through the use of zoning and other land use regulations, suburban municipalities have been able to ensure that the poor do not follow the middle classes in their "flight." Fiscal zoning (maintaining larger minimum lot sizes and higher levels of building standards and specifications) is used as a socioeconomic screen against unwanted minorities. Zoning can also be used to exclude from the municipality, land uses which are considered to be noxious (apartments, low-income housing, group homes, and other residential-care facilities). As Wolch pointed out, "By excluding service facilities on which many service dependents rely, externality zoning policies effectively exclude the service dependents themselves" [49].

Similar trends have been noted in Canadian cities. The poor have often been excluded from suburban areas due to high costs for serviced land (a function of larger lot sizes and the cost of services installed by the developer). Lately, however, the dynamic nature of neighborhood change has become notable with middle class professionals seeking older, renovated housing in the city centers ("Cabbagetown" in Toronto and the "Glebe" in Ottawa). As the nature of neighborhoods change, service provision also must change.

Although in many parts of Canada, planning and development controls may act to exclude the poor from suburbanizing areas, the same trend is not the case in Canada's Atlantic provinces (New Brunswick, Nova Scotia, Prince Edward Island, and Newfoundland). Land use planning and development control, particularly in the rural areas surrounding urban centers, has not been widespread. In one study, the author found the absence of these planning tools to be a primary factor in the incursion and spread of suburban shopping centres throughout the Atlantic region [20].

In New Brunswick, the provincial Equal Opportunities Program in 1967 established urban and rural equality in the provision of a number of services (education, health, hospitals, welfare, and the administration of justice). This program coupled with the general absence of development control regulations in the suburban areas has facilitated urban sprawl throughout the province. For example, within the city of Fredericton, developers are required to provide full services in their subdivision; outside the city, however, these same deelopers can sell less-serviced subdivided lots at much lower costs. Rural/suburban residents are guaranteed equal services in many sectors (education, street maintenance) at less cost (initial land price and subsequent lower property taxes). The result is that Fredricton is losing its young, middle-class families to the outlying suburban areas [8].

In summary, municipalities are able to use their planning and development control regulations to screen out "undesirables." This enables suburban municipalities to attract the affluent middle-class and avoid the costs of supporting the social services required by the poor. From a metropolitan perspective, therefore, service distribution may well vary in extreme amounts. In the Canadian context, one approach to reducing these forms of servicing extremes is to amalgamate the suburban units into the central city—such as the formation of Winnipeg as "Unicity" in 1971 (an amalgamation of twelve urban municipalities into one major city) or through the formation of regional (metropolitan) levels of government.

RESOLVING DISTRIBUTIONAL INEQUITIES

> The product we produce in local government is service. The service is not the activity or the level of activity but is the impact on the results achieved [48].

Inequities in the provision of municipal goods and services exist in varying degrees. Although such inequities may arise from the differing socioeconomic status of the recipients (underclass hypothesis) or from the age of the facilities and population density of the area (ecological hypothesis), it is apparent that the most pervasive cause lies in the discretionary decision-making powers of the municipal bureaucracy. Given this situation, steps should be taken to increase the ability of municipal administrators to better understand the distributional impact of their decisions and to modify them, if required.

Lucy and Mladenka, in their handbook on establishing equity in municipal servicing argue that careful consideration must be given to the bureaucratic deicsion-rules which guide service distribution [30]. They pose several key questions which should be addressed by both elected and appointed officials concerned with problems of equitable service distribution:

- Which equity concepts (need, demand, equality, preference, or willingness to pay) are most relevant to a particular service? To which aspect should they be applied?
- What decision-rules (efficiency, economy, equality, safety, convenience, established practice) are most important in the current distribution of a given service? How, if at all, should these decision-rules be changed?
- What is the present distribution of services? How can this distribution best be measured?
- Who should participate in making distributional decisions (bureaucrats, politicians, community groups, in-

dividuals)? What process should they go through in making these decisions?

Ensuring that municipal services fully meet the needs of their community is generally considered to be a political responsibility. However, the actual definition of needs often rests with the civic administration. Municipal government tends to work best if both the elected and the appointed officials respect one another's contribution to the effective operation of the municipal system. Elected councillors must acknowledge and respect the technical expertise provided by their staff, as the appointed bureaucracy can clearly establish both the limitations and capabilities of their departmental organizations. Similarly, the appointed administration must acknowledge and respect the political astuteness and community awareness capabilities of their council [19].

Enhancing the political process of municipal service distribution requires improved information. Both the quality and quantity of staff advice must be appropriate for the policy-making needs of council. It is the lack of suitable information which often leads to political forays into adminstrative "trivia," breeding distrust and suspicion among both staff members and councilors. As Leo aptly stated:

> As urban issues have become more controversial, it has become obvious that technical and managerial competence without political guidance can go just as far wrong as a political body operating without the support of good managers and technicians [27].

Achieving greater equity in the distribution of municipal services requires several steps. The first involves enhancing the political role of municipal elected officials. This "re-politization" process aids in ensuring that the municipal politicians are better able to represent the specific needs of their communities. However, care must be taken in this approach, for more politics in itself will be insufficient unless it is accompanied by a municipal political process which is more rational, informed, and accountable. A second step entails modifying the inolvement of the civic bureaucracy in the distribution process. As one of the primary sources of information for municipal council (in particular during the budgetary process), the civic bureaucracy must ensure that their advice is both timely and comprehensive.

Need for Bureaucratic Reform

> A bureaucratic decision-rule explanation of existing distribution . . . suggests that bureaucratic reform lies at the heart of service-distribution problems. Bureaucracies would have to be altered and internal incentive structures changed in order to effect different outcomes [43].

In New York City, an increased degree of distributional equity has been achieved by the decentralization of various civic agencies providing municipal services (police, fire, public works). Traditional neighborhood areas were clearly defined such that all service agencies used common boundaries for their decentralized activities (a concept of coterminality). The management and provision of specific services as well as budgeting for program activities was focussed on meeting the unique needs of each of the defined service areas. Decentralized neighborhood budgeting reinforced local accountability for the services provided [43]. Decentralizing service provision enables differential levels of activity to be provided.

Coterminality may not ensure equity. The defined service areas should include roughly homogeneous groups to avoid internal wrangling. Although equity concerns dominate, the question of efficient and economic service provision must still be addressed. In some instances these latter considerations may dominate the former.

Another aspect of bureaucratic reform is the need for municipal service departments to measure the outcome (or impact) of their programs and activities rather than merely focussing on their productivity and efficiency. Outcomes can be determined by matching service deliverly outputs to defined needs. What is requird is an objective survey method which aids in determining the needs of various neighborhoods and community groups for specific services. In the municipal public works sector, the current conditions of streets can be measured through varios techniques aimed at determining "road needs" [34]. This permits the allocation of scarce roadway maintenance funds to those streets in greatest need of repair. How well these needs are met (outcome) can be determined through futher roadway condition surveys. Similar approaches to determining community needs should be developed for all municipal service activities. Although these techniques imply the development of additional bureaucratic decision-rules, these at least, are more visible, accountable, and likely to be equitable.

Increasing Political Responsiveness

Enhancing the political responsiveness and sensitivity of elected officials requires a reversal of many of the "depolitization" mechanisms which have come to be institutionalized in modern local government. Local government political involvement can be increased by redirecting authority and responsibility back to the municipal council. This may be achieved by disbanding many of the existing independent boards and commissions which would strengthen the ability of elected officials to direct service distribution.

Political responsibility can also be reinforced through a more widespread recognition of the role of partisanship in local government. Increasingly, in major Canadian municipalities, local government political parties are fielding candidates. Party discipline tends to clarify accountability to the electorate since citizens can more readily identify voting patterns within a partisan system. A municipal partisan system enables the formation of a clearly established governing party and opposition. In turn, this will lead to more informed and meaningful debates on appropriate municipal policies particularly with regards to service distribution.

The political nature of local government can be further enhanced by encouraging ward representation. Ward councillors provide an easily identifiable access point to the municipal council for concerned citizens and community groups. A major criticism of ward politics is the tendency for these elected representatives to become parochial in their outlook—focussing primarily on their ward's needs and ignoring the longer term, major distribution issues affecting the entire urban area. Encouraging a partisan approach may reduce the parochialness found in some ward systems.

To remain politically responsible, municipal councillors require information about the service needs of their community. One approach to providing this information is through the use of service-oriented opinion polls conducted among a representative sample of citizens. A number of U.S. and Canadian cities have conducted these surveys to obtain feedback on the citizens' perceptions of the types and levels of service being provided [9,40]. Although these surveys are often of limited value as a measure of actual service delivery (given the respondent's lack of knowledge of all the municipal services being provided), they do provide general measures of citizen satisfaction with the current service level. In Fredericton, the results of the citizen survey were used as guidance by both council and staff in preparing their "program options" budget for 1985.

Shifting priorities in service provision generally results from the municipal budgetary process. Political responsibility for appropriate service distribution implies an increased role for the elected officials in the allocation of municipal funds among competing programs and service levels. To do this effectively, departmental budget estimates must be provided in a manner which enables the municipal politician to easily select appropriate service levels.

Conventional wisdom argues that the annual municipal budgeting process enables councillors to redirect fiscal resources to reinforce specific policies and service delivery programs. In reality, councillors have limited discretionary authority to modify the annual budget. They are constrained by past decisions, previous policies, and current trends; debts must be paid back, salaries at union negotiated rates must continue, various essential services must be maintained. Given these various constraints, many councils find their role relegated to one of allocating scarce fiscal resources among competing municipal departments. The actual distribution of services usually rests with the appointed bureaucracy. As pointed out by Crecine, the municipal budget:

> may have a lot to say about how many inspectors are employed . . . but very little to say about how many buildings are inspected, the order in which they are inspected . . . Budgeted dollar amounts provided many (severe) constraints on the equality of resources available to government units, but few constraints on how these resources are utilized [7].

Municipal council can monitor service distribution programs more closely through the use of more effective budgeting approaches. Various techniques exist (such as Program Budgeting and its derivative, Zero-Based Budgeting) to provide budget information to municipal council in a comprehensive manner. These techniques enable them to select both the services to be provided and the level of activity in each. In other words, council should be provided with a "budget menu" which includes costs from which they can select appropriate service activities to meet the needs of their community.

CONCLUSIONS

The whole question of appropriate municipal service provision appears to revolve around the three competing values: allocative (focussing on efficiency and economy), distributive (dealing with equality), and democracy. The latter value tends to be independent of both allocative and distributive values. The various concepts raised in this chapter can be examined within the context of these three values:

- Need surveys enhance efficiency and equity but reduce democracy by constraining some of the actions of the municipal politicians.
- Decentralized service delivery (coterminality) results in more equity, more democracy with likely, less efficiency.
- Improved budgeting systems provide improved efficiency, increased democracy, and possibly more equity.
- A reduced role for independent boards and commissions may lead to more democratic accountability, more efficiency (through less fragmentation), and possibly more equity.
- Partisan systems produce more efficient outcomes, maybe less equity, and possibly less democracy.
- Ward systems provide more democracy, more equity and, likely, less efficiency.

As North American society embraces the realities of fiscal restraint and cut-back management, many challenges lie before local governments. Increased pressures to cut back on existing services will result in growing conflict in the distribution of scarce funds. In an era of general prosperity, identified facility and service deficiencies could be rectified by altering the allocation of growing revenues. But in our current cut-back era, resolving the problem of service inequities becomes increasingly difficult.

This review of municipal service distribution has focussed on the need to modify the approaches used by the civic bureaucracy in their service provision process and on the need to enhance the political responsiveness of the elected politicians in municipal government. Increased complexity in our urban systems coupled with the growing need to carefully allocate scare resources implies a need for a strengthened political role in local government.

Modifying bureaucratic service delivery approaches requires a broader look at the impacts these services have on the recipient groups. One approach to determining effectiveness is to match service delivery to defined needs. Municipal engineers and other civic administrators must devise objective techniques both to measure servicing needs and to monitor the equity aspects of their service delivery.

Throughout the past decade, concerns have been raised about the need to improve equity in the distribution of municipal goods and services. The time has now come for both elected and appointed municipal officials to add an equity dimension to their traditional public administration goals of economy, efficiency, and effectiveness.

REFERENCES

1. Antunes, G. and K. Mladenka, "The Politics of Local Services and Service Distribution," in L. H. Massotti and R. L. Linebury (eds.), *The New Urban Politics,* Ballinger, Cambridge, MA, pp. 147–169 (1976).
2. Antunes, G. and J. P. Plumlee, "The Distribution of an Urban Public Service," in R. Lineberry (ed.), *The Politics and Economics of Urban Services,* Sage, Beverly Hills, CA, pp. 51–70 (1978).
3. Balmer, K. R. and D. S. Williams, "A Personal Guide to the Future of Leisure and Leisure Services in Canada," Balmer, Crapo & Associates, Waterloo, ON (1980).
4. Biles, S., *Training Manual for Setting Street Maintenance Priorities,* National Science Foundation, Washington, DC (1979).
5. Blishen, B. R. and W. K. Carroll, "Sex Differences in a Socio Economic Index for Occupations in Canada," *Canadian Review of Sociology and Anthropology, 15,* No. 3, pp. 352–371 (1978).
6. Blishen, B. R. and H. A. McRoberts, "A Revised Socio Economic Index for Occupations in Canada," *Canadian Review of Sociology and Anthropology, 13,* No. 1, pp. 71–79 (1976).
7. Crecine, J., *Government Problem Solving,* Rand McNally, Chicago, IL (1969).
8. Comay, E. and L. D. Feldman, *A Study on Sprawl in New Brunswick,* Department of Municipal Affairs, Fredericton, NB (1980).
9. "Community Needs Survey," Human Development Council, Saint John, NB (1983).
10. Cervero, R. B., et al., *Efficiency and Equity: Implications of Alternative Transit Fare Policies,* UMTA, Department of Transportation, Washington, DC (1980).
11. Foresta, R. A., "Elite Values, Popular Values, and Open Space Policy," *Journal American Planners Association, 46,* No. 4 (1980).
12. Frederickson, H. G., *New Public Administration,* University of Alabama Press, University, AL (1980).
13. Gallagher, M. A. and M. C. Ircha, "Transit Pricing Alternatives: Efficiency and Equity," *Proceedings,* Canadian Transportation Research Forum, Toronto, ON (May 1985).
14. Gold, S. D., "The Distribution of an Urban Government Service in Theory and Practise: The Case of Recreation in Detroit," *Public Finance Quarterly, 2,* No. 1 (1974).
15. Haas, R., *Pavement Management Systems,* McGraw-Hill, New York, NY (1978).
16. *Hawkins vs. Town of Shaw,* 461 F. 2d 1286, Fifth Circuit (1971).
17. Ircha, M. C., "Recreational Services: Resolving Distributional Inecquities," *Planning and Administration, 10,* No. 2 (1983).
18. Ircha, M. C. and D. Sundararajan, "Municipal Service Distribution: Equity Concerns," *Journal of Urban Planning and Development,* ASCE, Vol. 110, No. 1, pp. 34–40 (1984).
19. Ircha, M. C., "Working with Council: A Staff Perspective," *Municipal World, 93,* No. 10 (1983).
20. Ircha, M. C., "Shopping Centres: Their Development and Impact in Atlantic Canada," *Plan Canada, 22,* No. 2, pp. 35–44 (1982).
21. Ircha, M. C. and M. A. Gallagher, "Urban Transit: An Equity Perspective," *Proceedings,* 1984 Conference of Canadian Society for Civil Engineering, Halifax, NS.
22. Ircha, M. C., "The Allocation of Public Goods and Services: A Case Study of Kingston, Ontario," thesis presented to Queen's University at Kingston, ON, in partial fulfillment of the requirements for the degree of Master of Urban and Regional Planning (1973).
23. Katzman, M. T., *The Quality of Municipal Services: Central City Decline and Middle Class Flight,* Economc Development Administration, Washington, DC (1978).
24. Kemper, T. O., "Why Are the Streets So Dirty? Social and Psychological Stratification Factors in the Decline of Municipal Services," *Social Forces, 58,* pp. 422–442 (1979).
25. Krumholtz, N., "A Retrospective View of Equity Planning: Cleveland 1969–1979," *Journal American Planners Association, 48,* No. 2, pp. 163–174 (1982).
26. Lasswell, H., *Politics: Who Gets What, When and How,* World Publishing, Cleveland, OH (1958).
27. Leo, C., *The Politics of Urban Development: Canadian Urban Expressway Disputes,* Institute of Public Administration of Canada, Toronto, ON (1977).
28. Levy, F. S., A. J. Meltsner and A. Wildavsky, *Urban Outcomes: Street, Schools, and Libraries,* University of California Press, Berkeley, CA (1974).
29. Lineberry, R. L., *Equity and Urban Policy: The Distribution of Municipal Services,* Sage, Beverly Hills, CA (1977).
30. Lucy, W. and K. R. Mladenka, *Equity and Urban Services Distribution,* Department of Housing and Urban Development, Washington, DC (1977).
31. McDougall, G. S. and H. Bunce, "Urban Service Distributions: Some Answers to Neglected Issues," *Urban Affairs Quarterly, 19,* No. 3, pp. 355–371 (March 1984).
32. Merget, A. E., "Achieving Equity in an Era of Fiscal Constraint," in R. D. Burchill and D. Listokin (eds.), *Cities Under Stress,* Center for Urban Policy Research, New Brunswick, NJ (1981).

33. Merget, A. E. and W. M. Wolff, "The Law and Municipal Services: Implementing Equity," *Public Management, 58,* No. 8, pp. 2-8 (1976).
34. *Methods Manual: Municipal Road Needs Measurement,* Ministry of Transportation and Communication, Toronto, ON (1975).
35. Meyer, J. R., "Urban Transportation: Problems and Perspectives, in D. M. Gordon (ed.), *Problems in Political Economy: An Urban Perspective,* D. C. Heath, Lexington, MA, pp. 418-424 (1971).
36. Mladenka, K. R., "The Urban Bureaucracy and the Chicago Political Machine: Who Gets What and the Limits To Political Control," *American Political Science Review, 74,* pp. 991-998 (1980).
37. Mladenka, K. R. and K. Hill, "The Distribution of Benefits in an Urban Environment: Parks and Libraries in Houston," *Urban Affairs Quarterly, 13,* No. 1 (1977).
38. Mladenka, K. R., "The Distribution of Urban Public Services," thesis presented to the Rice University at Houston, TX, in partial fulfillment of the requirements for the degree of Doctor of Philosophy (1975).
39. Muller, T., *Growing and Declining Urban Areas: A Fiscal Comparison,* The Urban Institute, Washington, DC (1975).
40. "Multi-Service Citizen Survey," City of Fredericton, NB (1984).
41. *Report of the National Advisory Commission on Civil Disorders,* U.S. Government Printing Office, Washington, DC (1968).
42. Rich, R. C., "Introduction," in R. C. Rich (ed.), *The Politics of Urban Public Services,* D.C. Heath, Lexington, MA (1982).
43. Sanger, M. B., "Academic Models and Public Policy: The Distribution of City Services in New York," in R. C. Rich (ed.), *The Politics of Urban Public Services,"* D.C. Heath, Lexington, MA, pp. 37-51 (1982).
44. Seale, R., "A Study of Methods Used by Four New Brunswick Municipalities for Determining Road Needs," Senior Report presented to the Univerity of New Brunswick in partial fulfillment of the requirements for the degree of Bachelor of Science in Engineering (Civil) (1981).
45. Staley, E. J., "Determining Reaction Priorities: An Instrument," *Journal of Leisure Research* (Winter 1979).
46. Statistics Canada, *Household Facilities by Income and Other Characteristics,* Catalogue 13-567, Annual, Supply and Services, Ottawa, ON.
47. Tait, J. J., *Social Equity and the Automobile,* Working Paper no. 6, Role of the Automobile Study, Transport Canada, Ottawa, ON (1979).
48. Wise, F., "Towards Equity of Results Achieved: One Approach," *Public Management, 58,* No. 6, pp. 9-12 (1976).
49. Wolch, J. R., "Spatial Consequences of Social Policy," in R. C. Rich (ed.), *The Politics of Urban Public Services,* D.C. Heath, Lexington, MA, pp. 19-35 (1982).

CHAPTER 29

Engineering Economic Evaluation

PHILIP D. CADY*

BASIC CONCEPTS

Engineering, in common with other business (as well as personal) activities, requires frequent decision-making, i.e., the choice of an action from the group of possible actions related to the problem at hand. Often, the available choices in a given instance can be expressed in monetary terms providing a rational basis for making the selection (minimum cost or maximum profit). Engineering economic evaluation, then, is a tool for making rational decisions in those engineering situations where a choice must be made from a group of alternatives whose *differences* can be expressed in monetary terms. Notice that it is only the differences among alternatives that are pertinent to any rational selection process.

Time Value of Money

The major complication in using money as the basis for rendering rational decisions is the fact that it is not a static entity. In the real world situation, money works to produce more money, i.e., the real value (purchasing power) of a given monetary sum is expected to increase with time. Therefore, monetary evaluations of alternative courses of action that extend over periods of time must take into account the dynamic growth nature of money.

INTEREST

The money that money earns is called *interest*. It is the manifestation of the time value of money. *Interest rate* is the interest earned over a given period of time—the interest period (usually one year)—expressed as a percentage of the original amount of money. The original amount of money is called the *principal*. If interest is calculated by applying the interest rate to the principal only for each interest period and then summed, it is called *simple interest*. However, as noted previously, money makes money. Therefore, the interest earned during a previous interest period itself earns money during subsequent interest periods. This is called *compound interest*. Engineering economic evaluations are based on compound interest.

NOMINAL AND EFFECTIVE INTEREST RATES

In engineering economic evaluations, interest is usually assumed to be compounded on an annual basis. However, in the business world, especially banking, interest periods shorter than a year may be used. In these instances interest may be compounded semi-annually, quarterly, monthly, weekly, daily, or even continuously, corresponding to interest periods of 6 months, 3 months, 1 month, 1 week, 1 day, or zero, respectively. The interest rate in these cases is normally given as a *nominal annual* interest rate compounded semi-annually, quarterly, or whatever. Therefore, it is necessary to convert such interest rates to *effective annual* interest rates in engineering economic evaluations. The relationship between effective and nominal interest rates (except for continuous compounding) is given by Equation (1).

$$i = \left(1 + \frac{r}{m}\right)^m - 1 \qquad (1)$$

where

i = effective annual interest rate (decimal)
r = nominal annual interest rate (decimal)
m = number of compounding periods per year

For the case of continuous compounding, the relationship becomes:

$$i = e^r - 1 \qquad (2)$$

*Department of Civil Engineering, The Pennsylvania State University, University Park, PA

where

e = base of the Napierian (natural) logarithms

Notice that the interest rates (i and r) in Equation (1) and (2) must be expressed in decimal (not percent) form.

Example 1: For a nominal annual interest rate of 5%, determine the effective annual interest rates for (a) annual, (b) monthly, and (c) continuous compounding.

(a) For annual compounding ($m = 1$ in Equation (1)), the effective and nominal interest rates are synonymous; therefore, $i = 5\%$.

(b) $i = \left(1 + \dfrac{0.5}{12}\right)^{12} - 1 = 0.05116$, or 5.116%

(c) $i = e^{0.05} - 1 = 0.05127$, or 5.127%

In certain instances it is convenient to perform engineering economic evaluations based on the shorter compounding periods. In these cases the effective interest rate *per compounding period* (not year) is equal to the nominal annual interest rate divided by the number of compounding periods per year (except for continuous compounding). Notice that in performing the economic computations it is always the effective interest rate that must be used.

Example 2: For $1,000 deposited in a bank that pays a 6% nominal annual interest rate compounded quarterly, calculate the interest earned in one year using (a) the effective quarterly interest rate, and (b) the effective annual interest rate.

(a) The effective quarterly interest rate is:

$$\dfrac{6\%}{4} = 1.5\% \text{ per quarter}$$

Interest:

1st Quarter: (1000.00)(0.015) = 15.00
2nd Quarter: (1015.00)(0.015) = 15.22
3rd Quarter: (1030.22)(0.015) = 15.45
4th Quarter: (1045.67)(0.015) = 15.69
Total Interest   = 61.36

(b) effective annual interest rate,

$$i = \left(1 + \dfrac{0.06}{4}\right)^4 - 1 = 0.06136, \text{ or } 6.316\%$$

and the total interest = (1000) (0.06136) = $61.36. Notice that the same result is obtained as long as the effective interest rate per compounding period is used.

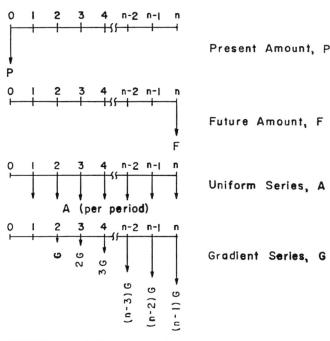

FIGURE 1. Cash flow diagrams defining the elements of cash flow.

MINIMUM ATTRACTIVE RATE OF RETURN

Minimum attractive rate of return (usually abbreviated *MARR*) is the interest rate, or earning power of money, that is generally used in engineering economic evaluations. The choice of value for *MARR* may be a management policy matter (typical in the manufacturing/industrial scenario), a matter of common usage (as in the public works sector), or it may be based on prevailing market or financial conditions. In any event, it should at least equal the "real" (inflation adjusted) interest rate for non-risk long-term investments. Furthermore, as the risk factor increases for enterprises being evaluated, *MARR* should be appropriately increased. While the choice of value for *MARR* is discretionary, it is not a matter that should be taken lightly. Decisions rendered in economic evaluations can, under certain conditions, vary with the value selected for *MARR*. In the public works area, it is common practice to use an *MARR* of 6%, this value being the approximate upper limit of the variation in the interest rate for "safe," long-term investments adjusted for the devaluating effect of inflation ("real" interest rate—typically ranges between 4% and 6%). This matter is covered in more detail under "Inflation."

Cash Flow

In order to perform engineering economic evaluations of alternative strategies, it is necessary to express all pertinent features of each of the alternatives in monetary terms. Further, the timing of the anticipated receipts and disbursements must be developed for each alternative. This process—the chronology of expected receipts and disbursements—is called cash flow.

CASH FLOW DIAGRAMS

A cash flow diagram is a time scale on which monetary quantities are presented as vectors (up for receipts; down for disbursements—by usual convention) and where time is represented in terms of compounding periods (usually annual). Each alternative in an engineering economic evaluation will be represented by a separate cash flow diagram. Cash flow diagrams should always be used in engineering economic evaluations because they greatly facilitate the organization of the data and the formulation of the solution.

ELEMENTS OF CASH FLOW

The elements of cash flow are defined as follows:

i = interest rate (in decimal form)
n = number of compounding periods
P = a present sum of money
F = a future sum of money
A = a uniform series of cash receipts or disbursements where the periodicity coincides with compounding frequency (i.e., usually annually).
G = a uniformly increasing series (gradient) of cash receipts or disbursements where the periodicity coincides with compounding frequency. The gradient series begins with a value of zero at the end of the first compounding period and increases by the amount G per compounding period.

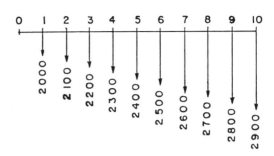

FIGURE 2. Cash flow diagram for operating and maintenance cost in Example 3.

The elements of cash flow are illustrated in cash flow diagrams on Figure 1. Notice that in this instance the cash flow elements are presented as disbursements (arrows down). The same elements may, of course, be receipts, in which case the arrows would be directed upward. It should be evident that one must adopt a consistent point of view in the matter of receipts and disbursements for each analysis, i.e., what is a disbursement by the lender is a receipt to the borrower. In terms of the results of the evaluation, it does not matter which viewpoint the analyst takes as long as he is consistent. Each alternative that is being evaluated will generally possess several cash flow elements, all of which are combined on a single cash flow diagram. Complex cash flow situations can generally be presented by combinations of the basic cash flow elements from Figure 1 using the principle of superposition.

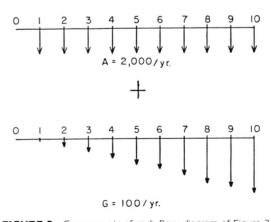

FIGURE 3. Components of cash flow diagram of Figure 2.

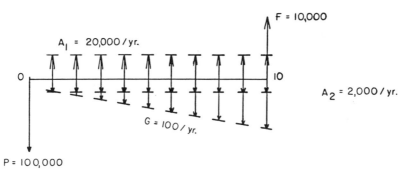

FIGURE 4. Cash flow diagram for Example 3.

Example 3: A contractor is evaluating various brands and models of equipment to perform a certain construction operation. Because of differences in construction and design, the various models have different costs and expected incomes (due to varying capacities). One of the models being considered has a first cost of $100,000 and an expected useful life of 10 years, after which it can be sold for salvage at an estimated value of $10,000. This model is expected to earn $20,000 per year for the contractor, based on its capacity. Maintenance and operating costs for this piece of equipment are estimated to be $2,000 in the first year and to increase by $100 per year thereafter due to wear.

Adopting the point of view of the contractor (which, of course, is the only one that makes sense here), this alternative involves:

1. A disbursement of $100,000 now (first cost), a *P* value
2. A receipt of $10,000 10 years in the future (salvage value), an *F* value at 10 years
3. A receipt of $20,000 per year for 10 years (income from the equipment), an *A* value over 10 years
4. Disbursements of $2,000 this year, $2,100 next year, etc., increasing by $100 per year for 10 years (operating and maintenance cost), combination of an *A* and a *G* over 10 years

The principle of superposition is illustrated by item 4, above. The cash flow diagram for this item is shown on Figure 2. This cash flow, itself, is not represented singularly by any of the cash flow elements shown on Figure 1. However, if an *A* of 2000 and a *G* of 100 are superimposed, as shown on Figure 3, the cash flow diagram for item 4 is produced. Notice that the definitions for *A* and *G*, as presented on Figure 1, are satisfied. The cash flow diagram for this alternative, then, is shown on Figure 4.

In the previous example, notice that the series elements (*A* and *G*), *by definition*, occur at the *end* of each year. This is referred to as *discrete* cash flow and it is the usual convention adopted, even though it is recognized that cash flows are generally more or less continuous throughout the year, especially for non-investment-type situations. Continuous cash flow convention may be adopted in conjunction with continuous compounding, but, in general, the added complexities are not warranted when the differences are compared with the uncertainties in predicting cash flows.

Two additional points regarding Example 3: notice that the gradient cash flow series (*G*) always starts at zero at the end of the first year. Secondly, notice that the cash flow diagram can be further simplified by applying the principle of superposition to *algebraically* combine the two uniform series (*A*-values), as shown on Figure 5.

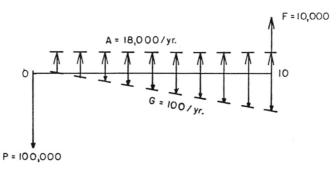

FIGURE 5. Simplified cash flow diagram for Example 3.

EQUIVALENCE

It should be evident that attempting to identify the most economical alternative simply by comparing cash flow diagrams would generally be difficult, if not impossible. In order to provide a simple, direct means of comparison, cash flows are converted to single, *equivalent* values based on the time-value of money. The cash flow elements usually used as equivalents are P ("present worth" or "present value") or A ("equivalent uniform annual amount"). To illustrate, the last cash flow diagram for Example 3 may be converted to a single A value (equivalent uniform annual amount) by adding (algebraically):

- (the uniform annual equivalent of a $100,000 present sum)
- (the uniform annual equivalent of a gradient of $100/yr for 10 years)
- $18,000 (already a uniform annual amount)
- (the uniform annual equivalent of $10,000 10 years in the future)

In the absence of time value of money (i.e., $i = 0$), the present worth of a given cash flow is simply the algebraic sum of all of the receipts and disbursements over the life of the alternative, and the equivalent uniform annual amount is the present worth divided by the life of the alternative. However, in the "real world" situation future receipts and disbursements are subject to discounting, as a reflection of the time value of money, when converting to equivalencies.

INTEREST FACTORS

Conversion of cash flow elements to equivalents is accomplished through the use of interest factors. The eight interest factors involved, along with their applications, symbols, and formulae, are presented in Table 1. Notice that the number and formats of the interest factors is such that calculation of equivalent values involves only multiplication. In order to facilitate calculations, tabled values of the interest factors are presented in the Appendix to this chapter for selected values of i and n. Notice that, due to the nonlinear nature of the interest factors, linear interpolation between tabled values will introduce errors. The errors are generally insignificant if the interpolation is done between immediate adjacent table values.

Example 4: The use of the interest factors to convert a

TABLE 1. Equivalency Conversions.

To Convert	Multiply by Interest Factor				
	To	Name	Symbol	Formula	
P	F	Single Payment Compound Amount Factor	(F/P,i,n)	$(i + 1)^n$	(3)
P	A	Capital Recovery Factor	(A/P,i,n)	$\dfrac{i(1 + i)^n}{(1 + i)^n - 1}$	(4)
F	P	Single Payment Present Worth Factor	(P/F,i,n)	$\dfrac{1}{(1 + i)^n}$	(5)
F	A	Sinking Fund Factor	(A/F,i,n)	$\dfrac{i}{(1 + i)^n - 1}$	(6)
A	P	Uniform Series Present Worth Factor	(P/A,i,n)	$\dfrac{(1 + i)^n - 1}{i(1 + i)^n}$	(7)
A	F	Uniform Series Compound Amount Factor	(F/A,i,n)	$\dfrac{(1 + i)^n - 1}{i}$	(8)
G	P	Gradient Present Worth Factor	(P/G,i,n)	$\dfrac{1}{i}\left[\dfrac{(1 + i)^n - 1}{i(1 + i)^n} - \dfrac{n}{(1 + i)^n}\right]$	(9)
G	A	Gradient Uniform Series Factor	(A/G,i,n)	$\dfrac{1}{i} - \dfrac{n}{(1 + i)^n - 1}$	(10)

cash flow to single equivalent values will be illustrated using the data from Example 3 and a minimum attractive rate of return (i) of 10%. Referring to the cash flow diagram on Figure 5, the equivalent present worth of this alternative at an *MARR* of 10% is:

$$P = -100,000 - (100) \overset{22.891}{(P/G, 10\%, 10)}$$
$$+ (18,000) \overset{6.1446}{(P/A, 10\%, 10)} + (10,000) \overset{0.3855}{(P/F, 10\%, 10)}$$
$$= +\$12,169$$

Likewise the equivalent uniform annual amount corresponding to this cash flow is:

$$A = -(100,000) \overset{0.16275}{(A/P, 10\%, 10)}$$
$$-(100) \overset{3.725}{(A/G, 10\%, 10)} + 18,000$$
$$+ (10,000) \overset{0.06275}{(A/F, 10\%, 10)} = +\$1,980/\text{yr}$$

which can also be obtained by multiplying the equivalent *P*-value (12,169) by the capital recovery factor (A/P, 10%, 10).

The positive (+) signs in Example 4 indicate net receipts, since receipts were designated positive. Most engineering applications involve comparisons of alternatives that consist entirely or primarily of disbursements. In these instances, it will be convenient to designate disbursements positive in order to reduce the number of minus signs in the calculations. Either convention is permissible provided that internal consistency is maintained.

COMPARING ALTERNATIVES

The first step in an engineering economic evaluation is to identify all technically feasible alternatives. A point of caution here: doing nothing often constitutes a technically feasible solution and, in such instances, must be included as an alternative. For instance, the alternatives for providing flood control in a river drainage basin might include a dam/reservoir, channel improvement, building levees, *and* doing nothing. The cost associated with doing nothing, in this case, is the expected average annual cost of flood damages.

After the technically feasible alternatives have been established, the various activities involved in each to which monetary values can be assigned must be identified and "costed-out," and cash flows developed. Finally, the alternatives are compared on the basis of their respective cash flows using one of four procedures: present worth, annual cost, rate of return, or benefit/cost. Each of these procedures will be discussed in the sections that follow, and the conditions that favor the use of one or another of the procedures will be presented. It is emphasized that all of the procedures will provide the same result in terms of the choice of alternative based on cost effectiveness.

Equivalent Uniform Annual Amount

GENERAL

The simplest and most widely used method for comparing alternatives in engineering economic evaluations is the equivalent uniform annual amount procedure. In this

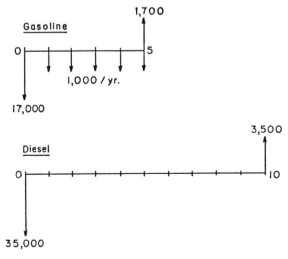

FIGURE 6. Cash flow diagrams for Example 5.

method, all elements of cash flow are converted to equivalent A values and algebraically summed for each alternative which permits ready comparison and selection on an economic basis. The major advantages of this method over the ones to be described subsequently are the ease with which it can be applied and the absence of problems from differing lengths of cash flow periods among the alternatives. These points are illustrated in Example 5.

Example 5: A contractor needs to purchase a new portable electrical generator for supplying field power requirements. A gasoline engine driven generator costs $17,000 and has an expected service life of 5 years. A diesel engine unit costs $35,000 and has an expected life of 10 years. Annual operating costs are expected to be $1,000/year more for the gasoline unit due to higher fuel and maintenance costs. Salvage values are estimated to be 10% of first costs for both machines. If the contractor's minimum attractive rate of return is 12%, which unit should he purchase?

The cash flow diagrams are presented on Figure 6.

$$EUAA \text{ (Gasoline)} = (17,000) \overset{0.27741}{(A/P, 12\%, 5)} + 1,000$$

$$- (1,700) \overset{0.15741}{(A/F, 12\%, 5)} = \$5,448/\text{yr}$$

$$EUAA \text{ (Diesel)} = (35,000) \overset{0.17698}{(A/P, 12\%, 10)}$$

$$- (3,500) \overset{0.05698}{(A/F, 12\%, 10)}$$

$$= \$5,995/\text{yr}$$

The gasoline engine driven unit has the lower equivalent uniform annual cost and, therefore, should be selected.

The life of the diesel unit is twice that of the gasoline unit. The *EUAA* for the gasoline unit over ten years (two life cycles), therefore, will be:

$$(17,000) [1 + \overset{0.5674}{(P/F, 12\%, 5)}] \overset{0.17698}{(A/P, 12\%, 10)}$$

$$- (1,700) [\overset{0.5674}{(P/F, 12\%, 5)} + \overset{0.3220}{(P/F, 12\%, 10)}]$$

$$\times \overset{0.17698}{(A/P, 12\%, 10)} + 1,000 = \$5,448/\text{yr}$$

i.e., exactly the same as the *EUAA* for 5 years. Since *EUAA* is a unit (annual) cost, it will be constant regardless of the number of life cycles. Incidently, notice that there are several ways—all equivalent and all giving the same answer (within the limits of the significant figures for the interest factors)—to perform the immediately preceding computation. For example, another way is:

$$(17,000) \overset{0.17698}{(A/P, 12\%, 10)} + 1000$$

$$+ [(17,000 - 1,700) \overset{1.7623}{(F/P, 12\%, 5)} - 1,700]$$

$$\times \overset{0.05698}{(A/F, 12\%, 10)} = \$5,448/\text{yr}$$

Notice here that the first cost for the initial cycle is converted to an equivalent A value ("capital recovery") over the entire 10 year cash flow for the two cycles. The $1,000/yr operating cost differential is already an A value over the entire cash flow period. The $17,000 initial expenditure for the second cycle and the $1,700 salvage value for the first cycle occur at the same point in time. Therefore, they are combined algebraically, converted to a future amount at the end of the 10-year cash flow, combined with the second cycle salvage value, and the sum is then converted to an equivalent A value over 10 years using the sinking fund factor.

The independence of *EUAA* from life cycles, demonstrated in Example 5, also applies to exactly repetitive cycles of elements contained within a single overall life cycle. This feature may serve to greatly reduce computational effort in many complex situations.

Example 6: What net annual income would a city have to obtain from the sale of tickets, rentals, and concessions to support a sports stadium having an initial cost of $20 million and an expected 60 year life? The stadium will have to be renovated every 15 years at a cost of $3 million. Turf replacement will have to be made every 5 years at a cost of $1 million and painting will have to be done every 3 years at a cost of $500,000. Annual operating and maintenance costs are expected to be $400,000/yr. Turf replacement and painting are not included in the renovation cost figure. The cost of money to the city is 6% per annum. The cash flow diagram (costs in millions) is presented on Figure 7. Notice that if "renovation," "turf replacement," and "painting" are considered to occur initially (i.e., are subtracted from the first cost), these three items exactly repeat their respective cycles over the entire 60 year cash flow. Therefore,

$$EUAA = (20 - 3 - 1 - 0.5) \overset{0.06188}{(A/P, 6\%, 60)}$$

$$+ (3) \overset{0.10296}{(A/P, 6\%, 15)} + (1) \overset{0.23740}{(A/P, 6\%, 5)}$$

$$+ (0.5) \overset{0.37411}{(A/P, 6\%, 3)} + 0.4$$

$$= \$2.092475 \text{ million/yr}$$

$$= \$2,092,475/\text{yr}$$

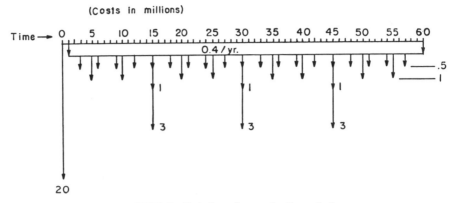

FIGURE 7. Cash flow diagram for Example 6.

It should be evident that this procedure is considerably easier than attempting to convert each of the disbursements shown on the cash flow diagram to an *A*-value over 60 years.

PERPETUAL SERVICE

Many types of civil engineering works have very long life expectancies—50, 100, or more years. Examples include highways, bridges, dams, major buildings, water supply and sewage collection systems and so forth. Since it is very difficult to predict future developments, it is often convenient to assume infinite life (perpetual service) in such instances. Table 2 shows that there are not significant differences between perpetual service and finite time spans of about 25 years for high interest rates (25%) or 100 years for low interest rates (5%).

The first question that comes to mind involves the matter of the capital recovery factor (*A/P*) for $n = \infty$ needed to convert present values to perpetual equivalent uniform annual cash flow. It can easily be demonstrated that the capital recovery factor for infinite life is equal to the interest rate (in decimal form). In mathematical terms:

$$(A/P, i, n = \infty) = i \qquad (11)$$

Example 7: In the situation described for Example 6, suppose that the land on which the stadium is to be built will have to be bought by the city at a cost of $3 million. The land is assumed to have infinite life. What, now, is the required net annual income for the project to be self-amortizing? From Example 6, *EUAA* = $2,092,475/yr not including the land.

$$\text{for the land, } EUAA = (3,000,000)\,(A/P, 6\%, \infty)$$
$$= (\$3,000,000)\,(0.06)$$
$$= \$180,000$$

Notice that the action of specifying infinite life is exactly the same thing as saying that the salvage value at any point in time equals first cost. For instance, in Example 7, it is stated that the salvage value of the land at the end of the expected 60 year stadium life is equal to its first cost; thus the *EUAA* for the land is:

$$\overset{0.06188}{(3,000,000)\,(A/P, 6\%, 60)} - (3,000,000)$$
$$\overset{0.00188}{\times\,(A/F, 6\%, 60)} = \$180,000$$

identical with the result in Example 7.

Present Worth

GENERAL

The present worth method is based on converting the cash flow for each alternative to an equivalent present value. The

TABLE 2. Percent Difference Between EUAAs of Present Values Based on Infinite Life Versus Finite Life Basis.

Finite Life (yr)	% Difference for Perpetual Service at Interest Rate i			
	i = 5%	i = 10%	i = 15%	i = 25%
25	−30.0	−9.2	−3.0	−0.4
30	−23.0	−5.7	−1.5	−0.1
40	−14.0	−2.2	−0.4	
50	−8.7	−0.8	−0.1	
80	−2.0			
100	−0.8			

economic evaluation involves the comparison of the respective equivalent present values. The major disadvantage of the present worth method is that *alternatives must be compared over the same time span*. Thus, if the cash flows for the various alternatives under consideration cover different lengths of time, the comparison must be made over the least common multiple of the time spans for the individual alternatives. In other words, varying numbers of cycles of each of the alternatives will be used depending on the respective cycle lengths.

Example 8: A classical situation in engineering economic evaluation is that of alternative equipment items to do a certain job where higher first cost alternatives have longer lives and/or lower operating/maintenance costs due to better efficiency or materials. Take, for example, the case of a recirculating pump in a water treatment plant. The engineer has a choice from among three available pumps having the proper operating characteristics. Pump A costs $8,000, has a 6 year estimated life, and is estimated to cost $1,500/yr for power and maintenance. Pumps B and C have progressively higher first costs at $12,000 and $20,000, but are expected to last longer (8 years and 12 years, respectively) and have lower power/maintenance costs ($1000/yr and $700/yr, respectively) due to higher efficiencies and better materials. Compare these pumps on a present worth basis at a minimum attractive rate of return of 10% compounded annually.

Since the present worth method is to be used, the pumps will have to be compared over a period of time evenly divisible by their respective service lives. The shortest period that satisfies this criterion is 24 years (i.e., four life cycles of pump A, three of B, or two of C). The respective cash flow diagrams are shown on Figure 8. The equivalent present worths for 24 years service are presented below. Because only disbursements are involved, sign convention shall be chosen positive for disbursements.

for A:

$$PW = (8000)\,[1 + \overset{0.5645}{(P/F, 10\%, 6)} + \overset{0.3186}{(P/F, 10\%, 12)}$$

$$+ \overset{0.1799}{(P/F, 10\%, 18)}] + (1500)\,\overset{8.9847}{(P/A, 10\%, 24)}$$

$$= \$29,981$$

for B:

$$PW = (12,000)\,[1 + \overset{0.4665}{(P/F, 10\%, 8)} + \overset{0.2176}{(P/F, 10\%, 16)}]$$

$$+ (1000)\,\overset{8.9847}{(P/A, 10\%, 24)} = \$29,194$$

for C:

$$PW = (20,000)\,[1 + \overset{0.3186}{(P/F, 10\%, 12)}]$$

$$+ (700)\,\overset{8.9847}{(P/A, 10\%, 24)} = \$32,661$$

Obviously, pump B represents the lowest cost alternative.

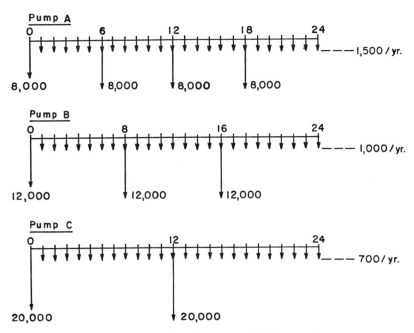

FIGURE 8. Cash flow diagram for Example 8.

Notice that if the time value of money had not been considered (i.e., $i = 0$), the present worths for 24 years service would have been:

$$(8000)(4) + (1500)(24) = 68,000 \text{ for A}$$

$$(12,000)(3) + (1000)(24) = 60,000 \text{ for B}$$

$$(20,000)(2) + (700)(24) = 56,800 \text{ for C}$$

which would have incorrectly selected pump C. Notice further that there is a simpler computational procedure that can be used when multiple life cycles occur in a single cash flow. That procedure is to convert the cash flow for one cycle to the equivalent uniform annual amount and then to convert that value to the present worth over the entire cash flow. For example, using the data for pump A in Example 5, *EUAA* for one life cycle of pump A is:

$$(8000)\overset{0.22961}{(A/P, 10\%, 6)} + 1500 = \$3,336.88$$

and

$$PW \text{ for 24 years} = (3336.88) \overset{8.9847}{(P/A, 10\%, 24)}$$

$$= \$29,981$$

which, of course, is the same result found in Example 8.

CAPITALIZED COST

Capitalized cost is the present worth of perpetual service. It is often used for comparing long lived alternatives, especially in the public works sector. As shown earlier, the ratio of a uniform annual amount in *perpetum* to its equivalent present value is equal to the interest rate. Therefore, the uniform series present worth factor for infinite time (P/A, i, $n = \infty$) must equal $1/i$ (i expressed in decimal form).

Example 9: One of several types of bridges that could be built at a certain location on a new highway project has a first cost of $250,000, an expected life of 50 years, and a negative salvage value of $50,000 (cost of removing the bridge when it is replaced by a new one). Annual maintenance costs are expected to average $500 per year. Annual repair costs are expected to be nil during the first ten years and $400 in the eleventh year, increasing by $100/yr thereafter. Since the bridge site (the road) is expected to be a long term installation, the alternatives are to be evaluated on the basis of capitalized cost. What is the capitalized cost for this alternative if $i = 6\%$? The cash flow diagram is presented in Figure 9.

$$\text{Capitalized Cost} = \overset{\frac{1}{0.06}}{(P/A, 6\%, \infty)}$$

$$\times \{[(250,000) \overset{0.06344}{(A/P, 6\%, 50)} + 500$$

$$+ (50,000) \overset{0.00344}{(A/F, 6\%, 50)}]$$

$$+ [(400) \overset{15.0463}{(P/A, 6\%, 40)}$$

$$+ (100) \overset{185.947}{(P/G, 6\%, 40)}]$$

$$\times \overset{0.5584}{(P/F, 6\%, 10)} \overset{0.06344}{(A/P, 6\%, 50)}\}$$

$$= \$290,066$$

Notice that the procedure first involved the determination of the equivalent uniform annual amount for the 50 year cycle — the computations between the brackets { }. As described in the preceding section, *EUAA* is a unit value that applies to any number of life cycles, including an infinite

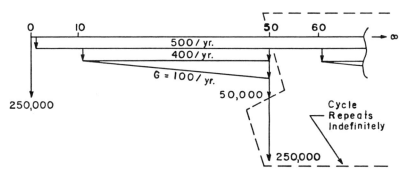

FIGURE 9. Cash flow diagram for Example 9.

number. Finally, the capitalized cost—the present worth of perpetual service—is determined by dividing the *EUAA* by *i* because the value of the appropriate interest factor, (*P/A, i, n* = ∞), is equal to 1/*i*. The physical significance of the $290,066 figure is that if this amount of money is in hand today, the bridge could be built for $250,000 and the interest earned on the remaining $40,066 at 6% compounded annually would be sufficient to cover all future costs (including bridge replacement every 50 years) *forever*. The effects of inflation on this situation are covered in a later section.

PROPERTY INVESTMENT VALUATION

The present worth method is most advantageous in those situations involving evaluation of a single investment opportunity where the alternative is not to invest.

Example 10: A certain piece of property is offered for sale at $100,000. It is estimated that in five years the property can be resold for $150,000. The property is expected to produce $10,000 per year in rental income. Annual taxes and maintenance costs are estimated to be $2,000 per year. If an investor must make at least 20% per year on his investments, should he purchase this property?

The cash flow diagram for this case is presented on Figure 10. Writing an equation to represent this cash flow diagram is analogous to summing the vertical loads on a beam, i.e.,

$$-P - (2{,}000)\,\overset{2.9906}{(P/A,\,20\%,\,5)}$$

$$+ (10{,}000)\,\overset{2.9906}{(P/A,\,20\%,\,5)}$$

$$+ (150{,}000)\,\overset{0.4019}{(P/F,\,20\%,\,5)}$$

from which, $P = \$84{,}210$.

This value ($84,210) represents the maximum amount that the investor can pay if he is to obtain a 20% return on his investment. Since the price of the property is $100,000, he should not purchase it.

One method for evaluating investments in municipal bonds (discussed in a later section) follows a similar procedure. The investment opportunity situation is also often evaluated, though with more difficulty, using the rate of return method discussed in the next section.

Rate of Return

The rate of return method of economic analysis is not commonly used in engineering. The primary reason for this is the rather considerably higher level of computational effort required for a given situation, compared with equivalent uniform annual amount (especially) or with present worth. Specifically, trial and error solutions are usually in-

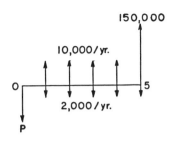

FIGURE 10. Cash flow diagram for Example 10.

volved, interpolation of the tabled interest factors is nearly always needed, and special procedures are required for cases involving multiple alternatives. Aside from the mechanistic disadvantages of the rate of return method, it is psychologically ill-suited to engineering applications. Where the *EUAA* and *PW* methods compare equivalent costs based on interest rates (or *MARR*s) that have been established by others as matters of policy, the rate of return method calculates rates of return which must be judged acceptable or not. Therefore, the rate of return method is more aligned with the activities of the world of business and finance, and, indeed that is where it is most commonly encountered.

SINGLE INVESTMENT

Rate of return on a single investment may be expressed in three mathematically equivalent ways:

(a) The interest rate at which the *PW* or *EUAA* of the cash flow equals zero.
(b) The interest rate at which the *PW* of future cash flow equals the initial investment.
(c) The interest rate at which the *PW* or *EUAA* of all receipts equals the *PW* or *EUAA*, respectively, of all disbursements.

An example of a single investment rate of return analysis, using definition (a) and the data of Example 10, is given in Example 11.

Example 11: Determine the rate of return in Example 10.

$$-100{,}000 - (2{,}000)\,(P/A,\,i,\,5) + (10{,}000)\,(P/A,\,i,\,5)$$

$$+ (150{,}000)\,(P/F,\,i,\,5) = 0$$

which can be simplified to:

$$100{,}000 + (8{,}000)\,(P/A,\,i,\,5) + (150{,}000)\,(P/F,\,i,\,5) = 0$$

Thus, we have an equation in one unknown, *i* (the rate of return), but it is an equation that must be solved by trial and error procedures.

Assume $i = 15\%$:

$$-100{,}000 + (8{,}000)\,\overset{3.3522}{(P/A,\ 15\%,\ 5)}$$

$$+ (150{,}000)\,\overset{0.4972}{(P/F,\ 15\%,\ 5)} \overset{?}{=} 0$$

$$+\ \$1397.60 \ne 0$$

Therefore, i is not 15%. However, since both P/A and P/F decrease with increasing values of i, i must be larger than 15%.

Assume $i = 20\%$:

$$-100{,}000 + (8{,}000)\,\overset{2.9906}{(P/A,\ 20\%,\ 5)}$$

$$+ (150{,}000)\,\overset{0.4019}{(P/F,\ 20\%,\ 5)} \overset{?}{=} 0$$

$$+\ 15{,}790.20 \ne 0$$

So, i is also not 20%. However, the sign change indicates that i must lie between these two interest rates. Therefore, by linear interpolation:

$$i = 15\% + (5\%)\,\frac{1{,}397.6}{1{,}397.6 + 15{,}790.2}$$

$$= 15.4\%$$

As stated in Example 10, the minimum attractive rate of return is 20%. Since the rate of return (15.4%) < MARR (20%), the investment should not be made. Notice that this is the same decision rendered in Example 10 (but with much greater effort!).

Since rate of return analysis almost invariably involves linear interpolation of nonlinear functions, two precautions must be observed:

(a) Interpolate only between *successive* interest tables.
(b) Do not calculate the rate of return beyond the first decimal place.

Notice that had the equation in Example 11 been set up according to definitions (b) or (c) (using *PW*), exactly the same relationship would have occurred. Had *EUAA* been used rather than *PW*, the equation would have been different, but the results the same (within the limits of the errors introduced by linear interpolation).

MULTIPLE ALTERNATIVES

Economic evaluations of situations involving multiple alternatives fall into one or the other of two categories:

(a) Mutually exclusive—when one alternative is selected, the others must be discarded.
(b) Capital budgeting—when the alternatives are independent and any combination of them may be selected, limited by the available investment budget.

Engineering economics evaluations invariably fall into category (a). Therefore, category (b) will not be discussed here. Using the *EUAA* and *PW* methods, any number of mutually exclusive alternatives can be evaluated simply by comparing the *EUAA* or *PW* values for each. However, this is not so for the rate of return method. It is *not* sufficient to merely compare the rates of return for the individual alternatives. The rates of return on the *incremental* investments between the alternatives must also be evaluated.

Alternatives Consisting Primarily of Disbursements

Most engineering economic evaluations involve mutually exclusive alternatives consisting primarily or entirely of disbursements, i.e., cost minimization situations. In these cases the internal rates of return for the individual alternatives are undefined, and the choice is based on the rate of return on the incremental first costs due to reduced incremental future costs.

Example 12: Solve the problem of Example 8 by the rate of return method. The minimum attractive rate of return is 10%. Notice that all three alternatives consist entirely of disbursements and that the object is to pick the lowest cost alternative. There is no way that rates of return can be calculated for each alternative individually, and even if some receipts were present to permit this, the values would be meaningless.

To begin, one alternative must be selected, so the one with the lowest first cost would be taken at the very least. The first step will be to compare the one with the lowest first cost (pump A) with the one having the next higher first cost (pump B). The question is: does the incremental (extra) first cost of pump B result in sufficiently reduced power and maintenance costs in the future to warrant the selection of pump B over pump A? In mathematical form:

B over A

$$[(12{,}000)\,(A/P,\ i,\ 8) - (8{,}000)\,(A/P,\ i,\ 6)]$$

$$+ [1{,}000 - 1{,}500] = 0$$

which can be simplified to:

$$24\,(A/P,\ i,\ 8) - 16\,(A/P,\ i,\ 6) - 1 = 0$$

try $i = 10\%$

$$24\,\overset{0.18744}{(A/P,\ 10\%,\ 8)} - 16\,\overset{0.22961}{(A/P,\ 10\%,\ 6)} - 1 \overset{?}{=} 0$$

$$-\ 0.1752 \ne 0$$

try $i = 12\%$

$$24 \overset{0.20130}{(A/P, 12\%, 8)} - 16 \overset{0.24323}{(A/P, 12\%, 6)} - 1 \overset{?}{=} 0$$

$$- 0.06048 \neq 0$$

But, it is obvious that the rate of return $(i) > 10\%$. Therefore, pump B is better than pump A (economically). The actual value of the rate of return on the incremental investment is not of importance—it only matters whether i is greater or less than $MARR$. If $i = MARR$ the alternatives are equal. Now, pump B has to be compared with the highest first cost alternative, pump C.

C over B

$[(20,000) (A/P, i, 10) - (12,000) (A/P, i, 8)]$

$$+ [700 - 1,000] = 0$$

$$200 (A/P, i, 10) - 120 (A/P, i, 8) - 3 = 0$$

try $i = 10\%$

$$(200) \overset{0.16275}{(A/P, 10\%, 10)} - (120) \overset{0.18744}{(A/P, 10\%, 8)} - 3 \overset{?}{=} 0$$

$$+ 7.0572 \neq 0$$

try $i = 12\%$

$$(200) \overset{0.17698}{(A/P, 12\%, 10)} - (120) \overset{0.20130}{(A/P, 12\%, 8)} - 3 \overset{?}{=} 0$$

$$+ 8.2400 \neq 0$$

Obviously, $i < 10\%$ and therefore, C is *not* better than B. The choice then is pump B, the same conclusion reached in Example 8.

Investment Scenarios

In the investment situation involving mutually exclusive alternatives, the objective is to pick the one that maximizes net receipts. This is not necessarily the same as the one that shows the highest internal *rate* of return on the investment. The internal rate of return on each investment must be determined for investment scenarios, but, as with the case of minimizing cost, the rate of return on the incremental investment must also be determined in order to assure the proper choice of alternative.

Example 13: Evaluate the following mutually exclusive alternatives using the rate of return method. The minimum attractive rate of return is 25%. All alternatives have an expected life of 10 years.

Proposal	Required Investment, P	Annual Savings, A
A	13,000	4,200
B	25,000	7,600
C	8,000	2,500
D	18,000	6,000
E	12,000	4,100
F	6,000	1,500

The general cash flow diagram that represents each alternative is shown on Figure 11.

The general equation for this cash flow diagram is:

$$-P + A (P/A, i, 10) = 0 \text{ or } P/A = (P/A, i, 10)$$

Notice that, with this simple cash flow situation, trial and error will not be necessary. Also, notice that with $MARR = 25\%$ the critical value of P/A, i.e., $(P/A, 25\%, 10)$, is 3.5705. Further examination of the tabled interest factors reveals that for $i < 25\%$, $(P/A) > 3.5705$.

The proposals are arranged in order of increasing capital investment (P) and the overall and incremental rates of return are determined in Table 3. The final decision is the one that appears in the last line, i.e., select proposal D. There are some important points illustrated in Table 3.

(a) Any alternative for which the overall rate of return is less than $MARR$ (e.g., Proposal F) is summarily dismissed.
(b) The highest overall internal rate of return (the alternative having the smallest value of P/A in Example 13, i.e., Proposal E) is not necessarily the most economical choice.
(c) Higher first cost proposals will be economically preferable as long as the *incremental* rate of return on the extra investment exceeds the $MARR$.

The selection of D as the economical choice can be verified by comparing the six alternatives at an interest rate equal to $MARR$, using either the $EUAA$ or PW methods. This is done in Table 4 using PW, and shows that Proposal D possesses the highest net worth (profit).

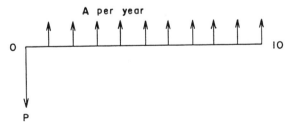

FIGURE 11. General cash flow diagram for all alternatives in Example 13.

TABLE 3. Overall and Incremental Rates of Return for Example 13.

Proposal	P	A	P/A	i	Incremental ΔP/ΔA	Return	Decision
F	6,000	1,500	4.0000	<25%	—	—	Drop F
C	8,000	2,500	3.2000	>25%	—	—	Keep C
E	12,000	4,100	2.9268	>25%	$\frac{12,000 - 8,000}{4,100 - 2,500} = 2.500$	>25%	Drop C; Keep E
A	13,000	4,200	3.0952	>25%	$\frac{13,000 - 12,000}{4,200 - 4,100} = 10,000$	<25%	Drop A; Keep E
D	18,000	6,000	3.0000	>25%	$\frac{18,000 - 12,000}{6,000 - 4,100} = 3.1579$	>25%	Drop E; Keep D
B	25,000	7,600	3.2859	>25%	$\frac{25,000 - 18,000}{7,600 - 6,000} = 4.3750$	<25%	Drop B; Keep D

Benefit/Cost Ratio

The benefit/cost ratio method of evaluating alternatives is based on the proposition that a project, in order to be acceptable, must show benefits that equal or exceed the costs associated with the project, i.e.,

$$B > C \quad (12)$$

where: B = benefits
C = costs

This is the same thing as:

$$B - C \geq 0 \quad (13)$$

or

$$B/C \geq 1 \quad (14)$$

TABLE 4. Present Worths of Proposals in Example 13 at a Minimum Attractive Rate of Return of 25%.

Proposal	PW of Net Cash Flow at i = 25%; (P/A, 25%, 10) = 3.5705
A	$-13,000 + (3.5705)(4,200) = +1,996$
B	$-25,000 + (3.5705)(7,600) = +2,136$
C	$-8,000 + (3.5705)(2,500) = +926$
D	$-18,000 + (3.5705)(6,000) = +3,423$
E	$-12,000 + (3.5705)(4,100) = +2,639$
F	$-6,000 + (3.5705)(1,500) = -644$

Any one of the three approaches defined by Equations (12), (13), and (14), respectively, can be employed in the analysis procedure, but common practice is to use the benefit/cost ratio [Equation (14)].

This method is commonly encountered in the public works area due to its appeal to the layman ("benefits" and "costs" are more meaningful to the lay public than are "present worth," "capitalized cost," etc.). However, it possesses some computational problems in common with the rate of return method.

SINGLE INVESTMENT

It is unusual to find B/C ratio used for evaluating investment-type situations. However, Example 14 shows how such a situation would be handled.

Example 14: Using the data of Example 10, evaluate the potential investment on the basis of benefit/cost ratio.

All receipts will be considered benefits and all disbursements costs.

$$PW \text{ Benefits} = (10,000) \overset{2.9906}{(P/A, 20\%, 5)}$$
$$+ (150,000) \overset{0.4019}{(P/F, 20\%, 5)}$$

$$= \$90,191$$

$$PW \text{ Costs} = 100,00 + (2,000) \overset{2.9906}{(P/A, 20\%, 5)}$$

$$= \$105,981$$

$$B/C = \frac{90{,}191}{105{,}981} = 0.85$$

0.85 < 1, therefore, do not buy the property.

MULTIPLE ALTERNATIVES

A typical situation in which benefit/cost procedures might be employed is multipurpose dam and reservoir construction. Almost invariably, in instances such as this, doing nothing constitutes one of the alternatives that have to be considered. Like rate of return analysis, benefit/cost evaluation of multiple alternatives *requires* the investigation of incremental benefit/cost values among alternatives.

Example 15: A certain stretch of river is prone to flooding. Several projects are being considered to remedy this situation and to provide additional benefits. The available alternatives are summarized below ($i = 6\%$):

Alternate A: Do nothing. Average annual flood damage = $500,000/yr

Alternate B: Channel improvement and levee construction. Cost $2,000,000; life 25 years; average annual flood damage = $200,000/yr; annual operating and maintenance cost = $25,000/yr

Alternate C: Low earth-fill dam for flood control and recreational reservoir. Cost $20,000,000; life 50 years; income from tourist trade $2,000,000/yr; flood damage = $0/yr; annual operating and maintenance cost = $200,000/yr

Alternate D: High concrete gravity dam for flood control, power generation, and recreational reservoir. Cost of dam $50,000,000 (100 year life); cost of power generating equipment $2,000,000 (20 year life); annual operating and maintenance costs = $500,000/yr; income from tourist trade = $2,000,000/yr; income from power generation = $2,000,000/yr; and annual flood damage = $0/yr

At the very least, nothing will be done (Alternate A). Therefore, the first incremental comparison is between A and B (B having the lowest capital investment).

B vs. A
Benefit:

savings in annual flood damage = 500,000 − 200,000
= $300,000/yr

Cost:

$$\overset{0.07823}{(2{,}000{,}000)\ (A/P,\ 6\%,\ 25)} + \$25{,}000 = \$181{,}460/\text{yr}$$

$$B/C = \frac{300{,}000}{181{,}460} = 1.65$$

1.65 > 1, therefore, drop A, keep B.

C vs. B
Benefit:

flood control: 200,000 − 0 = 200,000/yr
tourist income: = 2,000,000/yr
                                          ─────────────
                                          2,200,000/yr

Costs:

$$\text{Capital} = \overset{0.06344}{(20{,}000{,}000)\ (A/P,\ 6\%,\ 50)}$$

$$- \overset{.07823}{(2{,}000{,}000)\ (A/P,\ 6\%,\ 25)} = 1{,}112{,}340/\text{yr}$$

Oper. & Maint.: 200,000 − 25,000 = $\dfrac{175{,}000/\text{yr}}{1{,}287{,}340/\text{yr}}$

$$B/C = \frac{2{,}200{,}000}{1{,}298{,}340} = 1.71$$

1.71 > 1, therefore, drop B, keep C.

D vs. C
Benefits:

flood control: 0 − 0   = 0/yr
tourist income: 2,000,000 − 2,000,000 = 0/yr
power generation: 2,000,000 − 0   = 2,000,000/yr
                                         ─────────────
                                         2,000,000/yr

Costs:

$$\text{Capital (Dam)} = \overset{0.06018}{(50{,}000{,}000)\ (A/P,\ 6\%,\ 100)}$$

$$- \overset{0.06344}{(20{,}000{,}000)\ (A/P,\ 6\%,\ 50)} = 1{,}740{,}200$$

Capital (Power) = $\overset{.08718}{(2{,}000{,}000)(A/P,\ 6\%,\ 20)}$ = 174,360
Oper. & Maint. = 500,000 − 200,000 = 300,000
                                      ─────────
                                      2,214,560

$$B/C = \frac{2{,}000{,}000}{2{,}214{,}560} = 0.90$$

0.90 < 1, therefore, drop D, keep C.
And so, the decision is to go with the earth-fill dam.

DEPRECIATION AND TAXES

Many civil engineering projects are public works which are non-income producing, or if income producing, are exempt from federal and state income taxes. However, for those civil engineering applications in the private sector that generate income, income tax is a disbursement item that must be considered in the cash flows of the alternatives being compared. Further, income tax liability is reduced by the depreciation of capital equipment and facilities.

Depreciation

DEFINITIONS

Depreciation is the decrease in the value of equipment or facilities (assets) through wear, deterioration, or obsolescence. *Depreciation charge* is the assigned depreciation in the value of an asset in a given year. *Book value* is the first cost of an asset minus the accumulated depreciation charges at any given point in time in the life of the asset. Notice that the book value is a figure used in depreciation accounting for tax purposes and generally bears no relationship to the *market value* of the asset.

The nomenclature that will be used in this section is as follows:

D_m = depreciation charge in year m
P = first cost of asset
S = salvage value of asset
n = expected life of asset
B_m = book value after m years

METHODS OF DEPRECIATION

The appropriate method(s) of depreciation accounting are dictated by the Internal Revenue Service in accordance with applicable provisions of the Internal Revenue Code as established by Acts of the U.S. Congress. Accordingly, the methods change periodically and IRS Publication 534 – *Depreciation*, revised annually, should be consulted for specifics.

Accelerated Cost Recovery System (ACRS) Method

The *ACRS* method of depreciation accounting was established under changes in the tax code brought about by the Economic Recovery Tax Act of 1981. It involves the use of tabulated annual rates of depreciation published by the IRS for various classes of assets. Since legislated changes to the *ACRS* have occurred virtually every year since its inception, the annual depreciation charge rates will not be duplicated here. The use of this method merely involves the multiplication of the first cost of the asset by the prescribed annual depreciation charge rate for the year in the life of the asset in order to obtain the depreciation charge for that year. Asset life, expressed as "recovery period," is specified for classes of assets and salvage value is ignored in the *ACRS* method.

Three classical methods of depreciation accounting that have, at various times, been allowable under IRS regulations are presented in the following sections. They are the straight-line method, the declining-balance method, and the sum-of-the-year-digits method.

Straight-Line Method

In the straight-line method, the book value of an asset decreases linearly, from first cost to salvage value, throughout the life of the asset. Therefore, the annual depreciation charge is constant. In mathematical terms: the annual depreciation charge,

$$D_m = \frac{P - S}{n} \tag{15}$$

and the book value at the end of m years,

$$B_m = P - mD_m \tag{16}$$

Declining-Balance Method

The declining balance method is an accelerated depreciation procedure (depreciation greatest in early years) which, through greater reductions in tax liability in the early years of asset life, is economically more favorable than the straight-line method. In this method a *depreciation rate, d,* must be specified. Usually, the depreciation rate used is:

$$d = \frac{2}{n} \tag{17}$$

in which case the method is called the *double-rate-declining-balance* method. The depreciation charge in year m is:

$$D_m = Pd(1 - d)^{m-1} \tag{18}$$

and the book value at the end of m years is:

$$B_m = P(1 - d)^m \tag{19}$$

Notice that the salvage value does not enter into the calculation of the annual depreciation charge in this method. Therefore, since the book value is not allowed to fall below the salvage value, the annual depreciation charge may be zero in the later years of the asset life.

Sum-of-the-Year-Digits Method

The sum-of-the-years-digits method is another accelerated depreciation procedure. In using this method it is first necessary to determine the value of a parameter called the sum-of-the-years-digits, Y.

$$Y = \frac{n(n + 1)}{2} \tag{20}$$

Then, the annual depreciation charge,

$$D_m = \frac{n - m + 1}{Y}(P - S) \qquad (21)$$

and the book value at the end of m years,

$$B_m = P - \frac{m(n - m/2 + 0.5)}{Y}(P - S) \qquad (22)$$

Example 16: For an asset having a first cost of $10,000 and a salvage value of $2,000 at the end of a 5-year depreciation life, determine the annual depreciation charge and book value for each year of the depreciation life.

Straight-Line Method

$$D_m = \frac{P - S}{n} \qquad B_m = P - mD_m$$

m		
0	—	10,000
1	1,600	8,400
2	1,600	6,800
3	1,600	5,200
4	1,600	3,600
5	1,600	2,000

Double-Rate-Declining-Balance Method

$$d = \frac{2}{5} = 0.4$$

$$D_m = (10{,}000)(0.4)(1 - 0.4)^{m-1} = 4{,}000\,(0.6)^{m-1}$$

$$B_m = (10{,}000)(1 - 0.4)^m = 10{,}000\,(0.6)^m$$

m	$D_m =$ 4,000 $(0.6)^{m-1}$	$B_m =$ 10,000 $(0.6)^m$
0	—	10,000
1	4,000	6,000
2	2,400	3,600
3	1,440	2,160
4	160*	2,000*
5	0*	2,000*

*Book value cannot be less than salvage value.

Sum-of-the-Years-Digits Method

$$Y = \frac{5(5 + 1)}{2} = 15$$

$$D_m = \frac{5 - m + 1}{15}(10{,}000 - 2{,}000)$$

$$= (0.4 - m/15)(8{,}000)$$

$$B_m = 10{,}000 - \frac{m(5 - 0.5m + 0.5)}{15}(10{,}000 - 2{,}000)$$

$$= 1{,}000\,[10 - m(1.1 - 0.1\,m)(8/3)]$$

m	D_m	B_m
0	—	10,000
1	2,667	7,333
2	2,133	5,200
3	1,600	3,600
4	1,067	2,533
5	533	2,000

Taxes

Income taxes are disbursements. Therefore, alternatives in engineering economic evaluations which produce taxable income are liable for income tax disbursements which become part of the respective cash flows. Taxable income equals gross annual income minus annual expenses and annual depreciation charges. The tax for each year of cash flow, then, is equal to the taxable income times the tax rate.

Corporate income taxes are levied at the federal and state (and sometimes local) levels. Usually, for engineering economic analysis, the incremental rate of 46% for federal corporate income taxes on taxable incomes over $100,000 is used. Combining state and local income taxes, which are typically about 8% (total) and which are deductible from federal taxes, an *effective tax rate* of about 50 is obtained. The formula for effective tax rate is:

$$T_e = T_{s\&l} + T_f - (T_{s\&l})(T_f) \qquad (23)$$

where

T_e = effective tax rate (decimal)
$T_{s\&l}$ = state and local tax rate, combined (decimal)
T_f = federal tax rate (decimal)

Example 17: A certain piece of equipment having the first cost, salvage value, and life given in Example 16 is expected to produce income of $5,000/yr. Operating and maintenance expenses are estimated to be $1,000/yr. The MARR (after taxes) is 6% and the effective tax rate is 50%. The equipment is depreciated using the double-rate-declining-balance method. Calculate the after-tax present worth of this proposed investment.

From Example 16:

$$P = \$10{,}000$$
$$S = \$\,2{,}000$$
$$n = 5 \text{ years}$$

The cash flow is calculated in Table 5 and presented in diagram form on Figure 12.

$$PW \text{ (after taxes)} = (-10,000) + (4,000) \overset{0.9434}{(P/F, 6\%, 1)}$$

$$+ (3,200) \overset{0.8900}{(P/F, 6\%, 2)}$$

$$+ (2,720) \overset{0.8396}{(P/F, 6\%, 3)}$$

$$+ (2,080) \overset{0.7921}{(P/F, 6\%, 4)}$$

$$+ (4,000) \overset{0.7473}{(P/F, 6\%, 5)} = +\$3,542$$

Income and income tax liability are affected by factors other than depreciation. For example, net capital gains (capital gains minus capital losses) are taxed and should, therefore, be considered in after-tax evaluations. A capital gain (or loss) is the net selling price of an asset minus the book value of the asset at the time of sale. Also, current federal tax laws (1984) allow tax credits for investments in certain qualifying types of property, which reduces tax liability. Finally, taxes based on a percentage or millage of assessed values of property may be levied at local governmental levels. All such effects should be included in after-tax evaluations. Detailed discussion of these considerations is omitted here due to space limitations and the continually changing nature of the tax laws.

INFLATION

Until recent years, the effects of inflation were generally ignored in engineering economic evaluations. The primary reason for this was that annual inflation rates between the end of World War II and 1970 were generally quite low and money was considered as a fixed measure of worth. Furthermore, it was generally held that inflation affects all aspects of cash flow in similar fashions so that the net effect of inflation is negligible in the engineering economic comparison of alternatives. The latter point is generally true, but *only if cash flows and interest rates are consistently expressed in either fixed value or actual value terms.* "Fixed value" means constant purchasing power (i.e., corrected for inflation). "Actual" means then-current, or inflated, values. Furthermore, there are circumstances in which various elements of cash flow are affected in different ways by inflation and appropriate steps must be taken to avoid fallacious results.

Nature of Inflation

Examination of any of the several indices of price and cost variations over an extended period of time reveals that inflation is a geometric progression of the form:

$$I = B(1 + f)^m \qquad (24)$$

where

I = Cost (or price) index m years after the base index value of B
f = inflation rate (in decimal form)

Notice that Equation (24) is exactly analogous to the single payment compound amount relationship in compound interest calculations, Equation (3). Therefore, inflation rate is comparable to compound interest rate.

General Procedure

Handling the inflation situation in engineering economic evaluation, as emphasized earlier, *requires* that the cash flows be expressed entirely in either fixed value or actual value terms and that the interest rate appropriate to the respective expression of the cash flow be used in either case.

FIXED VALUE CONVENTION

The fixed value convention is the one most commonly used in engineering economic evaluations because the alternatives undergoing evaluation usually consist of cash flows based on today's costs. Even subsequent cycles of the same

TABLE 5. Calculation of Cash Flow for Example 17.

Year	Capital	Gross Income	Expenses	Depreciation Charges*	Taxable Income	Taxes @ 50%	Net Cash Flow
0	−10,000	—	—	—	—	—	−10,000
1		+5,000	−1,000	−4,000	0	0	+4,000
2		+5,000	−1,000	−2,400	+1,600	−800	+3,200
3		+5,000	−1,000	−1,440	+2,560	−1,280	+2,720
4		+5,000	−1,000	−160	+3,840	−1,920	+2,080
5	+2,000	+5,000	−1,000	0	+4,000	−2,000	+4,000

*See Example 16

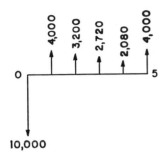

FIGURE 12. Cash flow diagram for Example 17.

cash flow within a given planning horizon or the repetition of an element within a given cash flow is usually presented as a constant value. Under these conditions and with inflation affecting the various elements within each cash flow being compared in the same manner, the usual methods of engineering economic evaluation apply. However, the interest rate used must be one that has been adjusted to remove the effects of inflation. This is called the *real interest rate*, i_r, (also, sometimes called inflation-free or constant dollar interest rate). During periods of inflation, the real interest rate will always be less than the prevailing interest rate, i_f (also called minimum attractive rate of return, or inflated, market, combined, current dollar, or actual interest rate). With inflation rate f, real interest rate can be calculated from the following equation (interest rates and inflation rate expressed in decimal form):

$$i_r = \frac{i_f - f}{1 + f} \quad (25)$$

Notice that when the inflation rate is zero, the real interest rate (i_r) is equal to the prevailing interest rate (i_f). Notice further that for most practical situations, the following approximation is sufficiently accurate:

$$i_r = i_f - f \quad (26)$$

As noted above, most engineering economic evaluations involve cash flows expressed in fixed dollars. Under inflationary conditions, the most common error in this situation involves the use of the prevailing interest rate (minimum attractive rate of return), while the real interest rate should be used. This can lead to erroneous decisions, as shown in Example 18.

Example 18: An engineer is evaluating the procurement of a piece of equipment needed for certain field operations. A new item can be purchased for $18,000 and will have an expected salvage value of $5,000 at the end of its useful life of 10 years. Operating and maintenance costs will be $850/yr. A used, but reconditioned item of the same equipment hav-

ing an estimated life of 5 years and $2,000 salvage value can be purchased for $10,000. The annual O&M cost for the used equipment is $1,000/yr. All costs are in today's dollars. The prevailing interest rate (*MARR*) is 10% and the inflation rate over the next 10 years is expected to average 5%. Compare the alternatives.

Cash flow diagrams for the two alternatives appear on Figure 13. Because the costs are all in constant (today's) dollars and inflation is expected to persist over the planning horizon, the correct interest rate to use in this case is the real interest rate, i_r.

$$i_r = \frac{i_f - f}{1 + f} = \frac{0.10 - 0.05}{1 + 0.05} = 0.0476 \ (4.76\%)$$

New:

$$\text{EUAA} = (18,000) \ \overset{0.12800}{(A/P, \ 4.76\%, \ 10)}$$

$$- (5,000) \ \overset{0.08040}{(A/F, \ 4.76\%, \ 10)} + 850 = \$2,752/\text{yr}$$

Used:

$$\text{EUAA} = (10,000) \ (A/P, \ 4.76\%, \ 5)$$

$$- (2,000) \ (A/F, \ 4.76\%, \ 5) + 1,000 = \$2,931/\text{yr}$$

Since the new machine has the lower equivalent annual cost, it should be selected.

If the *MARR* had been used as the interest rate (i.e., ignoring the effect of inflation):

New:

$$\text{EUAA} = (18,000) \ \overset{0.16275}{(A/P, \ 10\%, \ 10)}$$

$$- (5,000) \ \overset{0.06275}{(A/F, \ 10\%, \ 10)} + 850 = \$3,466/\text{yr}$$

Used:

$$\text{EUAA} = (10,000) \ \overset{0.26380}{(A/P, \ 10\%, \ 5)}$$

$$- (2,000) \ \overset{0.16380}{(A/F, \ 10\%, \ 5)} + 1,000 = \$3,310/\text{yr}$$

which indicates that the used machine has the lower equivalent uniform annual cost—an incorrect conclusion.

ACTUAL VALUE CONVENTION

Actual value convention is used when future receipts and disbursements are all expressed as actual (inflated) values. This situation is generally more common in investment evaluations than in engineering. The appropriate interest rate

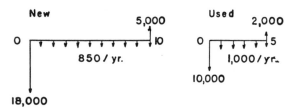

FIGURE 13. Cash flow diagrams for Example 18.

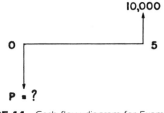

FIGURE 14. Cash flow diagram for Example 19.

here is the *prevailing* interest rate, i_f. The prevailing interest rate (or minimum attractive rate of return) is the most commonly expressed interest rate in all situations. Its relationship to the real interest rate and inflation can be obtained by solving Equation (25) for i_f:

$$i_f = i_r + f + i_r(f) \quad (27)$$

(all rates in decimal form)

Or, approximately:

$$i_f = i_r + f \quad (28)$$

Example 19: A promissory note for $10,000 due in 5 years is available to an investor. What is the maximum amount that he can pay for the note if he wishes to earn 6%, discounting the effect of a 5% inflation rate?

The cash flow diagram for this situation is shown on Figure 14.

The $10,000 figure is an "actual" value not changed by inflation. Therefore, the proper interest rate to use in this case is the prevailing interest rate, i_f. Since the investor wishes to earn 6% discounting inflation, this is a real interest rate, and

the "prevailing," or inflated, interest rate, i_f:

$$= i_r + f + i_r(f)$$

$$= 0.06 + 0.05 + (0.06)(0.05) = 0.113 \; (11.3\%)$$

and

$$P = (10{,}000) \, (P/F, 11.3\%, 5) = \$5{,}855$$
$$\phantom{P = (10{,}000) \,} ^{0.5855}$$

COMPARISON OF THE CONVENTIONS

Because of the exact analogy between inflation rate and compound interest, the relationship between fixed and actual future cash flow is as follows:

$$(\text{fixed value})(1 + f)^n = (\text{actual value}) \quad (29)$$

Or,

$$\text{fixed value} = (P/F, f, n)(\text{actual value}) \quad (30)$$

And,

$$\text{actual value} = (F/P, f, n)(\text{fixed value}) \quad (31)$$

Example 19, for instance, could have been solved by con-

TABLE 6. PW of Cash Flow for Example 20 (Using Fixed Values).

| End of Year | Capital (fixed) | Income (fixed) | Depreciation | | Income Tax (fixed)[b] | PW[c] |
			(actual)	(fixed)[a]		
0	−10,000	—	—	—	—	−10,000
1	0	+4,000	+1,600	+1,509	−1,245	+2,539
2	0	+4,000	+1,600	+1,424	−1,288	+2,304
3	0	+4,000	+1,600	+1,343	−1,329	+2,092
4	0	+4,000	+1,600	+1,267	−1,367	+1,901
5	+2,000	+4,000	+1,600	+1,196	−1,402	+3,059

[a] fixed value of depreciation = actual value/$(1.06)^n$
[b] income tax = −(0.5)(income−depreciation)
[c] at $i_r = \dfrac{i^f - f}{1 + f} = \dfrac{0.15 - 0.06}{1.06} = 0.0849 \; (8.49\%)$

TABLE 7. PW of Cash Flow for Example 20 (Using Actual Values).

End of Year	Capital (fixed)	Capital (actual)	Income (fixed)	Income (actual)[b]	Deprec. (actual)	Inc. tax (actual)[c]	PW[d]
0	−10,000	−10,000	—	—	—	—	−10,000
1			+4,000	+4,240	+1,600	−1,320	+2,539
2			+4,000	+4,494	+1,600	−1,447	+2,304
3			+4,000	+4,764	+1,600	−1,582	+2,092
4			+4,000	+5,050	+1,600	−1,725	+1,901
5	+2,000	+2,676[a]	+4,000	+5,353	+1,600	−1,877	+3,059
						Total PW =	+$1,895

[a] $(2,000)(1.06)^5 = 2,676$
[b] $(4,000)(1.06)^n$
[c] income tax = $-(0.5)$(income-deprecation)
[d] at i_f = MARR = 15%

verting the $10,000 actual dollar value to a fixed dollar (inflation free) value and using the real interest rate, as follows:

$$\text{fixed value} = (P/F, 5\%, 5)(10,000) = \overset{0.7835}{\$7,835}$$

This figure is the value, in terms of the purchasing power of today's dollar, of the $10,000 figure five years hence with a 5% inflation rate. Since this is a fixed dollar value the real interest rate must be used:

$$P = (7,835)(P/F, 6\%, 5)$$

= $5,855, exactly the same result as in Example 19.

Varying Inflationary Effects

In certain situations, specific elements of cash flow are affected differently by inflation than the remaining elements. In such instances, cash flows must be adjusted so that all elements are expressed in either fixed or actual values before economic evaluations can be made. The most common situation of this type is after-tax evaluation of a cash flow involving depreciable capital assets where the cash flow is expressed in terms of fixed values. However, the calculated depreciation charges are *actual* values. Therefore, the depreciation charges should be converted to fixed values, or the other elements of the cash flow should be converted to actual values.

Example 20: A certain asset has a first cost of $10,000, a salvage value of $2,000 at the end of its 5 year expected life, and it is expected to produce a net income of $2,000/yr, all values in today's (fixed) dollars. The asset is to be depreciated on the straight line basis. The minimum attractive rate of return is 15%, the expected inflation rate is 6%, and the effective tax rate is 50%. Calculate the after-tax net present worth of this prospective investment.

$$\text{Depreciation charges, } D_m = \frac{P - S}{n} = \frac{10,000 - 2,000}{5}$$

$$= 1,600/\text{yr}$$
$$(\text{constant})$$

For simplicity and clarity, calculations will be rendered in the form of cash flow tables, Table 6 using fixed values and Table 7 using actual values.

Notice that fixed and actual equivalencies are numerically equal only at the present because that is the only point in time at which inflation has no effect on cash flow.

REPLACEMENT ANALYSIS

Replacement analysis does not differ in principle from engineering economic evaluations in other situations. One of the alternatives, however, is an existing piece of equipment, building, or some other asset that is being considered for replacement. It is referred to as the *defender*. The other alternatives considered in a replacement analysis (the *challengers*) are possible candidates for replacing the defender. There could be any of a number of reasons for considering the replacement of an asset, including:

(a) The inability of the existing asset to continue to perform its intended duties without extensive repair and/or modifications.
(b) The inability of the existing asset to meet current and/or

predicted future requirements due to changes in demand.
(c) The appearance on the market of challengers that can perform the duties of the asset at a lower cost.

Since replacement analysis usually deals with depreciable assets (at least in the private, industrial, and commercial sectors), after-tax analysis is often required. This leads to the major difficulty in the application of replacement analysis methodology—the identification of the *appropriate* elements of cash flow, most particularly in the case of the defender. The cardinal rule is that *past receipts and disbursements are irrelevant*. The appropriate first cost for the defender is the *market* value at the present. However, the original first cost of the defender, which is always a past disbursement in replacement analysis, does enter into the calculation of depreciation charges and book values that may extend beyond the present time. While the latter two features are also not directly involved in cash flow, they have a bearing on income tax liability which *is* a cash flow disbursement item. Depreciation charges directly reduce income tax liability in proportion to the tax rate, and the differences between market values and book values at each end of the cash flow may result in increasing (or decreasing) tax liabilities due to capital gains (or losses).

Another difficulty in replacement analysis is that it is almost never appropriate (in the case of the defender), to assume the repeatability concept regularly used in economic analysis of alternatives over periods of time longer than the service life of the alternative. Rather, the time period for the replacement analysis (usually called "planning horizon") is based on the foreseen future need for the asset. If that period of time exceeds the expected remaining service life of the defender, the alternative involving the defender will include a deferred challenger. Obviously, in this case, a market value (salvage value) will have to be estimated for the added-on challenger at the termination of the planning horizon. For convenience and because of the difficulty in predicting future events, however, it is common practice to limit the planning horizon to the remaining life of the defender. Then it is necessary only to estimate the market values at the end of the planning horizon for each challenger.

Example 21: John Brown is the sole proprietor of a materials testing and consulting engineering firm. The coring rig that he purchased three years ago for $15,000 is a constant maintenance problem and has averaged $2,000 per year in maintenance and repair costs. It is being depreciated over a ten-year life using the straight-line method and a salvage value of $3,000. The present market value of the rig is $8,000. Mr. Brown is considering the purchase of a better quality rig that costs $20,000, but is expected to have maintenance and repair costs of only $500/yr. The new rig would have a ten-year depreciation life to a salvage value of $6,000 (using straight-line depreciation). Both rigs have comparable production rates and operating costs. Mr. Brown intends to retire and liquidate the business in five years. The market values for the present and the proposed rigs five years hence are estimated to be $2,000 and $12,000, respectively. Inflation is assumed to be zero, the effective income tax rate is 46%, and Mr. Brown's *MARR* is 6% after taxes. Should Mr. Brown retain the present equipment or purchase the new one?

Defender—If the defender is selected, the following cash flow elements exist:

(a) Planning horizon = 5 years
(b) First cost (present market value) = $8,000 (−)
(c) Annual maintenance costs = $2,000/yr (−)
(d) Salvage value (market value 5 years hence) = $2,000 (+)
(e) Reduction of income tax due to depreciation: annual depreciation charge =

$$\frac{15,000 - 3,000}{10} = 1,200/\text{yr}$$

reduction in inc. tx. = (0.46) (1,200) = $552/yr (+)
(f) Income tax due to capital gains (or losses): since the purchase-to-sale times are > 12 months, capital gains (losses) are long term and therefore one-half of the capital gains (losses) are subject to income tax. Present book value = 15,000 − (1,200) (3) = 11,400. Capital gain (loss) = market value − book value = 8,000 − 11,400 = − 3,400 (capital loss). Assuming that this can be used to off-set capital gains from other sources in the company, a tax *increase* results at the present (lost opportunity): (3,400) (0.46) (0.5) = 782 (−). Book value 5 years hence = 15,000 − (1,200) (8) = 5,400. Capital gain (loss) = 2,000 − 5,400 = − 3,400 (capital loss). Tax *reduction* 5 years hence = (3,400) (0.46) (0.5) = 782 (+).
(g) Reduction of income tax from annual maintenance cost: (2000) (0.46) = 920/yr (+). The cash flow diagram for the defender is shown on Figure 15.

PW of Cost = (8,000 + 782)

$$- (2,000 + 782) \overset{0.7473}{(P/F, 6\%, 5)}$$

$$+ (2,000 - 920 - 552) \overset{4.2124}{(P/A, 6\%, 5)}$$

$$= \$8,927$$

Notice that the initial cost, book values, and depreciation charges are *irrelevant* except for their effect on income taxes. Therefore, in before-tax analyses, these items do not appear at all. They are *sunk costs* stemming from previous

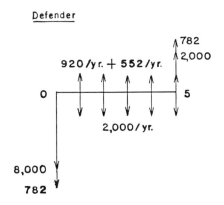

FIGURE 15. Cash flow diagram for defender—Example 21.

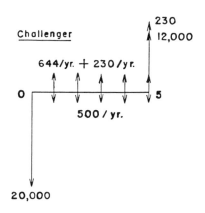

FIGURE 16. Cash flow diagram for challenger—Example 21.

decisions and have no bearing on the present nor the future (except for income taxes).

Challenger—Using the same procedure as was used for the defender, the following cash flow elements are appropriate:

(a) Planning horizon = 5 years
(b) First cost = 20,000 (−)
(c) annual maint. & repair cost = 500/yr (−)
(d) Salvage value (market value 5 yrs hence) = 12,000 (+)
(e) Reduction of income tax due to depreciation: annual depreciation charge =

$$\frac{20{,}000 - 6{,}000}{10} = 1{,}400/\text{yr}$$

reduction in inc. tx. = (0.46) (1,400/yr) = 644/yr (+)
(f) Income tax due to capital gains (losses): Book value 5 years hence = (20,000) − (5) (1,400) = 13,000. Capital gain (loss) = 12,000 − 13,000 = − 1,000 (loss) resulting tax *reduction* 5 years hence = (1,000) (0.46) (0.5) = 230 (+)
(g) Reduction of income tax from annual maintenance costs = (500) (0.46) = 230/yr (+). The cash flow diagram for the challenger is shown on Figure 16.

$$PW \text{ of Cost} = 20{,}000 - (12{,}000 + 230) \overset{0.7473}{(P/F, 6\%, 5)}$$

$$+ (500 - 644 - 230) \overset{4.2124}{(P/A, 6\%, 5)}$$

$$= \$9{,}285$$

Therefore, the defender, having the lower cost, should be retained.

The *before-tax* present worths of the defender and the challenger are, respectively:

$$8{,}000 - (2{,}000) \overset{0.7473}{(P/F, 6\%, 5)}$$

$$+ (2{,}000) \overset{4.2124}{(P/A, 6\%, 5)}$$

$$= \$14{,}930$$

and

$$20{,}000 - (12{,}000) \overset{0.7473}{(P/F, 6\%, 5)}$$

$$+ (500) \overset{4.2124}{(P/A, 6\%, 5)} = \$13{,}139$$

which reverses the decision in favor of the challenger.

MINIMUM COST/MAXIMUM PROFIT LIFE AND BREAKEVEN ANALYSES

Minimum Cost/Maximum Profit Analysis

In certain instances, the elements of the cash flow vary with time in such a way that a minimum cost (or a maximum profit, depending on circumstances) exists at some point in the service life of the asset. This situation commonly occurs in cases of an asset which produces net receipts that decrease as a function of time. Example 22 illustrates the point:

Example 22: The U.S. Army Corps of Engineers is considering the installation of a low-head turbine to generate power at one of its flood control dams. The power generated by the turbine would be sold to a public utility. The cost of

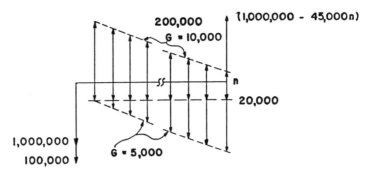

FIGURE 17. Cash flow diagram for Example 22.

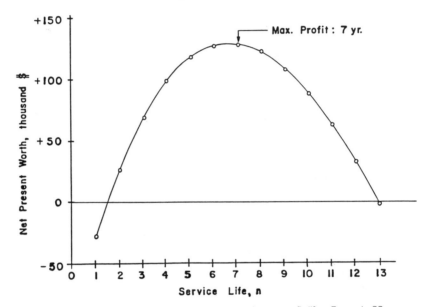

FIGURE 18. Profitable life range and maximum profit life—Example 22.

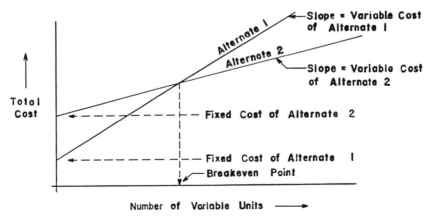

FIGURE 19. Illustration of breakeven involving two alternatives.

the turbine and attendant facilities is $1 million. The initial installation cost is $100,000. Annual operating and maintenance costs are estimated to be $20,000 in the first year of operation, increasing by $5,000 per year thereafter due to wear and tear. Receipts from the sale of power are expected to be $200,000 in the first year, but will decrease by $10,000 per year thereafter due to reduced efficiency and increased downtime with age. Since this is a governmental installation, no income taxes are paid. The life of the turbine is estimated to be 20 years at which time the salvage value (realizable market value) is expected to be $100,000. The market value of the turbine in other years is assumed to decrease linearly with time. All receipts and disbursements are expressed in today's dollars. The real interest rate (inflation free) is 6.%

(a) At what point in time should the turbine be replaced in order to maximize the return?
(b) What is the range of time during which the turbine realizes a profit?

It is clear that the effects of income taxes and inflation are to be ignored here. The cash flow diagram for this situation is shown on Figure 17. Notice that the annual decrease in salvage value is:

$$\frac{\$1,000,000 - \$100,000}{20 \text{ yr}} = \$45,000/\text{yr}$$

The general equation for the present worth of this cash flow is:

$$PW = -1,000,000 - 100,000 - (20,000)(P/A, 6\%, n)$$
$$- (5,000)(P/G, 6\%, n) + (200,000)(P/A, 6\%, n)$$
$$- (10,000)(PG, 6\%, n) + (1,000,000 - 45,000\ n)$$
$$\times (P/F, 6\%, n)$$

Using this equation, the range of time (n-values) over which PW is positive (a profit) and the year that the maximum PW occurs are determined (see Table 8). The results are also shown graphically on Figure 18. The maximum profit occurs for a turbine life of 7 years. A life between 2 to 12 years is profitable.

Example 22 could also have been handled by determining equivalent uniform annual amounts in lieu of present worths, producing the same results.

Breakeven Analysis

When one or more cost elements involved in comparing alternatives vary as functions of output or usage, there exist values of the output or usage at which pairs of the alter-

TABLE 8. Calculation of Profitable Life Range and Maximum Profit Life—Example 22.

End of Year	(P/A, 6%, n)	(P/F, 6%, n)	(P/G, 6%, n)	PW
1	0.9434	0.9434	0.000	− 29,241
2	1.8334	0.8900	0.890	+ 26,562
3	2.6730	0.8396	2.569	+ 68,859
4	3.4651	0.7921	4.946	+ 99,050
5	4.2124	0.7473	7.935	+118,365
6	4.9173	0.7050	11.459	+127,879
7	5.5824	0.6651	15.450	+128,676
8	6.2098	0.6274	19.842	+121,670
9	6.8017	0.5919	24.577	+107,832
10	7.3601	0.5584	29.602	+ 87,908
11	7.8869	0.5268	34.870	+ 62,626
12	8.3838	0.4970	40.337	+ 32,649
13	8.8527	0.4688	45.963	− 1,407

natives have the same total $EUAA$ or PW. This is called the "breakeven" point. The choice of alternative, then, is indicated by comparing the expected output or usage with the breakeven values. Breakeven is illustrated in its simplest form on Figure 19.

Notice that the higher fixed cost alternative must have the lower variable cost if a real (positive) result is to occur. Alternatives that do not meet this criterion can be summarily discarded. (Common sense dictates that in comparing alternatives one would not pay more for something that performs less well.)

Breakeven situations can be handled using a graphical approach, such as that shown on Figure 19, but the algebraic approach is more commonly used. In the algebraic method, all possible combinations of pairs of the alternatives are compared by equating their respective cash flows, expressed as functions of the variable whose value is being sought, and solving for the breakeven value.

Example 23: Sewer service must be supplied to a new subdivision development. Due to topographic limitations, gravity flow to the existing sewer system is not possible and so a pumping station and pressure sewer line must be used to service the new subdivision. The three types of pumps and their respective costs and service lives appear below.

Pump Type	First Cost, $	Operating & Maint. Cost, $/1000 gal of sewage	Est. Service Life, Years
Ejector	10,000	0.50	10
Positive Displacement	7,000	0.60	8
Diaphragm	6,000	0.65	7

The salvage values of all pumps at the ends of their respective service lives are considered to be negligible. $MARR = 6\%$. For what range of sewage flow rates, in thousands of gallons per day, does each of the three pumps present the most economical choice?

Let x = average sewage flow rate, 1,000 gal/day.

EUAA

Ejector:

$$(10{,}000)\,\overset{0.13587}{(A/P,\,6\%,\,10)} + (365)\,(0.50)\,x$$

$$= 1{,}358.70 + 182.5\,x$$

Pos. Displ.:

$$(7{,}000)\,\overset{0.16104}{(A/P,\,6\%,\,80)} + (365)\,(0.60)\,x$$

$$= 1{,}127.28 + 219.0\,x$$

Diaphragm:

$$(6{,}000)\,\overset{0.17914}{(A/P,\,6\%,\,7)} + (365)\,(0.65)\,x$$

$$= 1{,}074.84 + 237.25\,x$$

Ejector vs. Pos. Displ.:

$$1{,}358.70 + 182.5\,x = 1{,}127.28 + 219\,x$$

$$x = 6.340 \;(1{,}000\text{ gal/day})$$

Ejector vs. Diaphragm:

$$1{,}358.70 + 182.5\,x = 1{,}074.84 + 237.25\,x$$

$$x = 5.185 \;(1{,}000\text{ gal/day})$$

Pos. Displ. vs. Diaphragm:

$$1{,}127.28 + 219\,x = 1{,}074.84 + 237.25\,x$$

$$x = 2.873 \;(1{,}000\text{ gal/day})$$

Therefore, if the flow rate:

$x > 6.340$, 1,000 gal/day: use ejector

$6.340 > x > 2.873$, 1,000 gal/day: use Pos. Displ.

$x < 2.873$, 1,000 gal/day: use diaphragm

The graphical solution to this situation, shown on Figure 20, clearly demonstrates the logic involved. Notice that the "breakeven" intersection for the ejector vs. the diaphragm pump does not enter into the final solution because at the flow rate corresponding to that point the positive displacement pump has a lower total cost.

SENSITIVITY ANALYSIS

Engineering economic evaluation deals entirely with cash flows that extend from the present into the future. Uncertainties always exist regarding the lengths of service lives and the timing and amounts of future receipts and disbursements. Also, unanticipated expenditures or receipts may occur. Sensitivity analysis is the determination of the effect of variability in the elements of cash flow on the economic decision.

The general procedure involved in carrying out a sensitivity analysis consists of the following:

1. Determine which elements of the cash flow are most likely to vary from estimated values.
2. Estimate the probable range and choose the increment of variation for each of the selected elements.
3. Select an evaluation method (*PW, EUAA*, rate of return, etc.) to carry out the evaluations.
4. Carry out the computations of the selected evaluation method for each increment within the estimated range of variability for each of the variable elements.
5. For each element in question, plot the computed values (ordinate) against the respective increments of the element (abscissa).

The plots that result from the above procedures will immediately reveal the sensitivities of the elements in question. The more vertical the plot, the greater the sensitivity, i.e., a vertical line represents infinite sensitivity while a horizontal line depicts zero sensitivity.

Example 24: A certain piece of equipment costs $10,000. Its expected service life is 10 ± 2 years, salvage value $\$5{,}000 \pm \$1{,}000$, and annual operating and maintenance cost $\$1{,}000 \pm \$100$ per year. Further, the real interest rate over the life of the equipment is estimated to be $5 \pm 1\%$. Determine the sensitivity of cash flow to possible variations in:

(a) Interest rate
(b) Annual operating & maintenance cost
(c) Salvage value
(d) Life

Present worth will be used to carry out the evaluations. It will be sufficient to determine *PW* values for three points (minimum, maximum, and average) to establish sensitivity. The cash flow diagram appears on Figure 21.

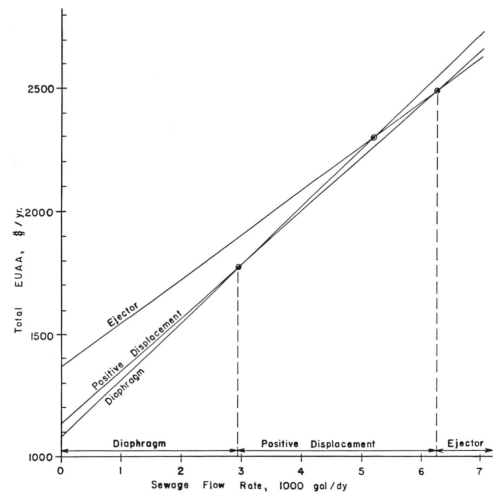

FIGURE 20. Graphical solution to Example 23.

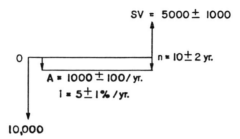

FIGURE 21. Cash flow diagram for Example 24.

General Equation:

$PW = 10,000 + A(P/A, i, n) - SV(P/F, i, n)$
(a) *Test i* for $A = 1,000$, $SV = 5,000$, $n = 10$

$i = 4\%$

$PW = 10,000 + (1,000) \overset{8.1109}{(P/A, 4\%, 10)}$

$ - (5,000) \overset{0.6756}{(P/F, 4\%, 10)} = 14,733$

$i = 5\%$

$PW = 10,000 + (1,000) \overset{7.7217}{(P/A, 5\%, 10)}$

$ - (5,000) \overset{0.6139}{(P/F, 5\%, 10)} = 14,652$

$i = 6\%$

$PW = 10,000 + (1,000) \overset{7.3601}{(P/A, 6\%, 10)}$

$ - (5,000) \overset{0.5584}{(P/F, 6\%, 10)} = 14,568$

Test A for $i = 5\%$, $SV = 5,000$, $n = 10$

$A = 900$

$PW = 10,000 + 900 \overset{7.7217}{(P/A, 5\%, 10)}$

$ - (5,000) \overset{0.6139}{(P/F, 5\%, 10)} = 13,880$

$A = 1,000$ (see above): $PW = 14,652$

$A = 1,100$

$PW = 10,000 + (1,100) \overset{7.7217}{(P/A, 5\%, 10)}$

$ - (5,000) \overset{0.6139}{(P/F, 5\%, 10)} = 15,424$

Test SV for $i = 5\%$, $A = 1,000$, $n = 10$

$SV = 4,000$

$PW = 10,000 + (1,000) \overset{7.7217}{(P/A, 5\%, 10)}$

$ - (4,000) \overset{0.6139}{(P/F, 5\%, 10)} = 15,266$

$SV = 5,000$ (see above): $PW = 14,652$

$SV = 6,000$

$PW = 10,000 + (1,000) \overset{7.7217}{(P/A, 5\%, 10)}$

$ - (6,000) \overset{0.6139}{(P/F, 5\%, 10)} = 14,038$

Test n for $i = 5\%$, $A = 1,000$, $SV = 5,000$

$n = 8$

$PW = 10,000 + (1,000) \overset{6.4632}{(P/A, 5\%, 8)}$

$ - (5,000) \overset{0.6768}{(P/F, 5\%, 8)} = 13,079$

$n = 10$ (see above): $PW = 14,652$

$n = 12$

$PW = 10,000 + (1,000) \overset{8.8633}{(P/A, 5\%, 12)}$

$ - (5,000) \overset{0.5568}{(P/F, 5\%, 12)} = 16,079$

The results are plotted on Figure 22 which shows that the equivalent cost is essentially unaffected by interest rate, moderately susceptible to variations in annual operating and maintenance cost and salvage value, and significantly affected by service life (within the respective expected ranges of variability). Notice that in this analysis the four elements were varied individually, one at a time while the others remained fixed at their respective average values. In actuality, two or more elements could vary simultaneously. Figure 22 suggests that the worst case (highest cost) would result from the combination of:

$$i = 4\%$$
$$A = \$1,100/\text{yr}$$
$$SV = \$4,000$$
$$n = 12 \text{ yr}$$

$PW = 10,000 + (1,100) \overset{9.3851}{(P/A, 4\%, 12)}$

$ - (4,000) \overset{0.6246}{(P/F, 4\%, 12)} = \$17,825$

This value is approximately 22% higher than the value obtained by using the average values of the variables ($14,652).

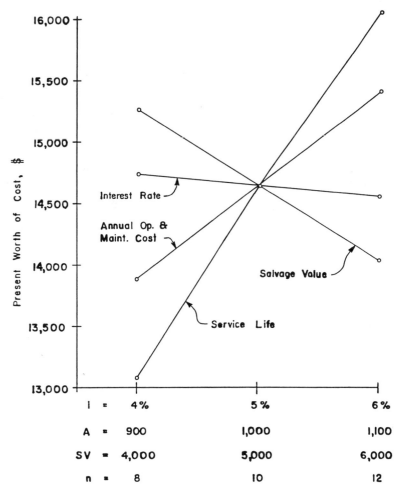

FIGURE 22. Sensitivity plots for each variable element (remaining elements held constant at mean values)—Example 24.

RISK ANALYSIS

In some cases, certain future costs are predicated on the occurrence of events that are, in turn, governed by the laws of probability. Examples include flood damage costs resulting from peak flows in excess of certain values, damage to crops from freezing temperatures, and other casualty losses (such as fire, wind, and theft). If the probability of the event occurring during any given year is known or can be estimated from past records, then the most probable *uniform annual cost* assignable to the event, *in the long run*, is merely the product of the probability and the cost of the consequence when the event occurs:

$$A = Cp \qquad (32)$$

where

A = average uniform annual cost of unfavorable consequence of risk over the long run
C = cost of the unfavorable event when it occurs
p = probability that the unfavorable event may occur in any given year (decimal)

Notice that probability must be within the range zero (no chance of the event occurring) to one (certainty of the event occurring).

Example 25: The storage capacity of the reservoir for a flood control dam and reservoir project is to be sized on the basis of minimum total costs (first cost plus expected average annual damage resulting from floods that exceed storage capacity). The estimated cost of damage if the emergency

TABLE 9. Data and Calculations for Example 25.

Flood Storage, acre-ft.	Probability of Exceeding Storage Capacity in any Given Year, p	First Cost Million $	EUAA* Million $
30,000	0.15	25	3.0
35,000	0.10	30	2.8
40,000	0.06	35	2.7
45,000	0.04	40	2.8
50,000	0.02	45	2.9
55,000	0.01	50	3.1

$= i = 0.06$
*EUAA (millions) = (first cost)(A/P, 6%, ∞) + (10)(p)

spillway is topped (reservoir capacity exceeded) is $10 million (each event). The interest rate is 6% and the reservoir is assumed to possess perpetual life. Six reservoir designs of different storage capacities, the probabilities of exceeding those capacities in any given year, and the estimated first costs are presented in Table 9 along with calculation of *EUAA* values. The minimum cost occurs with the 40,000 acre-ft storage reservoir.

MUNICIPAL FINANCING

Financing of Public Works

BACKGROUND

Projects in the public domain, such as highways, sewage treatment plants, flood control projects, schools, and so forth, must be financed using public funds. Privately funded projects are evaluated in terms of profits or return to the owner whereas public activities are evaluated in terms of both economic benefits and the general welfare of the public or user. Therefore, the evaluation approaches and methods of financing will differ somewhat.

There are three major approaches to financing public projects. These are through various taxes, charges for services, and borrowing of funds. Most state and local governments cannot finance extensive public improvements on a cash basis alone. Therefore, financing all or a portion of the improvement by borrowing is common. Borrowing for public works is usually done through the sale of bonds.

AUTHORITIES

The authority is a vehicle for the accomplishment of public purposes which are not possible or practical for the state or local governments. It is a corporate body authorized to acquire, construct, improve, maintain, and operate projects for the public, and to borrow money and issue bonds therefor. A municipality organizes an authority by adopting a resolution or ordinance specifying this intention, by giving due public notice of this resolution or ordinance, and by filing articles of incorporation with the state.

THE INDENTURE

In order for bonds to be salable, they must have some security behind them. In the case of authority bonds, this security takes the form of the assignment and pledge of the rentals and revenues of the authority from the project. The trustee (a bank) administers the money in accordance with the terms of a legally binding document, known as the *indenture*, as long as any of the authority's bonds are outstanding. This gives the buyer of the bonds assurance that the anticipated revenues of the authority will in fact be used during all the years the bonds are outstanding exactly in accordance with the intentions of the authority when the bonds were originally issued.

Municipal Bonds

TYPES

Three types of bonds may be used for municipal financing: general obligation, special assessment, and revenue bonds.

General obligation bonds are based on governmental credit and the taxing power of the community. As security, the full faith and credit of the governmental unit, backed by its taxing power, is pledged. The amount borrowed becomes a general obligation and must be paid without regard to any specific fund. General obligation bonds are not considered "risk" capital and generally carry a low interest rate. They usually require a referendum and may be rejected unless there has been a good educational campaign on the merits of the project for which they are intended.

Special assessment bonds are generally payable from assessments based on benefits to private property and become a lien on the property benefited. They are sometimes used in combination with general obligation or revenue bonds. Such a financing method has considerable merit because it permits the cost to be spread on a more equitable basis among users, property benefit, and the general public.

Financing through the sale of bonds on which principal and interest payments are paid for solely from the income derived from operation of the facility is referred to as revenue financing. Since the taxing power of the governmental unit is not used, revenue bonds do not constitute a municipal debt within the meaning of constitutional or statutory limitations. However, revenue bond financing should not be regarded as a device to overcome constitutional or statutory debt restrictions. This method should be based on the merits of the project and be able to withstand a critical financial analysis. There is a distinct advantage in using revenue financing for applicable public projects, such as water supply and sewage disposal, because it places these essential services on a businesslike basis. The service that these projects provide is ideally suited to revenue financing because a moderate periodic charge can be made to cover the financing of the project.

BOND NOMENCLATURE

Most bonds issued by municipal authorities are coupon interest bearing bonds. The bearer of such bonds submits coupons at specified time intervals (usually semi-annually) and receives interest payments in return. The interest payments are equal to the coupon interest rate (always expressed as a nominal interest rate), divided by the number of payments per year, times the face value (denomination) of the bond. Coupon interest is *not* compounded. At maturity (the stated life span of a bond), the bearer receives the face value of the bond by surrendering the bond certificate.

Municipal bonds are usually in denominations of $1,000 or multiples thereof. The price at which an authority sells bonds is usually not the same as the face value of the bonds. It must be borne in mind that the investment banker who buys the bonds does so only because he hopes to sell them at a profit. If he buys the bonds from the authority at a price of par (one hundred cents on the dollar) he must sell them at a premium (a price in excess of par) to make a profit. If he buys the bonds at less than par (a discount) he may then sell them at par and make his profit. Since it usually is easier to sell authority bonds to ultimate retail buyers at prices of par or near par, most sales of bonds by authorities are at discount. In order to make this possible, it is necessary for the authority to sell more bonds than are required to pay all other costs. The amount of this discount, which is a cost of financing, is considered, accounting wise, to be deferred interest. Normally, it is more economical for the authority to sell bonds at a discount even though this requires the selling of additional bonds. This occurs because interest coupons usually can be lower on bonds ultimately sold at a price of par to customers than on bonds sold at a premium. Investors generally have a reluctance to pay a premium for bonds, and therefore require greater attractiveness in the form of coupon income before they will buy them.

"Coverage" is an amount, usually equal to 10% or more, by which money provided for bond interest and retirement exceeds the actual debt service. It is rendered desirable because of the revenue nature of this financing, which does not have the strong security of reliance upon general taxing power, as in general obligation bonds. Without the coverage, the bonds would be salable to fewer buyers and they would have to bear a higher rate of interest.

TAX EXEMPTION

Interest earned on municipal bonds is not subject to federal income tax, a factor which renders them especially attractive to many investors, and at the same time explains why they bear rates of interest which are generally lower than those on corporate (taxable) obligations of equivalent rating or quality.

NET INTEREST COST

Most bond issues are composed of groups of bonds having varying maturity dates, with the largest number of bonds maturing at the end of the bond issue. The longer the term of the bond issue, the higher will be the average (or net) interest cost. This occurs because, generally speaking, the longer the period over which funds are loaned the higher the return the lender expects. This average interest cost is the weighted average of the coupons of interest which are affixed to the bond issue. This is expressed as a percentage by dividing the total interest to be paid on the bond issue (plus the amount of bond discount) by the total number of bond years involved. Bond years is simply the multiplication of each bond by the number of years it is to be outstanding.

Example 26: A municipal sanitary authority is planning a major addition to their sewage treatment plant. The project is expected to cost $5 million, including the engineering, legal, and financial costs for floating a bond issue and carrying out the construction. The project will be financed through the sale of $1,000 face value revenue bonds carrying 6% coupon interest, payable semi-annually, and maturing in 40 years. The bonds are to be sold at 1/2% discount. Calculate the debt service requirement based on uniform semi-annual payments to a sinking fund earning a nominal 7% interest rate, compounded semi-annually. The trust indenture requires a 10% coverage for debt service.

$$\text{The amount of the bond issue} = \frac{5,000,000}{1 - \text{discount rate}}$$

$$\frac{5,000,000}{0.995} = 5,025,125$$

However, since the bonds are $1,000 face value,

$$\text{Bond issue} = \$5,026,000$$

Semi-annual coupon interest payment =

$$(5,026,000)\frac{0.06}{2} = \$150,780$$

no. of interest periods to maturity = (40)(2) = 80
Semi-annual sinking fund deposit =

$$(1 + 0.1)\,[150{,}780 + (5{,}026{,}000)\,\overset{0.00238}{(A/F,\,7\%/2,\,80)}]$$

$$= \$179{,}016/6 \text{ months}$$

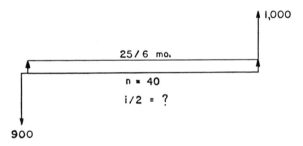

FIGURE 23. Cash flow diagram for Example 28.

Example 27: What is the net interest cost for Example 26?

Total coupon interest $= (2)(40)(150{,}780)$
$= 12{,}062{,}400$

plus discount:
$(5{,}026{,}000)(.005) = \underline{25{,}130}$
$\phantom{(5{,}026{,}000)(.005) = }12{,}087{,}530$

Bond years $= (40) \dfrac{5{,}026{,}000}{1{,}000} = 201{,}040$

net interest cost $= \dfrac{12{,}087{,}530}{201{,}040} = \$60.125/\$1{,}000$ bond

$= 6.0125\%$

Example 28: A \$1,000, 5% bond (compounded semi-annually) that matures in 20 years is being offered for sale at \$900 (i.e., 10% discount). What is the rate of return on this prospective investment?

Coupon Interest $= \dfrac{(1000)(.05)}{2} = \$25$ (every 6 months)

$n = (2)(20) = 40$ (coupon interest payment periods)

The cash flow diagram is shown on Figure 23.

$(-900) + (25)(P/A, i/2, 40) + (1000)(P/F, i/2, 40) = 0$

try $i/2 = 2\%$

$(-900) + (25) \overset{27.3555}{(P/A, 2\%, 40)}$

$+ (1000) \overset{0.4529}{(P/F, 2\%, 40)} \overset{?}{=} 0$

$+ 236.78 \neq 0$

try $i/2 = 3\%$

$(-900) + (25) \overset{23.1148}{(P/A, 3\%, 40)}$

$+ (1000) \overset{0.3066}{(P/F, 3\%, 40)} \overset{?}{=} 0$

$- 15.53 \neq 0$

Interpolating:

$i/2 = 2\% + (1\%)\dfrac{236.78}{236.78 + 15.53} = 2.94\%$

$i = (2)(2.94) = 5.88\%$ (nominal)

or

$\left(1 + \dfrac{0.0588}{2}\right)^2 - 1 = 0.0597$

$= 5.97\%$ (effective)

ACKNOWLEDGEMENTS

I wish to acknowledge the efforts of my son, Michael E. Cady, who wrote the computer program that generated the interest factor tables and who drafted the figure copy in this chapter.

APPENDIX

Interest Factor Tables

i = 0.5%

	SINGLE PAYMENT		UNIFORM SERIES				GRADIENT	
	COMPOUND AMOUNT	PRESENT WORTH	SINKING FUND	COMPOUND AMOUNT	CAPITAL RECOVERY	PRESENT WORTH	PRESENT WORTH	ANNUAL COST
n	F/P	P/F	A/F	F/A	A/P	P/A	P/G	A/G
1	1.0050	0.9950	1.00000	1.0000	1.00500	0.9950	0.000	0.000
2	1.0100	0.9901	0.49875	2.0050	0.50375	1.9851	0.990	0.499
3	1.0151	0.9851	0.33167	3.0150	0.33667	2.9702	2.960	0.997
4	1.0202	0.9802	0.24813	4.0301	0.25313	3.9505	5.901	1.494
5	1.0253	0.9754	0.19801	5.0503	0.20301	4.9259	9.803	1.990
6	1.0304	0.9705	0.16460	6.0755	0.16960	5.8964	14.655	2.485
7	1.0355	0.9657	0.14073	7.1059	0.14573	6.8621	20.449	2.980
8	1.0407	0.9609	0.12283	8.1414	0.12783	7.8230	27.176	3.474
9	1.0459	0.9561	0.10891	9.1821	0.11391	8.7791	34.824	3.967
10	1.0511	0.9513	0.09777	10.2280	0.10277	9.7304	43.386	4.459
11	1.0564	0.9466	0.08866	11.2792	0.09366	10.6770	52.853	4.950
12	1.0617	0.9419	0.08107	12.3356	0.08607	11.6189	63.214	5.441
13	1.0670	0.9372	0.07464	13.3972	0.07964	12.5562	74.460	5.930
14	1.0723	0.9326	0.06914	14.4642	0.07414	13.4887	86.583	6.419
15	1.0777	0.9279	0.06436	15.5365	0.06936	14.4166	99.574	6.907
16	1.0831	0.9233	0.06019	16.6142	0.06519	15.3399	113.424	7.394
17	1.0885	0.9187	0.05651	17.6973	0.06151	16.2586	128.123	7.880
18	1.0939	0.9141	0.05323	18.7858	0.05823	17.1728	143.663	8.366
19	1.0994	0.9096	0.05030	19.8797	0.05530	18.0824	160.036	8.850
20	1.1049	0.9051	0.04767	20.9791	0.05267	18.9874	177.232	9.334
21	1.1104	0.9006	0.04528	22.0840	0.05028	19.8880	195.243	9.817
22	1.1160	0.8961	0.04311	23.1944	0.04811	20.7841	214.061	10.299
23	1.1216	0.8916	0.04113	24.3104	0.04613	21.6757	233.677	10.781
24	1.1272	0.8872	0.03932	25.4320	0.04432	22.5629	254.082	11.261
25	1.1328	0.8828	0.03765	26.5591	0.04265	23.4456	275.269	11.741
26	1.1385	0.8784	0.03611	27.6919	0.04111	24.3240	297.228	12.220
27	1.1442	0.8740	0.03469	28.8304	0.03969	25.1980	319.952	12.698
28	1.1499	0.8697	0.03336	29.9745	0.03836	26.0677	343.433	13.175
29	1.1556	0.8653	0.03213	31.1244	0.03713	26.9330	367.663	13.651
30	1.1614	0.8610	0.03098	32.2800	0.03598	27.7941	392.632	14.126
32	1.1730	0.8525	0.02889	34.6086	0.03389	29.5033	444.762	15.075
34	1.1848	0.8440	0.02706	36.9606	0.03206	31.1955	499.758	16.020
36	1.1967	0.8356	0.02542	39.3361	0.03042	32.8710	557.560	16.962
38	1.2087	0.8274	0.02396	41.7354	0.02896	34.5299	618.105	17.901
40	1.2208	0.8191	0.02265	44.1588	0.02765	36.1722	681.335	18.836
42	1.2330	0.8110	0.02146	46.6065	0.02646	37.7983	747.189	19.768
44	1.2454	0.8030	0.02038	49.0788	0.02538	39.4082	815.609	20.696
46	1.2579	0.7950	0.01939	51.5758	0.02439	41.0022	886.538	21.622
48	1.2705	0.7871	0.01849	54.0978	0.02349	42.5803	959.919	22.544
50	1.2832	0.7793	0.01765	56.6452	0.02265	44.1428	1035.697	23.462
60	1.3489	0.7414	0.01433	69.7700	0.01933	51.7256	1448.646	28.006
70	1.4178	0.7053	0.01197	83.5661	0.01697	58.9394	1913.643	32.468
80	1.4903	0.6710	0.01020	98.0677	0.01520	65.8023	2424.645	36.847
90	1.5666	0.6383	0.00883	113.3109	0.01383	72.3313	2976.077	41.145
100	1.6467	0.6073	0.00773	129.3337	0.01273	78.5426	3562.793	45.361

i = 1%

	SINGLE PAYMENT		UNIFORM SERIES				GRADIENT	
	COMPOUND AMOUNT	PRESENT WORTH	SINKING FUND	COMPOUND AMOUNT	CAPITAL RECOVERY	PRESENT WORTH	PRESENT WORTH	ANNUAL COST
n	F/P	P/F	A/F	F/A	A/P	P/A	P/G	A/G
1	1.0100	0.9901	1.00000	1.0000	1.01000	0.9901	0.000	0.000
2	1.0201	0.9803	0.49751	2.0100	0.50751	1.9704	0.980	0.498
3	1.0303	0.9706	0.33002	3.0301	0.34002	2.9410	2.921	0.993
4	1.0406	0.9610	0.24628	4.0604	0.25628	3.9020	5.804	1.488
5	1.0510	0.9515	0.19604	5.1010	0.20604	4.8534	9.610	1.980
6	1.0615	0.9420	0.16255	6.1520	0.17255	5.7955	14.320	2.471
7	1.0721	0.9327	0.13863	7.2135	0.14863	6.7282	19.917	2.960
8	1.0829	0.9235	0.12069	8.2857	0.13069	7.6517	26.381	3.448
9	1.0937	0.9143	0.10674	9.3685	0.11674	8.5660	33.696	3.934
10	1.1046	0.9053	0.09558	10.4622	0.10558	9.4713	41.843	4.418
11	1.1157	0.8963	0.08645	11.5668	0.09645	10.3676	50.807	4.901
12	1.1268	0.8874	0.07885	12.6825	0.08885	11.2551	60.569	5.381
13	1.1381	0.8787	0.07241	13.8093	0.08241	12.1337	71.113	5.861
14	1.1495	0.8700	0.06690	14.9474	0.07690	13.0037	82.422	6.338
15	1.1610	0.8613	0.06212	16.0969	0.07212	13.8651	94.481	6.814
16	1.1726	0.8528	0.05794	17.2579	0.06794	14.7179	107.273	7.289
17	1.1843	0.8444	0.05426	18.4304	0.06426	15.5623	120.783	7.761
18	1.1961	0.8360	0.05098	19.6147	0.06098	16.3983	134.996	8.232
19	1.2081	0.8277	0.04805	20.8109	0.05805	17.2260	149.895	8.702
20	1.2202	0.8195	0.04542	22.0190	0.05542	18.0456	165.466	9.169
21	1.2324	0.8114	0.04303	23.2392	0.05303	18.8570	181.695	9.635
22	1.2447	0.8034	0.04086	24.4716	0.05086	19.6604	198.566	10.100
23	1.2572	0.7954	0.03889	25.7163	0.04889	20.4558	216.066	10.563
24	1.2697	0.7876	0.03707	26.9735	0.04707	21.2434	234.180	11.024
25	1.2824	0.7798	0.03541	28.2432	0.04541	22.0232	252.894	11.483
26	1.2953	0.7720	0.03387	29.5256	0.04387	22.7952	272.196	11.941
27	1.3082	0.7644	0.03245	30.8209	0.04245	23.5596	292.070	12.397
28	1.3213	0.7568	0.03112	32.1291	0.04112	24.3164	312.505	12.852
29	1.3345	0.7493	0.02990	33.4504	0.03990	25.0658	333.486	13.304
30	1.3478	0.7419	0.02875	34.7849	0.03875	25.8077	355.002	13.756
32	1.3749	0.7273	0.02667	37.4941	0.03667	27.2696	399.586	14.653
34	1.4026	0.7130	0.02484	40.2577	0.03484	28.7027	446.157	15.544
36	1.4308	0.6989	0.02321	43.0769	0.03321	30.1075	494.621	16.428
38	1.4595	0.6852	0.02176	45.9527	0.03176	31.4847	544.883	17.306
40	1.4889	0.6717	0.02046	48.8864	0.03046	32.8347	596.856	18.178
42	1.5188	0.6584	0.01928	51.8790	0.02928	34.1581	650.451	19.042
44	1.5493	0.6454	0.01820	54.9318	0.02820	35.4555	705.585	19.901
46	1.5805	0.6327	0.01723	58.0459	0.02723	36.7272	762.176	20.752
48	1.6122	0.6203	0.01633	61.2226	0.02633	37.9740	820.146	21.598
50	1.6446	0.6080	0.01551	64.4632	0.02551	39.1961	879.418	22.436
60	1.8167	0.5504	0.01224	81.6697	0.02224	44.9550	1192.806	26.533
70	2.0068	0.4983	0.00993	100.6763	0.01993	50.1685	1528.647	30.470
80	2.2167	0.4511	0.00822	121.6715	0.01822	54.8882	1879.877	34.249
90	2.4486	0.4084	0.00690	144.8633	0.01690	59.1609	2240.567	37.872
100	2.7048	0.3697	0.00587	170.4814	0.01587	63.0289	2605.776	41.343

$i = 1.5\%$

	SINGLE PAYMENT		UNIFORM SERIES				GRADIENT	
	COMPOUND AMOUNT	PRESENT WORTH	SINKING FUND	COMPOUND AMOUNT	CAPITAL RECOVERY	PRESENT WORTH	PRESENT WORTH	ANNUAL COST
n	F/P	P/F	A/F	F/A	A/P	P/A	P/G	A/G
1	1.0150	0.9852	1.00000	1.0000	1.01500	0.9852	0.000	0.000
2	1.0302	0.9707	0.49628	2.0150	0.51128	1.9559	0.971	0.496
3	1.0457	0.9563	0.32838	3.0452	0.34338	2.9122	2.883	0.990
4	1.0614	0.9422	0.24444	4.0909	0.25944	3.8544	5.710	1.481
5	1.0773	0.9283	0.19409	5.1523	0.20909	4.7826	9.423	1.970
6	1.0934	0.9145	0.16053	6.2296	0.17553	5.6972	13.996	2.457
7	1.1098	0.9010	0.13656	7.3230	0.15156	6.5982	19.402	2.940
8	1.1265	0.8877	0.11858	8.4328	0.13358	7.4859	25.616	3.422
9	1.1434	0.8746	0.10461	9.5593	0.11961	8.3605	32.612	3.901
10	1.1605	0.8617	0.09343	10.7027	0.10843	9.2222	40.368	4.377
11	1.1779	0.8489	0.08429	11.8633	0.09929	10.0711	48.857	4.851
12	1.1956	0.8364	0.07668	13.0412	0.09168	10.9075	58.057	5.323
13	1.2136	0.8240	0.07024	14.2368	0.08524	11.7315	67.945	5.792
14	1.2318	0.8118	0.06472	15.4504	0.07972	12.5434	78.499	6.258
15	1.2502	0.7999	0.05994	16.6821	0.07494	13.3432	89.697	6.722
16	1.2690	0.7880	0.05577	17.9324	0.07077	14.1313	101.518	7.184
17	1.2880	0.7764	0.05208	19.2014	0.06708	14.9076	113.940	7.643
18	1.3073	0.7649	0.04881	20.4894	0.06381	15.6726	126.944	8.100
19	1.3270	0.7536	0.04588	21.7967	0.06088	16.4262	140.508	8.554
20	1.3469	0.7425	0.04325	23.1237	0.05825	17.1686	154.615	9.006
21	1.3671	0.7315	0.04087	24.4705	0.05587	17.9001	169.245	9.455
22	1.3876	0.7207	0.03870	25.8376	0.05370	18.6208	184.380	9.902
23	1.4084	0.7100	0.03673	27.2251	0.05173	19.3309	200.001	10.346
24	1.4295	0.6995	0.03492	28.6335	0.04992	20.0304	216.090	10.788
25	1.4509	0.6892	0.03326	30.0630	0.04826	20.7196	232.631	11.228
26	1.4727	0.6790	0.03173	31.5140	0.04673	21.3986	249.607	11.665
27	1.4948	0.6690	0.03032	32.9867	0.04532	22.0676	267.000	12.099
28	1.5172	0.6591	0.02900	34.4815	0.04400	22.7267	284.796	12.531
29	1.5400	0.6494	0.02778	35.9987	0.04278	23.3761	302.978	12.961
30	1.5631	0.6398	0.02664	37.5387	0.04164	24.0158	321.531	13.388
32	1.6103	0.6210	0.02458	40.6883	0.03958	25.2671	359.691	14.236
34	1.6590	0.6028	0.02276	43.9331	0.03776	26.4817	399.161	15.073
36	1.7091	0.5851	0.02115	47.2760	0.03615	27.6607	439.830	15.901
38	1.7608	0.5679	0.01972	50.7199	0.03472	28.8051	481.595	16.719
40	1.8140	0.5513	0.01843	54.2679	0.03343	29.9158	524.357	17.528
42	1.8688	0.5351	0.01726	57.9231	0.03226	30.9941	568.020	18.327
44	1.9253	0.5194	0.01621	61.6889	0.03121	32.0406	612.496	19.116
46	1.9835	0.5042	0.01525	65.5684	0.03025	33.0565	657.698	19.896
48	2.0435	0.4894	0.01437	69.5652	0.02937	34.0426	703.546	20.667
50	2.1052	0.4750	0.01357	73.6828	0.02857	34.9997	749.964	21.428
60	2.4432	0.4093	0.01039	96.2147	0.02539	39.3803	988.168	25.093
70	2.8355	0.3527	0.00817	122.3638	0.02317	43.1549	1231.166	28.529
80	3.2907	0.3039	0.00655	152.7109	0.02155	46.4073	1473.074	31.742
90	3.8189	0.2619	0.00532	187.9299	0.02032	49.2099	1709.544	34.740
100	4.4320	0.2256	0.00437	228.8031	0.01937	51.6247	1937.451	37.530

i = 2%

	SINGLE PAYMENT			UNIFORM SERIES			GRADIENT	
	COMPOUND AMOUNT	PRESENT WORTH	SINKING FUND	COMPOUND AMOUNT	CAPITAL RECOVERY	PRESENT WORTH	PRESENT WORTH	ANNUAL COST
n	F/P	P/F	A/F	F/A	A/P	P/A	P/G	A/G
1	1.0200	0.9804	1.00000	1.0000	1.02000	0.9804	0.000	0.000
2	1.0404	0.9612	0.49505	2.0200	0.51505	1.9416	0.961	0.495
3	1.0612	0.9423	0.32675	3.0604	0.34675	2.8839	2.846	0.987
4	1.0824	0.9238	0.24262	4.1216	0.26262	3.8077	5.617	1.475
5	1.1041	0.9057	0.19216	5.2040	0.21216	4.7135	9.240	1.960
6	1.1262	0.8880	0.15853	6.3081	0.17853	5.6014	13.680	2.442
7	1.1487	0.8706	0.13451	7.4343	0.15451	6.4720	18.904	2.921
8	1.1717	0.8535	0.11651	8.5830	0.13651	7.3255	24.878	3.396
9	1.1951	0.8368	0.10252	9.7546	0.12252	8.1622	31.572	3.868
10	1.2190	0.8203	0.09133	10.9497	0.11133	8.9826	38.955	4.337
11	1.2434	0.8043	0.08218	12.1687	0.10218	9.7868	46.998	4.802
12	1.2682	0.7885	0.07456	13.4121	0.09456	10.5753	55.671	5.264
13	1.2936	0.7730	0.06812	14.6803	0.08812	11.3484	64.948	5.723
14	1.3195	0.7579	0.06260	15.9739	0.08260	12.1062	74.800	6.179
15	1.3459	0.7430	0.05783	17.2934	0.07783	12.8493	85.202	6.631
16	1.3728	0.7284	0.05365	18.6393	0.07365	13.5777	96.129	7.080
17	1.4002	0.7142	0.04997	20.0121	0.06997	14.2919	107.555	7.526
18	1.4282	0.7002	0.04670	21.4123	0.06670	14.9920	119.458	7.968
19	1.4568	0.6864	0.04378	22.8406	0.06378	15.6785	131.814	8.407
20	1.4859	0.6730	0.04116	24.2974	0.06116	16.3514	144.600	8.843
21	1.5157	0.6598	0.03878	25.7833	0.05878	17.0112	157.796	9.276
22	1.5460	0.6468	0.03663	27.2990	0.05663	17.6580	171.379	9.705
23	1.5769	0.6342	0.03467	28.8450	0.05467	18.2922	185.331	10.132
24	1.6084	0.6217	0.03287	30.4219	0.05287	18.9139	199.631	10.555
25	1.6406	0.6095	0.03122	32.0303	0.05122	19.5235	214.259	10.974
26	1.6734	0.5976	0.02970	33.6709	0.04970	20.1210	229.199	11.391
27	1.7069	0.5859	0.02829	35.3443	0.04829	20.7069	244.431	11.804
28	1.7410	0.5744	0.02699	37.0512	0.04699	21.2813	259.939	12.214
29	1.7758	0.5631	0.02578	38.7922	0.04578	21.8444	275.706	12.621
30	1.8114	0.5521	0.02465	40.5681	0.04465	22.3965	291.716	13.025
32	1.8845	0.5306	0.02261	44.2270	0.04261	23.4683	324.403	13.823
34	1.9607	0.5100	0.02082	48.0338	0.04082	24.4986	357.882	14.608
36	2.0399	0.4902	0.01923	51.9944	0.03923	25.4888	392.040	15.391
38	2.1223	0.4712	0.01782	56.1149	0.03782	26.4406	426.776	16.141
40	2.2080	0.4529	0.01656	60.4020	0.03656	27.3555	461.993	16.889
42	2.2972	0.4353	0.01542	64.8622	0.03542	28.2348	497.601	17.624
44	2.3901	0.4184	0.01439	69.5027	0.03439	29.0800	533.517	18.347
46	2.4866	0.4022	0.01345	74.3306	0.03345	29.8923	569.662	19.057
48	2.5871	0.3865	0.01260	79.3535	0.03260	30.6731	605.966	19.756
50	2.6916	0.3715	0.01182	84.5794	0.03182	31.4236	642.361	20.442
60	3.2810	0.3048	0.00877	114.0515	0.02877	34.7609	823.698	23.696
70	3.9996	0.2500	0.00667	149.9779	0.02667	37.4986	999.834	26.663
80	4.8754	0.2051	0.00516	193.7720	0.02516	39.7445	1166.787	29.357
90	5.9431	0.1683	0.00405	247.1567	0.02405	41.5869	1322.170	31.793
100	7.2446	0.1380	0.00320	312.2323	0.02320	43.0984	1464.753	33.986

i = 3%

	SINGLE PAYMENT		UNIFORM SERIES				GRADIENT	
	COMPOUND AMOUNT	PRESENT WORTH	SINKING FUND	COMPOUND AMOUNT	CAPITAL RECOVERY	PRESENT WORTH	PRESENT WORTH	ANNUAL COST
n	F/P	P/F	A/F	F/A	A/P	P/A	P/G	A/G
1	1.0300	0.9709	1.00000	1.0000	1.03000	0.9709	0.000	0.000
2	1.0609	0.9426	0.49261	2.0300	0.52261	1.9135	0.943	0.493
3	1.0927	0.9151	0.32353	3.0909	0.35353	2.8286	2.773	0.980
4	1.1255	0.8885	0.23903	4.1836	0.26903	3.7171	5.438	1.463
5	1.1593	0.8626	0.18835	5.3091	0.21835	4.5797	8.889	1.941
6	1.1941	0.8375	0.15460	6.4684	0.18460	5.4172	13.076	2.414
7	1.2299	0.8131	0.13051	7.6625	0.16051	6.2303	17.955	2.882
8	1.2668	0.7894	0.11246	8.8923	0.14246	7.0197	23.481	3.345
9	1.3048	0.7664	0.09843	10.1591	0.12843	7.7861	29.612	3.803
10	1.3439	0.7441	0.08723	11.4639	0.11723	8.5302	36.309	4.256
11	1.3842	0.7224	0.07808	12.8078	0.10808	9.2526	43.533	4.705
12	1.4258	0.7014	0.07046	14.1920	0.10046	9.9540	51.248	5.148
13	1.4685	0.6810	0.06403	15.6178	0.09403	10.6350	59.420	5.587
14	1.5126	0.6611	0.05853	17.0863	0.08853	11.2961	68.014	6.021
15	1.5580	0.6419	0.05377	18.5989	0.08377	11.9379	77.000	6.450
16	1.6047	0.6232	0.04961	20.1569	0.07961	12.5611	86.348	6.874
17	1.6528	0.6050	0.04595	21.7616	0.07595	13.1661	96.028	7.294
18	1.7024	0.5874	0.04271	23.4144	0.07271	13.7535	106.014	7.708
19	1.7535	0.5703	0.03981	25.1169	0.06981	14.3238	116.279	8.118
20	1.8061	0.5537	0.03722	26.8704	0.06722	14.8775	126.799	8.523
21	1.8603	0.5375	0.03487	28.6765	0.06487	15.4150	137.550	8.923
22	1.9161	0.5219	0.03275	30.5368	0.06275	15.9369	148.509	9.319
23	1.9736	0.5067	0.03081	32.4529	0.06081	16.4436	159.657	9.709
24	2.0328	0.4919	0.02905	34.4265	0.05905	16.9355	170.971	10.095
25	2.0938	0.4776	0.02743	36.4593	0.05743	17.4131	182.434	10.477
26	2.1566	0.4637	0.02594	38.5530	0.05594	17.8768	194.026	10.853
27	2.2213	0.4502	0.02456	40.7096	0.05456	18.3270	205.731	11.226
28	2.2879	0.4371	0.02329	42.9309	0.05329	18.7641	217.532	11.593
29	2.3566	0.4243	0.02211	45.2189	0.05211	19.1885	229.414	11.956
30	2.4273	0.4120	0.02102	47.5754	0.05102	19.6004	241.361	12.314
32	2.5751	0.3883	0.01905	52.5028	0.04905	20.3688	265.399	13.017
34	2.7319	0.3660	0.01732	57.7302	0.04732	21.1318	289.544	13.702
36	2.8983	0.3450	0.01580	63.2759	0.04580	21.8323	313.703	14.369
38	3.0748	0.3252	0.01446	69.1594	0.04446	22.4925	337.796	15.018
40	3.2620	0.3066	0.01326	75.4013	0.04326	23.1148	361.750	15.650
42	3.4607	0.2890	0.01219	82.0232	0.04219	23.7014	385.502	16.265
44	3.6715	0.2724	0.01123	89.0484	0.04123	24.2543	408.997	16.863
46	3.8950	0.2567	0.01036	96.5015	0.04036	24.7754	432.186	17.444
48	4.1323	0.2420	0.00958	104.4084	0.03958	25.2667	455.025	18.009
50	4.3839	0.2281	0.00887	112.7969	0.03887	25.7298	477.480	18.558
60	5.8916	0.1697	0.00613	163.0534	0.03613	27.6756	583.053	21.067
70	7.9178	0.1263	0.00434	230.5941	0.03434	29.1234	676.087	23.215
80	10.6409	0.0940	0.00311	321.3630	0.03311	30.2008	756.087	25.035
90	14.3005	0.0699	0.00226	443.3489	0.03226	31.0024	823.630	26.567
100	19.2186	0.0520	0.00165	607.2977	0.03165	31.5989	879.854	27.844

i = 4%

	SINGLE PAYMENT		UNIFORM SERIES				GRADIENT	
	COMPOUND AMOUNT	PRESENT WORTH	SINKING FUND	COMPOUND AMOUNT	CAPITAL RECOVERY	PRESENT WORTH	PRESENT WORTH	ANNUAL COST
n	F/P	P/F	A/F	F/A	A/P	P/A	P/G	A/G
1	1.0400	0.9615	1.00000	1.0000	1.04000	0.9615	0.000	0.000
2	1.0816	0.9246	0.49020	2.0400	0.53020	1.8861	0.925	0.490
3	1.1249	0.8890	0.32035	3.1216	0.36035	2.7751	2.703	0.974
4	1.1699	0.8548	0.23549	4.2465	0.27549	3.6299	5.267	1.451
5	1.2167	0.8219	0.18463	5.4163	0.22463	4.4518	8.555	1.922
6	1.2653	0.7903	0.15076	6.6330	0.19076	5.2421	12.506	2.386
7	1.3159	0.7599	0.12661	7.8983	0.16661	6.0021	17.066	2.843
8	1.3686	0.7307	0.10853	9.2142	0.14853	6.7327	22.181	3.294
9	1.4233	0.7026	0.09449	10.5828	0.13449	7.4353	27.801	3.739
10	1.4802	0.6756	0.08329	12.0061	0.12329	8.1109	33.881	4.177
11	1.5395	0.6496	0.07415	13.4864	0.11415	8.7605	40.377	4.609
12	1.6010	0.6246	0.06655	15.0258	0.10655	9.3851	47.248	5.034
13	1.6651	0.6006	0.06014	16.6268	0.10014	9.9856	54.455	5.453
14	1.7317	0.5775	0.05467	18.2919	0.09467	10.5631	61.962	5.866
15	1.8009	0.5553	0.04994	20.0236	0.08994	11.1184	69.735	6.272
16	1.8730	0.5339	0.04582	21.8245	0.08582	11.6523	77.744	6.672
17	1.9479	0.5134	0.04220	23.6975	0.08220	12.1657	85.958	7.066
18	2.0258	0.4936	0.03899	25.6454	0.07899	12.6593	94.350	7.453
19	2.1068	0.4746	0.03614	27.6712	0.07614	13.1339	102.893	7.834
20	2.1911	0.4564	0.03358	29.7781	0.07358	13.5903	111.565	8.209
21	2.2788	0.4388	0.03128	31.9692	0.07128	14.0292	120.341	8.578
22	2.3699	0.4220	0.02920	34.2480	0.06920	14.4511	129.202	8.941
23	2.4647	0.4057	0.02731	36.6179	0.06731	14.8568	138.128	9.297
24	2.5633	0.3901	0.02559	39.0826	0.06559	15.2470	147.101	9.648
25	2.6658	0.3751	0.02401	41.6459	0.06401	15.6221	156.104	9.993
26	2.7725	0.3607	0.02257	44.3117	0.06257	15.9828	165.121	10.331
27	2.8834	0.3468	0.02124	47.0842	0.06124	16.3296	174.138	10.664
28	2.9987	0.3335	0.02001	49.9676	0.06001	16.6631	183.142	10.991
29	3.1187	0.3207	0.01888	52.9663	0.05888	16.9837	192.121	11.312
30	3.2434	0.3083	0.01783	56.0849	0.05783	17.2920	201.062	11.627
32	3.5081	0.2851	0.01595	62.7015	0.05595	17.8736	218.792	12.241
34	3.7943	0.2636	0.01431	69.8579	0.05431	18.4112	236.261	12.832
36	4.1039	0.2437	0.01289	77.5983	0.05289	18.9083	253.405	13.402
38	4.4388	0.2253	0.01163	85.9703	0.05163	19.3679	270.175	13.950
40	4.8010	0.2083	0.01052	95.0255	0.05052	19.7928	286.530	14.477
42	5.1928	0.1926	0.00954	104.8196	0.04954	20.1856	302.437	14.983
44	5.6165	0.1780	0.00866	115.4129	0.04866	20.5488	317.870	15.469
46	6.0748	0.1646	0.00788	126.8706	0.04788	20.8847	332.810	15.936
48	6.5705	0.1522	0.00718	139.2632	0.04718	21.1951	347.245	16.383
50	7.1067	0.1407	0.00655	152.6671	0.04655	21.4822	361.164	16.812
60	10.5196	0.0951	0.00420	237.9907	0.04420	22.6235	422.997	18.697
70	15.5716	0.0642	0.00275	364.2905	0.04275	23.3945	472.479	20.196
80	23.0498	0.0434	0.00181	551.2450	0.04181	23.9154	511.116	21.372
90	34.1193	0.0293	0.00121	827.9833	0.04121	24.2673	540.737	22.283
100	50.5049	0.0198	0.00081	1237.6237	0.04081	24.5050	563.125	22.980

i = 5%

	SINGLE PAYMENT		UNIFORM SERIES				GRADIENT	
	COMPOUND AMOUNT	PRESENT WORTH	SINKING FUND	COMPOUND AMOUNT	CAPITAL RECOVERY	PRESENT WORTH	PRESENT WORTH	ANNUAL COST
n	F/P	P/F	A/F	F/A	A/P	P/A	P/G	A/G
1	1.0500	0.9524	1.00000	1.0000	1.05000	0.9524	0.000	0.000
2	1.1025	0.9070	0.48780	2.0500	0.53780	1.8594	0.907	0.488
3	1.1576	0.8638	0.31721	3.1525	0.36721	2.7232	2.635	0.967
4	1.2155	0.8227	0.23201	4.3101	0.28201	3.5460	5.103	1.439
5	1.2763	0.7835	0.18097	5.5256	0.23097	4.3295	8.237	1.903
6	1.3401	0.7462	0.14702	6.8019	0.19702	5.0757	11.968	2.358
7	1.4071	0.7107	0.12282	8.1420	0.17282	5.7864	16.232	2.805
8	1.4775	0.6768	0.10472	9.5491	0.15472	6.4632	20.970	3.245
9	1.5513	0.6446	0.09069	11.0266	0.14069	7.1078	26.127	3.676
10	1.6289	0.6139	0.07950	12.5779	0.12950	7.7217	31.652	4.099
11	1.7103	0.5847	0.07039	14.2068	0.12039	8.3064	37.499	4.514
12	1.7959	0.5568	0.06283	15.9171	0.11283	8.8633	43.624	4.922
13	1.8856	0.5303	0.05646	17.7130	0.10646	9.3936	49.988	5.322
14	1.9799	0.5051	0.05102	19.5986	0.10102	9.8986	56.554	5.713
15	2.0789	0.4810	0.04634	21.5786	0.09634	10.3797	63.288	6.097
16	2.1829	0.4581	0.04227	23.6575	0.09227	10.8378	70.160	6.474
17	2.2920	0.4363	0.03870	25.8404	0.08870	11.2741	77.140	6.842
18	2.4066	0.4155	0.03555	28.1324	0.08555	11.6896	84.204	7.203
19	2.5270	0.3957	0.03275	30.5390	0.08275	12.0853	91.328	7.557
20	2.6533	0.3769	0.03024	33.0660	0.08024	12.4622	98.488	7.903
21	2.7860	0.3589	0.02800	35.7193	0.07800	12.8212	105.667	8.242
22	2.9253	0.3418	0.02597	38.5052	0.07597	13.1630	112.846	8.573
23	3.0715	0.3256	0.02414	41.4305	0.07414	13.4886	120.009	8.897
24	3.2251	0.3101	0.02247	44.5020	0.07247	13.7986	127.140	9.214
25	3.3864	0.2953	0.02095	47.7271	0.07095	14.0939	134.228	9.524
26	3.5557	0.2812	0.01956	51.1135	0.06956	14.3752	141.259	9.827
27	3.7335	0.2678	0.01829	54.6691	0.06829	14.6430	148.223	10.122
28	3.9201	0.2551	0.01712	58.4026	0.06712	14.8981	155.110	10.411
29	4.1161	0.2429	0.01605	62.3227	0.06605	15.1411	161.913	10.694
30	4.3219	0.2314	0.01505	66.4388	0.06505	15.3725	168.623	10.969
32	4.7649	0.2099	0.01328	75.2988	0.06328	15.8027	181.739	11.501
34	5.2533	0.1904	0.01176	85.0670	0.06176	16.1929	194.417	12.006
36	5.7918	0.1727	0.01043	95.8363	0.06043	16.5469	206.624	12.487
38	6.3855	0.1566	0.00928	107.7095	0.05928	16.8679	218.338	12.944
40	7.0400	0.1420	0.00828	120.7998	0.05828	17.1591	229.545	13.377
42	7.7616	0.1288	0.00739	135.2318	0.05739	17.4232	240.239	13.788
44	8.5572	0.1169	0.00662	151.1430	0.05662	17.6628	250.417	14.178
46	9.4343	0.1060	0.00593	168.6852	0.05593	17.8801	260.084	14.546
48	10.4013	0.0961	0.00532	188.0254	0.05532	18.0772	269.247	14.894
50	11.4674	0.0872	0.00478	209.3480	0.05478	18.2559	277.915	15.223
60	18.6792	0.0535	0.00283	353.5837	0.05283	18.9293	314.343	16.606
70	30.4264	0.0329	0.00170	588.5285	0.05170	19.3427	340.841	17.621
80	49.5614	0.0202	0.00103	971.2288	0.05103	19.5965	359.646	18.353
90	80.7304	0.0124	0.00063	1594.6073	0.05063	19.7523	372.749	18.871
100	131.5013	0.0076	0.00038	2610.0251	0.05038	19.8479	381.749	19.234

i = 6%

	SINGLE PAYMENT			UNIFORM SERIES			GRADIENT	
	COMPOUND AMOUNT	PRESENT WORTH	SINKING FUND	COMPOUND AMOUNT	CAPITAL RECOVERY	PRESENT WORTH	PRESENT WORTH	ANNUAL COST
n	F/P	P/F	A/F	F/A	A/P	P/A	P/G	A/G
1	1.0600	0.9434	1.00000	1.0000	1.06000	0.9434	0.000	0.000
2	1.1236	0.8900	0.48544	2.0600	0.54544	1.8334	0.890	0.485
3	1.1910	0.8396	0.31411	3.1836	0.37411	2.6730	2.569	0.961
4	1.2625	0.7921	0.22859	4.3746	0.28859	3.4651	4.946	1.427
5	1.3382	0.7473	0.17740	5.6371	0.23740	4.2124	7.935	1.884
6	1.4185	0.7050	0.14336	6.9753	0.20336	4.9173	11.459	2.330
7	1.5036	0.6651	0.11914	8.3938	0.17914	5.5824	15.450	2.768
8	1.5938	0.6274	0.10104	9.8975	0.16104	6.2098	19.842	3.195
9	1.6895	0.5919	0.08702	11.4913	0.14702	6.8017	24.577	3.613
10	1.7908	0.5584	0.07587	13.1808	0.13587	7.3601	29.602	4.022
11	1.8983	0.5268	0.06679	14.9716	0.12679	7.8869	34.870	4.421
12	2.0122	0.4970	0.05928	16.8699	0.11928	8.3838	40.337	4.811
13	2.1329	0.4688	0.05296	18.8821	0.11296	8.8527	45.963	5.192
14	2.2609	0.4423	0.04758	21.0151	0.10758	9.2950	51.713	5.564
15	2.3966	0.4173	0.04296	23.2760	0.10296	9.7122	57.555	5.926
16	2.5404	0.3936	0.03895	25.6725	0.09895	10.1059	63.459	6.279
17	2.6928	0.3714	0.03544	28.2129	0.09544	10.4773	69.401	6.624
18	2.8543	0.3503	0.03236	30.9057	0.09236	10.8276	75.357	6.960
19	3.0256	0.3305	0.02962	33.7600	0.08962	11.1581	81.306	7.287
20	3.2071	0.3118	0.02718	36.7856	0.08718	11.4699	87.230	7.605
21	3.3996	0.2942	0.02500	39.9927	0.08500	11.7641	93.114	7.915
22	3.6035	0.2775	0.02305	43.3923	0.08305	12.0416	98.941	8.217
23	3.8197	0.2618	0.02128	46.9958	0.08128	12.3034	104.701	8.510
24	4.0489	0.2470	0.01968	50.8156	0.07968	12.5504	110.381	8.795
25	4.2919	0.2330	0.01823	54.8645	0.07823	12.7834	115.973	9.072
26	4.5494	0.2198	0.01690	59.1564	0.07690	13.0032	121.468	9.341
27	4.8223	0.2074	0.01570	63.7058	0.07570	13.2105	126.860	9.603
28	5.1117	0.1956	0.01459	68.5281	0.07459	13.4062	132.142	9.857
29	5.4184	0.1846	0.01358	73.6398	0.07358	13.5907	137.310	10.103
30	5.7435	0.1741	0.01265	79.0582	0.07265	13.7648	142.359	10.342
32	6.4534	0.1550	0.01100	90.8898	0.07100	14.0840	152.090	10.799
34	7.2510	0.1379	0.00960	104.1838	0.06960	14.3681	161.319	11.228
36	8.1473	0.1227	0.00839	119.1209	0.06839	14.6210	170.039	11.630
38	9.1543	0.1092	0.00736	135.9042	0.06736	14.8460	178.249	12.007
40	10.2857	0.0972	0.00646	154.7620	0.06646	15.0463	185.957	12.359
42	11.5570	0.0865	0.00568	175.9505	0.06568	15.2245	193.173	12.688
44	12.9855	0.0770	0.00501	199.7580	0.06501	15.3832	199.913	12.996
46	14.5905	0.0685	0.00441	226.5081	0.06441	15.5244	206.194	13.282
48	16.3939	0.0610	0.00390	256.5645	0.06390	15.6500	212.035	13.549
50	18.4202	0.0543	0.00344	290.3359	0.06344	15.7619	217.457	13.796
60	32.9877	0.0303	0.00188	533.1282	0.06188	16.1614	239.043	14.791
70	59.0759	0.0169	0.00103	967.9322	0.06103	16.3845	253.327	15.461
80	105.7960	0.0095	0.00057	1746.6000	0.06057	16.5091	262.549	15.903
90	189.4645	0.0053	0.00032	3141.0754	0.06032	16.5787	268.395	16.189
100	339.3021	0.0029	0.00018	5638.3684	0.06018	16.6175	272.047	16.371

i = 7%

	SINGLE PAYMENT		UNIFORM SERIES				GRADIENT	
	COMPOUND AMOUNT	PRESENT WORTH	SINKING FUND	COMPOUND AMOUNT	CAPITAL RECOVERY	PRESENT WORTH	PRESENT WORTH	ANNUAL COST
n	F/P	P/F	A/F	F/A	A/P	P/A	P/G	A/G
1	1.0700	0.9346	1.00000	1.0000	1.07000	0.9346	0.000	0.000
2	1.1449	0.8734	0.48309	2.0700	0.55309	1.8080	0.873	0.483
3	1.2250	0.8163	0.31105	3.2149	0.38105	2.6243	2.506	0.955
4	1.3108	0.7629	0.22523	4.4399	0.29523	3.3872	4.795	1.416
5	1.4026	0.7130	0.17389	5.7507	0.24389	4.1002	7.647	1.865
6	1.5007	0.6663	0.13980	7.1533	0.20980	4.7665	10.978	2.303
7	1.6058	0.6227	0.11555	8.6540	0.18555	5.3893	14.715	2.730
8	1.7182	0.5820	0.09747	10.2598	0.16747	5.9713	18.789	3.147
9	1.8385	0.5439	0.08349	11.9780	0.15349	6.5152	23.140	3.552
10	1.9672	0.5083	0.07238	13.8164	0.14238	7.0236	27.716	3.946
11	2.1049	0.4751	0.06336	15.7836	0.13336	7.4987	32.466	4.330
12	2.2522	0.4440	0.05590	17.8885	0.12590	7.9427	37.351	4.703
13	2.4098	0.4150	0.04965	20.1406	0.11965	8.3577	42.330	5.065
14	2.5785	0.3878	0.04434	22.5505	0.11434	8.7455	47.372	5.417
15	2.7590	0.3624	0.03979	25.1290	0.10979	9.1079	52.446	5.758
16	2.9522	0.3387	0.03586	27.8881	0.10586	9.4466	57.527	6.090
17	3.1588	0.3166	0.03243	30.8402	0.10243	9.7632	62.592	6.411
18	3.3799	0.2959	0.02941	33.9990	0.09941	10.0591	67.622	6.722
19	3.6165	0.2765	0.02675	37.3790	0.09675	10.3356	72.599	7.024
20	3.8697	0.2584	0.02439	40.9955	0.09439	10.5940	77.509	7.316
21	4.1406	0.2415	0.02229	44.8652	0.09229	10.8355	82.339	7.599
22	4.4304	0.2257	0.02041	49.0057	0.09041	11.0612	87.079	7.872
23	4.7405	0.2109	0.01871	53.4361	0.08871	11.2722	91.720	8.137
24	5.0724	0.1971	0.01719	58.1767	0.08719	11.4693	96.255	8.392
25	5.4274	0.1842	0.01581	63.2490	0.08581	11.6536	100.676	8.639
26	5.8074	0.1722	0.01456	68.6765	0.08456	11.8258	104.981	8.877
27	6.2139	0.1609	0.01343	74.4838	0.08343	11.9867	109.166	9.107
28	6.6488	0.1504	0.01239	80.6977	0.08239	12.1371	113.226	9.329
29	7.1143	0.1406	0.01145	87.3465	0.08145	12.2777	117.162	9.543
30	7.6123	0.1314	0.01059	94.4608	0.08059	12.4090	120.972	9.749
32	8.7153	0.1147	0.00907	110.2182	0.07907	12.6466	128.212	10.138
34	9.9781	0.1002	0.00780	128.2588	0.07780	12.8540	134.951	10.499
36	11.4239	0.0875	0.00672	148.9135	0.07672	13.0352	141.199	10.832
38	13.0793	0.0765	0.00580	172.5610	0.07580	13.1935	146.973	11.140
40	14.9745	0.0668	0.00501	199.6351	0.07501	13.3317	152.293	11.423
42	17.1443	0.0583	0.00434	230.6322	0.07434	13.4524	157.181	11.684
44	19.6285	0.0509	0.00376	266.1209	0.07376	13.5579	161.661	11.924
46	22.4726	0.0445	0.00326	306.7518	0.07326	13.6500	165.758	12.143
48	25.7289	0.0389	0.00283	353.2701	0.07283	13.7305	169.498	12.345
50	29.4570	0.0339	0.00246	406.5289	0.07246	13.8007	172.905	12.529
60	57.9464	0.0173	0.00123	813.5204	0.07123	14.0392	185.768	13.232
70	113.9894	0.0088	0.00062	1614.1342	0.07062	14.1604	193.519	13.666
80	224.2344	0.0045	0.00031	3189.0628	0.07031	14.2220	198.075	13.927
90	441.1030	0.0023	0.00016	6287.1857	0.07016	14.2533	200.704	14.081
100	867.7164	0.0012	0.00008	12381.6624	0.07008	14.2693	202.200	14.170

i = 8%

	SINGLE PAYMENT			UNIFORM SERIES			GRADIENT	
	COMPOUND AMOUNT	PRESENT WORTH	SINKING FUND	COMPOUND AMOUNT	CAPITAL RECOVERY	PRESENT WORTH	PRESENT WORTH	ANNUAL COST
n	F/P	P/F	A/F	F/A	A/P	P/A	P/G	A/G
1	1.0800	0.9259	1.00000	1.0000	1.08000	0.9259	0.000	0.000
2	1.1664	0.8573	0.48077	2.0800	0.56077	1.7833	0.857	0.481
3	1.2597	0.7938	0.30803	3.2464	0.38803	2.5771	2.445	0.949
4	1.3605	0.7350	0.22192	4.5061	0.30192	3.3121	4.650	1.404
5	1.4693	0.6806	0.17046	5.8666	0.25046	3.9927	7.372	1.846
6	1.5869	0.6302	0.13632	7.3359	0.21632	4.6229	10.523	2.276
7	1.7138	0.5835	0.11207	8.9228	0.19207	5.2064	14.024	2.694
8	1.8509	0.5403	0.09401	10.6366	0.17401	5.7466	17.806	3.099
9	1.9990	0.5002	0.08008	12.4876	0.16008	6.2469	21.808	3.491
10	2.1589	0.4632	0.06903	14.4866	0.14903	6.7101	25.977	3.871
11	2.3316	0.4289	0.06008	16.6455	0.14008	7.1390	30.266	4.240
12	2.5182	0.3971	0.05270	18.9771	0.13270	7.5361	34.634	4.596
13	2.7196	0.3677	0.04652	21.4953	0.12652	7.9038	39.046	4.940
14	2.9372	0.3405	0.04130	24.2149	0.12130	8.2442	43.472	5.273
15	3.1722	0.3152	0.03683	27.1521	0.11683	8.5595	47.886	5.594
16	3.4259	0.2919	0.03298	30.3243	0.11298	8.8514	52.264	5.905
17	3.7000	0.2703	0.02963	33.7502	0.10963	9.1216	56.588	6.204
18	3.9960	0.2502	0.02670	37.4502	0.10670	9.3719	60.843	6.492
19	4.3157	0.2317	0.02413	41.4463	0.10413	9.6036	65.013	6.770
20	4.6610	0.2145	0.02185	45.7620	0.10185	9.8181	69.090	7.037
21	5.0338	0.1987	0.01983	50.4229	0.09983	10.0168	73.063	7.294
22	5.4365	0.1839	0.01803	55.4568	0.09803	10.2007	76.926	7.541
23	5.8715	0.1703	0.01642	60.8933	0.09642	10.3711	80.673	7.779
24	6.3412	0.1577	0.01498	66.7648	0.09498	10.5288	84.300	8.007
25	6.8485	0.1460	0.01368	73.1059	0.09368	10.6748	87.804	8.225
26	7.3964	0.1352	0.01251	79.9544	0.09251	10.8100	91.184	8.435
27	7.9881	0.1252	0.01145	87.3508	0.09145	10.9352	94.439	8.636
28	8.6271	0.1159	0.01049	95.3388	0.09049	11.0511	97.569	8.829
29	9.3173	0.1073	0.00962	103.9659	0.08962	11.1584	100.574	9.013
30	10.0627	0.0994	0.00883	113.2832	0.08883	11.2578	103.456	9.190
32	11.7371	0.0852	0.00745	134.2135	0.08745	11.4350	108.857	9.520
34	13.6901	0.0730	0.00630	158.6267	0.08630	11.5869	113.792	9.821
36	15.9682	0.0626	0.00534	187.1021	0.08534	11.7172	118.284	10.095
38	18.6253	0.0537	0.00454	220.3159	0.08454	11.8289	122.358	10.344
40	21.7245	0.0460	0.00386	259.0565	0.08386	11.9246	126.042	10.570
42	25.3395	0.0395	0.00329	304.2435	0.08329	12.0067	129.365	10.774
44	29.5560	0.0338	0.00280	356.9496	0.08280	12.0771	132.355	10.959
46	34.4741	0.0290	0.00239	418.4261	0.08239	12.1374	135.038	11.126
48	40.2106	0.0249	0.00204	490.1322	0.08204	12.1891	137.443	11.276
50	46.9016	0.0213	0.00174	573.7702	0.08174	12.2335	139.593	11.411
60	101.2571	0.0099	0.00080	1253.2133	0.08080	12.3766	147.300	11.902
70	218.6064	0.0046	0.00037	2720.0801	0.08037	12.4428	151.533	12.178
80	471.9548	0.0021	0.00017	5886.9355	0.08017	12.4735	153.800	12.330
90	1018.9151	0.0010	0.00008	12723.9389	0.08008	12.4877	154.993	12.412
100	2199.7613	0.0005	0.00004	27484.5164	0.08004	12.4943	155.611	12.455

i = 9%

	SINGLE PAYMENT		UNIFORM SERIES				GRADIENT	
	COMPOUND AMOUNT	PRESENT WORTH	SINKING FUND	COMPOUND AMOUNT	CAPITAL RECOVERY	PRESENT WORTH	PRESENT WORTH	ANNUAL COST
n	F/P	P/F	A/F	F/A	A/P	P/A	P/G	A/G
1	1.0900	0.9174	1.00000	1.0000	1.09000	0.9174	0.000	0.000
2	1.1881	0.8417	0.47847	2.0900	0.56847	1.7591	0.842	0.478
3	1.2950	0.7722	0.30505	3.2781	0.39505	2.5313	2.386	0.943
4	1.4116	0.7084	0.21867	4.5731	0.30867	3.2397	4.511	1.393
5	1.5386	0.6499	0.16709	5.9847	0.25709	3.8897	7.111	1.828
6	1.6771	0.5963	0.13292	7.5233	0.22292	4.4859	10.092	2.250
7	1.8280	0.5470	0.10869	9.2004	0.19869	5.0330	13.375	2.657
8	1.9926	0.5019	0.09067	11.0285	0.18067	5.5348	16.888	3.051
9	2.1719	0.4604	0.07680	13.0210	0.16680	5.9952	20.571	3.431
10	2.3674	0.4224	0.06582	15.1929	0.15582	6.4177	24.373	3.798
11	2.5804	0.3875	0.05695	17.5603	0.14695	6.8052	28.248	4.151
12	2.8127	0.3555	0.04965	20.1407	0.13965	7.1607	32.159	4.491
13	3.0658	0.3262	0.04357	22.9534	0.13357	7.4869	36.073	4.818
14	3.3417	0.2992	0.03843	26.0192	0.12843	7.7862	39.963	5.133
15	3.6425	0.2745	0.03406	29.3609	0.12406	8.0607	43.807	5.435
16	3.9703	0.2519	0.03030	33.0034	0.12030	8.3126	47.585	5.724
17	4.3276	0.2311	0.02705	36.9737	0.11705	8.5436	51.282	6.002
18	4.7171	0.2120	0.02421	41.3013	0.11421	8.7556	54.886	6.269
19	5.1417	0.1945	0.02173	46.0185	0.11173	8.9501	58.387	6.524
20	5.6044	0.1784	0.01955	51.1601	0.10955	9.1285	61.777	6.767
21	6.1088	0.1637	0.01762	56.7645	0.10762	9.2922	65.051	7.001
22	6.6586	0.1502	0.01590	62.8733	0.10590	9.4424	68.205	7.223
23	7.2579	0.1378	0.01438	69.5319	0.10438	9.5802	71.236	7.436
24	7.9111	0.1264	0.01302	76.7898	0.10302	9.7066	74.143	7.638
25	8.6231	0.1160	0.01181	84.7009	0.10181	9.8226	76.926	7.832
26	9.3992	0.1064	0.01072	93.3240	0.10072	9.9290	79.586	8.016
27	10.2451	0.0976	0.00973	102.7231	0.09973	10.0266	82.124	8.191
28	11.1671	0.0895	0.00885	112.9682	0.09885	10.1161	84.542	8.357
29	12.1722	0.0822	0.00806	124.1354	0.09806	10.1983	86.842	8.515
30	13.2677	0.0754	0.00734	136.3075	0.09734	10.2737	89.028	8.666
32	15.7633	0.0634	0.00610	164.0370	0.09610	10.4062	93.069	8.944
34	18.7284	0.0534	0.00508	196.9823	0.09508	10.5178	96.693	9.193
36	22.2512	0.0449	0.00424	236.1247	0.09424	10.6118	99.932	9.417
38	26.4367	0.0378	0.00354	282.6298	0.09354	10.6908	102.816	9.617
40	31.4094	0.0318	0.00296	337.8824	0.09296	10.7574	105.376	9.796
42	37.3175	0.0268	0.00248	403.5281	0.09248	10.8134	107.643	9.955
44	44.3370	0.0226	0.00208	481.5218	0.09208	10.8605	109.646	10.096
46	52.6767	0.0190	0.00174	574.1860	0.09174	10.9002	111.410	10.221
48	62.5852	0.0160	0.00146	684.2804	0.09146	10.9336	112.962	10.332
50	74.3575	0.0134	0.00123	815.0836	0.09123	10.9617	114.325	10.430
60	176.0313	0.0057	0.00051	1944.7921	0.09051	11.0480	119.968	10.768
70	416.7301	0.0024	0.00022	4619.2232	0.09022	11.0844	121.294	10.943
80	986.5517	0.0010	0.00009	10950.5742	0.09009	11.0998	122.431	11.030
90	2335.5266	0.0004	0.00004	25939.1844	0.09004	11.1064	122.976	11.073
100	5529.0408	0.0002	0.00002	61422.6757	0.09002	11.1091	123.234	11.093

i = 10%

	SINGLE PAYMENT		UNIFORM SERIES				GRADIENT	
	COMPOUND AMOUNT	PRESENT WORTH	SINKING FUND	COMPOUND AMOUNT	CAPITAL RECOVERY	PRESENT WORTH	PRESENT WORTH	ANNUAL COST
n	F/P	P/F	A/F	F/A	A/P	P/A	P/G	A/G
1	1.1000	0.9091	1.00000	1.0000	1.10000	0.9091	0.000	0.000
2	1.2100	0.8264	0.47619	2.1000	0.57619	1.7355	0.826	0.476
3	1.3310	0.7513	0.30211	3.3100	0.40211	2.4869	2.329	0.937
4	1.4641	0.6830	0.21547	4.6410	0.31547	3.1699	4.378	1.381
5	1.6105	0.6209	0.16380	6.1051	0.26380	3.7908	6.862	1.810
6	1.7716	0.5645	0.12961	7.7156	0.22961	4.3553	9.684	2.224
7	1.9487	0.5132	0.10541	9.4872	0.20541	4.8684	12.763	2.622
8	2.1436	0.4665	0.08744	11.4359	0.18744	5.3349	16.029	3.004
9	2.3579	0.4241	0.07364	13.5795	0.17364	5.7590	19.421	3.372
10	2.5937	0.3855	0.06275	15.9374	0.16275	6.1446	22.891	3.725
11	2.8531	0.3505	0.05396	18.5312	0.15396	6.4951	26.396	4.064
12	3.1384	0.3186	0.04676	21.3843	0.14676	6.8137	29.901	4.388
13	3.4523	0.2897	0.04078	24.5227	0.14078	7.1034	33.377	4.699
14	3.7975	0.2633	0.03575	27.9750	0.13575	7.3667	36.800	4.996
15	4.1772	0.2394	0.03147	31.7725	0.13147	7.6061	40.152	5.279
16	4.5950	0.2176	0.02782	35.9497	0.12782	7.8237	43.416	5.549
17	5.0545	0.1978	0.02466	40.5447	0.12466	8.0216	46.582	5.807
18	5.5599	0.1799	0.02193	45.5992	0.12193	8.2014	49.640	6.053
19	6.1159	0.1635	0.01955	51.1591	0.11955	8.3649	52.583	6.286
20	6.7275	0.1486	0.01746	57.2750	0.11746	8.5136	55.407	6.508
21	7.4003	0.1351	0.01562	64.0025	0.11562	8.6487	58.110	6.719
22	8.1403	0.1228	0.01401	71.4028	0.11401	8.7715	60.689	6.919
23	8.9543	0.1117	0.01257	79.5430	0.11257	8.8832	63.146	7.108
24	9.8497	0.1015	0.01130	88.4973	0.11130	8.9847	65.481	7.288
25	10.8347	0.0923	0.01017	98.3471	0.11017	9.0770	67.696	7.458
26	11.9182	0.0839	0.00916	109.1818	0.10916	9.1609	69.794	7.619
27	13.1100	0.0763	0.00826	121.0999	0.10826	9.2372	71.777	7.770
28	14.4210	0.0693	0.00745	134.2099	0.10745	9.3066	73.650	7.914
29	15.8631	0.0630	0.00673	148.6309	0.10673	9.3696	75.415	8.049
30	17.4494	0.0573	0.00608	164.4940	0.10608	9.4269	77.077	8.176
32	21.1138	0.0474	0.00497	201.1378	0.10497	9.5264	80.108	8.409
34	25.5477	0.0391	0.00407	245.4767	0.10407	9.6086	82.777	8.615
36	30.9127	0.0323	0.00334	299.1268	0.10334	9.6765	85.119	8.796
38	37.4043	0.0267	0.00275	364.0434	0.10275	9.7327	87.167	8.956
40	45.2593	0.0221	0.00226	442.5926	0.10226	9.7791	88.953	9.096
42	54.7637	0.0183	0.00186	537.6370	0.10186	9.8174	90.505	9.219
44	66.2641	0.0151	0.00153	652.6408	0.10153	9.8491	91.851	9.326
46	80.1795	0.0125	0.00126	791.7953	0.10126	9.8753	93.016	9.419
48	97.0172	0.0103	0.00104	960.1724	0.10104	9.8969	94.022	9.500
50	117.3909	0.0085	0.00086	1163.9086	0.10086	9.9148	94.889	9.570
60	304.4816	0.0033	0.00033	3034.8165	0.10033	9.9672	97.701	9.802
70	789.7470	0.0013	0.00013	7887.4699	0.10013	9.9873	98.987	9.911
80	2048.4003	0.0005	0.00005	20474.0031	0.10005	9.9951	99.561	9.961
90	5313.0229	0.0002	0.00002	53120.2288	0.10002	9.9981	99.812	9.983
100	13780.6132	0.0001	0.00001	137796.130	0.10001	9.9993	99.920	9.993

i = 12%

n	SINGLE PAYMENT		UNIFORM SERIES				GRADIENT	
	COMPOUND AMOUNT F/P	PRESENT WORTH P/F	SINKING FUND A/F	COMPOUND AMOUNT F/A	CAPITAL RECOVERY A/P	PRESENT WORTH P/A	PRESENT WORTH P/G	ANNUAL COST A/G
1	1.1200	0.8929	1.00000	1.0000	1.12000	0.8929	0.000	0.000
2	1.2544	0.7972	0.47170	2.1200	0.59170	1.6901	0.797	0.472
3	1.4049	0.7118	0.29635	3.3744	0.41635	2.4018	2.221	0.925
4	1.5735	0.6355	0.20923	4.7793	0.32923	3.0373	4.127	1.359
5	1.7623	0.5674	0.15741	6.3528	0.27741	3.6048	6.397	1.775
6	1.9738	0.5066	0.12323	8.1152	0.24323	4.1114	8.930	2.172
7	2.2107	0.4523	0.09912	10.0890	0.21912	4.5638	11.644	2.551
8	2.4760	0.4039	0.08130	12.2997	0.20130	4.9676	14.471	2.913
9	2.7731	0.3606	0.06768	14.7757	0.18768	5.3282	17.356	3.257
10	3.1058	0.3220	0.05698	17.5487	0.17698	5.6502	20.254	3.585
11	3.4786	0.2875	0.04842	20.6546	0.16842	5.9377	23.129	3.895
12	3.8960	0.2567	0.04144	24.1331	0.16144	6.1944	25.952	4.190
13	4.3635	0.2292	0.03568	28.0291	0.15568	6.4235	28.702	4.468
14	4.8871	0.2046	0.03087	32.3926	0.15087	6.6282	31.362	4.732
15	5.4736	0.1827	0.02682	37.2797	0.14682	6.8109	33.920	4.980
16	6.1304	0.1631	0.02339	42.7533	0.14339	6.9740	36.367	5.215
17	6.8660	0.1456	0.02046	48.8837	0.14046	7.1196	38.697	5.435
18	7.6900	0.1300	0.01794	55.7497	0.13794	7.2497	40.908	5.643
19	8.6128	0.1161	0.01576	63.4397	0.13576	7.3658	42.998	5.838
20	9.6463	0.1037	0.01388	72.0524	0.13388	7.4694	44.968	6.020
21	10.8038	0.0926	0.01224	81.6987	0.13224	7.5620	46.819	6.191
22	12.1003	0.0826	0.01081	92.5026	0.13081	7.6446	48.554	6.351
23	13.5523	0.0738	0.00956	104.6029	0.12956	7.7184	50.178	6.501
24	15.1786	0.0659	0.00846	118.1552	0.12846	7.7843	51.693	6.641
25	17.0001	0.0588	0.00750	133.3339	0.12750	7.8431	53.105	6.771
26	19.0401	0.0525	0.00665	150.3339	0.12665	7.8957	54.418	6.892
27	21.3249	0.0469	0.00590	169.3740	0.12590	7.9426	55.637	7.005
28	23.8839	0.0419	0.00524	190.6989	0.12524	7.9844	56.767	7.110
29	26.7499	0.0374	0.00466	214.5828	0.12466	8.0218	57.814	7.207
30	29.9599	0.0334	0.00414	241.3327	0.12414	8.0552	58.782	7.297
32	37.5817	0.0266	0.00328	304.8477	0.12328	8.1116	60.501	7.459
34	47.1425	0.0212	0.00260	384.5210	0.12260	8.1566	61.961	7.596
36	59.1356	0.0169	0.00206	484.4631	0.12206	8.1924	63.197	7.714
38	74.1797	0.0135	0.00164	609.8305	0.12164	8.2210	64.239	7.814
40	93.0510	0.0107	0.00130	767.0914	0.12130	8.2438	65.116	7.899
42	116.7231	0.0086	0.00104	964.3595	0.12104	8.2619	65.851	7.970
44	146.4175	0.0068	0.00083	1211.8125	0.12083	8.2764	66.466	8.031
46	183.6661	0.0054	0.00066	1522.2176	0.12066	8.2880	66.979	8.082
48	230.3908	0.0043	0.00052	1911.5898	0.12052	8.2972	67.407	8.124
50	289.0022	0.0035	0.00042	2400.0183	0.12042	8.3045	67.762	8.160
60	897.5969	0.0011	0.00013	7471.6412	0.12013	8.3240	68.810	8.266
70	2787.7999	0.0004	0.00004	23223.3322	0.12004	8.3303	69.210	8.308
80	8658.4832	0.0001	0.00001	72145.6934	0.12001	8.3324	69.359	8.324
90	26891.9346	0.0000	0.00000	224091.122	0.12000	8.3330	69.414	8.330
100	83522.2671	0.0000	0.00000	696010.558	0.12000	8.3332	69.434	8.332

i = 15%

	SINGLE PAYMENT		UNIFORM SERIES				GRADIENT	
	COMPOUND AMOUNT	PRESENT WORTH	SINKING FUND	COMPOUND AMOUNT	CAPITAL RECOVERY	PRESENT WORTH	PRESENT WORTH	ANNUAL COST
n	F/P	P/F	A/F	F/A	A/P	P/A	P/G	A/G
1	1.1500	0.8696	1.00000	1.0000	1.15000	0.8696	0.000	0.000
2	1.3225	0.7561	0.46512	2.1500	0.61512	1.6257	0.756	0.465
3	1.5209	0.6575	0.28798	3.4725	0.43798	2.2832	2.071	0.907
4	1.7490	0.5718	0.20027	4.9934	0.35027	2.8550	3.786	1.326
5	2.0114	0.4972	0.14832	6.7424	0.29832	3.3522	5.775	1.723
6	2.3131	0.4323	0.11424	8.7537	0.26424	3.7845	7.937	2.097
7	2.6600	0.3759	0.09036	11.0668	0.24036	4.1604	10.192	2.450
8	3.0590	0.3269	0.07285	13.7268	0.22285	4.4873	12.481	2.781
9	3.5179	0.2843	0.05957	16.7858	0.20957	4.7716	14.755	3.092
10	4.0456	0.2472	0.04925	20.3037	0.19925	5.0188	16.979	3.383
11	4.6524	0.2149	0.04107	24.3493	0.19107	5.2337	19.129	3.655
12	5.3503	0.1869	0.03448	29.0017	0.18448	5.4206	21.185	3.908
13	6.1528	0.1625	0.02911	34.3519	0.17911	5.5831	23.135	4.144
14	7.0757	0.1413	0.02469	40.5047	0.17469	5.7245	24.972	4.362
15	8.1371	0.1229	0.02102	47.5804	0.17102	5.8474	26.693	4.565
16	9.3576	0.1069	0.01795	55.7175	0.16795	5.9542	28.296	4.752
17	10.7613	0.0929	0.01537	65.0751	0.16537	6.0472	29.783	4.925
18	12.3755	0.0808	0.01319	75.8364	0.16319	6.1280	31.156	5.084
19	14.2318	0.0703	0.01134	88.2118	0.16134	6.1982	32.421	5.231
20	16.3665	0.0611	0.00976	102.4436	0.15976	6.2593	33.582	5.365
21	18.8215	0.0531	0.00842	118.8101	0.15842	6.3125	34.645	5.488
22	21.6447	0.0462	0.00727	137.6316	0.15727	6.3587	35.615	5.601
23	24.8915	0.0402	0.00628	159.2764	0.15628	6.3988	36.499	5.704
24	28.6252	0.0349	0.00543	184.1678	0.15543	6.4338	37.302	5.798
25	32.9190	0.0304	0.00470	212.7930	0.15470	6.4641	38.031	5.883
26	37.8568	0.0264	0.00407	245.7120	0.15407	6.4906	38.692	5.961
27	43.5353	0.0230	0.00353	283.5688	0.15353	6.5135	39.289	6.032
28	50.0656	0.0200	0.00306	327.1041	0.15306	6.5335	39.828	6.096
29	57.5755	0.0174	0.00265	377.1697	0.15265	6.5509	40.315	6.154
30	66.2118	0.0151	0.00230	434.7452	0.15230	6.5660	40.753	6.207
32	87.5651	0.0114	0.00173	577.1005	0.15173	6.5905	41.501	6.297
34	115.8048	0.0086	0.00131	765.3654	0.15131	6.6091	42.103	6.371
36	153.1519	0.0065	0.00099	1014.3457	0.15099	6.6231	42.587	6.430
38	202.5433	0.0049	0.00074	1343.6222	0.15074	6.6338	42.974	6.478
40	267.8636	0.0037	0.00056	1779.0903	0.15056	6.6418	43.283	6.517
42	354.2495	0.0028	0.00042	2354.9970	0.15042	6.6478	43.529	6.548
44	468.4950	0.0021	0.00032	3116.6335	0.15032	6.6524	43.723	6.573
46	619.5847	0.0016	0.00024	4123.8978	0.15024	6.6559	43.878	6.592
48	819.4007	0.0012	0.00018	5456.0049	0.15018	6.6585	44.000	6.608
50	1083.6575	0.0009	0.00014	7217.7164	0.15014	6.6605	44.096	6.620
60	4383.9989	0.0002	0.00003	29219.9924	0.15003	6.6651	44.343	6.653
70	17735.7206	0.0001	0.00001	118231.470	0.15001	6.6663	44.416	6.663
80	71750.8819	0.0000	0.00000	478332.545	0.15000	6.6666	44.436	6.666

i = 20%

	SINGLE PAYMENT		UNIFORM SERIES				GRADIENT	
	COMPOUND AMOUNT	PRESENT WORTH	SINKING FUND	COMPOUND AMOUNT	CAPITAL RECOVERY	PRESENT WORTH	PRESENT WORTH	ANNUAL COST
n	F/P	P/F	A/F	F/A	A/P	P/A	P/G	A/G
1	1.2000	0.8333	1.00000	1.0000	1.20000	0.8333	0.000	0.000
2	1.4400	0.6944	0.45455	2.2000	0.65455	1.5278	0.694	0.455
3	1.7280	0.5787	0.27473	3.6400	0.47473	2.1065	1.852	0.879
4	2.0736	0.4823	0.18629	5.3680	0.38629	2.5887	3.299	1.274
5	2.4883	0.4019	0.13438	7.4416	0.33438	2.9906	4.906	1.641
6	2.9860	0.3349	0.10071	9.9299	0.30071	3.3255	6.581	1.979
7	3.5832	0.2791	0.07742	12.9159	0.27742	3.6046	8.255	2.290
8	4.2998	0.2326	0.06061	16.4991	0.26061	3.8372	9.883	2.576
9	5.1598	0.1938	0.04808	20.7989	0.24808	4.0310	11.434	2.836
10	6.1917	0.1615	0.03852	25.9587	0.23852	4.1925	12.887	3.074
11	7.4301	0.1346	0.03110	32.1504	0.23110	4.3271	14.233	3.289
12	8.9161	0.1122	0.02526	39.5805	0.22526	4.4392	15.467	3.484
13	10.6993	0.0935	0.02062	48.4966	0.22062	4.5327	16.588	3.660
14	12.8392	0.0779	0.01689	59.1959	0.21689	4.6106	17.601	3.817
15	15.4070	0.0649	0.01388	72.0351	0.21388	4.6755	18.509	3.959
16	18.4884	0.0541	0.01144	87.4421	0.21144	4.7296	19.321	4.085
17	22.1861	0.0451	0.00944	105.9306	0.20944	4.7746	20.042	4.198
18	26.6233	0.0376	0.00781	128.1167	0.20781	4.8122	20.680	4.298
19	31.9480	0.0313	0.00646	154.7400	0.20646	4.8435	21.244	4.386
20	38.3376	0.0261	0.00536	186.6880	0.20536	4.8696	21.739	4.464
21	46.0051	0.0217	0.00444	225.0256	0.20444	4.8913	22.174	4.533
22	55.2061	0.0181	0.00369	271.0307	0.20369	4.9094	22.555	4.594
23	66.2474	0.0151	0.00307	326.2369	0.20307	4.9245	22.887	4.647
24	79.4968	0.0126	0.00255	392.4842	0.20255	4.9371	23.176	4.694
25	95.3962	0.0105	0.00212	471.9811	0.20212	4.9476	23.428	4.735
26	114.4755	0.0087	0.00176	567.3773	0.20176	4.9563	23.646	4.771
27	137.3706	0.0073	0.00147	681.8528	0.20147	4.9636	23.835	4.802
28	164.8447	0.0061	0.00122	819.2233	0.20122	4.9697	23.999	4.829
29	197.8136	0.0051	0.00102	984.0680	0.20102	4.9747	24.141	4.853
30	237.3763	0.0042	0.00085	1181.8816	0.20085	4.9789	24.263	4.873
32	341.8219	0.0029	0.00059	1704.1095	0.20059	4.9854	24.459	4.906
34	492.2235	0.0020	0.00041	2456.1176	0.20041	4.9898	24.604	4.931
36	708.8019	0.0014	0.00028	3539.0094	0.20028	4.9929	24.711	4.949
38	1020.6747	0.0010	0.00020	5098.3735	0.20020	4.9951	24.789	4.963
40	1469.7716	0.0007	0.00014	7343.8579	0.20014	4.9966	24.847	4.973
42	2116.4711	0.0005	0.00009	10577.3554	0.20009	4.9976	24.889	4.980
44	3047.7184	0.0003	0.00007	15233.5918	0.20007	4.9984	24.920	4.986
46	4388.7144	0.0002	0.00005	21938.5722	0.20005	4.9989	24.942	4.990
48	6319.7488	0.0002	0.00003	31593.7439	0.20003	4.9992	24.958	4.992
50	9100.4382	0.0001	0.00002	45497.1913	0.20002	4.9995	24.970	4.995
60	56347.5151	0.0000	0.00000	281732.574	0.20000	4.9999	24.994	4.999

i = 25%

| | SINGLE PAYMENT | | | UNIFORM SERIES | | | GRADIENT | |
| | COMPOUND AMOUNT | PRESENT WORTH | SINKING FUND | COMPOUND AMOUNT | CAPITAL RECOVERY | PRESENT WORTH | PRESENT WORTH | ANNUAL COST |
n	F/P	P/F	A/F	F/A	A/P	P/A	P/G	A/G
1	1.2500	0.8000	1.00000	1.0000	1.25000	0.8000	0.000	0.000
2	1.5625	0.6400	0.44444	2.2500	0.69444	1.4400	0.640	0.444
3	1.9531	0.5120	0.26230	3.8125	0.51230	1.9520	1.664	0.852
4	2.4414	0.4096	0.17344	5.7656	0.42344	2.3616	2.893	1.225
5	3.0518	0.3277	0.12185	8.2070	0.37185	2.6893	4.204	1.563
6	3.8147	0.2621	0.08882	11.2588	0.33882	2.9514	5.514	1.868
7	4.7684	0.2097	0.06634	15.0735	0.31634	3.1611	6.773	2.142
8	5.9605	0.1678	0.05040	19.8419	0.30040	3.3289	7.947	2.387
9	7.4506	0.1342	0.03876	25.8023	0.28876	3.4631	9.021	2.605
10	9.3132	0.1074	0.03007	33.2529	0.28007	3.5705	9.987	2.797
11	11.6415	0.0859	0.02349	42.5661	0.27349	3.6564	10.846	2.966
12	14.5519	0.0687	0.01845	54.2077	0.26845	3.7251	11.602	3.115
13	18.1899	0.0550	0.01454	68.7596	0.26454	3.7801	12.262	3.244
14	22.7374	0.0440	0.01150	86.9495	0.26150	3.8241	12.833	3.356
15	28.4217	0.0352	0.00912	109.6868	0.25912	3.8593	13.326	3.453
16	35.5271	0.0281	0.00724	138.1085	0.25724	3.8874	13.748	3.537
17	44.4089	0.0225	0.00576	173.6357	0.25576	3.9099	14.108	3.608
18	55.5112	0.0180	0.00459	218.0446	0.25459	3.9279	14.415	3.670
19	69.3889	0.0144	0.00366	273.5558	0.25366	3.9424	14.674	3.722
20	86.7362	0.0115	0.00292	342.9447	0.25292	3.9539	14.893	3.767
21	108.4202	0.0092	0.00233	429.6809	0.25233	3.9631	15.078	3.805
22	135.5253	0.0074	0.00186	538.1011	0.25186	3.9705	15.233	3.836
23	169.4066	0.0059	0.00148	673.6264	0.25148	3.9764	15.362	3.863
24	211.7582	0.0047	0.00119	843.0329	0.25119	3.9811	15.471	3.886
25	264.6978	0.0038	0.00095	1054.7912	0.25095	3.9849	15.562	3.905
26	330.8722	0.0030	0.00076	1319.4890	0.25076	3.9879	15.637	3.921
27	413.5903	0.0024	0.00061	1650.3612	0.25061	3.9903	15.700	3.935
28	516.9879	0.0019	0.00048	2063.9515	0.25048	3.9923	15.752	3.946
29	646.2349	0.0015	0.00039	2580.9394	0.25039	3.9938	15.796	3.955
30	807.7936	0.0012	0.00031	3227.1743	0.25031	3.9950	15.832	3.963
32	1262.1774	0.0008	0.00020	5044.7098	0.25020	3.9968	15.886	3.975
34	1972.1523	0.0005	0.00013	7884.6090	0.25013	3.9980	15.923	3.983
36	3081.4879	0.0003	0.00008	12321.9517	0.25008	3.9987	15.948	3.988
38	4814.8249	0.0002	0.00005	19255.2995	0.25005	3.9992	15.965	3.992
40	7523.1638	0.0001	0.00003	30088.6554	0.25003	3.9995	15.977	3.995
42	11754.9435	0.0001	0.00002	47015.7740	0.25002	3.9997	15.984	3.996
44	18367.0993	0.0001	0.00001	73464.3969	0.25001	3.9998	15.990	3.998
46	28698.5926	0.0000	0.00001	114790.369	0.25001	3.9999	15.993	3.998
48	44841.5508	0.0000	0.00001	179362.203	0.25001	3.9999	15.995	3.999
50	70064.9232	0.0000	0.00000	280255.692	0.25000	3.9999	15.997	3.999

i = 30%

	SINGLE PAYMENT		UNIFORM SERIES				GRADIENT	
	COMPOUND AMOUNT	PRESENT WORTH	SINKING FUND	COMPOUND AMOUNT	CAPITAL RECOVERY	PRESENT WORTH	PRESENT WORTH	ANNUAL COST
n	F/P	P/F	A/F	F/A	A/P	P/A	P/G	A/G
1	1.3000	0.7692	1.00000	1.0000	1.30000	0.7692	0.000	0.000
2	1.6900	0.5917	0.43478	2.3000	0.73478	1.3609	0.592	0.435
3	2.1970	0.4552	0.25063	3.9900	0.55063	1.8161	1.502	0.827
4	2.8561	0.3501	0.16163	6.1870	0.46163	2.1662	2.552	1.178
5	3.7129	0.2693	0.11058	9.0431	0.41058	2.4356	3.630	1.490
6	4.8268	0.2072	0.07839	12.7560	0.37839	2.6427	4.666	1.765
7	6.2749	0.1594	0.05687	17.5828	0.35687	2.8021	5.622	2.006
8	8.1573	0.1226	0.04192	23.8577	0.34192	2.9247	6.480	2.216
9	10.6045	0.0943	0.03124	32.0150	0.33124	3.0190	7.234	2.396
10	13.7858	0.0725	0.02346	42.6195	0.32346	3.0915	7.887	2.551
11	17.9216	0.0558	0.01773	56.4053	0.31773	3.1473	8.445	2.683
12	23.2981	0.0429	0.01345	74.3270	0.31345	3.1903	8.917	2.795
13	30.2875	0.0330	0.01024	97.6250	0.31024	3.2233	9.314	2.889
14	39.3738	0.0254	0.00782	127.9125	0.30782	3.2487	9.644	2.969
15	51.1859	0.0195	0.00598	167.2863	0.30598	3.2682	9.917	3.034
16	66.5417	0.0150	0.00458	218.4722	0.30458	3.2832	10.143	3.089
17	86.5042	0.0116	0.00351	285.0139	0.30351	3.2948	10.328	3.135
18	112.4554	0.0089	0.00269	371.5180	0.30269	3.3037	10.479	3.172
19	146.1920	0.0068	0.00207	483.9734	0.30207	3.3105	10.602	3.202
20	190.0496	0.0053	0.00159	630.1655	0.30159	3.3158	10.702	3.228
21	247.0645	0.0040	0.00122	820.2151	0.30122	3.3198	10.783	3.248
22	321.1839	0.0031	0.00094	1067.2796	0.30094	3.3230	10.848	3.265
23	417.5391	0.0024	0.00072	1388.4635	0.30072	3.3254	10.901	3.278
24	542.8008	0.0018	0.00055	1806.0026	0.30055	3.3272	10.943	3.289
25	705.6410	0.0014	0.00043	2348.8034	0.30043	3.3286	10.977	3.298
26	917.3333	0.0011	0.00033	3054.4444	0.30033	3.3297	11.005	3.305
27	1192.5333	0.0008	0.00025	3971.7777	0.30025	3.3305	11.026	3.311
28	1550.2933	0.0006	0.00019	5164.3111	0.30019	3.3312	11.044	3.315
29	2015.3813	0.0005	0.00015	6714.6044	0.30015	3.3317	11.058	3.319
30	2619.9957	0.0004	0.00011	8729.9857	0.30011	3.3321	11.069	3.322
32	4427.7928	0.0002	0.00007	14755.9759	0.30007	3.3326	11.085	3.326
34	7482.9698	0.0001	0.00004	24939.8993	0.30004	3.3329	11.094	3.329
36	12646.2190	0.0001	0.00002	42150.7298	0.30002	3.3331	11.101	3.330
38	21372.1101	0.0000	0.00001	71237.0334	0.30001	3.3332	11.105	3.332
40	36118.8660	0.0000	0.00001	120392.896	0.30001	3.3332	11.107	3.332
42	61040.8936	0.0000	0.00000	203466.278	0.30000	3.3333	11.109	3.333
44	103159.093	0.0000	0.00000	343860.312	0.30000	3.3333	11.110	3.333
46	174338.867	0.0000	0.00000	581126.228	0.30000	3.3333	11.110	3.333
48	294632.687	0.0000	0.00000	982105.625	0.30000	3.3333	11.111	3.333

CHAPTER 30

Microprocessor-Based Data Acquisition Systems

CHANG-NING SUN* AND CARL D. TOCKSTEIN*

INTRODUCTION

All laboratory experiments and field tests involve data collection. A clear trend in the design of devices for data collection is to increase the utilization of digital technologies. Microprocessor-based data acquisition systems (DAS) have existed for many years, but major advances have been made in the past few years as a result of decreasing cost and broad availability of advanced microprocessors and other chips.

Microprocessor-based DAS is a part of the closed measurement and control loop as shown in Figure 1. The essential purpose of a data acquisition system is to make a measurement, convert the measurand to a predetermined form, and produce an output signal having a form appropriate for process control purpose or being an end result. An instrument that is designed around either a microprocessor or a microcomputer is referred to as a microprocessor-based instrument.

Dedicated instruments are used for specific experiments and no software or hardware modifications are possible. Dedicated microprocessor-based instruments are suitable for situations where continuous monitoring is required, such as patient-monitoring, weather monitoring, or stack emission monitoring systems. They are also used for special process control purposes such as the control unit for power plants, military testing facilities, load testing machines with servo control device, or shaking tables for earthquake simulation where the computer's cost is relatively insignificant. Simple dedicated microprocessor-based instruments which contain a microinterpreter is not sophisticated but adequate for low level A/D conversion. A microinterpreter is a microprocessor with a built-in BASIC interpreter.

For general civil engineering applications, either in the field or laboratory, all-purpose microprocessor-based data acquisition systems are more appropriate. The system should be able to handle data acquisition for static as well as dynamic testings and accept electric signals from all commonly used transducers.

Since almost all transducers produce analog electric signals, the main function of a microprocessor-based DAS is to digitize the analog signals. This leads to a conclusion that the analog-to-digital converter (ADC) is the heart of the system.

Microprocessor-based DAS have been used in the field [7,8,14,18,25,27,30,39,42,52,55,57,58], and in the laboratories [4,20,37,41,68]. Comprehensive discussions on this topic have also appeared in recent years [48,49,59,60]. Selected materials are discussed adequately in this chapter so that a civil engineer will be able to communicate easily and intelligently with a computer specialist or equipment supplier to specify and procure flexible and economical data acquisition systems. References, codes and standards, and glossary given at the end of this chapter make this chapter a self-contained source for further studies.

DATA ACQUISITION SYSTEM

Without a computer, a test in a civil engineering laboratory is manually controlled from a control panel where an operator starts and stops the test and performs all data acquisition and on-line calculations. The operator's tools for these tasks are switches and knobs, pocket calculator, pen, and clipboard. If only a few slowly changing physical parameters are involved, this method of data collection may be

---

*Geology and Geotechnical Engineering Group, Tennessee Valley Authority, 400 West Summit Hill Drive, Knoxville, TN

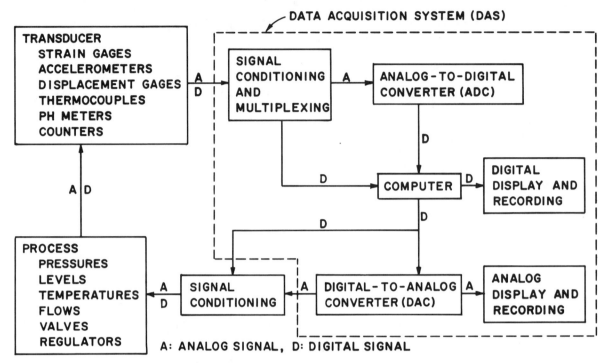

FIGURE 1. Closed measurement and control loop.

satisfactory. The potential problems with the manual method include the inability to accurately recreate the actual control and data collection sequence, no allowance for monitoring more than two or three parameters when there may actually be more of interest, and the inability to monitor dynamic tests. Fast changing analog signals can be recorded on analog tapes or strip chart for future analyses; but, without digitization, the analyses are very limited. Using a microprocessor-based data acquisition and control system resolves all of these problems and opens the door to more detailed numerical analysis and to a variety of ways of analyzing data graphically.

Signal Conditioning

Most transducers used in a civil engineering laboratory require a 10V DC supply, but the output voltage range varies greatly. Displacement gages are generally of the linear variable differential transformer type and have an output in the range −5V to +5V. Load cells and pressure transducers are generally of the strain gage type and have an output in the range 0 to +0.1V. On the other hand, process control requires the control signal to have a DC current in the range 4 to 20 mA or a voltage between 0 and 5V. Some form of signal conditioning is necessary to deal with these extremes.

Signal Multiplexing

If the input signals are more than one channel, switching from channel to channel is required. In some cases, the number of input channels may exceed 1,000. Some ADC can make more than 2 million conversions every second, but only one conversion can be made at any particular time. A time-division analog multiplexer is needed to switch the input signals consecutively or in a predetermined sequence. An analog multiplexer consists of a set of IC switches controlled by a counter and decoder. Output demultiplexing is usually handled by standard computer hardware such as the programmable peripheral interface chip or by software data manipulation. Figure 2 shows a combined amplifier/multiplexer unit.

Analog-To-Digital Conversion

Analog signals may be in the form of electric voltage, in the form of graphs, or even in the form of a continuously moving object. We encounter all these forms of analog signals in the field and laboratory.

The conversion of electric voltage signal into binary numbers is best performed by an ADC, the conversion may be performed on-line (actual signal input) or off-line (input from analog tape). Output is in parallel format.

For the conversion of analog signals in graphic form, a "digitizer" is used. This type of conversion is used widely in computer aided design. By tracing the curve, the coordinates of selected points of the curve become the input data of the computer. It is time consuming, but still widely used in works related to seismology studies and geological surveys. Transparent analog records may be made from ordinary graphic records and a flying spot scanner can be used for automatic digitization. This process reduces human errors.

A position encoder is used to measure the angular position of a shaft. A disc having concentric rings of alternate transparent and opaque segments is mounted on the shaft. A light source is installed on one side of the disc and a group of photo cells is used to monitor the light passing or not passing the disc as it rotates. A binary code is produced this way. The analog positions of the shaft are digitized. The angular position change of the shaft can be converted into linear position change by adding a pulley and a wire. This type of device is widely used in depth determination in geotechnical investigations. A magnetic trigger may be used instead of optical means.

Some of the input signals from transducers are in digital form. Examples are electronic pulses from a nuclear particle detector, raingage monitor, and possible from counting items on a conveyor belt. After signal conditioning, digit signals can be processed by computer directly.

Digital-To-Analog Conversion

Digital-to-analog conversion is a much simpler process. The conditioned signal from a digital-to-analog converter (DAC) is usually in the range of $-10V$ to $+10V$ suitable for graphic display (meter, oscilloscope, strip chart), analog tape recording, and process control. Many DAC may be used in a DAS.

Microprocessor

An ADC has limited ability and intelligence. A microprocessor is needed to handle the data transfer between ADC and storage devices (RAM, disk, tape) and display devices (monitor, printer, plotter). A microprocessor is also needed to initiate conversion and multiplexing sequence.

Software

System control programs are needed for all data acquisition functions such as calibration, zero-drift correction, engineering unit calculation, timing control, data format, controlling stepper motor, and reading optical position encoders. These functioning programs are better furnished by an EPROM (Erasable Programmable Read-Only Memory). Programs stored on tape or disk would have to be read in each time it is used. Mechanical failures of the disk drive or tape drive are common reasons of an unsuccessful field data acquisition program.

PLANNING MICROPROCESSOR-BASED DAS

Thirty years ago designers had little choice; everything was analog. Very few engineers heard of or thought about digital signal processing. The physical size of the vacuum tubes and their high cost and low reliability prevented signal digitization from reaching most laboratories. As recent as 15 years ago, microprocessor-based DAS was still in its infancy. The best available DAS was minicomputer-based, bulky, inefficient, and the cost was still prohibitive for most laboratories [14]. Ten years ago, 3-bit or 4-bit microprocessors were the hearts of many DAS; now, 16-bit microprocessors are very popular and 32-bit microprocessors are also available.

Interfacing a microprocessor to the real world is no small effort; understanding and planning for a DAS are not easy. But civil engineers very often are forced to plan and, with some help from computer specialists, to purchase or build a DAS. Some knowledge of how a microprocessor-based DAS functions will help us to obtain an efficient system.

Inevitable Decisions

A frustrating situation arises when an engineer or manager senses the genuine need for a new approach to data acquisition, but cannot immediately decide what capabilities the new system should provide. Unless we have the knowledge to do the planning and specifying, we have to rely on people such as vendors, consultants, and EE colleagues. We may depend on others, but we must be knowledgeable enough to convey our needs. A survey shows that the testing equipment at many institutions needs to be replaced and it is a serious situation [43].

Most civil engineers have mixed feelings towards DAS. The major concerns are the high cost and the performance

FIGURE 2. Amplifier/multiplexer.

of the systems. It is common to hear that expensive DAS have failed to function in the field, often for some quite simple reason. It is also common to see a complete, expensive DAS stored at a corner of a laboratory, not functioning at all. The warranty of the equipment has expired and the vendor is not willing to do the repair work. Things have changed for the better now. We have many choices to suit our needs. The problem is in knowing what to do and what not to do. Depending on our own ability, we may:

1. Build simple systems with electronic components
2. Assemble more complicated systems with plug-in boards
3. Assemble more advanced systems with modules
4. Buy complete system from vendors
5. Have a system custom-built

Initial Cost and Maintenance Cost

To produce and market a DAS, a commercial company must write a users' manual and service manual, maintain a service group, advertise and carry administrative overhead. The system must also be as versatile as possible to appeal to a reasonable wide range of users. The cost is partly compensated for by the economics of greater production. The market for complete DAS is small and the cost must be shared by a small number of users. On the other hand, the market for modular units is much larger and the cost is correspondingly lower. The market for the microprocessors, ADC chips, and other components is so huge that the costs are very low.

It is obvious that the DAS built in house is very easy to repair. The system with plug-in board or modules are not too difficult to repair. The maintenance cost for the complete system and the custom-built system is high.

What a Civil Engineer Should Not Do

A civil engineer, just like the modern chemical engineer or biomedical engineer, may want to learn and even build DAS, but he should not attempt to do mathematical analysis of electronic circuits. The improvement of basic electronic circuitries is a job better left to the electronic engineers. The number of commercially available components, assembled parts, plug-in cards and boards, modular devices, and systems is much more than we can choose and handle properly.

A civil engineer also should refrain from building complex (DAS and control) systems, no matter how much he understands digital electronics. These systems have a function generator, servo controller, and DAS. They have to be purchased as a complete system. Examples are the shaking tables for earthquake simulation, load testing machines with servo control device, and process control units. However, we can always add our own DAS to expand the data acquisition capability of these facilities.

A civil engineer should not build special equipment which can be obtained from commercial sources at a reasonable cost. Examples are oscilloscope, sound level meter and analyzer, and nuclear density and moisture gage. In short, we should apply our level of expertise to the limit that it is economically feasible.

LEVEL OF CONVERSION

The level of conversion is essentially determined by the required speed and accuracy of the conversion. In most cases a physical phenomenon which needs a higher conversion speed also needs a higher conversion accuracy. Slowly varying signals usually need lower level conversion. If two or more signals are involved in the conversion, then the system conversion speed will be slowed down proportionally.

Speed of Conversion

Digital meters, e.g., voltmeter and ammeter, are ADC of the simplest form. After rectification, AC input is treated as DC and only slow A/D conversion is needed. Examples of slow changing signal measurements are static load test, triaxial shear test, temperature measurement, and consolidation test. The conversion time is 2 or 3 seconds which is fast enough.

Higher speed conversions are best illustrated by the A/D conversion of the signals from seismometers. A seismometer is usually an accelerometer. Only acceleration time-history is monitored; the velocity and displacement time-histories are obtained by integration of the acceleration record. For earthquake source study, sensitive seismographs are used, which have a frequency content up to 100 Hz. For engineering purposes, less sensitive strong motion accelorographs are placed near a fault to monitor strong motions in the vicinity of the epicenter, which have a frequency content up to 40 Hz. If, according to the Nyquist sampling theory, a minimum of two samples per cycle of the highest frequency component is assumed, then a sampling rate of 100 per second per channel is adequate.

Accuracy of Conversion

The accuracy of conversion depends mainly on the resolution of the ADC. A 12-bit ADC can resolve to one part in 4096 (2^{12}). The accuracy of the entire system depends on the response characteristics of the transducer and stability of all electronic components; but these are not considered as digital problem related to A/D conversion.

The simplest 1-bit ADC is a Schmidt trigger gate. As shown in Figure 3, it has a built-in threshold level to determine when to change state. The low-to-high transition will not occur until the input signal voltage crosses the upper threshold voltage, and the high-to-low transition will not occur until the lower threshold voltage is crossed. Apparently, the accuracy of this simple ADC is not very good; we don't even know the maximum value of the input signal, nor the minimum value. On the other hand, the Schmidt trigger gate can be used as a triggering device to start some action when the signal strength is higher than a predetermined threshold.

Figure 4 shows the relationship between the analog and digital signal of a 4-bit ADC which can resolve to one part in 16 (2^4). Figure 5 shows the AD363 data acquisition module with a 12-bit resolution which can accommodate a mixing of single-ended (unipolar mode) and differential (bipolar mode) signals.

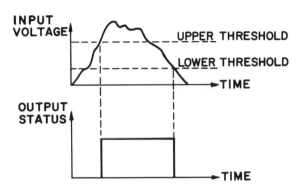

FIGURE 3. Schmidt Trigger Gate.

Throughput Rate and Throughout Rate

Throughput rate is the same as single-channel conversion rate. One conversion may involve up to 16 bits. The conversion rate not only depends on the ADC itself; it also depends on the characteristics of the microprocessor, the storage device, the multiplexer, and other digital and analog circuitries. The software for data transfer is the most influencing factor. If high speed conversion is required, assembly-language programming is a must.

Throughout rate is a simple way to measure the overall conversion capability of a system in terms of bits per second. A 3-channel digital strong motion acceloragraph with a 100 samples per second sampling rate and a 12-bit resolution needs a system throughout rate of 3,600 bits per second ($3 \times 100 \times 12$). If three independent ADC systems are used, then a throughout rate of only 1,200 bits per second is needed for each system. Systems with throughout rate up to 40 million bits per second are available in the U.S.

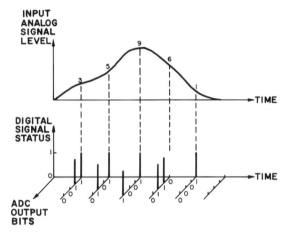

FIGURE 4. Analog vs. Digital Signal.

SYSTEM HARDWARE

The 8-bit microprocessors are slow and the 8-bit ADC do not provide good resolution for most of civil engineering works. A 12-bit ADC combined with a 16-bit microprocessor or an advanced 8-bit microprocessor is adequate for most of the work. The systems with a 16-bit microprocessor and a 16-bit ADC are the best choices.

Display and storage devices are less critical, but due considerations should be given for the most economical and efficient selection.

Microprocessor

If a civil engineer is not capable of building his DAS around a microprocessor from scratch, he should at least

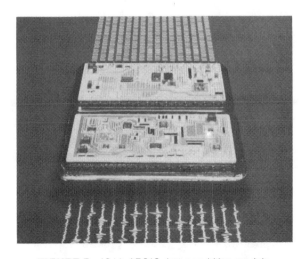

FIGURE 5. 12-bit AD363 data acquisition module.

FIGURE 6. A/D coverter modules.

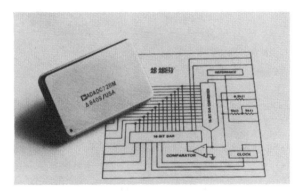

FIGURE 7. 16-bit A/D converter.

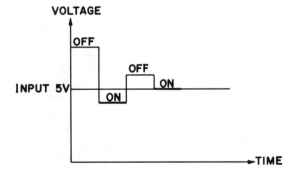

FIGURE 8. Process of successive approximation.

select a microcomputer with the proper microprocessor for efficient data acquisition work.

A true 16-bit microprocessor has 16 data lines, and every time data is moved a 16-bit computer word is moved. The size of the internal registers is invisible to the user and has very little meaning to him. A microprocessor having 16-bit data architecture is most likely to have a number of 32-bit internal registers; but, it is still a 16-bit microprocessor. The terminology of "32-bit technology" really has no meaning to the users.

An ADC generates a lot of digital data. The data should be stored in RAM, recorded on digital tape, or after D/A conversions, recorded on analog tape and displayed. Most 8-bit microprocessors have 16 address lines and can handle 65,536 bytes of data. Sampling at a 6,000 conversions per second rate with an 8-bit ADC, the RAM will be full in 10.92 seconds. The RAM is adequate for short duration sampling such as pile driving and blast monitoring. Most of the 16-bit microprocessors have 20 to 24 address lines and can handle 1 to 16 mega bytes RAM (2^{20}–2^{24}). With 32 address lines, the Motorola's MC68028 32-bit microprocessor can directly address 4 gigabytes.

A/D conversion rate not only depends on the ADC but also depends on the clock frequency of the microprocessor. High clock frequency means faster data transfer. Generally, an 8-bit microprocessor operates at 4 MHz, 16-bit microprocessor at 8 MHz, and 32-bit microprocessor at 16 MHz; but, the instruction processing rate of a microprocessor is much lower than the clock frequency. A low performance microprocessor will not accommodate a high performance ADC. Additionally, a microprocessor of less bit architecture must use additional hardware and software to link with a high-resolution ADC. An 8-bit microprocessor needs two input ports to handle the data from a 12-bit ADC and additional software routines are needed to read the 8-bit and 4-bit data alternately. The conversion rate will suffer. The newly available 32-bit microprocessors can interface to 32-, 16-, and 8-bit peripherals and allows transfer of 8-, 16-, and 32-bit data between various ports. A microprocessor of 32-bit architecture offers high performance and would be the ultimate choice for A/D conversion; except, the supporting chips are not readily available. This situation should be improved in two or three years. The following 32-bit microprocessors are available in the U.S.: Data General's Microeagle, Digital Equipment's Micro VAX 1, Hewlett-Packard's Focus, Inmos's Transputer, Intel's 80386, Motorola's MC68028, National Semiconductor's NS32032, NCR's NCR/32, Western Electric's WE3200, and Zilog's Z80000.

Analog-To-Digital Converter

Most input signals from transducers are in analog form and need to be converted to digital form. Hundreds of different approaches to A/D conversion are published in the literature, but only a few techniques have been successful commercially. Only one ADC is used in most DAS, but two or more may be used. The commonly used ADC are: successive approximation, dual-slope integration, voltage-to-frequency, flash (also called parallel), and servo (also called binary ramp) converters. The successive approximation converter, with its speed and stability, easily outperform all other types of converters. ADC with 16-bit resolution are available (see Figure 7).

All successive approximation ADC have an internal digital-to-analog converter. In the comparison process, the input voltage is compared with the analog voltage from the DAC until they match (see Figure 7). When an unknown voltage is to be converted, the ADC compares the input voltage with the voltage represented by the most significant bit of the ADC. The full range of a 4-bit ADC is 15V $(8 + 4 + 2 + 1)$. If at a certain instance an input voltage of 5V is being converted (see Figure 8), the most significant bit is compared with the 5V and set to zero (off) since 5 is less than 8. Next, the input voltage is compared with the voltage represented by the sum of the first two bits with the current status: in this case a decimal 4. This bit is set to one (on) since 5 is more than 4. This process continues until all bits are compared with the input voltage.

The dynamic range of an ADC needs to be considered too. The input voltage is usually between $-10V$ and $+10V$. A signal with a maximum voltage of 0.1V should be amplified and a signal with a maximum voltage of 100V should be attenuated before conversion. But, to handle a wide input voltage range of the analog signal (e.g., 0.1mV to 100V), an ADC must have enough resolution. Otherwise, the weak signals may be undetected and lost: strong signals may also be lost as a result of overloading the system. It is better to avoid selection of an ADC with a low resolution but supplemented by a so-called autoranging circuitry. An 8-bit ADC with an autoranging feature to extend the input range by 10 times is no better than a 12-bit ADC; the additional 4 bits can extend the input range by 16 times (2^4).

The analog input to a converter must be stable and constant for the entire length of time (may be as short as one microsecond) it takes to complete the conversion. This may be accomplished by using a sample and hold circuit. The usual method is to hold the voltage value in a high quality capacitor for the short duration of the conversion time and then reset the capacitor. An ADC with sample and hold unit and multiplexing unit as a modular package is a preferred choice for most data acquisition work.

Many suppliers offer single-board data acquisition systems. These boards are designed to be plugged into the mother board of a microcomputer or the backplane of a minicomputer. These boards usually have ADC, multiplexer, data acquisition system software in a ROM, analog and digital input/output ports, and sometimes, high speed Cache memory. Most of them are adequate for low level and median level data acquisition work; only a few of them have

FIGURE 9. High speed ADC board.

high level data acquisition capability. Figure 9 shows a high speed ADC board with a conversion rate of 300,000 per second, designed to be used with a minicomputer. Figure 10 shows a complete measurement and control system. Figure 11 shows a personal computer-based DAS. Figure 12 shows a data logger as a remote station of a personal computer-based DAS.

An ADC's cost and stability depend on the complexity of its circuit and its characteristic with regards to temperature variation. An 18-bit ADC, built with the best components available today, will either suffer greatly in speed or temperature stability, or be prohibitively expensive. Today's limit on resolution is probably reached at 16 bits. On the other hand, the auto-ranging circuitry can always be added to the 16-bit ADC if needed. Under certain conditions, two or more ADC with different input range settings may be used.

FIGURE 10. Measurement and control system.

FIGURE 11. Personal computer-based DAS.

Other types of ADC, such as the logarithmic ADC and the companding ADC are non-standard and will increase circuit and software complexity.

Bus Systems

Buses are conductors used for transmitting signals or power from sources to destinations. At the ends of the conductors, they have the forms of cable connectors (e.g., printer port), edge connectors (male), or expansion slots (female). Some bus systems have over 100 conductors. A microprocessor may be used in conjunction with different bus systems; on the other hand, one bus system may be used in conjunction with many microprocessors. The data bus, address bus, and control bus are laid out neatly on both sides of a printed circuit board. This will reduce noise, interference, and keep the interface work to a manageable magnitude.

A data acquisition card or board having the same bus system as the host computer can be inserted directly into the slot on the mother board. Bus compatibility should be determined early in the DAS selection process.

Signal Transmission

For transmitting the analog signals between the transducer and a DAS, the connection is restricted by the characteristics of the analog signals and the available power source. Weak signals may have to be amplified at the source and high frequency signals limit the distance of transmission.

For transmitting digital signals (other than on-off signals), cables and connectors must conform to certain standards. This is the case where the ADC is far away from the microprocessor or microcomputer. This is also the case when a single computer is used as a host controller with several DAS connected to it. Data transmission between the microcomputer and peripherals is also governed by these standards. The IEEE-488 Standard for parallel data transmission and the EIA RS-232-C Standard for serial data transmission are used in most of the data acquisition systems (see Codes and Standards). The IEEE-488 connector allows connection to be made piggyback with the connector since each connector has both a male and a female part. Fifteen instruments may be conveniently connected to a common IEEE-488 bus.

ADDITIONAL HARDWARE

In addition to the microprocessor, the ADC, and their supporting chips, many other components and devices are used in a DAS. Some of the more important devices are discussed below.

Data Logger and Signal Conditioner

A data logger is usually a multi-channel, low resolution, low speed DAS. Most of them have an alarm system, digital display, and a built-in small printer or plotter. Day and time are also recorded. Logging and scanning interval may vary from one second to 24 hours. They are used for monitoring temperature, pressure, and voltage at strategic locations of a building, a factory, or a mining environment. For medical usage, it is called physiological integrator/averager. When used to record brain waves of patients, it is called an alpha logger.

A signal conditioner consists of amplifier, filter, power

FIGURE 12. Data logger.

supply for the Wheatstone bridge circuit, and sometimes, multiplexer and sample and hold circuits. Other analog circuits for addition, subtraction, integration, and differentiation are sometimes included in a signal conditioner. A signal conditioner may be a part of a data logger. Special signals such as those from an eddy-current transducer [57] need special treatment.

Board, Card, and Module

A board is basically a printed circuit board with components on both sides. It usually has a standard pin connector at one edge to handle all connections required for data, address, control codes, and power source. A microcomputer is built around a large board which is called a motherboard. Smaller boards can be plugged into the slots on the motherboard. Apple II microcomputer has seven such slots, very convenient for DAS board adaption. A board may be relatively simple, e.g., the ADC board, signal conditioning board, or clock board.

Card and board are two terms used almost synonymously. Usually, a card is smaller than a board; a card with a front panel may be called a plug-in board; with a complete case, such a plug-in board may be called a module.

A module has its own case or container; components are not exposed (see Figures 6 and 7); and connectors are usually non-standard. Modules are designed to be plugged into a motherboard or a backplane and then rack mounted. A module may contain a complete microcomputer or may simply serve as a set of connectors. It is difficult to select a board, card, or a module on the basis of their exterior look and physical size; instructions must be read carefully.

Gate, Buffer, and Latch

A programmable peripheral interface chip is always used with a microprocessor to handle the input-output signals. The variety of digital signals that the user is likely to encounter makes it impossible to have one interface chip responding precisely to all signals. Usually, it is necessary to condition the signals from the input-output device to make them compatible with the interfacing chip. Gates, buffers, and latches are used for digital signal conditioning. A single chip usually contains a number of these devices of the same kind.

A gate, also called a logic gate, is nothing more than a device whose output status (0 or 1) depends on the input status arrangement. It responds to the input signals immediately and no clocking pulse is needed to start its action. These are the often heard AND, OR, NAND, NOR, . . . gates.

A buffer, also called a driver, is used for matching logic level (voltage level) among IC devices. Digital signals from a microprocessor and a programmable peripheral interface chip usually do not have the required voltage to drive a relay, an alarm, or other devices, a buffer must used to provide the required voltage and current. A buffer needs a clocking pulse to start its action.

A latch, also called a flip-flop, is used for time matching. The data from a microprocessor are short pulses unsuitable for other functions. A latch accepts the data and releases the data on request. Its function is to condition the digital signal so that it will have a proper time base.

Filter

Filters used in a signal conditioner are similar to the tone controls in an ordinary hi-fi audio system. Certain signals are filtered out (attenuated) and other signals are allowed to pass and amplified. An ideal low-pass filter will allow the passing of all signals below a certain frequency; a high-pass filter does the reverse. A band-pass filter allows only signals within a certain range of frequencies to pass. A notch filter is the complement of a band-pass filter. Digital filtering is accomplished by programming.

There are practical reasons to use a filter in a signal conditioner. Strong motion accelerographs usually use a 0.1 Hz high-pass filter to remove DC offsets and use a 40 Hz low-pass filter to remove ambient vibration signals of higher frequencies. For blast measurement, a 6 Hz high-pass filter is used to record signals of high frequency up to 500 Hz. For the monitoring of suspension bridges, a low-pass filter is used so that all frequencies higher than 3 Hz are filtered out. Generally, low-pass filters are used for static tests to monitor slow varying parameters and high-pass filters are used for dynamic tests. Filter circuits are simple: an operational amplifier, 2 or 3 resistors, and 2 or 3 capacitors.

Clock

Two entirely different clocks are used in a microprocessor-based DAS. A master clock is needed for any microprocessor, since all of the internal functions are performed synchronously. The clock signals are of square-wave form. The frequency of the clock signals are under 20 MHz and usually in the 2 to 6 MHz range. A crystal may or may not be used for the generation of the clock signals.

A real time clock is used for recording time, day, month, and year. It is similar to an ordinary digital watch. Digital time signals are generated for recording or printing. Many seismographs use self-contained real time clocks. Real time clocks are also used to generate digital time signals for field and laboratory works.

The radio synchronized digital clocks, gaining popularity in recent years, are also real time clocks. Through its radio receiver, the clock is automatically set by the standard radio time signal broadcasts, such as WWVB (US National Bureau of Standards time) at 60 kHz, Mainflingen, Germany at 77.5 kHz, Pragini, Switzerland at 75 kHz, and others. There are real time clocks synchronized to the eastern or western GOES satellites and OMEGA navigation signals. In the event of loss of signal, the internal oscillator takes over the clock operation until signal is restored.

SYSTEM SOFTWARE

System control programs are developed, verified, and then stored either in firmware (PROM or EPROM), on disk or tape for all data acquisition functions such as calibration, engineering unit calculation, timing control, data format, data transmission, etc. It is a routine matter to use a "programmer" for the programming of an EPROM. Many EPROM chips may be installed and selected by the DAS according to type of application.

The most important part of the software is the routine that controls the ADC and provides transfer of the binary data. At the beginning of a conversion, the interrupt logic is enabled and the computer is forced to stop (interrupt) whatever it is doing and branch to a predetermined memory location where the ADC is instructed to begin the conversion. The digitized signal is then loaded into the microprocessor's accumulator and then from there the data is transferred into a designated memory location. A return from interrupt command then forces the computer to return to the point in the main program where the interrupt request was detected. Between data conversions, the computer does other things such as calculation, calibration, zero drift setting, transferring data from memory to tape or disk, and display data on the monitor.

System control programs may be written in BASIC, FORTRAN, Pascal, C, PL/1, or preferable, assembly language. BASIC is the simplest, but the slowest. FORTRAN is usually 15 to 30 times faster than BASIC. Assembly language, 1,000 times faster than BASIC, is always needed for DAS with high conversion rate.

OVERALL CONSIDERATIONS

It is recognized that a microprocessor-based DAS is needed for every laboratory and field operation, but is it wise to use the system at all occasions? Many fundamental facts on data acquisition need to be considered.

Whether to Convert

The clear trend in DAS design is to increase the utilization of digital technologies. Engineers always hear the good points about microprocessor-based DAS. However, analog signals are sometimes best processed by analog devices and a conversion to digital form may not necessarily be an improvement. Improper conversions not only increase cost, complexity, and response time; it may even hamper the production rate in the field. Neither the analog nor the digital approach is clearly superior in every case. The proper choice depends on the type and use of the data being obtained.

On-Line or Off-Line Conversion

If a digital system is needed, then where to convert the signal from analog to digital form is an important issue. The following factors should be considered:

1. Analog recording devices are simple and the data recorded is complete. They are neither outdated nor inaccurate as claimed by many. The industrial grade tapes are suited for both analog and digital recordings.
2. If an analog signal can be converted into digital form on-line, it always can be recorded on an analog tape and then converted off-line. No signal is too fast for analog recording. Analog calibration signal recorded on analog tape is as good as, if not better than, the digital calibration signal stored in the RAM.
3. Analog tape speed varies from 15/16 to 240 inches per second. The variable time base is a special feature of analog tape recorders. Slow varying signals such as tip resistance and sleeve friction from an electric cone penetrometer may be recorded at low speed and played back at high speed for A/D conversion and analysis. Fast changing signals such as ground vibration due to pile driving [27] may be recorded at high speed and played back at lower speed. Best speed may be selected for direct plotting.
4. There are 4-, 7-, 14-, 28-, and 42-channel analog tape recorders. One reel can handle over 12,000 feet of continuous recording. Time marks and voice can be recorded as well.
5. Long analog record, such as vibration monitoring of structures subjected to random loads [30,58] may be searched for significant data and only a part of the data needs to be digitized. On-line A/D conversion is difficult because cut-off level of the signal is not known and a complete digital record contains too much data and offers no merit.
6. On-line A/D conversion is best used if long distance transmission of data, either through radio link or telephone line, is required; especially, when an array of instruments, such as seismograph network is involved. Analog signals transmitted over a distance of more than a few hundred feet will be distorted and noise becomes a problem.
7. On-line A/D conversion is a must if real-time control feedback is needed. For uranium mining application, the monitoring of the alpha particle concentration and temperature, control of fans, and sounding alarms are required. The number of stations monitored could be over 500. Only an on-line A/D conversion can fulfill such need. Dam instrumentation is another situation appropriate for using on-line A/D conversion technique.

Distributed System

A single computer can be used as a host controller with several DAS distributed from its serial-communication link. The EIA RS-232-C standard is generally used for defining the serial data transmission. Many remote stations can be controlled by only two wires. Each station has its own access code. Communication between the host computer and remote stations is handled by a Universal Asynchronous Receiver and Transmitter chip (UART). The stations are numbered by different DIP (Dual-Inline-Package) switch combination. Communication starts when the host computer sends out an access code followed by a command. All stations receive the same access code and command, but only the station with the correct code combination will respond. The UART is really a microprocessor to a lesser extent; it receives commands from the host computer and transmits the digitized data from the local ADC back to the host computer.

CONCLUSIONS

The information presented in this chapter will help civil engineers to intelligently select their DAS, to prepare technical specifications, and possibly to assemble and build some of the hardware. Writing data acquisition system control programs is a task most civil engineers are capable to do. Those who write their own system control programs will be awarded with unlimited usage of their microprocessor-based data acquisition systems.

Elements of microprocessor-based DAS such as signal conditioning, analog and digital multiplexing, A/D conversion, and level of conversion are discussed. Also covered are system considerations such as whether to convert, on-line or off-line conversion, distributed systems, system software, and limits of the microprocessors and ADC. A flexible data acquisition system incorporating a 16-bit microprocessor and a 16-bit ADC is recommended for use in a modern civil engineering laboratory. Field work, due to its harsh environment, requires additional planning.

Readers are directed to the references and codes and standards for further information. The glossary contains only important data acquisition terms; readers are directed to References 3, 21, 33, 34, 44, and 64 for other digital electronics terms.

REFERENCES

1. Allocca, J. A. and A. Stuart, *Electronic Instrumentation*, Reston Publishing Co., Inc. (1983).
2. Allocca, J. A. and A. Stuart, *Transducers: Theory and Applications*, Reston Publishing Co., Inc. (1984).
3. *American National Dictionary for Information Processing*, Computer and Business Equipment Manufactures Association, Washington, D.C. (1977).
4. Atkinson, J. H., J. S. Evans, and C. R. Scott, "Developments in Microcomputer Controlled Stress Path Testing Equipment for Measurement of Soil Parameters," *Ground Engineering*, Vol. 18 (1), 15–22 (January 1985).
5. Auslander, D. M. and P. Sagues, *Microprocessors for Measurement and Control*, McGraw-Hill Book Co. (1981).
6. "Automation and Instrumentation," *AWWA Manual No. M2*, American Water Works Association (1977).
7. Baker, V. A., N. A. Thomsen, C. R. Nardi, and M. J. Talbot, "Pile Foundation Design Using Pile Driving Analyzer," *Analysis and Design of Pile Foundations*, Proceedings of a Symposium sponsored by the ASCE Geotechnical Engineering Division, ASCE National Convention, San Francisco, California, pp. 350–373 (October 1–5, 1984).
8. Bell, F. G., *Foundation Engineering in Difficult Ground*, Newnes-Butterworths (1978).
9. Bordes, J. L. and C. Keryell, "Developments and Trends in Instrumentation," *Water Power & Dam Construction* (May 1983).
10. Carr, J. J., *Design Microprocessor-Based Instrumentation*, Reston Publishing Co., Inc. (1982).
11. Carr, J. J., *Digital Interfacing with an Analog World*, Tab Books, Inc. (1981).
12. Carr, J. J., *Interfacing Your Microcomputer to Virtually Anything*, Tab Books, Inc. (1984).
13. Chang, S. S. L., ed., *Fundamentals Handbook of Electrical and Computer Engineering*, John Wiley & Sons (1983).
14. Christiano, P. P., L. E. Goodman, and C. N. Sun, "Bridge Stress-Range History," Highway Research Record No. 382, *Loading History of Bridges* (1972).
15. Clayton, G. B., *Data Converters*, John Wiley & Sons (1982).
16. "Control and Monitoring Philosophy," Design Criteria P19F-50-D776, Tennessee Valley Authority (Nov. 1981).
17. "Cool Storage Instrumentation and Data Verification Program," EPRI Report No. EM-2485, Electric Power Research Institute (July 1982).
18. Cuthbert, L. G. and T. J. Poskitt, "Development of Instruments for Offshore Piles," *Ground Engineering*, Vol. 16 (1), 29–34 (January 1983).
19. "Data Processing System," Design Criteria P19F-IC-D775, Tennessee Valley Authority (August 1983).
20. Davison, L. R. and J. M. Hills, "A Data Acquisition System for Soil Mechanics," *Ground Engineering*, Vol. 16 (1), 15–17 (January 1983).
21. "Definition, Specification, and Analysis of Manual, Automatic, and Supervisory Station Control and Data Acquisition," ANSI/IEEE Standard C37.1 (1979).
22. "Digital Techniques for Control and Protection of Transmission Class Substations," EPRI Report No. WS-79-184, Electric Power Research Institute (August 1980).
23. Dorato, P. and D. Petersen, "Digital Control Systems," *Advances in Computers*, Marshall C. Yovits, ed., Vol. 3, Academic Press, Inc., pp. 177–252 (1984).
24. "Electrical Wiring Inputs and Outputs for Computers and In-

strumentation Systems," *Electrical Design Guide, DG-E20.3.4*, Tennessee Valley Authority (October 1983).
25. "Forced Vibration Tests of Dams Conducted by Caltech Researchers," *EERC News, Vol. 9* (1), Earthquake Engineering Research Center, University of California (March 1985).
26. Garrett, P. H., *Analog I/O Design*, Reston Publishing Co., Inc. (1981).
27. Giese-Koch, G. V. and C. N. Sun, "Experimental Study of Pile-Driving Effects on Nearby Nuclear Plant Structure," *International Symposium on Dynamic Soil-Structure Interaction*, Minneapolis, Minnesota, U.S.A., 133–138 (September 4–5, 1984).
28. Hallagen, R. C., *Interface Projects for the Apple II*, Prentice-Hall, Inc. (1982).
29. Hallagen, R. C., *Interface Projects for the TRS-80*, Prentice-Hall, Inc. (1982).
30. Hand, F. R., C. N. Sun, and J. M. Hoskins, "Full Scale Testing and Seismic Qualification of Cement Mortar Lined Carbon Steel Pipe," *Eighth World Conference on Earthquake Engineering*, San Francisco, California, VII 279–286 (July 21–28, 1984).
31. "Hierarchical Power Control Center Analyzer," EPRI Report No. EL-2835, Electric Power Research Institute (May 1983).
32. Houpis, C. H. and G. B. Lamont, *Digital Control Systems Theory, Hardware, and Software*, McGraw-Hill Book Co. (1985).
33. *IBM Data Processing Glossary*, Publication GC20-1699, IBM Corporation, White Plains, N.Y. (1977).
34. "IEEE Standard Dictionary of Electrical and Electronics Terms," ANSI/IEEE Standard 100 (1984).
35. "Instruments and Apparatus for Soil and Rock Mechanics," ASTM SPT 392, American Society for Testing and Materials (1965).
36. "Instrumentation and Control Complex," Design Criteria P19F-JM-D775, Tennessee Valley Authority (August 1983).
37. James, A. N. and A. J. Morris, "Automatic Recording and Data Processing of Laboratory Soil Tests," *Ground Engineering, Vol. 17* (3), 36–41 (April 1984).
38. Kuecken, J. A., *How to Measure Anything with Electronic Instruments*, Tab Books, Inc. (1981).
39. Law, K. T., "Computer-Aided Pressuremeter Tests," *Geotechnical Testing Journal, ASTM, Vol. 7* (2), 99–103 (June 1984).
40. Mellichamp, D. A., *Real Time Computing with Applications to Data Acquisition and Control*, Van Nostrand Reinhold Co. (1983).
41. Mindlin, H. and R. W. Landgraf, eds., "Use of Computers in the Fatigue Laboratory," ASTM SPT No. 613, American Society for Testing and Materials (1976).
42. Nigbor, R. L., "Full-Scale Ambient Vibration Measurements of the Golden Gate Suspension Bridge Instrumentation and Data Acquisition," *Proceedings of the Eighth World Conference on Earthquake Engineering*, San Francisco, California, Vol. VI, 63–70 (July 21–28, 1984).
43. "One-Fourth Academic Research Equipment Classified Obsolete," NSF 84-312, *Science Resources Studies Highlights*, National Science Foundation, Washington, D.C. (April 18, 1984).
44. Parker, S. P., *McGraw-Hill Encyclopedia of Electronics and Computer Technology*, McGraw-Hill Book Co. (1984).
45. Phillips, C. L. and H. T. Nagle, *Digital Control System Analysis and Design*, Prentice-Hall, Inc. (1984).
46. Poe, E. C. and J. C. Goodwin, *The S-100 & Other Micro Buses*, Howard W. Sams & Co., Inc. (1981).
47. "Revenue-Metering Device for HVDC Systems," EPRI Report No. EL-3505, Electric Power Research Institute (May 1984).
48. Rooney, M. F., "Data Acquisition Systems: Do It Yourself," *J. of Technical Topics in Civil Engineering, ASCE, Vol. 110* (1), 19–28 (May 1984).
49. Rooney, M. F., "Computer Hardware for Civil Engineers," *J. of the Technical Councils, ASCE, Vol. 107* (TC1), 153–168 (April 1981).
50. Schmitt, N. M. and R. F. Farwell, *Understanding Electronic Control of Automation Systems*, Texas Instruments, Inc. (1983).
51. Seitzer, D., G. Pretzl, and N. A. Hamdy, *Electronic Analog-to-Digital Converters*, John Wiley & Sons (1983).
52. Sheeran, C. T. and J. C. Franklin, "Microcomputer-Based Monitoring and Control System with Uranium Mining Application," Information Circulate 8981, U.S. Department of the Interior (1984).
53. Shen, C. K., X. S. Li, and Y. S. Kim, "Microcomputer Based Data Acquisition Systems for Centrifuge Modeling," *Geotechnical Testing Journal, Vol. 7* (4), 200–204 (Dec. 1984).
54. Sirohi, R. S. and H. C. R. Krishna, *Mechanical Measurements*, John Wiley & Sons (1980).
55. Srinivasan, M. G., C. A. Kot, and B. J. Hsieh, "Dynamic Testing of As-Built Civil Engineering Structures—A Review and Evaluation," NUREG/CR-3649, ANL-83-20, Argonne National Laboratory (1984).
56. Stout, D. F., *Micro-processor Applications Handbook*, McGraw-Hill, Inc. (1982).
57. Sun, C. N., "Pipe Lining Thickness and Thickness Gages," *J. of Transportation Engineering, ASCE, Vol. 110* (4), 447–450 (July 1984).
58. Sun, C. N., J. M. Hoskins, and Hunt, R. J., "Full Scale Testing of Cement-Mortar Lined Carbon Steel Pipe," Reprint No. 80-015, ASCE Spring Meeting (May 14–18, 1984).
59. Sun, C. N. and C. D. Tockstein, "Hardware for Data Acquisition Systems," 2nd International Conference on Computing in Civil Engineering, Hangzhou, China 665–682 (June 5–9, 1985).
60. Sun, C. N. and C. D. Tockstein, "Microprocessor-Based Data Acquisition Systems," ASCE Hydraulics Division Specialty Conference, Hydraulics and Hydrology in the Small Computer Age, Lake Buena Vista, Florida 1304–1309 (August 12–17, 1985).
61. Susskind, A. K., *Notes on Analog-Digital Conversion Techniques*, The MIT Press (1957).
62. Triebel, W. A. and A. E. Chu, *Handbook of Semiconductor and Bubble Memories*, Prentice-Hall, Inc. (1982).

63. Uffenbeck, J. E., *Hardware Interfacing with the Apple II Plus*, Prentice-Hall, Inc. (1983).
64. "Vocabulary for Information Processing," ANSI Standard X3.12 (1970).
65. Window, A. L. and G. S. Holister, *Strain Gauge Technology*, Applied Science Publishers (1982).
66. Wobsshall, D., *Circuit Design for Electronic Instrumentation*, McGraw-Hill Book Co. (1979).
67. Wojslaw, C. F., *Integrated Circuits Theory and Applications*, Reston Publishing Co., Inc. (1978).
68. Wonsiewics, B. C., ed., *Computer Automation of Materials Testing*, ASTM SPT 710, American Society for Testing and Materials (1980).
69. Yovits, M. C., *Advances in Computers,* Vol. 23, Academic Press, Inc.

APPENDIX

Codes and Standards

1. American Society for Testing and Materials (ASTM)
    E622-82, Standard Generic Guide for Computerized Systems.
    E623-83, Standard Guide Lines for Developing Functional Requirements for Computerized Laboratory Systems.
    E624-83, Standard Guidelines for Implementing Computerized Laboratory Systems.
    E625-83, Standard Guide for Training Users of Computerized Systems.
    E626-83, Standard Guidelines for Evaluating Computerized Systems.
    E627-82, Standard Guidelines for Documenting Computerized Laboratory Systems.
    E730-80, Standard Guide for Developing Functional Designs for Computerized Systems.
    E731-80, Standard Guide for Procurement of Commercially Available Computerized Systems.
    E792-81, Standard Guide for Computer Automation in the Clinical Laboratory.
    E919-83, Standard Specification for Software Documentation for a Computerized System.
2. American National Standards Institute (ANSI)
    X3.1-1969, Synchronous Signaling Rate for Data Transmission.
    X3.4-1968, Code for Information Interchange.
    X3.24-1968, Signal Quality at Interface Between Data Processing Terminal Equipment and Synchronous Data Communication Equipment for Serial Data Transmission.
3. American National Standards Institute/Institute of Electrical and Electronics Engineers (ANSI/IEEE)
    488-1978, IEEE Standard Digital Interface for Programmable Instrumentation.
    583-1982, Modular Instrumentation and Digital Interface System (CAMAC*).
    595-1982, Serial Highway Interface System (CAMAC).
    596-1982, Parallel Highway Interface System (CAMAC).
    675-1982, Multiple Controllers in a CAMAC Crate.
    683-1976, Block Transfers in CAMAC Systems (Reaffirmed 1981).
    726-1982, Real-Time BASIC for CAMAC.
    758-1979, Subroutines for CAMAC (Reaffirmed 1981).
    SHO8482, CAMAC Instrumentation and Interface Standards.
4. American National Standards Institute/Instrument Society of America (ANSI/ISA)
    RP55.1-1975, Recommended Practice, Hardware Testing of Digital Process Computers.
5. Electronic Industries Association (EIA)
    RS-247, Analog-to-Digital Conversion Equipment, 1961.
    RS-232-C, Interface Between Data Terminal Equipment and Data Communication Equipment Employing Serial Binary Data Interchange, 1969.
    RS-363, Standard for Specifying Signal Quality for Transmitting and Receiving Data Processing Terminal Equipment Using Serial Data Transmission at the Interface with Nonsynchronous Data Communication Equipment, 1969.
    RS-404, Standard for Start-Stop Signal Quality Between Data Terminal Equipment and Nonsynchronous Data Communication Equipment, 1973.
6. American Petroleum Institute (API)
    Manual on Installation of Refinery Instruments and Control Systems, 1982.

GLOSSARY

Accuracy—The deviation or error, by which an actual output varies from an expected ideal or absolute output.

Acquisition Time—The time the output of a sample/hold circuit takes to change from its previous value to a new value when the circuit is switched from the hold mode to sample mode.

Address—The label or number identifying the memory location where a unit of information is stored.

Alarm CRT—A CRT display that is dedicated solely to displaying alarms.

Alias Frequency—A false lower frequency component in a

*Computer Automated Measurement and Control

reconstructed analog signal due to inadequate sampling rate and/or low-pass filtering.

Alphanumeric—A set of symbols including letters, numbers, and other characters.

Analog—The representation of quantities by means of continuous physical signals.

Analog-To-Digital Converter (A/D, ADC)—A device or circuit that outputs a binary number corresponding to an analog voltage level at the input.

Aperture Time—When a sample/hold circuit is switched from sample to hold, a finite amount of time is required for the internal electronics to turn off. Aperture time is the time between the sample-to-hold command transition and the point at which the output ceases to follow the input.

Architecture—Characteristics of a computer system visible to the programmer (e.g., instruction set, registers, interrupt structure).

Arithmetic-Logic-Unit (ALU)—The part of a CPU where binary data is acted upon with mathematical operations.

ASCII—See Code.

Assembler—A program that translates assembly language instructions into machine language instructions.

Assembly Language—See Machine Language.

Asynchronous—See Synchronous.

Baud—The signaling speed, that is, keying rate of the modem. The signaling speed in baud is equal to the reciprocal of the shortest element duration in seconds to be transmitted. If the information transmitted per element duration is one-bit, then the baud and bit rate are equal; otherwise, they shall not be interchanged in usage. The element duration of binary train of signals is one-bit and that of the octal train of signals is three-bit.

Bidirectional—See Tristate.

Binary Coded Decimal—The representation of a decimal number (0 through 9) by means of a 4-bit binary nibble.

Bipolar Mode—Characterizing an ADC whose input range extends from negative to positive signal levels. Commonly, the input range is offset by one half the full-scale range.

Bit—Binary Digit.

Bit Rate—The number of bits transferred in a given time interval. Bits per second is a measure of the rate at which bits are transmitted.

Board—See Packaging Hierarchy.

Bootstrap—A technique or device designed to bring itself into a desired state, that is a computer routine whose first few steps provide the instructions necessary for continued operation.

Buffer—An isolating element used to prevent a circuit from influencing upstream elements, also called a driver, is used for matching logic level (voltage level) among IC devices. Also, a temporary storage register, used to accumulate data for subsequent processing.

Burn In—A period, usually prior to on-line operation, during which equipment is continuously energized for the purpose of stabilizing its characteristics and identifying early failures.

Bus—A group of conductors used for transmitting signals or power from one or more sources to one or more destinations.

Byte—Data word, generally 8 bits wide.

Cache—Very high speed buffer memory interposed between CPU and main memory, holding recently referenced information; used to improve system performance.

Calibration—Adjustment of a device such that the output is within a specified range for particular values of the input.

Card—See Packaging Hierarchy.

Cathode Ray Tube (CRT)—A display device in which controlled electron beams are used to present alphanumeric or graphical data on an electroluminescent screen.

Central Processing Unit (CPU)—Section of a computer system that includes ALU, control unit, and registers.

Chip—See Packaging Hierarchy.

Clock—Electronic circuitry for generating periodic timing signals.

Code—Bit pattern used to represent computer instruction, data character, etc.; "EBCDIC" (Extended Binary Coded Decimal Interchange Code) uses 8 bits to represent alphanumeric and control characters; "ASCII" (American National Standard Code for Information Interchange) is a different code, using 7 bits plus parity bit.

Common Mode Rejection—The ability of certain amplifiers to cancel a common mode signal while responding to an out-of-phase signal.

Common Mode Signal—A signal which appears at both input terminals of a differential amplifier.

Companding—Combination of compressing (in an ADC) and expanding (in an DAC) the analog signal range to yield large dynamic ranges.

Compiler—A program which translates instructions written in a user-oriented language to instructions in machine language, for example, a FORTRAN compiler.

Computer Word—An ordered set of bits which occupies one storage location and is treated by the computer circuits as a unit and transferred as such. A word is the smallest addressable unit of information in a programmable memory. A 32-bit microprocessor such as the NS32032 has 32-bit words and each 32-bit word occupies 4 bytes (32 bits) of memory.

Control Program—See Operating System.

Conversion Rate—The number of conversions per unit time.

Conversion Time—Time required for an ADC in the worst case to perform a complete conversion.

Cross Assembler—Assembler that runs on one computer and produces code to run on another computer.

Cross Compiler—Compiler that runs on one computer and produces machine language to run on another computer.

Crosstalk—The measure of the effect an off-channel signal has on the on-channel expressed in terms of dB of attenuation of the off-channel signal. The signal interference between channels usually is due to coupling between measurement channels occurring in some common element, e.g., a power supply having high internal impedance circuits.

Data—Any representation of a digital or analog quantity to which meaning has been assigned.

Data Acquisition—The collection of data.

Data Acquisition System (DAS)—A system processing one or more analog and/or digital signals.

Data Base—The entire body of data stored in the main memory or attached storage devices (e.g., tape file, desk file).

Data Highway—A serial communications link over which digitized status, alarm, and process signals are transmitted between the various components of a distributed control system.

Data Logging—The recording of selected data on suitable media.

Data Logger—A data logger is usually a multi-channel, low resolution low speed DAS.

Decoder—A synonym for digital-to-analog converter, a term used in communication engineering.

Demultiplexer—A device used to recover individual signals which have been combined for transmission over a single channel.

Device Controller—Hardware that attaches peripheral devices to I/O interface, generating control signals to the device, serializing and deserializing data, monitoring status, and performing other functions.

Diagnostic Routine—A program that locates malfunctions in computer hardware of software.

Digital filter—A computer program that separates digital data in accordance with a criterion such as signal frequency.

Digital-To-Analog Converter (D/A, DAC)—A device or circuit to convert a digital value to an analog voltage level.

Digitizer—Synonym for ADC.

Disable—A command or condition which prohibits some specific event from proceeding.

Disk Operating System (DOS)—Program used to control the transfer of information to and from a disk.

Distributed System—A method of distributing analog and digital control elements. These control elements communicate with the host computer via the data highway.

Droop—Voltage decay of a hold capacitor during the hold mode.

Droop Rate—Droop per unit of time.

Duplex—Penetrating to a simultaneous bidirectional independent data transmission.

Dynamic Accuracy—The total error of a converter operating at its maximum signal and/or sampling frequency.

Dynamic Range—The ratio of a converter's full-scale range to the smallest nominal difference the converter can resolve.

EBCDIC-See Code.

Emulation—The imitation of all or part of one computer system by another, primarily by hardware, so that the imitating computer system accepts the same data, executes the same programs, and achieves the same results as the imitated system.

Enable—A command or condition which permits some specific event to proceed.

EPROM—Erasable programmable read-only memory, a PROM that can be erased by the user, usually by exposing it to ultraviolet light.

Encoder—A communication engineering term for an ADC. Also, an electromechanical transducer that produces digital signals in accordance with mechanical angle or displacement.

Firmware—Programs stored in PROM or EPROM.

Flip-Flop—See Latch.

Gate—A gate, also called a logic gate, is a device whose output status (0 or 1) depends on the input status arrangement. It responds to the input signals immediately and no clocking pulse is needed to start its action. These are the often heard AND, OR NAND, NOR, etc., gates.

Half-Duplex—One-way-at-a-time communication. Both sides can transmit and receive data, but only one at a time.

Handshake—An interface procedure that is based on status/data signals that assure orderly data transfer as opposed to asynchronous exchange.

Interface—Hardware and software necessary to provide power source and signal compatibility between two systems or a system and a peripheral.

Interpreter—A program that translates and executes each user-oriented language instruction before translating and executing the next one, for example, a BASIC interpreter.

Interrupt—A suspension in the execution of a sequential program to permit processing of high priority data and performed in such a way that the execution of the program can be resumed. Examples are I/O interrupts, timer interrupts, hardware failure interrupts, and program interrupts.

K—Abbreviation for 1024 (2^{10}); often used in referring to memory size; M is abbreviation for 1,048,576 (2^{20}).

Latch—A latch, also called a flip-flop, is used for time duration matching. The data from a microprocessor are short pulses unsuitable for other functions. A latch accepts the data and releases the data on request. Its function is to condition the digital signal so that it will have a proper time base.

Log—A periodic printed summary of operating data.

Logger—An instrument that scans sensors and records or prints data on a chart.

Machine Language—Instructions written in binary form that a microprocessor can execute directly; also called object code and object language; the binary form instructions are very difficult to write. Assembly language instructions written in alphanumeric form (mnemonics) are in one-to-one correspondence with the machine language instructions.

Missing Code—A code word that never occurs at the output of an ADC when the input signal is varied over the full range.

Modem—Acronym for modulator/demodulator. A device that transforms digital signals into audio tones for transmission over telephone lines, and does the reverse for reception.

Module—See Packaging Hierarchy.

Motherboard—See Packaging Hierarchy.

Multiplexer (MUX)—A circuit designed to connect any of a number of input channels to an output load, switching from channel to channel in a consecutive, arbitrary, or changing sequence, in accordance with a digitally coded address instruction from the system program control source (PROM, ROM, RAM, tape, etc.).

Nyquist Frequency—The rate at which an analog signal must be sampled more than twice as fast as the highest frequency component to be expected in the signal.

Off-Line Conversion—Analog-to-digital conversion of the analog signals recorded on an analog tape.

On-Line Conversion—Analog-to-digital conversion of the analog signals from an instrument or transducer in its operational environment.

Operating System—A collection of computer programs that controls the overall operation of a computer and performs such tasks as assigning places in memory to programs and data, processing interrupts, scheduling jobs, and controlling the overall input/output of the system.

Packaging Hierarchy—Set of physical structures to rank hardware components and construct computer system; "chip," a semiconductor entity on which integrated circuits are fabricated; "board," also called "card," a printed circuit board with components on both sides, usually has standard pin connector at one edge to handle connections required for data, address, control codes, and power source; "mother board," a large board usually containing the most important portion of a microcomputer, smaller boards can be plugged into the mother board; "module," a device having its own case or container, components are not exposed, and connectors are usually non-standard, designed to be plugged into a backplane and then rack mounted.

Parallel Data Transmission—Sending all data bits simultaneously; eight wires are needed for an 8-bit data. Within a microprocessor, all data are transferred in a parallel fashion.

Parity Bit—A binary digit appended to an array of bits to make the sum of all the bits always odd or always even for the purpose of detecting transmission or recording errors.

Peripheral—A device that is external to the CPU and main memory, e.g., printer, modem, or terminal, but connected to it by electrical connections.

Pixels—Acronym for picture elements. Definable locations on a display screen that are used to form images on the screen. For graphics display, screens with more pixels provide higher resolution.

Polling—The process by which a DAS selectively requests data from one or more of its remote terminals.

Port—A signal input or output point on a computer.

PROM—Acronym for Programmable Read Only Memory. After it has been programmed, its contents cannot be changed by the computer operator and it behaves like a ROM.

Protocal—A formal definition that describes how data is to be exchanged.

RAM—Acronym for Random Access Memory.

ROM—Acronym for Read Only Memory.

Real-Time Clock—The circuitry that maintains time for use in program execution and event initiation.

Real-Time Operation—Computer monitoring, control, or processing functions performed at a rate compatible with the operation of physical equipment of processes.

Record—Basic block of data on input or output device.

Resolution—The smallest detectable increment of measurement, limited by the number of bits of an ADC. A 12-bit ADC can resolve to one part in 4096 (2^{12}).

Sample and Hold—A circuit used to hold a sampled value for a period of time after the signal itself has changed value, a necessary circuit for the sampling of high frequency events.

Scan—Collection of data from multiple sensors, usually through a MUX.

Serial Data Transmission—Sending one bit at a time on a single wire or data highway.

Signal Conditioner—A device consisting of amplifier, filter, power supply for the Wheatstone bridge circuit, and sometimes, multiplexer and sample and hold circuits to be used for conditioning the input signals so that they are acceptable to the microprocessor. Other analog circuits for addition, subtraction, integration, differentiation, and peak detection may be also a part of a signal conditioner.

Sign Bit—A single bit used to designate the algebraic sign of the information contained in the remainder of the word. A 13-bit ADC may have a 12-bit resolution and use one bit for the sign of the analog signals.

Simplex—One direction transmission of data.

Strobe—Same as enable.

Synchronous—Data transmission associated with clock having fixed cycle time; "asynchronous" is event not associated with clock.

Throughout Rate—System conversion rate in terms of bits; systems with throughout rate up to 40 million bits per second are available.

Throughput Rate—Same as conversion rate. One conversion may involve up to 16 bits.

Tristate—Logic device capable of offering "high," "low," and "disconnected" states.

UART—Acronym for Universal Asynchronous Receiver/Transmitter. A device for communicating along a serial bus or data highway, one bit at a time, with facilities for serializing and deserializing data.

Unipolar Mode—In this operating mode the range of an ADC has one polarity only such as 0 to 10 volts.

USART—Acronym for Universal Synchronous/Asynchronous Receiver/Transmitter. A device able to operate like either UART or USRT.

USRT—Acronym for Universal Synchronous Receiver/Transmitter. A device similar to UART, except that the clock pulses at both ends of the circuit must be matched.

CHAPTER 31

BASIC Programming for Civil Engineers on Micro-Computers

R. Hussein* and M. Morsi**

INTRODUCTION

Communicating with a micro-computer requires a special language. It is in the form of certain English words which can be interpreted by the computer. This chapter deals with the BASIC statements and instructions (or commands) and their applications to microcomputers.

Example:

```
10  PRINT 10*5
20  READ X,Y,Z
30  DATA 3,5,65
40  IF X=3, THEN 100
50  GO TO 90
60  FOR N=0 TO 20 STEP 2
RUN
LIST
NEW
GR
COLOR=5
SAVE PROGRAM1
LOAD PROGRAM1
```

BASIC STATEMENTS

A BASIC program consists of a sequence of numbered lines called statements. A statement may represent some information given to the computer.

Example:

```
10  DATA 1,2,3,4
20  END
30  REM A PROGRAM TO CALCULATE THE BENDING
    MOMENT
```

Also, a statement may instruct (or command) the computer to carry out certain operations.

Example:

```
40  PRINT "AXIAL FORCE"
50  READ A,B,C,D
60  GO TO 100
70  PRINT 5*4+3*9
```

Here, the words PRINT, READ, GO TO are the commands of the respective statements which instruct the computer to perform the actions.

Direct Instructions

These are the instructions which the computer accepts but are not included in a program. They are also termed as system commands.

Example:

```
RUN
LIST
LOAD PROGRAM1
DELETE PROGRAM2
```

Print

A PRINT statement may have numerical or string (words) output data. The variables in the PRINT statement may be separated from one another by semicolons or by commas.

---

*University of the District of Columbia, Washington, DC
**Arab Bureau for Design and Technical Consulting, Abbasseya, Cairo, Egypt

The PRINT statement can be used in the following two modes:

IMMEDIATE MODE

To perform simple tasks and a user gets the response immediately.

Example:

PRINT 4*10 − 5
35

When 4*10 − 5 is entered preceded by a PRINT command, the system displays the response immediately as 35.

Example:

PRINT "COMPUTER"
COMPUTER
PRINT "10*3 + 14"
10*3 + 14
PRINT 3 ^ 4
81

PROGRAM MODE

The statement in this case consists of a statement number, the command PRINT followed by an action to be carried out.

Example:

10 PRINT 10*5
20 PRINT 10/5
30 PRINT "ERROR"
40 PRINT "10 + 5"
50 PRINT

A system will not display any response to the above until the program is executed.

COMMAS AND SEMICOLONS IN A PRINT STATEMENT

Consider the following examples:

100 PRINT A,B,C
110 PRINT A;B;C
120 PRINT A;B;C,
130 PRINT D;E;F
140 PRINT

A comma, in general, sets the cursor to move to the first position in the next available zone. If there is not enough space on the line, then the cursor returns to start a new line. The screen on a computer is divided into three zones. If the PRINT statement has more than three items to be printed, the computer prints the first three items on the first line and the rest on the second line.

A semicolon, on the other hand, sets the cursor to move before a number or word (or letter). If there is not enough space on the line, then the cursor moves to the next line.

If the last item is also followed by a comma, it indicates that the PRINT statement is continued. In the above example, the last item C of the statement 120 is followed by a comma. In the next statement 130 PRINT is followed by D;E;F. Instead of typing in one line, the items are typed in two PRINT statements.

It should be noted that, if a PRINT command is not followed by an instruction, it produces a blank line.

Run

This command should be used to execute a program when it is entered.

Example: enter the following first

10 PRINT 10*4
20 PRINT 20/5
30 PRINT 10 + 5 ^ 4
40 PRINT "10 − 8"
50 PRINT "SHEAR FORCE"
60 PRINT A;B;C
70 PRINT "A;B;C"

Then, enter RUN. The system will display the following results:

RUN
40
4
635
10 − 8
SHEAR FORCE
000
A;B;C

List

This command can be used to print the program which is currently in the computer's memory. If LIST is entered, the entire program will be displayed on the screen.

LIST followed by either a line number or a range (the first and last statement numbers of the part desired) will result in either that line or that specified range displayed on the screen. The two items in the range should be separated by a hyphen.

Example: enter the following:

10 PRINT 10*5
20 PRINT 10/5
30 PRINT "MOMENT DISTRIBUTION"
40 PRINT
50 PRINT "RESULTS"

If LIST is entered at this stage, the user will see the LISTing of the entire program displayed on the screen.

LIST
10 PRINT 10*5
20 PRINT 10/5
30 PRINT "MOMENT DISTRIBUTION"
40 PRINT
50 PRINT "RESULTS"

New

This command insures that the existing program in the computer's memory is cleared. It is always advisable to use this command before a new program is entered in a computer.

Example:

```
10  PRINT 10*5
20  PRINT 20/5
30  PRINT "TORSION"
40  PRINT "X*Y"
```

If the user does not need this program anymore, it can be deleted from computer's memory as follows:

NEW

Read, Data

The READ statement specifies the variables whose values are to be entered into the computer. The format for this statement is:

Statement no. READ < a list of input variables (including subscripted variables) separated by commas >

The DATA statement provides the appropriate values to the variables listed in the READ statement. The format for DATA statement is:

Statement no. DATA < numbers or strings separated by commas >

The DATA statement can be placed either preceding a READ statement or succeeding it.

Example:

```
100   READ A$,B,C
 . . .
 . . .
 . . .
180   DATA JUNE 12, 1950
10    READ A,B,C,D
20    PRINT A*(B/C)+D ^ 3
30    DATA 2,8,4,2,
RUN
12
```

If the program has more than one DATA statement, the computer starts reading data from the first statement and continues sequentially through the remaining DATA statements.

Input

INPUT statement implements the same function as READ and DATA statements when put in action together. In some programs, INPUT command is preferred instead of READ, DATA command. When INPUT is used, the computer displays a question mark (?) requesting the data to be entered.

Example:

```
100   INPUT X
110   PRINT X â 3 + 15
RUN
?10
1015
```

The program stops execution at this stage. If there is more than one value for X, the user will be required to LIST and RUN it as many times as the number of X. To avoid this cumbersome procedure, either GO TO or FOR . . . To, NEXT statements can be used.

Go To

Statements in a program are executed one after the other in the given order. Sometimes it is required to alter the normal flow of operations. This is known as unconditional branching operation, where the sequence is transferred from one part of the program to the other. GO TO statement is very useful in such circumstances. GO TO statement can also be used to execute repititious programs.

Example:

```
100   X = 1
110   PRINT X
120   X = X + 2
130   GO TO 110
```

Example:

```
200   INPUT Y
210   PRINT Y*10 + 2
220   GO TO 200
RUN
?2
22
?3
32
?10
102
?
```

The question mark will be displayed every time the RETURN key is depressed unless the user tries to interrupt the execution of the program which can be done by pressing C key while holding down the CTRL key. The user then will see the response:

BREAK IN 200

On . . . Go To

This statement can be used to carry out multiple branching operations in a program. The format for this statement is:

Stmt. no. ON < numeric variable or formula > GO TO < statements where the control is to be transferred >

Example:

100   ON P GO TO 20,50,30,140

Example:

200   ON P+1 GO TO 180,220,60,310

Rem

Rem stands for REMARK. REM statements can be included at various stages of the program to explain the purpose of the program or the steps following the REM statement. These statements do not cause any execution by the computer.
Example:

100   REM A PROGRAM FOR THE DESIGN OF REINFORCED CONCRETE BEAMS
110   REM ******** DONE BY XYZ********
. . .
. . .
. . .
300   REM CHECK FOR SHEAR
. . .

Restore

RESTORE statement can be used if a set of data is to be read more than once in a program. Data can be read again starting at its beginning.
Example:

100   READ A,B,C
110   PRINT A;B;C
120   READ E,F,G
130   PRINT E;F;G
140   RESTORE
150   READ A,B,C
160   PRINT A;B;C
170   DATA 1,2,3,4,5,6,7,8,9
180   END
RUN
123
456
123

End

The END statement indicates the end of a program. It is used to terminate the RUN of a program, so it should always be the last line. The format is:
Statement no. followed by END.
Example:

120   END

Stop

STOP statement is used to terminate the execution at any point in the program. Whereas the END statement is always placed at the end of a program, the STOP statement can be placed anywhere except in the end. The format for the statement is:
Stmt. no. STOP.
Example:

100   STOP
. . .
. . .
. . .
230   END

If . . . Then

This is a conditional statement as the action carried out depends upon whether that condition is found true. The format of this statement is:
Statement no. IF <condition> THEN <the no. of statement to where sequence of operation goes>
Example:

100   IF I< >0 THEN 180

It explains that IF I is not equal to 0 (i.e., if this condition is true), skip to statement no. 180. If the condition is not met with (i.e., if it is false), execution continues.
The following are the conditions that are valid for use in IF . . . THEN statements:

DESIGNATION	MEANING
=	Equal to
>	Greater than
<	Less than
< >	Not equal to
< =	Less than or Equal to
> =	Greater than or Equal to

IF . . . THEN statement also permits the following conditions:

100   IF A >=0 THEN PRINT A
200   IF B <=0 THEN PRINT "THE VALUE IS NEGATIVE"
300   IF C >=0 THE PRINT "IS NOT THE SOLUTION":
       GO TO 200

The following example explains the use of STOP, END, INPUT, IF . . . THEN statements. This example solves any pair of simultaneous equations by using KRAMER'S rule.

10   PRINT "SOLUTION OF SECOND ORDER SIMULTANEOUS EQUATIONS"
20   PRINT "TYPE COEFFICIENTS OF X & Y"

```
30   PRINT "FIRST EQUATION"
40   INPUT A,B
50   PRINT "SECOND EQUATION"
60   INPUT C,D
70   REM EVALUATE THE VALUE OF DENOMINATOR, S
80   S = A*D − C*B
90   IF ABS(S) > 0.00001 THEN 120
100  PRINT "SOLUTION IMPOSSIBLE; DET =";S
110  STOP
120  PRINT "TWO VALUES FOR CONSTANTS"
130  INPUT P,Q
140  X = (P*D − Q*B)/S
150  Y = (Q*A − P*C)/S
160  PRINT "X =";X;"Y =";Y
170  PRINT "ANY MORE CONSTANTS? YES?"
180  INPUT A$
190  IF A$ = "YES" THEN 110
200  REM YOU COULD HAVE "STOP" HERE
210  END
```

And, Or

These are called statement modifiers. They can be used within an IF . . . THEN statement.
Example:

```
10   IF X > 21 and Y < 14 THEN 90
```

The control of the execution skips to line 90 only if the two conditions are satisfied.
Example:

```
10   IF X > 21 or Y < 14 THEN 90
```

Here, the cursor skips to line 90 if either or both the conditions are satisfied.

For . . . To . . . Step, Next

A loop can be built in BASIC programming by using either GO TO or IF . . . THEN statements as examined earlier. A simpler way to build a loop is by using FOR . . . TO, NEXT statement.

The FOR . . . TO statement specifies how many times a loop will be executed. The variable in this case will always increase by an increment of one. Using the STEP clause, we can increment the variable by any value other than one. The NEXT statement provides an end to the loop. The general format is:

FOR <variable> = <string variable> TO <ending value> STEP <increment>
. . .
. . .
NEXT <variable>

Example:
```
10   FOR X = 1 TO 10
20   Y = 3*X + 15
30   PRINT X,Y
40   NEXT X
50   END
```

Example:
```
110  FOR X = −20 TO 20 STEP 2
110  Y = X ^ 3 + 13
120  PRINT X,Y
130  NEXT X
140  END
```

Example:
```
200  FOR X = 10 TO 1
210  Y = 5*X ^ X − 3*X
220  PRINT X,Y
230  NEXT X
240  END
```

Print, Tab

This command facilitates the user to tabulate various quantities.
Example:

```
10   PRINT TAB (10); "MOMENT"; TAB (20); "SHEAR"; TAB(30); "AXIAL"
20   PRINT
30   PRINT TAB (20);"COMPUTER"
RUN
```

```
         MOMENT        SHEAR         AXIAL
                       COMPUTER
```

Gosub . . . Return

This statement facilitates the transfer of control in a program to skip to a line number where a SUBROUTINE begins. A subroutine is a program or an extended DEF function separated from the main program. Its execution may be repeated in the main program several times. The control will be transferred to the main program again when a RETURN statement is encountered within the subroutine. The format for this statement is:

Stmt. no. GOSUB <the line no. where the subroutine starts>
. . .
. . .
The subroutine starts here
. . .
. . .
Stmt. no. RETURN

Dim

DIM stands for DIMENSION. Microcomputers generally permit an array consisting of a limited maximum of rows and columns. If a larger array is required, it should be defined. This can be achieved by means of DIM statement. The format for DIM is:
Stmt. No. DIM <array>
Example:

10 DIM A(100)
20 DIM B$(200)
30 DIM C(40,50)
40 DIM D(50),E(60,78),F(20,20)

If there is more than one number of arrays in a DIM statement, they must be separated by commas. It is always preferable to place the DIM statement at the beginning of a program.

Clear

CLEAR statement can be used to delete variables including subscripted ones from the memory of the computer. This saves memory space.
Example:

80 CLEAR

Mathematical Functions

A user can implement many mathematical functions on microcomputers. The known functions are listed next.

ABS

ABS stands for ABSOLUTE. It can be used to find the ABSOLUTE value of a number as per the following principle:

ABS(X) = X   if X = 0
ABS(X) = −X (or −1*X) if X < 0

Example:

100 PRINT ABS (10.25)
110 PRINT ABS(0)
120 PRINT ABS (−.8)
RUN
10.25
0
0.8

SGN

The purpose of this function is to find out whether the variable is positive, negative or equal to zero. The computer executes the function as follows:

SGN(X) = 1   if X > =0
SGN(X) = 0   if X   =0
SGN(X) = −1   if X   <0

Example:

100 PRINT SGN(10)
110 PRINT SGN(−10)
120 PRINT SGN(0)
RUN
1
−1
0

SQR

SQR stands for SQUAREROOT. This function can be used to calculate the positive square root of a number.
Example:

100 PRINT SQR(9)
120 PRINT SQR(1001)

If a negative number is entered for execution, the system will display an error message.

SIN, COS, TAN, COT

These four trigonometric functions are used to find the sine, cosine, tangent, cotangent of an angle measured in radians, respectively.
Example:

5   INPUT X
10 A=(3.141592654/180)*X
20 PRINT SIN(A);COS(A);TAN(A);COT(A)

LOG

This function can be used to find the natural logarithm of a variable.
Example:

100 PRINT LOG(10)
120 PRINT LOG(1)
140 PRINT LOG(2.71828)
RUN
2.30258
0
1

EXP

The EXP function calculates the value of an exponential function (e to the power x, where $e = 2.71828$).
Example:

10 FOR X=2 TO 10
20 Y = EXP(X)

```
30  Z = LOG(X)
40  PRINT X,Y,Z
50  NEXT X
60  END
```

INT

The function Y = INT(X) can be used to assign to Y the largest integer that does not exceed X.

Example:

INT(10.57) = 10
INT(19.91) = 19
INT(−7.245) = −8

ATN

This function can be used to calculate the arc tangent of a variable which is measured in radians.

Example:

```
10  Y = ATN(X)
```

To calculate ARCSINE, ARCCOSINE, etc., the user should represent these functions in terms of ARCTANGENT.

RND

RND function can be used to generate pseudo-random numbers.

Example:

```
10  DIM A(200)
. . .
. . .
40  FOR I = 1 TO 200
50  A(I) = RND
60  NEXT I
```

RANDOMIZE

This statement makes a BASIC program to start the cycle at an unpredicatable place on each encounter. Its format is:
Stmt. no. RANDOMIZE
Example:

```
10  RANDOMIZE
20  X = INT(1 + 6*RND)
30  Y = INT(1 + 6*RND)
40  PRINT "THROW:";X;"AND";Y
50  END
```

DEF

In addition to the mathematical functions described previously, the user can DEFINE any other function by using DEF statement. This is very useful where a function is repeated in the same program several times.

Example:

```
10  DEF FNA(X) = (3*X − 15) (X − 3)
20  DEF FNA7(X) = (X â 3 − 2*X â 2 − 5*X + 11)/(X + 29)
```

The function name (i.e., FNA or FNA7) may be any alpha numeric variable that is accepted by the computer.

EDITING ON MICRO-COMPUTERS

The editing features of microcomputers are introduced, in brief, next.

Using the Backspace and Retype Keys

These keys can be used when a line is entered and before the RETURN key is depressed.
Example:

```
20  PRIMT "WHAT IS YOUR SELECTION"
```

If the user wants to change M in PRIMT to N:

(a) Depress the backspace key repeatedly until the cursor comes to the position of M where the correction is to be made.
(b) Then change M to N.
(c) Depress the retype key repeatedly until the end of the statement.
(d) Depress the RETURN key.

Using ESC Key in Conjunction with I, J, K and M Keys

When a program is entered entirely, corrections can be done, on some systems, using the four keys: I,J,K and M. First, the system has to be brought into the edit mode by depressing and releasing the ESC twice.

The cursor can now be moved freely by depressing one of the following keys:

I   key to move the cursor upward
J   key to move it to the left direction
K   key to move it to the right direction
M   key to move it downward

To move the cursor repeatedly, hold down one of the four keys. Consider the following example to elaborate on the edit features:

```
10  REM PROGRAM TO CONVERT DEGREES FAHRENHEIT
    TO DEGREES CENTIGRADE
20  PRINT "TEMPERATURE IN DEGREES FAHRENHEIT =";
30  INPUT F
40  C = 5*(F − 32)*9
50  PRINT DEGREES F =";F;"DEGREES C =";C
60  END
```

There is a mistake in statement 40. We have to change *9 to /9.

1. Depress ESC key twice and release it.
2. Depress the I and REPT KEYS concurrently and release them when the cursor is on the statement 40.
3. Depress either J or REPT keys to bring the cursor to the place where correction is to be made then depress / in place of *.
4. Check whether correction is done by LISTing the program.

Deleting Lines From a Program

(a) To delete a line currently under typing: Hold down the CTRL key and enter the letter X.

(b) To delete any particular line: Type the number of that line and depress RETURN.

(c) To delete a range of a program, use the command DEL.

Example:

DEL   110,180

Clearing the Screen

The screen can be cleared fully or partially using the following commands:

(a) Depress ESC key and enter an @, the entire screen will be cleared.
(b) Enter HOME and then depress the RETURN key, the entire screen will be cleared.
(c) Depress ESC and enter the letter F, all characters to the end of the screen will be cleared.
(d) Depress ESC and enter the letter F, all characters to the end of the line will be cleared.

Other Features

CTRL + S

CTRL + S can be used for temporarily stopping a LISTing. To resume LISTing, depress the S key while the CTRL key is held down.

CTRL + C

Entering C while the CTRL key is held down will cause a break in the execution of a program.

IF RESET KEY IS DEPRESSED INSTEAD OF RETURN

If RESET key is depressed instead of RETURN, this will cause the screen to light up. Coming back to normal can be achieved by entering RUN.

FINDING THE LENGTH OF A STRING

The length of a string can be found as follows:

A$ = BENDING MOMENT DIAGRAM
PRINT LEN(A$)
22

TO PRINT A PART OF A STRING

To PRINT a part of a string, consider the following example:

A$ = THE UNIVERSITY OF EVANSVILLE
PRINT LEFT$(A$,14)
THE UNIVERSITY
PRINT RIGHT$(A$,10)
EVANSVILLE
PRINT MID$(A$,13)
OF EVANSVILLE

DISKETTE OPERATION ON MICRO-COMPUTERS

A brief introduction about disk operating commands, and initializing a new diskette is presented here. When a new diskette is to be used, it has to be formatted. The disk operating system must be loaded into the computer from a previously formatted diskette. The SYSTEM MASTER on DOS can be used for this purpose.

Initializing a New Diskette

(a) Insert the operating system disk in the disk drive and turn on the computer. After a few seconds, messages will appear on the screen.

(b) Remove this diskette from the drive and insert the new diskette which is to be initialized. Type NEW, in some systems or follow proper instructions of others. An initializing program can be entered containing: the date on which the diskette is initialized, the name of the person who initialized it, etc. This is generally called a greeting program and the user can name it with a remarkable title such as HELLO.

Example: a typical initializing program

10   REM GREETING PROGRAM
20   PRINT "DISKETTE FOR STRUCTURES PROGRAMS"
30   PRINT "INITIATED BY"
40   PRINT "NAME OF THE PERSON"
50   PRINT "TODAY'S DATE"
60   END

Once the initialization has been completed, the diskette can be removed from the drive and labelled for identification. It should be noted that with systems having two drives, there is no need to use one of them. Both can be used simultaneously.

Save

The SAVE command can be used to store a program on a diskette on space called a file. Each file has to be given a

unique name. A file name is formed of alpha numerics characters with the first one being a letter. The format for this command is:
SAVE <name of the file>
Example:

SAVE STRUCTURES FROM COLLAPSE
SAVE CURRENT

Using the SAVE command more than one time with the same file name will overwrite on existing version. This can be prevented by using LOCK command. Once a program is saved, it can be executed using the command RUN <file name>. The statements for save and lock vary from one system to another.

Load

LOAD command is used to call a program saved on a diskette to the computer's memory. The format is:
LOAD <file name>
With modifications conducted on a program called from the memory space, the command RUN should only be used but not RUN <file name>. If the latter command is used, the older version of the program will be reloaded into the memory and executed whereas the new version of it will be deleted. On some microcomputers, LOAD is not required, i.e., only the file name can be used.

Catalog

This command can be used to display a list of all the files stored on the diskette. On some systems, DIR can be used instead of CATALOG.

Rename

This command can be used to change the name of a file saved on the diskette. The format for RENAME command is:
RENAME <old file name>, <new file name>
On some systems, a question has to be answered with regard to this function.

Delete

DELETE command can be used to remove from the diskette any file that is no more required. The command format is:
DELETE <name of file>
It should be noted that general four commands LOAD, CATALOG, RENAME and DELETE can be used from within a program as follows:

110 PRINT CHR$(4); "LOAD ANALYSES"

110 PRINT CHR$(4); "CATALOG"
120 PRINT CHR$(4); "RENAME ANALYSIS,DESIGN"
130 PRINT CHR$(4); "DELETE ANALYSIS"

Lock and Unlock

The LOCK command can be used to prevent a file from being deleted or erased accidentally from a diskette. The command format is:
LOCK <name of file>
A file that is LOCKED cannot be renamed, overwritten or deleted unless it is first UNLOCKED. To UNLOCK a file the user should use the command format:
UNLOCK <name of file>
Again, other statements are being used on different systems.

USEFUL SUBROUTINES IN CIVIL ENGINEERING

In the following, four subroutines are presented in BASIC. They are often used by civil engineers in many applications. With each routine, an example is introduced to elaborate on its use. The first routine can be used to invert a matrix. The second routine can be used to find the N roots of a polynomial. The third and fourth routines can be used to solve simultaneous equations.

REMARKS

This chapter has been written about BASIC programming for civil engineers on micro-computers. Because there are many trade marks available at present and the way to operate them is not unique, this chapter presents, generally, the basics by which the functions of any system can be implemented easily.

```
PROGRAM NO. 1

100  DIM X(50), A(51,51,),IR(50),JC(50),Y(50),JO(50)
101  PRINT"THIS PROGRAM CALCULATE INVERSE MATRIX AND SOLVE LINEAR
     EQUATIONS"
102  PRINT"N     : NUMBER OF ROW"
103  PRINT"INDIC : COMPUTATIONAL SWITCH"
104  PRINT"          <0 N BY N MATRIX COMPUTE INVERSE MATRIX"
105  PRINT"          =0 N BY N+1 MATRIX COMPUTE INVERSE MATRIX AND
                     LINEAR EQUATION"
106  PRINT"          >0 N BY N+1 MATRIX SOLVE LINEAR EQUATION"
107  PRINT"EPS   : MINIMUM ALLOWABLE MAGNITUDE FOR A PIVOT ELEMENT"
108  PRINT
109  PRINT
110  READ N,IN,EP
120  PRINT "N=     ",N
130  PRINT "INDIC= ",IN
140  PRINT "EPS=   ",EP
145  PRINT "THE STARTING MATRIX IS"
150  MX=N
160  IF IN<0 THEN 180
170  MX=N+1
180  FOR I=1 TO N
190  FOR J=1 TO MX
200  READ A(I,J)
210  PRINT A(I,J)
220  NEXT J
230  NEXT I
240  GOSUB 1000
245  DT=SI
```

```
250     IF IN>0 THEN 330
255     PRINT "DETER=   ",DT
260     PRINT "THE INVERSE MATRIX IS"
270     FOR I=1 TO N
280     FOR J=1 TO N
290     PRINT A(I,J)
300     NEXT J
310     NEXT I
320     GO TO 3000
330     PRINT "DETER=   ",DT
340     PRINT "THE SOLUTIONS X(1)...X(",N,") ARE"
350     FOR I=1 TO N
360     PRINT X(I)
370     NEXT I
380     IF IN <> 0 THEN 3000
400     PRINT "THE INVERSE MATRIX IS"
410     FOR I=1 TO N
420     FOR J=1 TO N
430     PRINT A(I,J)
440     NEXT J
450     NEXT I
460     GO TO 3000
1000    IF N<=50 THEN 1040
1010    PRINT "N TOO BIG"
1020    SI=0
1030    RETURN
1040    DT=1
1050    FOR K=1 TO N
1060    K1=K-1
1070    PI=0
1080    FOR I=1 TO N
1090    FOR J=1 TO N
1100    IF K=1 THEN 1170
1110    FOR W=1 TO K1
1120    FOR Z=1 TO K1
1130    IF I=IR(W) THEN 1200
1140    IF J=JC(Z) THEN 1200
1150    NEXT Z
1160    NEXT W
1170    IF (ABS(A(I,J)) <= ABS(PI)) THEN 1200
1175    PI=A(I,J)
1180    IR(K)=I
1190    JC(K)=J
1200    NEXT J
1210    NEXT I
1220    IF ABS(PI) >EP THEN 1250
1230    SI=0
1240    RETURN
1250    IZ=IR(K)
1260    JZ=JC(K)
1270    DT=DT*PI
1280    FOR J=1 TO MX
1290    A(IZ,J)=A(IZ,J)/PI
1300    NEXT J
1310    A(IZ,JZ)=1/PI
1320    FOR I=1 TO N
1330    AI=A(I,JZ)
1340    IF I=IZ THEN 1400
1350    A(I,JZ)=AI/PI
1360    FOR J=1 TO MX
1370    IF J=JZ THEN 1390
1380    A(I,J)=A(I,J)-AI*A(IZ,J)
1390    NEXT J
1400    NEXT I
1410    NEXT K
1420    FOR I=1 TO N
1430    II=IR(I)
1440    JI=JC(I)
1450    JO(II)=JI
1460    IF IN<0 THEN 1475
1470    X(JI)=A(II,MX)
1475    NEXT I
1480    I%=0
1490    N2=N-1
1500    FOR I=1 TO N2
1510    P1=I+1
1520    FOR J=P1 TO N
1530    IF JO(J) > JO(I) THEN 1580
1540    JP=JO(J)
1550    JO(J)=JO(I)
1560    JO(I)=JP
1570    I%=I%+1
1580    NEXT J
1590    NEXT I
1600    IF ((FIX(I%/2)*2) = I%) THEN 1620
1610    DT=-DT
1620    IF IN < = 0 THEN 1650
1630    SI=DT
1640    RETURN
1650    FOR J=1 TO N
1660    FOR I=1 TO N
1670    II=IR(I)
1680    JI=JC(I)
1690    Y(JI)=A(II,J)
1700    NEXT I
1710    FOR I=1 TO N
1720    A(I,J)=Y(I)
1730    NEXT I
1740    NEXT J
1750    FOR I=1 TO N
1760    FOR J=1 TO N
1770    IJ=IR(J)
1780    JJ=JC(J)
1790    Y(IJ)=A(I,JJ)
1800    NEXT J
1810    FOR J=1 TO N
1820    A(I,J)=Y(J)
1830    NEXT J
1840    NEXT I
1850    SI=DT
1860    RETURN
3000    PRINT "DO YOU WANT CONTINUE?(Y/N) ";
3010    INPUT I$
3020    IF I$="N" THEN 7000
3030    GO TO 110
4000    DATA 6, -1,1E-10
4010    DATA 10,20,30,40,50,60
4020    DATA 20,40,70,80,90,100
4030    DATA 30,70,50,110,120,130
4040    DATA 40,80,110,60,140,150
4060    DATA 50,90,120,140,70,160
4070    DATA 60,100,130,150,160,80
7000    END
RUN
THIS PROGRAM CALCULATE INVERSE MATRIX AND SOLVE LINEAR EQUATIONS
N       : NUMBER OF ROW
INDIC   : COMPUTATIONAL SWITCH
            <0 N BY N MATRIX COMPUTE INVERSE MATRIX
            =0 N BY N+1 MATRIX COMPUTE INVERSE MATRIX AND LINEAR EQUATION
            >0 N BY N+1 MATRIX SOLVE LINEAR EQUATION
EPS     : MINIMUM ALLOWABLE MAGNITUDE FOR A PIVOT ELEMENT

N=       6
INDIC=  -1
EPS=     1E-10
THE STARTING MATRIX IS
 10         20         30         40         50         60
 20         40         70         80         90         100
 30         70         50         110        120        130
 40         80         110        60         140        150
 50         90         120        140        70         160
 60         100        130        150        160        80
DETER=  5.588E+09
THE INVERSE MATRIX IS
 .5825      -.3084     -7.8418E-02 -1.0343E-02  2.3156E-02  4.9212E-02
 -.3084      .1148      5.4026E-02  1.5980E-02 -4.6349E-03 -2.0669E-02
 -7.8418E-02 5.4026E-02 -7.7129E-03  4.2770E-03 -2.3264E-04 -3.7401E-03
 -1.0343E-02 1.5980E-02  4.2770E-03 -1.2580E-02  1.5211E-03  1.3779E-03
  2.3156E-02 -4.6349E-03 -2.3263E-04  1.5211E-03 -9.2877E-03  4.5275E-03
  .04921    -2.0669E-02 -3.7401E-03  1.3779E-03  4.5275E-03 -4.1338E-03
DO YOU WANT CONTINUE?(Y/N) ? N


PROGRAM NO. 2

100     DIM A(100),B(100),C(100)
101     PRINT "THIS PROGRAM USES GRAEFFE'S ROOT SQUARING TECHNIQUE
        TO FIND"
102     PRINT"THE REAL AND DISTINCT ROOTS OF THE NTH DEGREE POLYNOMIAL"
103     PRINT"N     : DEGREE OF STARTING POLYNOMIAL"
104     PRINT"ITMAX : MAXIMUM NUMBER OF ITERATIONS ALLOWED"
105     PRINT"TOP   : UPPER LIMIT ON THE MAGNITUDE OF COEFFICIENTS B(I)"
106     PRINT"        PRODUCED BY THE ROOT-SQUARING PROCESS"
107     PRINT"EPS   : TEST TO DETERMINE ANSWER IS TO BE CONSIDERED A ROOT"
108     PRINT"IPRINT : PRINT CONTROL VARIABLE. IF NONZERO, COEFFICIENTS B(I)"
109     PRINT"         ARE PRINTED AFTER EACH ITERATION"
110     PRINT
111     PRINT
115     B(1)=1
120     C(1)=1
130     READ N,IX,TP,EP,PR
140     N1=N+1
150     FOR I=1 TO N1
160     READ A(I)
170     NEXT I
180     FOR I=2 TO N1
190     A(I)=A(I)/A(1)
200     C(I)=A(I)
210     NEXT I
220     A(1)=1
230     PRINT "N=      ",N
240     PRINT "ITMAX=  ",IX
250     PRINT "TOP=    ",TP
260     PRINT "EPS=    ",EP
270     PRINT "IPRINT= ",PR
280     PRINT "A(1)...A(",N1,")"
283     FOR I=1 TO N1
285     PRINT A(I)
287     NEXT I
290     FOR P=1 TO IX
300     FOR I=2 TO N1
310     B(I)=C(I)*C(I)
320     N2=I-1
330     FOR L=1 TO N2
340     L1=I+L
```

```
350     L2=I-L
360     IF L1>N1 THEN 390
370     B(I)=B(I)+(-1)^L*2*C(L1)*C(L2)
380     NEXT L
390     B(I)=(-1)^N2*B(I)
400     NEXT I
410     IF PR=0 THEN 470
420     PRINT "ITER=       ",P
430     PRINT "B(1)...B(",N1,")"
440     FOR I=1 TO N1
450     PRINT B(I)
460     NEXT I
470     FOR I=2 TO N1
480     IF ((ABS(B(I))>TP) OR (ABS(B(I))<1/TP) AND (B(I)<>0)) THEN 560
490     NEXT I
500     FOR I=2 TO N1
510     C(I)=B(I)
520     NEXT I
530     NEXT P
540     PRINT "ITER EXCEEDS ITMAX - - CALCULATION CONTINUES"
550     P=IX
560     PRINT "ROOT        POLYNORMAL VALUE     ITER       COMMENT"
570     FOR I=2 TO N1
575     P2=1/2^P
580     RT=ABS(B(I)/B(I-1))^P2
590     US=1
600     NS=1
610     FOR J=2 TO N1
620     US=US*RT+A(J)
630     NS=NS*(-RT)+A(J)
635     NEXT J
640     IF (ABS(US)>ABS(NS)) THEN 670
650     PV=US
660     GO TO 690
670     PV=NS
680     RT=-RT
690     IF (PV>=EP) THEN 720
700     PRINT RT,PV," ",P;
710     PRINT"PROBABLY A ROOT"
715     GO TO 740
720     PRINT RT,PV," ",P;
730     PRINT "PROBABLY NOT A ROOT"
740     NEXT I
750     PRINT "DO YOU WANT CONTINUE?(Y/N) ";
760     INPUT I$
770     IF I$="N" THEN 7000
780     GO TO 130
1000    DATA 4,25,1E15, 0.1,0
1010    DATA 15,30,10,5,35
1020    DATA 3,25,1E15,0.1,0
1030    DATA 27,18,10,50
1040    DATA 2,25,1E15,0.1,0
1050    DATA 4,2,10
7000    END
RUN
THIS PROGRAM USES GRAEFFE'S ROOT SQUARING TECHNIQUE TO FIND THE REAL
AND DISTINCT ROOTS OF THE NTH DEGREE POLYNOMIAL
N       : DEGREE OF STARTING POLYNOMIAL
ITMAX   : MAXIMUM NUMBER OF ITERATIONS ALLOWED
TOP     : UPPER LIMIT ON THE MAGNITUDES OF COEFFICIENTS B(I)
          PRODUCED BY THE ROOT-SQUARING PROCESS
EPS     : TEST TO DETERMINE ANSWER IS TO BE CONSIDERED A ROOT
PRINT   : PRINT CONTROL VARIABLE. IF NONZERO, COEFFICIENTS B(I)
          ARE PRINTED AFTER EACH ITERATION
N=      4
ITMAX=  25
TOP=        1E+15
EPS=        .1
IPRINT=     0
A(1)...A(   4       )
 1
 2
 .6666667
 .3333334
 2.333333
ROOT        POLYNORMAL VALUE     ITER       COMMENT
-1.604125    1.880025                6  PROBABLY NOT A ROOT
-1.574678    1.800808                6  PROBABLY NOT A ROOT
- .9362908   1.732581                6  PROBABLY NOT A ROOT
- .9865882   1.680193                6  PROBABLY NOT A ROOT
DO YOU WANT CONTINUE?(Y/N) ? Y
N=      3
ITMAX=  25
TOP=        1E+15
EPS=        .1
IPRINT=     0
A(1)...A(   4       )
 1
 .6666667
 .3703704
 1.851852
ROOT        POLYNORMAL VALUE     ITER       COMMENT
-1.375808    -2.026558E-06           6  PROBABLY A ROOT
-1.160767     .756197                6  PROBABLY NOT A ROOT
-1.159587     .7595733               6  PROBABLY NOT A ROOT
DO YOU WANT CONTINUE?(Y/N) ? Y
N=      2
ITMAX=  25
TOP=        1E+15
EPS=        .1
IPRINT=     0
A(1)...A(   3       )
 1
 .5
 2.5
ROOT        POLYNORMAL VALUE     ITER       COMMENT
-1.590867    4.235424                6  PROBABLY NOT A ROOT
-1.57147     4.183784                6  PROBABLY NOT A ROOT
DO YOU WANT CONTINUE?(Y/N) ? N


PROGRAM NO. 3

100     DIM A(50,51)
101     PRINT"THIS PROGRAM FIND M SOLUTION VECTORS CORRESPONDING TO A SET OF "
102     PRINT"N SIMULTANEOUS LINEAR EQUATIONS USING THE GAUSS-JORDAN REDUCTION"
103     PRINT"ALGORITHM"
104     PRINT"N   : NUMBER OF EQUATIONS"
105     PRINT"M   : NUMBER OF SOLUTION VECTORS"
106     PRINT"EPS : MINIMUM ALLOWABLE MAGNITUDE FOR A PIVOT ELEMENT"
107     PRINT
108     PRINT
110     READ N,M,EP
120     NM=N+M
130     PRINT "N=    ",N
140     PRINT "M=    ",M
150     PRINT "EPS=  ",EP
160     PRINT "A(1,1)...A(",N,",",NM,")"
170     FOR I=1 TO N
180     FOR J=1 TO NM
190     READ A(I,J)
200     PRINT A(I,J)
210     NEXT J
220     NEXT I
230     DT=1
240     FOR K=1 TO N
250     DT=DT*A(K,K)
260     IF ABS(A(K,K)) > EP THEN 290
270     PRINT "SMALL PIVOT--MATRIX MAY BE SINGULAR"
280     GO TO 2000
290     K1=K+1
300     FOR J=K1 TO NM
310     A(K,J)=A(K,J)/A(K,K)
320     NEXT J
330     A(K,K)=1
340     FOR I=1 TO N
350     IF (I=K) OR (A(I,K)=0 THEN 400
360     FOR J=K1 TO NM
370     A(I,J)=A(I,J)-A(I,K)*A(K,J)
380     NEXT J
390     A(I,J)=0
400     NEXT J
410     NEXT K
420     PRINT "DETER=   ",DT
430     PRINT "A(1,",",M,")...A(",N,",",NM,")"
440     FOR I=1 TO N
445     M1=N+1
450     FOR J=M1 TO NM
460     PRINT A(I,J)
470     NEXT J
480     NEXT I
2000    PRINT "DO YOU WANT CONTINUE?(Y/N) ";
2010    INPUT I$
2020    IF I$="N" THEN 5000
2030    GO TO 110
3000    DATA 6,1,1E-20
3010    DATA 10,20,30,40,50,60,200
3020    DATA 20,40,70,80,90,100,300
3030    DATA 30,70,50,110,120,130,400
3040    DATA 40,80,110,60,140,150,500
3050    DATA 50,90,120,140,70,160,600
3070    DATA 60,100,130,150,160,80,700
5000    END
RUN
THIS PROGRAM FIND M SOLUTION VECTORS CORRESPONDING TO A SET OF N
SIMULTANEOUS LINEAR EQUATIONS USING THE GAUSS-JORDAN REDUCTION ALGORITHM
N       : NUMBER OF EQUATIONS
M       : NUMBER OF SOLUTION VECTORS
EPS     : MINIMUM ALLOWABLE MAGNITUDE FOR A PIVOT ELEMENT

N=      6
M=      1
EPS=    9.999999E-21
A(1,1)...A(6    ,      7     )
 10      20      30      40      50      60      200
 20      40      70      80      90      100     300
 30      70      50      110     120     130     400
 40      80      110     60      140     150     500
 50      90      120     140     70      160     600
 60      100     130     150     160     80      700
SMALL PIVOT -- MATRIX MAY BE SINGULAR
DO YOU WANT CONTINUE?(Y/N) ? N
```

PROGRAM NO. 4

```
100     DIM A(20,20),X(20)
110     READ N,IX,EP
120     PRINT "SOLUTION OF SIMULTANEOUS LINEAR EQUATIONS BY GAUSS-SEIDEL"
125     PRINT "METHOD, WITH "
126     PRINT"N       : NUMBER OF EQUATIONS"
127     PRINT"ITMAX   : MAXIMUM NUMBER OF ITERATIONS ALLOWED"
128     PRINT"EPS     : MINIMUM ALLOWED MAGNITUDE FOR A PIVOT ELEMENT"
129     PRINT
131     PRINT
135     PRINT "N=       ",N
140     PRINT "ITMAX=   ",IX
150     PRINT "EPS=     ",EP
160     PRINT "THE COEFFICIENT MATRIX A(1,1)...A(N+1,N+1) IS"
170     N1=N+1
180     FOR I=1 TO N
190     FOR J=1 TO N1
200     READ A(I,J)
205     PRINT A(I,J)
210     NEXT J
220     NEXT I
225     PRINT "THE STARTING VECTOR X(1)...X(N) IS"
230     FOR I=1 TO N
240     READ X(I)
245     PRINT X(I)
250     NEXT I
260     FOR I=1 TO N
270     AS=A(I,I)
280     FOR J=1 TO N1
290     A(I,J)=A(I,J)/AS
300     NEXT J
310     NEXT I
320     FOR K=1 TO IX
330     F%=1
340     FOR I=1 TO N
350     XS=X(I)
360     X(I)=A(I,N1)
370     FOR J=1 TO N
380     IF I=J THEN 400
390     X(I)=X(I)-A(I,J)*X(J)
400     NEXT J
410     IF (ABS(XS-X(I) <=EP) THEN 430
420     F%=0
430     NEXT I
440     IF F% <> 1 THEN 510
450     PRINT "PROCEDURE CONVERSED, WITH ITER=      ",K
460     PRINT "SOLUTION VECTOR X(1)...X(N) IS"
470     FOR I=1 TO N
480     PRINT X(I)
490     NEXT I
500     GO TO 580
510     NEXT K
520     PRINT "NO CONVERGENCE"
530     PRINT"ITER=   ",K
540     PRINT "THE CURRENT VECTOR X(1)...X(N) IS"
550     FOR I=1 TO N
560     PRINT X(I)
570     NEXT I
580     PRINT "DO YOU WANT CONTINUE?(Y/N)   ";
590     INPUT I$
600     IF I$="N" THEN 7000
610     GO TO 110
1000    DATA 6,15,0.0001
1010    DATA 10,20,30,40,50,60,200
1020    DATA 20,40,70,80,90,100,300
1030    DATA 30,70,50,110,120,130,140
1040    DATA 40,80,110,60,140,150,500
1050    DATA 50,90,120,140,70,160,600
1060    DATA 60,100,130,150,160,80,700
1070    DATA 0,0,0,0,0,0
7000    END
RUN
SOLUTION OF SIMULTANEOUS LINEAR EQUATIONS BY GAUSS-SEIDEL
METHOD, WITH
N         : NUMBER OF EQUATIONS
ITMAX     : MAXIMUM NUMBER OF ITERATIONS ALLOWED
EPS       : MINIMUM ALLOWED MAGNITUDE FOR A PIVOT ELEMENT

N=       6
ITMAX=   15
EPS=     .0001
THE COEFFICIENT MATRIX A(1,1)...A(N+1,N+1) IS
 10      20      30      40      50      60      200
 20      40      70      80      90      100     300
 30      70      50      110     120     130     400
 40      80      110     60      140     150     500
 50      90      120     140     70      160     600
 60      100     130     150     160     80      700
THE STARTING VECTOR X(1)...X(N) IS
 0
 0
 0
 0
 0
 0
NO CONVERGENCE
INTER=        16
THE CURRENT VECTOR X(1)...X(N) is
-4.511715E+07
 4295324
 3019488
 4551110
-5239222
 2.550716E+07
DO YOU WANT CONTINUE?(Y/N) ? N
```

INDEX

ABC analysis 167
accelerometer misalignment 99
accelerometers 97
accounting model 161
accumulated strain 264
acquisition strategy 152
activity time computations 190, 192
aerial tramways 309
aggregate blending 241
air network 389
air pollution 325, 558
air transportation 365, 389
airborne laser 73
airborne laser ranging 71
Alaskan oil pipeline 92
alternative planning 583
analog-to-digital conversion 653
analysis manipulation 194
annual budget form 172
area computation 39, 43, 45
Army Corps of Engineers 571
arrow bonds 216
assets and liabilities 174
astronomic latitudes 51
atmospheric refraction 84
auto ownership 348
automation 182
automobile ownership 285
automobile trends 279

ballast 257, 267
ballast gradations 259
baseline precision 89
basic programming 669
Bay Area rapid transit system 328
beam splitting optics 74
blasting vibration 206
block-level validation 53

bond nomenclature 631
bottleneck situation 417
boundary offsets 44
brake check areas 467
branch exchange 487
branching distribution networks 473
branching layout 477
Brandywine River 572
Brazos River Basin 529
breakeven analyses 625
budget 157
budget economic control 163
budget making 159
building construction 125, 183
building works analysis 133
Bureau of Reclamation 520
bureaucratic reform 597
bus lanes on freeways 301
bus priority systems 300
bus systems 658
bus transit systems 273, 277
business enterprises 135

cadastral overlay 3
cadastral map 8
calculation of area 41
capacity analysis 413
car occupancies 341
car occupancy distribution 336
carpool assignment 335
cash flow analysis 126, 561, 603
chance constrained aggregate blending 241
chance constrained programming 245
channelizing devices 218
charrette 575
choice of redundants 17
citizen advisory 575
city parking 287

Clarke ellipsoid 49
closed water loop 517
coefficients of condition equations 15
computation 192
computation of areas 39
computer applications 183, 184
computer based techniques 575
computer calculation 418
computer language 182
condenser cooling 515
congestion on crosswalks 393
conservation 527
constitutive modelling 257
construction 211
construction joints 442
construction noise 208
construction phase 132
construction site 200
contract acquisition 156
contract budget 171
contract financial budget 172
contracts 126
contracts portfolio analysis 153
control survey summary 6
convective storage 540
cooling ponds 516
cooling systems 514, 517
cooling technology 519
cooling water 513
coordinating interactivities 135
corporate level 138
cost functions 476
cost model 559
CPM in construction 189
CPM network 191
credits 158
critical path method 130, 189
cross street indication 445
crustal movement 66

681

curvature changes 43
curve warning 447
curve warning sign 446
curved boundaries 39, 42

data acquisition 651
data base management 52
data logger 658
data preparation 17
Dead Sea 547
debts 158
Defense Mapping Agency 333
deformation 16
demand density in air transportation 366
demand elasticities 362
denied boardings 367
depreciation 616
detour control 426
dial-a-ride transportation 312
discriminator 81
distance observations 52
distribution reservoirs 486
distribution trees 484
distributional equity 591
distributional inequities 596
Doppler satellite 54
Doppler satellite positioning 52
downgrade warning 465
downgrades 457, 462
dredging 202
driver attributes 501
driver error 430
driver expectancy 429
driver information 500
driver information needs 212
driver performance 453
driver response criterion 226, 499
driving task 429, 430
dry cooling 516

earth centered datum 49
earth science data 4
earth's tectonic plates 71
economic analysis 155, 601
economic management 157
electric utilities 515
Electricité de France 519
electrostatic gyroscope 97
energy consumption 324, 513
energy facility planning 525
energy requirements 326
environment 126
environmental impact statements 197
environmental impacts 197, 325
environmental specification 197
equivalency conversions 605

erosion 198
error analysis 43
error model 109
error variances 78
escape facilities 462
escape ramps 459
estimation algorithm 77
evaluation of plants 585
evaporation 551
evaporative recirculation 516
executive committee 145
expectancy 429
expectancy checklist 449, 450
expectancy concept 433
expectancy violation 448
explosives method 207
expressway 407

fare structure 356
Ferranti intertial land surveyor 113
ferry boats 308
field leveling 65
finance committee 147
financial budget 173
financial performance 155
financial resources management 139
financial strategies 153
financing of public works 630
finite element method 16
firm's trends analysis 168
fishbowl planning 576
fixed value convention 618
flagmen 219
flight frequencies 365, 389
flight schedules 365
fluctuation of traffic 411
fluid temperature 514
fluidized bed boilers 557
forces in members 15
fouling of pond water 551
foundation blasting 205
free walking speed 394
freeway work sites 211
freeways 301
fresh surface water 519

gas network 489
gas price 563
Gaussian process 231
geoid-ellipsoid 103
geodetic application 92
geodetic control 49
geodetic reference system 49, 50
geodetic techniques 47, 54
geometric control 5
giant-type barchart 133

global positioning 67
goal programming 247
good height change 60
grade speed limits 458
gravity values 66
ground vibrations 204
groundwater 521
guidance devices 218
guideway transit 309
gyroscopes 95

hazard avoidance 216
head loss 476
heat extraction 539, 543, 547
heat insulation 544
heat insulator thickness 546
heat storage 547
heat storage experiment 550
heat trap 535
Helmert block adjustment 53
Helmert blocking strategy 54
high density living 327
highway construction 457
highway design 429
highway embankment 200, 201
highway networks 473
highway system 429
Honeywell IMU 102
Honeywell inertial survey system 114
horizontal control data sheet 55
horizontal curvature 463
household income 288
Housner's design spectra 238
human resources 159
hydrologic reliability 526

indenture 630
inertial measuring unit 99
inertial navigation 95
inertial survey system 104, 109
inertial surveying 95
inertial technology 120
inflation 618
inflation effects 621
information 165
information gathering 580
information handling 431
innovative 175, 178
interchange lane drops 436
interest cost 631
interest rate 601
Intermountain Power Project 522
investment scenarios 613
invoices 163
irrigation water 523

INDEX

Japan Highway Public Corporation 407, 423
Japanese expressway 407
joint displacements 21
joint releases 20
junction points 498

labor relations 357
land control 71
land data systems 3, 10
land survey control station 7
land title 3
land use 198
land use planning 595
land-related data banks 3
lane closures 217, 222
lane markings 440
lane split 440
Laplace stations 445
large networks 495
large-scale surveying 71
laser ranging 72, 85
laser retroreflectors 71
laser targets 89
laser transmitter 78
laws of inertia 95
layout optimization 490
light rail transit 273
lighting devices 218
linear goal programming 252
linear programming 389
link calculation 74
Litton auto-surveyor 112
Litton IMU 101
load factor 367
LORAN-aided INS 75
low altitude aircraft 74

macrofirm level 139
magnetic error modeling 65
maintenance 179, 180, 424
management 157
management budgets 158
management information system 125, 165
mapping 4
market strategy 156
mass transportation 273, 355
mathematical logic 415
mathematical models 76
mean sea level 62
measuring sediment loads 201
mechanical stairs 403
Meishin expressway 423
method of traffic survey 408
micro-computers 669

microprocessor 651
midblock signal 445
minimal spanning tree 478
model for the organization chart 147
model of firm structure 138
model usage 424
motor bus usage 276
mountainous terrain 457
moving ramps 403
multilateration 78
multiload inputs 337
municipal bonds 630
municipal financing 630
municipal service distribution 589
municipal services 590

Nagoya Operation Bureau 426
narrow bridges 441
National Geodetic Reference System 47
navigation data 75
navigation expectancies 449
navigation principles 95
navigation systems 104
navy transit satellites 50
network geometry 473
networks 473
node-labeling 480
noise regulations 209
North American Datum 58
North American Geodetic Networks 47

off-ramp 437
one lane bridge 443
open traverse 105
operating budget 167
operations level 138
optomechanical configuration 74
organization 128, 134, 137
organization and finances 314
organization chart 144
organizational model 125
organizing meetings 146

parallel roadside 439
paratransit modes 355
parking 287
parking and transit 286
passenger capacities 296
passenger volumes on rapid transit 307
pedestrian crossings 401
pedestrian flow 393, 397, 405
pedestrian movements 393, 401
pedestrian systems 404, 405
Pennsylvania Department of Transportation 458

Pennsylvania utilities 522
people and organization 143
PERT and CPM 130
photomultiplier 80
pile driving operations 205
pipe size optimization 488
plane coordinates 61
plane trilateration 13
planning 129, 473
planning construction 127
planning data bank 10
planning for traffic control 221
planning framework 573
planning schedule 131
pointing system 75
polar coordinate system 43
political responsiveness 597
pollutant emissions by vehicle type 327
polynomial approximation 234
population distribution 284
population trends 277
position observations 52
power plant cooling 516
power production 517
power spectral density 231
premission calibration 113
production committee 146
production cycle 128
profit analysis 170, 173, 623
profit analysis summary 171
profit program 170
program evaluation and review technique 130
programmable works 127
programming model 485
project scheduling 221
property boundary 5
property investment 611
property law 4
public information 222
public involvement 571
public transit 349
public transportation 273, 279
purchasing committee 146
purchasing strategy 156

Q-switched lasers 78, 86
quasifirm theory 141

radiation 536
rail rapid transit 273
rail transit systems 295
railroad track beds 257
random analysis 231
range data model 76
ranging system 73

RAP on microcomputers 119
rate of return 603, 611
rate traverse 106
real property 3
recreation services 593
redundant observation 14
reference frames 103
refraction correction 65
regional area adjustment 116
replacement analysis 621
reservoirs 520
residential roads 403
residential function 403
resilient modulus 264
resources management 139
revenues and expenses 320
riparian water rights 520
risk analysis 629
river basin planning 527
road condition 227
road expectancies 436
roadway attributes 501
roadway maintenance 595
route characteristics 501
route choice 501
rugged topography 457

safety and subcontractors 176
safety shape barrier 220
salient boundary 39
salient points 45
saline water 539
salt-gradient creation 536
sampling 242
satellite laser ranging 78, 87
satellite pass 87
sediment loads 201
sedimentation 197
seismic monitoring 208
sensitivity analysis 562, 626
separation function of traffic control 220
service attributes 501
service delivery 590
Sevier River Basin 522
shallow solar pond 535
shared street for vehicles 403
shear test 262
shock wave behavior 416
short pulse lasers 71
short-term parking 287
sieve data 242
sight-line restrictions 451
signal conditioning 652, 658
signal transmission 658
signalized crossings 401
signing 465
Simpson's rule 39, 45

site controls 149
site improvement form 452
site management 148
smoothing mode 115
socio-economic characteristics 500
software logic 118
soil erosion 197, 200
soils map 9
solar ponds 535
solar radiation 536
space per pedestrian 395
space-stable inertial systems 115
spatial expectancy 436
spatial reference framework 3
speaker bureau 578
speed density relationships 397
speed zone 214
speed-density relationships 395
squares of corrections 15
stable element diagram 100
state plane coordinate system 4
Stationary Gaussian processes 231
steep positive grades 457
Steiner points 490
stochastic network assignment 506
storage temperature 543, 544, 545
strategic decision areas 141
streets 403
stress strain response 263
structural analysis 231
structural design language 17
structural framework 13
subcontractors 194
subcontractors offers 160
suburbanization trend 345
subway excavation 204
subway systems 277
sulfation process 558
sunlight spectrum 536
surface water 519
survey mode 115
survey questionnaire 579
surveying 72
surveyors 3
Susquehanna River Basin 573
system delay calibration 85
system hardware 655, 660

tangential off-ramps 437, 439
target acquisition 75
target effects 83
tax changes 565
tax exemption 631
taxation policies 565
thermal convection 535
thermal efficiency 553
throughput rate 655

tidal information 66
time interval unit 82
timing diagram 83
trade receivables 164
traffic assignment 389
traffic control 211, 429
traffic control devices 226, 441
traffic control plan 221
traffic contrtol zone 223
traffic demand propagation 417
traffic demands 407
traffic density 412
traffic flow 457
traffic signals 441
traffic simulation 407, 418
traffic survey 407
traffic volume 411
traffic-restraint schemes 403
training 175, 222
training program 579
transcontinental traverse 52
transit coat 330
transit industry equipment 294
transit oriented cities 346
transit patronage 279
transit performance 345
transit performance concepts 351
transit revenues 319
transit ridership 348
transit riding habits 282
transit service 273, 593
transit subsidies 349
transit time jitter 81
transit use 282
transit vehicles 299
transportation constraints 481
transportation planning 473
travel time 413
traverse stations 5
triangulation 5
triaxial compression 262
trips by mode 292
trips by public transport 290
trip-making process 499
truck accident rate 464
truck escape 459, 467
turbidity control 202

urban characteristics 279
urban laboratory 180
urban transit 274
urban transit bus 298
urban transit routes 297
U.S. Coast and Geodetic Survey 5
U.S. Public Land Survey 5
U.S. Public Land Survey System 4

vector diagram of accelerometer 98
vehicle characteristics 501
vehicle design 295
vehicle occupancy 335
vehicle operation 326
vehicle ownership 288
vehicular level 393
vertical network 61
vibration criteria 206
vibrations 204, 207
vibrations on humans 207
video recording 410

violation analysis 448
virtual work 13, 15

walking distances 401
walking speed 394
warning devices 215
warning signs 466
water conservation 515, 518
water for power analysis 528
water fouling 551
water quality standards 197
water reuse 523

water right transfer 522
water supply 509, 513, 525
water supply assessments 525
water transport 524
water-for-energy 527
water-level transfers 66
work payment 161
work travel mode 286
work zone types 222

zenith distances 51
zoning for parking 287